Culinary Nutrition

Principles and Applications

atp AMERICAN TECHNICAL PUBLISHERS
Orland Park, Illinois 60467-5756

Linda J. Trakselis, MS
Eric M. Stein, MS, RD, CCE

Cover Photos: Barilla America, Inc.; Irinox USA

American Technical Publishers, Inc., Editorial Staff

Editor in Chief:
Jonathan F. Gosse
Vice President—Production:
Peter A. Zurlis
Director of Product Development:
Cathy A. Scruggs
Art Manager:
Jennifer M. Hines
Multimedia Manager:
Carl R. Hansen
Technical Editors:
Sara M. Marconi
Cathy A. Scruggs
Copy Editors:
Talia J. Lambarki
Amy B. Weissenburger
Catherine A. Mini

Cover Design:
Jennifer M. Hines
Illustration/Layout:
Melanie G. Doornbos
Nicholas W. Basham
Digital Media Development:
Hannah A. Swidergal
Adam T. Schuldt
Daniel Kundrat
Cory S. Butler
Nicole S. Polak
Robert E. Stickley

1 2 3 4 5 6 7 8 9 – 14 – 9 8 7 6 5 4 3 2 1

Printed in the United States of America

ISBN 978-0-8269-4221-0

 This book is printed on recycled paper.

Foreword

There is a growing need for professionals equipped with a deep understanding of general nutrition, the science behind ingredients, and an awareness of food and nutrition trends to create flavorful and healthy dining experiences. Researchers at Clemson University and The Pennsylvania State University surveyed 300 chefs about how portion sizes are determined. Of the survey respondents, 70% said the executive chef makes decisions about portion sizes, 22% cited the restaurant owner, and 18% said portions were decided at the corporate level. The study concluded that the attitudes of chefs are critically important to how much food is put in front of guests each time they dine out.

With the frequency of eating out continuing to rise, learning ways to improve the nutrient density of menu items and serving appropriate portion sizes of foods and beverages are strategic ways to improve the nutrition of communities across the country. Examples of culinary nutrition in practice are the pairing of professional chefs with nutrition educators or nutrition-trained chefs with culinary-trained dietitians. A chef can take raw food ingredients and create a memorable meal for guests, but a chef cannot change the nutritional composition of those ingredients.

Bridging the gap between nutrition and culinary knowledge can translate into healthy and sustainable menu practices within culinary arts and hospitality education, community food preparation, and health outreach programs. *Culinary Nutrition Principles and Applications* provides students with opportunities to combine nutrition principles and food science with culinary arts applications. This inclusive textbook provides excellent recipe-modification examples and food science experiments that bring nutrition science to the forefront of culinary nutrition and food preparation courses.

Culinary Nutrition Principles and Applications addresses key nutrition principles from macronutrients to micronutrients and culinary applications from beverages on the menu to desserts on the menu. By tapping this valuable resource, culinary professionals will be able to plan healthy menus and produce flavorful, healthy menu items for both commercial and non-commercial food segments.

Dr. Margaret D. Condrasky, RD, LD, CCE
Associate Professor, Food Nutrition and Packaging Sciences Department
Clemson University, College of Agriculture, Forestry & Life Sciences

Acknowledgments

The authors and publisher are grateful for the technical information and assistance provided by the following individuals:

Dr. Margaret D. Condrasky, RD, LD, CCE
Associate Professor, Food Nutrition and Packaging Sciences Department
Clemson University

Brent T. Frei
Editor and Marketing Director
Center for the Advancement of Foodservice Education

Paul Mendoza
Culinary Arts Program Director
Galveston College

Joseph Mitchell
Program Director, The Culinary and Hospitality Institute
Jefferson State Community College

The authors and publisher would like to thank the following companies, organizations, and individuals for providing images.

Agricultural Research Service, USDA
Alinea/Photo by Lara Kastner
All-Clad Metalcrafters
Alpha Baking Co., Inc.
American Egg Board
American Lamb Board
American Metalcraft, Inc.
Barilla America, Inc.
Basic American Foods
Beef Checkoff
Browne Foodservice
Bunn-O-Matic Corporation
California Fresh Apricot Council
California Strawberry Commission
Canada Beef Inc.
Cape Cod Cranberry Growers' Association
Carlisle FoodService Products
Charlie Trotter's
Chef Eric LeVine
Chef Gui Alinat

Chef's Choice® by EdgeCraft Corporation
CROPP Cooperative
Daniel NYC
D'Artagnan, Photography by Doug Adams Studio
Edward Don & Company
Eloma Combi Ovens
Emu Today and Tomorrow
Entourage
Florida Department of Agriculture and Consumer Services, Bureau of Seafood and Aquaculture Marketing
Florida Department of Citrus
Florida Tomato Committee
Fortune Fish Company
Frieda's Specialty Produce
Harbor Seafood, Inc.
Harvard School of Public Health
Henny Penny Corporation
HerbThyme Farms
House Foods
Idaho Potato Commission
Indian Harvest Specialtifoods, Inc./Rob Yuretich
Irinox USA
L'Auberge Carmel, C.S. White, photographer
L. Isaacson and Stein Fish Company
MacArthur Place Hotel, Sonoma
MacFarlane Pheasants, Inc.
Manitowoc Beverage Systems
McCain Foods USA
Melissa's Produce
Mushroom Council
National Cancer Institute
National Cherry Growers and Industries Foundation
National Garden Bureau Inc.
National Honey Board
National Oceanic and Atmospheric Administration/ Department of Commerce
National Onion Association
National Pasta Association
The National Pork Board
National Turkey Federation
New Zealand Greenshell™ Mussels
Oregon Raspberry & Blackberry Commission
Paderno World Cuisine
Pear Bureau Northwest
Perdue Foodservice, Perdue Farms Incorporated
Planet Hollywood International, Inc.
The Publican
Rishi Tea
Scripps Health
SelectWisely
Service Ideas, Inc.
Shenandoah Growers
The Spice House
Sullivan University
Tanimura & Antle®
Trails End Chestnuts
United States Department of Agriculture
United States Potato Board
U.S. Apple® Association
U.S. Fish & Wildlife Service
U.S. Highbush Blueberry Council
Venison World
Vita-Mix® Corporation
Vulcan-Hart, a division of the ITW Food Equipment Group LLC
Wisconsin Milk Marketing Board, Inc.

Contents

Contents

Contents

Contents

Digital Resources

- Quick Quizzes®
- Illustrated Glossary
- Flash Cards
- Knowledge Checks and Reviews
- Culinary Nutrition Recipe Modifications
- Menu Planning Challenge
- Digestion and Absorption Process
- Media Library

Culinary Nutrition Recipe-Modification Process

The Culinary Nutrition Recipe-Modification Process provides a step-by-step procedure for modifying an existing recipe into a healthier version that upholds the flavor and integrity of the original dish. Recipes can be modified to reduce the amounts of sugars, fats, sodium, and/or calories. Recipes can also be modified for specific health or dietary reasons. The six steps of the culinary nutrition recipe-modification process include evaluating the original recipe for sensory and nutritional qualities, establishing goals, identifying modifications or substitutions, determining the functions of identified modifications or substitutions, selecting appropriate modifications or substitutions, and testing the modified recipe to evaluate sensory and nutritional qualities.

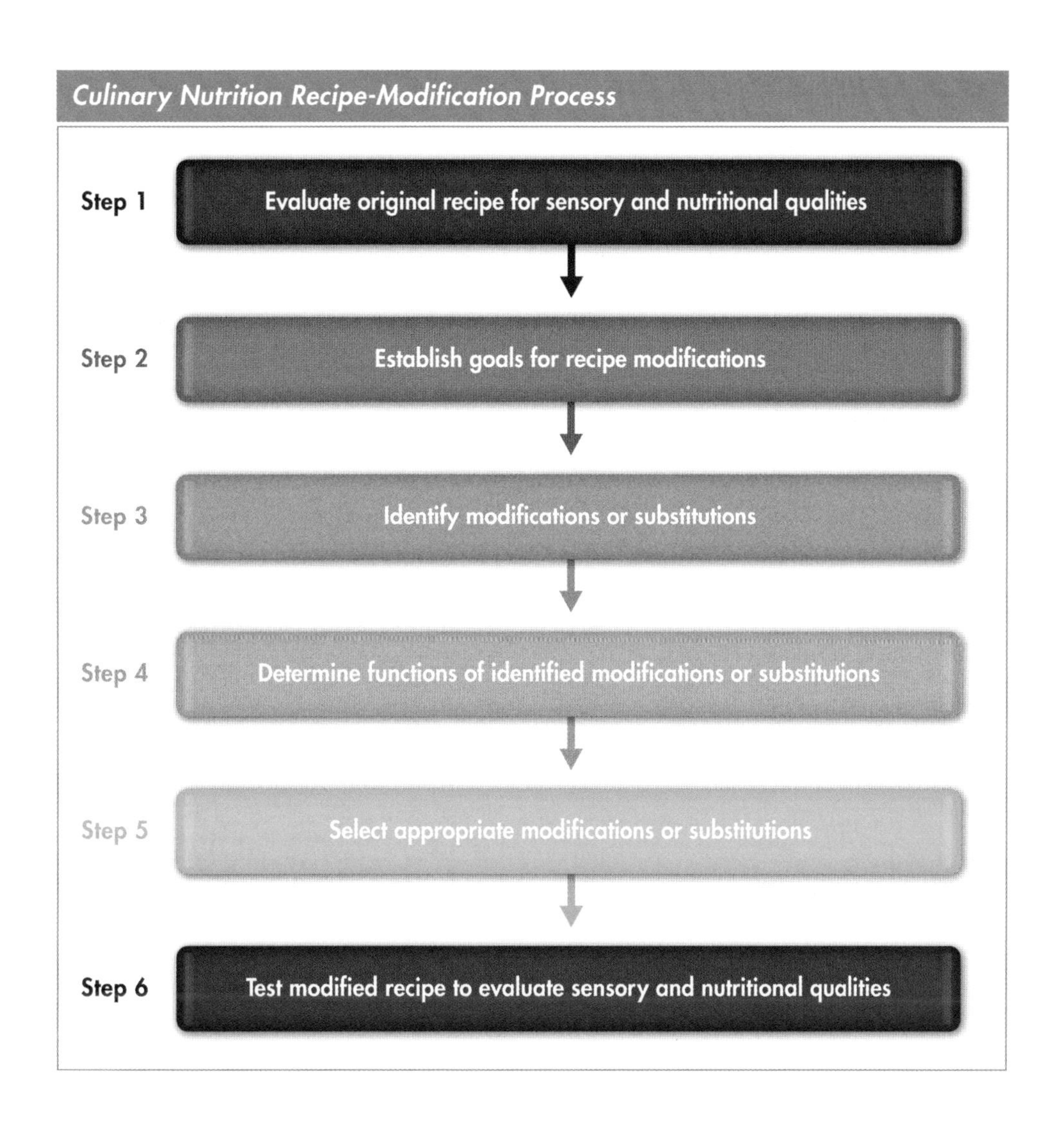

Culinary Nutrition Recipe-Modification Process Index

The Culinary Nutrition Recipe-Modification Processes listed below target specific food items covered in Chapters 7 through 15 and include the following recipes:

CHAPTER 7
Nutritious Menu Planning

CHAPTER 8
Beverages on the Menu

CHAPTER 9
Egg, Soy & Dairy Products on the Menu

CHAPTER 10
Poultry & Meats on the Menu

CHAPTER 11
Fish & Shellfish on the Menu

CHAPTER 12
Vegetables & Legumes on the Menu

CHAPTER 13
Fruits, Nuts & Seeds on the Menu

CHAPTER 14
Pastas, Grains & Breads on the Menu

CHAPTER 15
Desserts on the Menu

Book Features

Culinary Nutrition Principles and Applications combines the science of nutrition with the art of food preparation to provide learners with the knowledge base needed to plan and prepare healthy menu items. The first six chapters of this engaging textbook explain the functions and food sources of proteins, carbohydrates, lipids, water, vitamins, and minerals and their roles in the digestion and absorption process. The remaining nine chapters describe how to apply nutrition principles and healthy cooking techniques to create flavorful, nutrient-dense menu items that appeal to guests. A proven method for modifying recipes, called the Culinary Nutrition Recipe-Modification Process, is demonstrated for each menu category. Additionally, access is provided to a dynamic set of online digital resources that enhance chapter concepts and promote learning.

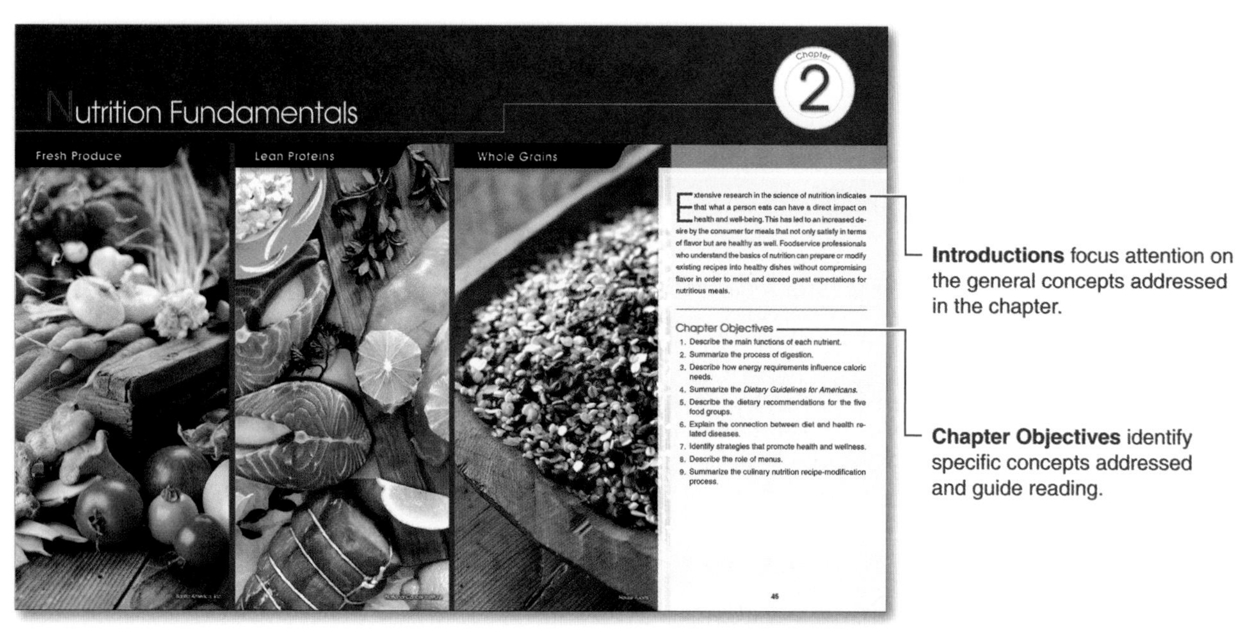

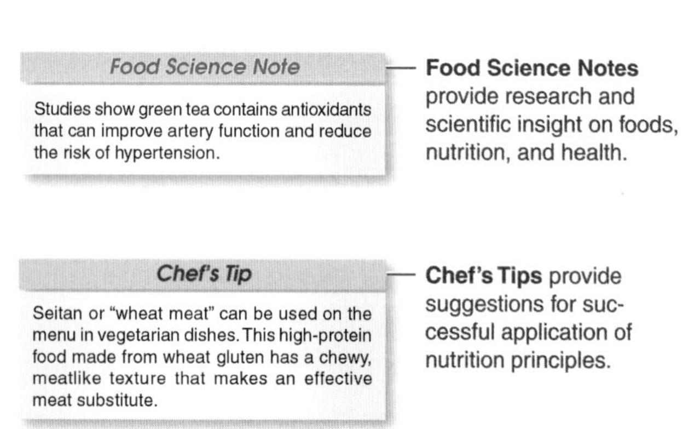

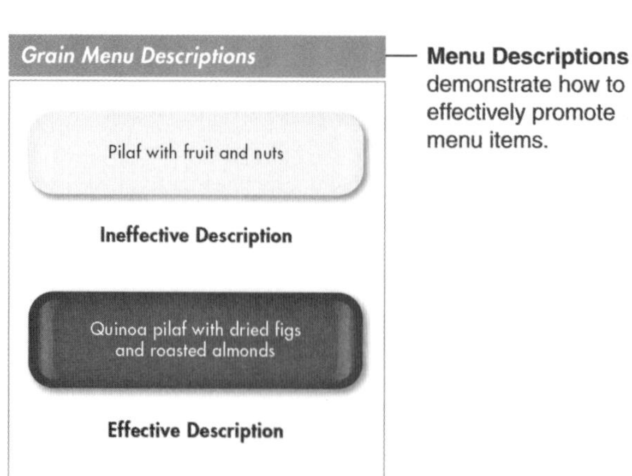

Digestion and absorption illustrations depict how essential nutrients are used in the body.

Numerous photos help learners visualize the concepts presented.

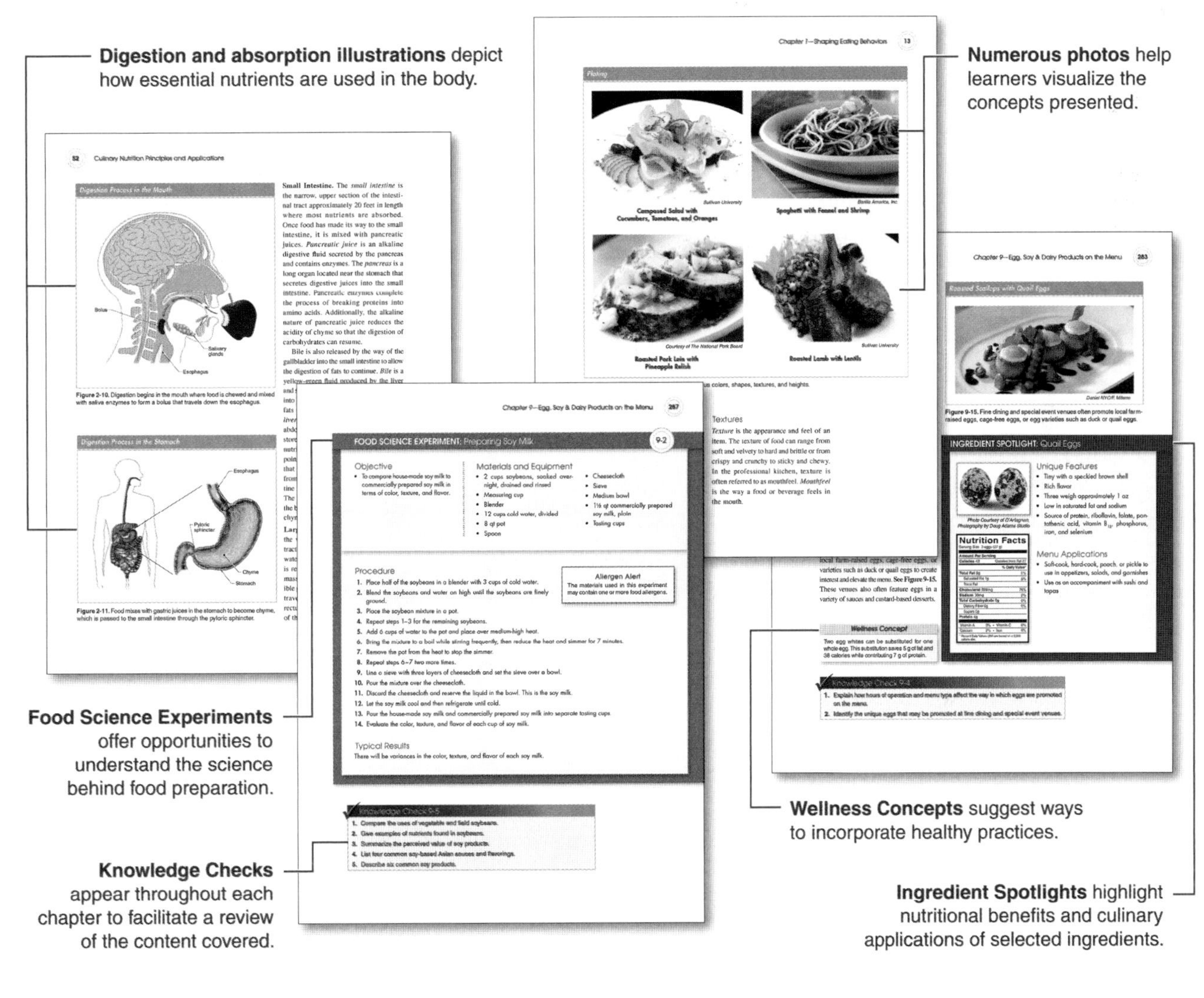

Food Science Experiments offer opportunities to understand the science behind food preparation.

Knowledge Checks appear throughout each chapter to facilitate a review of the content covered.

Wellness Concepts suggest ways to incorporate healthy practices.

Ingredient Spotlights highlight nutritional benefits and culinary applications of selected ingredients.

Recipe Modifications demonstrate how to modify recipes to create more nutrient-dense menu options.

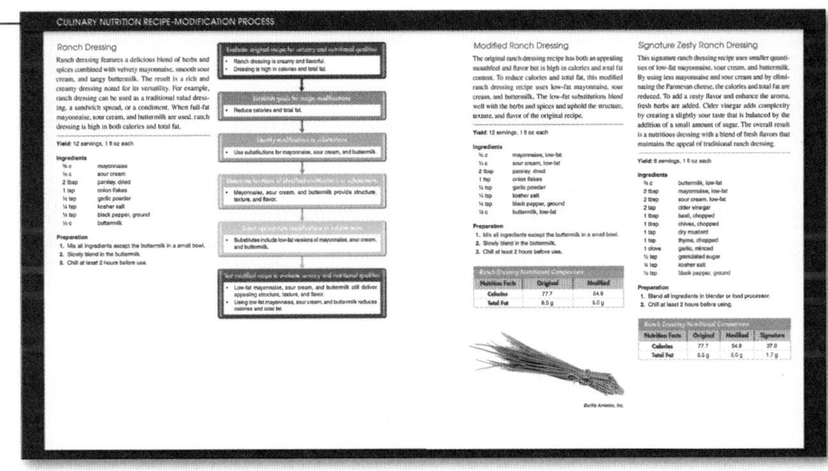

Digital Features

Culinary Nutrition Principles and Applications includes digital resources that enhance chapter concepts and promote learning. QuickLinks and Quick Response (QR) codes, located at the end of each chapter, offer easy access to digital resources.

Using QuickLinks

To access a QuickLink using an access code, follow these simple steps:

1. Key ATPeResources.com/QuickLinks into a browser.
2. Open the page and type in the access code provided on the last page of the chapter.
3. Instantly access the digital resources.

The Digital Resources include the following:

- **Quick Quizzes**® that provide 20 interactive questions for each chapter, with helpful links to textbook content and to the Illustrated Glossary
- An **Illustrated Glossary** that provides a helpful reference to commonly used terms, with selected terms linked to illustrations or media clips
- **Flash Cards** that provide a self-study tool for learning the terms and definitions used in the textbook
- **Knowledge Checks and Reviews** in PDF format that allow learners to demonstrate knowledge of chapter concepts
- **Culinary Nutrition Recipe Modifications** that provide a proven process for modifying traditional recipes to create healthier versions of menu items
- The **Menu Planning Challenge**, which provides interactive activities in which learners apply their nutrition knowledge to select healthy menu items
- The **Digestion and Absorption Process**, which simulates how food is digested and absorbed
- A **Media Library** that includes animations and video clips that enhance the textbook content
- **ATPeResources.com**, which links to online resources that support continued learning

Using Quick Response (QR) Codes

To access a QR code, follow these simple steps:

1. Download a QR code reader app to a mobile device. (Visit atplearning.com/QR for more information.)
2. Open the app and scan the QR code on the book page.
3. Instantly access the digital resources.

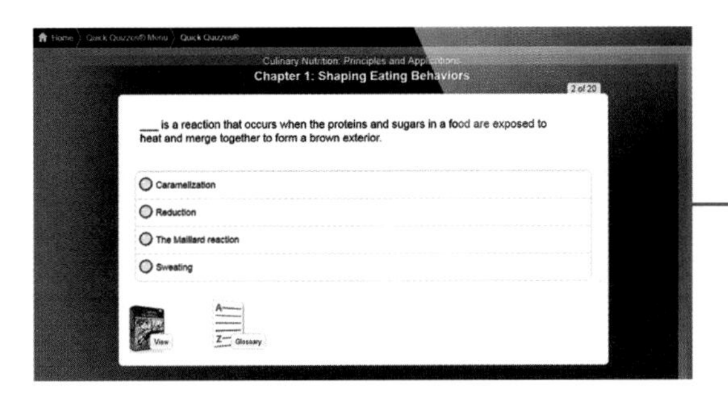

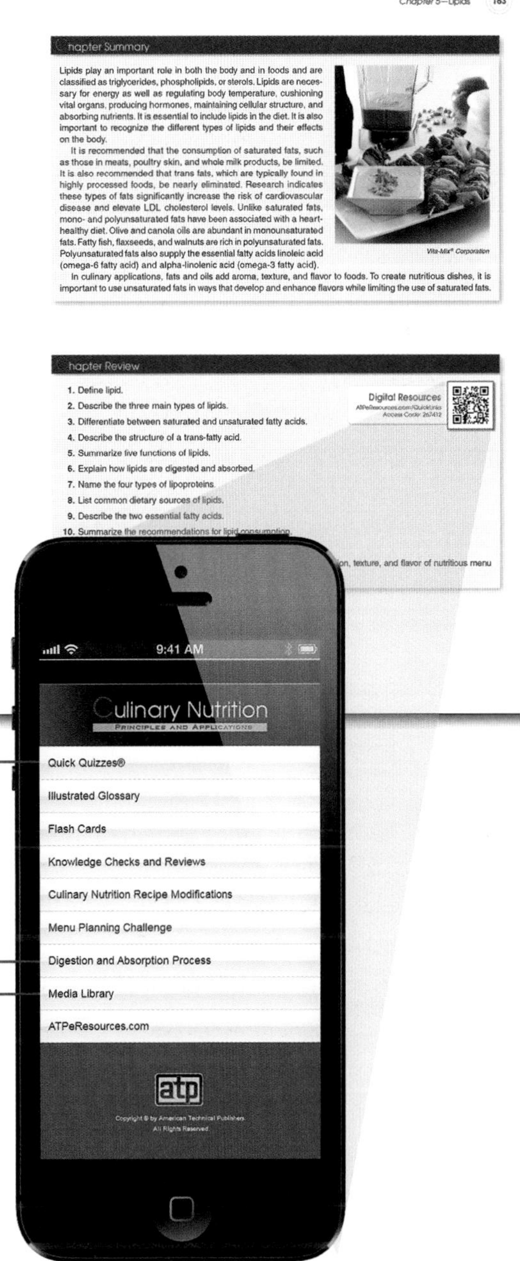

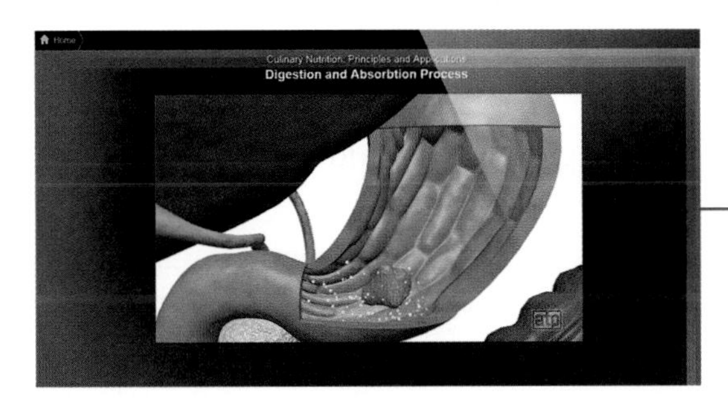

Study Guide

The *Culinary Nutrition Principles and Applications Study Guide* includes review questions and activities that reinforce and expand upon the information presented in the textbook.

Reviews for each chapter include true-false, multiple choice, completion, and critical thinking questions that reinforce comprehension.

Shaping Eating Behaviors 1 REVIEW

Name ______ Date ______

True-False

T F 1. Heat always decreases the nutritive value of foods.

T F 2. Layers of flavors are created when different ingredients are added at different times during the preparation and cooking processes.

T F 3. Research has shown that after taste, cost is the next most influential factor when choosing foods.

T F 4. Menu items will only sell on the basis that they are "healthy."

T F 5. Oil is derived from vegetables, nuts, seeds, and fruits.

T F 6. Research supports the idea that tastes can be categorized as sweet, salty, sour, bitter, or umami.

T F 7. Temperature affects the aroma of food.

T F 8. Bones are never used as a flavoring component in meat, poultry, and fish stocks.

T F 9. Spices are always added at the end of the cooking process.

T F 10. Heat typically flows from a cooler material to a warmer material.

T F 11. A gum is a thickening agent derived from plant flours.

T F 12. A rub is a blend of ingredients that is pressed onto the surface of uncooked foods to impart flavor.

T F 13. Sustainable agriculture is a method of food production that seeks to conserve land and water, reduce energy consumption, avoid the use of harmful fertilizers and pesticides, and limit air pollution.

T F 14. The Maillard reaction is a reaction that occurs when the lipids and sugars in a food are exposed to heat and merge together to form a brown exterior.

T F 15. Dry-heat cooking is any cooking technique that uses hot air, hot metal, a flame, or hot fat to conduct heat and brown food.

© 2014 American Technical Publishers, Inc. All rights reserved

CULINARY NUTRITION PRINCIPLES AND APPLICATIONS STUDY GUIDE

Multiple Choice

1. According to the Food and Drug Administration (FDA), milk, eggs, fish, crustacean shellfish, tree nuts, peanuts, wheat, and ___ account for 90% of all food allergies.
A. soybeans
B. rice
C. corn
D. quinoa

2. The Nutrition Facts label provides a standard for comparing foods based on serving size, calories, nutrients, and percent daily value for a ___ calorie diet.
A. 1000
B. 1500
C. 2000
D. 2500

3. A ___ is an individual who primarily consumes plant-based foods and only consumes animal-based foods, such as meat, dairy, and eggs, occasionally or in small amounts.
A. flexitarian
B. lacto vegetarian
C. vegetarian
D. vegan

4. A(n) ___ food is a food that has been altered by processes such as canning, cooking, freezing, dehydration, or milling.
A. whole
B. raw
C. organic
D. processed

5. Standard nutrition labels are required to include ___.
A. unsaturated fat and "trans" fat
B. saturated fat and unsaturated fat
C. the total grams of fat, saturated fat, and unsaturated fat
D. the total grams of fat, saturated fat, and "trans" fat

6. Sour tastes come from the ___ in foods.
A. glutamate
B. salts
C. acids
D. alkalies

7. Vinegar is a sour, acidic liquid made from fermented ___.
A. alcohol
B. oil
C. seeds
D. herbs

© 2014 American Technical Publishers, Inc. All rights reserved

Nutrition Fundamentals 2 ACTIVITIES

Name ______ Date

Activity 2-4

Macronutrient Calorie Contributions

Using the Nutrition Facts label provided, calculate the calorie contributions ... and protein.

1. ___ × ___ = ___ calories from fat
2. ___ × ___ = ___ calories from carbohydrates
3. ___ × ___ = ___ calories from protein

Nutritious Menu Planning 7 ACTIVITIES

Name ______ Date ______

Activity 7-4

Research: Seasoning Blends

Chefs often develop signature seasoning blends to help create flavorful menu items. These blends can be ... a variety of ways to enhance flavor while limiting added calories. Seasoning blends are typically prepared ... of time, which promotes efficiency during preparation.

Develop a signature seasoning blend and use the USDA Nutrient Database to prepare a 2-3 page report ... oral presentation to include:

1. Describe the name and ingredients of your signature seasoning blend.
2. List three nutritional reasons for the choices that you made.
3. Identify key nutrients found in the seasoning blend.
4. Explain the intended flavor profile and use.
5. Describe how the recipe was tested.
6. Explain any adjustments that were made to the original recipe after it was tested and the reason(s) ... change(s).
7. Describe three menu items the seasoning blend would complement.

Nutrition Project Rubric

Content:	
Written report	5
Cover page	5
Introduction	5
Quality content	20
Summary	5
Format:	
Typed, double spaced	5
Correct spelling	5
Correct grammar	5
Correct punctuation	5
Bibliography	5
Oral Presentation:	
Introduction	5
Quality content	10
Summary	5
Delivery	5
Visual aid(s)	10
Total	100

© 2014 American Technical Publishers, Inc. All rights reserved

Fruits, Nuts & Seeds on the Menu 13 ACTIVITIES

Name ______ Date

Activity 13-1

Fruit Classifications

The major classifications of fruits include berries, grapes, pomes, drupes, m... and exotic fruits. Fruits enhance menu items with varied colors, textures, an... assortment of vitamins and minerals, dietary fiber, phytonutrients, and antiox...

Match each fruit with its appropriate classification. Note: Classifications ma...

1. Currants
2. Lychees
3. Raisins
4. Pineapples
5. Star fruit
6. Mandarins
7. Peaches
8. Honeydews
9. Pomegranates
10. Mangoes
11. Grapefruits
12. Cherries
13. Couscou
14. Pears
15. Casabas
16. Limes
17. Raspberries
18. Apples

A. Berries
B. Grapes
C. Pomes
D. Drupes
E. Melons
F. Citrus fruits
G. Tropical fruits
H. Exotic fruits

Nutritious Menu Planning 7 ACTIVITIES

Name ______ Date ______

Activity 7-1

Menu Descriptions

Menu descriptions should be clear and appealing while communicating the main ingredients of a dish and the method of preparation. Effective menu descriptions promote clarity, add appeal, and increase the likelihood that an item will be ordered. Research indicates that terms such as "low-fat," "low-carb," and "cholesterol-free" may send a message of deprivation, while terms such as "whole grains," "natural," and "locally produced" accentuate the positive aspects of an item and can stimulate sales.

Example:
Ineffective Menu Description:
Veggie burger
Effective Menu Description:
Grilled portobello-lentil burger with chipotle-lime yogurt on a toasted whole grain bun

Rewrite the following ineffective menu descriptions to make them more appealing and complete in the description.

1. Oatmeal: ______
2. Chef's salad: ______
3. Chicken sandwich: ______
4. Steak fajitas: ______
5. Shrimp kebab: ______
6. Yogurt parfait: ______

Menu Description Grading Rubric	
Menu Description Criteria	**Points**
Accurate representation of menu item	0 1 2 3
Promotes clarity of menu item	0 1 2 3
Appealing portrayal of menu item	0 1 2 3
States unique characteristics of menu item	0 1 2 3
Conveys preparation/cooking technique(s)	0 1 2 3
Total Possible Points for Each Menu Description: 15	

© 2014 American Technical Publishers, Inc. All rights reserved

Activities help learners reinforce and apply chapter concepts by analyzing nutrition labels, conducting research, writing effective menu descriptions, and matching classifications of foods.

Instructor's Resource Guide

The *Culinary Nutrition Principles and Applications Instructor's Resource Guide* provides a comprehensive teaching resource that includes a detailed Instructional Guide, editable PowerPoint® Presentations, an interactive Image Library, multiple Assessments, and Answer Keys in addition to access to the digital resources used with the textbook.

- The **Instructional Guide** explains how to best use the learning resources provided and includes Instructional Plans for each chapter that identify images and media pieces that enhance the learning experience.
- **PowerPoint® Presentations** provide a review of key concepts and illustrations from each chapter of *Culinary Nutrition Principles and Applications* by addressing the chapter objectives and providing section knowledge checks and review questions. PowerPoint® Presentation notes are also provided.
- The **Image Library** provides all the numbered figures in a format that can be manipulated for maximum instructional use.
- **Assessments** include sets of questions based on objectives and key concepts from each chapter and consist of a pretest, posttest, and test banks. The test banks can be used with most test development software packages and learning management systems.
- **Answer Keys** list answers to pretest, posttest, textbook, and study guide questions.

To obtain information on related products, visit the American Technical Publishers website at atplearning.com.

The Publisher

Shaping Eating Behaviors

Presentation

Irinox USA

Flavor

Browne Foodservice

Nutrition

National Turkey Federation

Culinary arts is an exciting and dynamic field influenced by individual eating behaviors, industry trends, and the discovery of new foods, information, and practices. Chefs have the opportunity to help shape eating behaviors by using healthy preparation techniques and sound nutrition principles to combine flavors that excite and tempt all the senses. The result is meals that are visually appealing, flavorful, and nutritious.

Chapter Objectives

1. Identify primary influences that shape food choices.
2. Describe sustainable influences that affect food choices.
3. Differentiate between food allergies and food intolerances.
4. Identify the dominant senses that are used to detect flavor.
5. Identify common flavor profiles.
6. Explain how presentation, texture, taste, aroma, and temperature can influence food choices.
7. Describe the primary tastes recognized by the taste buds.
8. Explain how layers of flavors are created.
9. Identify common flavorings and seasonings.
10. Explain common techniques used to develop flavor.
11. Describe the heat transfer methods used to cook food.
12. Describe the main cooking techniques.

FOOD CHOICES

Whether it is a family dinner at home, a business luncheon at a local bistro, or a catered event, the foods and beverages that are chosen usually reflect the preferences of the individuals who are dining together. The primary factor influencing food choices and preferences is the palatability of food. Individuals seek foods with pleasing colors, textures, and flavors. **See Figure 1-1.**

Many food choices are also influenced by factors such as the media, the sustainability of foods, and purchasing considerations, such as affordability. In addition, individual influences, including health issues and nutrition knowledge, play a role in the food choices that determine eating behaviors.

Media Influences

Consumers are exposed to a variety of media influences that shape food choices. For example, both television and the Internet transmit information regarding the impact foods have on health as well as feature the preparation of recipes ranging from appetizers to elaborate desserts. Featured recipes often showcase ingredients once considered rare or exotic but have now become mainstream due in large part to the impact of the media. Evidence also supports the idea that consumer interest in food-related media continues to grow. For example, since its humble beginning in 1993, the Food Network has risen to a worth of over a billion dollars, brought celebrity status to many chefs, and helped popularize numerous cookbooks and cooking tools.

The popularity of food is also highlighted through social media outlets and thousands of blogs. Advertisements, videos, images, and reviews featured on these sites often help individuals connect with foodservice operations that meet their specific needs. For example, families with young children are more apt to be influenced by social media outlets that share information about family-style restaurants featuring child-friendly menus. Professionals often tap into sites reviewing fine dining restaurants and bistros to find an ideal venue for lunches and dinners with clients. Others connect to food blogs and foodservice operations featuring sustainable products.

Food Choices

Courtesy of The National Pork Board

Apricot Pecan Pork Salad

National Turkey Federation

Grilled Turkey atop Brown Rice with Mesclun Greens

Daniel NYC/B. Milne

Salmon and Vegetable Medley

Figure 1-1. Individuals seek foods with pleasing colors, textures, and flavors.

Sustainability Influences

Sustainable agriculture is a method of food production that seeks to conserve land and water, reduce energy consumption, avoid the use of harmful fertilizers and pesticides, and limit air pollution. A growing trend supported by sustainable agriculture is the practice of eating foods that are locally grown and produced. The main reasons for this practice are to conserve natural resources, support the local economy, and reduce the risk of food contamination due to transit time. Also, local foods are often harvested at the peak of freshness, offering maximum flavor.

Interest in sustainable practices has led some to incorporate organic foods into their diets. *Organic food* is food produced without the use of chemically formulated fertilizers, growth stimulants, antibiotics, pesticides, or spoilage-inhibiting radiation. The U.S. Department of Agriculture (USDA) sets standards for organic food products through the Organic Foods Production Act (OFPA) and the National Organic Program (NOP). **See Figure 1-2.** Some organic foods such as apples may be healthier than foods produced using conventional methods.

California Strawberry Commission

The purposes of sustainable agriculture are to conserve land and water, reduce energy consumption, avoid harmful chemicals, and limit air pollution.

Purchasing Influences

Choosing which foods to purchase requires individuals to make decisions based on a variety of factors. Some factors include the affordability of an item, whereas other factors may revolve around the availability of the item. Purchasing decisions may also be based on the nutritional qualities of a food or meal. Menus reflecting this trend often emphasize seasonal or local foods, whole grains, and smaller portion sizes. A *portion size* is the amount of a food or beverage item served to an individual.

Organic Foods

Classification	Requirements	May Display the Organic Seal (USDA ORGANIC — U.S. Department of Agriculture)
100% Organic	Contains 100% organic ingredients	Yes
Organic	Contains at least 95% organic ingredients; remaining ingredients must not be available in organic form	Yes
Made with Organic Ingredients	Contains at least 70% organic ingredients	No
Contains Organic Ingredients	Contains less than 70% organic ingredients	No

Figure 1-2. The U.S. Department of Agriculture (USDA) sets standards for organic food products through the Organic Foods Production Act (OFPA) and the National Organic Program (NOP).

Food labels may be used to market products in a manner that influences consumer purchasing decisions. This is observed on labels that include "fat-free," "light," or "extra lean" to convince consumers that certain products have benefits aside from fulfilling hunger. Many terms have precise meanings and can only be used to describe a product that meets specific guidelines. **See Figure 1-3.** Additionally, purchase decisions may be influenced by products that include Nutrition Facts labels, which inform consumers of the nutritional qualities of the product.

Affordability. The rising cost of food can affect food choices. Research has shown that after taste, cost is the next most influential factor when choosing foods. This often leads to purchasing products that are nutritionally inferior. For example, processed foods containing high amounts of sugars, fats, and/or sodium are often less expensive than foods containing quality, fresh ingredients. A *processed food* is food that has been altered by processes such as canning, cooking, freezing, dehydration, or milling. **See Figure 1-4.**

Interpreting Food Labels

Food Label Terms	Product Guidelines
Free, zero, no, without, trivial source of, negligible source of, insignificant source of	Food that contains no amount or a trivial amount of calories, fat, saturated fat, cholesterol, sodium, and/or sugars; for example, calorie-free means fewer than 5 calories per serving; sugar-free and fat-free both mean less than 0.5 g per serving
Low, little, few, contains a small amount of, low source of	Food where an individual can consume a large amount of the food without exceeding the daily value for the nutrient: • "low-calorie" means 40 calories or less per serving • "low-fat" means 3 g or less per serving • "low in saturated fat" means 1 g or less per serving • "low-cholesterol" means less than 20 mg per serving • "low-sodium" means less than 140 mg per serving
Percent fat-free	Food that is low-fat or fat-free; claim must reflect the amount of fat present in 100 g of the food; for example, if a food contains 2.5 g of fat per 50 g, the claim 95% fat-free would be based on 5 g of fat per 100 g of the food
Light, lite	Food containing one-third fewer calories, or half the fat, than the original product; low-calorie, low-fat food with a sodium reduction of 50%
Reduced	Food with 25% less calories, fat, saturated fat, cholesterol, sodium, or sugar than the original product
Less	Food containing 25% less calories, fat, saturated fat, cholesterol, sodium, or sugar than a reference food; for example, a bag of pretzels may claim "25% less fat than potato chips"
More	Food with at least 10% more of a nutrient than a reference food
Good source	Food with 10% to 19% of the daily value per serving
High	Food with 20% or more of the daily value per serving
Lean	Meat, poultry, or seafood with less than 10 g of fat, less than 4 g of saturated fat, and less than 95 mg of cholesterol per 100 g
Extra lean	Meat, poultry, or seafood with less than 5 g of fat, less than 2 g of saturated fat, and less than 95 mg of cholesterol per 100 g

Adapted from U.S. Food and Drug Administration

Figure 1-3. Many terms used on food labels have precise meanings and can only be used to describe a product that meets specific guidelines.

Examples of Processed Foods

- Bagged salad mixes
- Breads
- Breakfast cereals
- Canned foods
- Frozen fruits and vegetables
- Frozen meals
- Gelatins
- Ice creams
- Jarred pasta sauces
- Juices
- Milks
- Nut butters
- Pastas
- Rices
- Salad dressings
- Snack foods
- Soups
- Spice mixes
- Washed and packaged fruits and vegetables
- Yogurts

Figure 1-4. Processed foods have been altered by processes such as canning, cooking, freezing, dehydrating, or milling.

Farmers' Markets

Figure 1-5. Farmers' markets have a growing influence over the availability of items.

Availability. The availability of food often dictates the type of food purchased and consumed. For example, corn on the cob, watermelon, and strawberries are generally more available during the summer months, whereas a large variety of apples, pumpkins, and squashes are easily found in the fall.

Retail outlets and farmers' markets also have a growing influence over the availability of items. **See Figure 1-5.** Locally produced items ranging from fruits, vegetables, meats, and seafood to coffees, wines, cheeses, and honeys are giving both chefs and their guests the opportunity to experiment with new flavors while incorporating the freshest ingredients of the season.

Wellness Concept

Online resources can be helpful for locating farmers' markets. Some organizations, such as Sustainable Table and the Agricultural Marketing Service (AMS), a division of the U.S. Department of Agriculture (USDA), provide a comprehensive listing of farmers' markets on their websites.

Nutrition Facts Labels. In addition to packaging requirements, such as health claims, manufacturer information, and ingredient lists, the Nutrition Labeling and Education Act of 1990 requires food manufacturers to display a Nutrition Facts label on all packaged food and beverage items. The label provides a standard for comparing foods based on serving size, calories, nutrients, and percent daily value for a 2000 calorie diet. **See Figure 1-6.** This information can help shape nutritious eating behaviors by serving as a guide to identifying healthy foods.

- **Serving Size:** Listed at the top of all nutrition labels is information about serving sizes and the number of servings per container. The terms "portion size" and "serving size" are often used interchangeably. However, the terms have different meanings. While portion size refers to the amount of food served, serving size is an identifiable measurement. *Serving size* is a specific amount of food expressed using a unit of measurement such as a tablespoon, ounce, or cup. On the Nutrition Facts label, the values for calories, nutrients, and percent daily value are calculated based on the serving size.

- **Calories:** Total calories and calories from fat are listed directly below the serving size. Standard nutrition labels are required to include the total grams of fat, saturated fat, and "trans" fat. Some manufacturers choose to list monounsaturated and polyunsaturated fats because these types of fats are considered healthy.
- **Cholesterol and Sodium:** The amounts of cholesterol and sodium (in milligrams) appear after the calorie listing. Cholesterol and sodium are included because of their potential for causing adverse health effects.
- **Carbohydrates:** Total carbohydrates is listed in grams after cholesterol and sodium. Carbohydrates are also broken down into dietary fiber and sugars.
- **Protein:** Protein is listed in grams after total carbohydrates.

Interpreting Nutrition Facts Labels

Food Label Facts

Use this guide to help you better understand the nutrition facts on food labels and make better choices while shopping.

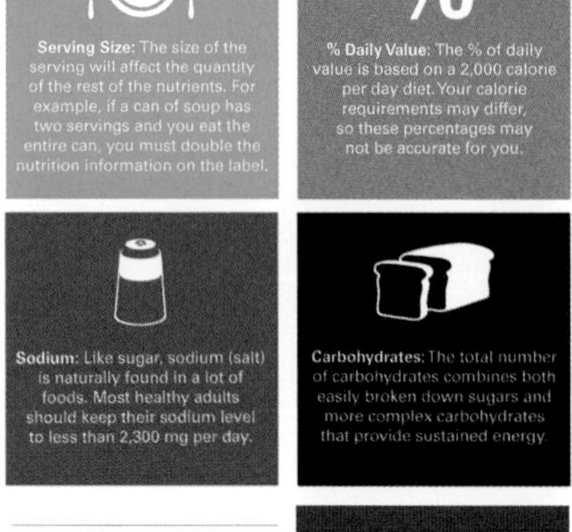

Nutrition Facts

Serving Size 1 cup
Servings per container 2

Amount Per Serving	
Calories 90	Calories from fat 20
	% Daily Value
Total Fat 2 g	3%
Saturated Fat 0 g	0%
Trans Fat 0 g	
Cholesterol 10 mg	3%
Sodium 890mg	37%
Total Carbohydrates 13 g	4%
Dietary Fiber 1 g	4%
Sugars 1 g	
Protein 6 g	

Ingredients: Chicken broth, carrots, cooked white chicken meat (white chicken meat, water, salt, sodium phosphate, isolates soy protein, modified cornstarch, cornstarch), potatoes, celery, rice, monosodium glutamate. Contains soy.

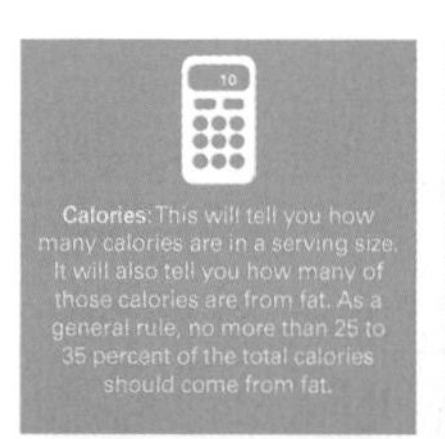

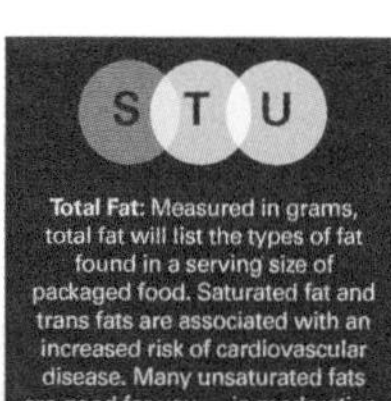

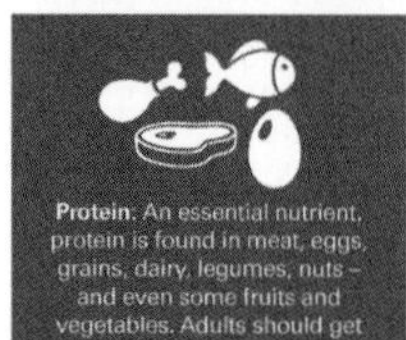

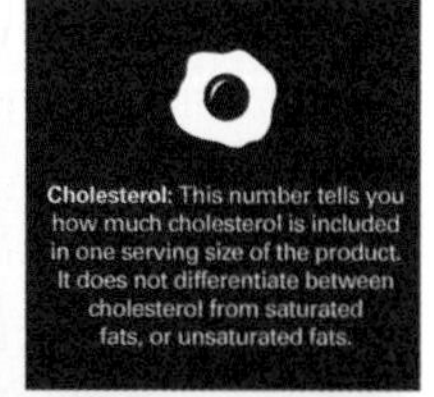

Hidden fats, sugars and salts

Sometimes fats, sugars and salts can be listed under different names, making it difficult to tell what you're eating. Below are some names of common ingredients that could be adding extra calories and sodium to your diet.

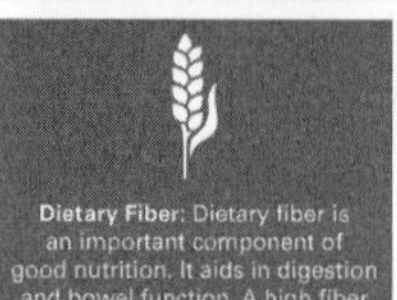

Types of added fat
Animal fat (pork, chicken and beef)
Butter
Cocoa butter
Coconut oil
Cream
Lard
Margarine
Milk solids
Palm kernel oil
Palm oil
Partially hydrogenated and hydrogenated oils
Shortening
Suet
Tallow
Vegetable oils (including avocado, olive, canola, peanut, sesame, soybean and sunflower)

Types of added sugar
Agave nectar
Anhydrous dextrose
Barley malt syrup
Brown sugar
Brown rice sugar
Corn sweetener
Corn syrup
Dextrin
Dextrose
Evaporated cane juice
Fructose
Glucose
High-fructose corn syrup
Honey
Invert sugar
Lactose
Maltodextrin
Maltose
Maple syrup
Molasses
Rice syrup
Saccharose
Sorghum or sorghum syrup
Sugar alcohol
Sucrose
Treacle
Xylose

Types of added sodium
Baking powder
Disodium phosphate
Iodized salt
Kosher salt
Monosodium glutamate (MSG)
Rock salt
Sea salt
Sodium benzoate
Sodium bicarbonate (Baking soda)
Sodium caseinate
Sodium citrate
Sodium propionate
Sodium saccharin
Sodium nitrite/nitrate
Sodium sulfite
Sodium phosphates
Sodium lactate

Scripps Health

Figure 1-6. The Nutrition Facts label provides a standard for comparing foods based on serving size, calories, nutrients, and percent daily value for a 2000 calorie diet.

- **Vitamins and Minerals:** The percentages of vitamin A, vitamin C, calcium, and iron are included on Nutrition Facts labels. It is at the discretion of the manufacturer to list other vitamins and minerals present in a food.
- **Percent Daily Value:** The percent daily value is located to the right of most nutrients on the Nutrition Facts label and is based on the recommended daily allowance for key nutrients. For example, if a food lists vitamin A as 20%, then one serving provides 20% of the vitamin A needed per day. A percent daily value of 5% or less is considered low, whereas a percent daily value of 20% or more is regarded as high. The percent daily value can be used to determine whether a food or beverage item is high or low in nutrients.

Individual Influences

Throughout life, individual influences shape eating behaviors. Some surrender to the desire to eat sweets, while others rely on convenience foods due to hectic schedules. A portion of the population cannot physically tolerate specific foods and ingesting them can lead to adverse reactions. Also, the more an individual knows about the effects of foods on the body, the more likely the individual is to choose nutritious foods. The major factors shaping individual eating behaviors are food preferences, food allergies and intolerances, and nutrition knowledge.

Food Preferences. Individual food preferences strongly influence food choices. For example, some people crave "comfort foods" that have a nostalgic appeal such as pot roast, meatloaf, macaroni and cheese, or apple pie. Other people prefer foods derived from plants and choose to avoid animal-based foods such as meat. A *vegetarian* is an individual who avoids the consumption of animal-based foods and consumes plant-based foods, such as vegetables, fruits, beans, grains, nuts, and seeds, and sometimes dairy products and/or eggs.

The number of individuals choosing to limit, but not completely avoid, animal-based foods is on the rise. **See Figure 1-7.** Many of these individuals follow a flexitarian approach to eating. Flexitarian is a combination of the words "flexible" and "vegetarian." A *flexitarian* is an individual who primarily consumes plant-based foods and consumes animal-based foods, such as meat, dairy, and eggs, only occasionally or in small amounts. Evidence suggests that a flexitarian diet can reduce the risk of cardiovascular disease and some cancers. The interest in plant-based meals has inspired movements such as "Meatless Mondays."

Plant-Based Meals

Tanimura & Antle®

Figure 1-7. The number of individuals choosing to eat more plant-based meals is steadily growing.

Food Allergies. A food allergy may limit what an individual eats. **See Figure 1-8.** A *food allergy* is an adverse reaction to a specific food that involves the immune system. Allergic reactions can be minor or extremely severe. *Anaphylaxis* is a severe allergic reaction in which breathing is prohibited due to narrowing of the airway. There are over 100 foods that can cause allergic reactions. However, eight specific foods are considered major allergens. According to the Food and Drug Administration (FDA), milk, eggs, fish, crustacean shellfish, tree nuts, peanuts, wheat, and soybeans account for 90% of all food allergies.

Food Allergy Cards

SelectWisely

Figure 1-8. A food allergy may limit what an individual eats.

Food Intolerances. A *food intolerance* is an adverse reaction to a specific food that does not involve the immune system. Symptoms of food intolerance usually include bloating, abdominal cramps, nausea, vomiting, and diarrhea. Some individuals may be able to tolerate the problem food in small quantities, while others may need to avoid the food.

Nutrition Knowledge. Individuals who are informed about the benefits of a nutritious diet are more likely to choose healthy foods. Nutrition knowledge is also an important influence for maintaining an appropriate weight and reducing the risk of diseases associated with diets high in sugar, fat, and sodium. When an individual is diagnosed with a dietary-related health issue, such as diabetes or obesity, nutrition knowledge can promote positive eating behaviors that help reduce disease-related symptoms and complications.

Courtesy of The National Pork Board

Colorful foods can be visually appealing, fresh, and flavorful.

Knowledge Check 1-1

1. Explain how social media can influence food choices.
2. Explain how organic food is produced.
3. Describe the product guidelines for "Light/Lite," "Good Source," and "Extra Lean" food label terms.
4. List five examples of processed food.
5. Summarize the information presented on a Nutrition Facts label.
6. Differentiate between vegetarian and flexitarian.
7. List the eight specific foods considered major allergens.

SENSORY PERCEPTIONS OF FOOD

Eating involves the senses, which unite to help detect flavors. *Flavor* is the combined sensory experience of taste and smell along with visual, textural, and temperature sensations. The senses of taste and smell are particularly significant because they are the dominant senses used by the brain to detect flavors. **See Figure 1-9.** Sometimes the scent of a particular food is so strong that simply its smell can generate an assessment of its flavor. For example, the scent of freshly chopped mint can generate a perception of the cool, refreshing taste to come.

Appearance, texture, and temperature can also affect flavor. Brightly colored foods offer visual appeal and are often perceived as being fresh and flavorful. Foods with a thick consistency may not taste as strong as foods with a thin consistency because thin foods easily move across the taste buds. Warm foods often have stronger smells and flavors than cold foods. This is why some foods are most flavorful at room temperature. The way the senses influence food choices requires an understanding of presentation, textures, tastes, aromas, and temperatures.

Chef's Tip

A dish should be evaluated in terms of sight (presentation), smell (aromas), taste (balanced and appealing), touch (textures), and even sound (sizzle or crunch) before presenting it to a guest.

Detecting Flavors

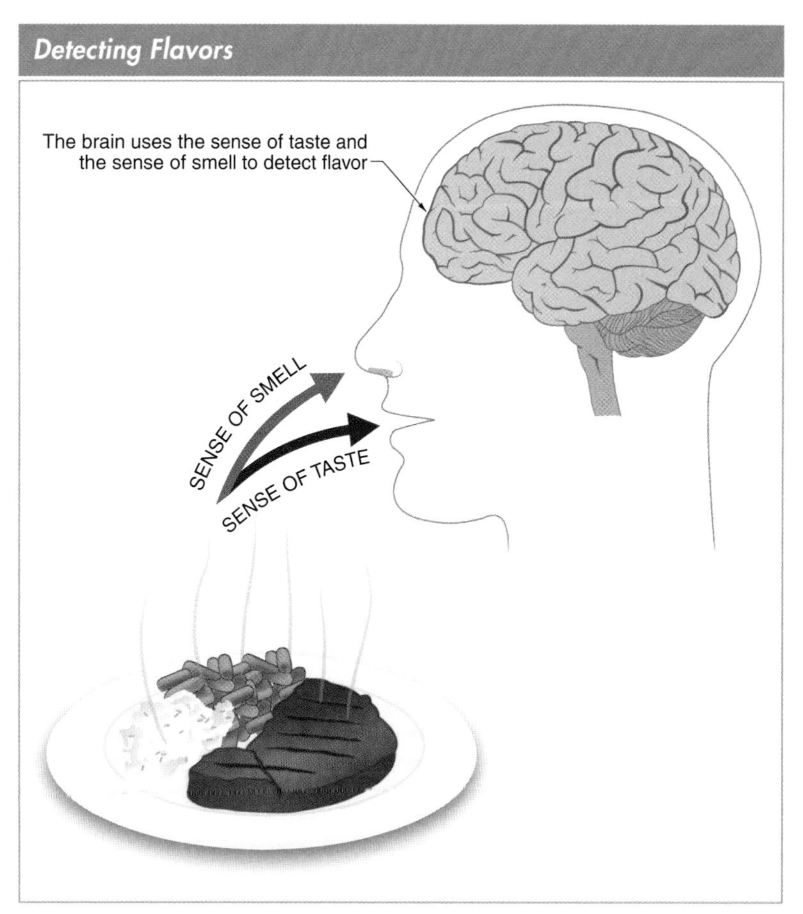

Figure 1-9. The senses of taste and smell are the dominant senses used to detect flavors.

Flavor Perceptions

Flavor perception is a preconceived idea about the taste of a food item. Flavor perception is subjective. Individuals respond differently to foods based on personal experiences such as exposure to various foods, the look, feel, and smell of foods, and even menu descriptions.

Flavor perceptions are also influenced by flavor profiles. A *flavor profile* is the combination of ingredients, such as herbs, spices, and seasonings, associated with a specific cuisine. **See Figure 1-10.** For example, a Mexican flavor profile often includes chile peppers, cumin, cilantro, and lime, whereas an Asian flavor profile may consist of ginger, garlic, lemongrass, and soy sauce. Understanding flavor profiles helps chefs create dishes that correlate to flavor perceptions guests have for certain items.

Presentation

The way food is presented gives a first impression about how the food will taste. If the presentation is appealing, a positive opinion of the dish is likely. If the presentation is not appealing, guests may immediately perceive the food as inadequate. Menu descriptions also create an image of food in the mind. Items should reflect their description in order to meet guest expectations.

The presentation of food can be enhanced with effective plating techniques. *Plating* is the process of arranging food in an appealing manner for presentation. Plating involves the use of various colors, shapes, textures, and heights. **See Figure 1-11.** For example, a dish of roasted vegetables consisting of broccoli florets, red onion wedges, and cubed sweet potatoes incorporates a variety of colors, shapes, textures, and heights to enhance presentation. A mound of wild rice with almonds presents contrast between the softness of the rice and the crunch of the almonds and complements the roasted vegetables. When a piece of meat is thinly sliced and fanned atop the mounded rice, height is added and a visually appealing plate can be presented. Using plates of various shapes can also enhance presentation.

Common Flavor Profiles

Cuisine	Common Ingredients
Asian	Anise, chile oil, chiles, coconut, garlic, ginger, green onions, lemongrass, sesame oil, soy sauce, star anise
Caribbean	Allspice, bay leaves, black pepper, cardamom, cayenne pepper, cilantro, cinnamon, cloves, coconut milk, coriander, cumin, curry, fenugreek, ginger, hot peppers, oregano, thyme, turmeric, vanilla
French	Bay leaves, black pepper, chervil, chives, fines herbes, garlic, green and pink peppercorns, marjoram, mustard, nutmeg, parsley, rosemary, shallots, tarragon, thyme
German	Allspice, caraway, cinnamon, dill, juniper berries, mustard, onions, vinegar, white pepper
Greek	Cinnamon, dill, garlic, lemon, mint, nutmeg, olives, onions, oregano, tomatoes, yogurt
Indian	Anise, cardamom, chiles, cilantro, cinnamon, coriander, cumin, curry, fennel, garlic, ginger, mint, mustard, nutmeg, saffron, turmeric, yogurt
Italian	Basil, bay leaves, fennel, garlic, marjoram, onions, oregano, parsley, pine nuts, rosemary, tomatoes
Mexican	Chiles, cilantro, cinnamon, cocoa, coriander, cumin, garlic, lime, onions, oregano, vanilla
North African	Cilantro, cinnamon, coriander, cumin, garlic, ginger, mint, red pepper, saffron, turmeric
Scandinavian	Cardamom, dill, lemons, mustard, nutmeg, white pepper

Figure 1-10. A flavor profile is the combination of ingredients associated with a specific cuisine.

Plating

Sullivan University

Composed Salad with Cucumbers, Tomatoes, and Oranges

Barilla America, Inc.

Spaghetti with Fennel and Shrimp

Courtesy of The National Pork Board

Roasted Pork Loin with Pineapple Relish

Sullivan University

Roasted Lamb with Lentils

Figure 1-11. Plating involves the use of various colors, shapes, textures, and heights.

Daniel NYC

Presentation can be enhanced by arranging food in an appealing manner.

Textures

Texture is the appearance and feel of an item. The texture of food can range from soft and velvety to hard and brittle or from crispy and crunchy to sticky and chewy. In the professional kitchen, texture is often referred to as mouthfeel. *Mouthfeel* is the way a food or beverage feels in the mouth.

The appearance of food gives the senses a preview of its texture. For example, the sight of a puréed butternut squash soup alerts the mouth to the thick, silky smooth texture to follow. A soup in which butternut squash is cut into cubes creates a chunky appearance and a different mouthfeel. Using a variety of textures can also add interest to the dining experience. For example, topping puréed butternut squash soup with toasted pumpkin seeds adds an element of crunch and an appealing contrast to the smooth soup.

Texture can also be evaluated when food comes in contact with a utensil or the fingers. **See Figure 1-12.** Meat that yields easily to a knife indicates the texture will be appealing, whereas meat requiring great effort to cut readies the mouth for a tough, chewy texture.

Tastes

Taste is a sense that is activated by receptor cells that make up the taste buds. Taste buds are located throughout the entire mouth but are visible as the small bumps, known as papillae, that cover the tongue. The sense of taste is triggered when food is moistened by saliva and comes into contact with taste buds. The number of taste buds varies among individuals, adding to the subjective nature of flavor. **See Figure 1-13.** For example, some people have a less than average number of taste buds and may find the flavor of foods bland. In contrast, some people, who are referred to as supertasters, have a greater than average number of taste buds, making the flavor of foods more pronounced.

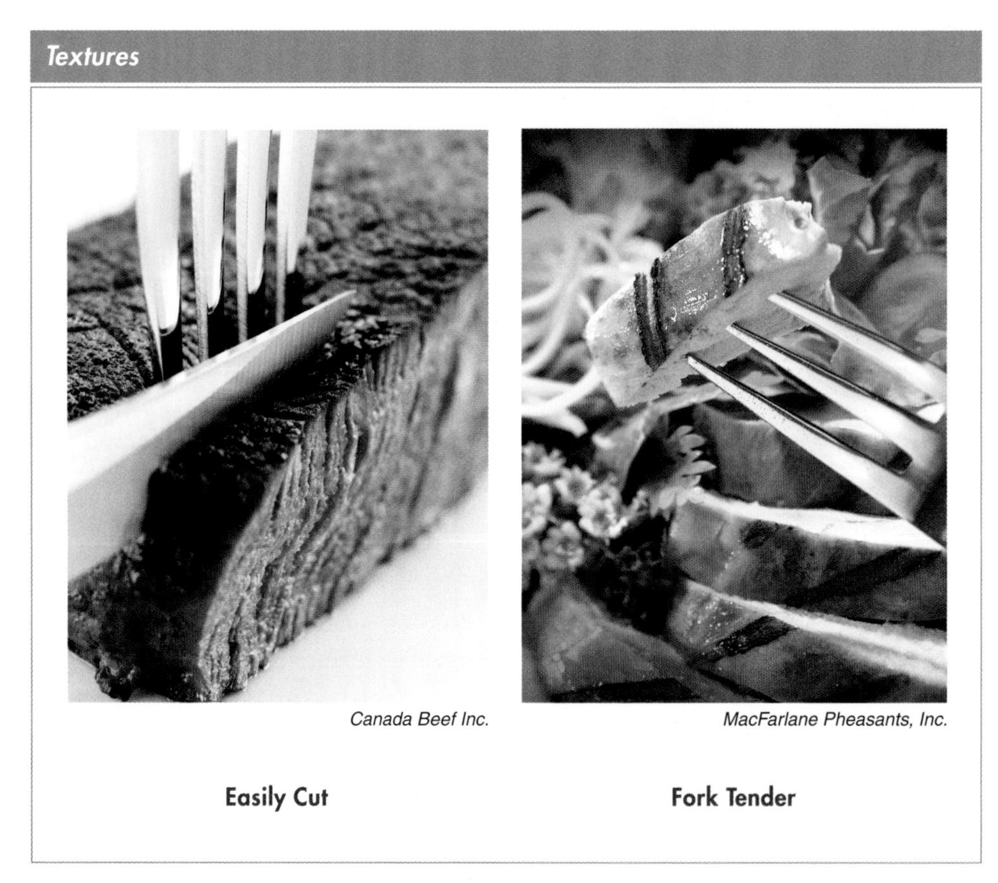

Figure 1-12. Texture can be evaluated when food comes in contact with a utensil.

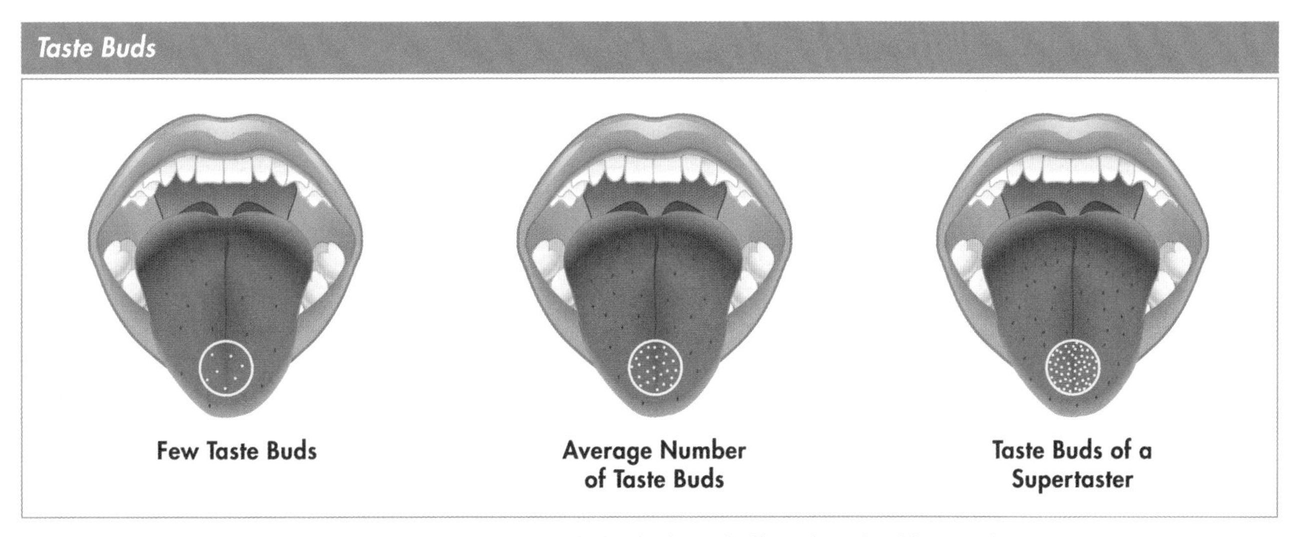

Figure 1-13. The number of taste buds varies among individuals and affects how food is tasted.

Research supports the idea that tastes can be categorized as sweet, salty, sour, bitter, or umami. **See Figure 1-14.** It was once thought that only specific regions of the tongue detected each of these tastes. For example, sweet tastes were believed to be sensed at the middle of the tongue, salty at the tip, sour at the sides, bitter near the back, and umami at the posterior of the tongue. While it is true that there are some areas of the tongue where certain tastes are more predominant, each taste is detectable wherever there are taste receptor cells, including the roof and sides of the mouth. Also, in addition to the five primary tastes, scientific studies theorize that receptor cells may be able to detect other tastes as well.

Sweet. Sweet tastes come from the sugars in foods. Foods that taste sweet are commonly regarded as desirable. Sugars are found naturally in foods such as fruits and milk. They are also commonly added to foods such as baked goods to enhance sweetness.

Salty. Salty tastes come from the salt (sodium chloride) in foods. Some foods, such as sea vegetables, have an inherently salty taste. Salt has also been used for centuries to enhance the flavor of food. The use of various products, such as iodized salt, sea salt, fish sauce, soy sauce, Worcestershire sauce, or Parmesan cheese, can heighten the salty taste of a dish.

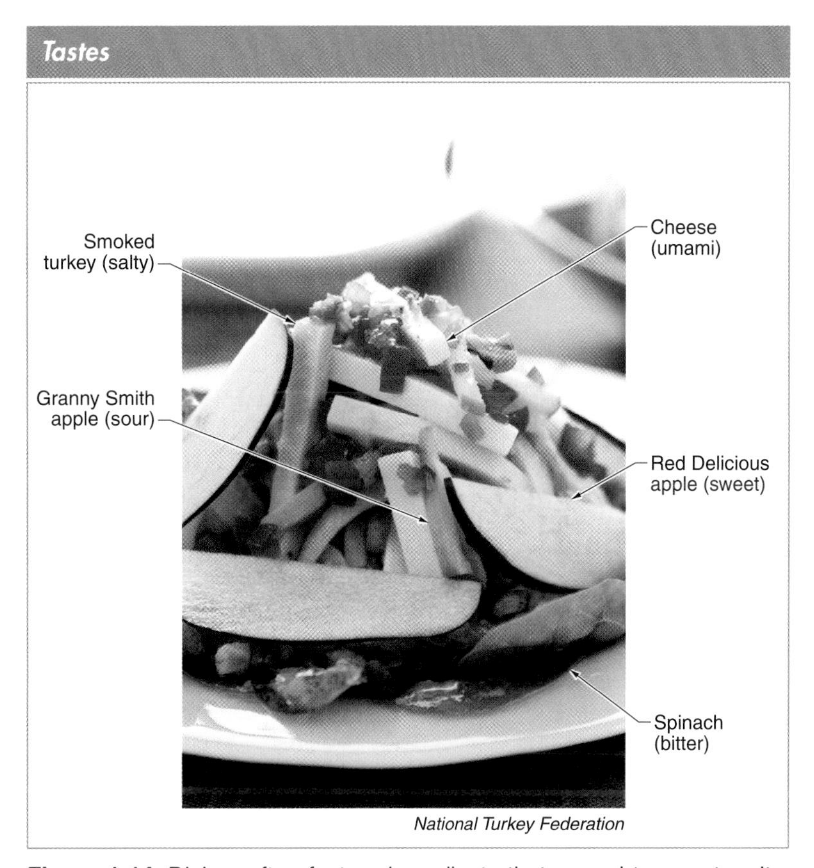

Figure 1-14. Dishes often feature ingredients that appeal to sweet, salty, sour, bitter, and umami tastes.

Sour. Sour tastes come from the acids in foods. Acidic foods have low pH levels. A *pH level* is the measurement of the acidity or alkalinity found in a substance based on a scale of 1 to 14. **See Figure 1-15.** The number seven represents a food with equal amounts of acidity and alkalinity. Lower numbers indicate higher acidity. Acidic items include citrus fruits, such as lemons and limes, and vinegars. Sour foods have the unique ability to naturally enhance salty tastes without adding salt.

Bitter. Bitter tastes come from the alkalies in foods. High alkaline foods have a pH level above seven on the pH scale. The higher the alkalinity, the higher the pH level. Many poisonous plants are high in alkalies. The extreme bitterness of poisonous plants is associated with a human survival mechanism to prevent the consumption of a toxic substance. Alkaline foods, such as spinach, chocolate, and coffee, contain bitter components but are safe to eat.

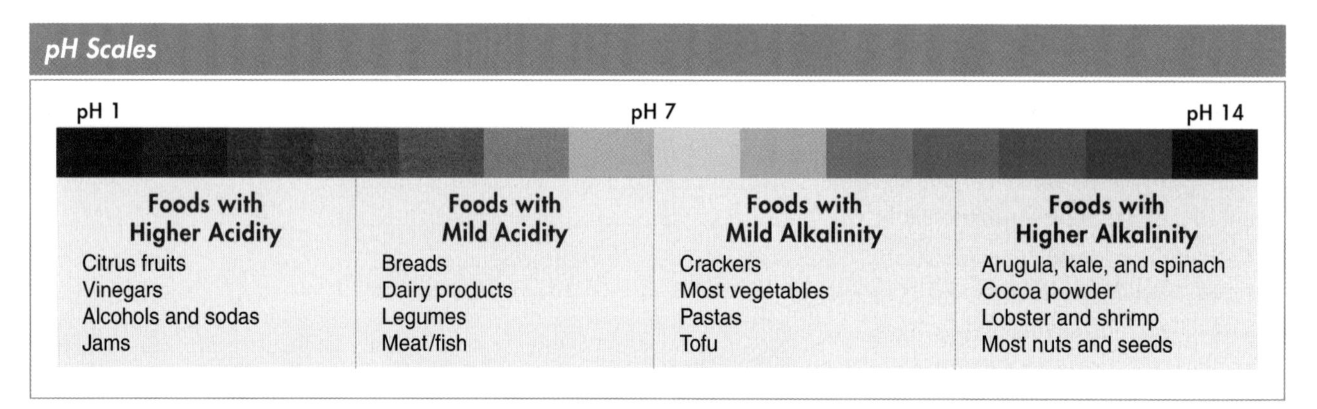

Figure 1-15. A pH scale can be used to measure the acidity or alkalinity of foods and beverages.

FOOD SCIENCE EXPERIMENT: Discovering Supertasters 1-1

Objective

- To determine if an individual is a supertaster.

Materials and Equipment

- Cotton swab
- Blue food coloring
- Reinforcement ring used for notepaper
- Magnifying glass

Procedure

1. Dip the cotton swab in blue food coloring and swab onto tip of the tongue.
2. Place a reinforcement ring on top of the blue-stained portion of the tongue.
3. Place a magnifying glass up to the tongue and count the pink dots that appear inside the reinforcement ring.

Typical Results

Supertasters will have 30 or more pink dots.

Food Science Note

Scientists are studying a sixth taste called kokumi. Although kokumi does not have an identifiable flavor, foods containing kokumi compounds, such as onions, garlic, and cheese, activate taste receptors on the tongue that make salty foods taste saltier and sweet foods taste sweeter.

Umami. *Umami* is the Japanese word for "deliciousness" and is a savory taste. Umami tastes come from glutamate found in foods. *Glutamate* is a substance derived from glutamic acid and is associated with the taste of umami. The savory taste of umami is described as earthy and robust. Mushrooms, tomatoes, cheeses, and seaweed have an umami taste.

In addition to occurring in foods naturally, glutamate is also a primary component of monosodium glutamate (MSG). MSG is commonly used to intensify the flavor of savory foods but lacks a pronounced flavor on its own. It is sold individually as MSG or included in various spice mixtures. MSG is also present in many processed foods. Some people experience headaches, dizziness, fatigue, and diarrhea as a result of consuming MSG.

Aromas

An *aroma* is a scent detected by the sense of smell. Flavor is dependent on the reaction between the sense of smell and the taste buds. The aromas greeting guests as they enter a foodservice operation can help set a tone for the dining experience. For example, when guests are welcomed by the aroma of freshly baked bread they often anticipate a great meal. Likewise, the aroma of certain foods prepares the mouth for what food will be eaten. The aroma of a newly peeled orange readies the taste buds for a sweet and slightly sour taste sensation while the aroma of lasagna prepares the taste buds for a savory experience.

Temperatures

Food temperature affects the way the senses detect flavor. Very hot and very cold foods tend to have less detectable flavors. Taste buds react more intensely to foods at room temperature or when slightly warm. For example, cheeses, fruits, and wines generally have more developed flavors at room temperature. **See Figure 1-16.**

Chef's Tip

Using contrasting temperatures, such as serving a chilled grain salad with a warm piece of seared fish, can add to the appeal of a dish.

Temperatures

Wisconsin Milk Marketing Board, Inc.

Figure 1-16. Taste buds react more intensely to foods, such as cheeses, fruits, and wines, at room temperature.

Temperature also affects the aroma of food. Warmer foods typically create a more powerful aroma. When pizza is fresh from the oven, the combination of warm cheese, sauce, seasonings, and toppings releases more aroma into the air than the same pizza in its cold state. The warm pizza has a more full-bodied flavor than cold pizza due to the stronger aroma and because of the taste buds' response to a warmer food.

FOOD SCIENCE EXPERIMENT: Detecting Flavor 1-2

Objective

- To observe the connection between the sense of taste and the sense of smell.

Materials and Equipment

- 5 jelly beans each with a common flavor such as cherry, orange, lemon, lime, or grape
- Blindfold

Procedure

1. Spread the jelly beans on a table.
2. Put the blindfold on and hold the nose closed with one hand.
3. Take a jelly bean with the free hand, place it in the mouth, and chew it while holding the nose.
4. Try to identify the flavor of the jelly bean while still holding the nose closed.
5. Release the hold on the nose and inhale while continuing to chew the jelly bean.
6. Try to identify the flavor of the jelly bean without holding the nose closed.

Typical Results

When the jelly bean is chewed while holding the nose closed, a slightly sweet taste may be detectable. However, after releasing the hold on the nose to allow for the sense of smell, the true flavor of the jelly bean should be identifiable.

Knowledge Check 1-2

1. Explain the meaning of a flavor profile.
2. Identify four elements included in effective plating techniques.
3. Name the term often used in the professional kitchen to refer to texture.
4. Identify two foods that have an umami taste.
5. Explain how the aroma of food can influence the dining experience.
6. Describe how food temperature affects the taste buds.

FLAVOR DEVELOPMENT

Menu items do not sell solely on the basis that they are "healthy." Dishes also need to appeal to the senses with outstanding colors, textures, and flavors. To accomplish this, it is imperative to develop flavors while incorporating sound nutritional principles.

Developing flavor involves the use of complementary ingredients to enhance the natural flavors and pleasing characteristics of food. Complementary ingredients provide balance to a dish. **See Figure 1-17.** For example, lemon pie provides a balance between the contrasting tastes of sour and sweet. The mild flavor of romaine lettuce can be enhanced by balancing complementary ingredients such as olive oil, balsamic vinegar, freshly ground pepper, and a touch of salt. Adding dried cranberries and pecans to the seasoned lettuce will further heighten the pleasing contrasts in color, texture, and flavor.

Balancing Tastes

- Balance sweet tastes with salty, sour, bitter, or spicy ingredients
- Balance salty tastes with sweet or sour ingredients
- Balance sour tastes with sweet, salty, or bitter ingredients
- Balance bitter tastes with sweet, salty, or sour ingredients
- Balance spicy tastes with sweet ingredients

Sweet Ingredients	Salty Ingredients	Sour Ingredients	Bitter Ingredients	Spicy Ingredients
Agave	Capers	Citrus juices	Arugula	Onions
Cinnamon	Celery	Citrus zests	Cocoa	Garlic
Fresh/dried fruits	Fish sauces	Cranberries	Coffee	Ginger
Fruit juices	Olives	Pickles	Cumin	Horseradish
Honeys	Salts	Vinegars	Kale	Hot peppers
Maple syrup	Soy sauces		Spinach	Mustards
Sugars	Worcestershire sauces			Wasabi

Figure 1-17. Complementary ingredients provide balance to a dish.

Developing flavor is also dependent on how a food is prepared. For example, raw onions impart a slight crunch and a pungent flavor. Quickly cooking onions until they are transparent yields a softer texture and milder onion flavor. Slowly cooked onions yield a rich brown color and a sweet and mild flavor.

Layering flavors is an essential component of developing flavor. Layers of flavors are created when different ingredients are added at different times during the preparation and cooking processes. Layering flavors can be as simple as adding freshly chopped cilantro to a bowl of chili just before service. The slight lemon taste of fresh cilantro adds a layer of flavor and a refreshing contrast to the spiciness of the chili.

More complex layering occurs when a variety of ingredients are added at specific stages of the cooking process. For instance, a recipe for red beans and rice might specify to cook onions until brown to incorporate sweetness. Adding garlic, celery, and bell peppers will impart a more savory layer of flavor. An additional layer of flavor can be developed by adding dried basil, oregano, and thyme. Combining this cooked mixture with the red beans, smoked sausage, and vegetable stock, and cooking to a thick consistency creates depth of flavor. Serving the red beans over rice and sprinkling chopped green onions on top appeals to all the senses with a variety of colors, a contrast of textures, and multiple layers of flavor.

The development of optimal flavor relies upon quality, fresh ingredients. The flavor of fresh ingredients can be enhanced through the use of flavorings and seasonings.

National Onion Association

The development of optimal flavor relies upon quality, fresh ingredients.

Flavorings

A *flavoring* is an item that alters or enhances the natural flavor of food. Flavorings often include herbs and spices that can be combined to complement specific foods and develop flavors. **See Figure 1-18.** Flavorings are a highly effective way to develop flavor without adding unwanted ingredients such as sugar, fat, and sodium.

The flavor of a dish can be altered depending on when flavorings are added. For example, herbs and spices are often added to uncooked dishes, such as salad dressings, several hours prior to service to let the flavors fully develop. Flavorings can also define regional cuisines. For example, the spices cardamom and turmeric are cultivated in India and found in many traditional Indian curry dishes. In addition to herbs and spices, flavorings include extracts and alcohols.

Herbs. An *herb* is a flavoring derived from the leaves or stem of a very aromatic plant. **See Figure 1-19.** For example, basil is a popular leaf herb and chives are common stem herbs. Most herbs are grown in temperate climates. Herbs are available fresh, dried, and sometimes frozen. Quality fresh herbs have a bright color, fresh smell, and offer exceptional flavor. Dried herbs have a stronger, more concentrated flavor and can impart a slightly different aroma and flavor than fresh herbs. For this reason, most chefs prefer to use fresh herbs.

Herbs should be added at the appropriate time during the cooking process. For optimal taste, fresh herbs are added at the end of the cooking process. Because dried herbs are not as delicate, they can be added earlier. If substituting dried herbs for fresh herbs, a smaller amount of dried herbs should be used because of their intense flavor.

Herbs should be stored properly to preserve freshness. To prevent browning and wilting, fresh herbs should be wrapped in a damp paper towel, placed in a sealable plastic bag, and refrigerated between 35°F and 45°F. They will typically remain fresh up to five days. Dried herbs should be stored away from light and in an airtight container between 50°F and 70°F. They should be used within six months for best quality.

Flavor Combinations

Food	Flavoring	Food	Flavoring
Beef	Bay leaf, marjoram, nutmeg, sage, thyme	**Potatoes**	Dill, paprika, parsley, sage
Carrots	Cinnamon, cloves, marjoram, nutmeg, rosemary, sage	**Poultry**	Ginger, marjoram, oregano, paprika, poultry seasoning, rosemary, sage, tarragon, thyme
Corn	Cayenne, chervil, chives, cumin, curry powder, paprika, parsley	**Salads**	Basil, celery seed, chervil, chives, cilantro, dill, oregano, rosemary, sage, tarragon, thyme
Eggs	Basil, cilantro, cumin, savory, tarragon, turmeric	**Soups**	Bay leaves, cayenne, chervil, chili powder, cilantro, cumin, curry powder, dill, marjoram, nutmeg, oregano, rosemary, sage, savory, thyme
Fish	Curry powder, dill, dry mustard, lemon zest and juice, marjoram, paprika		
Fruits	Cinnamon, ground cloves, ginger, mace, mint	**Summer squash**	Cloves, curry powder, marjoram, nutmeg, rosemary, sage
Green beans	Dill, curry powder, lemon juice, marjoram, oregano, tarragon, thyme	**Tomatoes**	Basil, bay leaves, dill, marjoram
Lamb	Curry powder, rosemary, mint	**Veal**	Bay leaves, curry powder, ginger, marjoram, oregano
Peas	Ginger, marjoram, parsley, sage	**Winter squash**	Cinnamon, ginger, nutmeg
Pork	Sage, oregano		

Figure 1-18. Flavorings often include herbs and spices that can be combined to complement specific foods and develop flavors.

Herbs

Type	Description	Common Use
Basil Shenandoah Growers	Leaf herb with pointy, green leaves; member of mint family; most common variety is sweet basil; some varieties have hints of lemon, garlic, cinnamon, clove, and chocolate flavor	Pasta, vegetable, seafood, chicken, and egg dishes; tomato sauces; pestos; infused oils
Cilantro Shenandoah Growers	Leaf herb from stem and leaves of coriander plant; added just before serving, as heat will destroy its flavor	Salsas, soups, salads, and sandwiches
Dill (dill weed) Shenandoah Growers	Leaf herb with feathery leaves; member of parsley family; slight licorice flavor	Salads, vegetables, meats, and sauces; pickling
Mint	General term used to describe family of similar leaf herbs, such as peppermint and spearmint; cool, refreshing flavor	Sweet and savory dishes; beverages
Oregano (wild marjoram) Shenandoah Growers	Leaf herb with small, dark-green, slightly curled leaves; member of mint family; pungent, peppery flavor	Italian and Greek dishes
Parsley Tanimura & Antle®	Leaf herb with curly or flat, dark-green leaves; tangy flavor (most predominant in stems)	Soups, salads, stews, sauces, and vegetable dishes; bouquets garnis; garnishing

Figure 1-19. (continued on next page)

Herbs

Type	Description	Common Use
Rosemary  *Shenandoah Growers*	Leaf herb with needlelike leaves; member of the evergreen family; slight mint flavor with scent resembling fresh pine	Grilled or roasted meats and poultry
Sage  *Shenandoah Growers*	Leaf herb with narrow, velvety leaves; member of mint family	Stews, sausages, and bean or tomato preparations
Tarragon  *Shenandoah Growers*	Leaf herb with smooth, slightly elongated leaves; slight licorice flavor	Seafood, poultry, tomato dishes, salads, salad dressings, béarnaise sauces
Thyme 	Leaf herb with very small leaves; member of mint family; some varieties have hints of nutmeg, lemon, mint, and sage flavor	Stews, chowders, poultry, and vegetable dishes; bouquets garnis
Chives 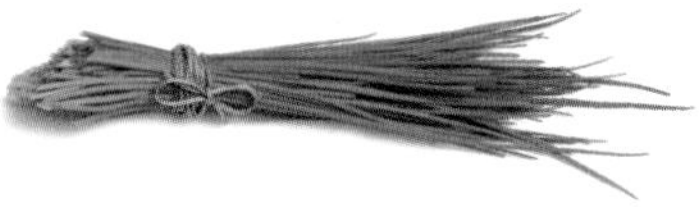 *Barilla America, Inc.*	Stem herb with hollow, grass-shaped sprouts; mild onion flavor	Teas, soups, and Thai dishes
Lemongrass (citronella)  *Shenandoah Growers*	Stem herb with long, thin, gray-green leaves and white scallionlike base; lemon flavor	Teas, soups, and Thai dishes

Figure 1-19. An herb is a flavoring derived from the leaves or stem of a very aromatic plant.

Spices. A *spice* is a flavoring derived from the bark, seeds, roots, flowers, berries, or beans of an aromatic plant. **See Figure 1-20.** For example, cinnamon is produced from bark, cumin from seeds, ginger from roots, capers from unopened flower buds, and peppercorns from berries. In addition, beans, such as cocoa beans and coffee beans, are commonly ground and used as spices. Spices are grown in tropical, humid climates and generally sold in dry form. They can be purchased whole or ground and should have a fresh, strong aroma. Ground spices lose aroma and flavor quicker than whole spices.

Spices are generally added at the beginning of the cooking process to help their flavors permeate the food. Heating spices in oil before adding liquid helps develop aromas and flavors. Like dried herbs, spices should be stored away from light and in an airtight container between 50°F and 70°F. They should also be used within six months.

Wellness Concept

Despite their heat, hot peppers help cool the body by causing perspiration, which has a cooling effect.

Spices

Type	Description	Common Use
Cassia (Chinese cinnamon) The Spice House	Bark spice from bark of small evergreen tree; most spices labeled as cinnamon and sold in United States are actually cassia; bittersweet, warm flavor stronger than cinnamon	Sweet and savory dishes
Cinnamon	Bark spice from dried, thin, inner bark of small evergreen tree; bark is rolled into quills (cinnamon sticks); slightly sweet, warm flavor	Sweet and savory dishes
Coriander	Seed spice from light-brown, ridged seed of coriander plant; combination of lemon, sage, and caraway flavor	Savory dishes and baked goods; global cuisines, especially Caribbean, Indian, Mexican, and North African
Cumin	Seed spice from crescent-shaped seed of plant in parsley family; slightly bitter, warm flavor	Chili; soups; tamales; rice and cheese dishes; curry powders and chili powders
Fennel seeds	Seed spice from oval seed of fennel plant; slight licorice flavor and aroma	Sausages; roasted duck, chicken, and pork dishes

Figure 1-20. (continued on next page)

Spices

Type	Description	Common Use
Mustard seeds  *The Spice House*	Extremely tiny seed from mustard plant; hot, pungent flavor when ground and mixed with hot water	Cabbage, beets, sauerkraut, sauces, salad dressings; pickling
Nutmeg 	Seed spice from oval seed found in yellow, nectarine-shaped fruit of large tropical evergreen; sweet, spicy, warm flavor	Soups, potato dishes, sauces, and desserts
Ginger 	Root spice from large, bumpy root of tropical plant; sold fresh, dried, powdered, crystallized, candied, or pickled; spicy, pungent, and warm flavor	Meat, poultry, and seafood dishes; fruits; desserts
Horseradish 	Root spice from large, brown-skinned root of shrub related to radish; commonly grated; hot, spicy, and pungent flavor	Grilled or roasted meats; seafood dishes; sauces
Capers 	Flower spice from unopened flower bud of shrub; only used after being pickled in strongly salted white vinegar	Fish and poultry dishes; sauces
Cardamom 	Berry spice from dried, immature fruit of tropical bush in ginger family; slight lemon flavor	Curries and pastries; pickling
Paprika	Berry spice from dried, ground sweet red pepper; comes in many varieties including sweet, smoked, and hot	Goulashes; chicken paprika and egg dishes; sauces

Figure 1-20. A spice is a flavoring derived from the bark, seeds, roots, flowers, berries, or beans of an aromatic plant.

Extracts. An *extract* is flavorful oil that has been combined with alcohol. Flavoring oils are obtained from natural sources such as spices, fruits, and nuts. Common extracts include vanilla, lemon, and almond. Extracts provide intense flavors, so typically only a small amount is needed. Adding extracts to an extremely hot or boiling liquid will cause the extract to lose flavor. For this reason, the hot liquid is removed from the heat and allowed to cool several minutes before the extract is incorporated. Extracts should be stored in tightly closed containers in a cool, dark location.

Alcohols. *Alcohol* is a liquid substance produced from fermented fruits or grains and is commonly used as a flavoring or beverage. Alcohols ranging from wines and beers to vodkas and rums are used to develop flavor in a variety of menu items including appetizers, entrées, and desserts. Some of the alcohol content evaporates during the cooking process, leaving the flavor of the alcohol behind. **See Figure 1-21.** However, the amount of alcohol that evaporates depends on factors such as the amount of alcohol used and the length of time the alcohol cooks. The flavor of the alcohol should complement and not overpower the dish. Fruit juice can often be substituted for alcohol if its flavor is appropriate for the dish.

Seasonings

A *seasoning* is an item that intensifies or improves the flavor of food. Salts and peppercorns are the most common seasonings, but seasonings also include oils, vinegars, citrus zests, and juices. Seasonings should be used in moderation so they will not be too obvious or overpowering. They are often added toward the end of the cooking process or at the time of service because their effect is fairly immediate. Acidic seasonings, such as lemon juice, lime juice, and vinegars, are essential for developing healthy dishes because they wake the taste buds and make food taste flavorful. This means less salt is needed to create dishes that are full of taste.

Alcohols

Courtesy of The National Pork Board

Figure 1-21. Some of the alcohol content evaporates during the cooking process, leaving the flavor of the alcohol behind.

Salts. *Salt* is a crystalline solid composed mainly of sodium chloride and is used as a seasoning and a preservative. Salt has the ability to make sweet foods taste sweeter, make sour foods taste more sour, and intensify the flavor of mild foods. It also has a more pronounced flavor in cold dishes, so less is needed. When salt is added prior to cooking or during the cooking process, it is absorbed into the food. Adding more salt after cooking is then often unnecessary because the salt has infused throughout the food.

Salt comes from evaporated sea water or is mined from oceans and dried up sea beds. Common types of salts include iodized salt, sea salt, kosher salt, and pickling salt. Specialty salts are also used in the professional kitchen. Specialty salts are typically mined and vary in color, texture, and flavor depending on where they were excavated.

Chef's Tip

To help impart superior flavor, the extracts and alcohols used for food preparation should be high quality.

Chef's Tip

Salt substitutes are made from potassium or calcium chloride instead of sodium chloride and can give an uncharacteristic flavor to food. To reduce sodium in a dish, a better alternative to salt substitutes is to use little or no salt and develop flavor with spices, herbs, and seasonings.

An abundance of salt is found in processed foods. When foods are made in the professional kitchen using less-processed, whole foods, the overall sodium content can be reduced. In addition, using a coarse grain salt, such as kosher salt, instead of iodized salt may help reduce sodium levels in a dish. Because of its coarse grain, a teaspoon of kosher salt contains less sodium than a teaspoon of iodized salt. **See Figure 1-22.** Salt used in moderation and combined appropriately with additional flavorings and seasonings can be part of nutritionally balanced, flavorful dishes.

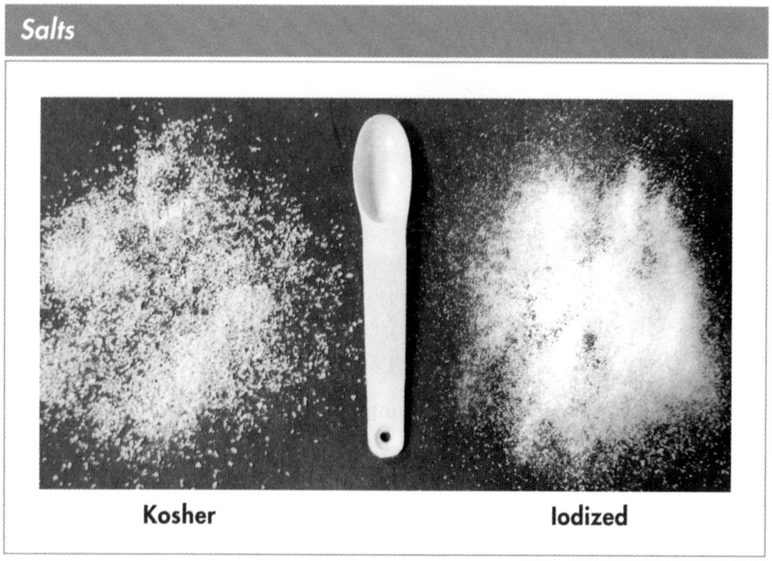

Salts

Kosher | Iodized

Figure 1-22. Because of its coarse grain, a teaspoon of kosher salt contains less sodium than a teaspoon of iodized salt.

Peppercorns. Pepper, derived from peppercorns, is the second most common seasoning found in the professional kitchen. A *peppercorn* is the dried berry of a climbing vine known as the Piper nigrum and is used whole, ground, or crushed. Peppercorns come in various colors such as green, white, and black. The color is dependent upon the ripeness of the peppercorns and how the peppercorns are processed. Colored peppercorns deliver assorted flavors ranging from mild and slightly sweet to spicy and pungent. **See Figure 1-23.**

The Spice House

Figure 1-23. Colored peppercorns deliver assorted flavors ranging from mild and slightly sweet to spicy and pungent.

Oils. Oil is derived from vegetables, nuts, seeds, and fruits. Because oils are plant based, they are often considered a healthy form of fat. The source of the oil determines its flavor. For example, corn oil has a mild corn flavor, whereas the flavor of sesame is apparent in sesame seed oil. Dishes often call for a particular type of oil depending on the flavors desired. Canola and corn oil are typically used when a neutral flavor is sought. Extra-virgin olive oil and walnut oil are commonly used in cold applications, such as salad dressings, where more flavor from the oil is preferred.

Different oils have different smoke points. A *smoke point* is the temperature at which oil begins to smoke and emit an odor. Oils with a high smoke point can tolerate high heat and are used in applications such as deep-frying. All oils have a limited shelf life and should be stored away from light and heat to prolong freshness.

Vinegars. *Vinegar* is a sour, acidic liquid made from fermented alcohol. Wine is commonly used to make vinegar, but other alcohols may be used. Vinegars provide mild to strong flavors that can enrich a dish without the need for salt. Common vinegars used in the professional kitchen include balsamic, Champagne, cider, distilled, malt, rice, sherry, and wine vinegar. Like oils, vinegars should be stored away from heat and light. They should always be clear with no signs of mold or discoloration.

Citrus Zests and Juices. *Zest* is the colored, outermost layer of a citrus fruit peel and contains a high concentration of oil. **See Figure 1-24.** Only the colored layer of the peel should be used because the white pith beneath the peel is bitter. The zest of citrus fruit is more flavorful than its juice and only a small amount is needed to impact flavor.

Citrus Zests

Browne Foodservice

Figure 1-24. Zest is the colored, outermost layer of a citrus fruit peel.

In addition to citrus juice, other types of juices such as apple, cranberry, and pomegranate are used in both sweet and savory applications. Juices can provide a natural sweetness to a dish without having to add extra sugar. Juices are often used to develop flavors in salad dressings and sauces.

Knowledge Check 1-3

1. Describe twelve herbs commonly used in professional kitchens.
2. Describe twelve spices commonly used in professional kitchens.
3. Give three examples of common extracts.
4. Identify the factors that determine alcohol evaporation during cooking.
5. Name the two most common seasonings used in professional kitchens.
6. Explain why a coarser salt may help reduce the amount of sodium in a dish.
7. Identify the flavors associated with green, white, and black peppercorns.
8. Identify four sources from which oils are derived.
9. Identify eight vinegars commonly used in professional kitchens.
10. Explain why only the colored layer of a citrus peel is used for zest.

FLAVOR DEVELOPMENT TECHNIQUES

Techniques that develop flavor rely upon the freshest ingredients available and skillful preparation. When ingredients are of the highest quality, their flavors alone can create a superior tasting dish without incorporating excessive amounts of sugars, fats, and sodium. Preparation techniques used to develop flavor include the Maillard reaction and caramelization as well as the use of rubs, marinades, reductions, stocks, and sauces. **See Figure 1-25.**

Chef's Tip

Cooking butter until it reaches a rich, golden brown color produces a deep, nutty aroma and a more flavorful product. Because the brown butter has more flavor, less needs to be used resulting in a lower fat dish.

Flavor Development Techniques

Type	Description	Nutritious Flavor Development Techniques
Maillard reaction	Reaction that occurs when proteins and sugars in food are exposed to heat and merge together to form brown exterior surface	Brown foods in small amount of oil to limit fats and increase flavor; toast nuts and spices for more intense flavor
Caramelization	Reaction that occurs when sugars are exposed to high heat and produce browning and change in flavor	Reduce added sugars by caramelizing foods to enhance their natural sweetness
Rubs	Blend of ingredients pressed onto surface of uncooked foods, such as meat, poultry, and fish, to impart flavor and sometimes to tenderize	Reduce salt in rubs by using ingredients such as herbs, spices, citrus zest/juice, mustards, cocoa powder, and ground coffee
Marinades	Flavorful liquid used to soak uncooked foods, such as meat, poultry, and fish, to impart flavor and sometimes to tenderize	Replace some oil in marinades with fruit juices, vinegars, or alcohols
Reductions	Gently simmering liquid until it reduces in volume and results in thicker liquid with more concentrated flavor	Reduce fats by making low-fat reduction from fruit juices, alcohols, or vinegars
Stocks	Unthickened liquid flavored by simmering seasonings with vegetables and often the bones of meat, poultry, or fish	Used in place of water to cook foods and enhance flavor
Sauces	Accompaniment served with food to complete or enhance flavor and/or moistness of dish	Replace high-fat butter or cream-based sauces with salsas, chutneys, coulis, and pestos that use fresh fruits, vegetables, flavorings, and seasonings

Figure 1-25. Flavor development techniques can create superior tasting dishes without incorporating excessive amounts of sugars, fats, and sodium.

Maillard Reaction

The *Maillard reaction* is a reaction that occurs when the proteins and sugars in a food are exposed to heat and merge together to form a brown exterior. For example, when meat comes in contact with a hot grill or nuts are toasted in a hot pan, the outside browns, creating color, texture, and a layer of flavor. Poultry, seafood, vegetables, fruits, grains, and seeds are also commonly browned to add flavor and complexity to a dish. **See Figure 1-26.**

Maillard Reaction

Figure 1-26. A variety of foods, including meats, are commonly browned as a result of the Maillard reaction.

Meat and poultry are often seared using the Maillard reaction. *Searing* is the process of using high heat to quickly brown the surface of a food. Searing is done with little or no fat in a very hot sauté pan, hot grill, or griddle. Some foods are seared before cooking, and others may be seared after they are cooked to add color. The entire cooking process may also consist of searing. Searing is an ideal technique for reducing the overall fat content of a dish.

Caramelization

Caramelization is a reaction that occurs when sugars are exposed to high heat and results in browning and a change in flavor. For example, as root vegetables, such as carrots, sweet potatoes, and parsnips, cook uncovered in a hot oven, their sugars caramelize and the vegetables develop a brown color and sweeter flavor.

Many dishes call for caramelized onions or fruits such as apples. This is often accomplished on a stovetop by first sweating the ingredient in a small amount of oil. *Sweating* is the process of slowly cooking food to soften its texture. As the item continues to cook, the sugars begin to darken and the item intensifies in aroma, color, and sweetness. **See Figure 1-27.**

Rubs

A *rub* is a blend of ingredients that is pressed onto the surface of uncooked foods, such as meat, poultry, and fish, to impart flavor and sometimes to tenderize. Rubs can be categorized as either dry rubs or wet rubs. Rub-coated foods should be refrigerated for several hours or overnight to let the flavors absorb fully.

A *dry rub* is a rub made by grinding herbs and spices into a fine powder. The mixture is then rubbed onto a protein prior to cooking. A mortar and pestle set or an electric spice grinder is often used to grind and combine dry rub ingredients. Spice and herb blends, such as Chinese five-spice powder, jerk seasoning, and poultry seasoning, are commonly used in dry rubs to add appealing aromas and flavors. **See Figure 1-28.**

A *wet rub* is a rub made by incorporating wet ingredients, such as Dijon mustard, flavored oils, puréed garlic, or honey, into a dry rub mixture. The mixture is then rubbed onto a protein prior to cooking. Wet rubs adhere to food more thoroughly than dry rubs, resulting in deeper flavors. Thickly coated wet rubs can either be wiped off prior to cooking or left on the food to form a crispy crust during cooking.

Figure 1-27. As the onion cooks, the sugars begin to darken and the onion intensifies in aroma, color, and sweetness.

Spice and Herb Blends

Name of Blend	Common Ingredients
Cajun seasoning	Zesty spice blend; ground chiles, fennel seeds, garlic, oregano, paprika, red and black pepper, salt, and thyme
Chili powder	Ground chiles, cloves, coriander, cumin, garlic, and oregano
Chinese five-spice powder	Equal proportions of ground cinnamon, cloves, fennel seeds, star anise, and Szechuan pepper
Curry powder	Mild to spicy blend; can consist of more than 20 spices including cardamom, cinnamon, cloves, coriander, cumin, fenugreek, ginger, mace, nutmeg, red and black pepper, and turmeric
Fines herbes	Chopped fresh chervil, chives, parsley, and tarragon; may include marjoram, savory, or thyme
Herbes de Provence	Dried basil, fennel seed, lavender, marjoram, rosemary, sage, savory, and thyme
Jerk seasoning	Spicy blend; ground allspice, chiles, cinnamon, cloves, garlic, and ginger
Pickling spice	Whole and coarsely ground allspice, bay leaves, cinnamon, cloves, dill, fennel seeds, ginger, mace, mustard, nutmeg, peppercorns, and red pepper
Poultry seasoning	Black pepper, marjoram, nutmeg, rosemary, sage, and thyme; may include celery salt, mustard powder, or oregano

Figure 1-28. Spice and herb blends are used in dry rubs to add appealing aromas and flavors.

Marinades

A *marinade* is a flavorful liquid used to soak uncooked foods, such as meat, poultry, and fish, to impart flavor and sometimes to tenderize. Marinades often consist of acidic bases to which flavorings and seasonings are added. **See Figure 1-29.** The acidic ingredient is what tenderizes the food. Oil is also a common ingredient in marinades. To promote optimal flavor development, the food item should be completely submerged in the marinade. Less tender cuts of meat can be submerged for several hours or overnight. The marinating time should be reduced to prevent a mushy texture if a delicate food item, such as fish, is to be marinated.

Food Safety

A nonaluminum container should be used to marinate foods because acidic ingredients react unfavorably with aluminum. Marinades should never be reused because of possible bacterial contamination.

Reductions

A *reduction* is a thick liquid with a flavor that has become concentrated as a result of being gently simmered until reduced in volume. **See Figure 1-30.** Many liquids can be reduced including stocks, soups, and sauces. Even if a reduction, such as a sauce, is high in sugar, fat, and/or sodium, it can still be used as a component of a healthy dish. This is because the intense flavor of a reduced sauce means only a small amount is needed to significantly impact the overall flavor of a dish. Alcohols, juices, and vinegars can also be made into flavorful reductions that are low in fat.

Stocks

A *stock* is an unthickened liquid that is flavored by simmering seasonings with vegetables, and often, the bones of meat, poultry, or fish. Stocks are composed of water, a flavoring component, mirepoix, and aromatics. Bones are the primary flavoring component of meat, poultry, and fish stocks. A *mirepoix* is a roughly cut mixture consisting of 50% onions, 25% celery, and 25% carrots. An *aromatic* is an ingredient added to a food to enhance its natural aromas and flavors.

Figure 1-29. Marinades often consist of acidic bases to which flavorings and seasonings are added.

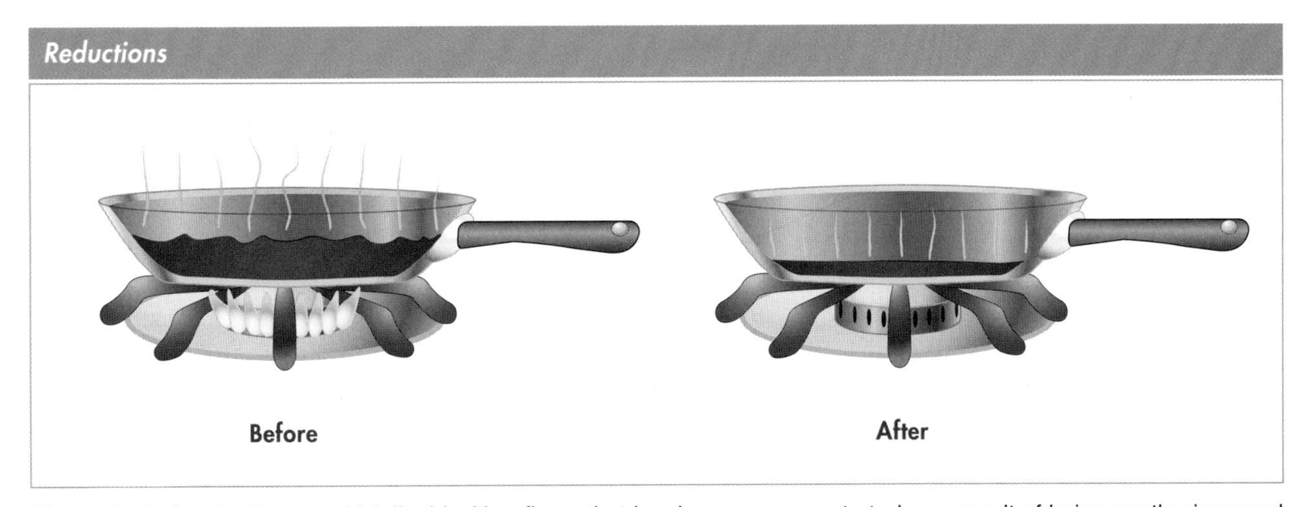

Figure 1-30. A reduction is a thick liquid with a flavor that has become concentrated as a result of being gently simmered until reduced in volume.

An aromatic is added to a stock with a bouquet garni or a sachet d'épices. A *bouquet garni* is a bundle of herbs and vegetables tied together with twine that is used to flavor stocks and soups. The bundle is tied to the handle of the stockpot so that it can be easily retrieved. A *sachet d'épices* is a mixture of spices and herbs placed in a piece of cheesecloth and tied with butcher's twine. **See Figure 1-31.** Tying or placing aromatics in cheesecloth allows them to be removed from a stock easily.

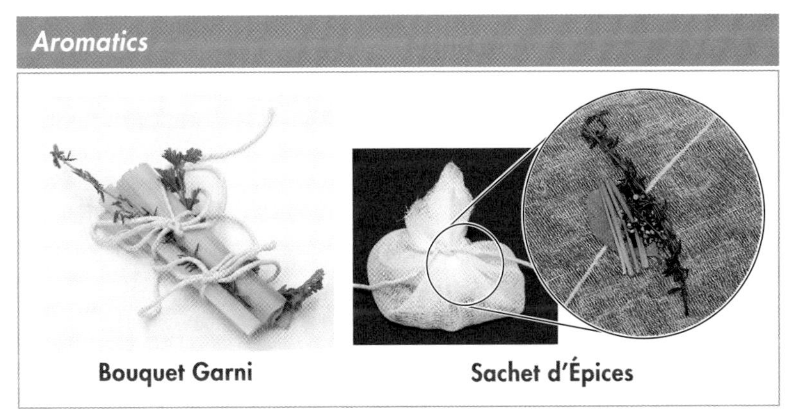

Figure 1-31. A bouquet garni and a sachet d'épices are made with aromatics that enhance the aroma and flavor of a stock.

Classical Sauces

A *sauce* is an accompaniment that is served with food to complete a dish or enhance the flavor and/or moistness of a dish. Sauces are used to create extraordinary flavor possibilities in dishes ranging from appetizers to desserts. However, many sauces are high in fat and calories. These sauces can still have a place in nutritious dishes when used in moderation.

Some sauces are based on mother sauces. A mother sauce is one of five classical sauces including béchamel (milk-based), espagnole (brown-stock-based), tomato (tomato-based), hollandaise (egg-based), and velouté (white-stock-based). Most mother sauces are composed of flavorful liquids, thickening agents (such as a roux) flavorings, and seasonings. **See Figure 1-32.**

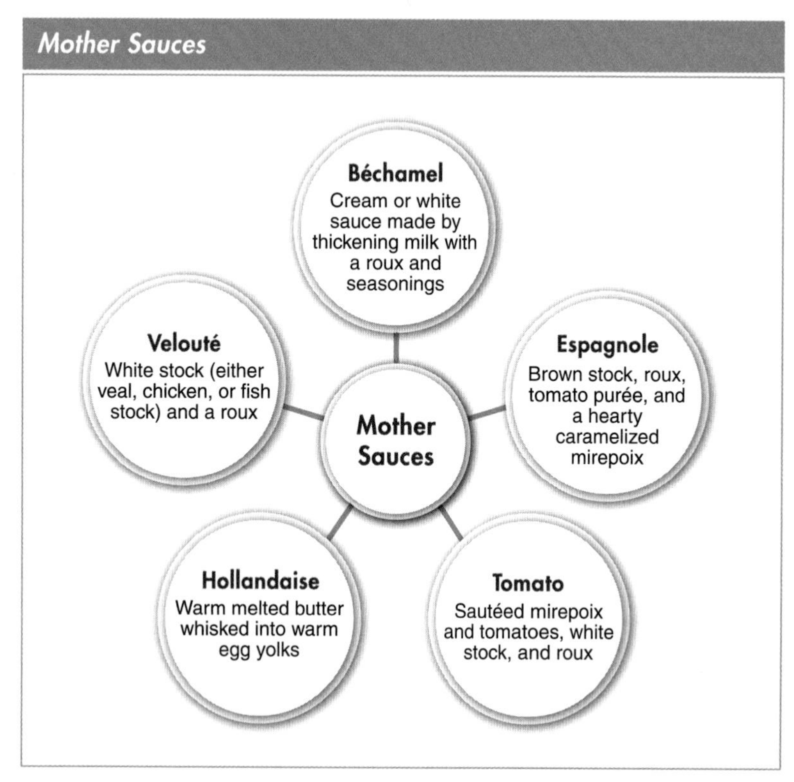

Figure 1-32. Most mother sauces are composed of flavorful liquids, thickening agents, flavorings, and seasonings.

Many sauces, including mother sauces, can be high in fat because they contain thickening agents. A *thickening agent* is a substance that adds body to a hot liquid. Classical thickening agents, as well as gums and stabilizers, can enhance the texture, viscosity, mouthfeel, and overall quality of a dish.

Classical Thickening Agents. Classical thickening agents are generally added to soups and sauces to develop a rich full-bodied consistency. Two classical thickening agents are rouxs and liaisons.

A *roux* is a thickening agent made by cooking a mixture of equal amounts, by weight, of flour and fat and is used to thicken soups and sauces. A *liaison* is a thickening agent that is a mixture of egg yolks and heavy cream used to thicken sauces. A liaison is typically used for cream- or milk-based dishes, such as custards, puddings, and rich cream sauces.

Gums and Stabilizers. A *gum* is a thickening agent derived from plant flours. Starches, such as cornstarch, arrowroot, tapioca starch, granulated tapioca, kudzu starch, and modified food starch, are derived from plants and, therefore, are classified as gums. When used as thickening agents, starches undergo gelatinization. *Gelatinization* is the process in which a heated starch absorbs moisture and swells, which thickens a liquid. Other gums commonly used include guar gum and xanthan gum. Gums, in addition to sauces, are used in many processed foods to improve their overall quality.

A *stabilizer* is a substance added to food to help maintain the suspension of one liquid in another. For example, eggs are used as a stabilizer to keep mayonnaise from separating. Like gums, stabilizers are used in many processed foods. Peanut butter often has stabilizers added to prevent oil separation. Common stabilizers include lecithin, carrageenan, and gelatin. **See Figure 1-33.**

Gums and Stabilizers

Type	Description	Common Use
Cornstarch	Powderlike starch derived from corn that produces smooth, shiny, and opaque appearance	Sauces
Arrowroot	Powderlike starch derived from roots of tropical tuber called arrowroot that produces smooth, shiny, and translucent appearance	Sauces
Tapioca starch	Powderlike starch derived from roots of cassava plant	Sauces
Granulated tapioca	Granulated starch derived from roots of cassava plant	Tapioca pudding; pie fillings; soups; sauces
Kudzu starch	Powderlike starch derived from roots of Asian tuber called kudzu	Soups; sauces
Modified food starch	Starch that has been physically or chemically altered from its original state	Protects processed foods from temperature changes; maintains food textures; increases or decreases viscosity of liquid
Guar gum	Powderlike substance derived from seeds of guar plant	Baked goods
Xanthan gum	Sticky substance derived from corn sugar	Processed salad dressings; condiments
Lecithin	Oil-like stabilizer derived from soybeans, legumes, peas, egg yolks, or yeast	Salad dressings; commercial nonstick oil sprays; prevents splattering when frying; prevents sugar crystallization in chocolates; increases shelf life of processed foods
Carrageenan	Powderlike stabilizer derived from seaweed	Fruit-based products; breading for frying; sweet doughs; vegetarian alternative to gelatin
Gelatin	Powderlike stabilizer, also available in sheet form, derived from animal bones and hide	Jellies; molded desserts and salads; cold soups; confectionery items

Figure 1-33. Gums and stabilizers are often used to improve the overall quality of a dish.

Contemporary Sauces

Contemporary sauces are often viewed as healthy alternatives to many high-fat sauces that rely on butter and cream. **See Figure 1-34.** This is because contemporary sauces are often composed of vegetables and fruits, making them low in fat and high in nutritional value. Contemporary sauces can be chunky mixtures to smooth purées and range in flavors from sweet to savory. Contemporary sauces include salsas, chutneys, coulis, and pestos.

Contemporary Sauces

Courtesy of The National Pork Board

Frittata with Tomato Salsa

Courtesy of The National Pork Board

Pork Chop with Cranberry Chutney

Harbor Seafood, Inc.

Shrimp with Pesto

Figure 1-34. Contemporary sauces, including salsas, chutneys, and pestos, are often healthy alternatives to sauces high in fats.

A *salsa* is a sauce made by mixing or puréeing diced vegetables or fruits, flavorings, and seasonings together. **See Figure 1-35.** Salsas can be made from raw or cooked ingredients that are left coarse or puréed. Their flavors range from sweet and mild to hot and pungent.

A *chutney* is a sauce typically made by cooking fruits with flavorings and seasonings. **See Figure 1-36.** Chutneys can be chunky or smooth in texture and range from mild to spicy in flavor.

A *coulis* is a sauce typically made from either raw or cooked puréed fruits or vegetables. Coulis can be warm or cold and served with various menu items including appetizers, entrées, and desserts. Vegetables are commonly puréed with stock to make a savory coulis, whereas a sweet coulis typically consists of frozen fruit puréed with a sugar-based liquid.

Salsas

Common Salsa Ingredients	
Vegetables and Fruits	Flavorings and Seasonings
Avocados	Basil
Bell peppers	Chili peppers
Black beans	Chili powder
Corn	Chives
Mangoes	Cilantro
Nectarines	Citrus zest/juice
Onions/shallots	Cumin
Oranges	Garlic
Papayas	Mint
Peaches	Olive oil
Pineapples	Parsley
Tomatillos	Red pepper flakes
Tomatoes	

Types of Salsas

- Black beans, corn, onions, lime juice, garlic, chipotle peppers, and parsley
- Mangoes, red and green bell peppers, onions, lime juice, jalapeno peppers, and mint
- Oranges, pineapples, shallots, orange zest, lemon zest and juice, basil, and olive oil
- Tomatillos, avocados, onions, lime zest and juice, garlic, cilantro, and chives
- Tomatoes, red and green bell peppers, onions, lime juice, garlic, jalapeno peppers, chili powder, and cilantro

Vita-Mix® Corporation

Figure 1-35. Salsas are made by mixing or puréeing diced vegetables or fruits, flavorings, and seasonings together.

Chutneys

Common Chutney Ingredients		Types of Chutneys
Fruits	Flavorings and Seasonings	
Apricots	Cardamom	• Apricots, raisins, pistachios, apple cider vinegar, and cardamom
Cherries	Cilantro	• Cherries, sherry vinegar, mustard seeds, and parsley
Cranberries	Cinnamon	• Cranberries, walnuts, red wine vinegar, and orange zest
Mangoes	Citrus zest/juice	• Oranges, pineapples, Champagne vinegar, cilantro, and mint
Oranges	Garlic	• Mangoes, onions, and rice vinegar
Pineapples	Mint	
Raisins	Mustard seeds	
	Nutmeg	
	Nuts	
	Onions	
	Poppy seeds	

Courtesy of The National Pork Board

Figure 1-36. Chutneys are typically made by cooking fruits with flavorings and seasonings.

A *pesto* is a sauce made from fresh ingredients that have been crushed with a mortar and pestle or finely chopped in a food processor before being mixed with oil. **See Figure 1-37.** Pestos are traditionally made with fresh basil, garlic, lemon juice, pine nuts, parmesan cheese, and olive oil. Chefs often modify pestos with an array of ingredients such as mint, cilantro, and walnuts. Although commonly served with pastas, pesto also makes an excellent accompaniment to seafood, poultry, and eggs.

In addition to being served as sauces, contemporary sauces may be used as condiments. A *condiment* is a savory, sweet, spicy, or salty accompaniment added to or served with a food to impart a particular flavor that will complement the dish. **See Figure 1-38.** Common condiments include cocktail sauces, ketchups, mayonnaises, mustards, soy sauces, tartar sauces, Worcestershire sauces, and yogurts.

Pestos

Common Pesto Ingredients		Types of Pestos
Almonds	Olive oil	• Basil leaves, garlic, lemon juice, pine nuts, Parmesan cheese, and olive oil
Arugula leaves	Olives	• Basil leaves, sun-dried tomatoes, garlic, Parmesan cheese, and olive oil
Asiago cheese	Nut oils	• Spinach, chives, lemon zest and juice, pine nuts, asiago cheese, and olive oil
Basil leaves	Parmesan cheese	• Cilantro, garlic, lime juice, almonds, and almond oil
Capers	Parsley	• Mint leaves, garlic, lemon juice, pistachios, and olive oil
Chile peppers	Pine nuts	
Chives	Pistachios	
Cilantro	Spinach	
Citrus zest/juice	Sun-dried tomatoes	
Garlic	Walnuts	
Mint leaves		

Figure 1-37. Pestos are made from fresh ingredients that have been crushed with a mortar and pestle or finely chopped in a food processor.

Condiments*

Type	Description	Total Fat	Sugar	Sodium
Cocktail sauces	Thick tomato-based product that usually contains vinegar, sugar, salt, spices, grated horseradish, hot sauce, Worcestershire sauce, and lemon juice	1 g	7 g	590 mg
Ketchups	Thick, tomato-based product that usually includes vinegar, sugar, salt, and spices	0 g	14 g	668 mg
Mayonnaises	Thick, uncooked emulsion formed by combining oil with egg yolks and vinegar or lemon juice	20 g	4 g	418 mg
Mustards	Pungent powder or paste made from the seeds of the mustard plant	3 g	1 g	707 mg
Soy sauces	Type of Asian sauce made from mashed soybeans, wheat, salt, and water	0 g	1 g	3594 mg
Tartar sauces	Mayonnaise-based sauce that contains finely chopped relish or pickles, capers, grated onions or minced chives, and fresh parsley	10 g	3 g	400 mg
Worcestershire sauces	Type of sauce traditionally made with anchovies, garlic, onion, lime, molasses, tamarind, and vinegar	0 g	7 g	674 mg
Yogurts	Semisolid food made from milk that has fermented due to the addition of bacteria	2 g	1 g	28 mg

* based on 2 oz serving size

Figure 1-38. Condiments impart a savory, sweet, spicy, or salty flavor to foods.

Care should be taken when using condiments since some contain high amounts of sugars, fats, and sodium. For example, ketchup-based sauces are often high in both sugars and sodium. Sour cream-based sauces are high in fats, and soy sauces and Worcestershire sauces are high in sodium. However, many condiments are available with reduced sugars, fats, and sodium, and are often used to replace higher sugar, fat, and sodium varieties.

Knowledge Check 1-4

1. Explain the Maillard reaction.
2. Describe the changes in color and flavor that result from caramelization.
3. Differentiate between a dry rub and a wet rub.
4. Identify three ingredients commonly found in marinades.
5. Explain why a reduction that is high in sugar, fat, and/or sodium can still be part of a healthy dish.
6. Describe two common ways in which aromatics are added to stocks.
7. Identify the five classical mother sauces.
8. Describe two classical thickening agents.
9. Describe eight common gums and three common stabilizers.
10. Describe four contemporary sauces.

COOKING FOOD

Cooking is the process of heating foods in order to make them taste better, make them easier to digest, and kill harmful microorganisms that may be present in the food. Heat is necessary to cook foods effectively and is transferred to food in different ways. The three different methods of heat transfer used with food are conduction, convection, and radiation.

Heat Transfer Methods

Heat is energy that is transferred between two objects or substances of different temperatures. Heat typically flows from a warmer material to a cooler material. When heat is transferred, the particles inside an object increase their movement as the temperature increases. Conduction, convection, and radiation heat transfer methods each move heat in a different manner and produce different results.

Conduction. *Conduction* is a type of heat transfer method in which heat passes from one object to another through direct contact. **See Figure 1-39.** The heat source warms the cookware, which transfers heat to the exterior of the food. The exterior molecules then transfer heat to the interior of the food to complete the cooking process. A range, flattop, and frying pan are examples of equipment that transfer heat directly to food.

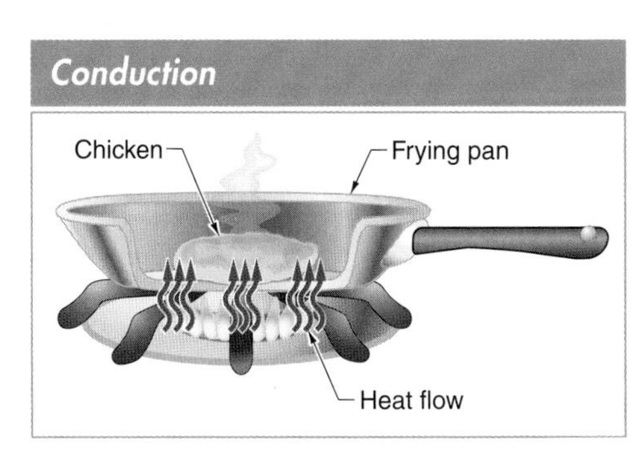

Figure 1-39. Conduction is a type of heat transfer method in which heat passes from one object to another through direct contact.

Convection. *Convection* is a type of heat transfer method that occurs due to the circular movement of a fluid or gas. **See Figure 1-40.** Heating a pot of water is an example of convection. The pot itself becomes heated by conduction. As the water heats, it expands and rises up through the surrounding cooler water, causing the cooler water to sink to the bottom of the pot where it will become heated. A fryer works using this same principle, but it uses fat instead of water as the cooking medium. This process provides a consistent temperature, which promotes even cooking.

Convection ovens also cook by convection. The hot circulating air in a convection oven transfers heat faster than the still air in a conventional oven. For this reason, convection ovens require a slightly reduced temperature than a conventional oven.

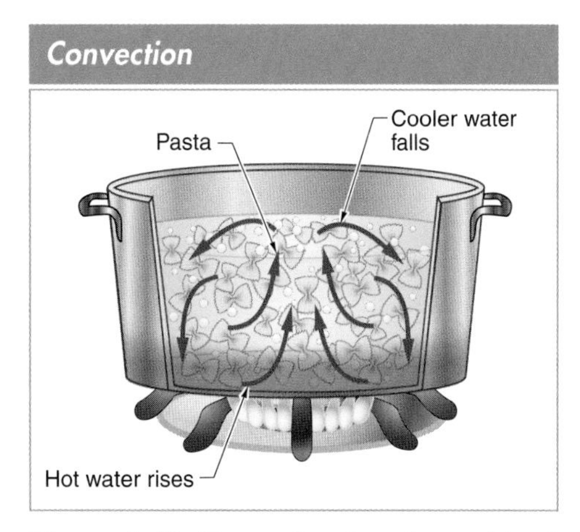

Figure 1-40. Convection is a type of heat transfer method that occurs due to the circular movement of a fluid or gas.

Radiation. *Radiation* is a type of heat transfer method in which electromagnetic waves transfer energy. When electromagnetic waves strike an object, the energy carried by the waves is transferred to the object. Types of radiation used to heat food include infrared, microwave, and induction radiation. **See Figure 1-41.**

Infrared radiation involves the use of an electric or ceramic heating element. The heating element becomes very hot and transfers the heat to food without direct contact. A broiler is an example of cooking equipment that uses infrared radiation.

Microwave radiation uses electromagnetic waves to heat the sugar, fat, and water molecules in food. The heated molecules vibrate, producing friction. Friction creates heat and cooks the food. Microwaves cook foods quickly and are efficient for reheating items. However, they are generally not effective for browning foods.

Induction radiation uses electromagnetic current to heat magnetic cookware. The cookware used on an induction cooktop must be flat-bottomed and made of a material that will conduct electricity, such as cast iron or magnetic stainless steel. The cookware heats the food and the cooktop remains cool. Induction cooking is rapid, and when the pan is removed from the cooktop, cooking slows quickly.

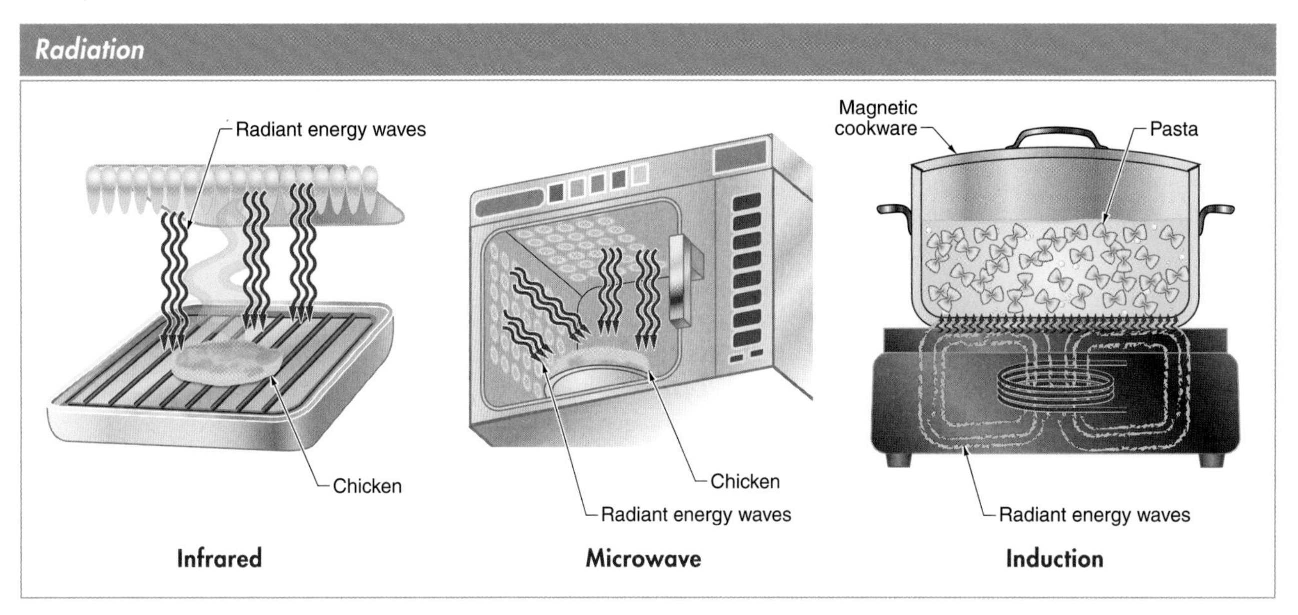

Figure 1-41. Radiation used to heat foods includes infrared, microwave, and induction.

Healthy Cooking Techniques

Healthy cooking techniques are those that preserve nutrients and limit added sugars, fats, and sodium. Some cooking techniques, such as grilling or steaming, are commonly regarded as healthier than other techniques that rely on fat to cook foods. However, all cooking techniques have a place in the development of nutritious dishes.

To cook nutritious dishes, exposure to heat and the cooking period should be considered. For example, when vegetables are cooked in a hot liquid, some vitamins can leach into the liquid. Heat can also increase the nutrient value of certain foods. For example, cooking tomatoes increases the compound lycopene, which is associated with reducing the risk of heart disease and some cancers.

The cooking medium can also significantly impact the nutritional outcome of a dish. **See Figure 1-42.** Deep-fried foods are cooked in a large amount of fat and are, therefore, typically higher in fat than grilled or roasted foods. However, deep-fried foods can be used to create healthy dishes when used in moderation. For example, a light sprinkling of deep-fried onions atop an otherwise healthy salad can still be nutritionally balanced because the portion of high-fat onions is limited.

With careful planning, quality ingredients, and an awareness of how dry-heat, moist-heat, and combination cooking will affect foods, chefs have the ability to shape positive eating behaviors by offering nutritious dishes filled with inspiring flavors.

Dry-Heat Cooking. *Dry-heat cooking* is any cooking technique that uses hot air, hot metal, a flame, or hot fat to conduct heat and brown food. **See Figure 1-43.** Tender meats, ground meats, poultry, seafood, and some fruits, vegetables, and breads are often cooked using dry-heat cooking techniques. Dry-heat cooking techniques include grilling, griddling, broiling, smoking, barbequing, roasting, baking, sautéing, stir-frying, pan-frying, and deep-fat frying.

Chef's Tip

A French griddle can be used to help produce a healthy dish. A French griddle has raised ridges that create grill marks where the food touches the ridges. The ridged design collects fat that drips from the food as it cooks, keeping fat away from the food.

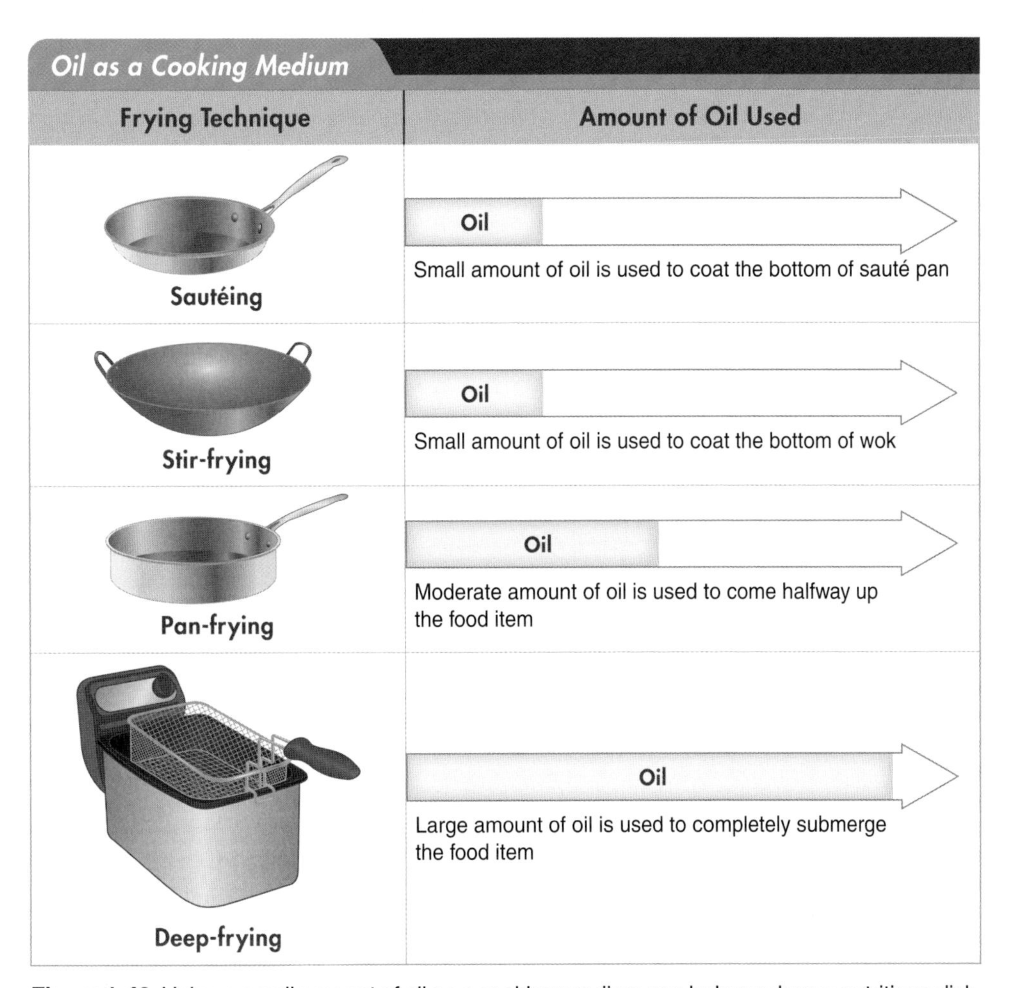

Figure 1-42. Using a small amount of oil as a cooking medium can help produce a nutritious dish.

Dry-Heat Cooking

Technique	Description	Common Foods Prepared Using Cooking Technique	Considered Healthy
Grilling	Food is cooked on open grates above a direct heat source to yield a smoky flavor with visually appealing char lines	• Meats • Poultry • Seafood • Some fruits and vegetables • Breads	**Yes;** small amount of fat is added to grill to prevent sticking
Griddling	Food is cooked on a solid metal cooking surface called a griddle	• Pancakes • French toast • Eggs • Hot sandwiches	**Yes;** small amount of fat is added to the griddle to prevent sticking *Note:* Griddling is often used with higher-fat foods, so consideration should be given to food item being griddled
Broiling	Food is cooked directly under or over a heat source	• Meats • Poultry • Seafood • Some fruits and vegetables • Breads • Foods topped with items, such as cheese, butter, and sugar, where melting is desired	**Yes;** fat drips away from food as it cooks *Note:* Using broiler to melt items can cause sugars, fats, and/or sodium to absorb into food or accumulate on surface

Figure 1-43. (continued on next page)

Dry-Heat Cooking

Technique	Description	Common Foods Prepared Using Cooking Technique	Considered Healthy
Smoking	Food is flavored, cooked, or preserved by exposing it to the smoke from burning or smoldering plant materials, most often wood	• Meats • Poultry • Seafood • Cheeses	**Yes;** small amount of fat may need to be added to smoking equipment to prevent sticking
Barbequing	Food is slowly cooked over hot coals or burning wood to yield a smoky flavor; sauces are commonly applied to barbequed foods	• Meats • Poultry	**Yes;** small amount of fat may need to be added to barbequing equipment to prevent sticking *Note:* Sauces can add extra sugars, fats, and/or sodium
Roasting	Food item that is typically savory is cooked uncovered at a high temperature in an oven or on a revolving spit over an open flame; roasted foods are commonly basted	• Large cuts of meat • Whole poultry • Some fruits • Root vegetables	**Yes;** small amount of fat may need to be added to roasting cookware to prevent sticking *Note:* Basting food as it roasts can add extra sugars, fats, and/or sodium
Baking	Food is cooked in an enclosed environment surrounded by dry, hot air	• Fabricated poultry • Seafood • Meatballs • Pastas • Casseroles • Breads • Cookies • Cakes • Pies and pastries	**Yes;** small amount of fat may need to be added to baking cookware to prevent sticking *Note:* Baking is healthiest when used to prepare lean meats, poultry, and seafood, whole grain bread products, and bakery items that limit sugars, fats, and sodium
Sautéing	Food is cooked quickly in sauté pan over direct heat using small amount of fat	• Meats • Poultry • Seafood • Fruits • Vegetables	**Yes;** small amount of fat needs to be added to pan to prevent sticking
Stir-frying	Food is cooked quickly in wok over direct heat using small amount of fat while constantly stirring items	• Meats • Poultry • Seafood • Vegetables	**Yes;** small amount of fat needs to be added to wok to prevent sticking *Note:* Quick cooking allows vegetables to retain maximum nutrients
Pan-frying	Food is often breaded and cooked in pan of hot fat; also known as shallow-frying	• Thin pieces of meat • Poultry • Seafood • Sliced vegetables	**No;** moderate amount of fat is used (enough to come halfway up thickness of item to be cooked)
Deep-frying	Food is completely submerged in hot oil heated to between 350°F and 375°F	• Meats • Poultry • Seafood • Some fruits • Vegetables • Donuts and fritters	**No;** large amount of fat is used (enough to completely submerge item to be cooked)

Figure 1-43. Dry-heat cooking techniques use hot air, hot metal, flames, or hot fat to conduct heat and brown food.

Chef's Tip

Pan drippings often remain after sautéing foods. These drippings can be reduced to make a flavorful sauce, which adds depth of flavor to food.

Moist-Heat Cooking. *Moist-heat cooking* is any cooking technique that uses liquid or steam as a cooking medium. **See Figure 1-44.** Flavoring the liquid with herbs and spices and using stocks or wine for the liquid can enhance both color and flavor. During the moist-heat cooking process, any accumulated fat can be skimmed from the top of the cooking liquid. Some foods can also be chilled to allow fat to raise and solidify on the surface where it can then be removed. Moist-heat cooking techniques, including poaching, simmering, boiling, blanching, and steaming, can produce nutritious dishes when lean food is cooked in a liquid that is low in sugar, fat, and sodium.

Moist-Heat Cooking

Technique	Description	Common Foods Prepared Using Cooking Technique	Considered Healthy
Poaching	• Food is cooked in liquid that is held between 160°F and 180°F • Small, motionless bubbles appearing on bottom of pan help indicate proper poaching temperature	• Eggs • Fish • Some fruits	**Yes;** fat is not required *Note:* Liquid used for poaching can add extra sugars, fats, and/or sodium
Simmering	• Food is gently cooked in liquid that is between 185°F and 205°F • Gently rising bubbles that do not break into full boil at top of pan indicates proper simmering temperature	• Meats • Poultry • Fish • Potatoes • Grains • Soups, stocks, and sauces	**Yes;** fat is not required *Note:* Liquid used for simmering as well as ingredients used to prepare soups, stocks, and sauces can add extra sugars, fats, and/or sodium
Boiling	• Food is cooked by heating liquid to its boiling point • Bubbles that rapidly rise from bottom of pan and break at surface indicate proper boiling temperature	• Pastas • Potatoes • Reductions	**Yes;** fat is not required *Note:* Reductions can add extra sugars, fats, and/or sodium
Blanching	• Food is briefly parcooked in boiling water and then shocked by being placed in ice-cold water to stop the cooking process • Reduces bitterness, preserves color and texture, and loosens skins on foods, such as tomatoes and nuts, for easy removal	• Vegetables • Fruits • Nuts	**Yes;** fat is not required *Note:* Quick cooking allows vegetables to retain maximum nutrients
Steaming	Food is placed in container that prevents steam from escaping	• Meats • Poultry • Seafood • Vegetables	**Yes;** fat is not required *Note:* Food is not submerged in liquid so nutrients remain inside food instead of leaching into cooking liquid

Figure 1-44. Moist-heat cooking uses liquid or steam as a cooking medium.

Chef's Tip

En papillote is an effective steaming technique in which food, such as seafood or poultry, is steamed in a parchment paper package as it bakes in an oven. Before the food item is enclosed in parchment paper, seasonings, vegetables, and a small amount of liquid, such as citrus juice or wine, are typically added.

Combination Cooking. *Combination cooking* is any cooking technique that uses both moist and dry heat. **See Figure 1-45.** In general, tougher cuts of meat, poultry, and seafood are cooked with a combination of moist and dry heat to help tenderize the food and develop flavors. Moist heat is used as foods are typically cooked in a liquid. Moist heat is also used in the form of steam. For example, when food is placed in a pot filled with liquid and a lid is placed on the pot, the steam created during cooking stays inside the pot and helps to cook and tenderize the food. Dry heat is used with combination cooking to brown foods, usually by searing the food. Foods can be browned before or after the moist-heat cooking process depending on the dish and the desired outcome. Combination cooking techniques include braising, stewing, poêléing, and sous vide.

Combination Cooking

Technique	Description	Common Foods Prepared Using Cooking Technique	Considered Healthy
Braising	• Food is seared in fat and then cooked, tightly covered, in small amount of liquid for a long period of time • Root vegetables and hearty greens, such as kale or cabbage, are often added • Liquid, such as stock, is added so that it comes halfway up food being braised	• Larger cuts of meat • Poultry	**Yes/No** *Note:* Healthy dish can be achieved by braising lean cuts of meat or poultry, removing fat that rises to surface, and using small amount of nutrient rich cooking liquid as sauce
Stewing	• Bite-sized pieces of food are barely covered with liquid and simmered for long period of time in tightly covered pot • Liquid, such as stock, is added so that it covers food being stewed	• Meats • Poultry • Seafood • Vegetables • Fruits	**Yes/No** *Note:* Healthy dish can be achieved by stewing lean cuts of meat or poultry, removing fat that rises to surface, and serving appropriate portion size
Poêléing	• Often referred to as butter roasting • Food item is placed atop aromatic vegetables in pot, brushed with butter, and then covered and placed in oven to steam • Food can be browned before or after it is poêléed	• Poultry	**No;** brushing food with butter contributes a significant amount of fat *Note:* Lower fat dish can be achieved by removing skin from poultry
Sous vide	• Process of cooking vacuum-sealed food by maintaining low temperature and warming food gradually to set temperature in water bath • Food is trimmed, marinated, or seared and cooled, and placed into pouch with desired flavorings before being sealed	• Meats • Poultry • Seafood • Eggs • Vegetables	**Yes;** fat is not required

Figure 1-45. Combination cooking uses both moist and dry heat.

Knowledge Check 1-5

1. Differentiate between conduction and convection.
2. Name the three types of radiation used with food.
3. Describe nine dry-heat cooking techniques that are considered healthy.
4. Explain why pan-frying and deep-frying are not considered healthy cooking techniques.
5. Describe five moist-heat cooking techniques that are considered healthy.
6. Summarize four combination cooking techniques.
7. Explain how braising and stewing can be used to prepare a healthy dish.

Chapter Summary

There are many factors contributing to the foods individuals choose to eat including demographic, sustainable, social, consumer, and individual influences. The way in which the senses perceive food also plays a significant role in the foods consumed. Pleasing all the senses with enticing presentations, textures, tastes, aromas, and temperatures creates interest and enhances the appeal of a dish. In addition, nutritious flavor development in conjunction with healthy cooking techniques can promote menu offerings filled with inspiring dishes that focus on healthy preparations and quality ingredients.

Eloma Combi Ovens

Chapter Review

Digital Resources
ATPeResources.com/QuickLinks
Access Code: 267412

1. Identify three primary influences that shape food choices.
2. Describe sustainable agriculture practices.
3. Differentiate between a food allergy and a food intolerance.
4. Identify the two main senses needed to detect flavor.
5. Give an example of a common flavor profile.
6. Explain how the presentation of food can influence a first impression.
7. Identify five textures associated with food.
8. Describe the five primary tastes recognized by the taste buds.
9. Explain why sour foods can help reduce the amount of sodium in a dish.
10. Explain how temperature can affect the aroma of food.
11. Explain how layers of flavors are created.
12. Differentiate between a flavoring and a seasoning.
13. Describe eight common flavor development techniques.
14. Describe the three heat transfer methods used to cook food.
15. Differentiate between dry-heat and moist-heat cooking techniques.
16. Describe the sous vide combination cooking technique.

Nutrition Fundamentals

Fresh Produce

Barilla America, Inc.

Lean Proteins

National Cancer Institute

Whole Grains

House Foods

Extensive research in the science of nutrition indicates that what a person eats can have a direct impact on health and well-being. This has led to an increased desire by the consumer for meals that not only satisfy in terms of flavor but are healthy as well. Foodservice professionals who understand the basics of nutrition can prepare or modify existing recipes into healthy dishes without compromising flavor in order to meet and exceed guest expectations for nutritious meals.

Chapter Objectives

1. Describe the main functions of each nutrient.
2. Summarize the process of digestion.
3. Describe how energy requirements influence caloric needs.
4. Summarize the *Dietary Guidelines for Americans.*
5. Describe the dietary recommendations for the five food groups.
6. Explain the connection between diet and health related diseases.
7. Describe the role of menus.
8. Summarize the culinary nutrition recipe-modification process.

NUTRITION

Nutrition is the science of how the body receives and uses the substances found in food. The science of nutrition is a relatively new field and evolves daily as scientists learn new details about how the nutrients in food affect the human body and overall health.

A *nutrient* is a substance that provides nourishment to the body. Nutrients are responsible for providing energy and help the body grow, maintain, and repair its bones, muscles, organs, and cells. The body requires six essential nutrients. An *essential nutrient* is a nutrient the body cannot make in sufficient amounts to meet its needs and must be obtained from food. The six essential nutrients are proteins, carbohydrates, lipids, water, vitamins, and minerals. **See Figure 2-1.**

Nutrients are classified as organic or inorganic based on whether they contain the element carbon. An *organic nutrient* is a nutrient that contains carbon. Proteins, carbohydrates, lipids, and vitamins are organic nutrients. An *inorganic nutrient* is a nutrient that does not contain carbon. Water and minerals are inorganic nutrients. Essential nutrients are also classified as either macronutrients or micronutrients depending on the amount that is needed by the body.

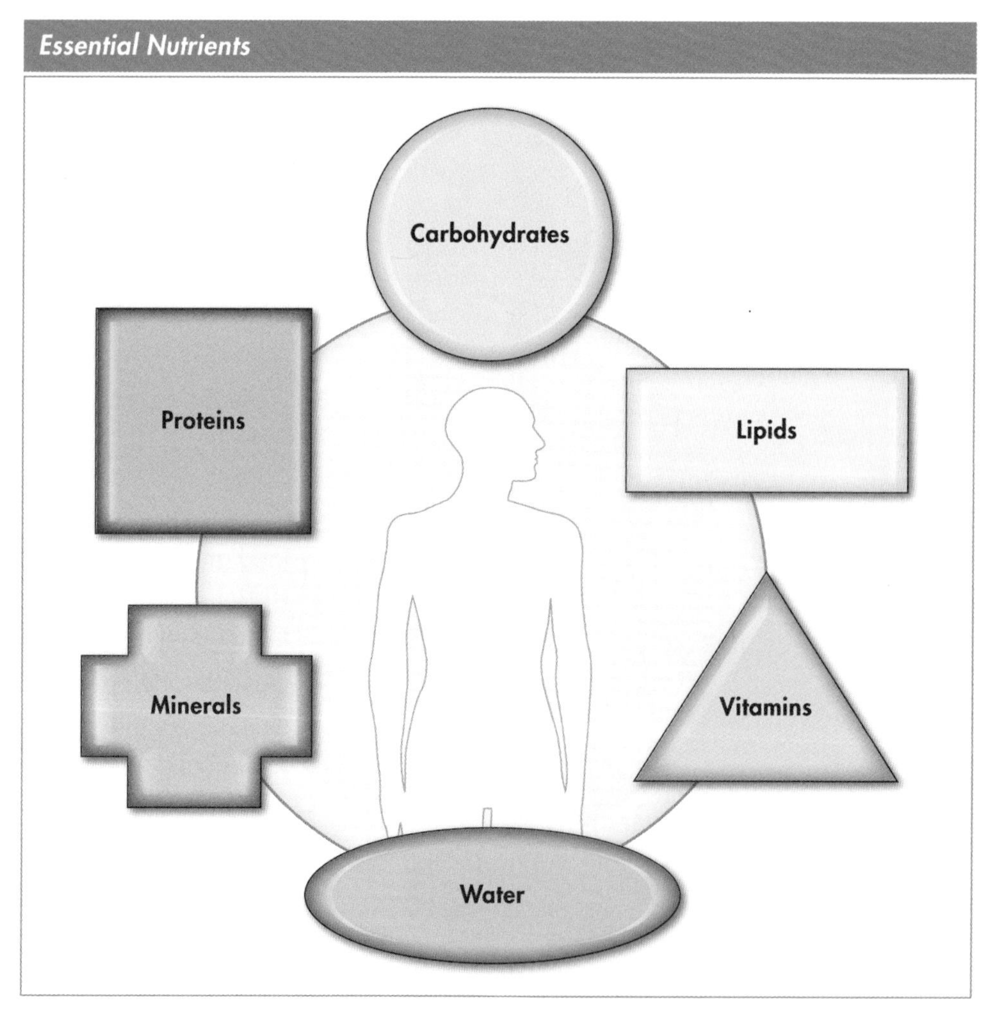

Figure 2-1. The body requires six essential nutrients in the form of proteins, carbohydrates, lipids, water, vitamins, and minerals.

Macronutrients

A *macronutrient* is a nutrient needed by the body in large amounts. Macronutrients include proteins, carbohydrates, lipids, and water. Proteins, carbohydrates, and lipids are needed in large amounts to supply energy for the body. According to the acceptable macronutrient distribution range (AMDR) set by the Food and Nutrition Board, a well-balanced diet should contain approximately 10%–35% proteins, 45%–65% carbohydrates, and 20%–35% lipids. **See Figure 2-2.** Although water does not supply energy, it is considered a macronutrient because the body requires a large amount (approximately 2½ qt daily) to function properly.

Proteins. A *protein* is an energy-providing nutrient made of carbon, hydrogen, oxygen, and nitrogen assembled in chains of amino acids. Amino acids are responsible for building and assembling specialized proteins within the body and are therefore commonly referred to as "the building blocks of proteins." Some of the key functions of protein include building muscle, providing cellular structure for organs and tissues, transporting molecules, and supplying energy. **See Figure 2-3.** Proteins are constantly broken down and rebuilt by the body, making it essential to consume protein foods daily.

Photo Courtesy of the Beef Checkoff

It is essential to consume protein foods, such as meat, daily.

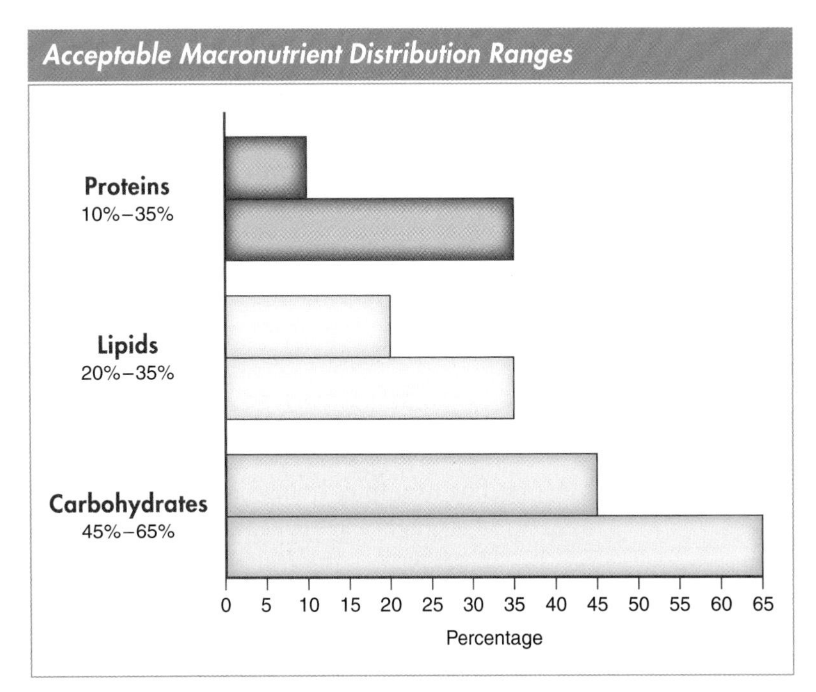

Figure 2-2. A well-balanced diet should contain approximately 10%–35% proteins, 45%–65% carbohydrates, and 20%–35% lipids.

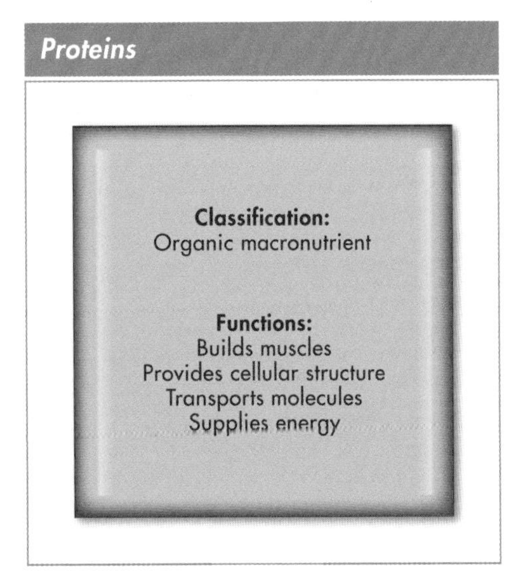

Figure 2-3. Some of the key functions of protein include building muscles, providing cellular structure, transporting molecules, and supplying energy.

Wellness Concept

Protein requirements may be higher during times of illness or for individuals who routinely participate in vigorous exercise regimens.

Carbohydrates. A *carbohydrate* is an energy-providing nutrient in the form of sugar, starch, or dietary fiber and is the main source of energy for the body. **See Figure 2-4.** *Dietary fiber* is an indigestible form of carbohydrate found in vegetables, fruits, and grains. Because carbohydrates are the preferred energy source of the body, they should make up the greatest percentage of the diet. The type of carbohydrate consumed determines how the body uses it for energy. For example, honey is a type of sugar that enters and exits the bloodstream very quickly, resulting in short-term energy. In contrast, lentils are a type of starch that take longer for the body to digest and absorb, resulting in sustained energy.

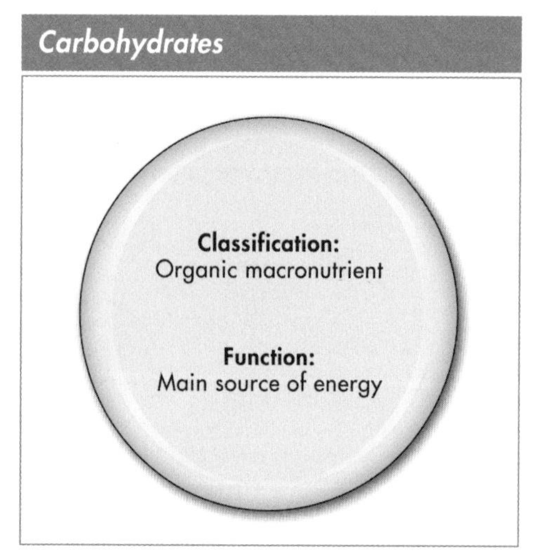

Figure 2-4. Carbohydrates are energy-providing nutrients in the form of sugar, starch, or dietary fiber and the main source of energy for the body.

Lipids. A *lipid* is an energy-providing nutrient made from fatty acids and includes solid fats and oils. Lipids come from both animal and plant sources, and they are also produced naturally within the human body. Lipids from animal sources are known as saturated fats and often referred to as solid fats because they are solid at room temperature. Lipids from plant sources are known as unsaturated fats and often referred to as liquid fats because they are liquid at room temperature. Unsaturated fats are considered a healthy form of lipids. Essential functions of lipids include the ability to provide insulation, manufacture hormones, and cushion organs. When the body has used its supply of carbohydrates for energy, lipids are the next energy source. **See Figure 2-5.** Unused lipids are stored as body fat.

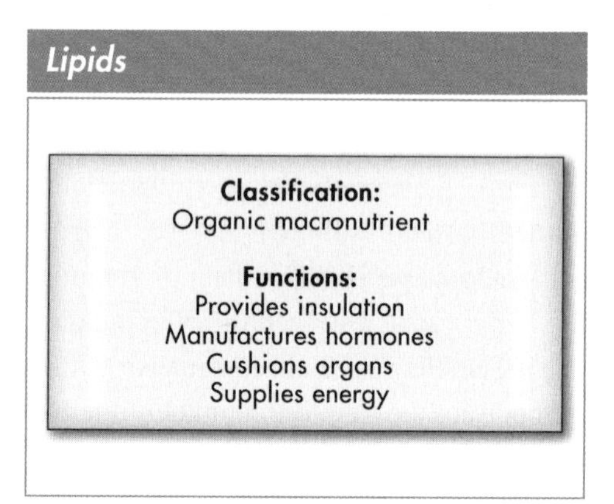

Figure 2-5. Essential functions of lipids include the ability to provide insulation, manufacture hormones, cushion organs, and supply energy when needed.

Water. Water is believed to be the most important nutrient. Water is necessary for just about every process that occurs within the body including transporting nutrients, carrying away waste, providing moisture, and normalizing body temperature. **See Figure 2-6.** Water is contained both inside and outside of the cells and comprises up to 60% of the human body.

Food Science Note

Most experts recommend consuming 8 to 10 cups of water daily to replenish fluids lost throughout the course of a day. Part of this goal can be met by eating fresh vegetables and fruits, which typically contain between 75% and 90% water.

Water

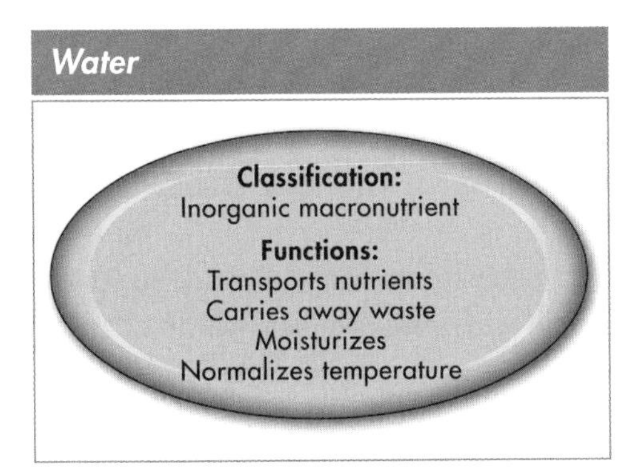

Figure 2-6. Water is necessary for just about every process that occurs within the body.

Micronutrients

A *micronutrient* is a nutrient needed by the body in small amounts. Micronutrients include vitamins and minerals. Micronutrients do not supply energy to the body, but they do play an essential role in maintaining and regulating body functions.

Vitamins. A *vitamin* is an organic nutrient that is required in small amounts to help regulate body processes. **See Figure 2-7.** Vitamins help the body process proteins, carbohydrates, and lipids; play an important role in regulating chemical reactions; and help build and maintain the structural components of the body. Because vitamins are non-energy-yielding nutrients, they are present in foods in much smaller quantities than macronutrients.

Vitamins

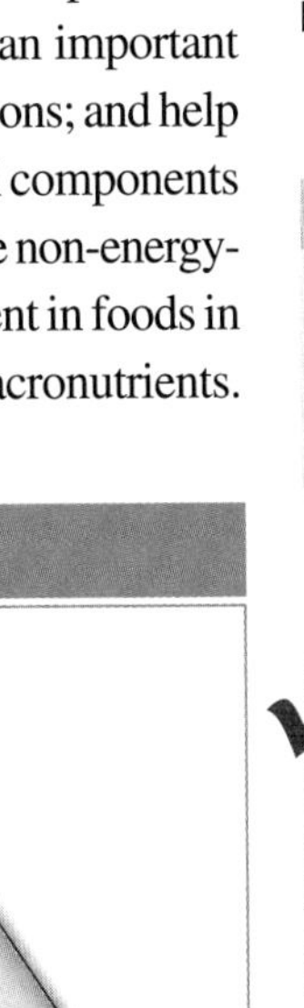

Figure 2-7. Vitamins are composed of organic substances and required in small amounts to help regulate body processes.

Minerals. A *mineral* is an inorganic nutrient that is required in small amounts to help regulate body processes. **See Figure 2-8.** Functions of minerals in the body include helping to regulate fluid balance, muscle contractions, and metabolism. Like vitamins, minerals are non-energy-yielding nutrients needed in smaller quantities than macronutrients.

Minerals

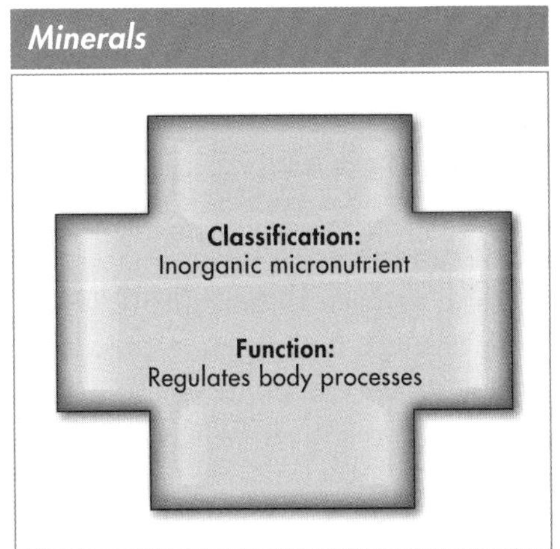

Figure 2-8. Minerals are inorganic substances required in very small amounts to help regulate body processes.

Wellness Concept

A well-balanced diet helps ensure that the body is supplied with an adequate amount of nutrients to function properly.

Knowledge Check 2-1

1. Differentiate between a macronutrient and a micronutrient.
2. Identify the four macronutrients.
3. Identify the AMDR for a well-balanced diet.
4. Describe two functions of proteins in the body.
5. Explain the main function of carbohydrates.
6. Identify the three main sources of dietary fiber.
7. Describe two functions of lipids in the body.
8. Describe two functions of water in the body.
9. Differentiate between a vitamin and a mineral.

DIGESTION

Digestion is the process the human body uses to break down food into a form that can be absorbed and used or excreted. Digestion begins in the mouth when food is ingested. As food passes through the digestive tract, it is broken down by both mechanical and chemical processes and changes form. Once nutrients are broken down, they can be used for energy production, cellular repair, and immune functions or stored for later use. The parts of food for which the body has no use are excreted as waste.

Digestive Tract

A main component of the digestive system is the digestive tract. The *digestive tract,* also known as the gastrointestinal (GI) tract or alimentary canal, is a hollow tube that serves as the passageway for food to move from the mouth through the esophagus, stomach, small and large intestines, rectum, and anus. **See Figure 2-9.** As food moves along this route, organs, such as the pancreas and liver, help convert nutrients into compounds that can be used for energy or to build and maintain cells.

Each part of the digestive tract performs a specific function. Some functions are mechanical such as chewing food to break it into smaller pieces. Some functions rely on chemical substances, such as saliva, to dissolve food into usable parts. Although all components of the digestive system are essential, the mouth, stomach, small intestine, and large intestine play a key role in the digestive process.

Mouth. Digestion begins in the mouth with the mechanical action of chewing along with the chemical action of enzymes present in saliva. An *enzyme* is a type of protein that acts as a catalyst to accelerate chemical reactions. *Saliva* is the watery fluid secreted by the salivary glands in the mouth that moistens food and contains digestive enzymes.

Saliva performs two key digestive functions. First, saliva moistens and helps compress food into a bolus. A *bolus* is a compacted mass of food that has been mixed with saliva and is ready to be swallowed. Second, saliva contains enzymes that initiate the digestion of carbohydrates. *Amylase* is an enzyme in saliva that breaks down carbohydrates into simple sugars. As each mouthful of bolus is swallowed, it travels down the esophagus and into the stomach. **See Figure 2-10.** The *esophagus* is a muscular tube that moves food from the mouth to the stomach.

Daniel NYC/E. Kheraj

The digestive process starts in the mouth as diners begin to chew food.

Stomach. After the bolus makes its way through the esophagus, it enters the stomach where it is mixed with gastric juices. The *stomach* is a sac-shaped organ located between the esophagus and small intestine where food is partially digested. *Gastric juice* is an acidic digestive fluid secreted by glands in the stomach and contains water, enzymes, and hydrochloric acid.

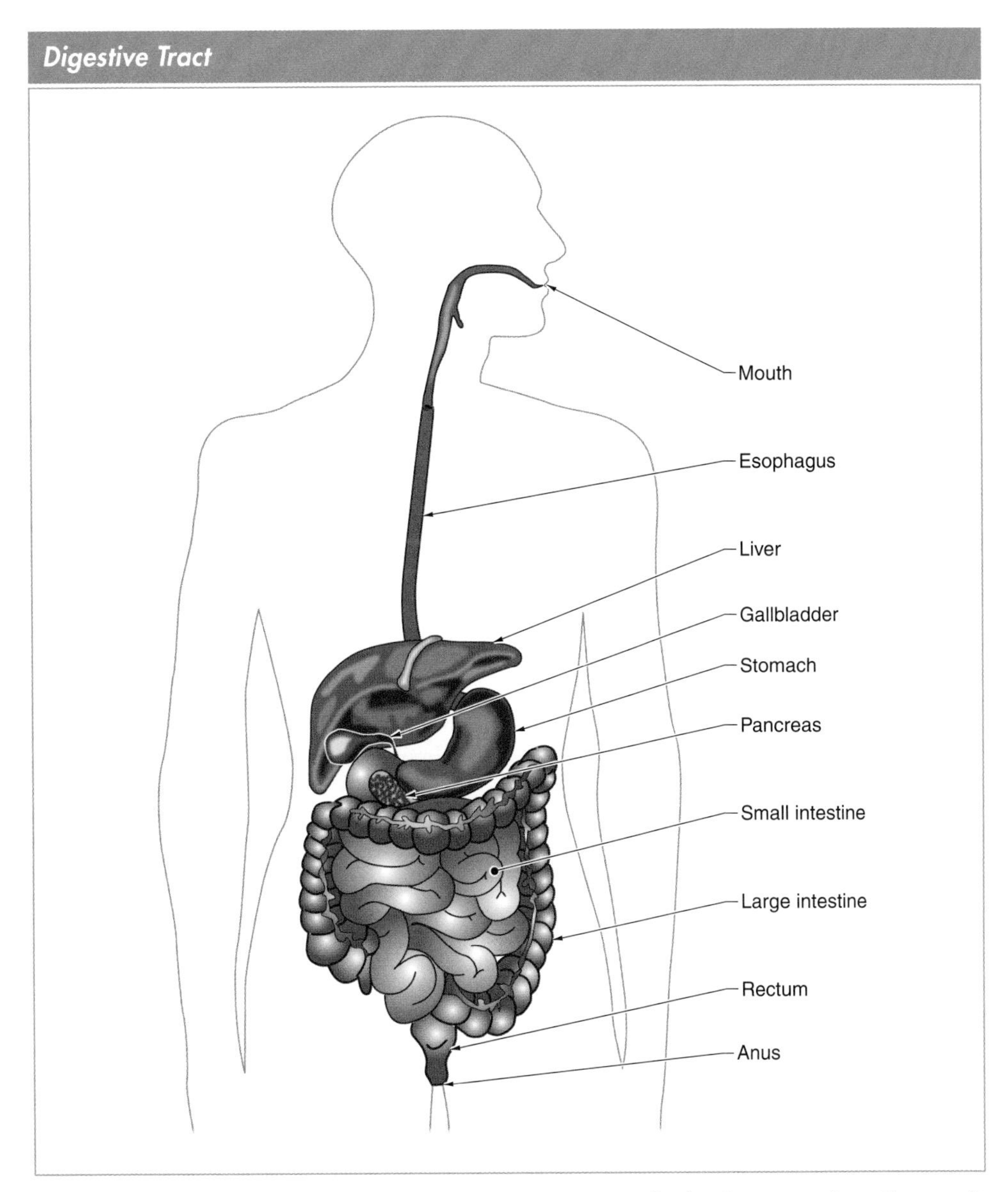

Figure 2-9. The digestive tract serves as the passageway for food to move from the mouth through the esophagus, stomach, small and large intestines, rectum, and anus.

Gastric juices begin the process of digesting proteins and lipids by separating them into their most basic form, amino acids and fatty acids. However, the digestion of carbohydrates comes to a temporary pause because the highly acidic gastric juices deactivate the amylase enzymes. The stomach and small intestine are shielded from this acidic mixture because they are lined with a thick layer of protective mucus.

Once the mass of food in the stomach has been mixed with gastric juices, it becomes chyme. *Chyme* is a thick, semifluid mass of partly digested food and gastric juices that originates in the stomach and is passed to the small intestine. Chyme is passed to the small intestine in tiny amounts through the pyloric sphincter. **See Figure 2-11.** The *pyloric sphincter* is a muscular ring located at the base of the stomach that enables chyme to pass from the stomach to the small intestine. This process can take up to 5 hours to complete depending on the contents of a meal.

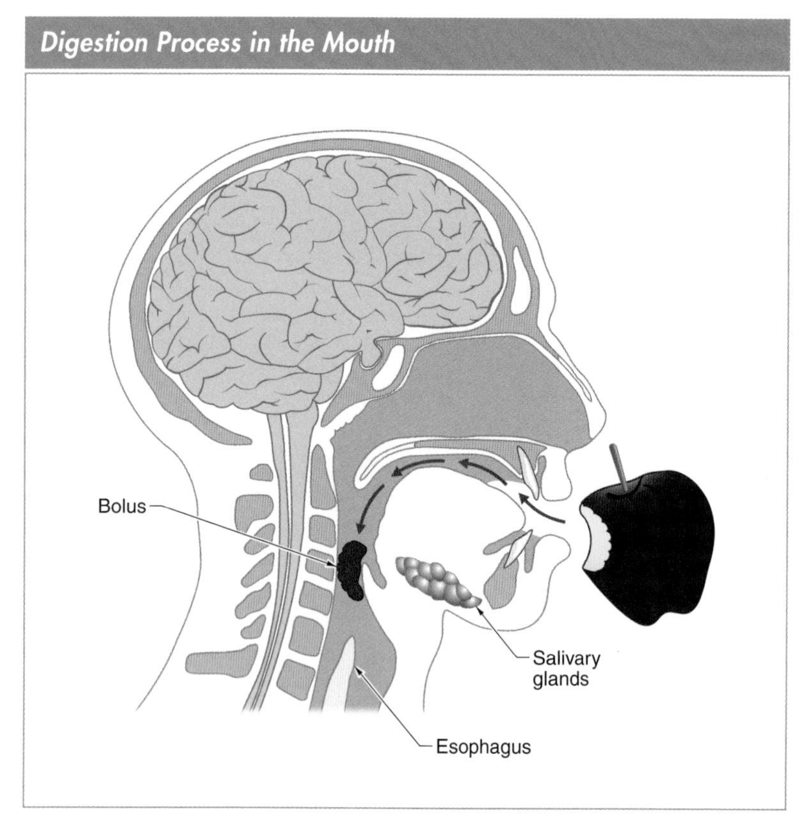

Figure 2-10. Digestion begins in the mouth where food is chewed and mixed with saliva enzymes to form a bolus that travels down the esophagus.

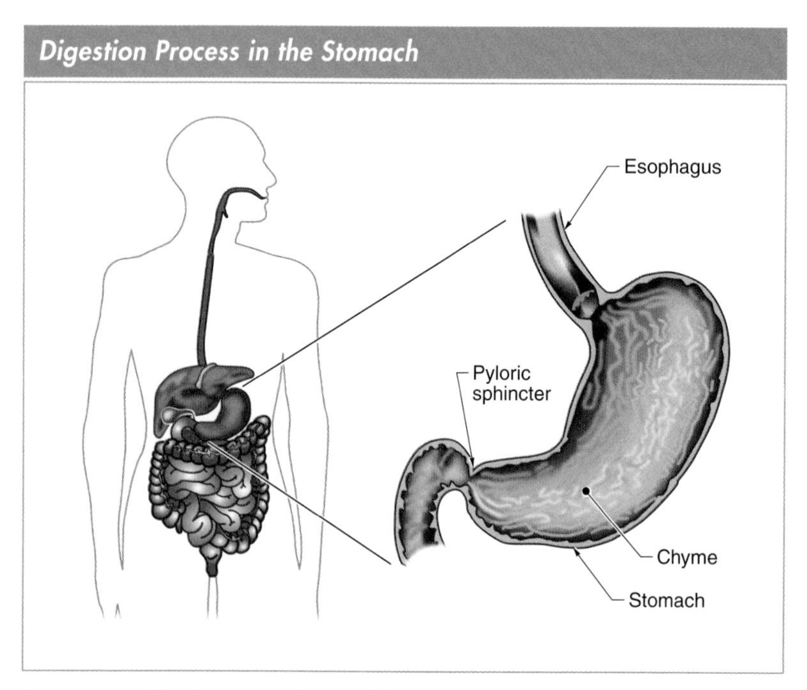

Figure 2-11. Food mixes with gastric juices in the stomach to become chyme, which is passed to the small intestine through the pyloric sphincter.

Small Intestine. The *small intestine* is the narrow, upper section of the intestinal tract approximately 20 feet in length where most nutrients are absorbed. Once food has made its way to the small intestine, it is mixed with pancreatic juices. *Pancreatic juice* is an alkaline digestive fluid secreted by the pancreas and contains enzymes. The *pancreas* is a long organ located near the stomach that secretes digestive juices into the small intestine. Pancreatic enzymes complete the process of breaking proteins into amino acids. Additionally, the alkaline nature of pancreatic juice reduces the acidity of chyme so that the digestion of carbohydrates can resume.

Bile is also released by the way of the gallbladder into the small intestine to allow the digestion of fats to continue. *Bile* is a yellow-green fluid produced by the liver and stored in the gallbladder until it passes into the small intestine where it enables fats to mix with water for digestion. The *liver* is a large organ located in the upper abdomen that produces bile for digestion, stores and filters blood, and helps convert nutrients into usable energy. From this point in the digestion process, any chyme that remains in the digestive tract moves from the small intestine to the large intestine through the ileum. **See Figure 2-12.** The *ileum* is a one-way valve located at the base of the small intestine that enables chyme to pass into the large intestine.

Large Intestine. The *large intestine* is the wide, lower section of the intestinal tract approximately 5 feet in length where water is extracted from chyme. After water is removed, chyme becomes a semisolid mass of waste comprised of the indigestible portions of food. This semisolid mass travels down the large intestine to the rectum. The *rectum* is the lowest section of the large intestine. The strong muscles of the rectum hold this waste until it is ready to be passed through the anus. The *anus* is the opening located at the end of the digestive tract where waste is excreted from the body. When the waste is passed through the anus, digestion is complete.

Digestive Processes

During digestion, food is transformed from a whole state in the mouth, to a state that is primarily liquid in the small intestine, and then into a semisolid waste product that is eliminated from the body through the anus. This transformation and movement of food is facilitated by a network of circular muscles that line the digestive tract. Food is moved through the digestive tract by a process called peristalsis. *Peristalsis* is the involuntary and wavelike contraction of muscles that transports food through the digestive tract. **See Figure 2-13.** Peristalsis begins once a mouthful of food has been swallowed and the bolus enters the esophagus. In addition to transporting food, the muscles of the digestive tract mix chyme with digestive enzymes and periodically squeeze the contents of the small intestine to remove water from the chyme.

Nutrient Absorption

The small intestine serves as the primary site for nutrient absorption after proteins have been broken down into amino acids, carbohydrates into sugars, and lipids into fatty acids. It is also the site where a majority of vitamins and minerals are absorbed. The small intestine has the surface area of almost 2700 square feet because it contains so many folds. These folds hold villi and microvilli. A *villus* is a fingerlike projection located in the small intestine that facilitates nutrient absorption. A *microvillus* is a hairlike projection on the villus of the small intestine that increases the surface area available for nutrient absorption. Villi and microvilli constantly move in wavelike motions that capture nutrients and absorb them into the cells of the small intestine. **See Figure 2-14.**

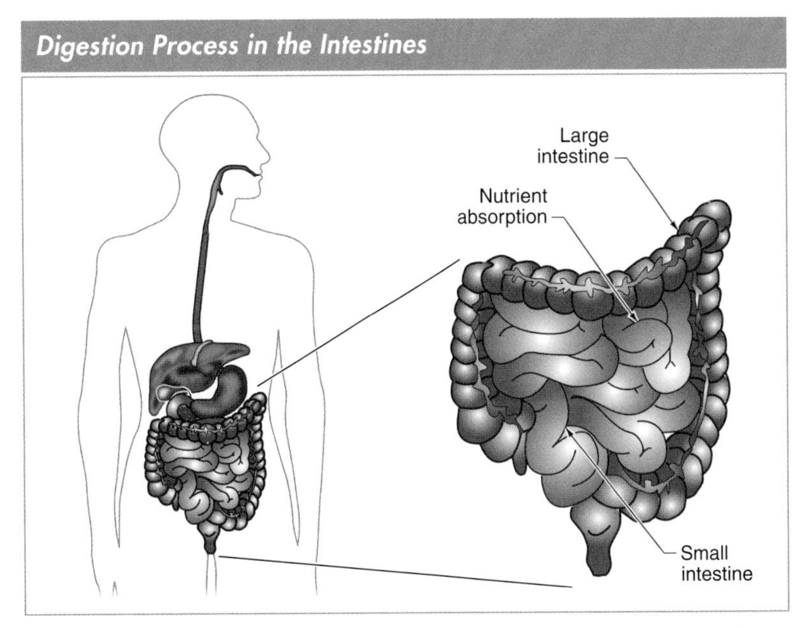

Figure 2-12. Most nutrients are absorbed in the small intestine, and any chyme that remains moves from the small intestine to the large intestine through the ileum.

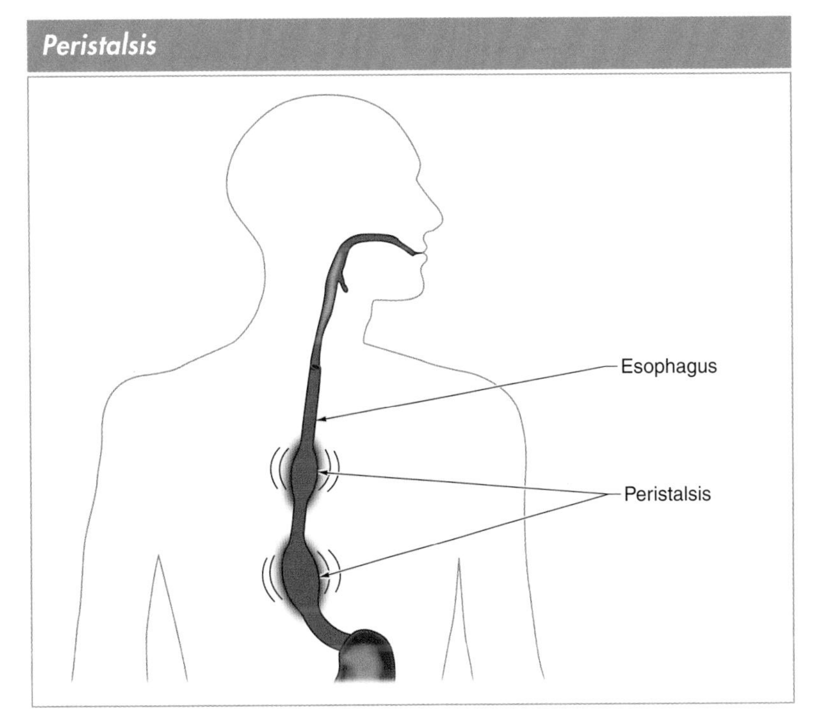

Figure 2-13. Peristalsis is the involuntary and wavelike contraction of muscles that transports food through the digestive tract.

Villi and Microvilli

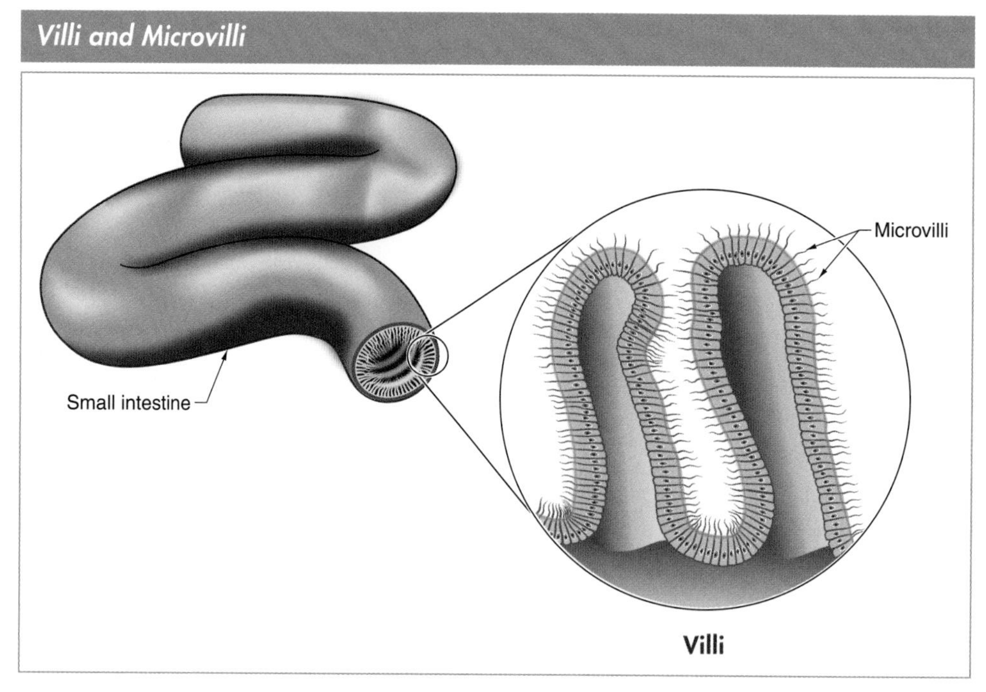

Figure 2-14. The villi and microvilli are constantly moving in wavelike motions to capture nutrients and absorb them into the cells of the small intestine.

After nutrients are absorbed into the cells of the small intestine, they are transported to the rest of the body to be used for energy, for structure, and to regulate body functions. The two main ways nutrients are delivered from the digestive tract to the rest of the body are through the portal system and the lymphatic system. The *portal system* is a system for transporting water-soluble nutrients from the small intestine to the liver and then into the bloodstream. The *lymphatic system* is a system for transporting fat-soluble nutrients from the small intestine into the bloodstream.

Food Science Note

The liver breaks down unwanted chemicals, such as alcohol, which are then eliminated by the body as waste.

Knowledge Check 2-2

1. Explain two main functions of saliva during digestion.
2. Explain the role of stomach gastric juices during digestion.
3. Describe chyme.
4. Identify the location in the body where most nutrients are absorbed.
5. Explain the process of digestion in the large intestine.
6. Describe the process of peristalsis.
7. Explain the roles of villi and microvilli in nutrient absorption.
8. Describe the two main ways nutrients are delivered from the digestive tract to the rest of the body.

ENERGY REQUIREMENTS

The body requires energy in the form of kilocalories to function properly. A *kilocalorie* is the amount of heat required to raise the temperature of 1 g of water by 1°C. Although the term "kilocalories" properly refers to the energy that foods provide, most resources, including Nutrition Facts labels, refer to kilocalories as calories.

Water, vitamins, and minerals do not supply energy, so they do not have calories. Proteins, carbohydrates, and fats (lipids) all supply energy and, therefore, have calories. Proteins and carbohydrates both provide 4 calories per gram, while fats (lipids) provide 9 calories per gram. **See Figure 2-15.** For example, a granola bar might contain 8 g of protein, 27 g of carbohydrates, and 5 g of total fat (lipids). Therefore, the amount of total calories is calculated as follows:

- protein calories = 4 calories/g × 8 g of protein = 32
- carbohydrate calories = 4 calories/g × 27 g of carbohydrates = 108
- fat (lipid) calories = 9 calories/g × 5 g of fats (lipids) = 45
- total calories = 32 protein calories + 108 carbohydrate calories + 45 fat (lipid) calories = **185**

National Turkey Federation

Meals that contain proteins, carbohydrates, and fats (lipids) supply energy and, therefore, have calories.

Calculating Calorie Contribution from Macronutrients

Nutrition Facts
Serving Size 1 bar

Amount Per Serving	
Calories 185	Calories from Fat 45
	% Daily Value*
Total Fat 5g	8%
Saturated Fat 2.5g	8%
Trans Fat 0g	
Cholesterol 0mg	0%
Sodium 210mg	9%
Total Carbohydrate 27g	9%
Dietary Fiber 5g	20%
Sugars 13g	
Protein 8g	11%
Vitamin A 0% • Vitamin C	15%
Calcium 20% • Iron	10%

* Percent Daily Values (DV) are based on a 2,000 calorie diet.

Nutrients	Calories per Gram	Grams in Granola Bar	Calories Supplied from Each Nutrient
Protein	4	8	32 (4 x 8 = 32)
Carbohydrates	4	27	108 (4 x 27 = 108)
Fat (lipids)	9	5	45 (9 x 5 = 45)
			Total Calories = 185

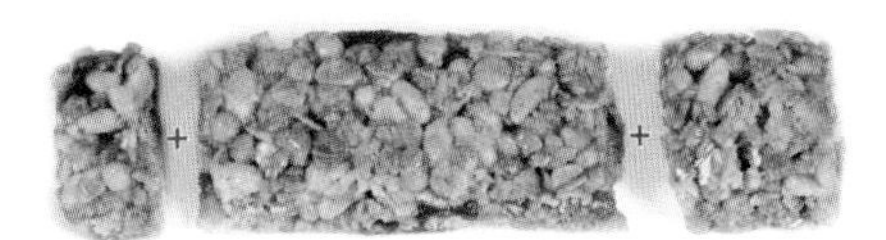

32 Calories from Protein + 108 Calories from Carbohydrates + 45 Calories from Fat (Lipids) = 185 Total Calories in Granola Bar (32 + 108 + 45 = 185)

Figure 2-15. Proteins and carbohydrates both provide 4 calories per gram, while fats (lipids) provide 9 calories per gram.

The total amount of calories required varies among individuals and is based upon factors such as gender, age, height, weight, activity level, and overall health. **See Figure 2-16.** Providing the body with an appropriate amount of calories per day promotes a healthy metabolism, fuels physical activity, encourages the consumption of nutritious foods, and contributes to healthy weight management.

Calorie Needs Based on Gender, Age, and Activity Level

Gender	Age	Sedentary	Active
Females	4 – 8	1200 – 1400	1400 – 1800
	9 – 13	1400 – 1600	1800 – 2200
	14 – 18	1800	2400
	19 – 30	1800 – 2000	2400
	31 – 50	1800	2200
	51+	1600	2000 – 2200
Males	4 – 8	1200 – 1400	1600 – 2000
	9 –13	1600 – 2000	2000 – 2600
	14 –18	2000 – 2400	2800 – 3200
	19 – 30	2400 – 2600	3000
	31 – 50	2200 – 2400	2800 – 3000
	51+	2000 – 2200	2400 – 2800

Adapted from USDA Dietary Guidelines for Americans, 2010

Figure 2-16. The total amount of calories required varies among individuals and is based on various factors, such as gender, age, and activity level.

Metabolism

Metabolism is the sum of all chemical reactions that occur within the body. Energy is necessary for chemical reactions to occur, and energy is obtained by the body through the calories in food. Metabolic chemical reactions occur by releasing energy through catabolism and by absorbing energy through anabolism.

Catabolism. *Catabolism* is the metabolic reaction by which the body breaks down the complex molecules in nutrients into simpler molecules in order to release useable energy. Energy released during catabolism is stored within adenosine triphosphate molecules. *Adenosine triphosphate (ATP)* is a molecule that captures the released energy generated during catabolism and uses that energy to power all cellular activity. For example, ATP energy generates heat for the body, allows muscles to contract, and creates fuel for anabolism.

Anabolism. *Anabolism* is the metabolic reaction by which the body absorbs the energy released during catabolism to build complex molecules from simpler molecules. **See Figure 2-17.** Complex molecules found in muscles, tissues, fat, and bones are built as a result of anabolism.

Basal Metabolic Rate. The ATP molecules also form the basis for what is referred to as basal metabolic rate. The *basal metabolic rate (BMR)* is the minimum amount of calories the body needs to support life. The BMR involves the most basic body functions, such as breathing, circulating blood, maintaining proper temperature, and building cells. A person with a high metabolism has a high BMR and, therefore, burns more calories at rest than a person with a low metabolism or low BMR. The majority of calories consumed each day are used to support the BMR. The calories not used to support the BMR provide energy for the digestion of food and physical activity.

Wellness Concept

The BMR typically slows down with age, but exercising regularly can help decrease the rate at which it slows.

Physical Activity

The body burns extra calories through movement and physical activity. **See Figure 2-18.** Taking part in moderate to vigorous physical activity for at least 30 minutes a day for three to five days per week has many health benefits including increased heart and lung function, stronger bones, and stress reduction. Physical activity is essential to overall well-being, but it is equally essential to consume the appropriate amount of calories with foods that are nutritious.

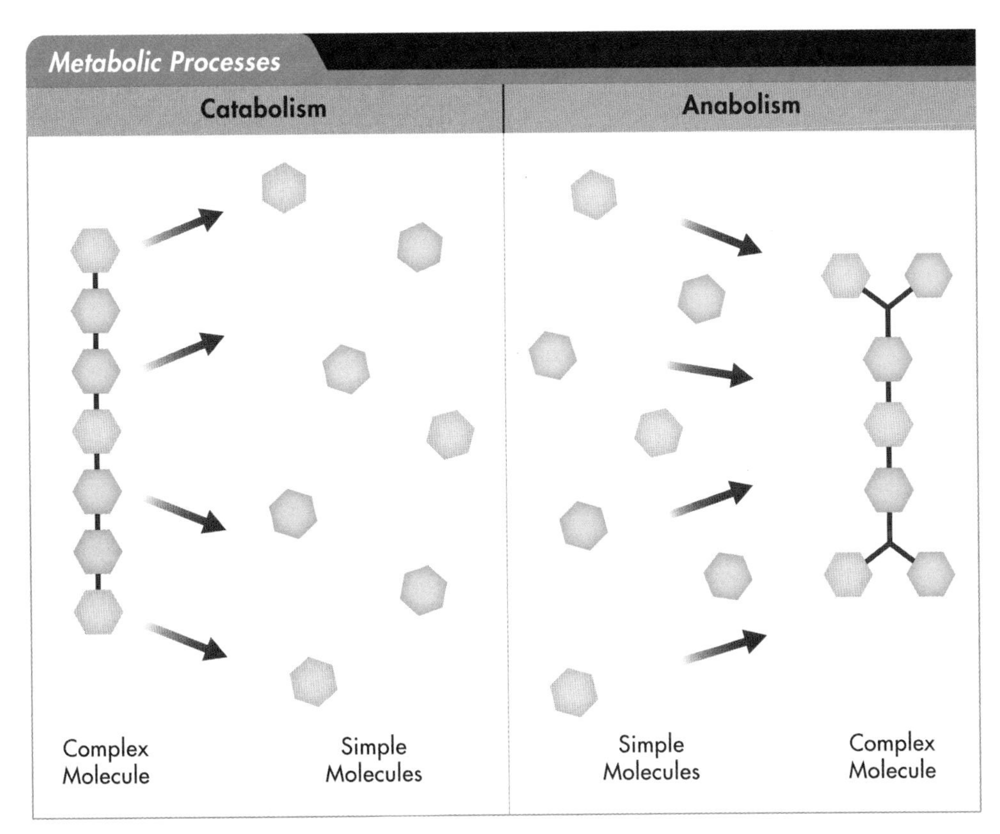

Figure 2-17. The metabolic process of catabolism breaks complex molecules into simpler molecules, whereas anabolism builds complex molecules from simpler molecules.

Calories Burned Through Physical Activities

Approximate Calories Used by a 154 Pound Adult	
Physical Activity	30 Minutes of Activity
Walking (3½ miles per hour)	140
Walking (4½ miles per hour)	230
Bicycling (less than 10 miles per hour)	145
Bicycling (more than 10 miles per hour)	295
Golf (walking and carrying clubs)	165
Light gardening/yard work	165
Heavy yard work	220
Weight lifting (vigorous)	220
Basketball (vigorous)	220
Swimming (laps)	255
Running/jogging (5 miles per hour)	295

Adapted from ChooseMyPlate.gov

Figure 2-18. The body burns extra calories through movement and physical activity.

Nutritious Choices

Foods are not all created equal. Some foods are highly nutritious and should be consumed regularly, while others can be detrimental to overall health if consumed frequently. Understanding how to make nutritious choices helps set the foundation for a healthy diet and can encourage the development of healthy recipes and menu offerings that promote nutrient-dense and functional foods.

Nutrient-Dense Foods. A *nutrient-dense food* is a food that is high in nutrients and low in calories. Foods such as fresh vegetables and fruit are prized for being nutrient dense because they are excellent sources of vitamins and minerals while being relatively low in calories. **See Figure 2-19.** For example, ⅓ cup of cooked spinach contains only 30 calories but supplies the body with 100% of its daily need for vitamin A. Studies have shown that a diet plentiful in nutrient-dense foods can help prolong life and prevent disease.

The opposite of a nutrient-dense food is an empty-calorie food. An *empty-calorie food* is a food that is high in calories and low in nutrients. Empty-calorie foods, such as soft drinks, chips, cookies, and candy, are often high in refined sugars, saturated fats, and/or sodium. Empty-calorie foods have been linked to weight gain and weight-associated health problems.

Functional Foods. A *functional food* is a food that provides potential health benefits when consumed on a regular basis as part of a varied diet. **See Figure 2-20.** The Academy of Nutrition and Dietetics identifies whole foods, fortified foods, and enriched foods as functional foods. A *whole food* is a food in its natural state such as fresh vegetables and fruit. A *fortified food* is a food that has nutrients added in order to increase nutritive value. For example, orange juice does not contain calcium but is commonly fortified with calcium to increase its health benefits. An *enriched food* is a food that has had nutrients added that were originally present in the food but were lost during processing. For example, when white bread is made, iron is removed during the processing of the wheat but is added back so that the final product contains iron. Consuming an assortment of functional foods in the appropriate amount, along with physical activity, promotes a healthy and effective way to manage weight.

Nutrient-Dense Foods

Vegetables

Fruits

National Honey Board

Figure 2-19. Fresh vegetables and fruits are nutrient dense because they are excellent sources of vitamins and minerals and low in calories.

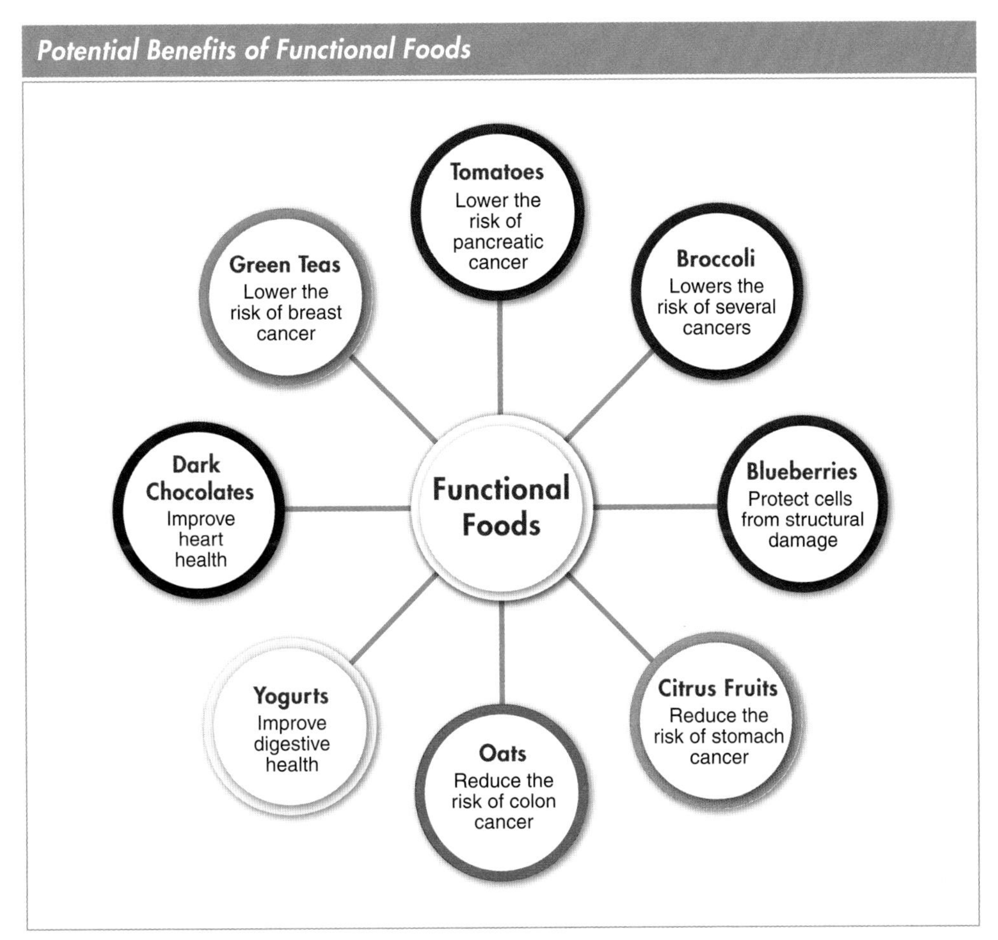

Figure 2-20. A functional food provides potential health benefits when consumed on a regular basis as part of a varied diet.

Weight Management

Weight management is dependent on energy intake and energy output. Energy intake is based on the calories consumed through food and beverages, and energy output is the result of the BMR, digestion, and physical activity. In order to maintain weight, the calories consumed (energy intake) and calories used (energy output) by the body need to be equal. **See Figure 2-21.** When more energy is used by the body than calories consumed, weight loss will occur. In contrast, when more calories are consumed than are used, weight gain will occur. Excess calories are stored as fat until the energy is used.

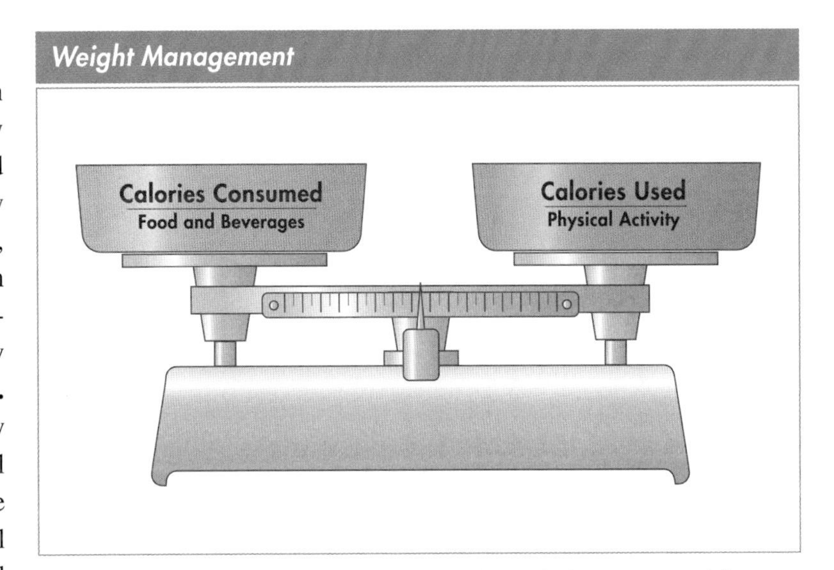

Figure 2-21. In order to maintain weight, the calories consumed (energy intake) and calories used (energy output) by the body must be equal.

For each individual, there is an ideal weight in which the body functions most efficiently. Effective methods used to determine a healthy body weight take into account the added skin, bones, fat, and fluids that accompany increases in height. One such method is referred to as the body mass index (BMI). The *body mass index (BMI)* is a calculation based on height and weight to determine body fat composition. **See Figure 2-22.** The Centers for Disease Control (CDC) uses BMI ranges to define the following:

- underweight—below 18.5
- normal weight—18.5 to 24.9
- overweight—25.0 to 29.9
- obese—30.0 and above

Although the BMI is a useful guide for determining a healthy body weight, it is not completely reliable. This is because the BMI does not account for muscle mass, which weighs more than fat. Therefore, a muscular individual may have a low body fat composition but weigh more due to muscle mass and be incorrectly categorized as overweight. In conjunction with the BMI, waist circumference is often used to help determine whether an individual is at risk for developing adverse health conditions associated with being overweight. Health risks increase for females with a waist circumference greater than 35 inches and for males with a circumference greater than 40 inches.

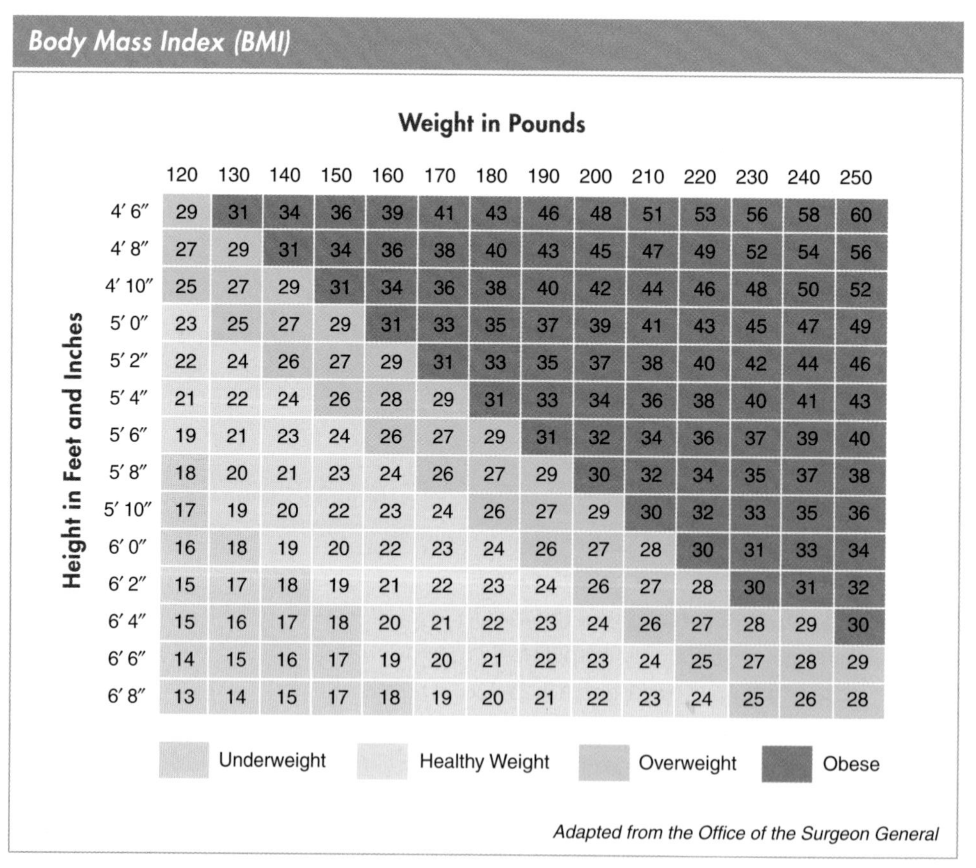

Body Mass Index (BMI)

Weight in Pounds

Height in Feet and Inches	120	130	140	150	160	170	180	190	200	210	220	230	240	250
4′ 6″	29	31	34	36	39	41	43	46	48	51	53	56	58	60
4′ 8″	27	29	31	34	36	38	40	43	45	47	49	52	54	56
4′ 10″	25	27	29	31	34	36	38	40	42	44	46	48	50	52
5′ 0″	23	25	27	29	31	33	35	37	39	41	43	45	47	49
5′ 2″	22	24	26	27	29	31	33	35	37	38	40	42	44	46
5′ 4″	21	22	24	26	28	29	31	33	34	36	38	40	41	43
5′ 6″	19	21	23	24	26	27	29	31	32	34	36	37	39	40
5′ 8″	18	20	21	23	24	26	27	29	30	32	34	35	37	38
5′ 10″	17	19	20	22	23	24	26	27	29	30	32	33	35	36
6′ 0″	16	18	19	20	22	23	24	26	27	28	30	31	33	34
6′ 2″	15	17	18	19	21	22	23	24	26	27	28	30	31	32
6′ 4″	15	16	17	18	20	21	22	23	24	26	27	28	29	30
6′ 6″	14	15	16	17	19	20	21	22	23	24	25	27	28	29
6′ 8″	13	14	15	17	18	19	20	21	22	23	24	25	26	28

Underweight — Healthy Weight — Overweight — Obese

Adapted from the Office of the Surgeon General

Figure 2-22. The BMI is a calculation based on height and weight to determine body fat composition.

Knowledge Check 2-3

1. Identify the term commonly used in place of kilocalories.
2. Identify the number of calories per gram supplied by proteins, carbohydrates, and lipids.
3. Differentiate between the metabolic processes of catabolism and anabolism.
4. Explain the difference between a high and a low basal metabolic rate.
5. Identify four benefits of physical activity.
6. Differentiate between nutrient-dense foods and empty-calorie foods.
7. Explain the meaning of functional foods.
8. Describe how functional foods can be classified.
9. Explain how to use energy intake and energy output to maintain weight.
10. Describe body mass index (BMI) including the four BMI ranges.

DIETARY RECOMMENDATIONS

Dietary recommendations are the result of scientific research focused on finding the optimum amounts of nutrients that should be consumed in order to prevent disease and promote overall wellness. These recommendations are intended to help individuals follow a nutritious diet and can be valuable tools for foodservice professionals when planning menus. Key dietary recommendations can be found in the *Dietary Guidelines for Americans,* U.S. Department of Agriculture (USDA) serving size recommendations, Harvard Healthy Eating Guides, and dietary reference intakes.

Dietary Guidelines for Americans

Every five years, the USDA publishes *Dietary Guidelines for Americans,* which includes key recommendations for healthy eating and physical activity. These recommendations stress the importance of the daily consumption of vegetables, fruits, whole grains, and nonfat or low-fat milk and dairy products. These recommendations also place emphasis on consuming protein from a variety of sources, including lean meats, poultry, fish, eggs, soy, and nuts, and reducing sugar, saturated fat, cholesterol, and sodium. The 2010 dietary guidelines include the following key recommendations:

- Choose foods that provide more potassium, dietary fiber, calcium, and vitamin D. These foods include vegetables, fruits, whole grains, and milk and milk products.
- Increase vegetable and fruit intake.
- Eat a variety of beans, peas, and vegetables, especially dark-green, red, and orange vegetables.
- Consume at least half of all grains as whole grains. Increase whole-grain intake by replacing refined grains with whole grains.
- Choose a variety of protein foods, including seafood, lean meat, poultry, eggs, beans, peas, soy products, and unsalted nuts and seeds.
- Increase the amount and variety of seafood consumed by choosing seafood in place of some meat and poultry.
- Replace protein foods that are high in solid fats with choices that are lower in solid fats, lower in calories, and/or sources of oils.
- Increase intake of nonfat or low-fat milk and milk products, such as yogurt and cheese.
- Use oils to replace solid fats when possible.

The USDA created the ChooseMyPlate website to help people understand how to meet the dietary guidelines and maintain their health and weight. The MyPlate icon is the symbol of the USDA ChooseMyPlate program. **See Figure 2-23.** The ChooseMyPlate program emphasizes the need to eat a variety from all food groups, including vegetables and fruits, whole grains, lean proteins, and low-fat dairy products.

MyPlate Icon

United States Department of Agriculture

Figure 2-23. The MyPlate icon is the symbol of the USDA ChooseMyPlate program.

Courtesy of The National Pork Board

The ChooseMyPlate program emphasizes eating a variety of foods from all food groups.

Serving Sizes

Serving size is a specific amount of food expressed using a unit of measurement, such as a tablespoon, ounce, or cup. Serving sizes differ for each food group because each group provides different amounts of calories and nutrients. Serving sizes also can vary within each food group. For example, a serving of fresh fruit is 1 cup, while the serving size for dried fruits is ½ cup. Serving sizes are not meant to be the amount of food on a plate at every meal but rather a guideline of how much food from each food group to consume daily. **See Figure 2-24.**

Vegetables. The daily recommended serving size for vegetables ranges from 1–3 cups depending on age, gender, and level of physical activity. In general, 1 cup of raw vegetables, cooked vegetables, or vegetable juice or 2 cups of raw leafy greens each equals 1 cup of vegetables.

Focus should be placed on consuming vegetables of every color, such as red beets and peppers, orange carrots and sweet potatoes, yellow corn, dark leafy greens, purple eggplant and cabbage, and white cauliflower and onions. **See Figure 2-25.** Vegetables contain a variety of vitamins and minerals, most notably vitamins A and C, potassium, folate, and magnesium. They are also a good source of dietary fiber.

Fruits. The recommended daily serving size of fruits is 1–2 cups daily. In general, 1 cup of fresh, canned, or frozen fruit, 1 cup of 100% fruit juice, and ½ cup of dried fruit are each the equivalent of a 1 cup serving. Emphasis should be placed on choosing a wide arrangement of colorful fruits in their fresh, canned, frozen, and dried forms as well as minimizing fruit juices because juices lose dietary fiber during processing. Like vegetables, fruits are nutrient-dense foods that provide a rich source of vitamins, minerals, and dietary fiber.

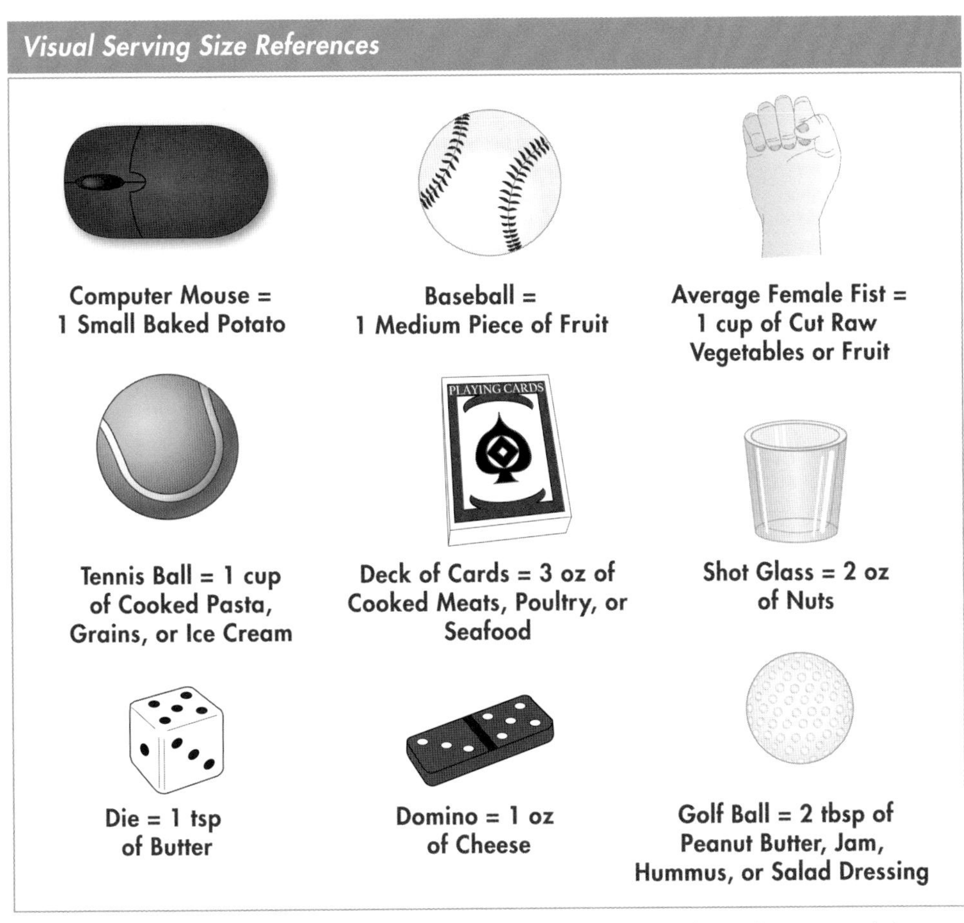

Figure 2-24. Visual references can be used as guidelines to help determine appropriate serving sizes.

Colorful Vegetables and Fruits

Reds	Oranges	Yellows	Greens	Blues/Purples	Whites
Oregon Raspberry & Blackberry Commission	*Melissa's Produce*		*Tanimura & Antle®*	*U.S. Highbush Blueberry Council*	*Tanimura & Antle®*
Radicchio	Butternut squash	Golden beets	Artichokes	Eggplant	Cauliflower
Radishes	Carrots	Sweet corn	Brussels sprouts	Purple asparagus	Bean sprouts
Red potatoes	Pumpkins	Wax beans	Endive	Purple cabbage	Jicama
Rhubarbs	Yams	Yellow bell peppers	Zucchinis	Purple potatoes	Parsnips
Cherries	Apricots	Kumquats	Avocados	Blackberries	Coconuts
Pomegranates	Cantaloupe melons	Lemons	Green apples	Blueberries	Plantains
Raspberries	Mangoes	Pineapples	Honeydew melons	Concord grapes	White nectarines
Watermelons	Tangerines	Yellow watermelons	Kiwifruit	Plums	White peaches

Figure 2-25. Consuming vegetables and fruits of every color provides optimal nutrients.

Grains. The recommended daily serving size of foods from the grain group is 3–8 oz, depending on age, gender, and level of physical activity. A slice of bread, 1 cup of ready-to-eat cereal, or ½ cup of cooked grains or pasta is considered a 1 oz equivalent for grains. Half of all grains consumed should be from whole grain sources, such as whole wheat breads and pastas, brown rice, wild rice, wheat berries, bulgur, oats, barley, rye, quinoa, millet, and spelt. **See Figure 2-26.** Grains are typically a rich source of B vitamins, iron, and dietary fiber.

Foods from the vegetable, fruit, and grain groups can supply the body with beneficial amounts of dietary fiber. Dietary fiber is a key nutrient because studies show diets rich in dietary fiber can help decrease the risk of developing heart disease, diabetes, and digestive disorders. Dietary fiber also plays an important role in weight management because it promotes satiety. *Satiety* is the state of feeling full. Research indicates that most individuals only get half of the total recommended daily dietary fiber intake of 21–38 g.

Protein Foods. The recommended daily serving size of protein foods is 2–6½ oz, depending on age, gender, and level of physical activity. Protein foods come from a variety of sources including both animal-based and plant-based foods. **See Figure 2-27.** Meat, poultry, seafood, eggs, and dairy products, such as milk and cheese, are excellent sources of animal-based proteins. Plant-based foods, such as beans, lentils, soy products, quinoa, nuts, and seeds, also supply protein and are alternatives to animal-based proteins. In general, ¼ cup of cooked beans, 1 egg, 1 tbsp of peanut butter, and 2 oz of tofu (soy) are each equivalent to a 1 oz serving of meat, poultry, or seafood. Protein foods contribute beneficial nutrients to the diet including B vitamins, iron, and zinc. Most plant-based proteins also provide a significant amount of dietary fiber.

Figure 2-26. Half of all grains consumed should be from whole grain sources, such as whole wheat breads and barley.

Figure 2-27. Protein foods come from a variety of sources including both animal-based and plant-based foods.

Dairy Foods. The recommended daily serving size of dairy foods is 2–3 cups per day, depending on age, gender, and level of physical activity. **See Figure 2-28.** In general, 1 cup of dairy is equal to 1½ oz of natural cheese or 1 cup of milk, yogurt, or calcium-fortified soy milk. Foods made from milk that have little to no calcium, such as butter, cream, and cream cheese, are not considered dairy foods. Low-fat and nonfat dairy products are considered the best choices for health and wellness. Dairy products provide protein, calcium, potassium, and vitamin D.

Fats. Although fats (lipids) are an essential part of a healthy diet, they do not contribute many vitamins and minerals for the large amount of calories they contain. Because foods containing fats can be highly caloric, it is easy to consume them in quantities greater than what is advised. The USDA recommends that individuals keep fat consumption in a range of 3–7 tsp daily, depending on age, gender, and level of physical activity. **See Figure 2-29.** In general, 1 oz of almonds, ½ of a medium avocado, and 1 tbsp of oil each equal 3 tsp of fat. Unsaturated fats from plant and seafood sources are considered healthier than the saturated fats found in animal products.

Figure 2-28. The recommended daily serving size of dairy foods is 2–3 cups per day.

Fats in Foods

Foods Containing Fat	Amount of Food	Amount of Fat in Food	Total Calories	Calories from Fat
Almonds	1 oz	3 tsp	170	130
Avocados	½ medium	3 tsp	160	130
Hazelnuts	1 oz	4 tsp	185	160
Mayonnaises	1 tbsp	3 tsp	100	100
Olive oils	1 tbsp	3 tsp	120	120
Olives	8 large	1 tsp	40	30
Peanut butter	2 tbsp	4 tsp	190	140
Sunflower seeds	1 oz	3 tsp	165	120

Adapted from ChooseMyPlate.gov

Figure 2-29. The USDA recommends that individuals keep fat consumption in a range of 3–7 tsp daily.

Harvard Healthy Eating Plate Icon

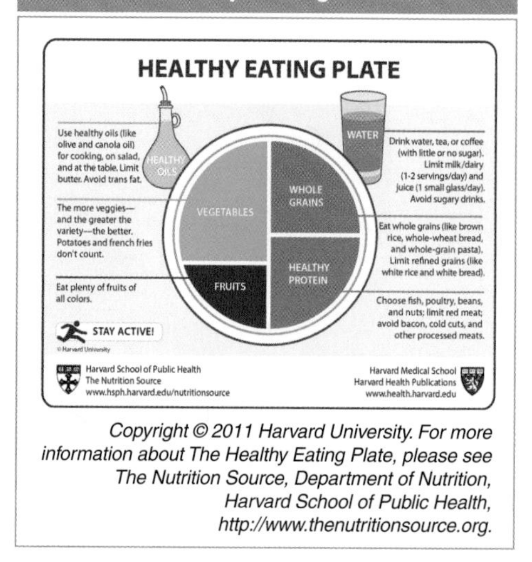

Figure 2-30. The Harvard Healthy Eating Plate represents what their scientific research shows to be a healthy eating approach.

Harvard Healthy Eating Guides

The Department of Nutrition at the Harvard School of Public Health created the Harvard Healthy Eating Plate to represent what their scientific research shows to be a healthy eating approach. **See Figure 2-30.** The Harvard Healthy Eating Plate icon differs from ChooseMyPlate in the following ways:

- Water, tea, or coffee is suggested instead of dairy as a beverage.
- Milk and dairy consumption is reduced from 2–3 cups per day to 1–2 cups per day.
- Vegetables make up a greater proportion of the plate, and potatoes do not count as a vegetable.
- Whole grains are emphasized versus grains in general.
- Fish, poultry, beans, and nuts are stressed as protein sources.
- A glass bottle entitled "Healthy Oils" is included as a reminder to use liquid oils (unsaturated), such as olive or canola oil.
- An image of a running figure represents the importance of physical activity.

Harvard also created a Healthy Eating Pyramid as a guide to choosing a healthy diet. **See Figure 2-31.** The foundation of the pyramid emphasizes physical activity, weight management, and balanced meals. The pyramid shows that most foods should be consumed from the bottom in the form of vegetables, fruits, whole grains, and oils. Moderate amounts of protein are represented on the next level with foods, such as nuts, seeds, beans, tofu, fish, poultry, and eggs. The pyramid narrows to suggest that dairy products be limited in the diet. At the peak, the pyramid includes foods that should be used sparingly, such as red meat, butter, white bread, soft drinks, and salt.

The general principles of the Harvard eating guides align with what many refer to as a Mediterranean diet. A *Mediterranean diet* is a diet that emphasizes a high consumption of vegetables, fruits, whole grains, olives, olive oils, nuts, and seeds; a moderate consumption of lean meats, fish, and wine; and a limited consumption of red meats and processed foods. Scientific studies indicate that a Mediterranean diet can reduce the risk of heart disease and some cancers.

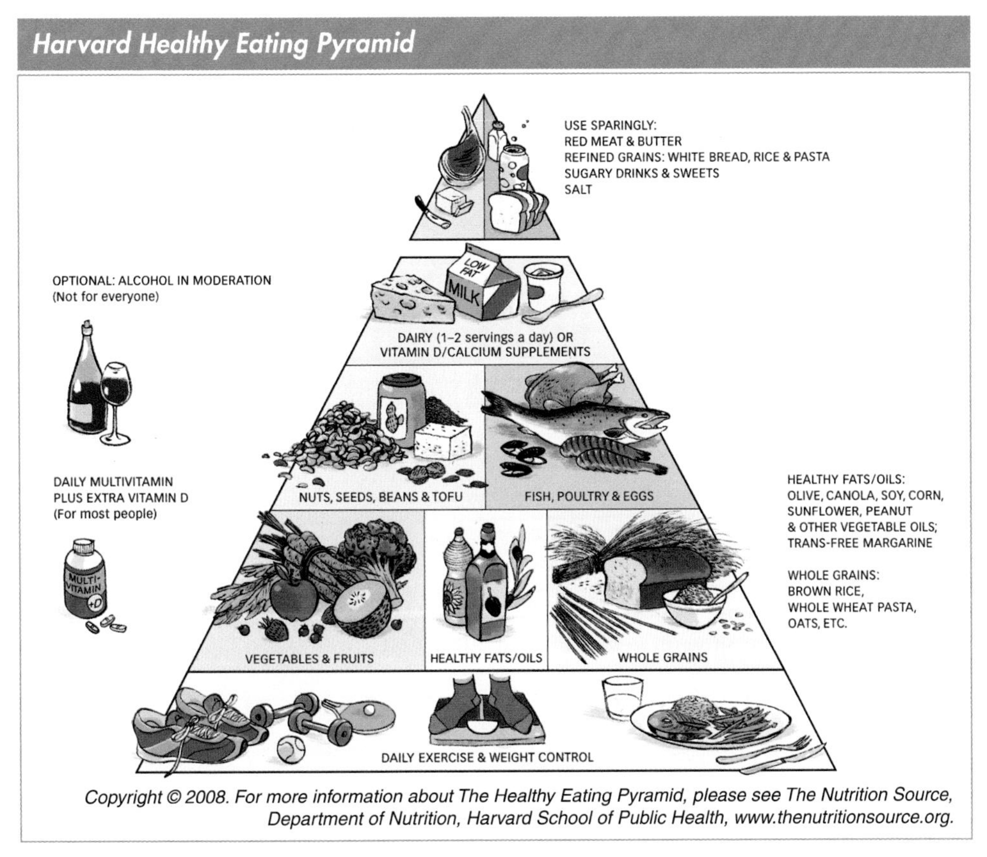

Figure 2-31. The Harvard Healthy Eating Pyramid shows that most foods should be consumed from the bottom part of the pyramid in the form of vegetables, fruits, whole grains, and oils.

Dietary Reference Intakes

A *dietary reference intake (DRI)* is a set of nutrient reference standards that include estimated average requirements, recommended dietary allowances, adequate intakes, tolerable upper intake levels, and estimated energy requirements. The Institute of Medicine developed DRIs for the United States and Canada. DRIs are often used by registered dietitians to help plan meals for individuals and for foodservice institutions, such as healthcare facilities and schools. A *registered dietician (RD)* is an individual who has specialized in the study of nutrition and met a stringent set of academic and professional standards.

Estimated Average Requirements. An *estimated average requirement (EAR)* is the daily nutrient intake level that is sufficient enough to meet the needs of half a healthy population group based on age and gender. An EAR is used in nutrition research to determine whether the diet of a specific group provides adequate amounts of nutrients.

Recommended Dietary Allowances. A *recommended dietary allowance (RDA)* is the daily nutrient intake level that is sufficient enough to meet the nutrient needs of approximately 98% of a healthy population group based on age and gender. The RDA for a nutrient is based on EAR data.

Adequate Intakes. *Adequate intake (AI)* is the estimated daily intake level for a nutrient for which no RDA has been set but nutrient intake appears to be adequate to meet the needs of a healthy population. The AI is intended to be used as a guide regarding the appropriate intake level for nutrients that have insufficient data to establish a requirement.

Tolerable Upper Intake Levels. A *tolerable upper intake level (UL)* is the maximum daily intake level of a nutrient before the body experiences an adverse effect. When daily intake levels rise above the UL value, health risks also rise.

Estimated Energy Requirements. An *estimated energy requirement (EER)* is the energy intake (calories) needed by healthy individuals to maintain their weight based on age, gender, weight, height, and activity level. Online templates allow individuals to calculate their EER by entering information regarding physical characteristics, a list of activities performed over a 24-hour period, and the length of time spent on each activity.

Knowledge Check 2-4

1. Describe the key recommendations of the *Dietary Guidelines for Americans.*
2. Give an example of a serving size for vegetables.
3. Give an example of a serving size for fruits.
4. Name four sources of whole grains.
5. Describe three benefits of eating foods high in dietary fiber.
6. Give three examples of protein foods aside from meat, poultry, and seafood.
7. Explain why butter, cream, and cream cheese are not considered dairy foods.
8. Identify the recommended daily serving range for fat consumption.
9. Differentiate between Harvard Healthy Eating Plate and ChooseMyPlate.
10. Describe the Harvard Healthy Eating Pyramid.

DIETARY CONNECTIONS

The field of nutrition is based on compelling scientific evidence showing a connection between diet and health. There is no such thing as a perfect diet. However, eating nutrient-dense foods, controlling portion sizes, and participating in daily physical activity are effective strategies to minimize the chances of developing nutrient-related illnesses. See Figure 2-32. Diets that are rich in plant-based foods, such as vegetables, fruits, and whole grains, as well as lean meats, seafood, eggs, and low-fat dairy products have been linked to increased health and well-being.

Nutrient-Dense Choices

Courtesy of The National Pork Board

Figure 2-32. Choosing nutrient-dense foods, such as mixed greens with fruits and lean pork, can help minimize the chances of developing nutrient-related illnesses.

When nutrient intake is poor, health and well-being often decline. For example, hypertension, obesity, cardiovascular disease, diabetes, and certain types of cancers have been linked to dietary factors. In addition, there has been a sharp increase in the number of individuals whose health can be adversely affected by the food they eat due to food allergies and intolerances.

Wellness Concept

Evidence suggests that diets high in sugar and fat can cause individuals to feel sluggish and irritable, while nutrient-dense diets can provide sustained energy and help fight stress and disease.

Hypertension

Hypertension is a condition characterized by high blood pressure. *Blood pressure* is the pressure of blood within the arteries. Blood pressure is measured using systolic pressure and diastolic pressure. *Systolic pressure* is the pressure of blood within the arteries after the heart contracts. *Diastolic pressure* is the pressure of blood within the arteries before the heart contracts. Blood pressure varies throughout the day. However, hypertension results when blood pressure remains elevated. If uncontrolled, hypertension can lead to heart and kidney disease, as well as blindness.

Factors such as high levels of sodium in the diet and being overweight have been shown to adversely affect blood pressure. The latest studies reveal that individuals consume an average of 3300 mg of sodium per day instead of the recommended dietary guideline of less than 2300 mg per day. Some organizations, such as the American Heart Association (AHA), recommend limiting daily sodium intake to less than 1500 mg per day. **See Figure 2-33.**

Research indicates an effective approach to limiting sodium and encouraging a healthy blood pressure is the Dietary Approaches to Stop Hypertension (DASH) eating plan promoted by the U.S. National Heart, Lung, and Blood Institute. The DASH eating plan involves limiting sodium intake to 1500 mg per day by consuming vegetables, fruits, whole grains, poultry, fish, nuts, and low-fat dairy products.

Wellness Concept

According to the American Heart Association, a normal systolic blood pressure reading should be less than 120 and diastolic pressure should be less than 80, commonly stated as 120 over 80. Prehypertension is in the range of 120–139 (systolic pressure) over 80–89 (diastolic pressure).

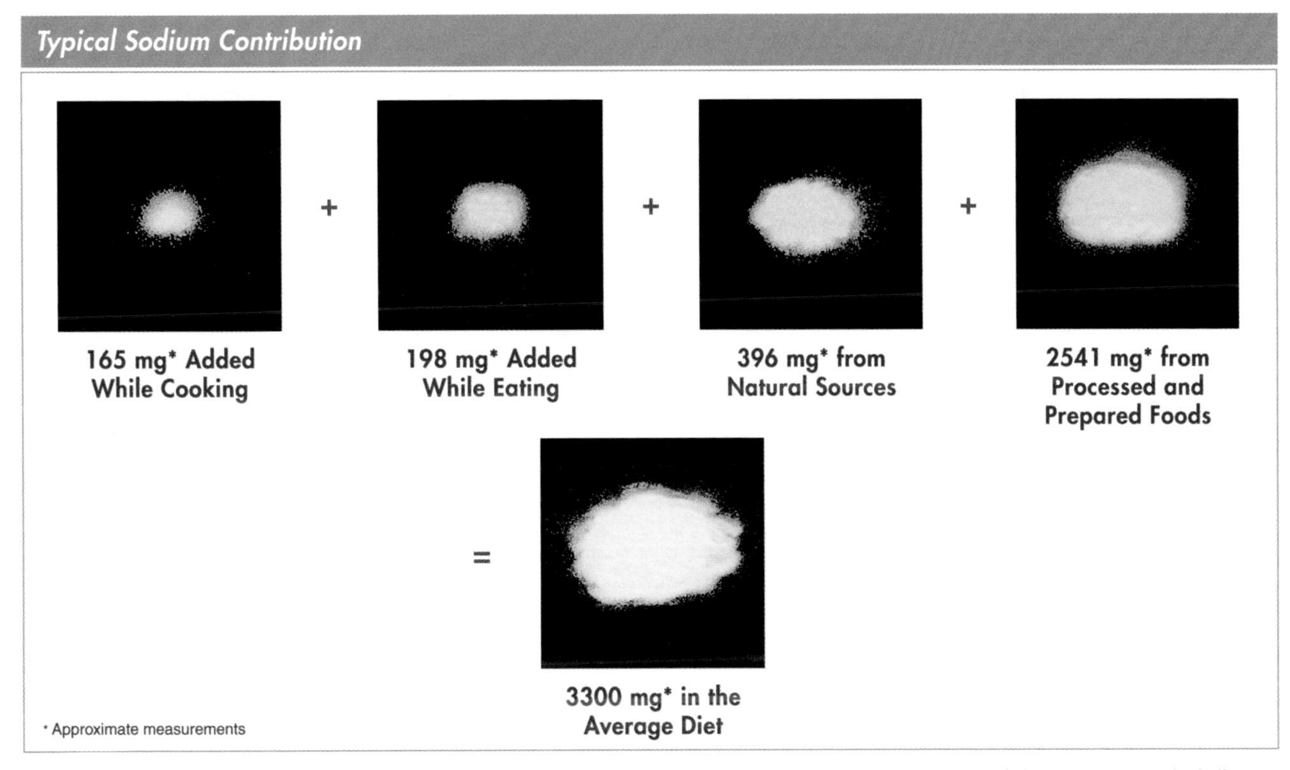

Figure 2-33. Individuals typically consume an average of 3300 mg of sodium per day instead of the recommended dietary guideline of less than 2300 mg per day.

FOOD SCIENCE EXPERIMENT: Sodium Sources

2-1

Objective

- To discover how food preparation can affect sodium levels.

Materials and Equipment

- 2 small containers
- Kosher salt
- Measuring spoons
- Pen and paper

Sodium by Measurement
⅛ tsp kosher salt = 250 mg
¼ tsp kosher salt = 500 mg
½ tsp kosher salt = 1000 mg
¾ tsp kosher salt = 1500 mg
1 tsp kosher salt = 2000 mg

Sodium in Common Foods Based on 3 oz Serving Size

Food	Sodium	Food	Sodium	Food	Sodium
Beef, ground	47 mg	Chicken breast	70 mg	Pasta	5 mg
Broccoli, cooked	10 mg	Egg, whole, raw	74 mg	Pork, lean	65 mg
Butter, unsalted	8 mg	Lemon juice	1 mg	Pork, sausage	900 mg
Carrots	40 mg	Lettuce	9 mg	Tomato, fresh	3 mg
Cheese, Cheddar	620 mg	Milk	50 mg	Tomato, canned	130 mg
Cheese, Parmesan	460 mg	Mustard, yellow	1252 mg	Tomato, ketchup	1042 mg

Note: The recommended dietary guideline for sodium is less than 2300 mg per day.
Based on the DASH eating plan, sodium intake should be limited to 1500 mg per day.

Procedure

1. Fill a small container with kosher salt.
2. Take a "pinch" of salt from the container in the amount commonly used to season a dish, and place in the second container.
3. Measure the pinch of salt in the second container using a ½ tsp measure and record the results.
4. Create a dish using the list of common foods, recording the ingredients and sodium level of each ingredient. Include the sodium in the pinch or pinches of salt that may be used for seasoning.
5. Add the sodium level from each ingredient to determine the total amount of sodium in the dish.
6. Compare the total sodium level in the dish to the sodium recommended by dietary guidelines and DASH.

Allergen Alert
The materials used in this experiment may contain one or more food allergens.

Typical Results

Note: In this example, ⅛ tsp represents a pinch of salt.

Ingredients in Chicken and Pasta Dish	Sodium
Chicken breast, 3 oz	70 mg
Pasta, 3 oz	5 mg
Tomato, canned, 3 oz	130 mg
Parmesan cheese, 1½ oz	230 mg
Two pinches of kosher salt for seasoning	500 mg
	TOTAL SODIUM: 935 mg

Obesity

Obesity is a medical condition characterized by an excess of body fat. According to the CDC, approximately 35% of adults and 17% of children and adolescents are considered obese. Obesity increases the risk of developing hypertension, cardiovascular disease, type 2 diabetes, and certain types of cancer. **See Figure 2-34.** Obesity is also the major form of malnutrition in the United States. *Malnutrition* is an adverse health condition caused by an excess or deficiency of nutrients.

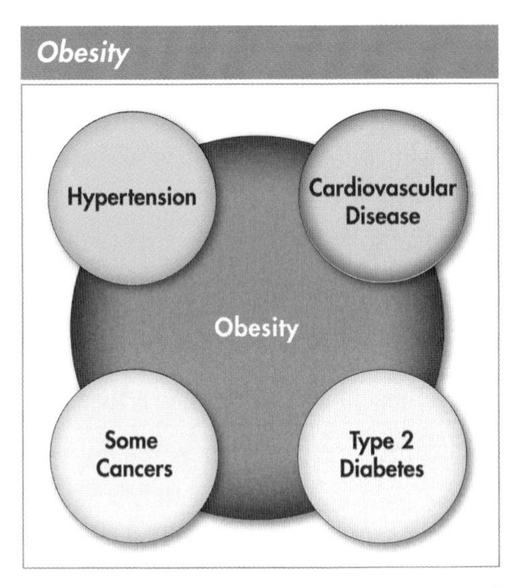

Figure 2-34. Obesity increases the risk of developing hypertension, cardiovascular disease, type 2 diabetes, and certain types of cancer.

Cardiovascular Disease

A *cardiovascular disease,* also known as heart disease, is a class of medical conditions that affect the heart and blood vessels. Cardiovascular disease is the primary cause of death in the United States. Eating a diet rich in dietary fiber and low in saturated fats has shown to reduce the risk of certain types of cardiovascular disease. Similarly, a diet rich in unsaturated fats, such as those found in avocados, olives, and nuts, can also promote heart health. Maintaining an appropriate weight, taking part in physical activity, and abstaining from smoking are additional ways to diminish the threat of cardiovascular disease.

Diabetes

Diabetes, scientifically known as diabetes mellitus, is a chronic disease characterized by elevated blood sugar levels and ineffective insulin production or absorption. *Insulin* is a hormone produced by the pancreas that is necessary for regulating blood sugar levels. Uncontrolled blood sugar levels can lead to serious ramifications such as amputations, blindness, heart and kidney disease, and premature death. There are two types of diabetes: type 1 diabetes and type 2 diabetes.

Type 1 Diabetes. *Type 1 diabetes* is a form of diabetes in which the pancreas does not produce enough insulin to normalize blood sugar levels. Because the body does not produce enough insulin, blood sugar levels remain elevated. Type 1 diabetes is the least common form of diabetes and generally occurs during childhood or young adulthood. Individuals with type 1 diabetes usually receive insulin through injections or via an insulin pump. It is also important for those with type 1 diabetes to monitor their diets to help manage blood sugar levels.

Type 2 Diabetes. *Type 2 diabetes* is a form of diabetes in which the body resists absorbing enough insulin to normalize blood sugar levels. Individuals with type 2 diabetes usually produce enough insulin, but because the insulin is not absorbed properly, blood sugar levels remain high. Type 2 diabetes accounts for approximately 80%–90% of all diabetes cases in the United States. Obesity is a major risk factor for developing type 2 diabetes. Weight reduction using sound nutrition principles and physical activity has a major influence in helping control blood sugar.

Foods with a lower glycemic index have been found to help control blood sugar levels for individuals with diabetes. The *glycemic index (GI)* is a measure indicating the rate at which an ingested food causes blood sugar levels to rise. Foods with a low GI, such as vegetables, have less effect on blood sugar levels than white rice, white bread, and most snack foods, which have a higher GI. **See Figure 2-35.**

Cancers

Cancer is a group of diseases characterized by abnormal cell growth. Cancers can disrupt the normal functioning of the organs and tissues due to the rapid multiplication of cancerous cells. Factors influencing the formation of cancer include obesity, physical inactivity, diets high in saturated fats, cigarette smoking, and a family history of cancer.

Food Allergies

A food allergy involves an abnormal response by the immune system after exposure to the offending food. An estimated 2% of children and about 6% of adults are affected by food allergies. The symptoms of food allergies are very specific and include swelling of the lips, itching, hives, rashes, vomiting, diarrhea, and breathing problems. Anaphylaxis is the most severe type of allergic reaction and can often be fatal because the airways close and block the ability to breathe. Allergic reactions can occur immediately upon contact with food, or several hours after exposure.

In 2004, the FDA developed the Food Allergen Labeling and Consumer Protection Act. This act states that food labels must identify the eight foods that account for 90% of all food allergies: milk, eggs, fish, shellfish, tree nuts, peanuts, wheat, and soy. **See Figure 2-36.** In addition to reading the menu, some guests may request a list of ingredients to help identify menu items they must avoid. This essential information helps individuals with food allergies prevent an allergic reaction by avoiding the food that triggers it.

Glycemic Index (GI)

Low-GI Foods (GI of 55 or less)		Medium-GI Foods (GI between 56 and 69)		High-GI Foods (GI of 70 or more)	
Yogurts, low-fat	14	Mangoes	56	Bagels, white	72
Spinach	15	Apricots	57	Watermelon	72
Tomatoes	15	Potatoes, new	57	Roll, kaiser	73
Cherries	22	Rice, wild	57	Graham crackers	74
Lentils	29	Bread, rye	64	Vanilla wafers	77
Spaghetti, whole wheat	37	Raisins	64	Pretzels	81
Apple juice	41	Oatmeal, instant	66	Honeys	87
Barley	48	Pineapples	66	Parsnips	97
Rice, brown	55	Cornmeal	68	Dates	103

Figure 2-35. Foods with a low GI have less of an effect on blood sugar levels than foods with a high GI.

Sources of Common Food Allergens

Allergen	Potential Food Sources
Milk	• Many nondairy products contain casein (a milk derivative) • Some brands of tuna fish contain casein • Butter melted on top of grilled steaks (often not visible) or foods cooked in butter • Many baked products contain dry milk
Eggs	• Egg whites are often used to create the foam topping on specialty coffee drinks • Egg wash is sometimes used on pretzels before they are dipped in salt • Some brands of egg substitutes contain egg whites • Most processed cooked pastas (including those in prepared soups) contain eggs or are processed on equipment shared with egg-containing pastas
Fish	• Imitation fish or shellfish • Salad dressings, Worcestershire sauce, and sauces made with Worcestershire
Shellfish	• Many Asian dishes include fish sauce • Shellfish protein can become airborne in the steam released during cooking
Tree nuts	• Salads and salad dressings, barbeque sauce, and honey • Meat-free burgers, fish dishes, pancakes, and pasta • Piecrust and the breading on meat, poultry, and fish
Peanuts	• Cereals, cookies, crackers, granola, and muffins • Cakes, ice creams, pies, and puddings • Sauces and marinades
Wheat	• Some brands of hot dogs, imitation crabmeat, and ice cream • Some Asian dishes contain wheat flour shaped to look like beef, pork, or shrimp
Soy	• Baked goods, cereals, crackers, infant formula, soups, and sauces • Some brands of tuna and peanut butter

Figure 2-36. Eight foods account for 90% of all food allergies. Those foods include milk, eggs, fish, shellfish, tree nuts, peanuts, wheat, and soy.

Food Intolerances

Unlike food allergies, a food intolerance does not involve the immune system. Symptoms of food intolerances are less severe and may include bloating, abdominal cramps, nausea, and diarrhea. Like food allergies, there are a variety of foods that can cause an adverse reaction. However, the most common food intolerances are gluten intolerance and lactose intolerance.

Gluten Intolerances. *Gluten* is a type of protein found in grains such as wheat, rye, barley, and some varieties of oats. Gluten is prevalent in flour and flour-based products such as breads, cereals, pastas, and baked goods. It can also be found in other foods such as bouillon cubes, soy sauce, and deli meats. **See Figure 2-37.** Gluten-free grains include quinoa, rice, and corn.

Foods Containing Gluten

- Bouillon cubes
- Brown rice syrup
- Candies
- Croutons
- Gravies
- Hot dogs
- Malt vinegars
- Panko
- Processed deli meats
- Roux
- Soup bases and soups
- Soy sauce
- Sausages

Figure 2-37. Gluten is found in flour-based products, such as breads, cereals, pastas, and baked goods, as well as other types of foods.

Indian Harvest Specialtifoods, Inc./Rob Yuretich

Rice is considered a gluten-free grain.

Celiac disease is a condition in which gluten damages the ability of the small intestine to absorb nutrients. When gluten is ingested, an inflammatory reaction in the small intestine causes the villi to flatten. This can result in malnutrition due to the decreased ability of the villi to absorb nutrients. It is, therefore, essential for a person with celiac disease to abstain from eating gluten. Gluten-free menu offerings can provide guests with safe and healthy alternatives.

Lactose Intolerances. Lactose is a sugar found in milk and dairy products. In addition to milk, yogurts, and cheeses, lactose can be found in many processed foods such as salad dressings, soups, and breads. Lactose intolerance is an inability to properly digest lactose. Alternatives such as lactose-free milk, almond milk, or soy products can be offered for those who are lactose intolerant.

Knowledge Check 2-5

1. Name the eating plan found to help promote a healthy blood pressure.
2. Identify four health risks associated with obesity.
3. Summarize what fats in the diet have been shown to reduce the risk of heart disease.
4. Differentiate between type 1 and type 2 diabetes.
5. Identify three factors that may influence the formation of cancer.
6. Explain why anaphylaxis is the most severe food allergy reaction.
7. Identify the eight food allergens required to be identified on all food labels.
8. Compare and contrast the symptoms associated with food allergies and food intolerances.
9. Identify five products that contain gluten.
10. Name three gluten-free grains.
11. Name three lactose-free products that may be alternatives to products with lactose.

ROLE OF MENUS

A *menu* is a document that markets a foodservice operation to guests. A menu has two primary functions. First, a menu functions as a communication tool, providing information such as item availability, preparation techniques, ingredients, and costs. Second, a menu is an essential sales tool and can be used to promote high profit items. The menu should reflect the ambience of the operation, be visually appealing, and have enticing descriptions.

Menu planning is critical to the success of a foodservice operation because it helps define the needs of the operation. The recipes used to create a menu play a vital role in the layout of tools and equipment, inventory of items, and staffing needs. Menus also help identify target markets, determine advertising and marketing strategies, and are useful for analyzing profits and losses.

The role of menus in regard to offering nutritious choices has become increasingly important as individuals are more aware of how food contributes to overall well-being. Menus can help shape healthy eating behaviors with recipes that focus on high quality ingredients, such as fresh produce, whole grains, and lean meats, poultry, and seafood. Developing and modifying recipes that focus on appropriate portion sizes and healthy preparation techniques result in visually appealing, flavorful, and nutritious dishes.

Portion Sizes

Portion sizes have steadily increased over time and often create a misconception between what is served and what is healthy. **See Figure 2-38.** According to the CDC, individuals eat up to 30% more food when presented with larger portions. This often leads to the consumption of too many calories, resulting in weight gain.

Chef's Tip

Reducing the portion sizes of animal-based foods and increasing the amounts of whole grains and fresh produce can often be a cost-effective way to create more nutritious dishes.

Portion Sizes and Calorie Comparisons

Common Portion Sizes	Calories in Common Portion Sizes	Appropriate Portion Sizes	Calories in Appropriate Portion Sizes
Chicken stir-fry: 4½ cups	865	Chicken stir-fry: 2 cups	540
Cheesecake: 7 oz	640	Cheesecake: 3 oz	260
Chocolate chip cookie: 5-inch diameter	170	Chocolate chip cookie: 2-inch diameter	70
French fries: 7 oz	610	French fries: 2.5 oz	210
Hamburger: 6 oz on kaiser roll	690	Hamburger: 3 oz on whole wheat bun	310
Muffin: 4 oz	500	Muffin: 1.5 oz	210
Spaghetti: 2 cups with 1 cup marinara sauce and four 1 oz meatballs	1000	Spaghetti: 1 cup with ½ cup marinara sauce and four ½ oz meatballs	500

Figure 2-38. Portion sizes have steadily increased over time.

Guests have become accustomed to seeing plates overfilled with food and often perceive larger portions as a better value. Presenting food on smaller plates is one way to alter this perception. For example, a 12-inch plate can be used instead of a 14-inch plate, and an 8-inch plate can be used instead of a 10-inch plate. **See Figure 2-39.**

In addition to value, recent surveys show that individuals desire to see a push toward health and nutrition when dining out. There are several ways to capitalize on this trend using portion sizes. For example, offering lower priced entrées that are one-quarter to one-half the size of traditional menu items encourages guests to choose appropriately sized meals. Another option is to promote larger-sized appetizer selections that are specifically designed for sharing. Finally, a small sampling of mini desserts can help promote moderation. In addition to portion size, recipes and ingredients should be examined and suitable modifications should be made so that meals are both flavorful and highly nutritious.

Culinary Nutrition Recipe-Modification Process

In order to include more nutritious foods on the menu, it is important to understand how to modify recipes. The purpose of the culinary nutrition recipe modification is to change an existing recipe into a healthier one, while maintaining the texture, flavor, and overall appeal of the original dish.

Recipes can be modified to reduce the amounts of sugars, fats, sodium, and/or calories. Recipes can also be modified for specific health or dietary reasons. For example, a traditional recipe can be made gluten-free or a meat-based recipe can be modified to become a vegetarian option. The six steps of the culinary nutrition recipe-modification process include evaluating the original recipe for sensory and nutritional qualities, establishing goals, identifying modifications or substitutions, determining the functions of identified modifications or substitutions, selecting appropriate modifications or substitutions, and testing the modified recipe to evaluate sensory and nutritional qualities. **See Figure 2-40.**

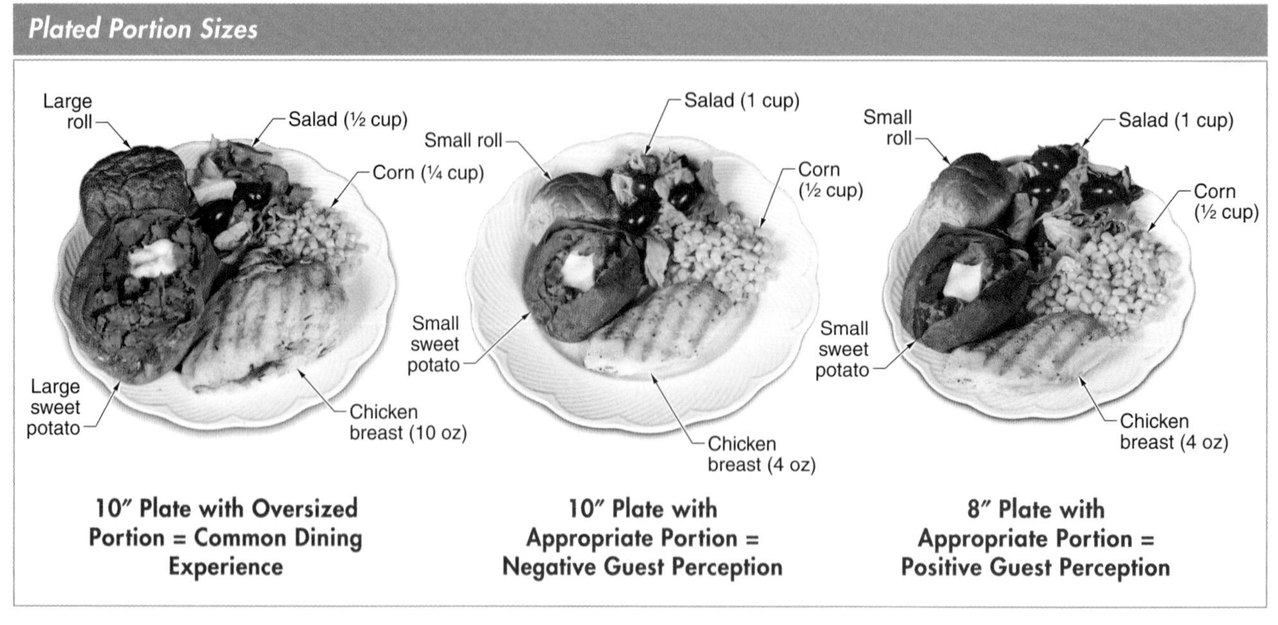

Figure 2-39. Using a smaller plate with an appropriate portion size can result in a nutritious dish and a positive experience for the guest.

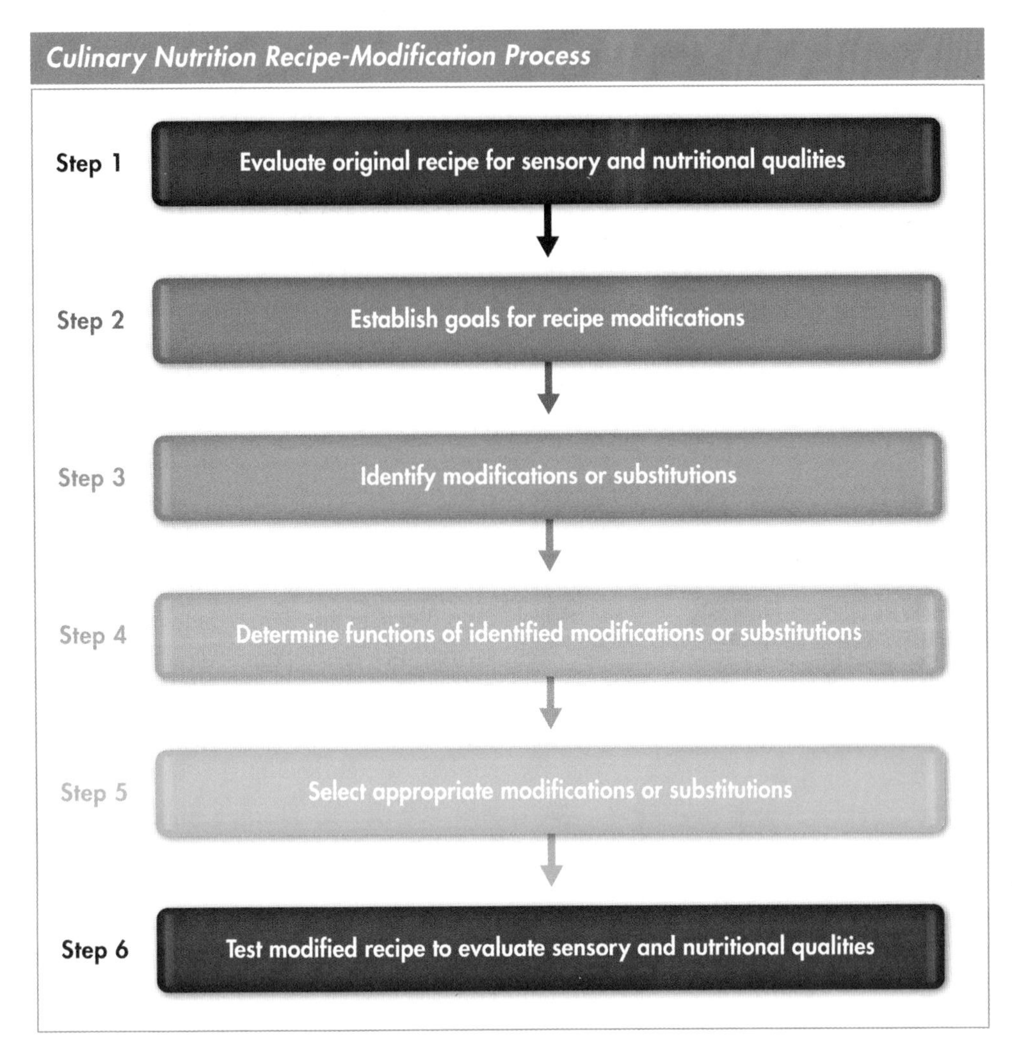

Figure 2-40. Following the six steps in the culinary nutrition recipe-modification process can help transform an existing recipe into a healthier dish.

Evaluating Recipes. The first step in the culinary nutrition recipe-modification process is to evaluate a recipe for sensory and nutritional qualities. Sensory qualities include components such as presentation, textures, tastes, aromas, and temperatures. These qualities should be upheld or enhanced in the modified recipe.

When analyzing a recipe for nutritional qualities, sugars, fats, sodium, and calories should be evaluated. It is equally important to consider how to add ingredients that will increase vitamins, minerals, and dietary fiber. Recipe modifications often entail the following:

- reducing the amount of sugars, fats, and/or high-sodium ingredients
- adding nutrient-dense foods including vegetables, fruits, and whole grains
- adding herbs and spices to enhance flavor
- changing from a high-fat to a low-fat preparation technique
- using rubs and marinades to tenderize and add depth of flavor
- using vegetable- or fruit-based reductions and sauces

Establishing Goals. The next step in the culinary nutrition recipe-modification process is to establish goals. This requires a decision regarding what ingredients or preparation techniques should be changed. Reducing sugars, fats, sodium, or calories may be the goal. In contrast, replacing dairy products with alternative ingredients that uphold texture and flavor may be the goal for creating a lactose-free dish.

Courtesy of The National Pork Board

Recipes can be modified to create healthier versions of a dish.

Identifying Modifications or Substitutions. Once a goal has been set for modifying a recipe, the next step in the culinary nutrition recipe-modification process is to identify the ingredients or preparation techniques to be altered. **See Figure 2-41.** If sodium reduction is the goal, all sources of sodium should be identified to see where it can best be minimized. This may include modifying or making substitutions for iodized salt, soy sauce, stocks, and convenience products.

Determining Functions. After identifying the ingredients to be modified, the next step in the culinary nutrition recipe-modification process is to determine the function of those ingredients in the recipe. Functions include factors such as structure, color, texture, and flavor. The identified ingredients often contribute flavor along with higher amounts of sugars, fats, and sodium. For example, sugars and salts are commonly added in recipes to enhance flavor but can be replaced by flavor enhancers such as herbs, spices, and extracts. Fats help spread flavor throughout a dish, but should be used in moderation. For example, by adding a finely shredded high-fat cheese, less cheese is needed to incorporate flavor.

Identifying Modifications or Substitutions

Meatloaf Recipe Modification
Goal: Reduce fats and sodium

Amount	Ingredients
1 tbsp	vegetable oil
3 oz	celery, small dice
4 oz	onion, small dice
2 fl oz	milk
2	eggs, beaten
1 c	bread crumbs
1 tbsp	salt
1½ tsp	black pepper
1½ tsp	thyme
3 lb	ground beef

Photo Courtesy of the Beef Checkoff

Figure 2-41. Once a goal has been set for modifying a recipe, the next step is to identify the ingredient(s) to be altered.

Structure adds shape and form to a dish and comes from the main ingredients. Flours create the structure for breads, whereas grains, such as rice, provide structure on their own. Color can come from a variety of different ingredients, such as vegetables, fruits, and meats. One of the functions of using vegetables, such as red peppers, peapods, and carrots, in a stir-fry is to enhance color for presentation. All foods provide texture. For example, nuts and seeds add a crunchy texture, dried fruits add an appealing chewiness, and puréed vegetables can create an enticing texture that is silky and smooth.

Selecting Appropriate Modifications or Substitutions. The fifth step of the culinary nutrition recipe-modification process is to select appropriate modifications or substitutions. An appealing final product can often be produced simply by reducing or eliminating the identified ingredient. If this is not the case, it is important to determine whether the ingredient can be modified or substituted for a more nutritious alternative. **See Figure 2-42.** For example, ground beef adds both fat and structure to meatloaf. Appropriate modifications include lower-fat varieties of ground beef, ground chicken, ground turkey, or substituting some of the meat with another ingredient, such as lentils or beans, that would provide structure.

Chef's Tip

For a lactose-free dessert, the soft and creamy texture of silken tofu can often be used as an effective substitute for dairy products in smoothies, puddings, mousses, and cream pies.

Ingredient Modifications

Original Ingredient	Ingredient Options
Bacon	Turkey bacon, Canadian bacon
Bread crumbs	Oat flour
Cheeses (1 cup coarsely shredded)	½ cup finely shredded, low-fat varieties
Cream	Half-and-half, evaporated skim milk
Eggs	Two egg whites for each whole egg
Ground beef	Extra-lean or lean ground beef or poultry
Iceberg lettuce	Mixed greens, Romaine lettuce, spinach
Marinades with oil	Wine, vinegars, fruit juice
Nuts (1 cup coarsely chopped)	½ cup finely chopped and toasted nuts
Ricotta or cottage cheese	Low-fat ricotta or cottage cheese
Salt	Herbs, spices, citrus zest/juice, vinegars
Sour cream	Low-fat sour cream, Greek yogurt
White rice	Brown or wild rice, barley, bulgur, quinoa
Whole milk	2%, 1%, or skim milk

Figure 2-42. Ingredients can often be modified or substituted for more nutritious alternatives.

Testing Modified Recipes. The last step in the culinary nutrition recipe-modification process is to test the modified recipe to evaluate sensory qualities and nutritional values. At this point, it is important to evaluate how the modified recipe compares in presentation, texture, taste, aroma, temperature, and nutrient value to the original recipe. It is not uncommon to make additional changes to the recipe before it is deemed appropriate for the menu.

Knowledge Check 2-6

1. Explain how menus can help shape healthy eating behaviors.
2. Give an example of how to effectively use plate sizes to control portions.
3. List the six steps in the culinary nutrition recipe-modification process.
4. Identify five qualities to consider when evaluating a traditional recipe for modification.
5. Give an example of a recipe modification goal.
6. Name three ingredients that can potentially increase the sodium content of a dish.
7. Identify four ways ingredients can function in a dish.
8. Give three examples of ingredient substitutions.
9. Identify three items to compare when evaluating the modified recipe against the traditional.

Chapter Summary

Eating a varied diet provides the body with essential nutrients in the form of proteins, carbohydrates, lipids, water, vitamins, and minerals. The process of digestion breaks down nutrients so they can be used to support functions necessary for energy production, cellular repair, and body maintenance.

The body receives energy from food in the form of calories. Research indicates that consuming the appropriate amount of calories from nutrient-dense foods promotes health and well-being. In addition to eating a nutritious diet, physical activity encourages an appropriate weight while reducing the risk of diet-related ailments.

Menus can promote healthy eating behaviors by offering nutrient-dense dishes, which often entails modifying an existing recipe. The culinary nutrition recipe-modification process allows foodservice professionals the opportunity to follow a step-by-step procedure that modifies an existing recipe into a healthier version that upholds the flavor and integrity of the original dish.

Photo Courtesy of Perdue Foodservice, Perdue Farms Incorporated

Chapter Review

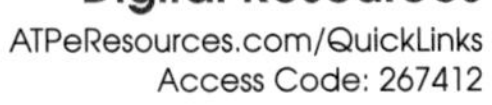

1. Summarize the six essential nutrients.
2. Explain the process of digestion as food moves through the mouth, stomach, and small and large intestines.
3. Identify the average number of calories needed per day by males and females.
4. Give six examples of factors that influence caloric needs.
5. Summarize the *Dietary Guidelines for Americans*.
6. Identify the recommended daily serving sizes from each of the five food groups.
7. Describe two diseases that have a connection to diet.
8. Explain the two primary functions of a menu.
9. Explain the purpose of the culinary nutrition recipe-modification process.

Proteins

Complete Proteins

American Lamb Board

Incomplete Proteins

Barilla America, Inc

Complementary Proteins

Alpha Baking Co., Inc.

Protein is an essential macronutrient necessary for maintaining a variety of body functions. Current research has increased the awareness of the benefits of protein as well as health concerns associated with excess protein consumption. As a result of this information, many individuals are adopting healthier eating habits and foodservice operations are reflecting this change. The oversized portions of animal-based proteins that once occupied the center of the plate are often being reduced in size or replaced with plant-based proteins to reflect the increased demand for more nutritious dishes.

Chapter Objectives

1. Describe how amino acids contribute to the structure of proteins.
2. Summarize the functions of proteins.
3. Explain how proteins are digested and absorbed by the body.
4. Identify dietary sources of proteins.
5. Describe how protein consumption affects health.
6. Explain how preparation and cooking processes change proteins.
7. Describe how proteins can be showcased on a menu.

PROTEIN STRUCTURE

The word "protein" is derived from the Greek word "proteos," which means "prime importance." Proteins are of prime importance because they build, repair, and maintain all cells and tissues throughout the body and support vital internal functions. For example, without protein, muscles, teeth, hair, and nails would cease to grow; infections, cuts, and burns would not heal; and internal chemical reactions necessary for life would not occur.

There are thousands of different proteins, and each protein has its own specific structure and function. However, all proteins are made of the same basic elements. A *protein* is an energy-providing nutrient made of carbon, hydrogen, oxygen, and nitrogen assembled in chains of amino acids. The three energy-providing nutrients, which are proteins, carbohydrates, and lipids, all contain carbon, hydrogen, and oxygen. Nitrogen is the element that differentiates proteins from carbohydrates and lipids. **See Figure 3-1.** The presence of nitrogen also contributes to the structure and functions of protein by helping form amino acids.

Amino Acids

An *amino acid* is a chemical compound consisting of carbon and hydrogen with a nitrogen-containing amino group at one end, an acid group at the other end, and a distinct side chain. It is the presence of nitrogen that allows amino acids to build and assemble specialized proteins throughout the body. This is why amino acids are commonly referred to as "the building blocks of protein."

There are 20 amino acids found in nature. Each type of amino acid has the same "backbone." The backbone of an amino acid consists of carbon and hydrogen atoms situated between the nitrogen-containing amino group on one side and the acid group on the other side. The backbone does not change, but the side chain can vary in shape, size, and electrical charge to give each amino acid its unique structure and function. **See Figure 3-2.**

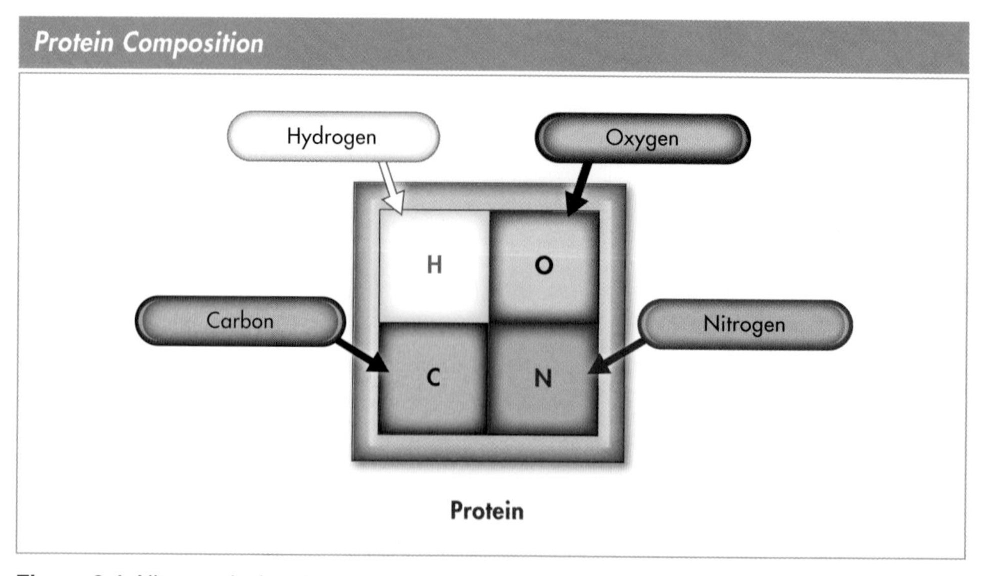

Figure 3-1. Nitrogen is the element that differentiates proteins from carbohydrates and lipids.

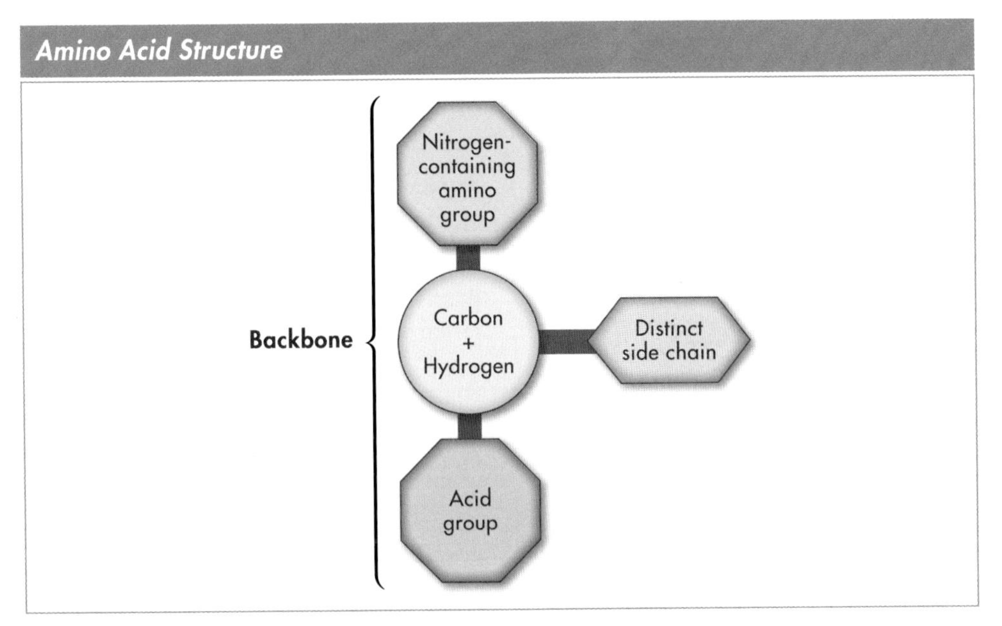

Figure 3-2. The backbone of an amino acid does not change, but the side chain can vary in shape, size, and electrical charge and gives each amino acid its unique structure and function.

The body has the ability to produce some amino acids, while others must be obtained from food. An *essential amino acid* is an amino acid the body cannot make in sufficient quantities and must be obtained from dietary sources. A *nonessential amino acid* is an amino acid that can be made by the body in sufficient quantities. **See Figure 3-3.**

Amino Acids Required by Adults

Essential	Nonessential
• Histidine	• Alanine
• Isoleucine	• Arginine
• Leucine	• Asparagine
• Lysine	• Aspartic Acid
• Methionine	• Cysteine
• Phenylalanine	• Glutamic Acid
• Threonine	• Glutamine
• Tryptophan	• Glycine
• Valine	• Proline
	• Serine
	• Tyrosine

Figure 3-3. An essential amino acid must be obtained from dietary sources, whereas a nonessential amino acid is made by the body.

Wellness Concept

Phenylketonuria (PKU) is a genetic disorder in which the body is unable to convert the protein phenylalanine into tyrosine, which is a nonessential amino acid. A screening test for PKU is performed on infants to help prevent brain damage.

Sometimes a nonessential amino acid is considered a conditionally essential amino acid. A *conditionally essential amino acid* is a nonessential amino acid that the body cannot make in sufficient quantities. This is usually a temporary situation and occurs during times of stress, injury, physical exertion, or illness.

Protein Formation

To form a protein, the backbones of amino acids must connect in a peptide bond. A *peptide bond* is a bond that connects a nitrogen-containing amino group of one amino acid to the acid group of another amino acid to form a link in a protein chain. **See Figure 3-4.** Protein chains can contain as few as 35 peptide bonds or up to several hundred.

When amino acids first link together, they form a straight chain known as the primary structure of protein. A *primary structure* is a long, straight chain of amino acids linked by peptide bonds.

Immediately after a primary structure is formed, the strand of protein changes from its straight shape into a secondary structure, tertiary structure, or quaternary structure. **See Figure 3-5.** A *secondary structure* is a coiled protein chain that has the appearance of a spring. A *tertiary structure* is a protein chain that folds back in an irregular pattern to form a compact spherical or globular shape. A *quaternary structure* is a complex protein chain that results from globular structures combining with each other.

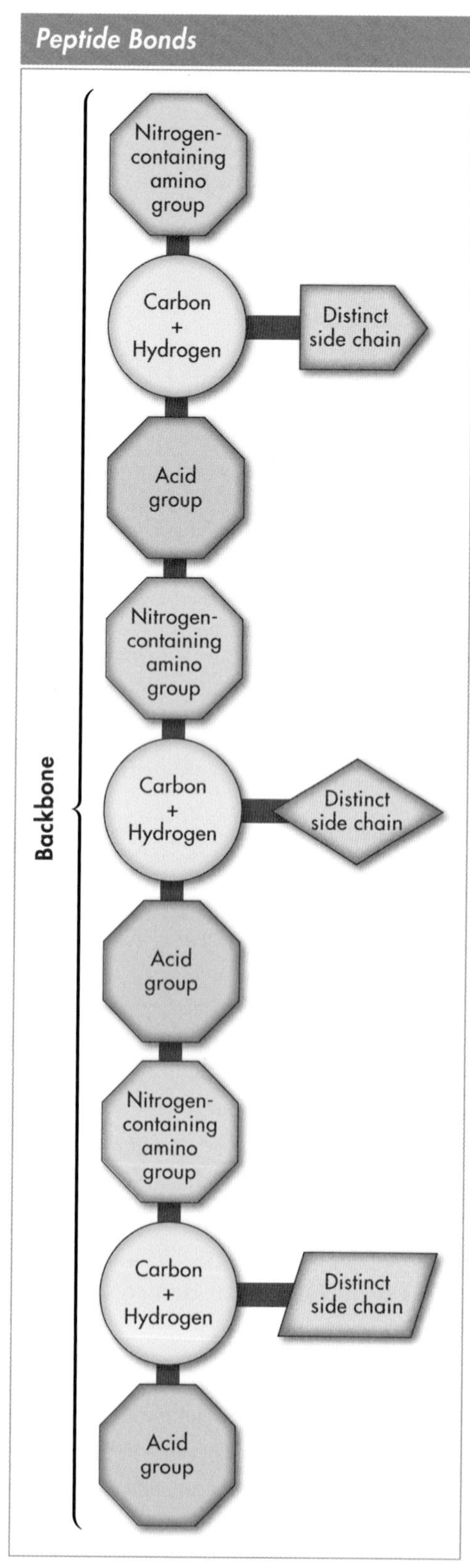

Figure 3-4. A peptide bond connects a nitrogen-containing amino group of one amino acid to the acid group of another amino acid to form a link in a protein chain.

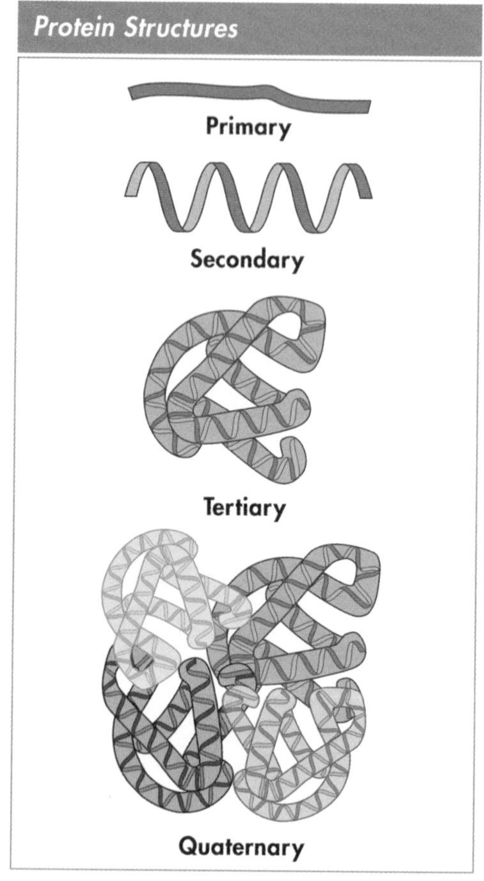

Figure 3-5. Immediately after the primary structure is formed, the strand of protein changes from its straight shape into a secondary structure, tertiary structure, or quaternary structure.

Knowledge Check 3-1

1. Name the element that differentiates proteins from carbohydrates and lipids.
2. Identify the number of amino acids found in nature.
3. Describe the "backbone" of an amino acid.
4. Identify the part of an amino acid that gives it a unique structure and function.
5. Differentiate between an essential and nonessential amino acid.
6. Describe a conditionally essential amino acid.
7. Describe a peptide bond.
8. Summarize the four structures associated with protein formation.

PROTEIN FUNCTIONS

Different amino acids can combine in countless ways to produce an almost infinite number of sequences. For example, one cell in the body can contain thousands of different proteins with different sequences of amino acids. The amino acid sequence enables each protein to perform a specific function. Some proteins work alone, while other proteins work together to carry out functions. The most important functions of proteins are to build, repair, and maintain the cells and tissues of the body. However, proteins are involved in other vital functions, such as regulating fluid and acid-base balances, transporting substances in and out of cells, and blood clotting. Also, protein can be used as energy if necessary and is involved in the formation and production of hormones, enzymes, and antibodies. **See Figure 3-6.**

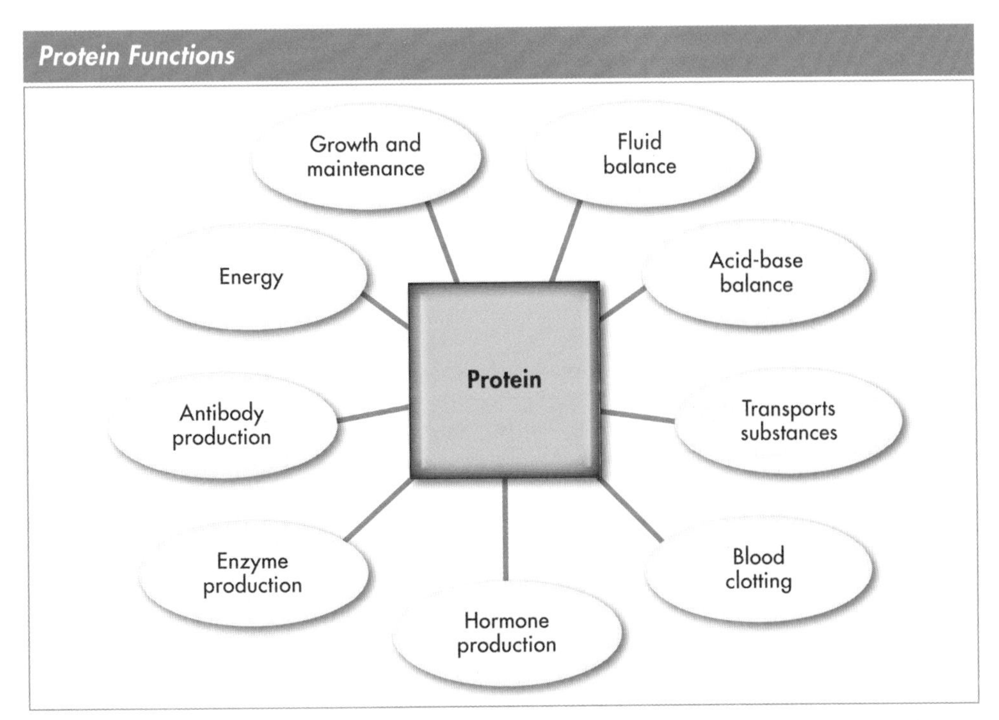

Figure 3-6. Proteins support tissue growth and maintenance, promote fluid and acid-base balance, transport substances in and out of cells, promote blood clotting, form hormones, act as enzymes and antibodies, and produce energy.

Growth and Maintenance

Proteins are essential for the growth and maintenance of living cells. They are found in every living cell throughout the body, including muscle fibers, tendons and ligaments, skin, bones, and blood. Cells are continually broken down, and proteins make replacing these cells possible. For example, proteins provide the necessary material for new cells to grow underneath the skin. Therefore, old skin cells that die and rub away are replaced. Red blood cells have a life span of three to four months and are replaced with new cells produced by bone marrow proteins. Even the proteins found within cells are continually replaced with new proteins.

Fluid Balance

Proteins keep optimal fluid balance between and within cells. Albumin, a protein in the blood, helps to push or attract fluid from the blood vessels. Without proper fluid balance, fluid retention takes place and causes edema. *Edema* is a condition in which body tissues swell due to fluid retention. **See Figure 3-7.** This condition is most evident in the hands, feet, and abdomen of protein deficient individuals.

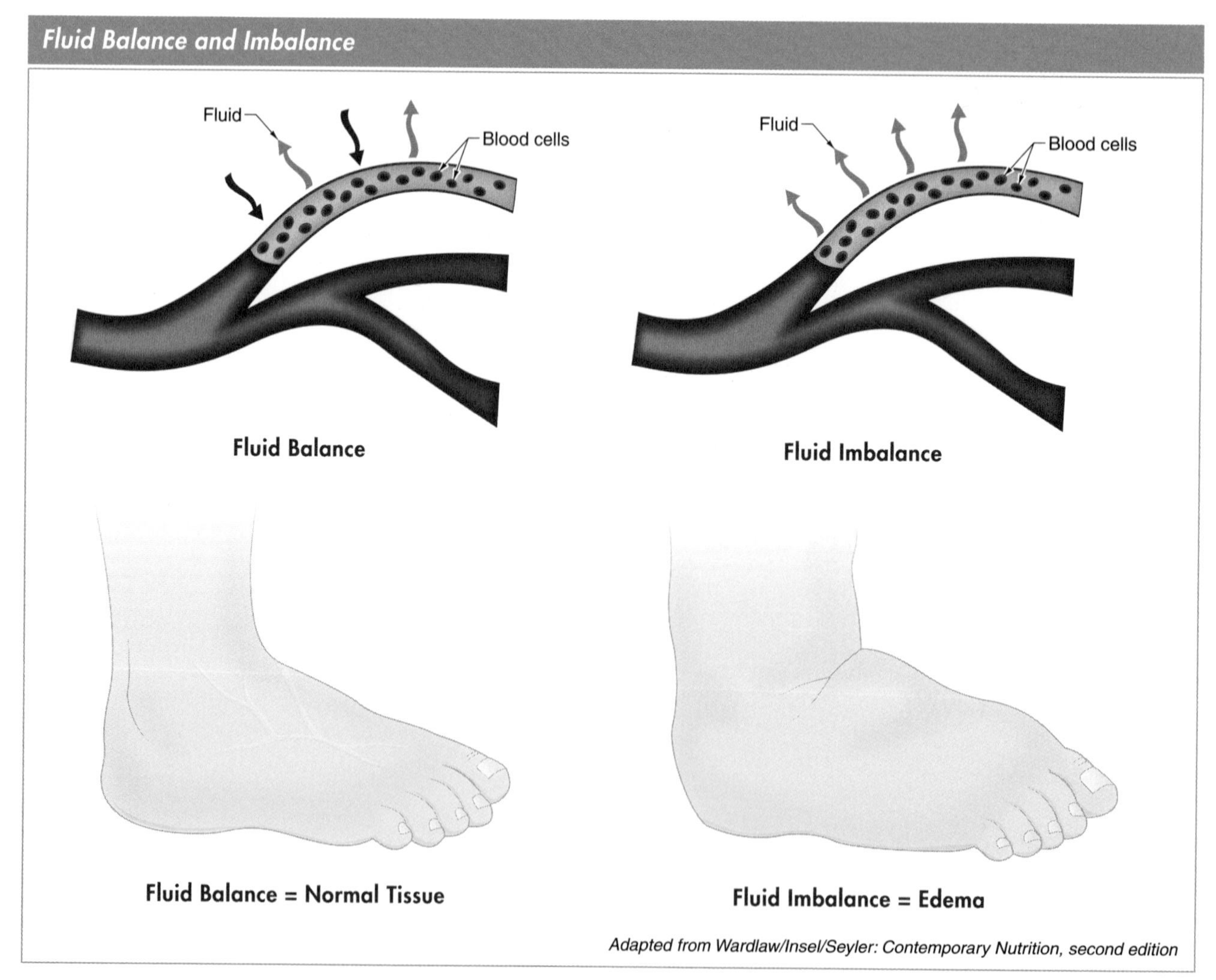

Figure 3-7. Edema is a condition in which body tissues swell due to fluid retention.

Acid-Base Balance

The body constantly produces acids and bases that are transported through the blood to excretion organs, such as the kidneys and lungs. For this to occur, the blood must maintain an acid-base balance. An *acid-base balance* is the state of equilibrium between acids and bases in body fluids. Acids and bases are opposites. An *acid* is a compound that releases hydrogen ions and produces a solution with a pH less than 7. A *base* is a compound that receives hydrogen ions and produces a solution with a pH greater than 7.

If the pH of blood varies from an optimal level of 7.4, proteins neutralize the acids and bases by releasing hydrogen ions when the blood is too acidic or accepting hydrogen ions when the blood is too basic. An acid-base imbalance can cause severe illness, such as acidosis or alkalosis, or death. *Acidosis* is a harmful condition in which the blood contains excess acids. *Alkalosis* is a harmful condition in which the blood contains excess bases.

Transportation

Proteins help transport lipids (fats) and fat-soluble vitamins throughout the body. In addition to transporting these nutrients, the protein hemoglobin transports oxygen in the bloodstream. *Hemoglobin* is a protein in red blood cells that transports oxygen throughout the body.

Blood Clotting

Fibrin is a protein fiber that forms a clot to stop bleeding. Protein fibrins interweave and form traps to help prevent the blood loss that results from cuts or wounds.

Hormone Production

A *hormone* is a chemical messenger secreted by an organ and sent to another destination in the body to deliver a message. Hormones help the internal environment of the body stay in balance. For example, when blood sugar is too high, the pancreas secretes the hormone insulin to regulate and balance blood sugar levels.

Enzyme Production

An *enzyme* is a type of protein that acts as a catalyst to accelerate chemical reactions. Enzymes cause chemical changes to occur, but the enzymes themselves are not changed chemically. There are thousands of different enzymes in the body that affect every process. Some enzymes, such as those involved with the digestion of food, break compounds apart. Other enzymes build compounds, such as teeth. There are also enzymes that synthesize compounds from other compounds. **See Figure 3-8.** Without enzymes, chemical reactions would occur only randomly.

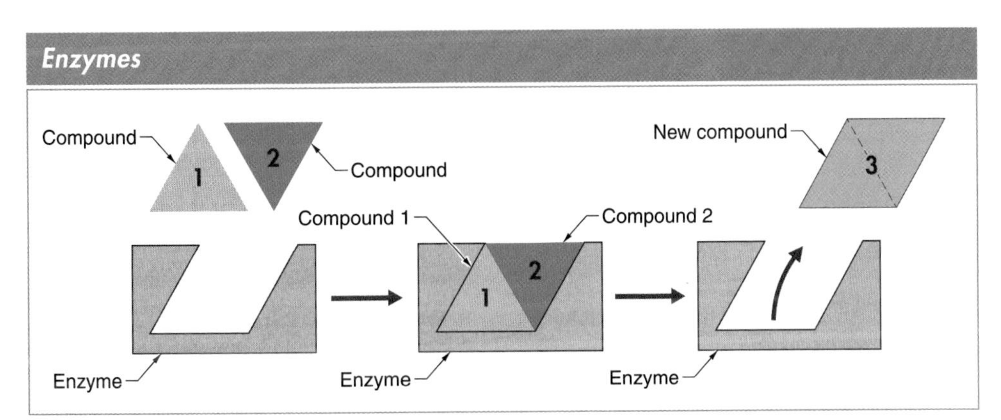

Figure 3-8. Some enzymes can synthesize compounds from other compounds.

Antibody Production

An *antibody* is a type of protein produced by the immune system that destroys or deactivates antigens. An *antigen* is a foreign substance the body considers harmful. Antigens include viruses, bacteria, and chemicals. Each antibody is unique and protects the body against a specific antigen. For example, in order to fight a specific cold virus, the body will produce a unique antibody to combat only that cold virus. If that cold virus invades again, the body will be more efficient at making that particular antibody and will, therefore, destroy the virus faster.

Energy

If the body lacks a sufficient supply of carbohydrates and lipids (fats) for energy, proteins will sacrifice their vital functions to provide 4 calories of energy per gram. When proteins are used for energy, nitrogen is removed. The remaining protein fragment is composed of carbon, hydrogen, and oxygen, which are the same energy-producing elements found in carbohydrates and lipids.

PROTEIN DIGESTION AND ABSORPTION

Most dietary protein is digested and absorbed over the course of several hours. During digestion, proteins must first be broken down into amino acids. This process begins in the stomach and continues in the small intestine. The strong acids in the stomach denature proteins to start digestion. *Denaturing* is the process of structurally changing protein from its natural state. **See Figure 3-9.**

As proteins become denatured, they start to untangle so that the peptide bonds can be broken by pepsin. *Pepsin* is a digestive enzyme in the stomach that breaks the peptide bonds connecting amino acids.

Knowledge Check 3-2

1. Identify nine functions of protein.
2. Give an example of how protein is used for growth and maintenance.
3. Describe the condition that occurs if proteins are unable to maintain fluid balance.
4. Differentiate between an acid and a base.
5. Explain how proteins help keep body pH levels at an optimal level.
6. Describe how proteins help prevent blood loss.
7. Describe the function of a hormone.
8. Summarize the role of enzymes.
9. Explain how antibodies protect the body.
10. Identify what happens to the main functions of proteins if they are used for energy.
11. Describe what happens to nitrogen when proteins are used for energy.

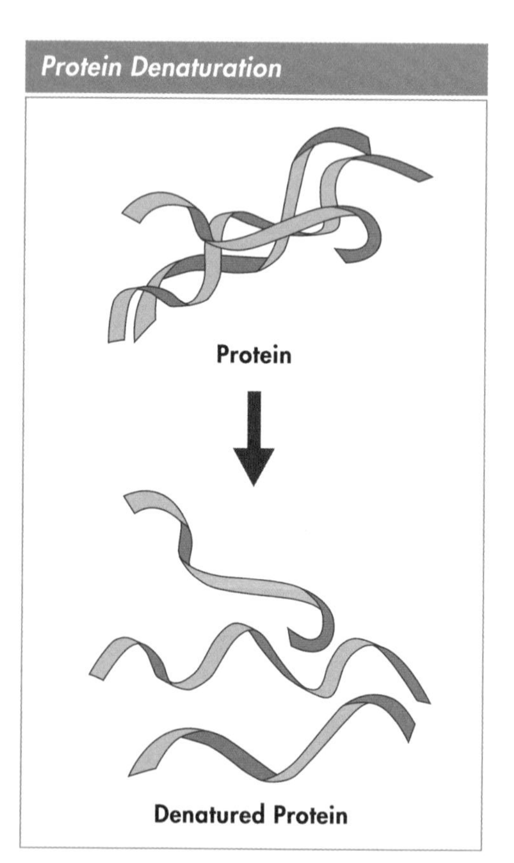

Figure 3-9. Denaturing is the process of structurally changing protein from its natural state.

As proteins travel from the stomach to the small intestine, some are single amino acids, but most are long strands of peptide bonds called polypeptides. A *polypeptide* is a protein fragment consisting of more than 10 peptide bonds. Unlike the acidic environment of the stomach, the small intestine has a neutral pH that allows enzymes called proteases to further break the amino acid bonds.

Protease is a digestive enzyme in the small intestine that breaks the peptide bonds connecting amino acids. The bonds are broken until the strands are single amino acids, dipeptides, or tripeptides. A *dipeptide* is a protein fragment consisting of two peptide bonds. A *tripeptide* is a protein fragment consisting of three peptide bonds. The amino acids are then absorbed by the cells of the small intestine, which break the remaining dipeptide and tripeptide bonds into single amino acids.

Once amino acids are absorbed into the cells of the small intestine, they travel in the blood to the liver. The liver determines where the amino acids are needed and sends them into the bloodstream to the required cells. **See Figure 3-10.** The cells can then link amino acids together to build proteins for their own use or send the proteins where they will be used in the body. The blood, liver, and body tissues constantly interchange amino acids depending upon need.

A short-term supply of amino acids is also maintained in the blood, organs, and cells by an amino acid pool. An *amino acid pool* is a collection of amino acids broken down from their original protein structure and distributed through the body to provide the material for cells to build new proteins. Amino acids broken down from dietary protein as well as disassembled proteins within the body supply the amino acid pool. This constant recycling of amino acids is part of protein turnover. *Protein turnover* is the continuous breakdown, rebuilding, and recycling of amino acids. To finally complete the process of digestion, waste is collected in the large intestine and excreted.

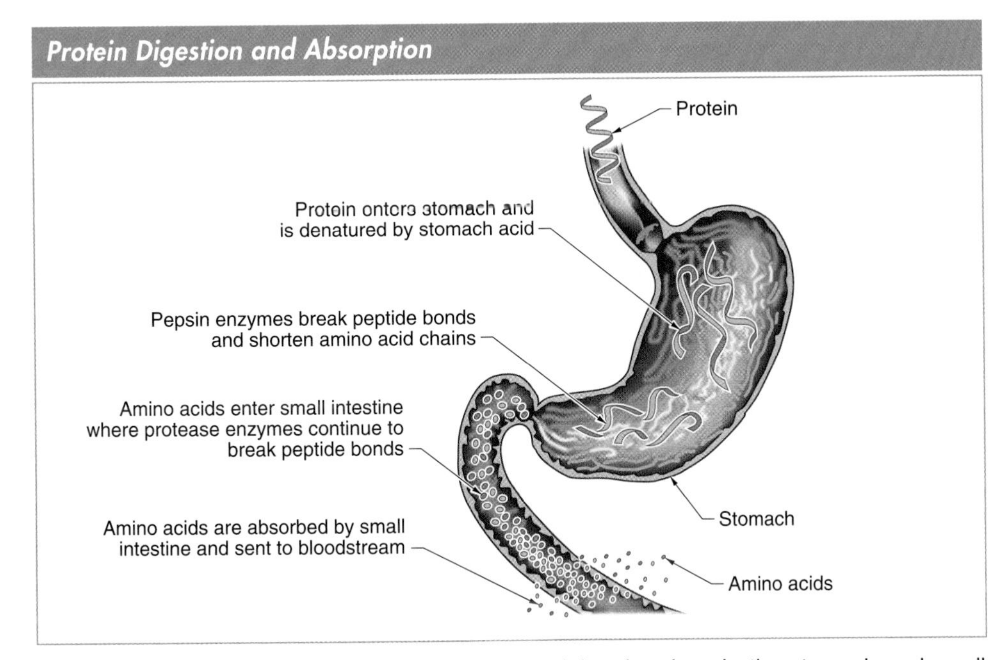

Figure 3-10. Protein digestion and absorption mainly take place in the stomach and small intestine.

Knowledge Check 3-3

1. Describe the process of protein digestion in the stomach.
2. Identify the stomach enzyme that breaks peptide bonds during digestion.
3. Differentiate between the pH levels in the stomach and the small intestine.
4. Identify the small intestine enzyme that breaks peptide bonds during digestion.
5. Differentiate between a polypeptide, dipeptide, and tripeptide.
6. Identify the location amino acids are initially sent once absorbed by the small intestine.
7. Summarize the digestive process after amino acids are sent to the liver.
8. Explain an amino acid pool.
9. Describe protein turnover.

PROTEIN SOURCES

Proteins can be found in both animal- and plant-based foods. They supply the body with the essential amino acids the body cannot make. Some protein foods supply the body with all the essential amino acids, and the foods are, therefore, complete protein foods. Other protein foods are considered incomplete and lack one or more of the essential amino acids.

Complete Proteins

A *complete protein* is a dietary source of protein that contains all the essential amino acids in the amounts needed by the body. The body uses the essential amino acids supplied by complete proteins to make nonessential amino acids. Most animal-based foods, such as meats, poultry, seafood, eggs, and dairy products, are considered complete proteins. Complete proteins are rarely found in plant-based foods. However, soybeans and quinoa contain complete proteins. Foods that contain all essential amino acids are typically considered high-quality protein sources. **See Figure 3-11.**

Incomplete Proteins

An *incomplete protein* is a dietary source of protein that lacks one or more of the essential amino acids. The majority of plant-based foods including vegetables, fruits, grains, legumes, nuts, and seeds are incomplete proteins. Although vital to a healthy diet, plant-based foods are often deemed low-quality protein sources.

California Strawberry Commission

The majority of plant-based foods are incomplete proteins.

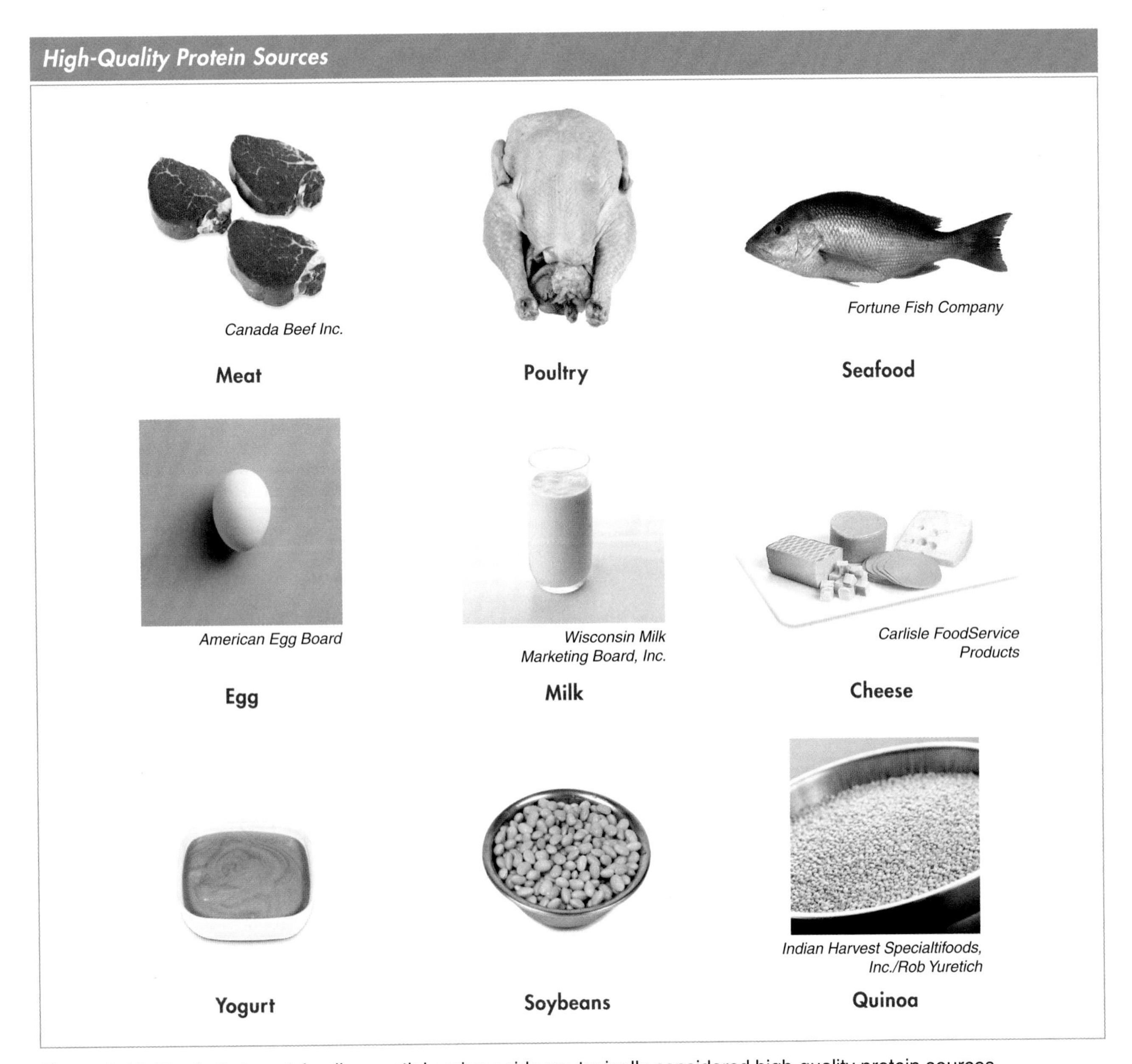

Figure 3-11. Foods that contain all essential amino acids are typically considered high-quality protein sources.

Incomplete proteins may be abundant in some amino acids but lack others. A *complementary protein* is a combination of two or more incomplete proteins that supplies sufficient amounts of all the essential amino acids. Although dairy products are complete proteins, they are often considered complementary proteins for some vegetarian diets. **See Figure 3-12.** An example of complementary proteins includes beans and rice. Beans are low in the essential amino acid methionine, whereas rice is low in lysine. By consuming beans with rice, a complete protein is created. It was once believed that complementary proteins had to be eaten together to form a complete protein. However, studies indicate that protein needs can be met by consuming a variety of incomplete proteins throughout the course of a day.

Complementary Proteins

Food Source	Complementary Food Source	Complete Protein Examples
Whole grains (Barilla America, Inc.)	Legumes	• Barley and lentil burgers • Whole grain pita chips with hummus • Brown rice and red beans
Whole grains (Barilla America, Inc.)	Nuts and seeds	• Spelt flour pumpkin spice muffins with pumpkin seeds • Whole grain couscous and pine nut-stuffed bell peppers • Granola with oats, coconut, and macadamia nuts
Nuts and seeds	Legumes	• Black bean chili with almonds • Lentil cakes with pecans • Mung bean noodles with sesame seeds

Figure 3-12. A complementary protein is a combination of two or more incomplete proteins that supplies sufficient amounts of all the essential amino acids.

Food Science Note

Although gelatin is an animal-based protein, it lacks all essential amino acids and is, therefore, an incomplete protein.

Knowledge Check 3-4

1. Differentiate between a complete and incomplete protein.
2. Identify five foods considered complete proteins.
3. List three plant-based foods considered complete proteins.
4. Identify six foods considered incomplete proteins.
5. Explain the meaning of a complementary protein.
6. Describe how a complete protein can be formed from the combination of incomplete protein foods.

PROTEIN CONSUMPTION

Because essential amino acids can only be replenished from food, it is important to consume protein daily. If an amino acid is not available in sufficient quantities, it is considered a limiting amino acid. Proteins that rely on that amino acid cannot form and carry out necessary functions. A *limiting amino acid* is an amino acid that is missing or unavailable in an adequate amount. Because amino acids contain nitrogen, a limiting amino acid can affect the nitrogen balance of the body. *Nitrogen balance* is the difference between nitrogen intake and nitrogen excretion.

Most healthy adults are in a state of nitrogen equilibrium in which the amount of protein ingested and excreted is the same. When protein needs are greater during periods of growth, injuries, and minor illnesses, nitrogen intake is higher than the level excreted. This creates a positive nitrogen balance because the body is adding protein. If the body is in negative nitrogen balance, more nitrogen is excreted than what is taken in and puts the body under extreme stress. Negative nitrogen balance can affect individuals recovering from surgery, with severe illnesses, and who fast for a prolonged period of time. **See Figure 3-13.**

Protein Recommendations

The recommended dietary allowance (RDA) for protein is approximately 56 g per day for an adult male and 46 g per day for an adult female. However, protein requirements vary throughout the life cycle and are affected by gender, age, activity level, and overall health. **See Figure 3-14.** To determine individual protein needs more accurately, 0.36 g of protein should be consumed per pound of body weight. For example, an individual weighing 150 lb needs approximately 54 g of protein each day (150 lb × 0.36 g = 54 g). Protein should also make up 15%–20% of the total daily caloric intake for most individuals. For example, in a diet of 2000 calories per day, between 300 and 400 calories should come from protein (2000 calories × 15% = 300 calories; 2000 calories × 20% = 400 calories).

Nitrogen Balance

Balance	Nitrogen Intake and Excretion	Conditions Affecting Nitrogen Balance
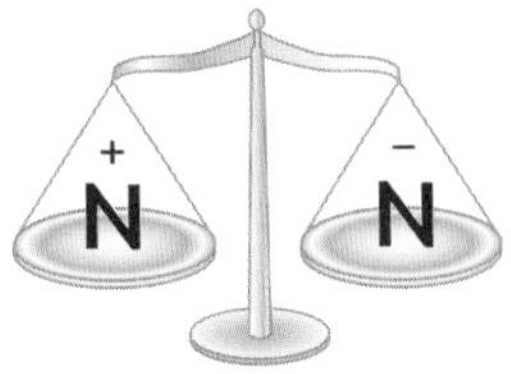 Nitrogen Equilibrium	Nitrogen intake equals nitrogen excreted	Most healthy adults are in a state of nitrogen equilibrium
 Positive Nitrogen Balance	Nitrogen intake is greater than nitrogen excreted	When protein needs are greater during periods of growth, injuries, and minor illnesses, individuals may experience a positive nitrogen balance
 Negative Nitrogen Balance	Nitrogen intake is less than nitrogen excreted	Negative nitrogen balance can affect individuals recovering from surgery, with severe illnesses, and who fast for a prolonged period of time

Figure 3-13. Nitrogen balance is the difference between nitrogen intake and nitrogen excretion.

Protein Recommended Dietary Allowances (RDAs)

Gender	Age	Daily Protein RDA
Males and females	1–3	13 g
Males and females	4–8	19 g
Males and females	9–13	34 g
Females	14–18	46 g
Males	14–18	52 g
Females	19–70+	46 g
Males	19–70+	56 g

Figure 3-14. Protein requirements vary throughout the life cycle.

Protein requirements are fairly easy to meet. Most people consume an adequate amount of protein throughout the day. **See Figure 3-15.** For example, 1 egg during breakfast, a 3 oz chicken breast during lunch, and 3 oz of salmon with 8 fl oz of milk during dinner yields approximately 59 g of protein. Likewise, consuming 1 cup of cooked oats topped with 1 oz of walnuts during breakfast, ½ cup of cooked black beans and 6 oz of Greek yogurt during lunch, and 1 cup of cooked quinoa and ½ cup of cooked soybeans (edamame) during dinner yields 56 g of plant-based protein.

Since most people receive enough protein through their usual diet, the Federal Drug Administration (FDA) does not require a percent daily value for protein on food labels. However, if a claim is made for the protein content of a particular food or if the food is designed for children less than four years of age, a percent daily value must be listed.

Protein Content of Common Foods

Food	Protein Content
Meat (3 oz cooked)	
• Top sirloin steak	26 g
• Flank steak	24 g
• Ground beef	22 g
• Pork tenderloin	26 g
Poultry (3 oz cooked)	
• Chicken breast	26 g
• Turkey breast	26 g
Seafood (3 oz cooked)	
• Cod	20 g
• Salmon	19 g
• Tuna	25 g
• Shrimp	18 g
Eggs	
• Egg, large, whole	6 g
• Egg, large, white	4 g
Legumes (½ cup cooked)	
• Black beans	8 g
• Chickpeas	8 g
• Soybeans (edamame)	11 g
Nuts and Seeds (1 oz)	
• Almonds	6 g
• Walnuts	7 g
• Sunflower seeds	6 g
Vegetables (1 cup cooked)	
• Asparagus	5 g
• Broccoli	5 g
• Brussels sprouts	4 g
• Spinach	5 g
Grains (1 cup cooked)	
• Brown rice	4 g
• Oats	6 g
• Pasta	7 g
• Quinoa	6 g
Dairy	
• Cheddar cheese, 1 oz	7 g
• Greek yogurt, 6 oz	12 g
• Milk, 8 fl oz	8 g

Figure 3-15. Protein requirements are fairly easy to meet. Most people consume an adequate amount of protein throughout the day.

Wellness Concept

There is a growing body of evidence that suggests plant-based protein sources are better for long-term health than animal-based proteins.

Protein Excess

High-protein diets are popular among some groups of people. However, medical research indicates that a diet containing excessive amounts of protein can have adverse health effects. For example, excess dietary protein can lead to weight gain, dehydration, gout, osteoporosis, and ketosis.

Weight Gain. Many protein foods that come from animal-based sources are high in saturated fats and calories. Diets high in saturated fats have been linked to cardiovascular disease and certain cancers. If not burned for energy, the excess calories will be stored for later use and lead to weight gain.

Dehydration. *Dehydration* is a condition in which the body does not have an adequate amount of water to function properly. High levels of protein intake can put a strain on the kidneys as they try to filter the extra protein from the body. As the kidneys labor to eliminate excess protein from the body, water is also excreted. If too much water is lost, the body can go into a state of dehydration and exhibit symptoms ranging from a dry mouth and sleepiness to extreme thirst or unconsciousness. **See Figure 3-16.**

Symptoms of Dehydration

Moderate Dehydration	Severe Dehydration
• Thirst • Slightly dry/sticky mouth • Sleepiness • Decreased urination • Few tears when crying • Dry skin • Headache • Dizziness	• Extreme thirst • Very dry mouth • Irritability/confusion • Very little/no urination • No tears • Rapid heartbeat/ breathing • Fever • Unconsciousness

Figure 3-16. If too much water is lost, the body can go into a state of dehydration and exhibit symptoms ranging from a dry mouth and sleepiness to extreme thirst or unconsciousness.

Gout. *Gout* is a condition in which an excess of uric acid builds up within the joints and results in joint pain and swelling. *Uric acid* is a product of nitrogen waste found in the urine and blood. Uric acid is typically eliminated from the body. When protein intake is excessive, the kidneys cannot process uric acid efficiently and the build-up leads to gout.

Osteoporosis. *Osteoporosis* is a condition characterized by a loss of bone density that results in porous, brittle bones that are susceptible to breakage. **See Figure 3-17.** Diets high in protein increase the excretion of calcium, which is a nutrient necessary for healthy bones. A lack of calcium can place the body at risk of developing osteoporosis.

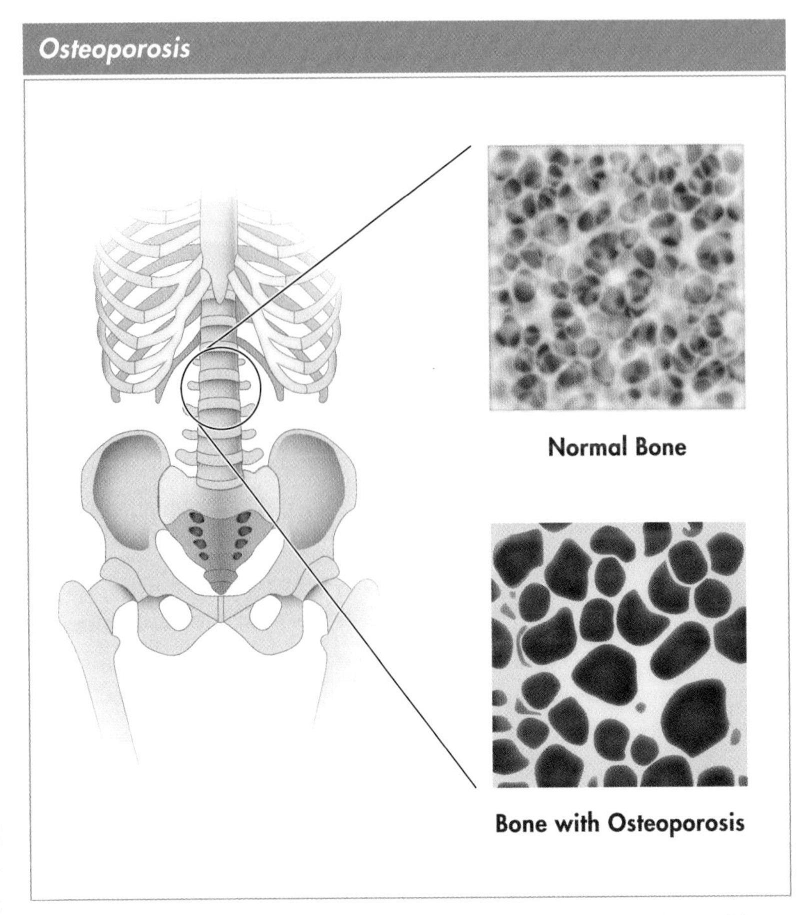

Figure 3-17. Osteoporosis is a condition characterized by a loss of bone density that results in porous, brittle bones that are susceptible to breakage.

Ketosis. *Ketosis* is a condition in which ketones build up in the body because lipids are improperly metabolized for energy. A *ketone* is an acidic substance that forms in the blood when lipids are metabolized for energy instead of carbohydrates. Ketosis is a result of diets in which excessive amounts of high-fat protein foods replace carbohydrate-rich foods, such as fruits, vegetables, and grains. When protein foods, which are high in lipids, are consistently metabolized for energy instead of carbohydrates, ketosis can lead to kidney failure.

Photo Courtesy of The National Pork Board

An appropriate portion size of protein can help promote good health.

Protein Deficiencies

When there is not a sufficient supply of protein, the body breaks down muscles to use as an energy source. This protein deficiency causes wasting of the muscles, organs, and tissues. A lack of protein can also impair the digestion and absorption of vitamins and minerals as well as cause fatigue, hair loss, hormone irregularities, and reduce the ability of the immune system to fight diseases. In severe cases, protein deficiencies lead to death.

Protein energy malnutrition (PEM), also known as protein-calorie malnutrition, is a general term used to describe various forms of malnutrition that result from a body deficient of protein.

PEM is found throughout the world and can affect any age group, but it is most prevalent in developing countries that lack adequate protein supplies. This can greatly impact the development of children because proper growth is dependent upon protein. In developed countries where PEM is rare, it is common among the economically disadvantaged, elderly, individuals participating in long-term fasting or starvation diets, and patients with serious medical conditions, such as cancer. The two basic forms of PEM include kwashiorkor and marasmus.

Kwashiorkor. *Kwashiorkor* is a form of PEM characterized by a diet deficient in proteins but adequate in caloric intake. The term "kwashiorkor" translates to "sickness of the weaning" because it most often affects infants when they are taken off protein-rich breast milk and fed a diet lacking protein. Kwashiorkor develops quickly and is characterized by moderate muscle wasting, which results in extremely thin arms and legs. Additional symptoms include a swollen abdomen, edema in the feet, stunted growth, a lowered immune system, dry skin, skin lesions, and changes in skin and hair color.

Marasmus. *Marasmus* is a form of PEM characterized by a diet deficient in both proteins and caloric intake. The term is derived from the Greek word "marasmos," which means "to wither away." Marasmus develops slowly and is characterized by severe muscle wasting and a skeletal appearance since the body breaks down its own tissues to use for energy. Growth and metabolism slows, body temperature drops, and lethargy can set in as energy stores are depleted.

Wellness Concept

A protein deficiency can result when individuals routinely consume 50%–75% of the recommended amount of daily protein.

Knowledge Check 3-5

1. Describe what happens if an amino acid is not available in sufficient quantities.
2. Differentiate between nitrogen equilibrium, positive nitrogen balance, and negative nitrogen balance.
3. Identify the adult male and adult female RDA for protein.
4. Explain how to determine individual protein needs.
5. List five potential consequences of consuming excess proteins.
6. Explain how protein excess can lead to dehydration.
7. Describe the relationship between consuming excess proteins and ketosis.
8. List five potential consequences for the body when protein supplies are insufficient.
9. Describe protein energy malnutrition (PEM).
10. Differentiate between kwashiorkor and marasmus.

PREPARING PROTEINS

During the preparation and cooking process, proteins are exposed to agents that can cause them to become denatured. This change in structure generally means that the amino acid chains unfold and realign, and the original characteristics or properties of the protein are altered in some way. Denatured proteins lose their biological activity as enzymes but not their nutritional value. This process is irreversible and usually improves the digestibility of protein. When preparing proteins, exposure to agents such as air, heat, moisture, and pH affect protein denaturation.

Reactions to Air

Air affects proteins due to the presence of oxygen, which can change the color of meat products. *Myoglobin* is an iron-containing protein that binds oxygen in muscle tissues. The color of meat is dependent upon how much myoglobin is present. For example, meat is somewhat purple in color when it is initially fabricated. When a meat becomes exposed to oxygen, some myoglobin is released and the meat turns red. After prolonged exposure to oxygen, more myoglobin is released and the meat turns reddish-brown. **See Figure 3-18.** Reddish-brown meat does not mean the meat is spoiled, only that the condition of the myoglobin has changed the color of the meat. Acidity, temperature, and salt also destabilize protein and cause color changes.

Photo Courtesy of D'Artagnan, Photography by Doug Adams Studio

The cooking process usually improves the digestibility of protein.

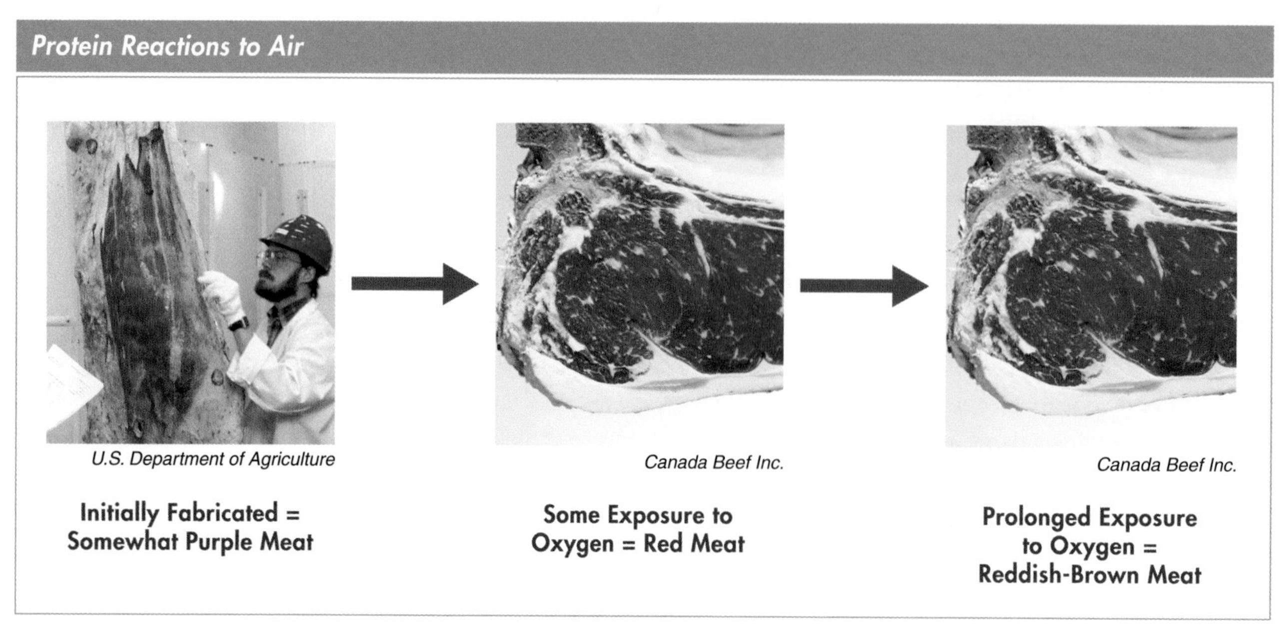

Figure 3-18. The color of meat is dependent upon how much myoglobin is present.

Eggs are a unique protein food in which the incorporation of air is often desirable. For example, air is incorporated when whipping egg whites to form peaks for foods such as meringues. The whipping action causes the protein bonds to break and re-form. As a result, a wall of protein strands is produced that hold air bubbles. When egg whites are overbeaten, the protein bonds accumulate and cluster tightly together. This causes air to be released, and the egg foam breaks down and collapses.

Reactions to Heat

Exposure to heat causes proteins to change in terms of color, texture, and flavor. In general, these changes become more pronounced the longer a protein is cooked. Coagulation and the Maillard reaction are examples of changes that occur when proteins react with heat.

Coagulation. When proteins are subjected to heat, they begin to coagulate. *Coagulation* is the process of a protein changing from a liquid to a semisolid or a solid state when heat or friction is applied. Coagulation takes foods from a high-moisture state to a low-moisture state, such as when a raw egg begins to coagulate as it is poached. Egg whites also coagulate due to friction as they are beaten into soft or firm peaks.

As proteins cook and coagulate, the bonds holding amino acid chains break apart and the protein strands unwind to become more linear. The unwound strands of protein then bump into each other and bind together in a different pattern, forming pockets that trap water. **See Figure 3-19.** As the protein continues to cook or denature, the protein bonds become tighter and squeeze the water from the pockets. This produces shrinkage and can lead to a tough and dry product if overcooked. *Shrinkage* is the loss of volume and weight of food as it cooks.

As a general rule, protein foods should be prepared using low to moderate heat to reduce shrinkage and uphold tenderness. Protein foods are sometimes cooked briefly at high heat to sear the outside. The heat is then reduced to complete the cooking process.

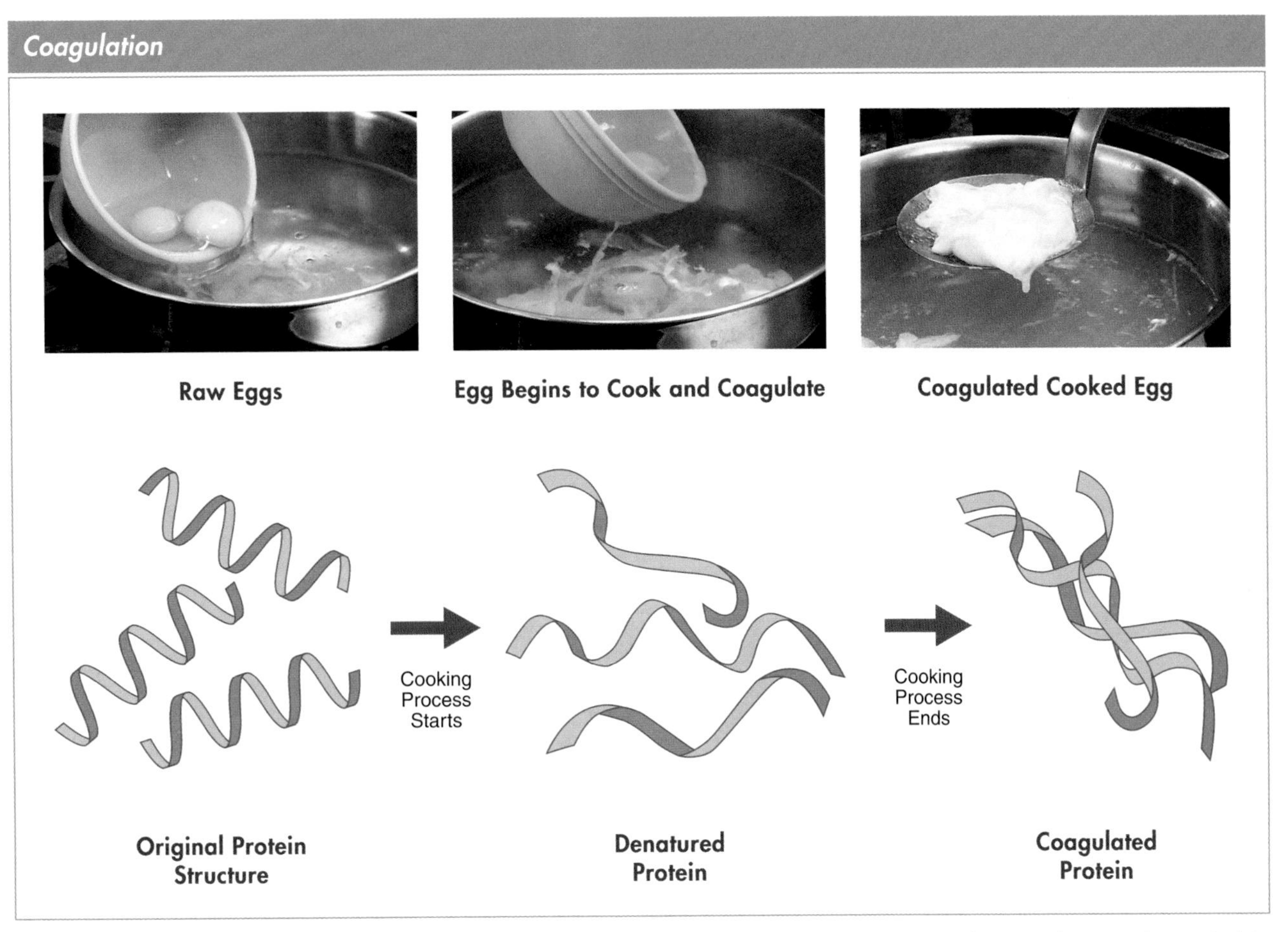

Figure 3-19. When proteins coagulate, the bonds holding amino acid chains break apart, the protein strands unwind to become more linear, and then the strands bind together in a different pattern.

Maillard Reaction. Heat also changes the color, texture, and flavor of proteins because of the Maillard reaction. The *Maillard reaction* is a reaction that occurs when the proteins and sugars in a food are exposed to heat and merge together to form a brown exterior. Searing, grilling, frying, and toasting foods are examples of cooking techniques that cause the Maillard reaction. **See Figure 3-20.** As food browns, aromatic compounds are released and flavor is heightened.

Reactions to Moisture

Protein molecules are surrounded by water and have the potential to absorb and hold water. This trait allows proteins to form a gel with a firm structure that can be used as a thickener. For example, the flesh of animal-based proteins, such as meat, poultry, and seafood, contains collagen. *Collagen* is a soft, white connective tissue that breaks down into gelatin when heated. When protein foods containing collagen are cooked in the presence of moisture, such as during braising, the dissolved collagen forms a gelatin that helps thicken the liquid. **See Figure 3-21.**

Wellness Concept

Research suggests charred foods that result from cooking over an open flame contain compounds that may increase the risk of certain cancers.

Maillard Reaction

Photo Courtesy of D'Artagnan, Photography by Doug Adams Studio

Seared Veal Chop

Venison World

Grilled Bison Ribeye

All-Clad Metalcrafters

Fried Chicken Cutlet

National Cancer Institute

Toast

Figure 3-20. Searing, grilling, frying, and toasting foods are examples of cooking techniques that cause the Maillard reaction.

Protein Reactions to Moisture

Figure 3-21. When protein foods containing collagen are cooked in the presence of moisture, such as during braising, the dissolved collagen forms a gelatin that helps thicken the liquid.

Reactions to pH

Acidic ingredients, such as citrus juices, vinegars, and wines, change protein foods from their original state. Acidic ingredients can produce desirable changes in foods, such as tenderizing meats. Other times, acidic ingredients can cause changes in textures and flavors that are unappealing. Reactions to pH include protein denaturation, changes in the ability of foods to brown, and the curdling of liquids that contain proteins.

Denaturation. Acids cause denaturation to occur in protein foods by changing the texture of the foods. For example, the acidic base of a marinade helps tenderize meats by breaking down the protein structure of the meat. Some protein foods are "cooked" in an acidic marinade. An example of this is ceviche, which is a dish containing raw fish that has been marinated in citrus juice. The acids from the citrus juice unravel the protein molecules in the fish and change its properties. The result is a firm flesh that is opaque. If allowed to marinate in an acid too long, protein molecules continue to become denatured, and foods will have a mushy texture.

Browning. Acids also prevent foods from browning. Marinades and wet rubs containing an acid often include brown sugar or honey because sugars help with the browning process.

Curdling. In liquids such as milk, acids can cause proteins to curdle, which may or may not be desirable. For example, if lemon juice or vinegar is added to milk, it causes the milk to curdle, which is important for making cheese and for baking purposes. **See Figure 3-22.** If a sauce curdles by mistake, the appearance and flavor of a dish may be ruined.

> ***Food Safety***
>
> Although acids in marinades help prevent the growth of bacteria, all foods need to be handled and prepared properly to ensure safety.

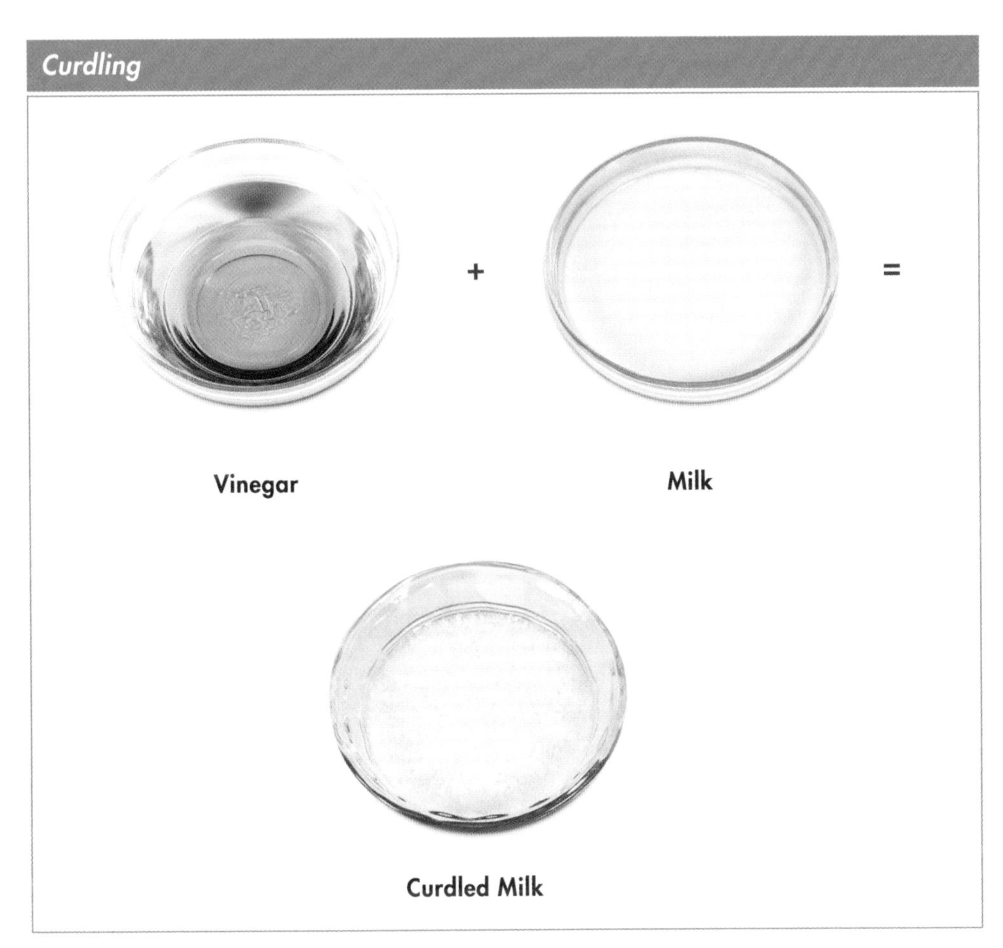

Figure 3-22. If an acidic ingredient, such as vinegar, is added to milk, it causes the milk to curdle.

FOOD SCIENCE EXPERIMENT: Using Acids to Tenderize Meats

3-1

Objective

- To compare meats that have been tenderized using a vinaigrette, tomato juice, and lemon juice in terms of tenderness, texture, and flavor.

Materials and Equipment

- 3 pieces of meat, 6 oz each (same grade and thickness)
- 3 sealable plastic bags, 1 gal. each
- Measuring cup
- 3 oz vegetable oil
- 1 oz vinegar
- Mixing bowl
- Whisk
- 4 oz tomato juice
- 4 oz lemon juice
- Sheet pan
- Instant-read thermometer
- 3 plates
- Knives and forks

Procedure

1. Place one piece of meat in each of the three plastic bags.
2. Make a vinaigrette by placing 3 oz of oil in a bowl and slowly whisking in 1 oz of vinegar until thoroughly combined.
3. Add the vinaigrette to the first bag and seal the bag.
4. Add tomato juice to the second bag and seal the bag.
5. Add lemon juice to the third bag and seal the bag.
6. Refrigerate the three plastic bags for 1–3 hours.
7. Remove each piece of meat and place it on the sheet pan.
8. Roast the meat in a 350°F oven until an internal temperature of 160°F is reached.
9. Place each piece of meat on a separate plate and allow the meat to rest for 10 minutes.
10. Evaluate the tenderness, texture, and flavor of each piece of meat.

Typical Results

There will be variances in the tenderness, texture, and flavor of each piece of meat.

Knowledge Check 3-6

1. Explain what generally happens to an amino acid as it denatures during preparation and cooking.
2. List four agents that affect protein denaturation during protein preparations.
3. Explain how myoglobin affects the color of meat.
4. Describe coagulation.
5. Describe the relationship between heat and shrinkage as food cooks.
6. Explain the Maillard reaction.
7. Explain what happens when proteins containing collagen are cooked in a liquid.
8. Describe the result of adding an acidic ingredient to a liquid protein such as milk.

PROTEINS ON THE MENU

The signature dishes that draw customers to a particular foodservice operation stabilize menu sales and profits. These dishes often feature animal-based proteins at the center of the plate, and their flavors cannot be drastically changed without affecting customer expectations. However, many restaurants now offer signature dishes in smaller portion sizes to create healthier menu options and customers are responding favorably. A smaller portion of protein can also be paired with additional vegetables and healthy grains to offer a more nutritious version of the meal. Revised menu options featuring healthier choices not only help preserve existing clientele, but also have the potential to attract new customers looking for restaurants that offer more balanced options.

Animal-based proteins will always have a significant place on most menus because their flavors and essential nutrients help meet guest expectations. The unique shapes and colors of proteins can be used to enhance plated presentations, while the textures and flavors of proteins can inspire the creation of exceptional dishes. Chefs are also considering the increased demand for vegetarian options. Having the ability to form complete proteins from plant-based foods, while incorporating enticing and complementary flavors, is essential for professional chefs.

Presentation

Animal-based proteins have often been the focal point of the plate, but due to nutritional concerns regarding excess protein, fat, cholesterol, and calories, many individuals are taking a healthier approach to eating. Menu changes due to consumer concerns and the increased demand for nutritious dishes have given chefs the opportunity to focus on presenting a greater abundance of colorful produce and whole grains paired with smaller portions of animal-based proteins.

Even though the portion of animal-based proteins might be smaller, the unique shapes and colors of the proteins can still be showcased on the plate in such a way that customer perceptions of value and quality are preserved. For example, thinly slicing and fanning cooked meats, poultry, or seafood gives the appearance of a larger portion size. In addition, elevating proteins on or near vegetables or grains draws attention to the overall composition of the plate and away from the portion size of each individual item. **See Figure 3-23.**

Protein Presentation

National Turkey Federation

Figure 3-23. Proteins can be showcased on plates in such ways that customer perceptions of value and quality are preserved.

Chef's Tip

Meat from bison (buffalo), elk, deer, and antelope can inspire unique and appealing dishes that are healthy alternatives to lean beef.

Texture and Flavor

Just as color, shape, and temperature affect foods, texture makes a meal more interesting. Some proteins such as meats, poultry, and whole grains can add a chewy texture to a dish, whereas eggs can offer a soft and smooth mouthfeel. Other proteins such as nuts and seeds can provide an appealing complement to a dish with their firm, crunchy textures.

Protein foods add both texture and flavor to meals in different ways. Animal-based proteins that are served raw, such as beef for carpaccio or tuna for sashimi, provide unique textures and flavors on their own. However, textures and flavors become even more complex when a protein is slightly seared to produce a combination of raw and cooked properties. **See Figure 3-24.** Cooking proteins also allows aromas to become more noticeable. Since flavor is dependent on the sense of smell, an enticing aroma can add to the overall dining experience.

Protein Textures and Flavors

Daniel NYC/E. Kheraj

Figure 3-24. Textures and flavors become complex when a protein is slightly seared to produce a combination of raw and cooked properties.

Knowledge Check 3-7

1. Describe how signature dishes can be revised to be more nutritious.
2. Identify ways to present smaller portions of animal-based proteins while preserving customer perceptions of value.
3. Describe three different textures proteins can provide to a dish.
4. Explain how cooking proteins can enhance flavor.

Chapter Summary

National Turkey Federation

Proteins are composed of amino acids that combine in various ways to produce thousands of different proteins. Each protein is unique and performs a specific function, such as growing and maintaining body tissues and supplying energy when needed. For proper functioning, proteins are digested in the stomach and broken down into amino acids. Then, amino acids are absorbed by the small intestine and sent to the liver where they are dispersed throughout the body for new proteins to be built and functions to be carried out.

Some amino acids are classified as essential and must be obtained from food. Most animal-based foods contain all of the essential amino acids and are considered complete proteins. Most plant-based foods are considered incomplete proteins because they lack essential amino acids. However, plant-based foods can be combined to create complete proteins. In general, people consume an adequate amount of protein, but diets in which protein consumption is too high or too low can lead to adverse health effects. When preparing proteins, exposure to such agents as air, heat, moisture, and acids cause proteins to become denatured. This change in structure usually makes proteins more digestible. As a result, chefs have many opportunities to showcase the appealing textures and flavors of both animal- and plant-based proteins.

Chapter Review

Digital Resources
ATPeResources.com/QuickLinks
Access Code: 267412

1. Explain how proteins are formed.
2. Describe the main functions of proteins.
3. Identify when proteins are used for energy.
4. Describe the path of proteins as they are digested and absorbed.
5. Differentiate between a high-quality and a low-quality protein source.
6. Give six examples of complementary proteins.
7. Explain the meaning of a limiting amino acid.
8. Summarize how the body is affected by a diet that lacks protein.
9. Give four examples of cooking techniques that cause the Maillard reaction.
10. Describe the role of an acidic ingredient in a marinade.
11. Give two examples of how proteins can be showcased on a plate.

Carbohydrates

Sugars

National Honey Board

Starches

National Cancer Institut

Chapter

4

Carbohydrates are an essential component of a nutritious diet and an integral part of every menu. They can make an everyday meal special and create varied and unique menu items. Nutrient-dense carbohydrates such as fruits, vegetables, legumes, and whole grains are rich in vitamins, minerals, and dietary fiber and fuel the body with optimal energy. It is essential for chefs to understand the nutritional profile of carbohydrates as well as proper techniques for preparation.

Chapter Objectives

1. Explain the process plants use to make carbohydrates.
2. Differentiate between simple and complex carbohydrates.
3. Summarize the functions of carbohydrates.
4. Describe how carbohydrates are digested and absorbed.
5. Explain how glucose is used for energy.
6. Identify sources of simple and complex carbohydrates.
7. Compare whole grains and refined grains.
8. Summarize the recommendations for consuming carbohydrates.
9. Describe the ways in which carbohydrates react to air, light, heat, moisture, and pH.
10. Describe how to use carbohydrates on the menu to enhance presentation, texture, and flavor.

CARBOHYDRATE CLASSIFICATIONS

A *carbohydrate* is an energy-providing nutrient in the form of sugar, starch, or dietary fiber and is the main source of energy for the body. Carbohydrates supply the body with an ideal form of fuel. For example, the brain and central nervous system rely solely on carbohydrates for energy and cannot function properly without them. Carbohydrates also play a role in digesting and metabolizing other nutrients in order to ensure that the energy needs of the body are met.

Carbohydrates, commonly abbreviated as CHO, are made from the elements carbon (C), hydrogen (H), and oxygen (O). Because carbohydrates contain carbon, they are classified as an organic compound. Carbohydrates are the most abundant organic compounds found in living things. For example, plants are mainly comprised of carbohydrates. Plants are able to make their own carbohydrates through photosynthesis. *Photosynthesis* is the process by which plants turn carbon dioxide and water into carbohydrates by using chlorophyll to capture energy from sunlight. *Chlorophyll* is the green pigment plants use to capture sunlight for photosynthesis.

In the process of photosynthesis, leaves absorb carbon dioxide from the air, which contributes carbon and oxygen to the plant. The roots of the plant absorb water, which adds hydrogen and oxygen. When they combine, these elements create carbohydrates, which provide energy for plant growth and repair. **See Figure 4-1.** Humans and animals are incapable of making carbohydrates. Therefore, they rely on plant sources to supply the carbohydrates necessary to meet their energy needs. Although plant-based foods are the primary source of carbohydrates, a significant amount of carbohydrates are also found in milk and some milk products.

Photosynthesis

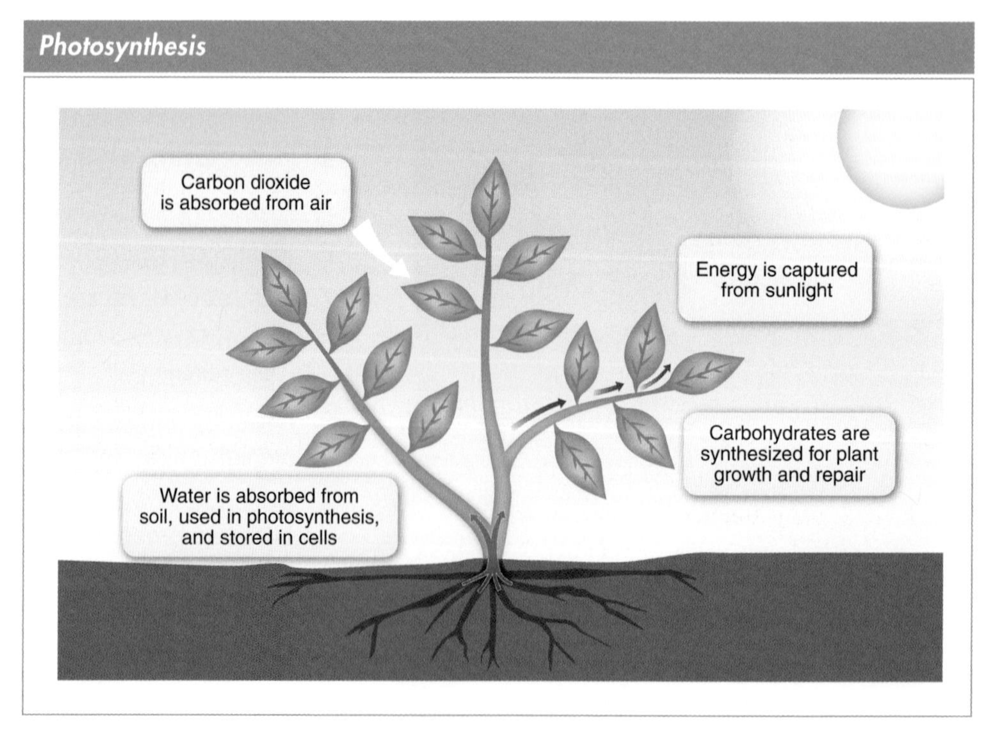

Figure 4-1. During the process of photosynthesis, leaves absorb carbon dioxide from the air, and roots absorb water to create carbohydrates that provide energy for plant growth and repair.

Carbohydrates are divided into two main groups: simple carbohydrates and complex carbohydrates. Simple and complex carbohydrates are both composed of saccharides. A *saccharide* is a unit of sugar. Simple carbohydrates contain less saccharide units than complex carbohydrates. Simple and complex carbohydrates occur naturally in foods, but their effect on the body differs.

Simple Carbohydrates

A *simple carbohydrate,* also known as simple sugar, is a type of carbohydrate made from one or two units of sugar. Simple carbohydrates have a simple chemical structure compared to complex carbohydrates. Because of their chemical structure, simple carbohydrates can be broken down quickly and used for energy faster than complex carbohydrates. Simple carbohydrates are categorized as monosaccharides or disaccharides, based on the number of sugar units (saccharides) they contain.

Monosaccharides. A *monosaccharide* is a simple carbohydrate made of one sugar unit. Monosaccharides are considered the simplest forms of sugar and are sometimes called the "building blocks of carbohydrates." There are three types of monosaccharides: glucose, fructose, and galactose. **See Figure 4-2.**

Glucose, also known as dextrose, is a monosaccharide that is the primary source of energy in plants and animals. All carbohydrates must be converted to glucose before they can be used for energy. Glucose circulates throughout the body in the bloodstream and is frequently called blood glucose or blood sugar. The amount of glucose in the blood can vary. A *blood glucose level* is the amount of glucose circulating in the blood. Blood glucose levels affect the functioning of all body cells. For example, if the brain does not have enough glucose, dizziness, weakness, and nausea may result.

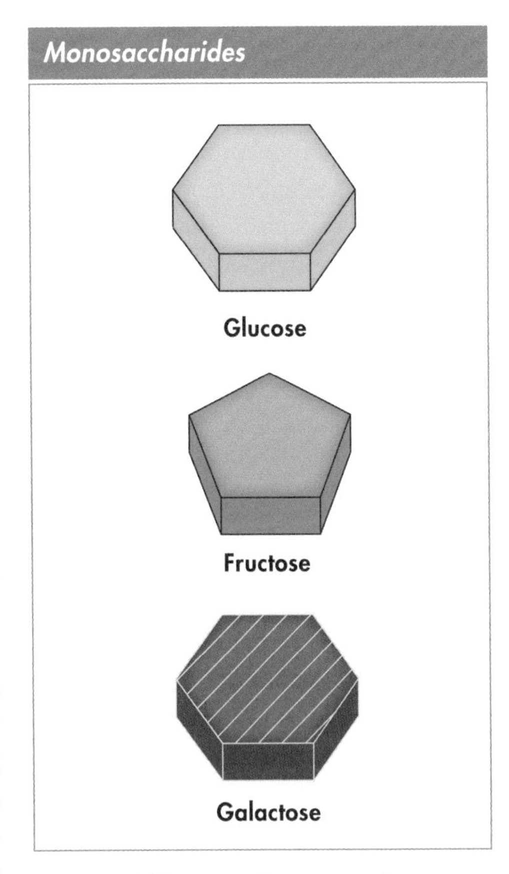

Figure 4-2. There are three types of monosaccharides: glucose, fructose, and galactose.

Fructose is a monosaccharide that is the sweetest of all sugars. Fructose is found primarily in fruits and honey. Fructose is metabolized more slowly than glucose and causes a slower rise in blood glucose levels.

Galactose is a monosaccharide that is most often attached to glucose to form lactose (milk sugar). Galactose stays connected to glucose until it is broken down during digestion. Galactose is also found in beets and seaweed and is not as sweet as glucose.

Food Science Note

In addition to the most common monosaccharides (glucose, fructose, and galactose), ribose is a monosaccharide that is found in veal. Research indicates that ribose plays a role in cellular metabolism, acts as a mood stabilizer, and can relieve pain associated with muscle diseases.

Disaccharides. A *disaccharide,* also known as a double sugar, is a simple carbohydrate made of two attached sugar units. Disaccharides contain glucose in combination with itself or in combination with fructose or galactose. Disaccharides must be broken down during digestion into the monosaccharide glucose before they can be absorbed and used for energy. Disaccharides include sucrose, lactose, and maltose. **See Figure 4-3.**

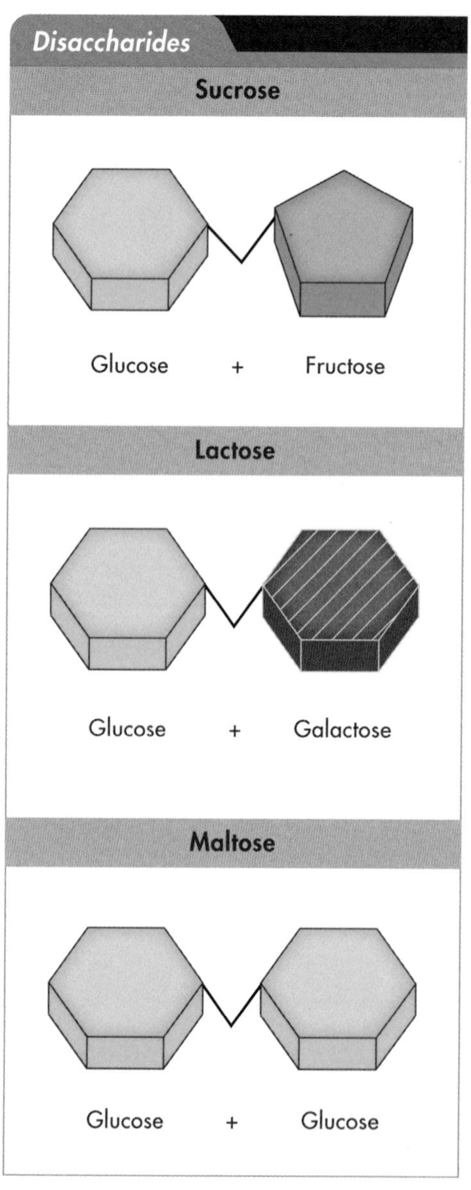

Figure 4-3. Disaccharides include sucrose, lactose, and maltose.

Sucrose is a disaccharide comprised of glucose and fructose. Sucrose is the second sweetest of the sugars, after fructose. The most common form of sucrose is granulated sugar, which is made by extracting sucrose from sugar cane or sugar beets during processing. A small amount of sucrose is also found in a variety of vegetables and fruits.

Lactose, also known as milk sugar, is a disaccharide comprised of glucose and galactose. In contrast to most carbohydrates, which are associated with plants, lactose is found only in milk and milk products such as cheese, yogurt, and ice cream.

Maltose, also known as malt sugar, is a disaccharide comprised of two units of glucose. Maltose is found in the germinating seeds of plants such as wheat, barley, and rye and is commonly used to add color and flavor to beers.

Complex Carbohydrates

A *complex carbohydrate* is a type of carbohydrate made from more than two attached sugar units. Complex carbohydrates are found in vegetables, fruits, grains, and grain products such as breads, cereals, and pastas. Complex carbohydrates must be broken down and converted to glucose before they can be used for energy. Because complex carbohydrates take more time to digest than simple carbohydrates, they are a longer lasting energy source. Complex carbohydrates include oligosaccharides and polysaccharides.

An *oligosaccharide* is a complex carbohydrate with a chain that generally ranges from three to eleven sugar units. Oligosaccharides, such as raffinose (glucose + fructose + galactose) and stachyose (glucose + fructose + 2 galactoses), are found mainly in potatoes, beans, peas, and beets. Oligosaccharides are not easily broken down during digestion and often produce gas in the intestine.

A *polysaccharide* is a complex carbohydrate with a chain usually consisting of hundreds to thousands of sugar units. Polysaccharides are classified as starches and dietary fiber. **See Figure 4-4.**

Starches. A *starch* is a polysaccharide made by plants to store long chains of glucose molecules. After energy needs are met for growth and repair, plants store extra glucose in the form of starch. Starch is abundant in foods such as corn, carrots, potatoes, legumes, and grains.

Amylose is an insoluble component of starch that forms an unbranched chain. *Amylopectin* is a soluble component of starch that forms a branched chain. Amylose typically comprises 20%–30% of the total structure of starch, while the remaining 70%–80% of starch is comprised of amylopectin. Starches, such as potato starch, are high in amylopectin and are effective thickening agents. Starches with a lower percentage of amylopectin, such as cornstarch, form a firmer gel and are less effective thickening agents. **See Figure 4-5.**

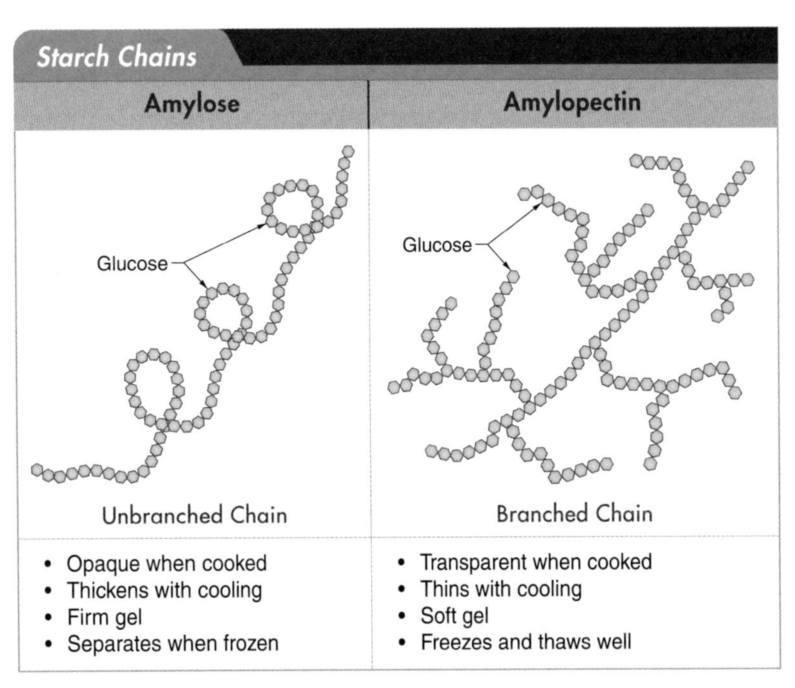

Figure 4-5. Amylose, an unbranched chain of starch, and amylopectin, a branched chain of starch, are commonly used as thickeners and gelling agents.

Chef's Tip

Arborio rice is a short-grain rice that is high in amylopectin. This makes it an ideal rice to use for risotto because the amylopectin released during cooking creates a rich and creamy texture.

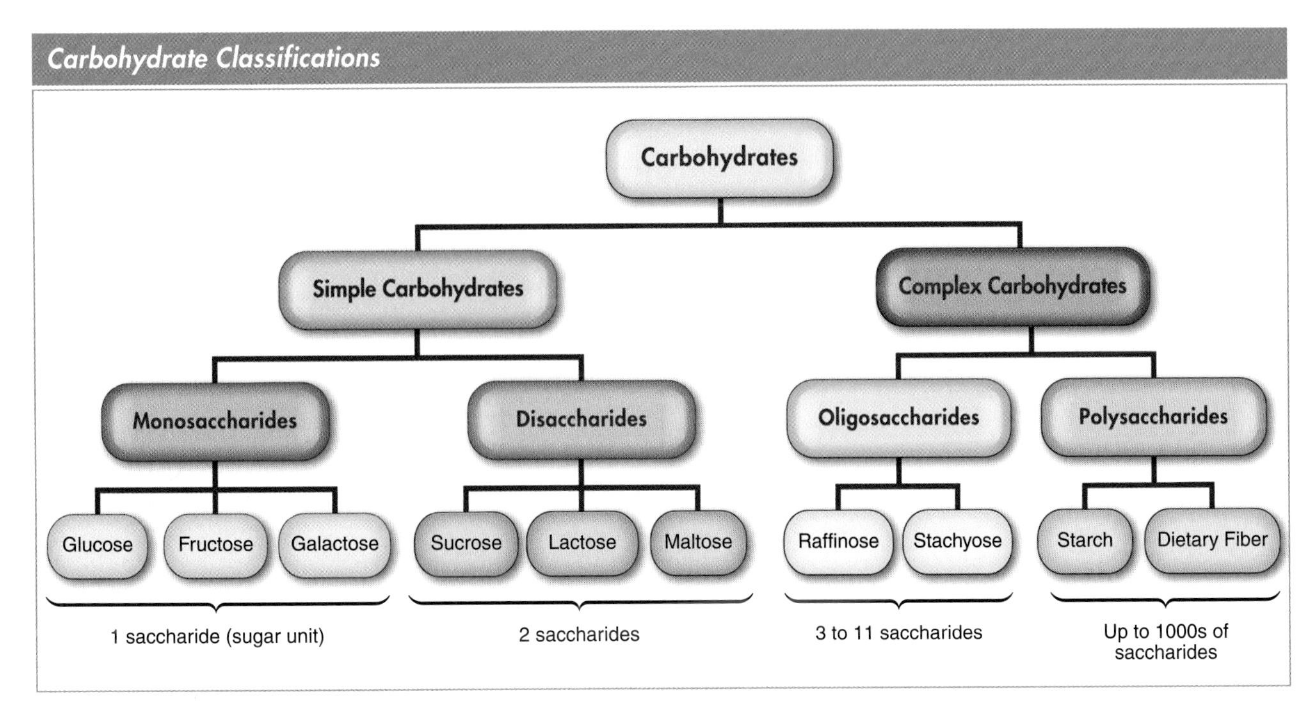

Figure 4-4. Carbohydrates are classified according to the number of saccharides, or sugar units, they contain.

Dietary Fiber. *Dietary fiber* is an indigestible form of carbohydrate found in vegetables, fruits, and grains. Dietary fiber consists of long polysaccharide chains of glucose and lignin. *Lignin* is a nonpolysaccharide form of fiber that provides rigidity and support for the woody cell walls of plants. Lignin is the major component of wood, but dietary fiber is also found in the leaves, stems, and seeds of plants. Dietary fiber helps eliminate waste from the body and also helps prevent digestive disorders.

Barilla America, Inc.

Dietary fiber is an indigestible form of carbohydrate found in vegetables, fruits, and grains.

The chains of glucose in dietary fiber are held together by bonds that human enzymes cannot break. Therefore, most dietary fiber passes through the body without being digested. A small amount of dietary fiber is metabolized for energy by bacteria in the large intestine and often produces gas as a by-product. Most complex carbohydrates contain two types of dietary fiber: soluble fiber and insoluble fiber.

Soluble fiber is dietary fiber that dissolves in water. Consuming soluble fiber produces a viscous material with a gel-like quality that is resistant to flow, binds with water, and helps slow digestion. Soluble fiber is commonly found in vegetables, fruits, legumes, and whole grains. A common type of soluble fiber used in the foodservice industry is known as pectin. *Pectin* is a soluble fiber found in and around plant cell walls. Pectin is often used to thicken jams, jellies, and salad dressings. It is also used to add texture to processed foods.

Insoluble fiber is dietary fiber that does not dissolve in water. Insoluble fiber is a nonviscous material that passes through the digestive system with very little change. As insoluble fibers move through the intestines, they absorb water and swell. This increases stool weight and volume, which promotes excretion. The most common form of insoluble fiber is cellulose. *Cellulose* is an insoluble fiber that is the main component of plant cell walls. Cellulose gives plants strength and rigidity for proper growth and is easily seen on the stalks of celery and kernels of corn. **See Figure 4-6.**

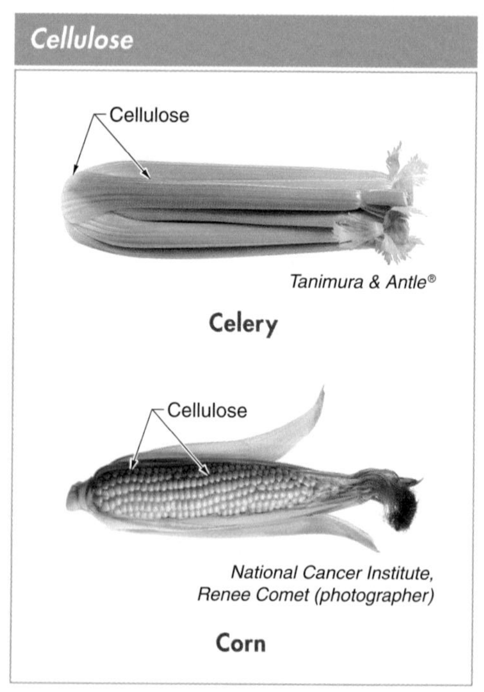

Figure 4-6. Cellulose gives plants strength and rigidity for proper growth and is easily seen on the stalks of celery and kernels of corn.

Knowledge Check 4-1

1. Define carbohydrate.
2. List the elements that compose carbohydrates.
3. Explain the process of photosynthesis.
4. Compare glucose, fructose, and galactose.
5. Compare sucrose, lactose, and maltose.
6. Identify two classifications of polysaccharides.
7. Differentiate between amylose and amylopectin.
8. Describe the two types of dietary fiber.

CARBOHYDRATE FUNCTIONS

The main function of carbohydrates is to supply the body with energy. Carbohydrates contribute 4 calories of energy per gram and are the preferred energy source over proteins and lipids. For example, when proteins are used for fuel, vital functions that rely on proteins cannot be performed. Therefore, carbohydrates are sometimes described as protein sparing. *Protein sparing* is the process by which carbohydrates and lipids supply energy so that the main functions of proteins can be fulfilled. Although skeletal muscles can use lipids for energy, the brain and central nervous system cannot use lipids for energy. Thus, carbohydrates are a crucial energy source for all cells.

Glucose is the only form of carbohydrate that the body can use for energy. This requires that all digestible forms of simple and complex carbohydrates be converted to glucose for energy use. Because complex carbohydrates are digested more slowly than simple carbohydrates, they can provide the body with a steady flow of energy. Although the main purpose of carbohydrates is to supply energy, carbohydrates are vital nutrients involved in many essential body functions. **See Figure 4-7.** For example, carbohydrates do the following:

- help regulate blood glucose levels to supply needed energy to cells while maintaining a 24-hour reserve of energy
- help eliminate unwanted materials from the body
- provide nutrients for bacteria in the large intestine
- supply raw materials used to build compounds such as amino acids
- affect the structure and functions of proteins
- influence communication among cells

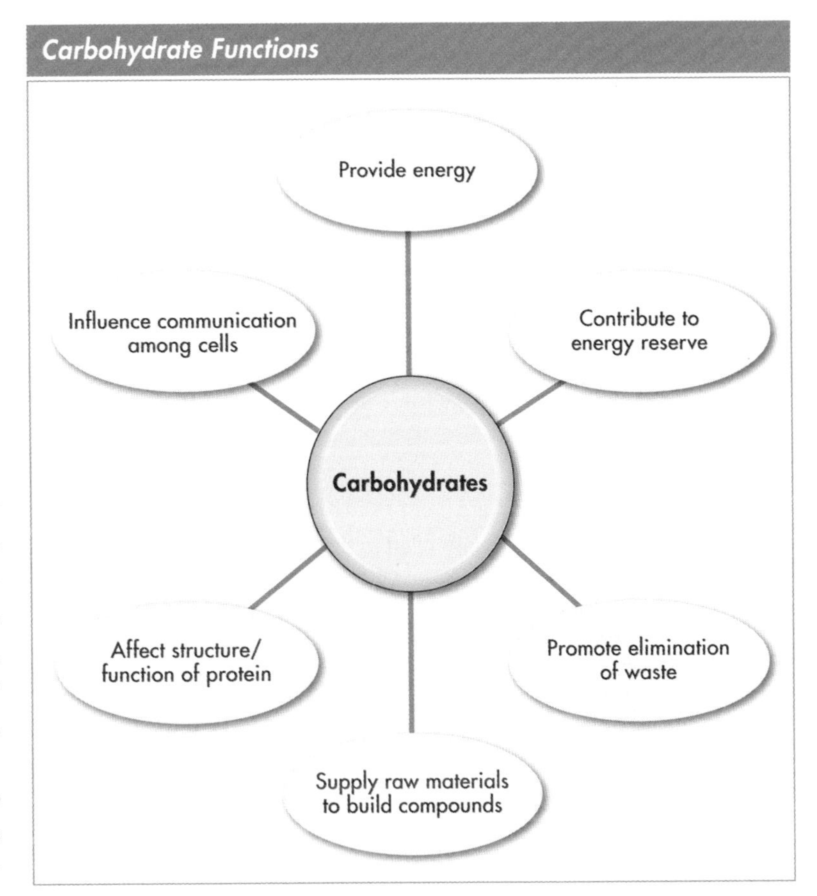

Figure 4-7. In addition to providing energy, carbohydrates perform a variety of functions necessary for health.

Knowledge Check 4-2

1. Identify the main function of carbohydrates.
2. Define protein sparing.
3. Name the only form of carbohydrate that can be used for energy.
4. Summarize the essential body functions that carbohydrates facilitate.

CARBOHYDRATE DIGESTION AND ABSORPTION

The consumption of carbohydrates is vital in order to supply the body with energy. However, carbohydrates from food cannot provide energy until they are broken down into their most basic form, which is glucose.

The digestion of carbohydrates begins in the mouth where amylase enzymes in saliva begin breaking down carbohydrates. Once carbohydrates reach the stomach, the acidic stomach environment deactivates the amylase enzymes. In the more neutral environment of the small intestine, enzymes released by the pancreas easily break down simple carbohydrates into glucose. In contrast, complex carbohydrates require more extensive breakdown before they are a viable energy source for cells.

Simple Carbohydrates

Carbohydrates need to be broken down into monosaccharides (glucose, fructose, and galactose) in order to be absorbed through the cells that line the small intestine. Because disaccharides (sucrose, lactose, and maltose) are comprised of two monosaccharides, they only need to be split once in order to be absorbed. Glucose, fructose, and galactose are then sent into the bloodstream to the liver.

The liver converts fructose and galactose to glucose, which is sent to the cells to be used for energy, stored by the liver and muscles as glycogen, or converted to fat. *Glycogen* is a polysaccharide made and used by the liver and muscles to store long chains of glucose. Glycogen can be converted back to glucose by the liver when the supply of glucose is low. The amount of glucose used as energy or stored as glycogen is regulated by the hormone insulin. *Insulin* is a hormone produced by the pancreas that is necessary for regulating blood sugar levels.

United States Potato Board

The consumption of carbohydrates is vital in order to supply the body with energy.

Complex Carbohydrates

Because complex carbohydrates contain long chains of sugars, they take longer to digest and absorb than simple carbohydrates. In the small intestine, complex carbohydrates are broken down into disaccharides and smaller polysaccharides. Enzymes continue to split disaccharides into monosaccharides that can be transported to the liver for conversion to glucose. Polysaccharides are broken into disaccharides and eventually to monosaccharides for absorption. Dietary fiber moves from the small intestine to the large intestine. **See Figure 4-8.**

Dietary Fiber

Dietary fiber cannot be broken down by enzymes and is mostly excreted as waste. However, some bacteria found in the large intestine digest a portion of soluble fiber through the process of fermentation. *Fermentation* is the breakdown of carbohydrates due to the actions of bacteria, microorganisms, or yeast. Because it is a viscous material, soluble fiber is easily fermented by the bacteria in the small intestine.

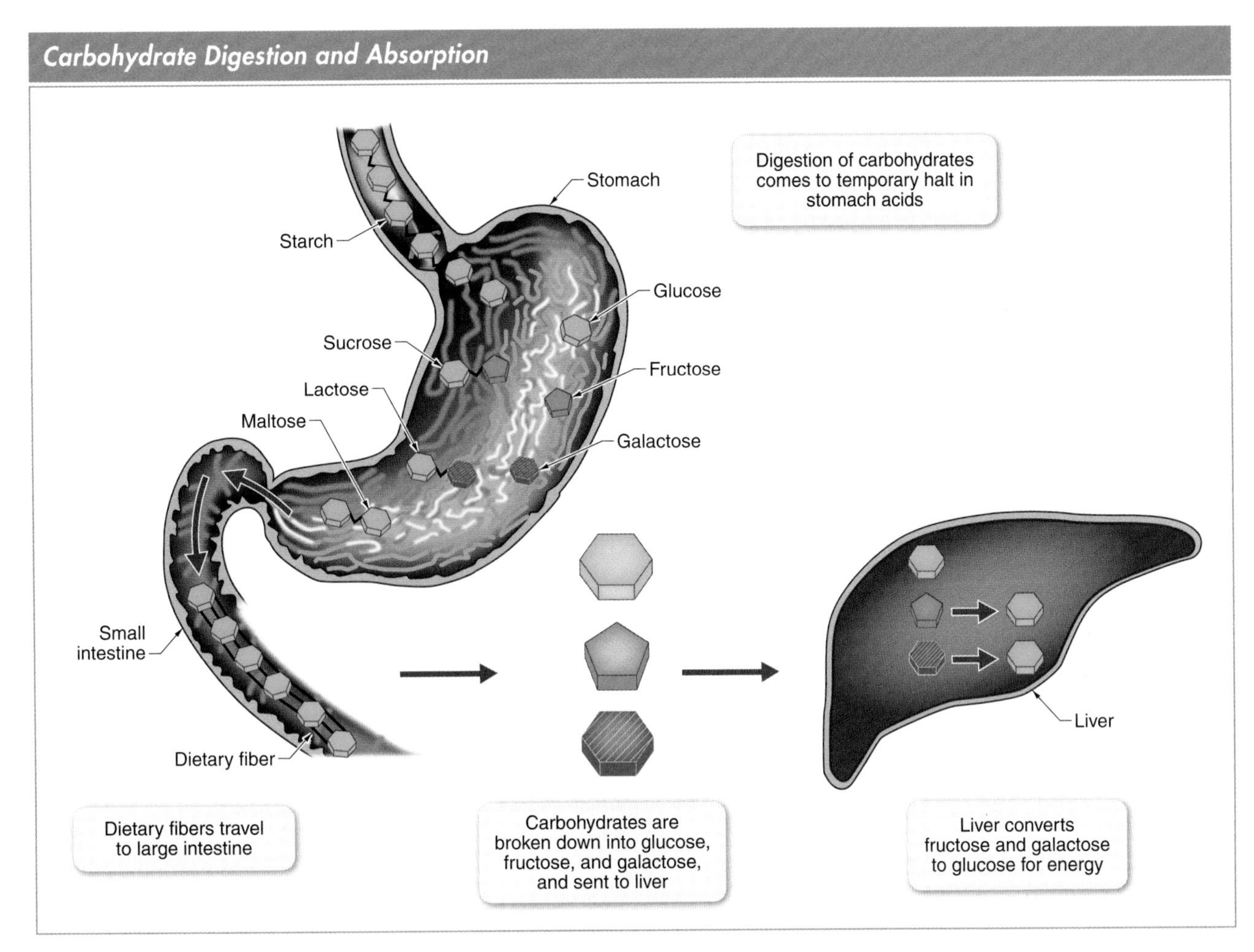

Figure 4-8. Carbohydrates from food cannot provide energy until they are broken down into their most basic form, which is glucose.

Knowledge Check 4-3

1. Identify the forms carbohydrates are broken down into so that they can be absorbed through the cells of the small intestine.
2. Explain the role of the liver during carbohydrate digestion and absorption.
3. Define insulin.
4. Describe the digestion of complex carbohydrates.
5. Describe the digestion of dietary fiber.
6. Define fermentation.

THE ROLE OF GLUCOSE

When glucose is used as energy, enzymes split the bonds between the six carbon atoms in the glucose molecule. This split results in two fragments, each with three carbons. A three-carbon fragment can attach to another three-carbon fragment to form glucose and provide energy to cells. The three-carbon fragments can also be further broken down.

When carbon fragments are further broken down, chemical reactions generate heat and adenosine triphosphate (ATP) molecules. ATP molecules capture energy to fuel all cellular activity. ATP can be produced with or without the presence of oxygen. In the presence of oxygen, glucose is converted to energy to yield heat, ATP, water, and carbon dioxide (a waste product).

In red blood cells, which do not contain oxygen, the conversion of glucose to energy yields heat, ATP, and lactic acid. *Lactic acid* is a type of organic, colorless acid made by red blood cells and muscle tissues. If oxygen levels in muscle cells fall low, these cells have the ability to convert glucose to energy without oxygen. This often occurs during strenuous exercise, and the lactic acid produced may lead to a feeling of heaviness in the muscles. When oxygen levels are restored back to normal, lactic acid is eliminated from the muscles.

Barilla America, Inc.

Carbohydrate-rich foods are broken down by the body to help form the ATP molecules necessary to fuel all cellular activity.

Storing Glucose

Just as plants store extra glucose in the form of starch, the body stores extra glucose in the form of glycogen. Glycogen is structurally similar to amylopectin, but its chains are longer and more branched. When the body has more glucose than it can store as glycogen, the excess is converted to fat. In contrast, if the body runs out of glycogen, fat stores are broken down for energy.

When carbohydrates are consumed and broken down into glucose, the pancreas releases insulin. Insulin acts as a messenger to cells and tissues to take up excess glucose so that it can be stored as glycogen. Glycogen is primarily stored in the liver and muscles. The brain also stores a tiny amount of glycogen to protect the body in times of severe carbohydrate deprivation. The majority of the glycogen produced is stored and used only by the muscles. The remaining glycogen is collected in the liver and dispersed to the tissues of the body and the brain when energy needs arise. **See Figure 4-9.**

Low Glucose Levels

Blood glucose levels are maintained when the body receives an adequate supply of carbohydrates at regular intervals throughout the day. When blood glucose levels drop, an individual may feel faint or weak. However, the body tries to protect itself from low glucose levels by storing glucose as glycogen. When glucose levels fall, the pancreas secretes a hormone called glucagon. *Glucagon* is a pancreatic hormone sent to the liver to trigger the conversion of glycogen to glucose for energy. **See Figure 4-10.**

High Glucose Levels

When the liver and muscles have reached their storage capacity for glycogen and glucose is still circulating throughout the body, excess glucose is burned for energy. When this situation occurs, excess lipids are stored in body tissues as fat. The liver also breaks the excess glucose into components that are reconstructed into lipids to be released into the blood and stored as fat.

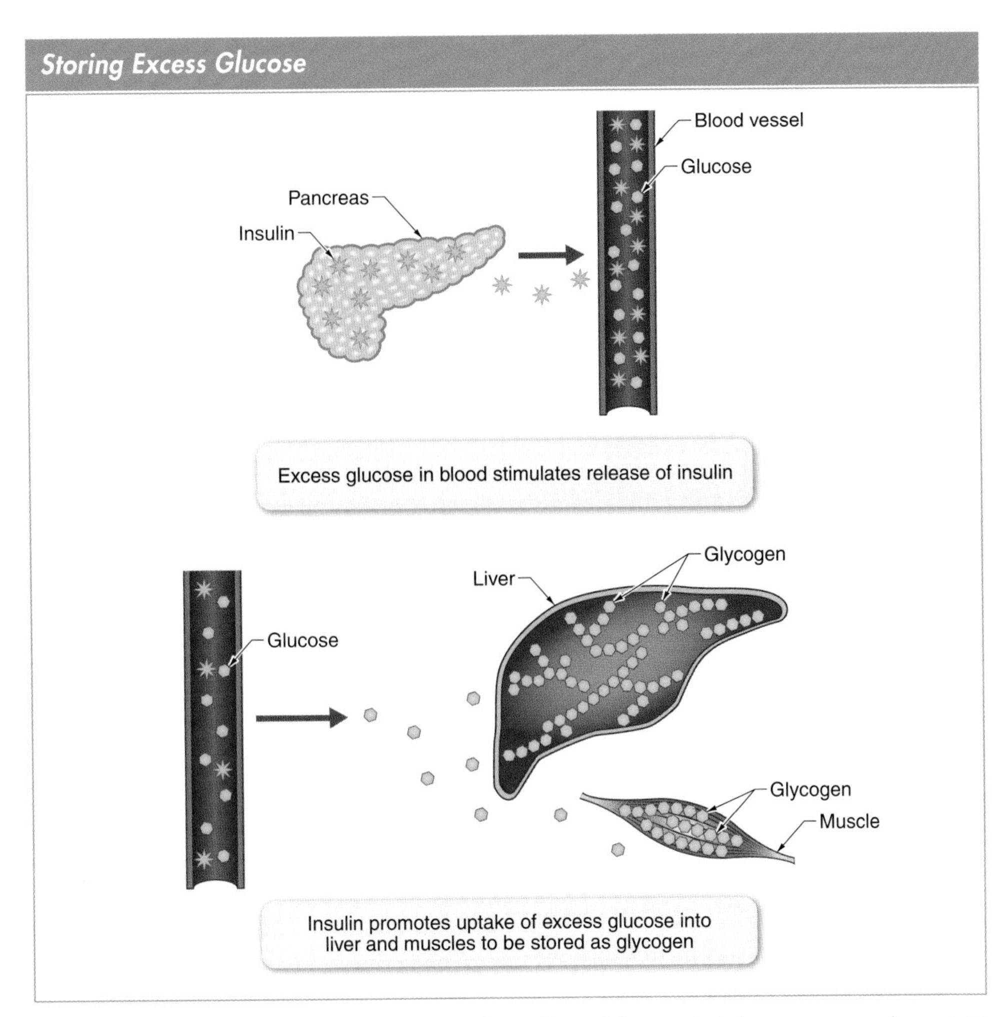

Figure 4-9. Insulin acts as a messenger to the cells and tissues to take up excess glucose so that it can be stored as glycogen.

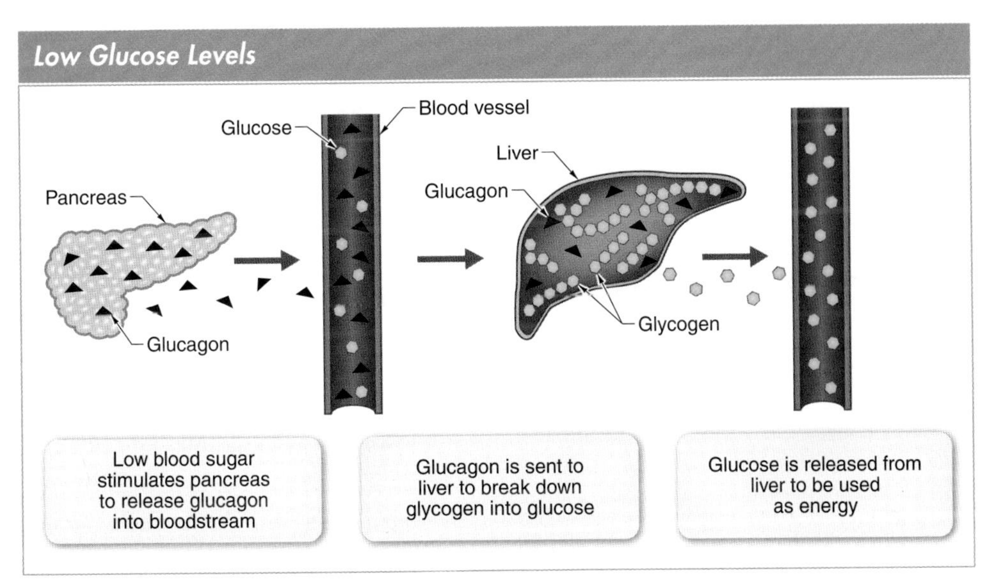

Figure 4-10. When glucose levels fall, the pancreas secretes glucagon to trigger the conversion of glycogen to glucose for energy.

Knowledge Check 4-4

1. Explain the function of adenosine triphosphate (ATP) molecules.
2. Define lactic acid.
3. Describe glycogen.
4. Identify where glycogen is stored.
5. Explain how the body protects itself from low glucose levels.
6. Describe what occurs when the body has high glucose levels.

CARBOHYDRATE SOURCES

Vegetables, fruits, whole grains, and milk provide a natural source of simple carbohydrates and complex carbohydrates. Most adults require a minimum of 130 g of carbohydrates daily. A serving of vegetables such as asparagus, broccoli, cauliflower, spinach, or tomatoes typically supplies the body with approximately 5 g of carbohydrates. A serving of starchy vegetables like potatoes, corn, peas, beans, or squash as well as most fruits contributes an average of 15 g of carbohydrates. In general, a serving of grains also provides 15 g of carbohydrates. However, the nutritional quality of grain can vary significantly depending on the way it has been processed. An 8 oz serving of milk includes 11 g of carbohydrates, but milk products such as yogurt often have more carbohydrates because sugars are usually added for sweetness.

National Cancer Institute, Daniel Sone (photographer)

Fruits provide a natural source of simple carbohydrates.

Simple Carbohydrates

Simple carbohydrates have a sweet flavor and occur naturally in foods such as fruits, agave nectar, honey, molasses, maple syrup, and milk. Simple carbohydrates are often used in desserts and are also abundant in processed foods. In addition to naturally occurring sugars, food manufacturers often use sugar replacers and artificial sweeteners to provide a sweet flavor in food. Added sugars, sugar replacers, and artificial sweeteners also affect the texture, caramelization, fermentation, and shelf life of foods.

Added Sugars. An *added sugar* is any sugar not naturally occurring in a food and added during processing or preparation. Sugars are often added to foods to heighten sweetness, promote browning in baked goods, or increase shelf life. Added sugars provide calories but little nutrients, so they are considered "empty calories." Processed foods can be a significant source of empty calories. If the ingredient list contains items that end in "ose," such as sucrose, maltose, dextrose, fructose, or lactose, then the product contains added sugars. Cane sugar, brown sugar, honey, molasses, nectars, corn syrup, and high-fructose corn syrup are also added sugars found on ingredient lists. **See Figure 4-11.**

Examples of Added Sugars on Ingredient Lists

- Agave nectar
- Brown sugar
- Corn syrup
- Dextrose
- Fructose
- Fruit juice concentrate
- High-fructose corn syrup
- Honey
- Invert sugar
- Lactose
- Maltose
- Malt syrup
- Maple syrup
- Molasses
- Sucrose
- Sugar

Figure 4-11. Added sugars can be found on ingredient lists.

High-fructose corn syrup (HFCS) is corn syrup that has been treated with an enzyme to convert part of its glucose to fructose. The conversion of glucose to fructose yields a very sweet product that is inexpensive to produce. This makes high-fructose corn syrup an attractive ingredient for food manufacturers and contributes to its wide use in beverages, baked goods, fruit spreads, syrups, and candies. In addition to flavor, high-fructose corn syrup can help prevent mold growth and spoilage in fruit spreads as well as promote moisture retention and caramelization in baked goods.

Stevia is also added to foods to make them sweeter. *Stevia* is a natural sweetener derived from a plant native to South America. Stevia is often mistakenly considered an artificial sweetener because it is intensely sweet (300 times sweeter than sucrose) and has no calories. In 2008, the FDA approved rebaudioside A, which is a substance extracted from the stevia plant, as a food additive. A *food additive* is a substance added to food to preserve or improve its appearance, texture, flavor, and/or nutritional qualities.

Sugar Replacers. A *sugar replacer,* also known as sugar alcohol or polyol, is a naturally occurring carbohydrate that is only partially digestible and has a chemical structure that resembles both sugar and alcohol. Sugar replacers are typically less sweet than sucrose (granulated sugar) and provide 2 to 3 calories of energy per gram versus a typical carbohydrate at 4 calories per gram. **See Figure 4-12.** Because sugar replacers are only partially digested and absorbed, they provide less energy and have less impact on blood glucose levels.

Sugar replacers can be found in items such as chewing gum, breath mints, candies, fruit spreads, baked goods, and ice cream. In addition to being low in calories, sugar replacers help foods stay moist, prevent browning, and do not promote tooth decay.

Sugar Replacers

Sugar Replacer	Sweetness Compared to Sucrose (100%)
Xylitol	100%
Hydrogenated starch hydrolysates	40% – 90%
Maltitol	75%
Mannitol	50% – 70%
Isomalt	45% – 65%
Sorbitol	50%
Lactitol	30% – 40%

Figure 4-12. Sugar replacers are typically less sweet than sucrose (granulated sugar).

Artificial Sweeteners. An *artificial sweetener* is an intensely sweet, nonnutritive synthetic substance with zero to almost no calories. Artificial sweeteners are considered nonnutritive because they are not absorbed by the body and do not contribute nutrients. Five types of artificial sweeteners have FDA approval. These include neotame, saccharin, sucralose, acesulfame K, and aspartame. **See Figure 4-13.** Other artificial sweeteners are currently seeking FDA approval, and some sweeteners, such as cyclamate, have approval in other countries but are banned in the United States.

Artificial Sweeteners

Artificial Sweetener	Sweetness Compared to Sucrose
Neotame	8000 times sweeter
Saccharine	200–700 times sweeter
Sucralose	600 times sweeter
Acesulfame K	200 times sweeter
Aspartame	200 times sweeter

Figure 4-13. An artificial sweetener is an intensely sweet, nonnutritive synthetic substance with zero to almost no calories.

Food Science Note

An adequate intake (AI) has been established for each of the approved artificial sweeteners and is the maximum amount considered safe to consume each day over the course of a lifetime.

Since artificial sweeteners have relatively no calories, they are primarily used to replace higher calorie sweeteners, such as granulated sugar, brown sugar, and honey, in order to lower the calorie content of foods. Individuals who must limit sugar, such as diabetics, often find artificial sweeteners appealing because they can help control blood glucose levels.

Artificial sweeteners are commonly found in chewing gum, cookies, candies, syrups, dairy products, desserts, and beverages. However, adjustments typically have to be made to recipes when artificial sweeteners are used because they do not provide the same structural, textural, caramelization, or fermenting capabilities. Artificial sweeteners also leave an aftertaste and can develop an "off" flavor when heated.

Wellness Concept

Because they do not promote tooth decay, sugar replacers and artificial sweeteners are effective ingredients in products that remain in the mouth for long periods of time such as chewing gum and breath mints.

Complex Carbohydrates

Complex carbohydrates are mainly found in vegetables, fruits, legumes, and whole grains. The body prefers foods containing complex carbohydrates because they take longer to digest. As a result, glucose is released slowly and evenly into the bloodstream.

In addition to being a rich source of dietary fiber, complex carbohydrates contain beneficial vitamins, minerals, and phytonutrients. *Phytonutrients,* also known as phytochemicals, are naturally occurring substances found in plants that have been found to protect against disease. Complex carbohydrates are also naturally low in added sugar, fat, and calories, and they promote a long lasting feeling of satiety.

Starches. Starchy foods, such as carrots, corn, winter squash, plantains, potatoes, yams, and dried beans and peas, come in a variety of colors, textures, and flavors. Although starchy foods can be somewhat higher in calories than other vegetables and fruits, they are rich sources of vitamins and minerals that are essential to health.

National Honey Board

Starchy foods such as carrots are a rich source of vitamins and minerals.

Dietary Fiber. In addition to vegetables, fruits, legumes, and whole grains, dietary fiber is also found in nuts and seeds. Most foods high in dietary fiber contain a combination of soluble and insoluble fiber. **See Figure 4-14.** Rich sources of soluble fiber include broccoli, Brussels sprouts, kidney beans, barley, oat bran, and psyllium seeds. Fruits such as apples, prunes, and citrus fruits also contain a significant amount of soluble fiber. Like soluble fiber, insoluble fiber is present in a variety of vegetables, fruits, legumes, nuts, seeds, and whole grains.

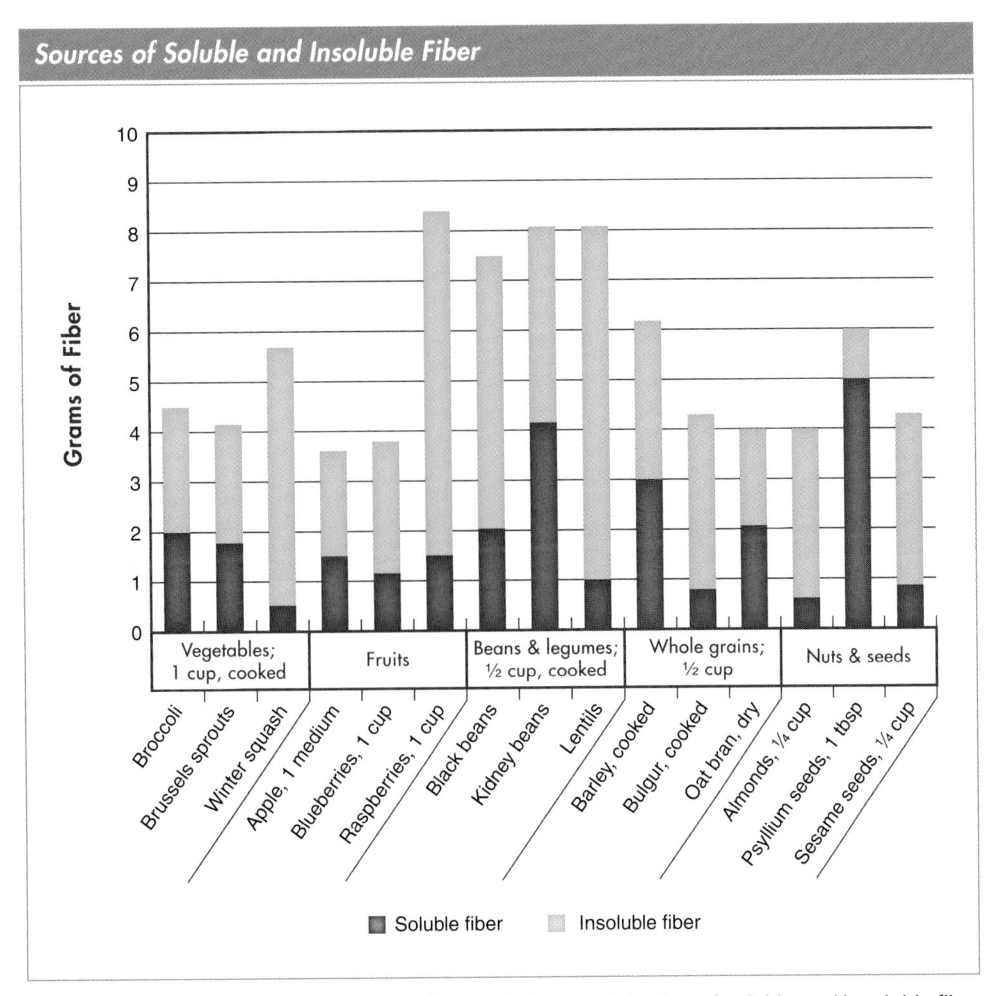

Figure 4-14. Most foods high in dietary fiber contain a combination of soluble and insoluble fiber.

Grains

A *grain* is the edible fruit, in the form of a kernel or seed, of a grass. Grains include cereal grains and pseudocereals. A *cereal grain* is a grain that is derived from plants in the grass family. Examples of cereal grains include wheat, barley, rice, oats, millet, corn, and teff. A *pseudocereal* is a seed that is classified as a grain but is derived from broadleaf plants instead of grasses. Pseudocereals include quinoa, amaranth, and buckwheat. Like cereal grains, pseudocereals are often processed into flours to be used in products such as packaged cereals, pastas, and baked goods.

When selecting grain-based products, the ingredient list is useful in determining whether the item is made with whole grains or refined grains. The term "whole grain" is the only term that ensures a product contains a whole grain. Terminology such as "wheat flour," "whole wheat flour," and "unbleached flour" do not guarantee a whole grain product. The amount of grain included in a product can also be inferred from the ingredient list. For example, when "whole grain" is placed near the beginning of the list, the product contains a higher percentage of whole grains.

Grain Composition. Grains come in a variety of shapes and forms. Most grains are hard and fairly indigestible in their natural form. It is necessary to process grains to some extent to make them easier to digest. However, the amount of processing directly affects the nutritional values of a grain as well as how the grain is used in culinary applications. A kernel of grain is composed of a husk, bran, endosperm, and germ. **See Figure 4-15.**

- A *husk,* also known as a hull, is the inedible, protective outer covering of grain.
- A *bran* is the tough outer layer of grain that covers the endosperm. While it is often removed during processing, bran provides necessary dietary fiber, complex carbohydrates, vitamins, and minerals.
- An *endosperm* is the largest component of a grain kernel and consists of carbohydrates and a small amount of protein. It is milled to produce flours and other grain products.
- A *germ* is the smallest part of a grain kernel and contains a small amount of natural oils as well as vitamins and minerals.

Whole Grains. A *whole grain* is a grain that has only had the husk removed. Because the germ and bran remain intact, whole grains may require more time to cook than processed grains. Whole grains include brown rice, wild rice, wheat berries, corn, bulgur, oats, barley, rye, millet, and spelt. A *cracked grain* is a whole kernel of grain that has been cracked by being placed between rollers. Bulgur wheat is a cracked grain.

Whole grains are rich in dietary fiber, low in fat, and contain vitamins, minerals, phytonutrients, and some protein. Consuming whole grains instead of refined grains has been shown to improve cardiovascular and digestive health, lower the risk of cancer, reduce cholesterol, and promote stable blood glucose levels.

Chef's Tip

Menus that feature global dishes, such as tabbouleh, not only create interest but also expose guests to meals rich in whole grains. Tabbouleh is a traditional Middle Eastern dish featuring bulgur, tomatoes, onions, olive oil, lemon juice, parsley, and mint.

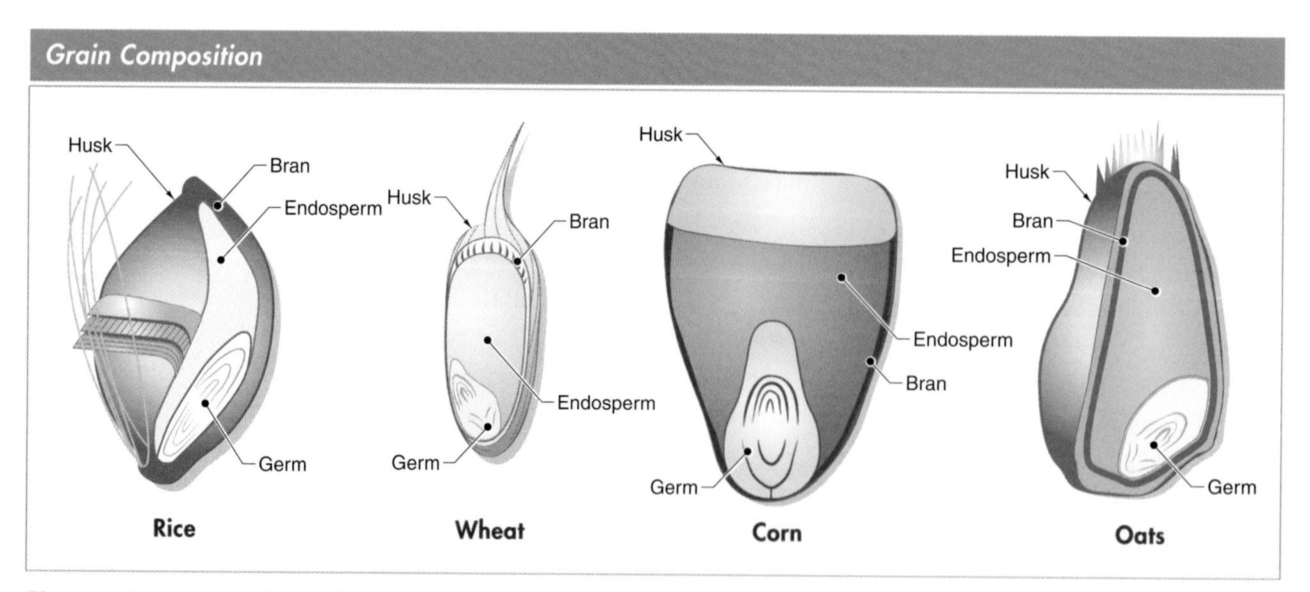

Figure 4-15. A kernel of grain is composed of a husk, bran, endosperm, and germ.

Refined Grains. A *refined grain* is a grain that has been processed to remove the bran, germ, or both. **See Figure 4-16.** Refined grains are easier to digest, have a longer shelf life, and take less time to cook. Processing often goes beyond removing the husk and can include removing the nutrient-rich bran and germ. The bran is removed from all-purpose flour to make the product white in color instead of brown. Removing the germ can help preserve the product so that it does not spoil quickly. However, grains lose vitamins, minerals, and dietary fiber during processing.

Sullivan University

Pearled barley is often used to create side dishes and soups.

Refined Grains

United States Department of Agriculture

Figure 4-16. A refined grain is a grain that has been processed to remove the germ, bran, or both.

Knowledge Check 4-5

1. Identify the four main sources of carbohydrates.
2. Differentiate between an added sugar, a sugar replacer, and an artificial sweetener.
3. Describe high-fructose corn syrup.
4. Describe stevia.
5. Identify the benefits of sugar replacers.
6. List the five types of FDA-approved artificial sweeteners.
7. Define phytonutrient.
8. Differentiate between cereal grains and pseudocereals.
9. Describe the four parts of a grain kernel.
10. Identify the benefits of whole grains.

Refined grains have been milled, pearled, or flaked. A *milled grain* is a refined grain that has been ground into a fine meal or powder. Meal, such as cornmeal, and all varieties of flour are milled grains. A *pearled grain* is a refined grain with a pearl-like appearance that results from having been scrubbed and tumbled to remove the bran. Barley is often refined into a pearled grain. A *flaked grain,* also known as a rolled grain, is a refined grain that has been rolled to produce a flake. Oatmeal is a flaked grain.

CARBOHYDRATE CONSUMPTION

Research indicates that most individuals consume an adequate amount of carbohydrates but are lacking in fiber-rich complex carbohydrates. For example, an individual who consumes a 12 oz can of soda typically takes in 35 g of carbohydrates and 140 calories. However, the soda lacks nutrients and, therefore, the simple carbohydrates in soda provide the body with nothing more than empty calories and short-term energy. In contrast, a cup of cooked oats supplies 27 g of carbohydrates and 160 calories. The oats are nutrient dense with 4 g of dietary fiber as well as vitamins, minerals, and protein. These complex carbohydrates also provide sustained energy and promote satiety.

Vegetables, fruits, grains, milk, and milk products also contribute to carbohydrate intake. With the exception of milk, complex carbohydrates also provide dietary fiber to the diet. Nutrient-dense, fiber-rich complex carbohydrates provide the body with optimal energy throughout the day. **See Figure 4-17.**

Carbohydrate Recommendations

Because they are classified as macronutrients, the body requires a large amount of carbohydrates to function properly. According to the Institute of Medicine, a healthy diet consisting of approximately 45%–65% carbohydrates helps ensure that energy needs are met without causing proteins to be burned as energy. For an individual following a 2000 calorie diet, this translates to between 225 g and 325 g (900 calories to 1300 calories) of carbohydrates per day. However, carbohydrate requirements can differ among individuals depending upon height, weight, age, gender, health, and physical activity levels.

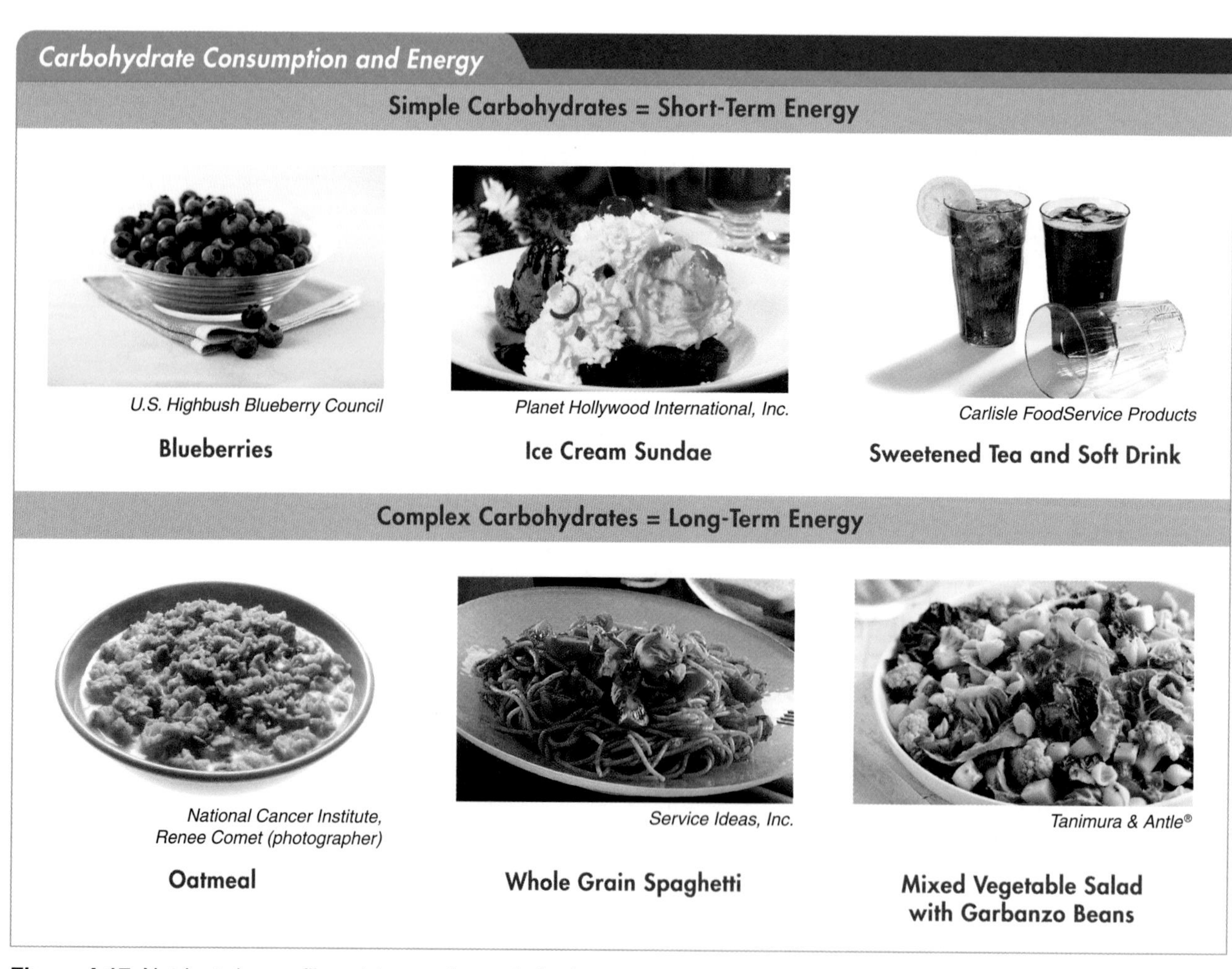

Figure 4-17. Nutrient-dense, fiber-rich complex carbohydrates provide the body with optimal energy throughout the day.

It is also recommended that the majority of carbohydrates come from nutrient-dense sources, such as vegetables, fruits, and whole grains, while added or refined sugars be limited. To help achieve these recommendations, the following is suggested:

- at least half of all grains consumed should be whole grains
- fiber-rich fruits and vegetables, including legumes, should be chosen
- whole fruits, due to their dietary fiber content, should be chosen over fruit juice
- the use of added sugars, sugar replacers, and artificial sweeteners should be minimized

The Dietary Reference Intake (DRI) Committee recommends that added sugars not exceed 25% of total carbohydrate intake. This recommendation is based on the possibility that when large amounts of added sugars are consumed other essential nutrients could be sacrificed for high-sugar foods. The committee has established a recommended dietary allowance for carbohydrates and an adequate intake for dietary fiber.

Recommended Dietary Allowances. The recommended dietary allowance (RDA) for carbohydrates is a minimum of 130 g per day and includes sugars and starches but not dietary fiber. This RDA was established to ensure that enough glucose reaches the brain for proper functioning. However, the amount required by the brain does not reflect the amount necessary to maintain and promote optimal health.

Adequate Intakes. The adequate intakes (AIs) for dietary fiber are based on 14 g of dietary fiber per 1000 calories consumed. **See Figure 4-18.** Fiber-rich foods have been found to help with weight management and may reduce the risk of cardiovascular disease, colon cancer, and digestive disorders such as diverticulosis. Fiber-rich foods are associated with a lower incidence of colon cancer because dietary fiber promotes regularity, which reduces the time that potentially harmful substances remain in the colon.

Adequate Intakes (AIs) for Dietary Fiber

Gender	Age	Dietary Fiber (grams per day)
Females	1–3	19 g
	4–8	25 g
	9–13	26 g
	14–50	25 g
	50+	21 g
Males	1–3	19 g
	4–8	25 g
	9–13	31 g
	14–50	38 g
	50+	30 g

Figure 4-18. AIs for dietary fiber are based on 14 g of dietary fiber per 1000 calories consumed.

Carbohydrate Excess

When carbohydrates with added sugars and empty calories are consumed in excessive amounts, individuals may be at greater risk for obesity, diabetes, cardiovascular disease, and tooth decay. In contrast, diets high in complex carbohydrates where dietary fiber is plentiful and saturated fats are low can reduce these risks. A sensible eating plan focused on nutrient-dense foods is an integral component of disease prevention.

Obesity. Obesity is the result of eating more calories than the body uses. Because foods and beverages with added sugars are often high in both saturated fats and calories, there can be a greater tendency to gain weight when these foods are consumed often. If a healthy weight is not maintained, the chance for diabetes, heart disease, high blood pressure, and some forms of cancer also increases.

Diabetes. Diabetes is characterized by abnormal blood glucose levels and insufficient or ineffective insulin production. Diabetics must control the number of carbohydrates they consume in order to maintain a safe blood glucose level. If uncontrolled, diabetes leads to complications such as blindness, heart disease, kidney disease, and death.

Foods rich in dietary fiber generally cause less fluctuation in blood glucose levels, which results in a low glycemic response. A *glycemic response* is a measure indicating the rate at which blood sugar rises after eating. Generally, a low glycemic response is preferable to a high response. The *glycemic index (GI)* is a measure indicating the rate at which an ingested food causes blood sugar levels to rise. **See Figure 4-19.** It is important to note that the GI is only a guide since the glycemic response varies based on the type and amount of food, the preparation technique, the amount of processing, metabolism, and individual differences in height, weight, age, and health.

Glycemic Index (GI) Levels of Food

Low	Medium	High
• Broccoli • Cashews • Kidney beans • Peaches • Zucchini	• Basmati rice • Corn on the cob • Couscous • Pita bread • Muesli	• Doughnuts • Jelly beans • Puffed wheat • Rice cakes • Soda crackers

Figure 4-19. The GI is a measure indicating the rate at which an ingested food causes blood sugar levels to rise.

Cardiovascular Disease. Studies have found that diets high in fructose and sucrose can increase cholesterol levels, which, in turn, can increase the risk of cardiovascular disease. When dietary fiber was added to the diet on a consistent basis, the risk of contracting cardiovascular disease was found to decrease.

Tooth Decay. *Tooth decay,* also known as dental caries, is a condition in which teeth decay due to acids produced by bacterial growth and improper dental care. **See Figure 4-20.** Bacteria in the mouth thrive on sugars and, to a lesser extent, starches. When a sugar is consumed, bacteria feed on the simple carbohydrates and produce acids for approximately 30 minutes. The acids destroy tooth enamel and give bacteria a location to attach and multiply. This promotes plaque buildup. *Plaque* is the accumulation of bacteria and food residue on teeth that can cause tooth decay and gum disease.

Tooth Decay

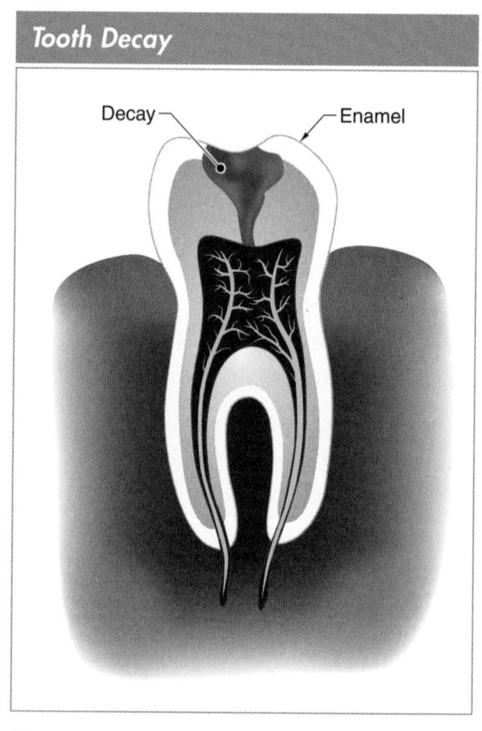

Figure 4-20. Tooth decay, also known as dental caries, is a condition in which teeth decay due to acids produced by bacterial growth and improper dental care.

Carbohydrate Deficiencies

It is important to consume adequate amounts of carbohydrates daily to meet energy needs and promote health. Carbohydrate deficiencies cause a decrease in energy, which can trigger dizziness, confusion, digestive problems, and weight loss. An insufficient amount of carbohydrates can also lead to hypoglycemia and ketosis.

Hypoglycemia. *Hypoglycemia* is a condition that results from low blood glucose levels. Hypoglycemia most often affects individuals with diabetes who have taken in too much insulin, have had too much exercise, or have not eaten enough. Although rare, hypoglycemia can occur in individuals who do not have diabetes, but it usually signifies an underlying disease.

Symptoms of hypoglycemia may include dizziness, weakness, sweating, shakiness, anxiety, and difficulty with speech. To prevent carbohydrate deficiencies, such as hypoglycemia, it is recommended that individuals eat a well-balanced meal that includes complex carbohydrates, proteins, and unsaturated fats at regular intervals throughout the day. **See Figure 4-21.**

Preventing Carbohydrate Deficiencies with Well-Balanced Meals

U.S. Apple® Association

Figure 4-21. To prevent carbohydrate deficiencies, it is recommended that individuals eat well-balanced meals at regular intervals throughout the day.

Wellness Concept

Soluble fibers slow digestion and absorption and can help counteract the low blood glucose levels associated with hypoglycemia.

Ketosis. Carbohydrates are necessary for lipids (fats) to be burned for energy. When carbohydrates are deficient, lipids combine with one another to form ketones. A *ketone* is an acidic substance that forms in the blood when lipids are metabolized for energy instead of carbohydrates. When ketone bodies accumulate, ketosis can occur. *Ketosis* is a condition in which ketones build up in the body because lipids are improperly metabolized for energy. Ketosis can cause the blood to become too acidic, which can be fatal in severe cases.

Knowledge Check 4-6

1. Identify the type of carbohydrate that provides optimal energy.
2. Identify the percentage of the diet that should consist of carbohydrates.
3. Explain the recommended dietary allowance (RDA) for carbohydrates.
4. Summarize the benefits of fiber-rich foods.
5. Describe four health problems associated with carbohydrate excess.
6. Define hypoglycemia.
7. Describe ketosis.

PREPARING CARBOHYDRATES

Although they both add texture and flavor to dishes, simple and complex carbohydrates often function differently during food preparation. For example, simple carbohydrates, or sugars, promote moisture retention, caramelization, preservation, crystallization, and fermentation. In contrast, complex carbohydrates, or starches, are more commonly used as thickening agents.

Carbohydrates react to air, light, heat, moisture, and changes in pH. Some of these reactions are necessary to produce a specific result that is essential to a successfully finished dish.

Reactions to Air

Carbohydrates are hygroscopic, which means they absorb moisture from the air. Hygroscopic properties are desirable for some foods because the foods retain moisture and stay fresher longer. However, other foods must be stored in airtight containers to prevent the absorption of moisture and preserve texture. For example, soft cookies can be stored in a loosely covered container to maintain a tender, chewy texture, while crisp cookies should be stored in an airtight container to preserve the crunchy texture.

Reactions to Light

Light affects carbohydrates through photosynthesis. This is especially important in the storage of potatoes. Potatoes should always be stored in a dark, cool location because sunlight begins the process of photosynthesis. **See Figure 4-22.** If potatoes undergo photosynthesis, they will begin to turn green and eventually develop a toxic compound called solanine. *Solanine* is a bitter, poisonous alkaloid that most commonly develops in potatoes and tomatoes.

Potato Reaction to Light

Figure 4-22. Potatoes should always be stored in a dark, cool location because sunlight begins the process of photosynthesis.

Reactions to Heat

Heat affects sugars and starches in different ways. Sugars are more stable and tolerate heat, whereas starches break down into smaller chains of glucose molecules when heated. If heat is applied to carbohydrates, caramelization, the Maillard reaction, dehydration, crystallization, and fermentation can occur.

Caramelization. When exposed to heat, sugars will begin to brown and caramelize. Caramelizing foods enhances color and adds a sweet flavor to the foods. This is evident in roasted vegetables, sautéed onions, and desserts such as flan and crème brulée.

Maillard Reaction. Sugars brown as a result of the Maillard reaction when the proteins and sugars in food are exposed to heat and merge to form a brown exterior surface. The Maillard reaction is evident in foods that have been seared, grilled, broiled, or toasted.

Dehydration. *Dehydration* is the process of removing moisture from food. Dehydration is an effective way to preserve foods and can be accomplished using low heat. Vegetables and fruits commonly dehydrated include beans, tomatoes, mushrooms, apples, apricots, grapes, and plums. As foods dehydrate, water is extracted and causes the product to shrink and concentrate in flavor. **See Figure 4-23.** Dehydrated foods are often cost effective, easy to store, and cook quickly, which promotes nutrient retention.

Dehydration

Frieda's Specialty Produce

Figure 4-23. Tomatoes are commonly dehydrated, resulting in a concentrated flavor.

Crystallization. *Crystallization* is the process in which crystals precipitate from a solution to form a solid in the presence of heat. Candy making is based upon this principle. Sugar is highly soluble in water, and the water when heated absorbs more sugar than could be absorbed in cold water. At high temperatures, the sugars remain suspended in the solution even as the water evaporates. However, as the mixture cools, there is more sugar than can be absorbed. When this occurs, the solution is supersaturated. A *supersaturated solution* is a solution in which a substance has been dissolved into another substance in an amount greater than what is usually possible as a result of the substances being heated and cooled.

Supersaturated solutions of sugar and water are unstable. As a result of being unstable, the sugar molecules begin to crystallize into a solid. **See Figure 4-24.** Candy often includes ingredients such as corn syrup, butter, egg whites, and cream of tartar because these ingredients interfere with crystal growth to help create a product of higher quality.

Fermentation. Temperature control is essential to the fermentation process for both baking and making alcohols. In baking, fermentation is accomplished by using yeast to convert sugars into carbon dioxide and alcohol. Warm water is generally added to activate the yeast and encourage fermentation. Fermentation increases volume, adds texture to dough, and improves flavor.

Fermentation is also essential in making wine, beer, and spirits. Wine is typically made from grapes, and beer is made using grains such as barley, wheat, or rye. Spirits are made from fruits, sugars, such as molasses, and starches, such as grains and potatoes. After the carbohydrates have been pulped into a mash and mixed with hot water, yeast is added. The yeast feeds on the sugars present in the mash and converts it to carbon dioxide and alcohol. Environments that are too cold or too hot will inactivate or destroy the yeast and create an inferior product.

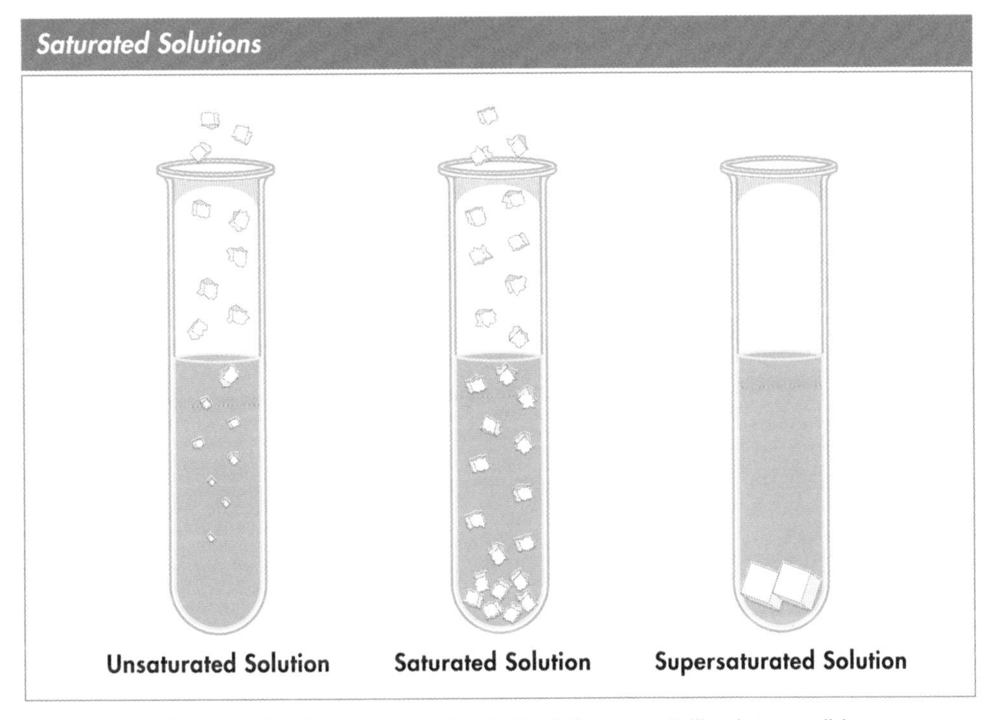

Figure 4-24. Sugar molecules in supersaturated solutions crystallize into a solid.

Reactions to Moisture

Many starches are prepared using both moisture and heat to make them more digestible. For example, potatoes, pasta, and grains are often boiled or simmered in a liquid to transform a hard food into one that is soft and appealing. Starches are also affected by heat and moisture as they undergo some reactions, such as gelatinization, retrogradation, and hydrolysis, which can affect the thickness and texture of gravies, sauces, soups, and desserts.

Gelatinization. Gelatinization occurs as heated starch granules absorb moisture and swell, which thickens liquids and adds structure to baked products. **See Figure 4-25.** When starch granules undergo gelatinization, their hydrogen bonds break, which allows water to enter the starch granules and causes them to swell. Starch granules can swell up to 30 times their original size, which makes starch an effective thickening agent. Flours, cornstarch, arrowroot, and tapioca are starches commonly used as thickeners.

Gelatinization is an irreversible process that generally occurs between 133°F and 167°F. The thickness, or viscosity, of the starch and liquid mixture increases during the heating process. However, if the liquid is heated and held at very high temperatures, the starch granules will collapse and the viscosity of the liquid will decrease. Viscosity also decreases when starches and sugars are used together in recipes such as puddings and pie fillings. Sugar competes with starch, which makes less water available for starch to reach its thickening capacity.

Retrogradation. *Retrogradation* is a process in which a gelatinized starch causes a thickened liquid to turn into a gel as it cools. Retrogradation is what causes instant puddings to gel as they cool. In starches with a high concentration of amylose, such as cornstarch, the gel will be firm. In contrast, potato starch with a high concentration of amylopectin produces a softer gel. If retrogradation is not desired, the product can usually be restored to an ideal viscosity by applying gentle heat.

Figure 4-25. Gelatinization occurs as heated starch granules absorb moisture and swell, which thickens a liquid.

FOOD SCIENCE EXPERIMENT: Gelatinization

4-1

Objective

- To determine whether the consistency of gelatinization is altered by changing the ratio of ingredients.

Materials and Equipment

- Masking tape
- Marker
- 3 small microwave-safe bowls
- Measuring spoons
- Measuring cups
- Wooden spoon
- Cornstarch
- Water
- Sheet pan lined with wax paper

Procedure

1. Use the masking tape and marker to label one bowl "10%," the second bowl "20%," and the third bowl "30%."
2. In the appropriate bowl, mix the following amounts of cornstarch in water:

Ingredient	10%	20%	30%
Cornstarch	4 tsp	8 tsp	12 tsp
Water	6 oz	6 oz	6 oz

3. Heat the "10%" bowl in the microwave on high for 30 seconds. Remove the bowl from the microwave and stir the mixture. Return the bowl to the microwave and heat again for 30 seconds. Remove the bowl from the microwave and stir the mixture. Heat the mixture a third time for 30 seconds. Remove the bowl and stir the mixture.

   Safety Note: The mixture can become extremely hot and cause a burn. Take caution when stirring.
4. Repeat step 3 using the bowl labeled "20%."
5. Repeat step 3 using the bowl labeled "30%."
6. Allow all three suspensions to cool to room temperature. Observe any changes as each suspension cools.
7. Measure the thickness (viscosity) of each of the three suspensions using the "ooze" method. Line a sheet pan with wax paper. Gently place each suspension onto the wax paper, and allow the suspension to spread out undisturbed.
8. After each suspension stops spreading, evaluate the differences in the three suspensions.

Typical Results

There will be variances in the gelatinization of each cornstarch suspension.

Variations

This experiment used starch made from corn. However, starches made from other sources, such as potatoes, rice, and tapioca, can also be used to observe gelatinization properties.

As they age, gels may experience syneresis. *Syneresis,* also known as weeping, is the separation of a liquid from a gel. Evidence of syneresis is the droplets of water that often form on a pudding that has been left in the refrigerator for some time. It is also evident in a soggy pie crust that has a starch-thickened filling such as chocolate cream pie.

Hydrolysis. *Hydrolysis* is the process of splitting a substance into smaller parts by the addition of water. Enzymes, acids, and heat can be used to help speed the process of hydrolysis. When it is hydrolyzed, a polysaccharide or disaccharide is broken down into simpler sugars. For example, sucrose is often hydrolyzed to yield the two monosaccharides that combine to form it, which are glucose and fructose. **See Figure 4-26.** The end result is a very sweet product known as invert sugar syrup, which is commonly used in baking to help maintain moisture and inhibit crystallization.

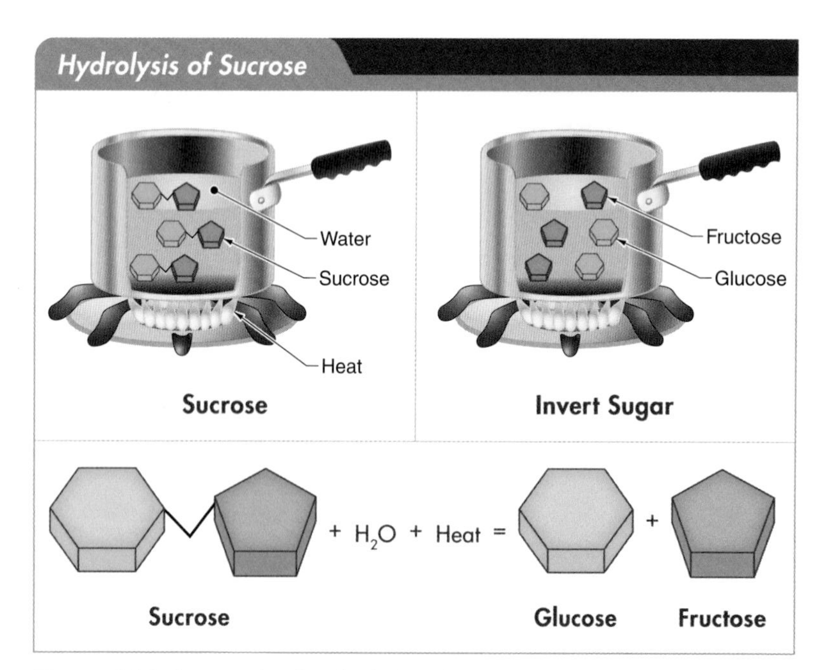

Figure 4-26. Sucrose is often hydrolyzed to yield the two monosaccharides that combined to form it: glucose and fructose.

Reactions to pH

The pH level of ingredients can affect the texture and flavor of food. For example, water is an important ingredient in baking bread. To encourage the rising action of the yeast and to help preserve the bread after it is baked, water with a pH level between 4 and 5 is ideal. Acidic and alkali ingredients can also impact the color and texture of vegetables. In addition, acids can affect the gelatinization of starches.

Vegetables. An acidic ingredient, such as lemon juice, or an alkaline ingredient, such as baking soda, can affect both the color and texture of vegetables during cooking. Acidic and alkaline ingredients can also alter the natural pigments present in vegetables. Vegetables high in chlorophyll, such as broccoli, will turn a drab green color but retain their firmness when an acidic ingredient is added to the cooking liquid. Adding an alkali will cause green vegetables to become brighter in color but mushy in texture. In addition to the chlorophyll found in green vegetables, natural pigments include carotenoids and flavonoids. **See Figure 4-27.**

A *carotenoid* is an organic pigment found in orange or yellow vegetables. Neither acidic nor alkaline ingredients affect the color of carotenoids. However, an alkali will cause orange and yellow vegetables to become mushy.

A *flavonoid* is an organic pigment found in purple, dark red, and white vegetables. Acidic ingredients cause purple and dark red vegetables to turn bright red. In contrast, alkaline ingredients cause purple and dark red vegetables to turn blue and white vegetables to turn yellow. In addition, alkaline ingredients cause a mushy texture in purple, dark red, and white vegetables.

Starches. The gelatinization of starches is also affected by pH levels. For example, when acidic ingredients are added to a starch-thickened product as it cooks, the product can decrease in viscosity. Therefore, acidic ingredients such as vinegars, wine, or citrus juices should be added after gelatinization is complete to prevent an adverse impact on viscosity.

When acidic ingredients are used in pie fillings, barbecue sauces, or glazes, modified starches are often used to help maintain the desired viscosity. A *modified starch* is a starch that has been chemically or physically altered from its original state. Modified starches are also used due to their ability to withstand extreme heat or freezing temperatures, improve textures, and increase shelf life.

Acid and Alkali Reactions

Pigment	Cooked Vegetables*	Acid Added	Alkali Added
Chlorophyll	Broccoli	Color Loss	Mushy texture
Carotenoids	Carrots	Little or no effect	Mushy texture
Flavonoids	Beets	Brighter red	Turns blue; mushy texture

* no acidic or alkaline ingredients used

Figure 4-27. Acidic and alkaline ingredients can impact the color and texture of vegetables.

Knowledge Check 4-7

1. Identify five functions of simple carbohydrates in food preparation.
2. Identify a common function of complex carbohydrates in food preparation.
3. Define hygroscopic.
4. Explain what happens to potatoes if stored in sunlight.
5. List five reactions that can occur in carbohydrates when exposed to heat.
6. Summarize gelatinization.
7. Describe retrogradation.
8. Explain the process of hydrolysis.
9. Describe how acidic and alkaline ingredients can affect the color and texture of vegetables.
10. Explain the purpose of modified starches.

CARBOHYDRATES ON THE MENU

Nutrient-dense carbohydrates come in an assortment of stunning colors, textures, and flavors, giving chefs the opportunity to take advantage of their versatility. For example, an array of grilled vegetables with vivid colors and varied textures can be a flavorful and nutritious start to a meal. **See Figure 4-28.**

Chef's Tip

Replacing granulated sugar with maple syrup enhances flavor and provides essential minerals not found in granulated sugar.

Carbohydrates on the Menu

Figure 4-28. Grilled vegetables can be a flavorful and nutritious start to a meal.

Carbohydrates can be the basis for a smooth and silky puréed soup without the need to add cream or the focal point of a healthy entrée such as a whole-grain pasta dish. When carbohydrates are served as an accompaniment to a protein such as meat, poultry, or seafood, their complementary flavorings and contrasting textures can be highlighted. Mini desserts, such as a bite-size trio of ice creams topped with fresh berries, a whole-grain bread pudding, or a tartlet featuring caramelized grilled peaches can be an enticing end to a meal.

Food costs can often be reduced by increasing the amount of carbohydrates on the plate while reducing the portion size of protein. Fruits, vegetables, grains, and legumes are typically low in cost, which is a significant consideration when analyzing food costs.

Food Safety

Vegetables and fruits must be carefully washed before using in food preparations.

Presentation

Carbohydrates can offer great diversity in terms of presentation. For example, a potato can be served mashed, roasted, or lightly fried to offer a soft, firm, or crisp texture. Carbohydrates can also be used to add height, dimension, and color to plated dishes to create an eye-appealing menu offering. Instead of serving white rice, presenting red rice or farro can build interest as well as introduce guests to grains that are both nutrient dense and flavorful.

Texture and Flavor

Carbohydrates are often cooked to enhance their innate characteristics and improve digestibility. Carbohydrates offer a range of textures and flavors from soft and sweet to firm and crisp. For example, sautéing broccoli with garlic-infused olive oil and lightly seasoning it with salt and pepper not only preserves the appealing crisp texture of broccoli but also develops flavor and makes the broccoli easier to digest. Appetizers, salads, sides, entrées, and desserts can showcase nutrient-dense carbohydrates by highlighting their many colors, textures, and flavors. Carbohydrates can be used to offer guests nutritious and innovative menu choices that complement any cuisine. **See Figure 4-29.**

Versatility of Carbohydrates on Menus

Courtesy of The National Pork Board

Moroccan Pork with Garbanzo Beans and Couscous

Photo Courtesy of the Beef Checkoff

Beef Stir-Fry with Jasmine Rice

Idaho Potato Commission

Potato Pesto Chicken Salad with Haricots Verts

Figure 4-29. Carbohydrates can be used to offer guests nutritious and innovative menu choices that complement any cuisine.

Knowledge Check 4-8

1. Summarize how carbohydrates can be used to help keep food costs low.
2. Describe how carbohydrates can be used to enhance presentation.
3. Give an example of how to enhance the texture and flavor of a carbohydrate.

Chapter Summary

Carbohydrates are macronutrients that fuel both the body and the brain and are an essential part of a healthy diet. Carbohydrates are classified as simple carbohydrates (sugars), complex carbohydrates (starches), or dietary fiber. Simple carbohydrates are easily broken down into glucose and supply short-term energy, whereas complex carbohydrates take longer to digest and deliver sustained energy. Dietary fiber works to keep the digestive system functioning properly. Choosing fiber-rich complex carbohydrates has been shown to significantly improve health and help prevent serious illness.

The public interest in healthy eating has led to an increased demand for flavorful and nutrient-dense menu offerings. When preparing carbohydrates, simple changes such as using whole grains instead of refined grains or grilling vegetables instead of deep-frying them can positively impact the nutrient content of a dish. Due to their wide range of colors, textures, and flavors, carbohydrates can be paired with other ingredients to develop complex flavors and nutrient-dense meals that appeal to a variety of guests.

Florida Tomato Committee

Chapter Review

1. Define photosynthesis.
2. List three monosaccharides and three disaccharides.
3. Compare starches and dietary fiber.
4. List seven functions of carbohydrates.
5. Summarize the digestion of carbohydrates.
6. Describe how glucose is stored.
7. Describe common sources of simple carbohydrates.
8. Differentiate between whole grains and refined grains.
9. Identify the recommended dietary allowance (RDA) for carbohydrates and the adequate intakes (AI) for dietary fiber.
10. Identify the consequences of carbohydrate excess and carbohydrate deficiency.
11. Describe the reactions of carbohydrates to heat.
12. Give an example of how carbohydrates can be used to enhance presentation.

Lipids

Fats

Oils

Barilla America, In

Chapter 5

Essential Fatty Acids

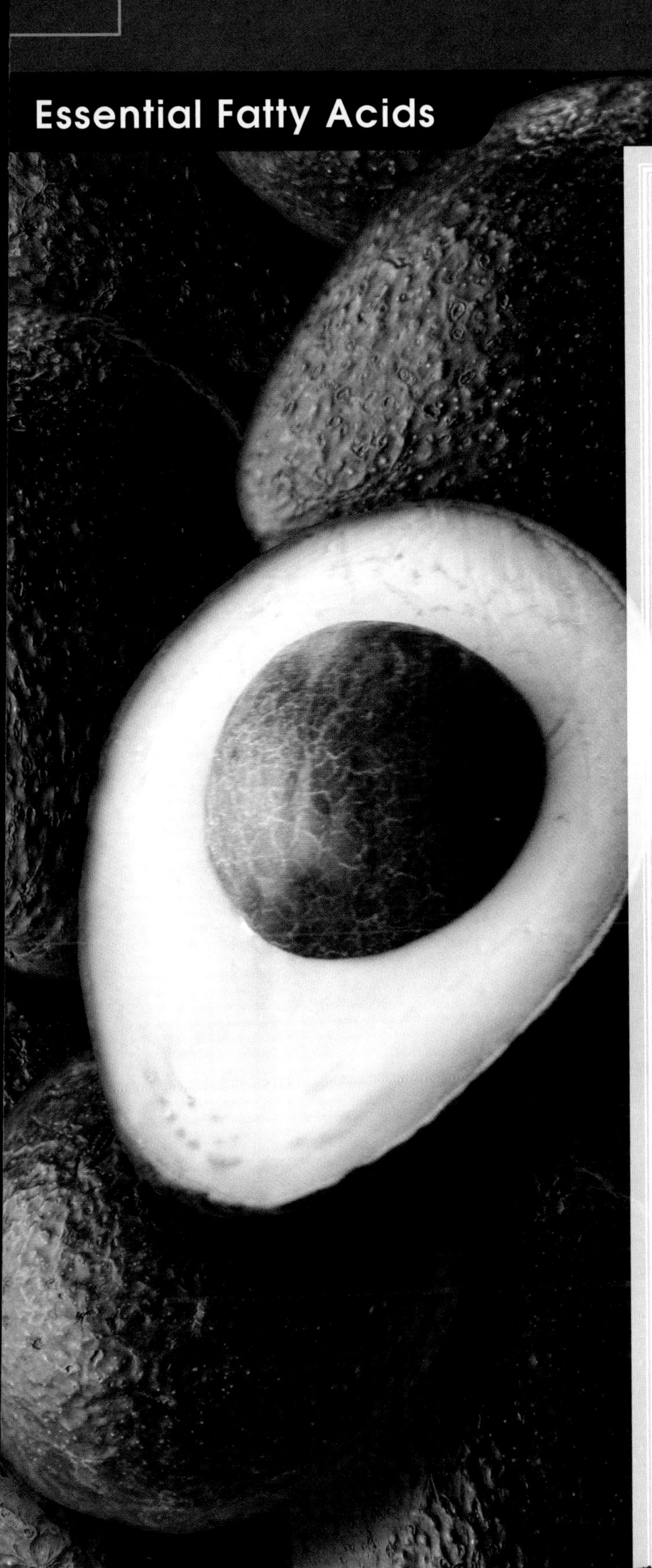

Lipids, commonly known as fats, are a large and structurally diverse group of naturally occurring compounds made by the body and found in virtually all food groups. Lipids are the chief form of energy storage for the body and are vital for normal growth and development. A varied diet that includes unsaturated fats provides essential nutrients and promotes health, whereas saturated and trans fats have been linked to cardiovascular disease. Healthy lipids, such as those found in olive oil, fish, avocados, and nuts, can be used to create exceptional dishes filled with pleasing colors, textures, and flavors.

Chapter Objectives

1. Summarize the role of lipids.
2. Differentiate between a triglyceride, phospholipid, and sterol.
3. Compare saturated fats and unsaturated fats.
4. Describe a trans fat.
5. Describe the main functions of lipids.
6. Summarize lipid digestion and absorption.
7. Differentiate between low-density lipoproteins and high-density lipoproteins.
8. Identify sources of unsaturated and saturated fats.
9. Name the essential fatty acids.
10. Describe how lipid consumption can affect health.
11. Explain how lipids react to air, light, heat, and moisture.
12. Identify ways lipids can be used to create flavorful, nutritious menu items.

THE ROLE OF LIPIDS

A *lipid* is an energy-providing nutrient made from fatty acids and includes solid fats and oils. **See Figure 5-1.** Lipids enable the body to function properly and are an essential part of a healthy diet. Lipids are calorie-rich, supply 9 calories of energy per gram, and play a significant role in providing the body with a sense of satiety, or fullness. Individuals often have the misconception that consuming lipids will result in weight gain. This is not necessarily the case. Whether calories are from proteins, carbohydrates, or lipids, the primary cause of weight gain is consuming more calories than the body burns through physical activity.

Lipids are found throughout the body in cell membranes, organs, and muscles. They provide natural oils that add shine to skin, hair, and nails. Lipids are also stored in the body as either visceral fat or subcutaneous fat. *Visceral fat,* also called intra-abdominal fat, is fat that is stored within the abdomen and surrounds the internal organs. *Subcutaneous fat* is fat that is stored just below the surface of the skin. **See Figure 5-2.** Too much visceral fat is associated with cardiovascular disease and is typically harder to lose than subcutaneous fat because it is deeply embedded within the body. A person can maintain a healthy weight and still have a fair amount of visceral fat.

In addition to their role in the body, lipids play a vital role in culinary applications such as lubricating foods that are to be grilled or sautéed. In cooking, lipids enhance the aroma, texture, and flavor of foods. For example, lipids help keep foods tender and juicy, provide breads with a soft crumb, and provide pastries with a flaky crust. Lipids also impart a smooth and rich mouthfeel that helps distribute flavors across the taste buds.

Figure 5-1. A lipid is an energy-providing nutrient made from fatty acids and includes solid fats and oils.

Stored Lipids

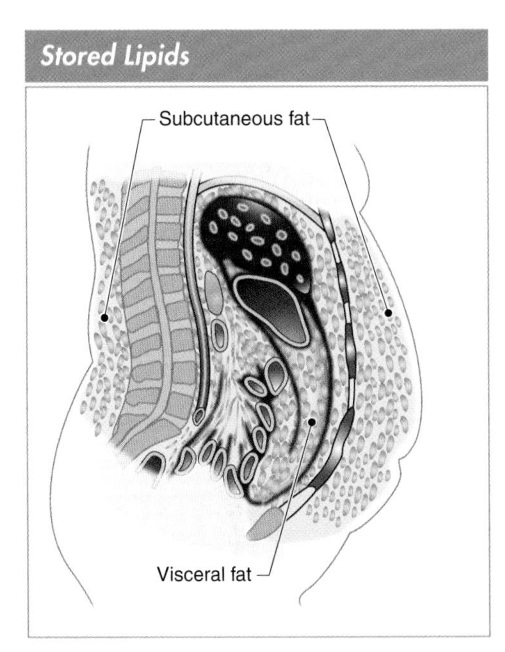

Figure 5-2. Visceral fat is stored within the abdomen and surrounds the internal organs, whereas subcutaneous fat is stored just below the surface of the skin.

Although lipids are commonly thought of as fats and oils, lipids are categorized as triglycerides, phospholipids, and sterols. Some lipids are found in both animal- and plant-based foods, whereas others are made by the body.

Triglycerides

Triglycerides make up 95% of all lipids and are found in the body as well as in food. A *triglyceride* is a lipid that consists of a glycerol backbone with three fatty acids attached and is the main form of fat in the body and foods. *Glycerol* is an organic compound that contains three carbon atoms. A *fatty acid* is an organic compound consisting of a chain of carbon and hydrogen atoms attached to an acid group on one end. To form a triglyceride, each carbon atom of glycerol attaches to one of the three fatty acids. **See Figure 5-3.**

In the body, triglycerides are stored and used for energy. In food, triglycerides are known as fats and oils. *Fat* is a triglyceride that remains solid at room temperature. Fats are commonly found in animal-based foods such as meats, cheeses, butter, and shortening. *Oil* is a triglyceride that remains liquid at room temperature. Oils are commonly derived from plant-based foods and include olive, canola, corn, and peanut oils among others. Solid fats are often associated with health problems, such as cardiovascular disease, whereas many oils are associated with a "heart-healthy" diet.

The fatty acid chains of a triglyceride vary in length. Short-chain fatty acids consist of fewer than 6 carbon atoms, medium chains consist of 6 to 10 carbon atoms, and long chains consist of 12 to 24 carbon atoms. Typically, the longer the fatty acid chain, the more solid the lipid (fat) will be at room temperature. In contrast, the shorter the fatty acid chain, the more liquid the lipid (oil) will be at room temperature.

Triglyceride Structure

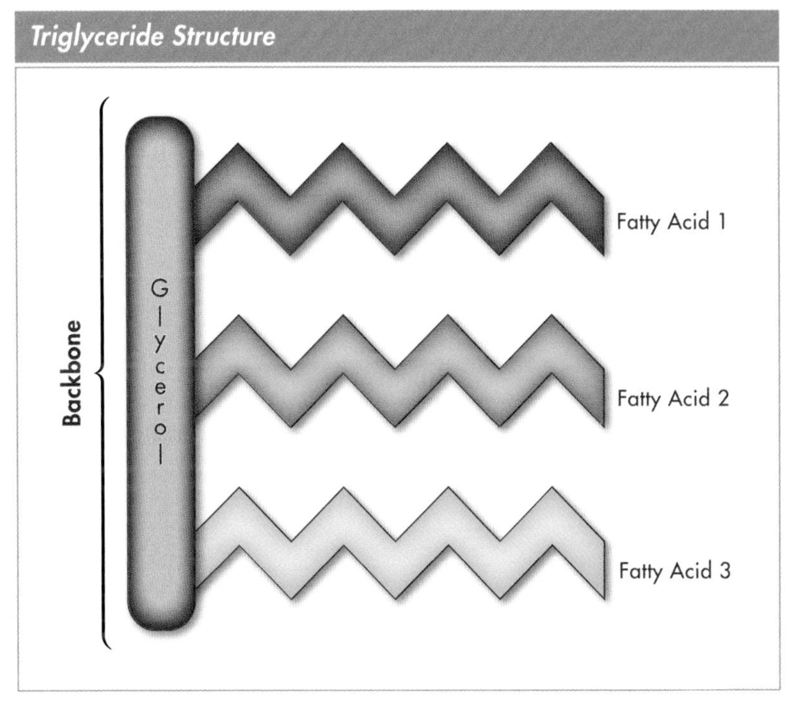

Figure 5-3. In a triglyceride, each carbon atom of glycerol is attached to one of the three fatty acids.

The degree of saturation is another factor that affects whether a lipid is solid or liquid at room temperature. Solid fats are considered saturated fats and oils are considered unsaturated fats.

Saturated Fats. A *saturated fat* is a triglyceride made primarily of saturated fatty acids. The more saturated fatty acids a fat contains, the more solid the fat will be at room temperature. **See Figure 5-4.** For example, butter contains more saturated fatty acids than corn oil and is, therefore, firmer at room temperature. Saturated fats are commonly found in animal-based foods and have a lower smoke point than unsaturated fats.

Saturated Fats

Entourage

Figure 5-4. Butter-rich cake and frosting are abundant in saturated fats.

The saturation of fats relates to the number of hydrogen atoms in a fatty acid chain. A *saturated fatty acid* is a fatty acid chain in which each carbon atom is filled to capacity with hydrogen atoms. Therefore, the fatty acid is entirely "saturated" with hydrogen atoms and forms a straight chain. When saturated fatty acids attach to glycerol, a saturated fat is made.

Unsaturated Fats. An *unsaturated fat* is a triglyceride made primarily of unsaturated fatty acids. In general, unsaturated fats are found in vegetable and olive oils, fish, avocados, nuts, and seeds. Pasta tossed with a pesto made with olive oil and walnuts is an example of a dish rich in unsaturated fats.

Unsaturated fats are typically categorized as a monounsaturated or polyunsaturated fat. A *monounsaturated fat* is a triglyceride made primarily of monounsaturated fatty acids. A *polyunsaturated fat* is a triglyceride made primarily of polyunsaturated fatty acids. Pesto made with olive oil and walnuts is not only rich in unsaturated fats but also provides both mono- and polyunsaturated fats. The olive oil is abundant in monounsaturated fats, and polyunsaturated fats are plentiful in the walnuts.

Unsaturated fats remain liquid at room temperature, are commonly found in plant-based foods, and have a higher smoke point. For example, canola oil is a plant-based oil with a smoke point of 468°F, which makes it ideal for high-heat applications such as stir-frying. In contrast, butter, which is an animal-based fat, does not tolerate heat well and begins to smoke between 300°F and 350°F.

An *unsaturated fatty acid* is a fatty acid chain that lacks some hydrogen atoms. The point on the fatty acid chain where hydrogen is missing is replaced with a double carbon bond. In contrast to the straight structure of a saturated fatty acid chain, an unsaturated fatty acid chain is typically bent. **See Figure 5-5.** When unsaturated fatty acids attach to the glycerol backbone, an unsaturated fat is made.

A double carbon bond prevents saturation from occurring and is known as the point of unsaturation. The *point of unsaturation* is the point on a fatty acid chain where a missing hydrogen atom is replaced with a double carbon bond. Fatty acids are classified according to the number of points of unsaturation they contain. A *monounsaturated fatty acid* is a type of fatty acid that contains one point of unsaturation. A *polyunsaturated fatty acid* is a type of fatty acid that contains two or more points of unsaturation. **See Figure 5-6.**

Saturated and Unsaturated Fatty Acid Comparison

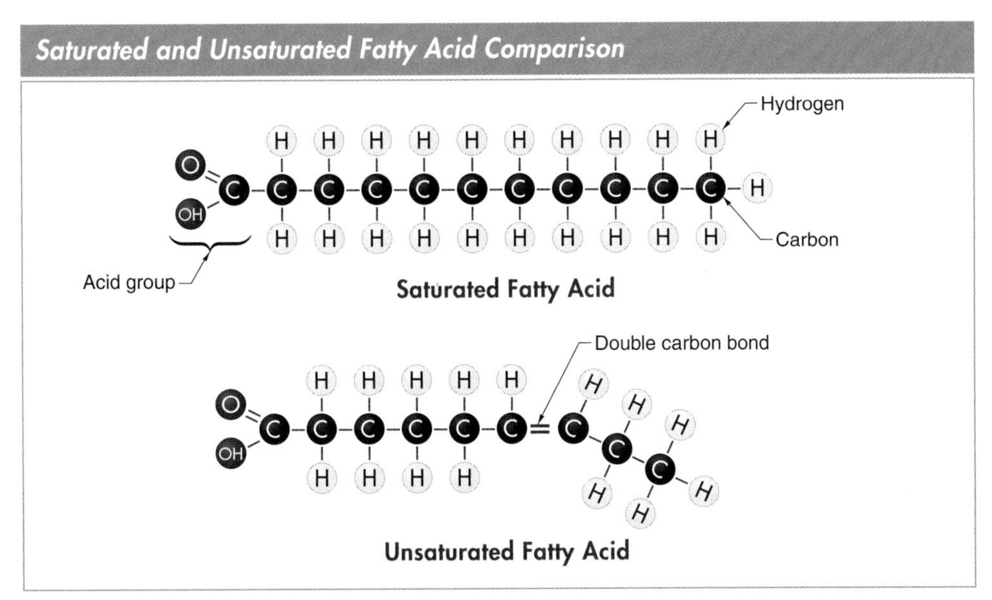

Figure 5-5. A saturated fatty acid chain is straight and filled to capacity with hydrogen atoms, whereas an unsaturated fatty acid chain is bent and lacks some hydrogen atoms.

Points of Unsaturation

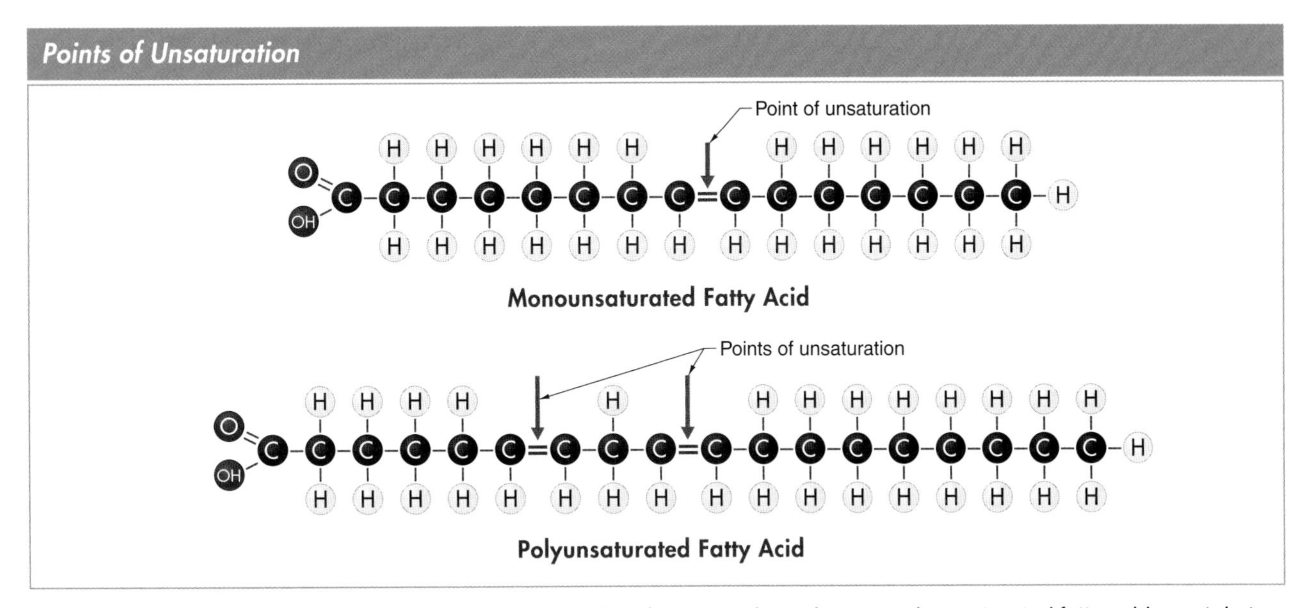

Figure 5-6. Monounsaturated fatty acids contain one point of unsaturation, whereas polyunsaturated fatty acids contain two or more points of unsaturation.

Trans Fats. A *trans fat,* also known as a trans-fatty acid, is an unsaturated fatty acid with hydrogen atoms on opposite sides of a double carbon bond, which makes its structure similar to a saturated fat. **See Figure 5-7.** Trans fats are rare in natural foods but abundant in processed foods due to hydrogenation. *Hydrogenation* is the chemical process of using heat to force hydrogen atoms into an unsaturated fatty acid to make it similar in structure to a saturated fat. Hydrogenation turns a liquid fat into a solid fat at room temperature. Vegetable shortening, which is 10% air by weight, is an example of a product that has undergone hydrogenation. Solid fats have a longer shelf life than liquid fats because they are more resistant to becoming rancid. Fats that are rancid have an undesirable odor and taste because they have decomposed and should not be used.

Trans-Fatty Acid Configuration

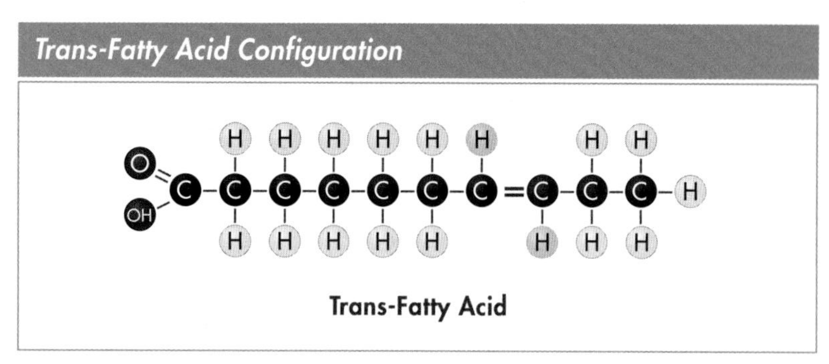

Figure 5-7. A trans fat is an unsaturated fatty acid with hydrogen atoms on opposite sides of a double carbon bond, which makes its structure similar to a saturated fat.

In addition to an increased shelf life, trans fats are less expensive than animal-based fats. However, trans fats affect the body similarly to saturated fats and can increase the risk of cardiovascular disease. Therefore, they should only be a small part of the diet.

Phospholipids

A *phospholipid* is a type of lipid consisting of two fatty acids and a phosphate molecule. Phospholipids are naturally made by the body and play an important role in the structure of cell membranes. The structure of phospholipids is similar to the structure of a triglyceride, except a phospholipid contains only two fatty acid chains. The third chain is replaced by the phosphate molecule. **See Figure 5-8.**

The two fatty acid chains of a phospholipid are soluble in fat, whereas the phosphate molecule is soluble in water. This characteristic allows a phospholipid to act as an emulsifier by keeping fats suspended in water. An *emulsifier* is a substance that mixes with two unlike liquids to produce an emulsion. An *emulsion* is a combination of two unlike liquids that have been forced to bond with each other. For example, mayonnaise generally contains vinegar and oil. Without an emulsifier, the oil would float on top of the vinegar. Mayonnaise also contains egg yolks, which contain the phospholipid lecithin. The lecithin acts as an emulsifier by keeping the vinegar and oil from separating. Lecithin is also found in soy and is often used as an emulsifier in prepared salad dressings, soups, and sauces.

The emulsifying properties of phospholipids also keep nutrients, including fats, suspended in the blood. This allows nutrients to be transported through lipid-containing cell membranes to the watery fluid in and around cells.

Sterols

A *sterol* is a type of lipid that is a waxy, insoluble substance composed of natural steroid alcohols derived from plants or animals. A steroid alcohol includes a group of fat-soluble organic compounds in a ringlike structure that consists of single or double carbon bonds and an acid group at one end. The most well-known type of sterol is cholesterol. **See Figure 5-9.**

Cholesterol. *Cholesterol* is a type of sterol in the lipid family that is a soft, waxy substance made by the body and found in every cell and in foods of animal origin. Sterols are found in both plant- and animal-based foods, whereas cholesterol is found only in animal-based foods such as egg yolks, meat, poultry, seafood, and dairy products. The body needs cholesterol to maintain cell membranes, produce a variety of hormones, synthesize vitamin D, and make the bile necessary for digestion. However, consuming excess cholesterol can contribute to plaque buildup in the arteries and increase the risk of cardiovascular disease.

Ergosterol. *Ergosterol* is a type of sterol present in fungi. Ergosterol is a necessary component of fungi cell membranes, such as those found in mushrooms, and is also a component of yeast. In the presence of ultraviolet light, ergosterol is converted to ergocalciferol, which is known as vitamin D_2. Because vitamin D is primarily found in animal-based foods, ergosterol-containing foods like mushrooms can serve as an important source of vitamin D for vegetarians.

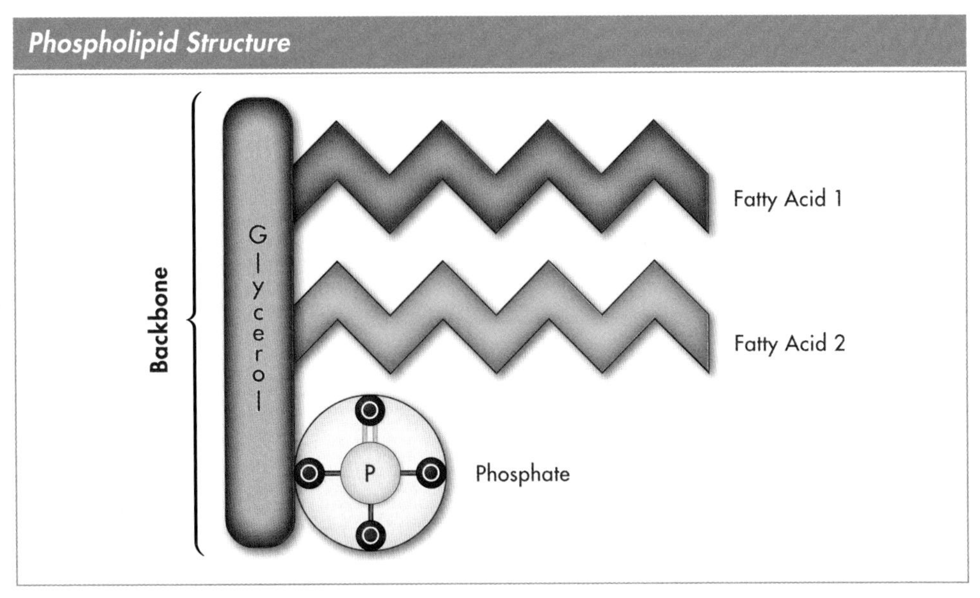

Figure 5-8. A phospholipid is a type of lipid consisting of two fatty acids and a phosphate molecule.

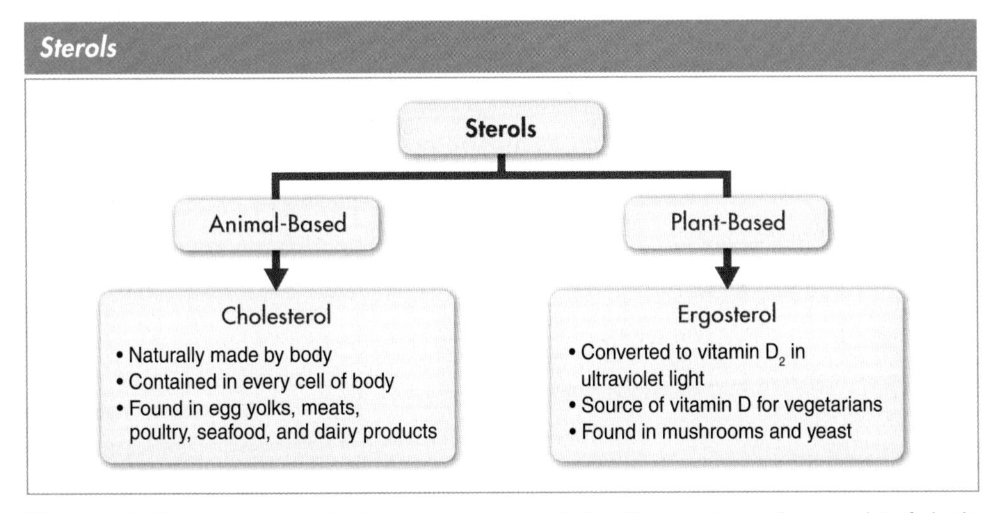

Figure 5-9. Sterols, such as cholesterol, are made of ring-like structures that consist of single or double carbon bonds and an acid group.

Knowledge Check 5-1

1. Differentiate between visceral fat and subcutaneous fat.
2. Identify two functions of lipids in cooking.
3. Describe how a triglyceride is formed.
4. Compare a fat with an oil.
5. Define a saturated fatty acid.
6. Explain the point of unsaturation.
7. Summarize the process hydrogenation.
8. Describe the structure of phospholipids.
9. Identify four functions of cholesterol.

LIPID FUNCTIONS

The healthy body needs lipids to function properly. Lipids supply energy to the body, provide insulation, and help protect organs. Lipids are also involved in the production of hormones, give structure to cells, and are necessary for the proper absorption of nutrients. **See Figure 5-10.**

Adipose cells also have the unique ability to expand. The more fat that an adipose cell stores, the larger it becomes. It can expand up to 20 times in weight. If adipose cells are not used for energy, they keep expanding and ultimately lead to weight gain. **See Figure 5-11.**

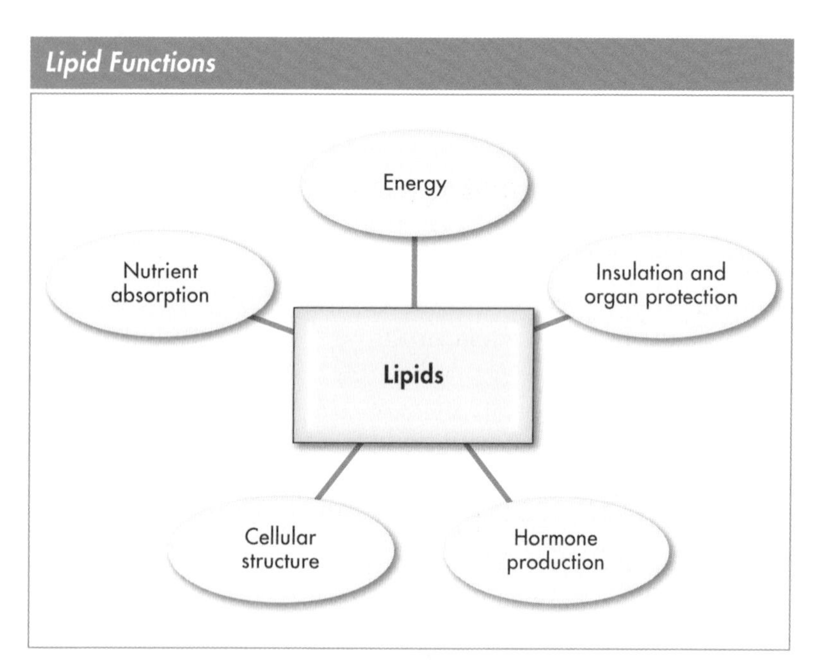

Figure 5-10. Lipids supply energy, provide insulation, protect organs, help produce hormones, give structure to cells, and are necessary for nutrient absorption.

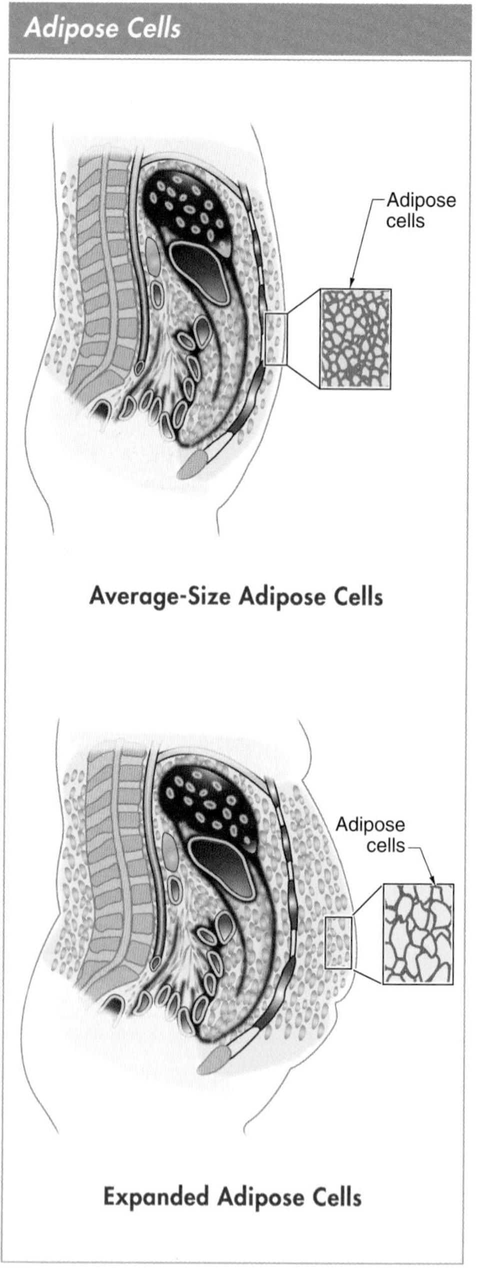

Figure 5-11. Adipose cells keep expanding and ultimately lead to weight gain if they are not used for energy.

Energy

Lipids are the most calorie-dense of the macronutrients. Lipids supply 9 calories of energy per gram compared to 4 calories per gram from proteins and carbohydrates. The body also has the ability to store excess fat for later energy use. Excess fat is stored in specialized "fat cells" called adipose cells. An *adipose cell* is a cell with the primary function of storing excess fat for later energy use. Unlike most cells, adipose cells lack water, which allows them the ability to store a large amount of fat in a small space.

Insulation and Organ Protection

Lipids form a layer of "fat" under the skin that insulates the body from temperature extremes. This layer of insulation helps regulate the internal temperature of the body. Fats also cushion the organs such as the heart, kidneys, and intestines. The fat surrounding these organs protects them from injury, such as shock, and helps hold the organs in place.

Hormone Production

Lipids are involved in the production of hormones that help regulate many functions throughout the body. Hormones work within the body as messengers by sending signals that influence hormonal responses, insulin regulation, and metabolism.

Cellular Structure

Cellular structure depends on lipids because cell membranes are made of lipids. Lipids coat cell membranes and provide a protective barrier that helps regulate the materials entering and exiting through cell walls. Lipid-coated cell membranes also enable information to be transmitted across nerves more efficiently.

Courtesy of The National Pork Board

Mediterranean pasta salad tossed with olive oil provides unsaturated fats that are necessary for nutrient absorption.

Nutrient Absorption

Lipids help the body absorb fat-soluble nutrients such as vitamins A, D, E, and K. With the exception of vitamin D, fat-soluble vitamins are also found in many vegetables and fruits. In order to absorb these vitamins, it is important to consume lipids.

Knowledge Check 5-2

1. List five functions of lipids.
2. Identify the calories of energy per gram supplied by lipids, proteins, and carbohydrates.
3. Define adipose cell.
4. Explain how lipids protect the body from temperature extremes.
5. Summarize the role lipids play in cellular structure.
6. Identify the nutrients that lipids help the body absorb.

LIPID DIGESTION AND ABSORPTION

Lipids are digested differently than both proteins and carbohydrates. The majority of lipid digestion takes place in the small intestine, with only a minimal amount occurring in the mouth and stomach. In the mouth, an enzyme known as lingual lipase helps digest lipids found in milk. *Lingual lipase* is an enzyme in saliva that helps break down the lipids found in milk for digestion. This enzyme plays a minor role for adults but is critical for infants whose diets primarily consist of milk.

As food travels down the esophagus and into the stomach, lipids separate from the water-based fluids and float to the top. Once lipids reach the small intestine, they are exposed to bile that was manufactured in the liver and secreted into the small intestine by the gallbladder.

The Role of Bile

Bile works to emulsify lipids and water-based fluids by transforming large lipid globules into smaller drops that have the ability to repel one another. Once lipids are emulsified, pancreatic lipase can break lipids into digestible parts. *Pancreatic lipase* is an enzyme released by the pancreas that breaks lipids into glycerol, fatty acids, and monoglycerides. A *monoglyceride* is a large lipid fragment consisting of a glycerol molecule with one fatty acid chain attached. **See Figure 5-12.** The smaller fatty acid chains and glycerol are mixed with bile, so they can be transported to the villi and absorbed into the bloodstream. Bile is then either sent back to the liver where it is recycled or trapped by dietary fiber and excreted as waste.

Lipid Digestion and Absorption

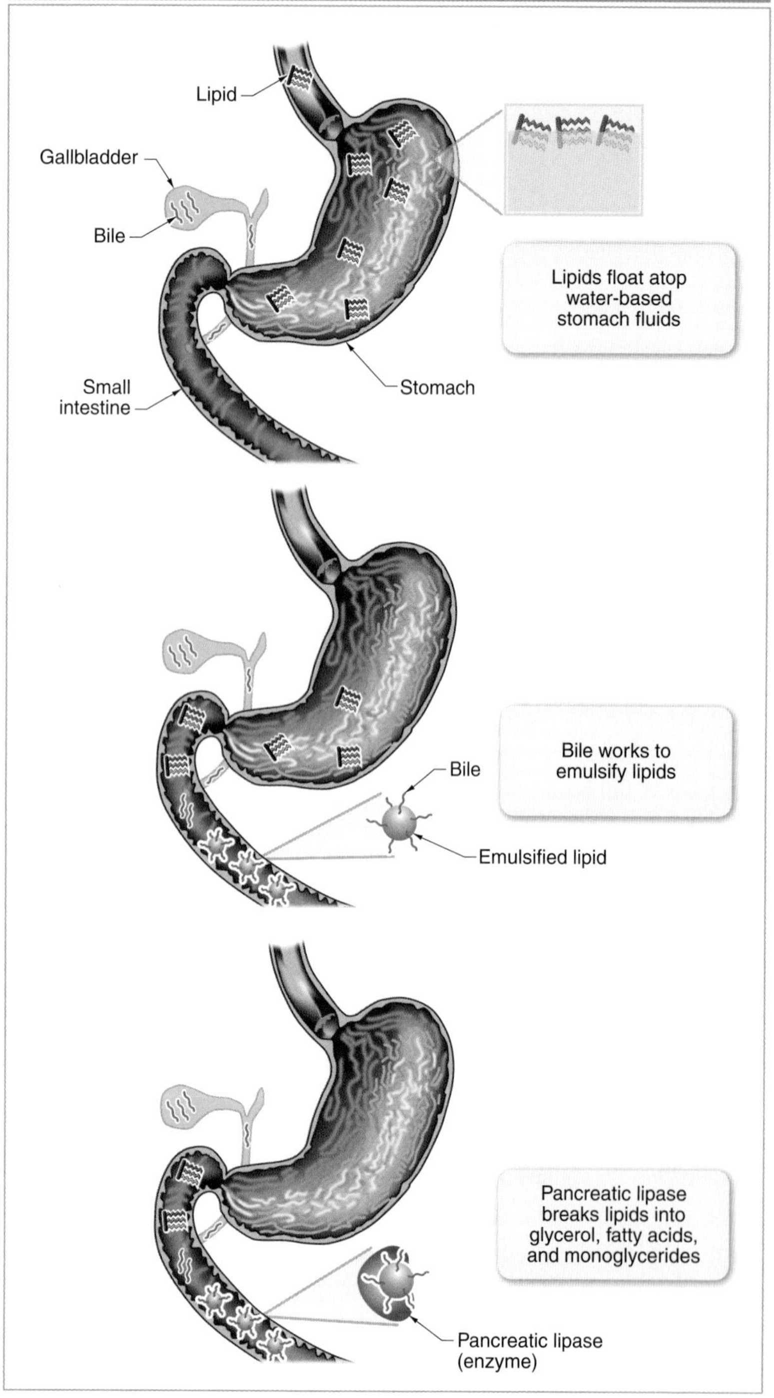

Figure 5-12. Bile works as an emulsifier so that lipids can be digested and absorbed.

Monoglycerides and long fatty acid chains are too large to remain emulsified. Therefore, they are unable to be transported in their natural state. For digestion and absorption to occur, monoglycerides and long fatty acid chains are bound by protein, which makes them water soluble and ready to be transported. These protein-coated lipids are known as lipoproteins.

The Role of Lipoproteins

A *lipoprotein* is a protein-coated lipid that acts as an emulsifier for larger lipid fragments and cholesterol to facilitate their transport into the lymph and bloodstream for absorption. *Lymph* is a type of fluid drained from tissues that can transport material from the bloodstream into tissues. Lipoproteins are made of four components: triglycerides, protein, cholesterol, and phospholipids. **See Figure 5-13.** Lipoproteins are classified based on their weight and density. The four types of lipoproteins include chylomicrons, very-low-density lipoproteins, low-density lipoproteins, and high-density lipoproteins.

Chylomicrons. A *chylomicron* is a type of lipoprotein that transports large fragments of partially digested triglycerides and cholesterol from the small intestine to the liver and other tissues for absorption. To reach the liver, chylomicrons are absorbed through the lymphatic system and then into the bloodstream. **See Figure 5-14.** Once the chylomicron reaches the bloodstream, the triglycerides are broken down into single fatty acid chains and glycerol by lipoprotein lipase. *Lipoprotein lipase* is an enzyme that breaks down the triglycerides transported by a chylomicron into fatty acids and glycerol that can be absorbed. The protein and cholesterol that remain from the chylomicron are metabolized by the liver.

Lipoprotein Structure

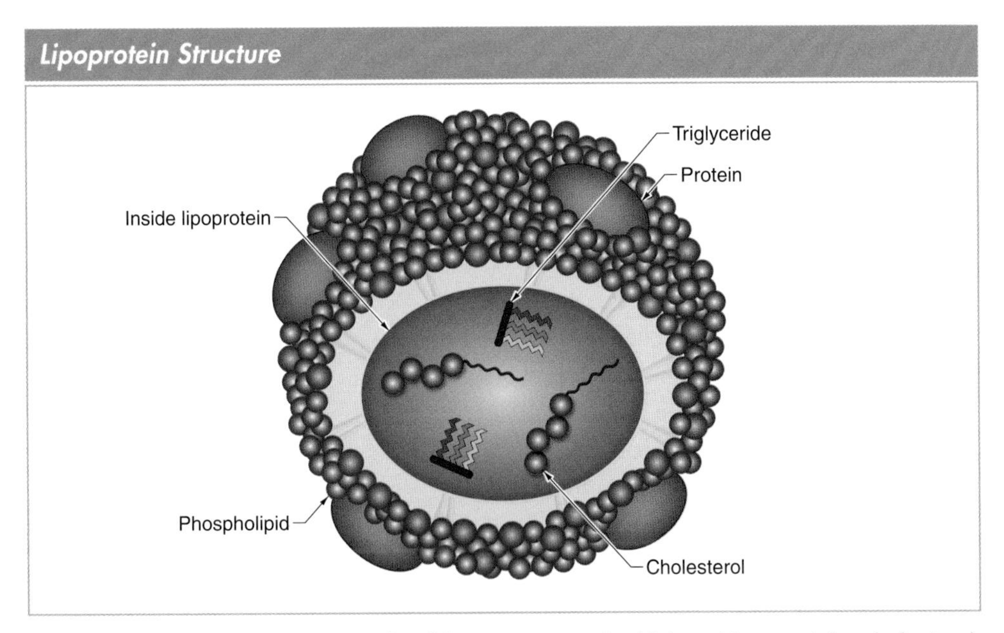

Figure 5-13. Lipoproteins are made of four components: triglycerides, protein, cholesterol, and phospholipids.

Lipid Absorption

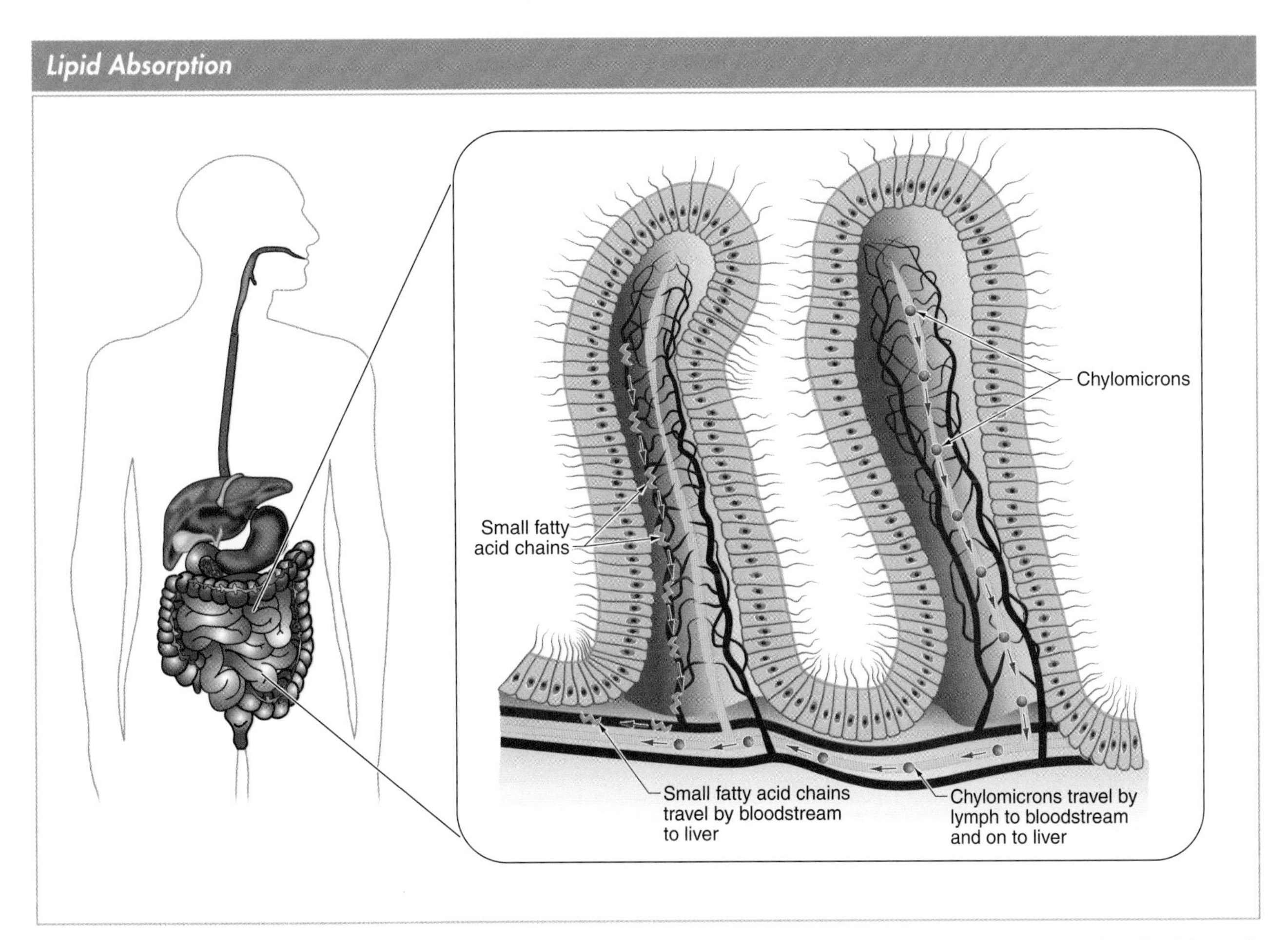

Figure 5-14. Small fatty acid chains are sent through the bloodstream to the liver, whereas chylomicrons are absorbed through the lymphatic system and then into the bloodstream.

Very-Low-Density Lipoproteins. Triglycerides and cholesterol can be made by almost every cell in the body but are most efficiently made by the liver and then transported through the body by very-low-density lipoproteins. A *very-low-density lipoprotein (VLDL)* is a type of lipoprotein used to transport triglycerides and cholesterol made by the liver through the body for absorption. VLDLs supplement the cholesterol made by cells. This is especially important in areas of the body that use more cholesterol such as the sex hormones testosterone and estrogen.

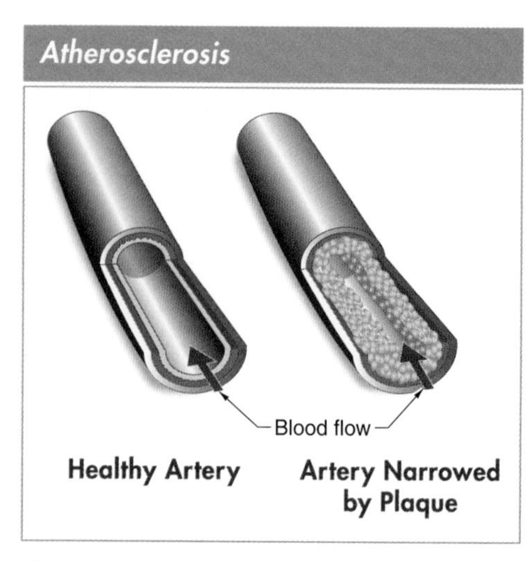

Figure 5-15. Atherosclerosis is a condition characterized by excess cholesterol that causes plaque to build up in artery walls and restrict blood flow.

Low-Density Lipoproteins. A *low-density lipoprotein (LDL)* is a type of lipoprotein used to transport cholesterol to the cells and tissues of the body. LDLs contain a higher ratio of lipids than protein and are commonly referred to as bad cholesterol because they are most responsible for depositing cholesterol into the arteries. An *artery* is a blood vessel that carries oxygen-containing blood from the heart to the rest of the body.

If the level of LDL is high, too much cholesterol can be deposited into the arteries and atherosclerosis may develop. *Atherosclerosis* is a condition characterized by excess cholesterol that causes plaque to build up in artery walls and restrict blood flow. Atherosclerosis is the most common cause of cardiovascular disease. **See Figure 5-15.** Studies have found that a high level of LDL is often related to the amount of saturated fat and cholesterol in the diet.

High-Density Lipoproteins. A *high-density lipoprotein (HDL)* is a type of lipoprotein used to collect cholesterol and bring it to the liver so that it can be dismantled and recycled in bile or excreted. HDLs contain a higher ratio of protein than lipids and are commonly referred to as good cholesterol because they lower the risk of cardiovascular disease by removing cholesterol from the body. **See Figure 5-16.** Unlike a high level of LDL, a high level of HDL is considered beneficial to health.

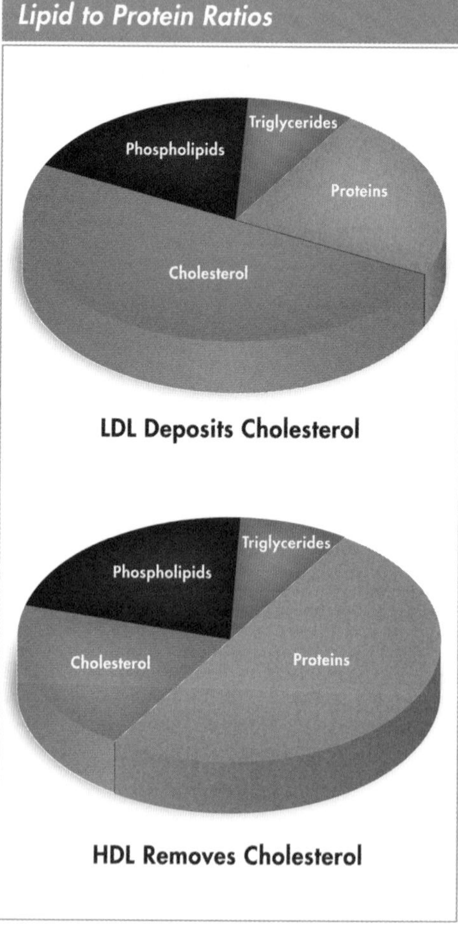

Figure 5-16. LDLs contain a higher ratio of lipids than protein, whereas HDLs contain a higher ratio of protein than lipids.

Knowledge Check 5-3

1. Identify the primary location of lipid digestion.
2. Summarize the role bile plays in lipid digestion.
3. Name the four components of lipoproteins.
4. Describe a chylomicron.
5. Differentiate between LDL and HDL.
6. Describe the relationship between LDL and atherosclerosis.

LIPID SOURCES

Lipids are naturally found in virtually all food groups and used for many cooking techniques. **See Figure 5-17.** For example, some vegetables and fruits, such as soybeans, avocados, and olives, naturally contain lipids. Despite being low in fat, some foods, such as potatoes, are often fried and become high-fat foods. Dairy products including whole milk, cheese, and butter are higher fat items. However, most dairy products are available in lower fat options. Protein foods such as meats, poultry, seafood, eggs, and nuts naturally contain lipids, although the amount and type can vary greatly. For example, some pork products are very high in fat, such as bacon, and some pork products are low in fat, such as pork tenderloin.

Lipid sources include both plant- and animal-based foods. However, it is the type of lipids found in foods that can have the greatest impact on health and well-being. Foods high in saturated fats, trans fat, and cholesterol should be a limited part of the diet but do not have to be entirely eliminated. Instead, foods that contain saturated fats, trans fat, and cholesterol should be eaten in moderation. Foods high in monounsaturated and polyunsaturated fats are healthier choices.

Monounsaturated Fats

Monounsaturated fats are mainly found in plant-based foods such as avocados, olives, and nuts. Oils derived from plants, including olive and canola oils, are also high in monounsaturated fats.

Lipid Sources

Figure 5-17. Lipids are naturally found in virtually all food groups and used for most cooking techniques.

Polyunsaturated Fats

Foods high in polyunsaturated fats include a variety of vegetable oils, fatty fish, nuts, and seeds. Polyunsaturated fats also contain the two fatty acids that are considered essential fatty acids: linoleic acid (omega-6 fatty acid) and alpha-linolenic acid (omega-3 fatty acid).

Essential Fatty Acids

An *essential fatty acid* is a fatty acid that cannot be made by the body and must be obtained from dietary sources. The essential fatty acids, linoleic acid and alpha-linolenic acid, are found primarily in plant-based foods and fatty fish. **See Figure 5-18.** Essential fatty acids help maintain cellular structure, reduce inflammation, regulate blood pressure, improve blood clotting, and increase immune functions.

Essential Fatty Acids

Figure 5-18. Nuts are a rich dietary source of essential fatty acids.

Linoleic Acid. *Linoleic acid,* commonly referred to as omega-6 fatty acid, is an essential polyunsaturated fatty acid and is part of the omega-6 fatty acid family. An *omega-6 fatty acid* is a polyunsaturated fatty acid with a final double bond that is six carbons from the end of its carbon chain. Dietary linoleic acid enables the body to produce omega-6 fatty acids, such as arachidonic acid, which is essential to regulating body functions.

The primary contribution of dietary linoleic acid is vegetable oils such as corn, cottonseed, safflower, soybean, and sunflower oil. Linoleic acid is also found to a lesser degree in whole grains and vegetables.

Alpha-Linolenic Acid. *Alpha-linolenic acid,* commonly referred to as omega-3 fatty acid, is an essential polyunsaturated fatty acid and is part of the omega-3 fatty acid family. An *omega-3 fatty acid* is a polyunsaturated fatty acid with a final double bond that is three carbons from the end of its carbon chain. The body can convert alpha-linolenic acid into other omega-3 fatty acids such as eicosapentaenoic acid (EPA) and docosahexaenoic acid (DHA). **See Figure 5-19.** EPA and DHA have been linked to a heart-healthy diet.

Fatty coldwater fish such as anchovies, bluefish, halibut, herring, mackerel, salmon, sardines, trout, and tuna are rich sources of alpha-linolenic acid. Alpha-linolenic acid is also abundant in wheat germ, flaxseeds, pumpkin seeds, and walnuts.

Saturated Fats

Saturated fats are mainly found in animal-based foods such as meats, poultry with skin, eggs, and whole milk dairy products like cream, cheese, and sour cream. Butter, lard, and shortening are also high in saturated fats, and food items made using these ingredients will also be higher in saturated fats. Plant-based foods are generally lower in saturated fats. However, tropical oils such as coconut, palm, and palm kernel oil are exceptions. **See Figure 5-20.**

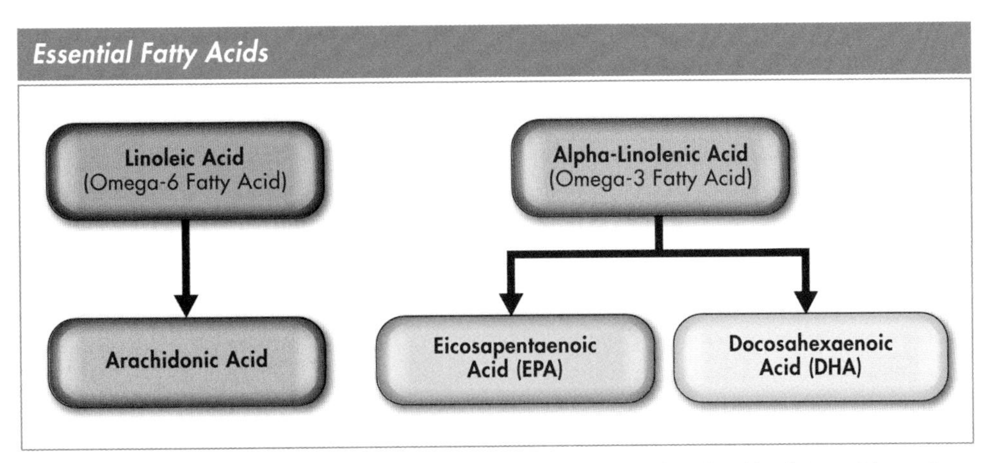

Figure 5-19. The body can convert essential fatty acids into other healthy fatty acids such as arachidonic acid, EPA, and DHA.

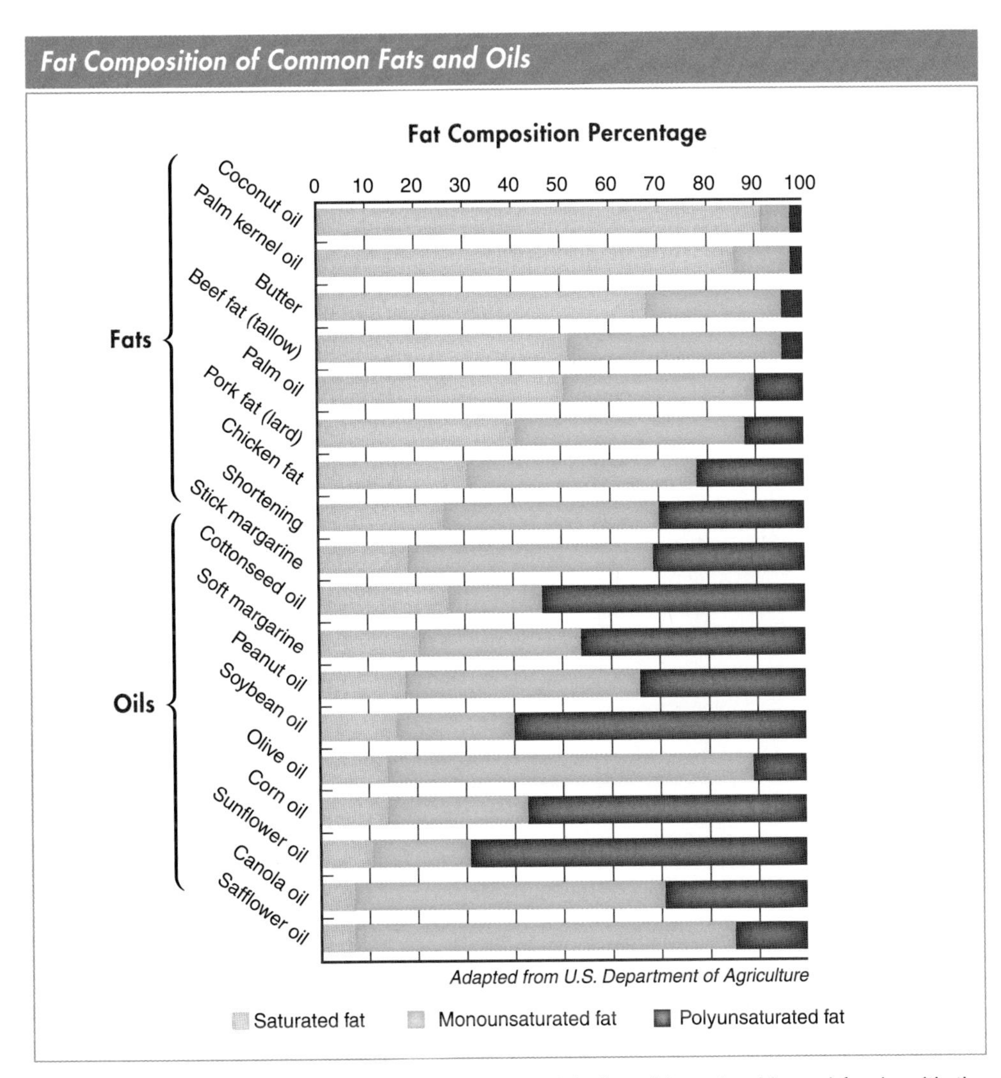

Figure 5-20. High levels of saturated fats are mainly found in animal-based foods with the exception of coconut, palm, and palm kernel oil.

In 2012, the USDA enacted new foodservice industry guidelines requiring Nutrition Facts labels for 40 cuts of meat as well as ground meats and poultry. These labels enable individuals to easily identify the amount of saturated fat in fresh and ground meats. In addition, the labels on ground meats often specify the ratio of lean meat to fat. For example, ground beef labeled "96/4" means that the meat is 96% lean and 4% fat based on weight.

Trans Fats

Small amounts of trans fats are found naturally in meats and dairy products. They are most commonly found in processed foods, such as margarine and snack foods, due to hydrogenation. **See Figure 5-21.** According to the USDA, approximately 80% of trans fats found in the diet are supplied through processed foods and 20% from natural sources.

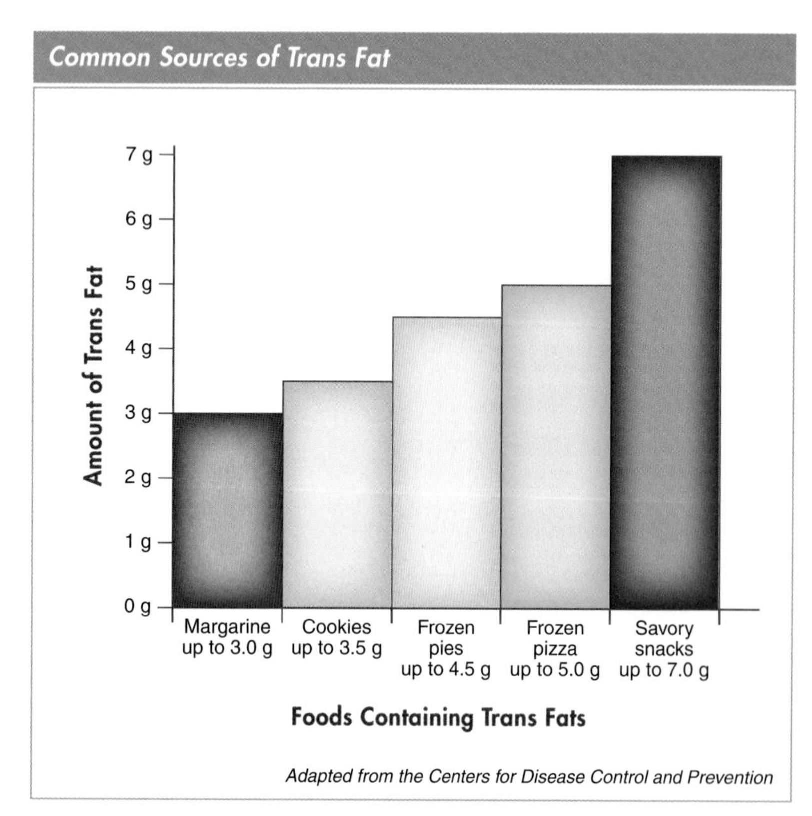

Figure 5-21. Trans fats are commonly found in processed foods.

Products containing less than 0.5 g of trans fat per serving can be labeled as having 0 g of trans fat. Therefore, it is important to look at the ingredient list for the words "hydrogenated" and "partially hydrogenated" to determine whether the product contains trans fat.

Wellness Concept

In an effort to reduce trans fat, many foodservice operations have voluntarily banned the use of hydrogenated and partially hydrogenated oils in favor of cooking oils that are free of trans fat. Food manufacturers are responding to this movement by offering blends of unsaturated oils that provide the same flavor and function of hydrogenated oils.

Cholesterol

Foods containing cholesterol raise blood cholesterol levels but to a lesser extent than foods high in saturated fats and trans fats. Dietary cholesterol is only found in animal-based foods including meat, poultry, seafood, eggs, and dairy products as well as dishes made using these ingredients. **See Figure 5-22.**

Common Sources of Dietary Cholesterol

Food	Cholesterol*
Beef liver	396 mg
Whole eggs	372 mg
Pound cake	221 mg
Butter	215 mg
Shrimp	211 mg
Bacon	113 mg
Cheddar cheese	105 mg
Vanilla ice cream	92 mg
Roasted chicken	88 mg
Halibut	60 mg

* based on 100 g serving size

Figure 5-22. Dietary cholesterol is only found in animal-based foods including meat, poultry, seafood, eggs, and dairy products as well as dishes made using these ingredients.

FOOD SCIENCE EXPERIMENT: Total Fat Content in Foods

5-1

Objective

- To measure the fat content of common food items.

Materials and Equipment

- Nutrition Facts label listing "Total Fat" grams for one food item from each of the following categories:
  - Raw protein (e.g., meat, poultry, seafood, eggs, or nuts)
  - Dairy product
  - Processed snack food
  - Frozen food entrée
  - Fast food item
- Marker
- 5 paper plates
- Gram scale
- Spoon
- Vegetable shortening (amount will vary)

Procedure

1. Read the Nutrition Facts label and identify the "Total Fat" content for the raw protein item.
2. Write the name of the raw protein and its total fat grams on one of the paper plates.
3. Place the paper plate on the gram scale.
4. Spoon vegetable shortening onto the plate so that the weight in grams is equal to the total fat grams in the raw protein.
5. Repeat steps 1–4 for each of the remaining four food items (dairy product, processed snack food, frozen food entrée, and fast food item).
6. Compare the total fat content of the five food items.

Typical Results

There will be variances in the fat content of each common food item.

Knowledge Check 5-4

1. List three foods high in monounsaturated fats.
2. Name the two polyunsaturated fats considered essential fatty acids.
3. Identify five types of fish that are rich sources of alpha-linolenic acid (omega-3 fatty acid).
4. Name three animal-based and three plant-based foods that are high in saturated fats.
5. Identify the primary source of trans fats.
6. List five sources of dietary cholesterol.

LIPID CONSUMPTION

Many health organizations, such as the Institute of Medicine (IOM), recommend that fat intake for healthy adults fall within the range of 20% to 35% of total caloric intake. **See Figure 5-23.** This range has been found to reduce the risk of cardiovascular disease while providing an adequate intake of beneficial nutrients. However, the type of fat consumed has been found to be a more influential factor in raising cholesterol levels and the risk of cardiovascular disease than the total amount of fat in the diet. For this reason, it is recommended that the consumption of saturated fats and trans fats be limited and more emphasis placed on the heart-healthy benefits of consuming monounsaturated and polyunsaturated fats. To reap the benefits of healthy fats, consuming foods such as olive and canola oils, fatty fish, nuts, and seeds is recommended.

Recommended Fat Intake Based on Calorie Needs

Total Calories in Daily Diet	Recommended Fat Intake*
1600	36–62 g
1800	40–70 g
2000	44–78 g
2200	49–86 g
2400	53–93 g

* based on 20% to 35% range

Figure 5-23. Many health organizations recommend that the fat intake for healthy adults fall within the range of 20% to 35% of total caloric intake.

Cholesterol is necessary for health, but the body makes an adequate amount to function properly. Therefore, cholesterol does not need to be incorporated into the diet. Some studies have shown that dietary cholesterol moderately raises LDL cholesterol levels but only by a minimal amount if the saturated fat intake is low. Keeping cholesterol intake to less than 300 mg per day has also been found to help maintain healthy blood cholesterol levels.

Although it is not advisable to get a large percentage of calories from fat, lipids are essential to health. Similar to a diet with excessive fat intake, a diet lacking fat can have negative effects on health and well-being.

Wellness Concept

Research suggests that a diet plentiful in dietary fiber, nuts, soy protein, and margarine enriched with plant sterols can lower cholesterol levels.

Lipid Recommendations

A recommended dietary allowance (RDA) does not currently exist for fat intake. However, based on scientific research, many experts recommend limiting saturated fats to less than 10% and trans fats to 1% of the total calories consumed in a day. Further reducing saturated fats to 7% of total caloric intake has been found to be even more effective at lowering the risk of cardiovascular disease.

While there is no RDA set for total fat intake, adequate intakes have been established for the essential fatty acids, which include linoleic acid (omega-6 fatty acid) and alpha-linolenic acid (omega-3 fatty acid). **See Figure 5-24.** There is significant scientific evidence that linoleic acid and alpha-linolenic acid help maintain structural components of cells, aid in optimal immune functions, and reduce symptoms associated with cardiovascular disease such as high blood pressure and inflammation.

Adequate Intakes for Linoleic and Alpha-Linolenic Acids

Gender	Age	Linoleic Acid	Alpha-Linolenic Acid
Females	19 – 50	12 g	1.1 g
	50+	11 g	1.1 g
Males	19 – 50	17 g	1.6 g
	50+	14 g	1.6 g

Food and Nutrition Board, Institute of Medicine, National Academies

Figure 5-24. Adequate intakes have been established for linoleic acid and alpha-linolenic acid.

Linoleic acid is typically not lacking in the diet due to its prevalence in margarine, commercially produced salad dressings, and common frying oils such as corn oil. If linoleic acid intake is high in proportion to alpha-linolenic acid intake, the body cannot effectively use omega-3 fatty acids. In order to remedy an imbalance between linoleic acid and alpha-linolenic acid, it is recommended that wheat germ, fatty fish, walnuts, and canola, soybean, and flaxseed oils be incorporated into the diet.

Lipid Excess

Whether a diet contains saturated or unsaturated fats, lipids contain 9 calories per gram versus the 4 calories per gram found in proteins and carbohydrates. If more calories are consumed than are burned through physical activity, body fat accumulates. This leads to weight gain and the potential of becoming overweight or obese. There are a variety of health problems associated with being overweight, such as high blood pressure, type 2 diabetes, cardiovascular disease, and increased joint pain. A diet in which fat consumption is excessive may also place the body at risk of being nutrient deficient if foods such as vegetables, fruits, and whole grains are sacrificed for high-fat and processed foods. An individual whose diet is nutrient deficient may exhibit a lack of energy, an inability to concentrate, and decreased immunity.

Even if caloric intake is appropriate, consistently consuming dishes high in saturated fats has been shown to increase LDL cholesterol and reduce HDL cholesterol. **See Figure 5-25.** This heightens the risks associated with cardiovascular disease. Additionally, a diet high in saturated fats has been linked to certain forms of cancer, including breast and prostate cancer.

Common Dishes High in Saturated Fats

Nutrition Facts
Serving Size 2 slices
Amount Per Serving
Calories 650 Calories from Fat 261
Total Fat 29 g
Saturated Fat 12 g

Pepperoni Pizza

Nutrition Facts
Serving Size 1 rack
Amount Per Serving
Calories 495 Calories from Fat 360
Total Fat 40 g
Saturated Fat 15 g

Barbeque Ribs

Nutrition Facts
Serving Size 1 slice
Amount Per Serving
Calories 257 Calories from Fat 162
Total Fat 18 g
Saturated Fat 8 g

Cheesecake

Nutrition Facts
Serving Size 2 thighs
Amount Per Serving
Calories 720 Calories from Fat 486
Total Fat 54 g
Saturated Fat 12 g

Fried Chicken with Skin

Figure 5-25. Consistently consuming dishes high in saturated fats has been shown to increase LDL cholesterol and reduce HDL cholesterol.

Lipid Deficiencies

Dietary fats serve an important role in many of the natural processes of the human body. For example, a diet too low in fat can interfere with the absorption of the fat-soluble vitamins A, D, E, and K. Because these vitamins are fat soluble, the body requires dietary forms of fat in order to use them. Fats are also responsible for cushioning vital organs, helping to regulate body temperature, and storing energy. A diet that includes too little fat can disrupt these body functions. Signs of lipid deficiency include the inability to regulate body temperature, muscle weakness, and dry skin, hair, and lips.

A diet lacking the essential fatty acids, linoleic acid (omega-6 fatty acid) and alpha-linolenic acid (omega-3 fatty acid), can potentially have an adverse impact on mental health and well-being. For example, studies show that the essential fats play a role in producing chemicals and hormones that influence mood and behavior. If the diet is deficient in essential fatty acids, depression and irritability may result. The essential fatty acids also raise HDL cholesterol levels. Therefore, a lipid deficiency can lower HDL cholesterol levels and raise LDL cholesterol, which contributes to cardiovascular disease.

Knowledge Check 5-5

1. Identify the recommended percentage range for the daily caloric intake of lipids.
2. Explain why cholesterol does not need to be incorporated into the diet.
3. Identify the recommendations for saturated and trans fat intake.
4. Explain how to remedy an imbalance between linoleic and alpha-linolenic acids.
5. List four health problems associated with being overweight.
6. Identify three signs of lipid deficiency.

PREPARING LIPIDS

Like carbohydrates, lipids are composed of carbon, hydrogen, and oxygen. The main difference is that lipids contain more carbon and hydrogen atoms and less oxygen atoms than carbohydrates. This structural difference allows lipids to perform their many vital functions and causes them to react to such agents as air, light, heat, and moisture differently than other nutrients during the food preparation and cooking process. Various types of lipids also have different reactions to these agents because of their structural differences.

For example, hydrogenation involves heat and changes the structure of an unsaturated fat into one that resembles a saturated fat. Hydrogenation not only changes absorption within the body but also how the fat is used in culinary applications. For example, hydrogenated fats are often used instead of oil for frying because they have a higher smoke point and can be used longer.

Reactions to Air

When fats or oils are exposed to air for prolonged periods of time, they can become rancid. A rancid fat or oil is typically identified by its strong, unpleasant odor and uncharacteristic flavor. The oil may also have a darker color than normal. Due to their structural composition, unsaturated fats are more sensitive to air. Therefore, they are more susceptible to rancidity than saturated fats. **See Figure 5-26.** Consuming a rancid fat or oil usually does not result in immediate illness, but some evidence suggests that rancid fats and oils contain substances that damage cells over time. To prevent the use of rancid fats and oils, foods such as oils, nuts, and seeds should be stored in airtight containers and used by their expiration date.

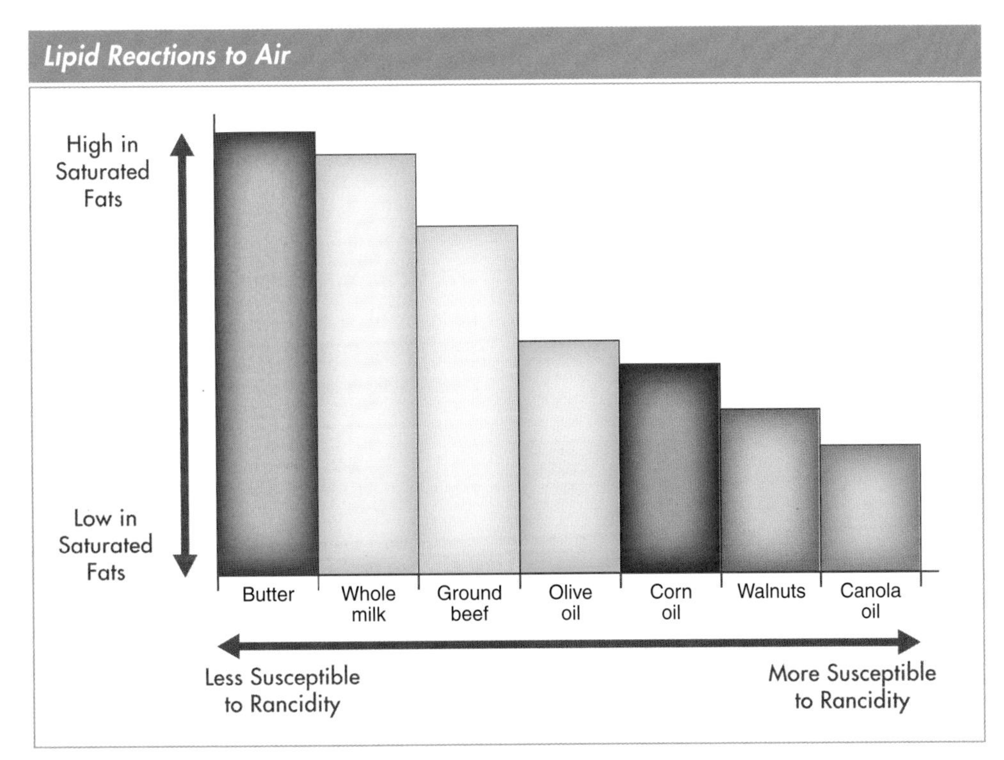

Figure 5-26. Due to their structural composition, unsaturated fats are more sensitive to air and more susceptible to rancidity than saturated fats.

Reactions to Light

In addition to air, exposure to light can cause fats and oils to become rancid. Light can also destroy some nutrients such as the alpha-linolenic acids found in walnuts and flax seeds. Most light-sensitive foods are packaged in containers that block light to minimize nutrient loss.

Reactions to Heat

Like air and light, heat also promotes rancidity in fats and oils. Therefore, fats and oils should be stored in a cool location or refrigerated based on safety and sanitation guidelines. When heat is applied during the cooking process, various fats and oils react differently because of their smoke points. The smoke point of a fat or oil is determined by the temperature that causes the fat or oil to break down, smoke, and emit an odor. Once reaching its smoke point, a fat or oil will begin to give food an unpleasant taste. If heated above its smoke point, the fat or oil can catch on fire.

The smoke point of a fat or oil is determined by the temperature that causes the fat or oil to break down, smoke, and emit an odor.

Smoke point is a key consideration in determining the type of fat or oil to use when cooking. For example, rice bran oil with a smoke point of 490°F and peanut oil with a smoke point of 470°F can withstand high-heat cooking techniques such as deep-frying. In contrast, extra-virgin olive oil is more appropriately used for quickly sautéing foods because its smoke point is low (330°F). **See Figure 5-27.**

Smoke Points

Oil	Smoke Point
Rice bran oil	490°F
Peanut oil	470°F
Canola oil	468°F
Sunflower oil	465°F
Corn oil	453°F
Safflower oil	446°F
Sesame oil	410°F
Extra-virgin olive oil	330°F

Figure 5-27. Smoke point is a key consideration in determining the type of fat or oil to use when cooking.

Reactions to Moisture

Water is an important part of many cooking processes and a major component of most foods. However, fats and oils do not mix with water (or water-based fluids) unless they are properly incorporated to produce an emulsion. For example, salad dressings are often made by whisking oil and vinegar to create a temporary emulsion with a creamy, smooth appearance. If allowed to rest, the mixture will separate into a pool of oil floating on a pool of vinegar. **See Figure 5-28.** The oil and vinegar mixture can be re-emulsified by shaking or stirring it vigorously. Sometimes an emulsion cannot be recovered, such as a sauce that has separated after plating.

Chef's Tip

In addition to mayonnaise and vinaigrettes, various liquid-to-fat ratios are used to create emulsions in hollandaise sauce, beurre blanc, aiolis, and heavy whipping cream.

Lipid Reactions to Moisture

Oil Separated from Vinegar

Oil Emulsified with Vinegar

Figure 5-28. Fats and oils do not mix with water (or water-based fluids) unless they are properly incorporated to produce an emulsion.

Knowledge Check 5-6

1. Name the three elements that compose lipids.
2. Explain what happens to fats or oils when exposed to air for a prolonged period.
3. Name the nutrient in fats and oils that is destroyed by light.
4. Explain why the smoke point of a fat or oil should be considered when cooking.
5. Describe the appearance of an emulsified vinegar and oil that has been left to rest.

LIPIDS ON THE MENU

It is often said that "fat is flavor," but this statement only provides a glimpse into the usefulness of lipids in culinary applications. While some fats and oils, such as butter, extra-virgin olive oil, and nut oils, do provide flavor, lipids also play an essential role in food preparation. For example, some cooking techniques, such as sautéing, stir-frying, and grilling, rely on fats and oils to help prevent foods from sticking. In soups and sauces, fats and oils help develop and spread flavors throughout the entire product. The texture of baked goods and desserts, such as ice cream, is also enhanced through the use of fats and oils. In addition, fats and oils lubricate the taste buds and allow flavors to be dispersed throughout the palate.

Presentation

The presentation of certain foods can be hindered by the misuse of fats and oils. For example, without a properly oiled pan, meats may cook unevenly and appear dry, rubbery, and unappealing. Likewise, a pasta dish sitting in a pool of oil is received poorly by guests since the excessive use of oil can overpower or mask the true flavors of the meal. Finding the right balance between the amount and types of fats and oils used is important to the presentation, textures, and flavors of a dish.

Meats and dairy products, such as whole milk, cream, sour cream, and cheese, are primary sources of saturated fats. Properly cooking lean cuts of meat, removing skin and fat from poultry, and using low-fat dairy products can lead to dishes that not only look appealing but also are flavorful and nutrient dense. An effective way to incorporate unsaturated fats into the menu is to replace solid fats, such as butter and shortening, with olive, canola, or soybean oils when possible. For example, bruschetta, a sliced Italian bread that has been toasted, lightly rubbed with garlic, and finished with a drizzle of olive oil, is a nutritious alternative to garlic bread coated with butter. The presentation of bruschetta can be further enhanced by topping it with a burst of color from fire-roasted tomatoes along with a small garnishing of cheese and fresh basil. **See Figure 5-29.**

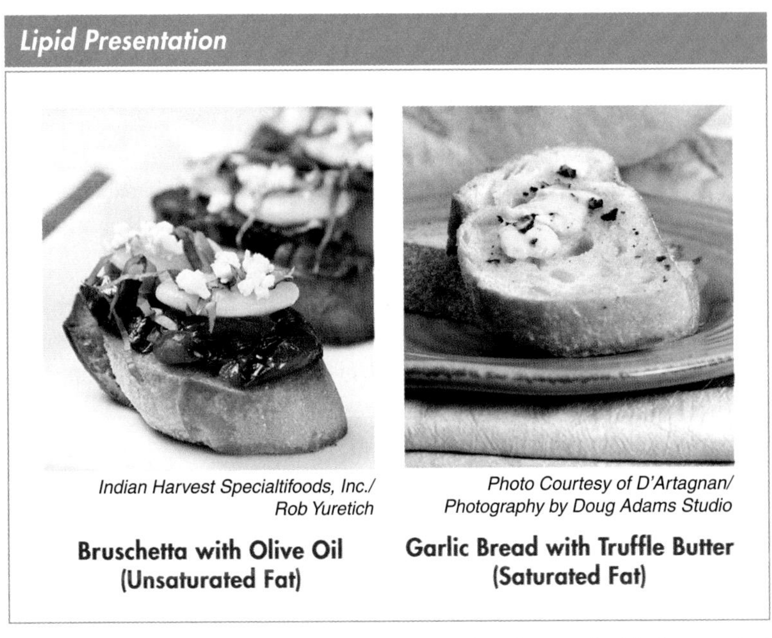

Figure 5-29. Bruschetta drizzled with olive oil and topped with fire-roasted tomatoes is a nutritious alternative to garlic bread coated with butter.

Texture and Flavor

Fats and oils enhance both the texture and flavor of foods. Without some fat or oil, muffins may become rubbery, a salad of mixed greens can be bland, and meats are often chewy and flavorless. However, there are many ways to maximize the texture and flavor of foods while lowering the total fat content. For example, some of the oil in quick breads can be replaced with puréed bananas without sacrificing texture or flavor. This technique also creates a product enriched with more vitamins, nutrients, and dietary fiber.

A unique approach to reducing the saturated fat content of a meat dish is to use alternatives to beef, such as bison or ostrich, that have textures and flavors similar to beef but are significantly lower in fat. An effective way to replace butter, mayonnaise, or cream cheese is to use avocados or puréed chickpeas (hummus) to create a spread that is not only creamy, rich, and flavorful but also abundant in heart-healthy unsaturated fats. **See Figure 5-30.** Nuts and seeds are also beneficial sources of unsaturated fats. They can provide contrasting textures and rich flavors to dishes ranging from soups and salads to entrées and desserts. With an emphasis on using unsaturated fats, high-quality ingredients, and techniques that develop textures and flavors, chefs can use fats and oils to create inspiring dishes that surpass guest expectations for healthy, flavorful meals.

Nutritional Comparison of Spreads*

Spread	Calories	Total Fat	Saturated Fat	Cholesterol
Butter	200	25 g	14 g	60 mg
Mayonnaise	110	10 g	1 g	10 mg
Cream cheese	100	10 g	6 g	30 mg
Fresh avocado	50	5 g	1 g	0 mg
Hummus	50	5 g	0 g	0 mg

* based on 1 oz serving size

Figure 5-30. An effective way to replace butter, mayonnaise, or cream cheese is to use avocados or puréed chickpeas (hummus) to create a spread that is not only creamy, rich, and flavorful but also abundant in heart-healthy unsaturated fats.

Knowledge Check 5-7

1. Explain how fats or oils can be used for cooking techniques such as sautéing, stir-frying, and grilling.
2. Describe two scenarios where fats and oils can hinder presentation.
3. Name an effective way to incorporate unsaturated fats into a menu.
4. Identify three ways to reduce the saturated fat content of various dishes.

Chapter Summary

Lipids play an important role in both the body and in foods and are classified as triglycerides, phospholipids, or sterols. Lipids are necessary for energy as well as regulating body temperature, cushioning vital organs, producing hormones, maintaining cellular structure, and absorbing nutrients. It is essential to include lipids in the diet. It is also important to recognize the different types of lipids and their effects on the body.

It is recommended that the consumption of saturated fats, such as those in meats, poultry skin, and whole milk products, be limited. It is also recommended that trans fats, which are typically found in highly processed foods, be nearly eliminated. Research indicates these types of fats significantly increase the risk of cardiovascular disease and elevate LDL cholesterol levels. Unlike saturated fats, mono- and polyunsaturated fats have been associated with a heart-healthy diet. Olive and canola oils are abundant in monounsaturated fats. Fatty fish, flaxseeds, and walnuts are rich in polyunsaturated fats. Polyunsaturated fats also supply the essential fatty acids linoleic acid (omega-6 fatty acid) and alpha-linolenic acid (omega-3 fatty acid).

Vita-Mix® Corporation

In culinary applications, fats and oils add aroma, texture, and flavor to foods. To create nutritious dishes, it is important to use unsaturated fats in ways that develop and enhance flavors while limiting the use of saturated fats.

Chapter Review

Digital Resources
ATPeResources.com/QuickLinks
Access Code: 267412

1. Define lipid.
2. Describe the three main types of lipids.
3. Differentiate between saturated and unsaturated fatty acids.
4. Describe the structure of a trans-fatty acid.
5. Summarize five functions of lipids.
6. Explain how lipids are digested and absorbed.
7. Name the four types of lipoproteins.
8. List common dietary sources of lipids.
9. Describe the two essential fatty acids.
10. Summarize the recommendations for lipid consumption.
11. Describe the reactions of lipids to air, light, heat, and moisture.
12. Give an example of how lipids can be used to enhance the presentation, texture, and flavor of nutritious menu items.

Water, Vitamins & Minerals

Chapter 6

Minerals

Paderno World Cuisine

Water, vitamins, and minerals are vital components of a nutrient-dense diet. The body depends on water more than any other nutrient for proper functioning. Water is primarily supplied to the body by consuming beverages, but foods also provide water. Foods like vegetables and fruits are high in water content as well as essential vitamins and minerals that regulate body processes. Incorporating a variety of nutritious foods on the menu allows guests to experience an array of colors, textures, and flavors while providing the nutrients needed for proper hydration, metabolic functions, and overall well-being.

Chapter Objectives

1. Summarize the role of water in the body.
2. Identify sources and recommendations for water intake.
3. Explain how the body maintains fluid balance.
4. Summarize the role of antioxidants.
5. Describe the characteristics of water-soluble vitamins.
6. Describe the functions, sources, and recommendations for water-soluble vitamins.
7. Identify toxicities and deficiencies associated with water-soluble vitamins.
8. Describe the characteristics of fat-soluble vitamins.
9. Describe the functions, sources, and recommendations for fat-soluble vitamins.
10. Identify toxicities and deficiencies associated with fat-soluble vitamins.
11. Describe the functions, sources, and recommendations for major minerals.
12. Identify toxicities and deficiencies associated with major minerals.
13. Describe the functions, sources, and recommendations for trace minerals.
14. Identify toxicities and deficiencies associated with trace minerals.
15. Explain vitamin and mineral reactions to air, light, heat, moisture, and pH.
16. Describe how to effectively include vitamins and minerals on menus.

THE ROLE OF WATER

A person can live for weeks without food, but when deprived of water, a person can only survive a few days. Despite its simple composition, just two hydrogen atoms attached to a single oxygen atom, water is regarded as the most important of all the nutrients. It is so abundant in the body that the average weight of a healthy adult is composed of approximately 50%–60% water.

Water is a major component of lungs, which are comprised of 90% water. Blood that circulates through the body consists of approximately 83% water. Muscle and brain tissue contain roughly 75% water. Bone consists of nearly 22% water. Even body fat contains 10% water. **See Figure 6-1.** Water is indispensable to life because of the role it plays in the digestion and absorption of nutrients; the regulation of body temperature; the health of tissues, organs, and joints; and maintaining fluid balance.

Aiding Digestion and Absorption

Water is an essential solvent that aids in the digestion and absorption of nutrients. A *solvent* is a substance that has the ability to dissolve other substances. Beginning with saliva in the mouth, water acts as a solvent to break down and transport food through the digestive tract. Because blood is primarily made of water, it is an ideal vehicle for transporting oxygen and water-soluble nutrients such as proteins, carbohydrates, water-soluble vitamins, and minerals throughout the body. Lipids and fat-soluble vitamins rely on water to be chemically modified through emulsification so that they too can be transported throughout the body via the bloodstream. Metabolic reactions that supply energy to the body would cease to exist without water. In the final stages of the digestive process, water helps flush waste and toxins from the body. **See Figure 6-2.**

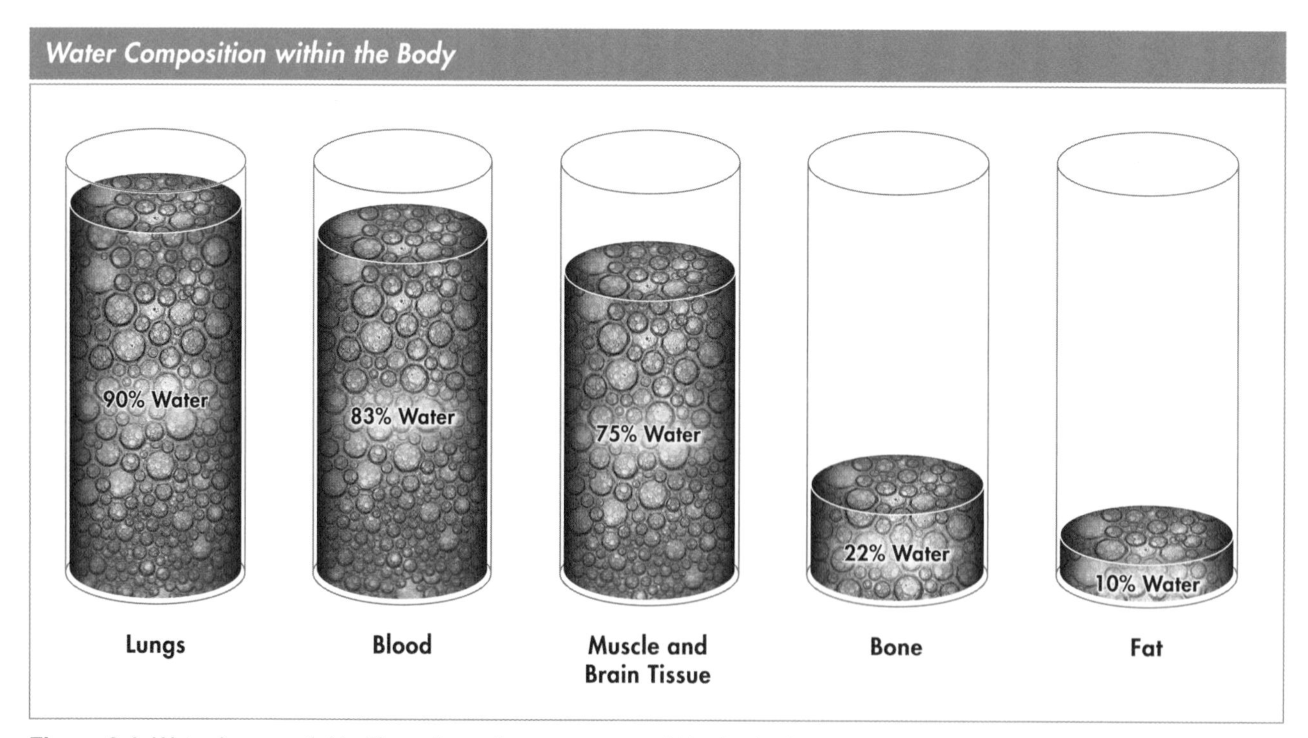

Figure 6-1. Water is essential to life and a major component within the body.

Role of Water in Digestion and Absorption

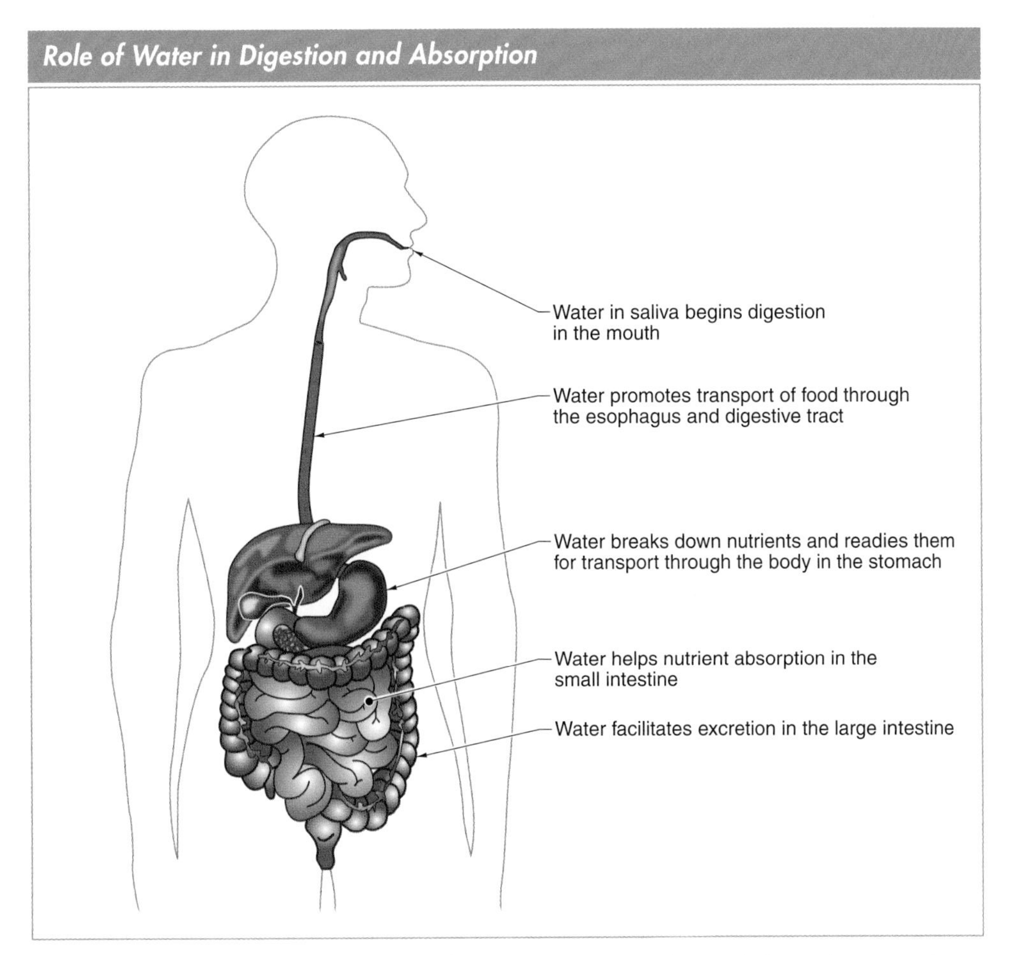

Figure 6-2. Water is an essential solvent that aids in the digestion and absorption of nutrients.

Regulation of Body Temperature

Water plays a critical role in helping the body maintain an ideal temperature due in part to its heat capacity. *Heat capacity* is the amount of energy needed to raise 1 g of a substance 1°C. Because water has a high heat capacity, it takes a significant amount of energy to raise or lower its temperature. For example, a pot of cold water placed on the stove requires high heat over an extended time period for the water to reach its boiling point. **See Figure 6-3.** Similarly, it takes an extended amount of time for that water to return to room temperature once the pot is removed from the heat. Because the body is primarily made of water, it takes prolonged exposure to heat to increase body temperature as well as prolonged exposure to cold to decrease body temperature. The heat capacity of water protects the body from extreme temperature changes and helps maintain a consistent body temperature.

Heat Capacity of Water

Figure 6-3. Because of the high heat capacity of water, it takes a significant amount of energy or heat over an extended time period to raise the temperature of water.

Water also has the ability to act as a coolant for the body. When the body senses that its temperature is rising, it reacts by increasing the flow of blood from the internal core of the body to the capillaries just under the skin. The sweat glands then emit sweat, which begins to evaporate off the skin. As the sweat evaporates, heat is released and both the skin and blood lying beneath the skin are cooled. The cooled blood then flows back to the core of the body to maintain the appropriate internal temperature.

Health of Tissues, Organs, and Joints

Water provides a moist environment that supports the health and functions of tissues, organs, and joints. For example, water acts as a shock absorber by surrounding tissues and organs to protect and cushion them during sudden movements that can occur during physical activity. Water also provides lubrication for joints to facilitate smooth, friction-free movement. **See Figure 6-4.** It also lubricates the lungs to enable them to expand and contract. The stomach and intestines are lined with watery mucous to help transfer nutrients through the digestive tract. The skin, nose, and eyes also rely on water for proper hydration. In addition, watery tears are a cleansing agent for the eyes.

Water Lubricates Joints

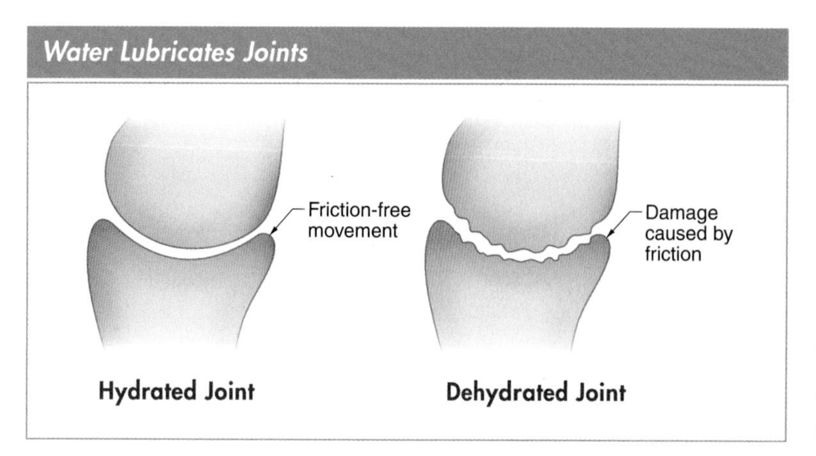

Figure 6-4. Water provides lubrication for joints to facilitate smooth, friction-free movement.

Maintaining Fluid Balance

To maintain a healthy body, there must be a balance between intracellular fluid and extracellular fluid. *Intracellular fluid* is fluid that is contained within the cells of the body. *Extracellular fluid* is fluid that surrounds the cells of the body. **See Figure 6-5.** Approximately 66% of the water in the body is intracellular and can be found in the cells of muscles, organs, and skin. The remaining extracellular fluid is found in the blood, lymph, saliva, and body excretions such as sweat and urine.

Intracellular and Extracellular Fluids

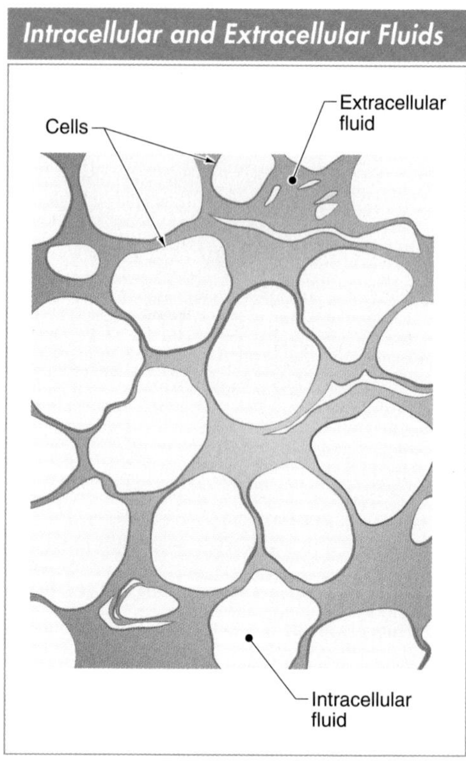

Figure 6-5. Intracellular fluid is contained within the cells of the body, and extracellular fluid surrounds the cells of the body.

Water and Electrolytes. The crucial balance of fluids in and around cells is aided by water and electrolytes. An *electrolyte* is a mineral compound that forms positively or negatively charged ions when dissolved in body fluids. For example, potassium and sodium are electrolytes with positively charged ions, whereas chloride is an electrolyte with negatively charged ions. The electric charge is necessary for functions such as muscle contractions and nerve impulses.

Electrolytes are also crucial to balancing fluids in the body. There must be the correct mix of electrolytes and water both in and around cells for the body to function properly. For example, potassium ions are found inside the cells of the body, while sodium and chloride ions are found outside the cells. If fluid balance were to become unstable and sodium ions accumulated inside the cells, the cells would not survive. This is because sodium attracts water, and if too much water builds up in the cells, they will burst and die. In contrast, drawing water out of cells will cause cells to shrink. This would create too much water in the spaces surrounding the cells resulting in a buildup of fluids that would cause the body to swell.

The Process of Osmosis. Water follows the movement of electrolytes through cell membranes to maintain fluid balance through a process called osmosis. *Osmosis* is the movement of water across cell membranes from a dilute solution to a more concentrated solution. Cell membranes are permeable to water. This allows water to flow freely in and out of cells. However, cell membranes are only semi-permeable to electrolytes, so they either stay inside or outside of the cells unless they are transported by specialized proteins. For example, when electrolytes are more concentrated on one side of the cell membrane, osmosis causes water to travel from the dilute side of the cell to the concentrated side of the cell. **See Figure 6-6.**

The concept of osmosis can be demonstrated by slicing eggplants and sprinkling them with salt. Eggplants contain a high percentage of water, which can move freely around the cells of the eggplant. Salt sprinkled on the eggplant slices creates a high concentration of electrolytes on the surface of the eggplant. In response, beads of water form on the eggplant slices as water moves from the dilute area of the eggplant to the salted area of high electrolyte concentration.

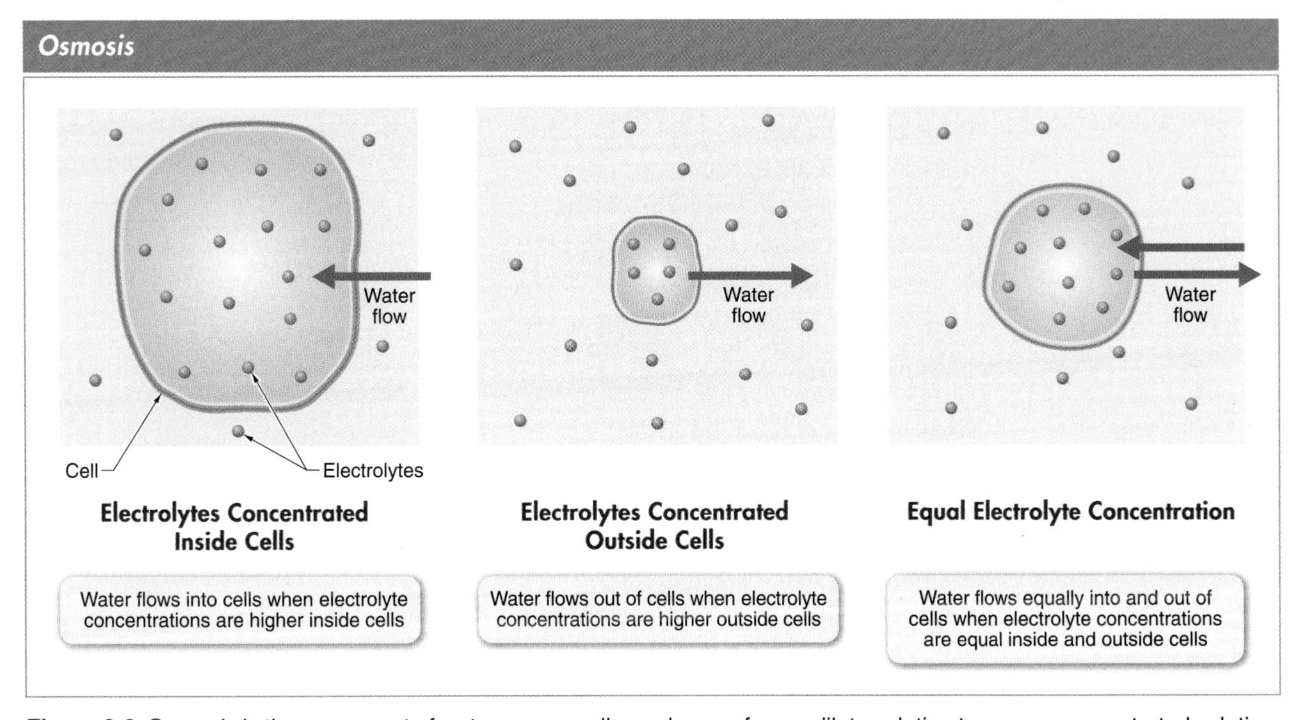

Figure 6-6. Osmosis is the movement of water across cell membranes from a dilute solution to a more concentrated solution.

Knowledge Check 6-1

1. Identify the percentages of water typically found within the body.
2. Define solvent.
3. Explain how water helps regulate body temperature.
4. Summarize the function of water in tissue, organ, and joint health.
5. Differentiate between intracellular and extracellular fluid.
6. Explain why the balance between electrolytes and water is important.
7. Describe the process of osmosis.

WATER INTAKE

Fluid balance is maintained by the body through processes that monitor water intake and water output. When excess water is consumed, the body increases urine production to eliminate the excess and restore balance. When fluids are lost, the kidneys conserve water by reducing the output of urine. The body is simultaneously prompted to drink liquids as the result of feeling thirst.

Thirst is an important indication that the body has entered a state of mild dehydration. The feeling of thirst is caused by the functioning of a gland at the base of the brain known as the hypothalamus. For example, a midmorning snack of salted peanuts followed by a salad topped with salty bacon for lunch increases the sodium level in the blood and creates a fluid imbalance. The intake of water is required to bring body fluids back into balance. This causes the hypothalamus to activate the thirst mechanism of the body. Thirst is also triggered when there is a reduction in blood volume and blood pressure due to the loss of fluids that can occur from vomiting, diarrhea, excessive sweating, or low fluid intake.

Although thirst increases the desire to drink liquid, the amount of liquid consumed may not always be adequate. An individual tends to drink only enough to suppress the feeling of thirst. This amount of water intake may not be enough to restore fluid balance. Therefore, it is important to drink liquids at regular intervals throughout the day.

Food Science Note

The caffeine in beverages, such as coffee, tea, and soft drinks, has been shown to cause insomnia, irritability, stomach upset, rapid heartbeat, and elevated blood pressure.

Water from Beverages

Approximately 80% of the water needed by the body comes from the beverages consumed. **See Figure 6-7.** Water in its pure form is considered the best source of hydration for the body. Other beverages, such as milk, juice, coffee, tea, soft drinks, and alcohol, also contribute water to the body but often contain added ingredients that also supply extra sugar or fat and calories. In addition, alcohol acts as a diuretic. A *diuretic* is a substance that causes an increased output of urine. Because diuretics increase urine production, they also increase the risk of dehydration.

Wellness Concept

Sports drinks are often used to replenish electrolytes that are lost through excessive sweating. However, water is a healthier choice because it lacks the added sugars and chemicals generally found in sports drinks.

Water from Beverages

Florida Department of Citrus

Figure 6-7. Approximately 80% of the water needed by the body comes from the beverages consumed.

Water from Foods

The remaining 20% of water intake comes from water that is produced during metabolic processes and from water found in foods. Many fruits and vegetables are exceptionally high in water content. For example, watermelons, grapefruits, tomatoes, zucchinis, cucumbers, and spinach are composed of roughly 90% water. Meats, poultry, seafood, legumes, and eggs also supply water to the body. For example, a serving of cod and an egg both contain approximately 75% water. Other foods that contribute significant amounts of water to the body include soups, yogurts, and frozen desserts such as ice cream. Dry foods, such as nuts, toast, crackers, and pretzels, contain very little water.

Dietary Recommendations. Because water needs vary due to factors such as age, weight, and activity level, only an adequate intake (AI) for water has been established. For adult males, the AI for water is 3.7 L per day (about 15½ cups). For adult females, the AI for water is 2.7 L per day (about 11½ cups). This recommendation takes into account all sources of daily water consumption including foods, beverages, and drinking water. Obtaining the recommended amount of water on a daily basis helps ensure proper hydration and enables proper metabolism. **See Figure 6-8.**

Water Toxicity. *Toxicity* is the degree to which a substance can harm living organisms. Water toxicity is relatively uncommon, but its effect on the body can be severe. When water is consumed to the point of excess, electrolyte balance is disturbed and the concentration of sodium in the blood drastically falls. This can create a sodium deficiency that leads to hyponatremia. *Hyponatremia* is a sodium deficiency characterized by muscle cramps, nausea, dizziness, seizures, and coma. Without proper medical attention, hyponatremia can be fatal.

Water Deficiency. A *deficiency* is the lack of a substance necessary to maintain health. A water deficiency can lead to dehydration. *Dehydration* is a condition in which the body does not have an adequate amount of water to function properly. Physical activity and exposure to extreme heat can accelerate dehydration due to the water lost through sweat. Dehydration reduces the amount of oxygen the blood can transport throughout the body, which can lead to a variety of adverse conditions. For example, early symptoms of dehydration include thirst, irritability, headache, and slight nausea. Severe dehydration can result in complications such as an elevated heart rate, increased body temperature, muscle spasms, impaired mental function, kidney failure, coma, and even death.

Water			
Characteristics	Solvent; high heat capacity		
Functions	Necessary for aiding digestion and absorption; regulating body temperature; sustaining health of tissues, organs, and joints; and maintaining fluid balance		
Sources	• Water (1 cup): 237 g • Canned tomato juice (1 cup): 228 g • 2% milk (1 cup): 218 g	• Low-fat yogurt (1 cup): 208 g • Tomatoes (1 cup, sliced): 170 g • Watermelon (1 cup, diced): 139 g	• Cod (5 oz): 114 g • Lean ground beef (5 oz): 103 g • Egg (1 large): 38 g
DRI	• Adult males: 3.7 L • Adult females: 2.7 L		
Toxicity	No UL		
Deficiency	Causes dehydration		

Figure 6-8. Obtaining the recommended amount of water on a daily basis helps ensure proper hydration.

Knowledge Check 6-2

1. Identify the body part that stimulates the feeling of thirst.
2. Explain why diuretics increase the risk of dehydration.
3. List fruits and vegetables that contain a high percentage of water.
4. Identify the water adequate intake (AI) established for males and females.
5. Define toxicity.
6. Describe the potential effects of dehydration on the body.

WATER OUTPUT

The body does not store water. Therefore, water needs to be replenished daily in order to replace the amount excreted through normal body functions such as the exhalation of breath, sweating, evaporation through the skin, feces, and urine. **See Figure 6-9.** Approximately 1 L of water is typically lost each day through urine. This accounts for the majority of fluid loss from the body. Replacing the water lost by the body each day can vary dramatically based on factors such as body size and composition, physical activity, and even the weather. For example, exercising strenuously outdoors in hot weather causes heavy sweating and results in more fluid loss than a casual stroll on a cool day. In general, heavy exercise and heat require an individual to drink more water in order to maintain or restore fluid balance.

Daily Water Loss

Source of Loss	Amount of Loss
Urine	500–1000 ml
Skin	450–1900 ml
Lungs	250–350 ml
Feces	100–200 ml

Adapted from Institute of Medicine

Figure 6-9. Water needs to be replenished daily in order to replace the amount excreted through normal body functions.

When water output is high or water intake is low, sodium levels rise in the body. The body reacts to this fluid imbalance by signaling the brain to release an antidiuretic hormone. An *antidiuretic hormone (ADH)* is a hormone secreted by the body that signals the kidneys to conserve water rather than eliminate it. The release of ADH returns water to the bloodstream, thus diluting sodium levels to help restore fluid balance. The kidneys also release enzymes that cause blood vessels to tighten and constrict the flow of fluids in an effort to retain water. The same time the body is trying to conserve water, it is also alerting the body that it needs water by creating a feeling of thirst.

Edward Don & Company

Water in the body needs to be replenished daily in order to replace the water excreted through normal body functions.

Knowledge Check 6-3

1. Explain why water must be replenished daily.
2. Identify factors that influence the amount of water the body needs to replenish.
3. Define antidiuretic hormone (ADH).

VITAMINS AND MINERALS

Commonly referred to as non-energy-yielding nutrients, vitamins and minerals are a powerful group of substances necessary for health and well-being. A *vitamin* is an organic nutrient that is required in small amounts to help regulate body processes. According to the Linus Pauling Institute, organic compounds are considered vitamins when a lack of that compound has an adverse impact on health. A *mineral* is an inorganic nutrient that is required in small amounts to help regulate body processes. Minerals occur naturally in soil and the water that is absorbed by plants. **See Figure 6-10.** Therefore, consuming plant-based foods is an effective way to supply the body with minerals. Because animals eat plants as well, animal-based foods also supply the body with minerals.

Minerals

HerbThyme Farms

Figure 6-10. Minerals occur naturally in soil and the water that is absorbed by plants.

Although vitamins and minerals do not contribute calories, they are essential components of a nutritious diet because they assist in keeping the body healthy. For example, vitamins and minerals help facilitate chemical reactions necessary for digestion, metabolism, growth, repair, and maintenance.

Dietary Recommendations

Vitamins and minerals are classified as micronutrients because they are needed by the body in small amounts. Vitamins and minerals are commonly measured using small units such as micrograms (mcg) and milligrams (mg). Despite the small amount required by the body, vitamins and minerals have a significant impact on how the body functions. To ensure the proper intake of vitamins and minerals, it is recommended to nourish the body with a variety of foods.

In addition to eating a variety of foods, the dietary reference intakes (DRIs), established by the Institute of Medicine, can help quantify the specific amount of a vitamin or mineral that should be consumed daily. DRIs are based on the bioavailability of a nutrient. *Bioavailability* is the extent to which a substance is absorbed by the body. The bioavailability of vitamins and minerals depends on multiple factors including age, gender, and overall health. The most common DRIs used to measure vitamin and mineral intake include the recommended dietary allowance (RDA), adequate intake (AI), and tolerable upper intake level (UL). **See Figure 6-11.**

Dietary Measurements

Dietary Reference Intake (DRI)	Description
Recommended dietary allowance (RDA)	Daily nutrient intake level sufficient enough to meet the nutrient needs of approximately 98% of healthy population based on age and gender
Adequate intake (AI)	Estimated daily intake level for nutrient for which no RDA has been set but nutrient intake appears to be adequate to meet the needs of a healthy population
Tolerable upper intake level (UL)	Maximum daily intake level of a nutrient before the body experiences an adverse effect

Figure 6-11. The DRIs can help quantify the specific amount of a vitamin or mineral that should be consumed daily.

Vitamin and Mineral Toxicities

A vitamin or mineral toxicity can occur when a given nutrient is ingested at levels that exceed recommendations. Vitamin and mineral toxicities rarely occur from food sources. Most micronutrient toxicities come from taking supplements that contain vitamins and minerals in amounts that exceed the DRI.

Depending on the type of nutrient involved, vitamin and mineral toxicities can happen very quickly or they can build over time. In addition, consuming excessive amounts of one nutrient can cause a deficiency of another. For example, high levels of iron and zinc in the blood can interfere with the absorption of copper.

Vitamin and Mineral Deficiencies

Adverse symptoms that range in severity may occur when the body is deficient in a vitamin or mineral. For example, a mild deficiency of a vitamin or mineral may result in dry skin, whereas a more serious deficiency can result in coma or even death. A deficiency of a vitamin or mineral is less likely to occur when ample amounts of diverse foods are consumed on a daily basis.

Antioxidants

Some vitamins and minerals are referred to as antioxidants and are valued for their potential in fighting cellular damage. An *antioxidant* is a substance that may protect cells from damage caused by free radicals. A *free radical* is an unstable compound produced as a byproduct of metabolism. Free radicals are also found in some foods and can be formed when the body is exposed to pollutants. Free radicals are unstable because they contain an odd number of electrons in their outer shell.

Antioxidants act as electron donors in order to stabilize free radicals and prevent them from "oxidizing" or taking electrons from healthy cells. **See Figure 6-12.** Antioxidants are made by the body and found in a variety of substances. Vitamin antioxidants include vitamins A, C, and E, and mineral antioxidants include zinc, selenium, copper, and manganese. Research indicates that antioxidants may help prevent certain cancers, reduce cholesterol levels, and increase immune functions.

Food Science Note

Studies show green tea contains antioxidants that can improve artery function and reduce the risk of hypertension.

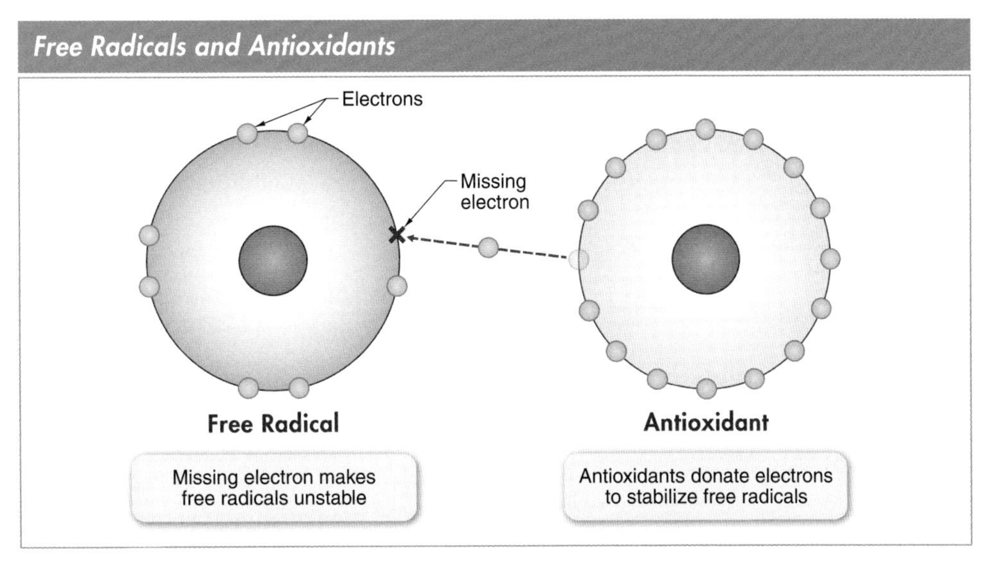

Figure 6-12. Antioxidants act as electron donors in order to stabilize free radicals and prevent them from "oxidizing" or taking electrons from healthy cells.

Knowledge Check 6-4

1. Differentiate between vitamins and minerals.
2. Define bioavailability.
3. Explain how most vitamin and mineral toxicities occur.
4. Describe the function of antioxidants.
5. Identify vitamins and minerals that are considered antioxidants.

WATER-SOLUBLE VITAMINS

A *water-soluble vitamin* is a vitamin that dissolves in water. Water-soluble vitamins can be easily broken down and absorbed by the intestines, sent to the bloodstream, and transported to cells and tissues where they are used. Water-soluble vitamins include vitamin C and a group of vitamins known as the B vitamins.

Being water-soluble also means that these vitamins are excreted daily and not stored for future use. Because water-soluble vitamins are excreted daily, toxicities are rare. In contrast, deficiencies may be more likely because water-soluble vitamins need to be replenished daily in order to ensure the body has an adequate supply to function properly.

Water-soluble vitamins help regulate body processes by acting as coenzymes. A *coenzyme* is a compound that attaches to an enzyme to activate or enhance the functions of the enzyme. As coenzymes, water-soluble vitamins are mainly involved in energy-producing metabolic reactions.

Vitamin C

Vitamin C, also known as ascorbic acid, is a water-soluble vitamin with antioxidant properties that protect the body from disease by improving immune functions. Research suggests that the antioxidant properties of vitamin C may provide some protection against cardiovascular disease, certain types of cancers, and diseases of the eye. Vitamin C also plays a significant role in producing collagen, the connective tissue found in skin, bones, teeth, ligaments, and tendons. In addition, vitamin C helps to improve the absorption of iron from foods.

The best dietary sources of vitamin C are vegetables and fruits. **See Figure 6-13.** In addition to citrus fruits and their juices, key sources of vitamin C include strawberries, broccoli, bell peppers, cantaloupe, and tomatoes.

Vitamin C Sources

California Strawberry Commission

Figure 6-13. Strawberries are among the best dietary sources of vitamin C.

Vitamin C has an RDA of 90 mg for males and 75 mg for females. However, needs may be higher during times of infection or wound healing. Individuals who smoke require an additional 35 mg of vitamin C per day. Because cigarette smoke causes the oxidation of vitamin C in the blood, the additional vitamin C is needed to repair damage caused by free radicals.

The UL for vitamin C is 2000 mg per day. In general, consuming large amounts of vitamin C will not result in toxicity because any excess is passed from the body through urine. However, regular intake above 2000 mg per day may cause stomach cramps, diarrhea, and vomiting.

Although rare, individuals who intake less than 10 mg of vitamin C daily for a prolonged time period may develop scurvy. *Scurvy* is a disease caused by a vitamin C deficiency with symptoms that include swollen or bleeding gums, loosening of the teeth, impaired wound healing, and general weakness. **See Figure 6-14.**

Wellness Concept

Research suggests pregnant women, burn victims, and individuals recovering from surgery may need extra vitamin C.

Vitamin C			
Alternative term	Ascorbic acid		
Characteristics	Water-soluble; antioxidant; improves absorption of iron; smokers require an additional 35 mg per day		
Functions	Aids immune functions, helps produce collagen, and protects against cardiovascular disease, certain cancers, and eye disease		
Sources	• Strawberries (1 cup, sliced): 84.7 mg • Broccoli (1 cup, chopped): 81.2 mg • Kale (1 cup, chopped): 80.4 mg	• Brussels sprouts (1 cup): 74.8 mg • Bell peppers (1 cup, sliced): 74.0 mg • Orange (1 medium): 67.9 mg	• Cantaloupe (1 cup, cubed): 58.7 mg • Cauliflower (1 cup): 46.4 mg • Tomatoes (1 cup, sliced): 22.9 mg
DRI	• Adult males: 90 mg • Adult females: 75 mg		
Toxicity	UL: 2000 mg		
Deficiency	Causes scurvy		

Figure 6-14. Vitamin C is a water-soluble vitamin with antioxidant properties that protect the body from disease.

FOOD SCIENCE EXPERIMENT: Vitamin C Content

6-1

Objective

- To observe the vitamin C content in various types of orange juice.

Materials and Equipment

- Measuring spoons
- Cornstarch
- Saucepan
- Spoon
- Measuring cup
- Water
- Bowl
- Eyedropper
- 2% iodine solution
- Masking tape
- Marker
- 3 test tubes
- Juice from fresh orange
- Bottled orange juice
- Orange juice concentrate prepared according to package directions
- Plain white paper

Procedures

To make the indicator solution, apply the following procedure:

1. Place 1 tbsp of cornstarch in the saucepan.
2. Stir enough water into the cornstarch to make a paste.
3. Add 1 cup of water to the paste.
4. Place the saucepan on a stovetop and bring to a boil, stirring occasionally.
5. Boil for 5 minutes, then remove the saucepan from heat.
6. Place 2½ oz of water in the bowl.
7. Use the eyedropper to add 10 drops of the cornstarch solution to the bowl of water.
8. Use the eyedropper to add enough iodine to the bowl with the water-cornstarch solution to produce a dark purple-blue color.

To conduct the experiment, apply the following procedure:

1. Use the masking tape and marker to label one test tube "Fresh," the second test tube "Bottled," and the third test tube "Concentrate."
2. Add 1 tsp of indicator solution to each of the three test tubes.
3. Use a clean eyedropper to add each of the following:
   A. Add 10 drops of fresh orange juice to the test tube labeled "Fresh."
   B. Add 10 drops of bottled orange juice to the test tube labeled "Bottled."
   C. Add 10 drops of prepared orange juice concentrate to the test tube labeled "Concentrate."
4. Line up test tubes against a plain white paper background.
5. Compare the color of the juice in each test tube.

Typical Results

Vitamin C causes the purple indicator solution to lose its color. Therefore, the test tube with the lightest color contains the greatest amount of vitamin C. In contrast, the test tube with the darkest color contains the least amount of vitamin C.

Variations

This experiment can also be used to compare the vitamin C content in beverages such as apple juice, pineapple juice, grapefruit juice, and juice blends.

B Vitamins

B vitamins are a group of vitamins that include thiamin, riboflavin, niacin, pantothenic acid, vitamin B_6, biotin, folate, and vitamin B_{12}. The reason for the break in the sequence of numbers is that some substances originally identified as B vitamins are no longer considered essential for the body. The B vitamins play a key role in energy production, nervous system function, and immunity. They have been found to be so essential to health that B vitamins are commonly added to processed grains. **See Figure 6-15.** As with most nutrients, consuming a variety of foods from all food groups will help ensure that the body receives a healthy supply of B vitamins.

Foods Enriched with B Vitamins

National Pasta Association

Figure 6-15. Processed grains are commonly enriched with B vitamins.

Thiamin. *Thiamin,* also known as vitamin B_1, is a water-soluble vitamin necessary for metabolizing carbohydrates, regulating nerve impulses, and the proper functioning of the heart and muscles. Thiamin can be found in abundance in pork, tuna, legumes, seeds, and whole grains. Enriched foods, such as breads and cereals, are also high in thiamin. Although fruits, vegetables, and milk and dairy products contain only a small amount of thiamin, they can be good sources of thiamin when they are plentiful in the diet.

The thiamin RDA for males age 14 and older is 1.2 mg per day. For females age 19 and older, the RDA is 1.1 mg of thiamin per day. Recommended levels of thiamin are typically met through a well-balanced diet. A UL does not exist for thiamin because thiamine is considered one of the safest of all the vitamins and toxicity is unlikely.

In industrialized countries, a thiamin deficiency is most likely to occur among individuals who abuse alcohol. Frequently consuming large amounts of alcohol inhibits the absorption of thiamin and increases the amount excreted from the body. A thiamin deficiency may also occur if calorie intake is inadequate. If left untreated, a thiamin deficiency can lead to a disease known as beriberi. *Beriberi* is a disease caused by a thiamin deficiency with symptoms that include muscle weakness, fatigue, nerve damage, edema, and heart damage. Beriberi can ultimately lead to heart failure. **See Figure 6-16.**

Riboflavin. *Riboflavin,* also known as vitamin B_2, is a water-soluble vitamin necessary for energy production, growth, red cell production, and the activation of vitamin B_6. Beef liver, milk and dairy products, clams, nuts, eggs, and green leafy vegetables are among the best dietary sources of riboflavin. Additional sources include foods such as breads and grain products that have been enriched with riboflavin.

The riboflavin RDA for adult males is 1.3 mg per day. The RDA for adult females is 1.1 mg per day. There are no known adverse effects associated with consuming large amounts of riboflavin due to its daily excretion from the body. Therefore, no UL has been set.

Thiamin			
Alternative term	Vitamin B_1		
Characteristics	Water-soluble; absorption inhibited and amount excreted from body increases when large amounts of alcohol are frequently consumed		
Functions	Metabolizes carbohydrates, regulates nerve impulses, and aids in heart and muscle functions		
Sources	• Pork loin (5 oz): 1.0 mg • Tuna (5 oz): 0.5 mg • Green peas (1 cup): 0.4 mg	• Sunflower seeds (¼ cup): 0.4 mg • Corn (1 cup): 0.3 mg • Oats (¼ cup): 0.3 mg	• Kaiser roll (3½ inch): 0.3 mg • Kidney beans (¼ cup): 0.25 mg • Brown rice (¼ cup): 0.2 mg
DRI	• Adult males: 1.2 mg • Adult females: 1.1 mg		
Toxicity	No UL		
Deficiency	Causes beriberi		

Figure 6-16. Thiamin is a water-soluble vitamin necessary for proper metabolic, nerve, heart, and muscle functions.

Since riboflavin is found in a wide variety of foods, riboflavin deficiency is uncommon. Riboflavin deficiency usually does not occur in isolation; rather it is symptomatic of a diet that is lacking in multiple nutrients. A lack of riboflavin can cause cracked lips, skin disorders, inflammation of the tongue, and red, itchy eyes. **See Figure 6-17.**

Niacin. *Niacin,* also known as vitamin B_3, is a water-soluble vitamin necessary for skin, nerve, and digestive system functions. Niacin also performs an integral role in energy production. Dietary sources rich in niacin include meats, poultry, seafood, legumes, whole grains, nuts, and seeds. Niacin is also found in milk and dairy products as well as green leafy vegetables and enriched grain products.

The niacin RDA for adult males is 16 mg per day, and it is 14 mg per day for adult females. Tryptophan, an amino acid, can be converted by the body into niacin. For this reason, the RDA for niacin refers to niacin equivalents. A niacin equivalent reflects the intake of niacin from food and sources of tryptophan that are converted into niacin.

Riboflavin			
Alternative term	Vitamin B_2		
Characteristics	Water-soluble; helps sustain supply of other B vitamins		
Functions	Aids in energy production, growth, and red cell production; activates vitamin B_6		
Sources	• Beef liver (5 oz): 4.0 mg • Plain nonfat yogurt (1 cup): 0.6 mg • Clams (5 oz): 0.5 mg	• Almonds (¼ cup): 0.35 mg • Nonfat milk (1 cup): 0.3 mg • Oat bran bagel (3 oz): 0.3 mg	• Egg (1 large): 0.2 mg • Banana (1 medium): 0.1 mg • Spinach (1 cup): 0.1 mg
DRI	• Adult males: 1.3 mg • Adult females: 1.1 mg		
Toxicity	No UL		
Deficiency	Uncommon		

Figure 6-17. Riboflavin is a water-soluble vitamin necessary for energy production, growth, red cell production, and vitamin B_6 activation.

For adult males and females, the UL of niacin is 35 mg per day. Niacin toxicity is likely to occur in patients who are prescribed nicotinic acid, a form of niacin, as part of treatment to help regulate cholesterol levels. This form of niacin is often given in levels higher than the RDA. Common symptoms of niacin toxicity include flushing in the face, tingling in the extremities, itchy rashes, and blurred vision. Liver damage and elevated blood glucose levels are also potential complications that arise from niacin toxicity.

A niacin deficiency is likely to affect a variety of normal body functions. For example, a niacin deficiency may result in a loss of appetite, muscle weakness, confusion, and skin rashes. If left untreated, a niacin deficiency can develop into a disease known as pellagra. *Pellagra* is a disease caused by a niacin deficiency with symptoms referred to as "the three Ds": dementia (mental impairment), dermatitis (inflamed skin), and diarrhea. **See Figure 6-18.**

Pantothenic Acid. *Pantothenic acid,* also known as vitamin B_5, is a water-soluble vitamin necessary for producing hormones and metabolizing carbohydrates and lipids. Pantothenic acid is sometimes called the "antistress" vitamin due to the role it plays in manufacturing stress-reducing hormones in the adrenal glands. It also helps the body manufacture red blood cells and synthesize cholesterol. Pantothenic acid is found in a wide variety of foods including meats, poultry, seafood, fruits, vegetables, dairy products, eggs, legumes, and whole grains. Seeds also contain some pantothenic acid. **See Figure 6-19.**

Beef Liver

Figure 6-19. Beef liver is one of the best dietary sources of B vitamins, including pantothenic acid.

Niacin			
Alternative term	Vitamin B_3		
Characteristics	Water-soluble; tryptophan, an amino acid found in food, can be converted by the body into niacin		
Functions	Necessary for skin, nerve, and digestive system functions; aids in energy production		
Sources	• Beef liver (5 oz): 18.5 mg • Chicken breast (5 oz): 15.5 mg • Salmon (5 oz): 12.0 mg	• Peanuts (¼ cup): 4.4 mg • Shrimp (5 oz): 3.5 mg • Green peas (1 cup): 3.0 mg	• White button mushrooms (1 cup, sliced): 2.5 mg • Barley (¼ cup): 2.3 mg • Pine nuts (¼ cup): 1.4 mg
DRI	• Adult males: 16 mg • Adult females: 14 mg		
Toxicity	UL: 35 mg Toxicity may occur in patients taking nicotinic acid to regulate cholesterol levels		
Deficiency	Causes pellagra		

Figure 6-18. Niacin is a water-soluble vitamin necessary for skin, nerve, and digestive system functions.

The pantothenic AI for adult males and females stands at 5 mg per day. Studies have shown that most adults easily consume between 5 mg and 6 mg per day. A UL does not exist for pantothenic acid since there is little risk of toxicity.

Pantothenic acid deficiency is rare and has been observed only in cases of malnutrition. A deficiency of pantothenic acid can cause fatigue, disturbances in sleep patterns, irritability, vomiting, muscle cramps, and tingling, burning, or numbness in the feet. **See Figure 6-20.**

Food Science Note

Pantothenic acid is a component of coenzyme A, which plays an essential role in reactions that support life. Coenzyme A is required for metabolizing proteins, carbohydrates, and lipids, and for synthesizing essential fatty acids, cholesterol, and various hormones.

Vitamin B_6. *Vitamin B_6* is a water-soluble vitamin necessary for metabolic functions, nervous system and immune system functions, and the formation of hemoglobin, which is the protein in red blood cells that transports oxygen throughout the body. Vitamin B_6 is available in three forms: pyridoxal, pyridoxine, and pyridoxamine, which are converted to active coenzymes. Key sources of vitamin B_6 include a variety of fruits and vegetables. Animal-based protein foods, legumes, nuts, and whole and enriched grains are also sources of vitamin B_6.

Food Science Note

Vitamin B_6 is necessary to convert the amino acid tryptophan into niacin.

The RDA for both males and females aged 19 through 50 is 1.3 mg of vitamin B_6 per day. Research indicates needs increase slightly after age 50. Vitamin B_6 has a UL of 100 mg per day for adults. When vitamin B_6 levels rise above the UL for a year or more, irreversible nerve damage is possible. This most often occurs as the result of overconsuming the supplemental form of vitamin B_6 pyridoxine.

If caloric intake is inadequate or the majority of a diet is comprised of highly processed foods, a vitamin B_6 deficiency may occur. Signs of a deficiency include nausea, depression, confusion, inflammation of the tongue, and skin rashes. **See Figure 6-21.**

Pantothenic Acid			
Alternative term	Vitamin B_5		
Characteristics	Water-soluble; commonly called the "antistress" vitamin due to role in manufacturing stress-reducing hormones		
Functions	Produces hormones, metabolizes carbohydrates and lipids, helps manufacture red blood cells, and helps synthesize cholesterol		
Sources	• Beef liver (5 oz): 10.0 mg • Avocado (1 cup, cubed): 2.1 mg • Corn (1 cup): 1.2 mg	• Chicken breast (5 oz): 1.0 mg • Mackerel (5 oz): 1.0 mg • Nonfat milk (1 cup): 0.8 mg	• Chickpeas (¼ cup): 0.8 mg • Egg (1 large): 0.7 mg • Wheat germ (¼ cup): 0.65 mg
DRI	• Adult males: 5 mg • Adult females: 5 mg		
Toxicity	No UL		
Deficiency	Uncommon		

Figure 6-20. Pantothenic acid is a water-soluble vitamin necessary for producing hormones and metabolic functions.

Vitamin B_6			
Alternative term	Pyridoxal; pyridoxine; pyridoxamine		
Characteristics	Water-soluble; pyridoxal, pyridoxine, and pyridoxamine are converted to active B_6		
Functions	Helps produce energy, form hemoglobin, and aids in nervous system and immune system functions		
Sources	• Tuna (5 oz): 1.5 mg • Cod (5 oz): 0.5 mg • Avocado (1 cup, cubed): 0.4 mg	• Banana (1 medium): 0.4 mg • Sweet potato (1 medium): 0.3 mg • Pistachios (¼ cup): 0.5 mg	• Brown rice (¼ cup): 0.25 mg • Lentils (¼ cup): 0.25 mg • Cauliflower (1 cup): 0.2 mg
DRI	• Adult males: 1.3 mg • Adult females: 1.3 mg		
Toxicity	UL: 100 mg Toxicity associated with supplemental form of vitamin B_6 pyridoxine		
Deficiency	Uncommon		

Figure 6-21. Vitamin B_6 is a water-soluble vitamin necessary for metabolic functions, nervous system and immune system functions, and the formation of hemoglobin.

Biotin. *Biotin,* also known as vitamin B_7, is a water-soluble vitamin made in small amounts by the body that is necessary for energy production, the synthesis of fatty acids, proper nervous system function, and healthy skin, hair, and nails. Small amounts of biotin are found in a wide array of foods. Some of the best sources of biotin include peanuts, almonds, Swiss chard, dairy products, and eggs (especially the yolk).

The biotin AI for adults is 30 mcg per day. Because biotin is found in so many different foods, a well-balanced diet will most likely meet the biotin needs of the body. A UL does not exist for biotin, and there is no toxicity associated with biotin intake.

The body requires such a small amount of biotin that deficiencies are rare. However, the risk of developing a biotin deficiency increases for individuals who are fed long-term through the veins or have a condition that inhibits the absorption of nutrients. Some medications also may interfere with the ability of the body to manufacture and absorb biotin, which puts the body at risk for deficiency. Fatigue, depression, tingling and numbness in the hands and feet, hair loss, and a scaly, red facial rash are symptoms associated with a biotin deficiency. **See Figure 6-22.**

Biotin			
Alternative term	Vitamin B_7		
Characteristics	Water-soluble; made in small amounts by the body		
Functions	Helps produce energy, synthesizes fatty acids, assists in nervous system function, promotes healthy skin, hair, and nails		
Sources	• Peanuts (¼ cup): 26.3 mcg • Almonds (¼ cup): 24.2 mcg • Swiss chard (1 cup): 10.5 mcg	• Plain low-fat yogurt (1 cup): 7.3 mcg • Egg (1 large): 7.0 mcg • Salmon (5 oz): 5.6 mcg	• Nonfat milk (1 cup): 4.9 mcg • Walnuts (¼ cup): 4.8 mcg • Banana (1 medium): 3.1 mcg
DRI	• Adult males: 30 mcg • Adult females: 30 mcg		
Toxicity	No UL		
Deficiency	Uncommon		

Figure 6-22. Biotin is a water-soluble vitamin necessary for energy production, the synthesis of fatty acids, nervous system function, and healthy skin, hair, and nails.

Folate. *Folate,* also known as vitamin B_9, is a water-soluble vitamin that supports red blood cell production, the nervous system, and the formation of DNA. *DNA* is the genetic building blocks of the body. Folate also promotes healthy blood circulation by preventing a buildup of the amino acid homocysteine. High levels of homocysteine are associated with atherosclerosis and cardiovascular disease. Folate occurs naturally in foods, whereas folic acid is added to fortified foods and supplements. *Folic acid* is a synthetic form of folate. Folic acid is more easily absorbed by the body than folate.

Folate, derived from the Latin word meaning "leaf," is abundant in green leafy vegetables. Folate is also plentiful in legumes and chicken liver. Folate has been widely acknowledged for its importance during pregnancy in the prevention of birth defects. Therefore, the FDA mandates that certain grain products such as breads, flour, pastas, rice, and some cereals be fortified with folic acid to help meet dietary needs. It is estimated that foods fortified with folic acid help increase folic acid intake by 100 mcg per day.

To take into account different levels of absorption, the RDA for folate is expressed in dietary folate equivalents (DFE). One DFE is the equivalent of 1 mcg of folate or 0.6 mcg of folic acid from fortified foods or supplements. The RDA for adults is 400 mcg DFE. During pregnancy, the RDA for folate increases to 600 mcg DFE. The UL for folate is based upon 1000 mcg DFE from folic acid. Excess folate in the diet may hide symptoms of a vitamin B_{12} deficiency. However, folate toxicity is rare because excess is passed from the body through the urine.

A deficiency of folate can affect many components of health. For example, a lack of folate during pregnancy can cause neural tube defects in which the brain and spine do not develop properly. A lack of folate also can lead to anemia. *Anemia* is a blood condition in which red blood cells do not supply an adequate amount of oxygen to the body. Additional symptoms of a folate deficiency include digestive disorders, decreased immunity, fatigue, weakness, irritability, disturbed sleep, and impaired mental function. **See Figure 6-23.**

Folate			
Alternative term	Vitamin B_9		
Characteristics	Water-soluble; folic acid, the synthetic form of folate, is better absorbed by the body, helps prevent birth defects, and is required in some fortified grain products; excess may hide symptoms of a vitamin B_{12} deficiency		
Functions	Supports red blood cell production, assists in forming DNA, helps support nervous system, and promotes blood circulation		
Sources	• Chicken liver (5 oz): 825 mcg • Pinto beans (¼ cup): 253 mcg • Lentils (¼ cup): 230 mcg	• White beans (¼ cup): 196 mcg • Beets (1 cup): 148 mcg • Orange juice (1 cup): 74.4 mcg	• Romaine lettuce (1 cup, shredded): 63.9 mcg • Spinach (1 cup): 58.2 mcg • Flour tortilla (8 inch): 47.8 mcg
DRI	• Adult males: 400 mcg DFE • Adult females: 400 mcg DFE • During pregnancy: 600 mcg DFE		
Toxicity	UL: 1000 mcg DFE		
Deficiency	Causes neural tube defects during pregnancy		

Figure 6-23. Folate is a water-soluble vitamin that supports red cell production, the nervous system, and the formation of DNA.

Vitamin B_{12}. *Vitamin B_{12}* is a water-soluble vitamin that helps support the formation of red blood cells and DNA, promotes proper nerve development, and plays a role in energy production. The absorption of vitamin B_{12} depends on a protein known as intrinsic factor. *Intrinsic factor* is a protein produced in the stomach that promotes vitamin B_{12} absorption. Vitamin B_{12} differs from other B vitamins because it can be stored by the body and is only found naturally in animal-based foods such as meats, seafood, eggs, and dairy products. **See Figure 6-24.** Plant-based foods that have been fortified also contain vitamin B_{12}.

Starting at age 14 and continuing through adulthood, the RDA for males and females is 2.4 mcg of vitamin B_{12} per day. A UL does not exist for vitamin B_{12}, and toxicity is not associated with this vitamin.

Vitamin B_{12} Sources

Fortune Fish Company

Figure 6-24. Vitamin B_{12} differs from other B vitamins in that it is only found in animal-based foods such as meats, seafood, eggs, and dairy products.

Individuals whose diet is void of animal-based foods and vitamin-B_{12}-fortified foods may be at greater risk for developing a vitamin B_{12} deficiency. A vitamin B_{12} deficiency is also more likely to occur later in life because less intrinsic factor is produced with age, which reduces the absorption of vitamin B_{12}. Individuals with digestive disorders may have trouble absorbing this nutrient as well.

Service Ideas, Inc.

Cereals are often fortified with vitamin B_{12}.

When vitamin B_{12} is not properly absorbed, nerves can become damaged, which leads to impaired balance, mental confusion, and a tingling sensation in the extremities. Anemia may also develop. A vitamin B_{12} deficiency can be mistaken for a folate deficiency due to their similar symptoms. If this happens, and folate is supplemented in the diet, folate will help treat anemia but not the adverse effects a vitamin B_{12} deficiency can have on the nervous system. Long-term damage to the nervous system may be irreversible. **See Figure 6-25.**

Vitamin B_{12}			
Characteristics	Water-soluble; stored by the body; occurs naturally only in animal-based foods; fortified in plant-based foods; depends on intrinsic factor for absorption		
Functions	Helps support formation of red blood cells, promotes proper nerve development, and assists in energy production		
Sources	• Beef liver (5 oz): 83 mcg • Crab (5 oz): 12.5 mcg • Trout (5 oz): 11 mcg	• Lamb (5 oz): 4 mcg • Scallops (5 oz): 2 mcg • Plain low-fat yogurt (1 cup): 1.4 mcg	• Nonfat milk (1 cup): 0.9 mcg • Egg (1 large): 0.6 mcg • Provolone cheese (1 oz): 0.4 mcg
DRI	• Adult Males: 2.4 mcg • Adult Females: 2.4 mcg		
Toxicity	No UL		
Deficiency	Typically uncommon		

Figure 6-25. Vitamin B_{12} is a water-soluble vitamin that helps support red blood cells and DNA, proper nerve development, and energy production.

Knowledge Check 6-5

1. Summarize the characteristics of water-soluble vitamins.
2. List the functions of vitamin C.
3. Identify dietary sources of vitamin C.
4. List the eight B vitamins.
5. Describe the functions of each of the eight B vitamins.
6. Identify common dietary sources of each of the eight B vitamins.
7. Define folic acid.

FAT-SOLUBLE VITAMINS

A *fat-soluble vitamin* is a vitamin that dissolves in fat. Fat-soluble vitamins play an essential role in a variety of functions including maintaining healthy skin, teeth, and bones and are important to proper immune and nervous system functioning. Fat-soluble vitamins include vitamins A, D, E, and K. It is necessary to consume dietary fats and oils in order to digest and absorb fat-soluble vitamins. Fat-soluble vitamins must also be emulsified during digestion before they can be transported and used by the body.

Unlike water-soluble vitamins, which are eliminated daily, fat-soluble vitamins are stored by the body. Fat-soluble vitamins are primarily stored in the liver and fatty tissues, and therefore they do not need to be consumed as frequently as water-soluble vitamins. **See Figure 6-26.** Because they are stored in the body, fat-soluble vitamins have a greater chance of accumulating to toxic levels. Individuals with conditions that interfere with fat absorption may be at risk of developing a fat-soluble vitamin deficiency.

Comparing Vitamins

	Water-Soluble Vitamins	Fat-Soluble Vitamins
Absorption	Directly into blood	Lymphatic system and blood
Transportation	Freely transported	May require transport proteins
Storage	Circulate in watery cells	Stored in liver and fatty tissues
Excretion	Filtered in kidneys; passed in urine	Remain in fat stores
Requirement	Daily	Every 2–3 days

Figure 6-26. Water-soluble and fat-soluble vitamins are used differently within the body.

Vitamin A

Vitamin A is a fat-soluble vitamin that plays a key role in promoting healthy vision and proper growth and development and is vital for a healthy immune system. Vitamin A is obtained from preformed retinoid or provitamin carotenoid. *Preformed retinoid* is a collective term for the most active forms of vitamin A, including retinol, retinal, and retinoic acid. *Provitamin carotenoid* is a collective term for compounds that are converted to vitamin A, including beta-carotene, alpha-carotene, and beta-cryptoxanthin. During digestion and absorption, preformed retinoid is converted to retinol that is transported, along with fat, to the liver for storage. Provitamin carotenoid is not absorbed as readily as preformed retinoid. Of the provitamin carotenoid compounds, beta-carotene has the highest absorption rate.

Preformed retinoid is mainly found in animal-based foods such as beef liver, eggs, and butter. Several foods, including breakfast cereals and milk, are commonly fortified with vitamin A. Provitamin carotenoid is predominantly found in orange or yellow vegetables and fruits such as carrots, sweet potatoes, butternut squash, and cantaloupe. **See Figure 6-27.** Some foods that are high in provitamin carotenoid do not have a bright yellow or orange pigment because it is masked by the green pigment chlorophyll. Green vegetables high in provitamin carotenoid include spinach, kale, and collard greens.

Because preformed retinoid and provitamin carotenoid differ in their bioavailability, the RDA is measured in retinol activity equivalents (RAEs). Retinol is used as the measurement because all dietary sources of vitamin A are converted to retinol. The vitamin A RDA for adult males is 900 mcg RAE and 700 mcg RAE for adult females. Most nutritionists rely on RAEs. However, vitamin A is most often expressed in international units (IU) on food and supplement labels. An *international unit* is a globally accepted amount of a substance, such as a vitamin, that is necessary to facilitate a specific body response.

Provitamin Carotenoid

National Honey Board

Figure 6-27. Provitamin carotenoid is predominantly found in orange or yellow vegetables and fruits.

A prolonged intake of vitamin A that rises above the UL of 3000 mcg per day can result in hypervitaminosis A. *Hypervitaminosis A* is an adverse health condition caused by the excessive intake of preformed vitamin A. Symptoms of hypervitaminosis A include nausea, fatigue, headache, painful bones and joints, dry itchy skin, peeling skin, and liver damage. In addition, the prolonged and excessive consumption of carotene-rich foods results in discoloration of the skin. The skin takes on an alarming, yet harmless, deep orange appearance.

Most cases of a vitamin A deficiency occur in developing countries where malnutrition is prevalent. Early symptoms of a vitamin A deficiency include diarrhea, a greater susceptibility to infections such as measles, respiratory disorders, and dry eyes. A prolonged deficiency in vitamin A can lead to the inability to see in dim light (night blindness) and eventually may result in total blindness. **See Figure 6-28.**

Vitamin A			
Alternative terms	• Preformed retinoid including retinol, retinal, and retinoic acid • Provitamin carotenoid including beta-carotene, alpha-carotene, and beta-cryptoxanthin		
Characteristics	Fat-soluble; antioxidant; preformed retinoid is converted to retinol for digestion and absorption; beta-carotene has highest absorption rate of provitamin carotenoid compounds; preformed retinoid is mainly found in animal-based foods; provitamin carotenoid is mainly found in orange and yellow fruits and vegetables		
Functions	Promotes healthy vision, proper growth and development, and healthy immune system		
Sources	• Beef liver (5 oz): 6955 mcg RAE • Carrots (1 cup, chopped): 1069 mcg RAE • Pumpkin (½ cup, canned): 953 mcg RAE	• Sweet potato (1 medium): 922 mcg RAE • Butternut squash (1 cup, cubed): 745 mcg RAE • Cantaloupe (1 cup, cubed): 270 mcg RAE	• Spinach (1 cup): 141 mcg RAE • Butter (1 oz): 130 mcg RAE • Egg (1 large): 70 mcg RAE
DRI	• Adult males: 900 mcg RAE • Adult females: 700 mcg RAE		
Toxicity	UL: 3000 mcg Causes hypervitaminosis A		
Deficiency	Uncommon		

Figure 6-28. Vitamin A is a fat-soluble vitamin that promotes healthy vision, growth and development, and a healthy immune system.

Vitamin D

Vitamin D is a fat-soluble vitamin that is required for proper kidney functioning, the health of teeth and bones, and for the absorption of vitamin A, calcium, phosphorus, magnesium, iron, and zinc. Vitamin D can be synthesized by the body after exposure to sunlight. The skin interacts with energy from the sun to produce vitamin D_3, which is further synthesized by the liver and kidneys to produce vitamin D. Because exposure to sunlight helps the body acquire vitamin D, individuals who live where there is little sunlight or are unable to go outside must get vitamin D through food or supplements. The best dietary sources of vitamin D include fatty fish and vitamin-D-fortified foods such as soy milk, orange juice, and breakfast cereals.

For adults between the ages of 19 and 70, the AI for vitamin D is set at 15 mcg. The body acquires 80%–90% of the vitamin D it needs through exposure to sunlight. However, the AI is formed on the assumption of minimal sun exposure.

Starting at age 9 and continuing through adulthood, the UL for vitamin D has been set at 100 mcg per day. Vitamin D toxicity occurs almost exclusively from overusing supplements. Excess vitamin D can raise the levels of calcium in the blood and cause muscle weakness, irregular heartbeat, and confusion. This condition also causes calcium deposits to form in the blood, kidneys, and other soft tissues. Additional signs of vitamin D toxicity include nausea, vomiting, excessive thirst, constipation, and weight loss.

Vitamin D is strongly linked with calcium absorption in the body. Therefore, a deficiency can lead to problems with bone development. In children, a vitamin D deficiency can cause rickets. *Rickets* is a childhood disease caused by a vitamin D deficiency characterized by bones that soften and bend. In adults, vitamin D deficiency is known as osteomalacia. *Osteomalacia* is an adulthood disease caused by a vitamin D deficiency characterized by bones that soften, resulting in bone pain and muscle weakness. If there is a long-term deficiency of vitamin D, osteomalacia will likely develop into the bone disease osteoporosis. **See Figure 6-29.**

Vitamin D			
Characteristics	Fat-soluble; synthesized by the body with exposure to sunlight		
Functions	Helps kidney functioning; necessary for absorption of vitamin A, calcium, phosphorus, magnesium, iron, and zinc; promotes healthy teeth and bones		
Sources	• Cod liver oil (1 tbsp): 34.0 mcg • Tuna (5 oz): 8.1 mcg • Herring (5 oz): 6.0 mcg	• Trout (5 oz): 5.5 mcg • Fortified soy milk (1 cup): 2.7 mcg • Fortified orange juice (1 cup): 2.5 mcg	• Fortified low-fat milk (1 cup): 2.4 mcg • Bran cereal (1 cup): 1.3 mcg • Egg (1 large): 1.0 mcg
DRI	• Adult males: 15 mcg • Adult females : 15 mcg		
Toxicity	UL: 100 mcg		
Deficiency	• Causes rickets in children • Causes osteomalacia in adults		

Figure 6-29. Vitamin D is a fat-soluble vitamin required for kidney functioning, healthy teeth and bones, and the absorption of vitamin A, calcium, phosphorus, magnesium, iron, and zinc.

Vitamin E

Vitamin E is a fat-soluble vitamin that strengthens the immune system and widens blood vessels, which reduces the risk of blood clots. Vitamin E also acts as an antioxidant. The cell membrane is at the highest risk for oxidation, which can eventually lead to cancer. Vitamin E works to limit the amount of free radicals that are exposed to the cell, thus helping cells stay healthy. Foods most abundant in vitamin E include oils such as wheat germ, canola, and olive oils. Nuts, especially almonds, and seeds are regarded as rich sources of vitamin E. Grains, fruits, and vegetables also contribute small amounts of vitamin E to the diet as well as foods, such as breakfast cereals, fruit juices, and spreads, that are fortified with vitamin E.

Alpha-tocopherol is the most active form of vitamin E. For this reason, the RDA for vitamin E is expressed in milligrams of alpha-tocopherol. An RDA of 15 mg per day of vitamin E applies to both males and females.

Eating foods rich in vitamin E poses no health risks. However, a UL exists for vitamin E obtained from supplements. Supplements that contain vitamin E in its natural form have a UL of 1500 IU per day. Supplements that contain synthetic vitamin E have a UL of 1000 IU per day. Although rare, vitamin E toxicity can reduce the ability of vitamin K to assist with blood clotting. This can result in excessive bruising and prolonged bleeding.

Due to the widespread use of vegetable oils, a vitamin E deficiency is rare among the general population. A vitamin E deficiency is most often linked to diseases where fat is not properly absorbed and may also affect individuals whose long-term diet is dangerously low in fat. Symptoms associated with a vitamin E deficiency include muscle weakness and loss of motor skills. Over time, a vitamin E deficiency results in the breakdown of cell membranes and causes nerve tissue damage. **See Figure 6-30.**

Vitamin E			
Alternative terms	Tocopherol; tocotrineol		
Characteristics	Fat-soluble; antioxidant; most active form is alpha-tocopheral		
Functions	Promotes cellular health, widens blood vessels, and fights invading bacteria and viruses		
Sources	• Wheat germ oil (1 tbsp): 20.2 mg • Almonds (¼ cup): 9.3 mg • Sunflower seeds (¼ cup): 3.8 mg	• Avocado (1 cup, cubed): 3.1 mg • Canola oil (1 tbsp): 2.4 mg • Olive oil (1 tbsp): 1.9 mg	• Pomegranate (1 medium): 1.7 mg • Mustard greens (1 cup, chopped): 1.1 mg • Papaya (1 cup, cubed): 1.0 mg
DRI	• Adult males: 15 mg • Adult females: 15 mg		
Toxicity	UL: 1500 IU for natural form UL: 1000 IU for synthetic form		
Deficiency	Uncommon		

Figure 6-30. Vitamin E is a fat-soluble vitamin that strengthens the immune system and widens blood vessels.

Vitamin K

Vitamin K is a fat-soluble vitamin that plays an essential role in blood clotting and bone formation. Of the 13 proteins needed for blood clotting to occur, vitamin K helps synthesize four of them. Like vitamin D, vitamin K is unique because it can be obtained from a nonfood source. Although bacteria in the intestines can synthesize some vitamin K, it is not in amounts sufficient to meet the needs of the body. Therefore, it is important to obtain vitamin K from dietary sources such as green leafy vegetables. Herbs and spices can also contribute vitamin K to the diet. **See Figure 6-31.**

The AI set for vitamin K is 120 mcg for men and 90 mcg for women. In general, approximately half of the vitamin K a person needs is produced by bacteria in the intestines. The remaining vitamin K must be provided by the diet and is reflected in the AI. There is no UL for vitamin K. Few adverse reactions have been attributed to an excess of vitamin K.

Although relatively rare, there are certain medications that can increase the likelihood of a vitamin K deficiency. For example, some antibiotics can destroy bacteria in the intestines. This limits the ability of the body to synthesize vitamin K. Taking medications that prevent blood clotting can also increase the risk of developing a deficiency. Additionally, individuals with conditions that impair fat absorption may be lacking vitamin K. A vitamin K deficiency interferes with blood clotting and results in undue bruising and bleeding. Low bone density may also be a sign of a vitamin K deficiency. **See Figure 6-32.**

Vitamin K Sources

Figure 6-31. Herbs such as parsley can be a key source of vitamin K.

Vitamin K			
Characteristics	Fat-soluble; some can be synthesized by bacteria in intestines		
Functions	Plays essential role in blood clotting, and helps form bone structure		
Sources	• Kale (1 cup, chopped): 547 mcg • Swiss chard (1 cup): 299 mcg • Brussels sprouts (1 cup): 156 mcg	• Spinach (1 cup): 145 mcg • Broccoli (1 cup, chopped): 92.5 mcg • Cabbage (1 cup, chopped): 67.6 mcg	• Fresh parsley (1 tbsp): 61.5 mcg • Soybeans (¼ cup): 21.8 mcg • Dried basil (1 tsp): 8.6 mcg
DRI	• Adult males: 120 mcg • Adult females: 90 mcg		
Toxicity	No UL		
Deficiency	Uncommon		

Figure 6-32. Vitamin K is a fat-soluble vitamin that plays an essential role in blood clotting and bone formation.

Knowledge Check 6-6

1. List the fat-soluble vitamins.
2. Explain how fat-soluble vitamins are absorbed by the body.
3. Identify the functions of each of the four fat-soluble vitamins.
4. Describe the two main forms of vitamin A.
5. Name the vitamin synthesized by the body after exposure to sunlight.
6. List dietary sources of vitamin E.
7. Identify signs of a vitamin K deficiency.

MAJOR MINERALS

A *major mineral* is a mineral with a recommended dietary allowance that exceeds 100 mg per day. Major minerals are involved in vital body functions such as providing the structure for teeth and bones and maintaining fluid balance. The major minerals include calcium, phosphorus, magnesium, potassium, sodium, and chloride.

Calcium

Calcium is a major mineral required for blood clotting and the development of healthy teeth and bones. Calcium is the most abundant mineral in the body. Approximately 99% of the calcium found in the body is in the teeth and bones. The remaining calcium circulates through the blood and is necessary for muscles to contract properly and for nerves to carry messages from the brain to all body parts.

The richest dietary sources of calcium include dairy products such as milk, cheese, and yogurt. Many dairy alternatives, such as rice milk and soy milk, as well as fruit juices are fortified with calcium. Other nondairy foods that contain calcium include fish with soft bones, sesame seeds, almonds, dried figs, and green leafy vegetables.

For males and females between the ages of 19 and 50, the RDA for calcium is 1000 mg. However, calcium needs tend to fluctuate throughout the life-cycle. In general, the greatest needs for calcium are during adolescence, pregnancy, and for females over the age of 50.

Most people do not exceed the calcium UL of 2500 mg from food sources. Rather, calcium excess typically results from the use of calcium supplements. When calcium is ingested in large doses it can interfere with the absorption of other nutrients and encourage the development of calcium deposits that form into kidney stones. The long-term intake of calcium above 2500 mg per day can also result in an irregular heartbeat and even death.

Wisconsin Milk Marketing Board, Inc.

Cheese is one of the richest dietary sources of calcium.

The body has the ability to maintain calcium levels by increasing the absorption of calcium from the blood or decreasing the amount of calcium that is excreted as waste. Therefore, calcium deficiencies do not produce obvious symptoms in the short term. If a person suffers from a long-term calcium deficiency, the body will pull calcium stores from the bones, which results in osteoporosis. Other symptoms of a calcium deficiency can include muscle cramps, numbness in the fingers, and convulsions. **See Figure 6-33.**

Food Science Note

Minerals account for approximately 5 lb of body weight, and calcium contributes 2 lb of those 5 lb.

Phosphorus

Phosphorus is a major mineral that works with calcium to build strong teeth and bones, helps synthesize proteins, plays an essential role in energy production and storage, filters waste from the kidneys, and assists in the formation of DNA. Phosphorus is the second most abundant mineral in the body. Approximately 85% of phosphorus in the body is found in the teeth and bones, while the remainder is present in cells and tissues.

The best sources of phosphorus are found in protein-rich foods such as meats, poultry, seafood, eggs, and dairy products. Phosphorus is also found in whole grains, but animal-based proteins are a better source because they are more easily absorbed by the body.

The RDA for phosphorus is 700 mg per day and applies to adult males and females 19 years of age and older. This recommendation is typically met with limited effort because phosphorus is plentiful in the food supply.

Calcium			
Characteristics	Major mineral; most abundant mineral in the body; may reduce risk of developing high blood pressure		
Functions	Required for healthy teeth and bones, necessary for muscle contraction, helps carry messages from the brain to all body parts, and assists in blood clotting		
Sources	• Sardines, with bones (5 oz): 535 mg • Fortified orange juice (1 cup): 500 mg • Plain low-fat yogurt (1 cup): 448 mg	• Sesame seeds (¼ cup): 351 mg • Nonfat milk (1 cup): 301 mg • Cheddar cheese (1 oz): 202 mg	• Dried figs (½ cup): 120 mg • Almonds (¼ cup): 94.5 mg • Spinach (1 cup): 29.7
DRI	• Adult males: 1000 mg • Adult females: 1000 mg • Calcium needs are greatest during adolescence, pregnancy, and for females over the age of 50		
Toxicity	UL: 2500 mg		
Deficiency	Causes osteoporosis		

Figure 6-33. Calcium is a major mineral required for blood clotting and healthy teeth and bones.

For adults up to the age of 70, the UL for phosphorus intake is set at 4000 mg per day. Excessive amounts of phosphorus can be toxic. Some symptoms of phosphorous toxicity include diarrhea, hardening of the organs and tissues, and an inhibition of the absorption of calcium, magnesium, iron, and zinc. In addition, if phosphorus levels become greater than calcium levels, the body will extract calcium from bones, which leads to an increased risk of developing osteoporosis.

Because of the wide variety of dietary sources rich in phosphorus, deficiencies are unlikely. However, health conditions that interfere with the absorption of nutrients can result in low levels of phosphorus in the body. Symptoms associated with a phosphorous deficiency include bone pain and weakness, decreased appetite, fatigue, irritability, and joint stiffness. **See Figure 6-34.**

Magnesium

Magnesium is a major mineral that is needed to keep bones strong and helps facilitate hundreds of chemical reactions in the body. For example, magnesium helps support a healthy immune system, assists in proper nerve and muscle functioning, helps regulate blood glucose levels, and promotes normal blood pressure. In addition, magnesium is involved in metabolizing proteins and lipids for energy. Magnesium is predominantly found in the bones, cells of body tissues, and organs.

Because magnesium is found in chlorophyll, green leafy vegetables are significant sources of magnesium. The bran and germ of whole grains also contain magnesium. Additional sources of magnesium include fish, legumes, nuts, and seeds.

For males between the ages of 19 and 30, the RDA for magnesium is 400 mg and increases to 420 mg for males over 30. For females between the ages of 19 and 30, the RDA for magnesium is 310 mg and increases to 320 mg for females over 30.

There is not a UL for magnesium from dietary sources. The magnesium UL for both males and females over the age of 19 from supplements is 350 mg. However, toxicity from dietary magnesium is extremely rare because excess magnesium is easily excreted from the body as waste. Magnesium toxicity may develop if magnesium-containing pharmaceuticals such as laxatives and antacids are abused. Symptoms of magnesium toxicity include nausea, vomiting, and diarrhea. **See Figure 6-35.**

Phosphorus			
Characteristics	Major mineral; second most abundant mineral in the body; phosphorus from animal-based foods is most easily absorbed		
Functions	Works with calcium to build strong teeth and bones, helps synthesize proteins, assists in energy production and storage, filters waste from kidneys, and helps form DNA		
Sources	• Sardines, with bones (5 oz): 685 mg • Pumpkin seeds (¼ cup): 405 mg • Halibut (5 oz): 311 mg	• Bison (5 oz): 262 mg • Nonfat milk (1 cup): 247 mg • Chicken breast (5 oz): 243.5 mg	• Lentils (¼ cup): 216.5 mg • Swiss cheese (1 oz): 159 mg • Sunflower seeds (¼ cup): 76 mg
DRI	• Adult males: 700 mg • Adult females: 700 mg		
Toxicity	UL: 4000 mg		
Deficiency	Uncommon		

Figure 6-34. Phosphorus is a major mineral that helps build strong teeth and bones, synthesize proteins, produce and store energy, filter waste, and form DNA.

Magnesium			
Characteristics	Major mineral; found in bones, tissues, and organs		
Functions	Keeps bones strong; helps facilitate many chemical reactions that are necessary for healthy immune system, proper nerve and muscle functioning, and maintaining normal blood glucose and normal blood pressure; helps metabolize proteins and lipids		
Sources	• Pumpkin seeds (¼ cup): 184.5 mg • Firm tofu (1 cup): 146.2 mg • Almonds (¼ cup): 96 mg	• Black beans (¼ cup): 83 mg • Cashews (¼ cup): 81.8 mg • Millet (¼ cup): 57 mg	• Salmon (5 oz): 40.5 mg • Swiss chard (1 cup): 29.2 mg • Spinach (1 cup): 23.7 mg
DRI	• Adult males to age 30: 400 mg • Adult males 30+: 420 mg • Adult females to age 30: 310 mg • Adult females 30+: 320 mg		
Toxicity	UL: 350 mg (from supplements)		
Deficiency	Uncommon		

Figure 6-35. Magnesium is a major mineral that keeps bones strong and helps facilitate hundreds of chemical reactions in the body.

Research suggests that some individuals may not have adequate body stores of magnesium because of a diet lacking in fruits, vegetables, and whole grains. Having adequate body stores of magnesium may provide protection against cardiovascular disease. Signs of a magnesium deficiency are similar to magnesium toxicity and include nausea, vomiting, and fatigue. As the deficiency worsens, muscle cramping, seizures, and abnormal heart rhythms can occur.

Potassium

Potassium is a major mineral that helps regulate normal heart functioning and muscle contraction and is necessary for energy production and protein synthesis. Potassium along with sodium and chloride are three electrolytes that are crucial to balancing the amount of water in and around cells. Electrolytes are also vital for maintaining an acid-base balance of 7.4 by neutralizing acids and bases.

The richest sources of potassium are fresh fruits and vegetables. Many people think that bananas are the best source of dietary potassium. However, russet potatoes contain more than twice as much potassium as bananas. **See Figure 6-36.** Dairy products and legumes are also considered good sources of potassium. It is best to choose foods that are unprocessed because foods often lose potassium during processing.

Potassium Sources

Idaho Potato Commission

Figure 6-36. Russet potatoes contain more than twice as much potassium as bananas.

The AI for potassium is 4700 mg per day and applies to males and females 14 years of age through adulthood. A UL for potassium does not exist because large amounts of dietary potassium have not been shown to adversely affect healthy individuals. However, health conditions such as kidney failure may lead to a condition known as hyperkalemia. *Hyperkalemia* is a condition characterized by an elevated concentration of potassium in the blood. Hyperkalemia results when potassium intake surpasses the ability of the kidneys to eliminate it. Symptoms of hyperkalemia include muscular weakness, tingling or numbness in the hands and feet, and an abnormal heart rhythm.

A deficiency in potassium can cause hypokalemia. *Hypokalemia* is a condition characterized by a low concentration of potassium in the blood. Hypokalemia can result from a diet low in potassium, excessive vomiting or diarrhea, and from the use of diuretics, which increases urination and, therefore, the excretion of potassium. Common symptoms of a potassium deficiency include weakness, fatigue, muscle cramps, and abnormal heart rhythm. If left untreated, a long-term potassium deficiency can be life threatening. **See Figure 6-37.**

Sodium

Sodium is a major mineral that helps maintain the volume of fluid surrounding the cells of the body and regulate the acid-base balance of the body. Maintaining fluid balance is critical for proper cellular structure, transmission of nerve impulses, muscle contraction, and cardiovascular function. Although naturally found in most foods, the majority of sodium in the U.S. diet comes from salt added to processed foods. For example, one medium-size pear contains 2 mg of sodium, whereas some varieties of canned chicken noodle soup contain as much as 1400 mg of sodium due to the addition of salt.

The AI for sodium is set at 1500 mg per day. This number is based on the amount needed to replace the sodium a moderately active person loses through sweat and to provide adequate amounts of other essential nutrients. Although the UL for sodium is 2300 mg daily, most adults in the U.S. exceed this recommendation by taking in approximately 3300 mg of sodium per day.

Potassium			
Characteristics	Major mineral; electrolyte		
Functions	Helps regulate normal heart functioning and muscle contraction; necessary for energy production and protein synthesis; helps maintain fluid and acid-base balance		
Sources	• Russet potato (1 medium): 888 mg • Dried apricots (½ cup): 755 mg • Raisins (½ cup): 618 mg	• Acorn squash (1 cup, cubed): 486 mg • Beets (1 cup): 442 mg • Banana (1 medium): 422 mg	• Nonfat milk (1 cup): 407 mg • Honeydew melon (1 cup, diced): 388 mg • Lima beans (¼ cup): 182 mg
DRI	• Adult males: 4700 mg • Adult females: 4700 mg		
Toxicity	No UL		
Deficiency	Causes hypokalemia		

Figure 6-37. Potassium is a major mineral that helps regulate heart functions and muscle contractions and is necessary for energy production and protein synthesis.

A diet consistently in excess of sodium recommendations has been found to increase the risk of developing high blood pressure. High blood pressure has been linked to other chronic conditions such as heart disease, kidney disease, and stroke. Research suggests that consuming potassium-rich foods can help minimize the rise in blood pressure associated with excess sodium consumption.

In addition to consuming too much sodium, certain conditions such as dehydration can create a fluid imbalance that results in excess sodium in the blood. *Hypernatremia* is a condition characterized by abnormally high levels of sodium in the body usually due to excessive water loss. Symptoms of hypernatremia and dehydration are similar including dizziness, low blood pressure, difficulty breathing, and decreased urine production.

In most cases, there is little risk of sodium deficiency due to the prevalence of salt in the diet. However, conditions contributing to a loss of sodium from the body, such as severe vomiting or diarrhea, excessive long-term sweating, or some forms of kidney disease, can result in hyponatremia. Hyponatremia is a sodium deficiency characterized by muscle cramps, nausea, dizziness, seizures, and coma. **See Figure 6-38.**

Wellness Concept

Even though a food lacks a salty taste, it may still be high in sodium. For example, commercially prepared white bread does not taste salty compared to potato chips, yet 2 slices of white bread typically contain 300 mg of sodium compared to 15 potato chips that contain 170 mg of sodium.

Chloride

Chloride is a major mineral that helps maintain the acid-base balance of the body as well as proper blood volume and blood pressure. As an electrolyte, chloride assists sodium in maintaining the balance of fluids surrounding the cells of the body. Chloride is found in the hydrochloric acid of the stomach. As part of hydrochloric acid, chloride supports the acidic environment of the stomach, which is necessary for digestion.

The most common dietary source of chloride is salt, also known as sodium chloride. Salt is comprised of approximately 40% sodium and 60% chloride. **See Figure 6-39.** In addition to being attached to the sodium in salt, chloride also accompanies potassium in many foods. Therefore, foods that contain sodium or potassium will most often include chloride. Foods that contain chloride include olives, capers, celery, lettuce, tomatoes, sea vegetables such as kelp, cheese, and preserved meats such as ham and bacon. Processed foods that contain salt also include chloride.

Sodium			
Characteristics	Major mineral; electrolyte		
Functions	Helps maintain fluid and acid-base balance		
Sources	• Iodized salt (1 tsp): 2300 mg • Corn chips (7 oz bag): 1220 mg • Soy sauce (1 tbsp): 902 mg	• Canned tuna (5 oz): 473 mg • Bacon (2 oz): 466 mg • Parmesan cheese (1 oz): 449 mg	• Cottage cheese (½ cup): 409.5 mg • Celery (1 cup, chopped): 80.8 mg • Halibut (5 oz): 75.5 mg
DRI	• Adult males: 1500 mg • Adult females: 1500 mg		
Toxicity	UL: 2300 mg Causes hypertension		
Deficiency	Uncommon		

Figure 6-38. Sodium is a major mineral that helps maintain the fluid and acid-base balance of the body.

Figure 6-39. Salt is comprised of approximately 40% sodium and 60% chloride.

The AI for chloride stands at 2.3 g per day for adult males and females. Chloride needs typically decrease with age because the body tends to be more sensitive to spikes in blood pressure from salt intake. In general, a diet that contains ample levels of sodium will also provide adequate levels of chloride.

Chloride has a UL of 3.6 g per day. Since excess chloride is typically excreted by the body, chloride toxicity is rare. However, high salt intake also results in higher levels of chloride in the body. This may cause the body to exhibit symptoms similar to those associated with excess sodium intake including high blood pressure and fluid imbalance.

Because salt is prevalent in a variety of foods, chloride deficiency is unlikely. However, the loss of body fluids through excessive sweating, vomiting, diarrhea, or overuse of diuretics can result in a deficiency of chloride. When the body is deficient of chloride, there may be a loss of appetite, weakness, fatigue, and dehydration.

Chloride is found in various types of salts.

Knowledge Check 6-7

1. Define major mineral.
2. Identify six major minerals.
3. Describe the functions of each of the six major minerals.
4. Explain how the body maintains calcium levels.
5. Identify the effects excess phosphorus can have on the body.
6. Explain why some individuals lack adequate body stores of magnesium.
7. Compare hyperkalemia and hypokalemia.
8. Describe hypernatremia.
9. List dietary sources of chloride.

TRACE MINERALS

A *trace mineral* is a mineral with a recommended dietary allowance of less than 100 mg per day. Trace minerals are just as essential as major minerals even though the amount required is less. Trace minerals assist in chemical reactions, act as coenzymes, and work in conjunction with hormones in the body. Trace minerals are also needed to maintain proper immune functions and for cellular functions such as growth and repair. Essential trace minerals include iron, zinc, iodine, selenium, copper, manganese, fluoride, and chromium.

Iron

Iron is a trace mineral that is required to make the proteins that distribute oxygen throughout the body. Iron is necessary to make two oxygen-carrying proteins: hemoglobin and myoglobin. *Hemoglobin* is a protein in red blood cells that transports oxygen throughout the body. *Myoglobin* is an iron-containing protein that binds oxygen in muscle tissues. Myoglobin is essential to the functioning of skeletal muscles as well as the heart.

Foods containing iron are found in two main forms: heme iron and nonheme iron. *Heme iron* is a type of dietary iron found only in animal-based foods and is easily absorbed by the body. *Nonheme iron* is a type of dietary iron found primarily in plant-based foods and is not easily absorbed by the body. A very small amount of nonheme iron can also be found in animal-based foods.

Research shows that consuming foods containing nonheme iron with foods containing vitamin C helps increase nonheme iron absorption. **See Figure 6-40.** For example, nonheme iron is found in black beans and vitamin C is found in tomatoes. Making a salsa of black beans (nonheme iron) and tomatoes (vitamin C) increases iron absorption.

An RDA of 8 mg per day of iron applies to males 19 years and older and to females 51 years and older. For females between the ages of 19 and 50, the RDA for iron is 18 mg per day. Iron needs for females are much higher than males in this age group due to reproductive cycles.

A UL has been set for iron at 45 mg per day and includes males and females 14 years and older. Iron toxicity can occur quickly from ingesting a high concentration of iron in supplemental form. If toxicity occurs, the body will naturally rid itself of the excess iron through vomiting and diarrhea. Iron toxicity can also occur in individuals with certain genetic disorders that cause the body to absorb excess iron. **See Figure 6-41.**

Figure 6-40. Consuming foods containing nonheme iron with foods containing vitamin C helps increase nonheme iron absorption.

Iron			
Alternative terms	Heme iron; nonheme iron		
Characteristics	Trace mineral; iron needs are higher for females due to reproductive cycles; heme iron is found only in animal-based foods; nonheme iron is primarily in plant-based foods; heme iron is better absorbed by the body		
Functions	Required to make oxygen-carrying proteins hemoglobin and myoglobin		
Sources	• Clams (5 oz): 19.5 mg • Oysters (5 oz): 9.5 mg • Firm tofu (1 cup): 6.8 mg	• Pumpkin seeds (¼ cup): 5.18 mg • Unsweetened chocolate (1 oz): 4.9 mg • Lean ground beef (4 oz): 2.8 mg	• Lamb chops (5 oz): 2.5 mg • Raisins (½ cup): 1.5 mg • Pistachios (¼ cup): 1.3 mg
DRI	• Adult males: 8 mg • Adult females to age 50: 18 mg • Adult females 51+: 8 mg		
Toxicity	UL: 45 mg		
Deficiency	Causes anemia		

Figure 6-41. Iron is a trace mineral that is required to make the proteins that distribute oxygen throughout the body.

A deficiency of iron is one of the most common nutrient deficiencies. Adolescent females and women of child-bearing age are at a high risk of deficiency due to a low intake of iron in the diet coupled with the loss of iron in the blood during menstruation. Symptoms of an iron deficiency include depression, lethargy, and irritability. If iron stores become severely depleted, iron deficiency anemia can develop resulting in a pale appearance and impaired cognitive and muscle function.

Zinc

Zinc is a trace mineral that is considered a primary protector of the immune system, plays an essential role in regulating genetic material, and assists in wound healing, protein synthesis, energy metabolism, and insulin storage. Hearing, sight, taste, and smell are also dependent on zinc for proper functioning. As an antioxidant, zinc is contained in every cell of the body and works as a coenzyme to facilitate over 200 enzyme reactions. Key sources of zinc include oysters, calf liver, red meat, poultry, dairy products, legumes, nuts, seeds, and whole grains.

The RDA for adult males is 11 mg per day of zinc and 8 mg per day for adult females. Because the highest concentrations of zinc are found in animal-based foods, individuals following a plant-based diet need to be more conscious of meeting the zinc RDA.

A UL of 40 mg per day is set for zinc. Excess zinc is excreted daily from the body. However, zinc can accumulate to unhealthy levels if supplements containing a high concentration of zinc are taken over an extended time period. Excess zinc can result in nausea, vomiting, lower immunity, and an impaired ability to absorb copper.

Although zinc deficiencies are rare, vegetarians, pregnant women, and the elderly may be more susceptible to developing a zinc deficiency. Symptoms of a zinc deficiency include reduced immune functions, a decreased sense of taste and smell, and poor growth in children. Hair loss, white spots on fingernails, impaired wound healing, and infertility in males may also signify a zinc deficiency. **See Figure 6-42.**

Zinc			
Characteristics	Trace mineral; antioxidant; contained in all cells; coenzyme that facilitates over 200 reactions		
Functions	Helps protect immune system; role in regulating genetic material; promotes proper functioning of senses; assists in wound healing, protein synthesis, energy metabolism, and insulin storage		
Sources	• Oysters (5 oz): 127 mg • Calf liver (5 oz): 17 mg • Strip steak (5 oz): 5 mg	• Sesame seeds (¼ cup): 2.8 mg • Lentils (¼ cup): 2.3 mg • Plain low-fat yogurt (1 cup): 2.2 mg	• Cashews (¼ cup): 1.9 mg • Turkey breast (5 oz) 1.5 mg • Oats (¼ cup): 1.5 mg
DRI	• Adult males: 11 mg • Adult females: 8 mg		
Toxicity	UL: 40 mg		
Deficiency	Uncommon		

Figure 6-42. Zinc is a trace mineral that protects the immune system, helps regulate genetic material, and assists in wound healing, protein synthesis, metabolism, and insulin storage.

Iodine

Iodine is a trace mineral that is required to make hormones in the thyroid gland. The *thyroid gland* is a gland found in the neck that secretes hormones necessary for regulating metabolism. Thyroid hormones assist in the regulation of body temperature, the production of red blood cells, and the proper functioning of nerves and muscles. Since iodine is plentiful in seawater, seafood is a significant source of iodine. Plants grown in iodine-rich soil, which is often near the ocean, produce fruits and vegetables that also contain iodine. Iodized salt contains iodine as well.

The RDA for iodine has been established at 150 mcg for adult males and adult females. In general, iodine levels in the average diet are above the recommended amounts but do not exceed the iodine UL of 1100 mcg.

Because iodine is needed to produce hormones in the thyroid gland, an excess of iodine can lead to a condition known as goiter. A *goiter* is an enlarged thyroid gland that appears as a protruding growth in the neck. High intakes of iodine can also cause minor swelling of the thyroid gland and thyroid cancer. A diet lacking iodine causes an insufficient production of thyroid hormones. Like iodine excess, a diet deficient of iodine can also lead to goiter or an enlarged thyroid gland. **See Figure 6-43.**

Food Science Note

Processed foods may contain a large amount of sodium but not iodine. Iodized salt is not used in the production of processed foods because it can react unfavorably when mixed with minerals found in the local water supply.

Iodine			
Characteristics	Trace mineral; plentiful in seawater		
Functions	Required to make hormones in the thyroid gland		
Sources	• Iodized salt (1 tsp): 400 mcg • Haddock (5 oz): 241 mcg • Plain low-fat yogurt (1 cup): 87 mcg	• 2% milk (1 cup): 39 mcg • Shrimp (5 oz): 35 mcg • White bread (1 slice): 35 mcg	• Egg (1 large): 26 mcg • Cottage cheese (½ cup): 26 mcg • Cheddar cheese (1 oz): 23 mcg
DRI	• Adult males: 150 mcg • Adult females: 150 mcg		
Toxicity	UL: 1100 mcg Causes thyroid gland to develop goiter		
Deficiency	Causes thyroid gland to develop goiter		

Figure 6-43. Iodine is a trace mineral that is required to make hormones in the thyroid gland.

Selenium

Selenium is a trace mineral that helps maintain the function of thyroid hormones and a healthy immune system. Selenium is also incorporated into proteins to make antioxidant enzymes that protect cells from damage during energy metabolism. Selenium is found in a variety of foods including meats, poultry, seafood, eggs, nuts, seeds, and whole grains. The selenium content of a food is strongly influenced by the soil and geographical area where it is grown. For example, selenium deficiency has been reported in some parts of Asia where the soil is low in selenium. In contrast, Brazil nuts that are grown in selenium-rich soil may contain far more selenium than the RDA guidelines suggest and should only be eaten occasionally. **See Figure 6-44.**

Selenium Sources

Figure 6-44. Brazil nuts may contain far more selenium than the RDA guidelines suggest and should only be eaten occasionally.

The RDA for selenium for both adult males and females is 55 mcg. In general, most diets that are rich in a variety of foods provide an adequate amount of selenium. A UL for selenium is set at 400 mcg to prevent a condition called selenosis. *Selenosis* is a toxicity of selenium with symptoms that include brittle hair and nails, nerve damage, and breath with a garlic odor.

In North America, a selenium deficiency is rare due to a food delivery system that transports foods from all areas of the country. This enables individuals living in an area with low-selenium soil to be exposed to selenium-rich foods. Selenium deficiency may be a contributing factor to thyroid disorders, cardiovascular disease, and weakened immune function. Selenium deficiency is also linked to a disease known as Keshan disease. *Keshan disease* is a disease associated with selenium deficiency that results in an enlarged heart that works inefficiently. **See Figure 6-45.**

Copper

Copper is a trace mineral that works with iron to form red blood cells. Copper is also an antioxidant involved in a variety of enzymatic reactions that keep tissues, nerves, and bones healthy. Organ meats, such as calf liver and kidney, as well as shellfish are considered good sources of copper. Nuts, legumes, whole grains, green leafy vegetables, and crimini mushrooms are additional dietary sources of copper.

The RDA for copper is 900 mcg per day and applies to both adult males and adult females. Research suggests that most individuals consume between 2000 mcg and 5000 mcg of copper daily.

For adults, the UL for copper is 10,000 mcg daily. Copper toxicity can cause gastrointestinal distress, vomiting, diarrhea, and liver damage. The toxic effects of copper are most often associated with Wilson's disease. *Wilson's disease* is a genetic disorder characterized by the accumulation of copper in various organs. Treatment for Wilson's disease involves the avoidance of foods and supplements rich in copper.

Copper deficiencies are rare since it is found in a large variety of foods, and a minimal amount is required by the body. However, when copper is lacking in the diet, iron cannot be absorbed properly, which can lead to iron deficiency anemia. Other symptoms that may indicate a copper deficiency include osteoporosis, poor immune functions, the loss of skin pigment, and neurological impairment. **See Figure 6-46.**

Selenium			
Characteristics	Trace mineral; antioxidant; incorporated into proteins to make antioxidant enzymes		
Functions	Helps maintain thyroid hormone functions; assists in keeping immune system healthy		
Sources	• Brazil nuts (¼ cup): 637 mcg • Cod (5 oz): 46.5 mcg • Clams (5 oz): 34 mcg	• Turkey breast (5 oz): 34 mcg • Crimini mushrooms (1 cup, sliced): 18.7 mcg • Egg (1 large): 15.8 mcg	• Brown rice (¼ cup): 10.8 mcg • Veal breast (5 oz): 9 mcg • Sunflower seeds (¼ cup): 6.1 mcg
DRI	• Adult males: 55 mcg • Adult females: 55 mcg		
Toxicity	UL: 400 mcg Causes selenosis		
Deficiency	Causes Keshan disease		

Figure 6-45. Selenium is a trace mineral that helps maintain the function of thyroid hormones and a healthy immune system.

Copper			
Characteristics	Trace mineral; antioxidant		
Functions	Helps form red blood cells; involved in enzymatic reactions that keep tissues, nerves, and bones healthy		
Sources	• Calf liver (5 oz): 16,500 mcg • Oysters (5 oz): 6000 mcg • Cashews (¼ cup): 800 mcg	• Kidney beans (¼ cup): 450 mcg • Crimini mushrooms (1 cup, sliced): 400 mcg • Prunes (½ cup): 250 mcg	• Quinoa (¼ cup): 250 mcg • Kale (1 cup, chopped): 200 mcg • Sweet potato (1 medium): 200 mcg
DRI	• Adult males: 900 mcg • Adult females: 900 mcg		
Toxicity	UL: 10,000 mcg		
Deficiency	Uncommon		

Figure 6-46. Copper is a trace mineral that works with iron to form red blood cells.

Manganese

Manganese is a trace mineral that facilitates protein and carbohydrate metabolism and bone formation. Evidence suggests that manganese also helps maintain the health of nerves, promotes the synthesis of fatty acids and cholesterol, and may also work as an antioxidant to protect tissues from free radical damage. When taken in conjunction with calcium, zinc, and copper, manganese has been used in the treatment of osteoporosis. Manganese is commonly found in a variety of fruits, whole grains, nuts, seeds, and green leafy vegetables. Tea and spices, such as cinnamon, also supply manganese to the diet.

The AI for manganese is 2.3 mg per day for adult males and 1.8 mg per day for adult females. A manganese UL of 11 mg has been established for both males and females. Manganese toxicity is rare because it is readily excreted by the body on a daily basis. Because the liver plays a significant role in eliminating excess manganese from the body, manganese toxicity is most likely to occur in individuals with liver disease. A toxicity of manganese can result in fatigue, irritability, and mental confusion. **See Figure 6-47.**

Manganese			
Characteristics	Trace mineral; antioxidant		
Functions	Works as coenzyme to facilitate protein and carbohydrate metabolism and bone formation; helps maintain nerve functions; promotes synthesis of fatty acids and cholesterol		
Sources	• Pineapple (1 cup, cubed): 2.6 mg • Amaranth (¼ cup): 1.6 mg • Pecans (¼ cup): 1.2 mg	• Chickpeas (¼ cup): 1.1 mg • Blackberries (1 cup): 0.9 mg • Flaxseeds (2 tbsp): 0.6 mg	• Black tea (1 cup): 0.5 mg • Ground cinnamon (1 tsp): 0.4 mg • Spinach (1 cup): 0.3 mg
DRI	• Adult males: 2.3 mg • Adult females: 1.8 mg		
Toxicity	UL: 11 mg		
Deficiency	Uncommon		

Figure 6-47. Manganese is a trace mineral that facilitates protein and carbohydrate metabolism as well as bone formation.

Because manganese is involved in a variety of enzymatic functions, a manganese deficiency can impact the body in a variety of ways. Symptoms associated with a manganese deficiency include fatigue, high blood glucose levels, bone loss, and nervous system abnormalities.

Fluoride

Fluoride is a trace mineral that provides protection against tooth decay and aids the formation of healthy teeth and bones. The main source of dietary fluoride in the United States is fluoridated (fluoride-added) tap water. Therefore, using fluoridated water to prepare or cook foods provides fluoride to the diet. **See Figure 6-48.** Fluoride also occurs naturally in foods such as meats, seafood, and tea.

The AI for fluoride is 4 mg per day for males age 19 through adulthood and 3 mg per day for females age 14 through adulthood. A fluoride UL is set at 10 mg per day for adults. When consumed in excessive amounts, fluoride can reach toxic levels. Symptoms of severe fluoride toxicity usually involve nausea, vomiting, abdominal pain, diarrhea, and sweating. An excess of fluoride may also cause fluorosis. *Fluorosis* is a condition caused by an excess of fluoride that results in white or brown marks on tooth enamel and damage to bones.

Fluoridated Water

Figure 6-48. Using fluoridated water to prepare or cook foods provides fluoride to the diet.

Because fluoride is primarily responsible for healthy teeth and bones, a fluoride deficiency increases the risk of tooth decay and weak bones. **See Figure 6-49.** Water and toothpaste with added fluoride have been shown to considerably reduce the incidence of tooth decay.

Fluoride			
Characteristics	Trace mineral; commonly added to water and toothpaste		
Functions	Protects teeth against tooth decay and helps with formation of healthy teeth and bones		
Sources	• Black tea (8 oz): 884 mcg • White wine (5 oz): 297 mcg • Tap water (8 oz): 169 mcg	• Beef hot dog (2 oz): 27.2 mcg • Canned tuna (5 oz): 26 mcg • Grapes (1 cup): 11.8 mcg	• Strawberries (1 cup, sliced): 7.3 mcg • Apple (1 medium): 6.0 mcg • Tomato (1 cup, sliced): 4.1 mcg
DRI	• Adult males: 4.0 mg (4000 mcg) • Adult females: 3.0 mg (3000 mcg)		
Toxicity	UL: 10 mg Causes fluorosis		
Deficiency	Uncommon		

Figure 6-49. Fluoride is a trace mineral that helps protect against tooth decay and aids the formation of healthy teeth and bones.

Rishi Tea

Fluoride occurs naturally in tea.

Chromium

Chromium is a trace mineral that is needed to regulate blood glucose levels. For example, chromium helps transfer glucose from the bloodstream into the cells of the body where it can be used for energy. Because of its role in regulating blood glucose levels, chromium has been used successfully to help treat type 2 diabetes. Chromium is found in a variety of animal- and plant-based foods. However, the chromium content in foods is influenced by agricultural and processing practices. Some of the better sources of chromium include sweet potatoes, onions, and broccoli.

The AI for chromium is 35 mcg per day for males and 25 mcg per day for females. Research suggests that most individuals consume an adequate amount of chromium in their daily diet. A UL has not been established for chromium. Chromium toxicity is extremely rare because less than 10% of ingested chromium is absorbed by the body.

Because chromium enhances the ability of insulin to help regulate blood glucose levels, a chromium deficiency can increase the risk of developing type 2 diabetes. Symptoms of a chromium deficiency may also include fatigue, weight loss, and hypoglycemia. *Hypoglycemia* is a condition that results from low blood glucose levels. Symptoms of hypoglycemia may include dizziness, weakness, sweating, shakiness, anxiety, and difficulty with speech. **See Figure 6-50.**

Chromium			
Characteristics	Trace mineral; less than 10% of ingested chromium is absorbed		
Functions	Helps regulate blood glucose levels		
Sources	• Sweet potato (1 medium): 35 mcg • Onions (1 cup, chopped): 24.8 mcg • Broccoli (1 cup): 22.0 mcg	• Turkey leg (5 oz): 17.3 mcg • Romaine lettuce (1 cup, shredded): 7.85 mcg • Grape juice (1 cup): 7.5 mcg	• Ham (5 oz): 6 mcg • English muffin (1 each): 3.6 mcg • Orange juice (1 cup): 2.2 mcg
DRI	• Adult males: 35 mcg • Adult females: 25 mcg		
Toxicity	No UL		
Deficiency	Uncommon		

Figure 6-50. Chromium is a trace mineral that is needed to regulate blood glucose levels.

Knowledge Check 6-8

1. Define trace mineral.
2. Identify eight trace minerals.
3. Describe the functions of each of the eight trace minerals.
4. Identify common dietary sources of each of the eight trace minerals.
5. Name the trace mineral in which adolescent females and women of child-bearing age are commonly deficient.
6. Summarize the role of the thyroid gland.

PREPARING VITAMINS AND MINERALS

During the preparation of foods, vitamins and minerals react to air, light, heat, moisture, and changes in pH. These reactions affect how well each nutrient is absorbed in the body. For example, heat causes many different reactions when applied to nutrients. Some nutrients are destroyed by heat, some become more easily absorbed, and some are unaffected. Knowing how nutrients interact with air, light, heat, moisture, and pH will help provide a better understanding of how to preserve nutrient content while preparing foods.

Reactions to Air

Water-soluble vitamin C and fat-soluble vitamins A and E react to air. Vitamin C is highly unstable and therefore deteriorates when exposed to air. Vitamins A and E serve as antioxidants in the body but can be destroyed if subjected to air for too long. Minerals are generally not lost in the presence of air. To prevent nutrient loss from exposure to air, foods should be cut for preparation as close to use as possible and stored in airtight containers or wrapped tightly with plastic wrap. **See Figure 6-51.**

Protecting Nutrients from Air

Sullivan University

Figure 6-51. To prevent nutrient loss from exposure to air, foods should be cut for preparation as close to use as possible and stored in airtight containers or wrapped tightly with plastic wrap.

Reactions to Light

Exposure to light can cause nutrient loss in water-soluble vitamins such as vitamin C, riboflavin, vitamin B_6, and vitamin B_{12}. Light can also cause the oxidation of vitamin E, which depletes vitamin E of its antioxidant health benefits. In addition, a slight loss of vitamins A and K occurs when these vitamins are exposed to light. For most minerals, there is a low risk of nutrient loss from light exposure. Foods sensitive to light are generally packaged in light-blocking containers to preserve nutrients.

Reactions to Heat

Heat plays a strong role in reducing the potency of vitamin C, thiamin, pantothenic acid, and vitamin B_6. Folate also loses strength but only when subjected to extreme heat. The remaining B vitamins can tolerate heat exposure for short periods of time. Most minerals hold up well to heat and exhibit only minimal loss. Although heat does not adversely affect some vitamins and minerals, the overall quality of food may decline when exposed to heat for too long.

Chef's Tip

Covering the pot when steaming foods helps preserve nutrients. When the pot is covered, steam circulates more consistently and cooks the food faster. Therefore, more nutrients are retained.

Reactions to Moisture

Although water is an essential component of many cooking processes, it can negatively impact vitamins and minerals. For example, large amounts of water-soluble vitamins and most minerals can be lost in a hot cooking liquid. To prevent these losses, foods can be served with the cooking liquid or steamed instead of boiled to retain more vitamin and mineral content. **See Figure 6-52.** In general, fat-soluble vitamins are not affected by moisture.

Steaming Foods

Vulcan-Hart, a division of the ITW Food Equipment Group LLC

Figure 6-52. Steaming foods promotes nutrient retention.

Reactions to pH

The pH of cooking water has little effect on vitamins and minerals. It is more likely for vitamins and minerals to be lost as a result of exposure to heat or to food additives that are high in acids or alkalis. Vitamin B_{12}, biotin, and vitamin K are at greatest risk of being destroyed when in contact with a strong acid such as lemon juice or vinegar. Vitamins affected by highly alkali solutions, such as baking soda, include thiamin, riboflavin, pantothenic acid, vitamin B_6, and vitamin K.

Knowledge Check 6-9

1. Identify the vitamins that react to air.
2. Explain how light can affect vitamins and minerals.
3. List the vitamins that lose strength due to heat.
4. Explain how to prevent vitamin and mineral loss due to cooking liquids.
5. Identify vitamins destroyed by acids.
6. Identify vitamins destroyed by alkalis.

VITAMINS AND MINERALS ON THE MENU

A well-rounded menu includes an abundance of vitamins and minerals. To ensure that a variety of these essential nutrients is available to guests, lean animal proteins, vegetables, fruits, legumes, whole grains, nuts, and seeds should be represented in dishes ranging from appetizers to desserts.

There is a variety of foods rich in different vitamins and minerals, and they can be combined in ways that are beneficial to health and wellness. For example, incorporating potassium-rich potatoes into a higher sodium soup encourages the kidneys to rid the body of excess sodium. By learning the sources of foods most abundant in different vitamins and minerals, chefs can be confident about putting the most nutritious choices on the menu.

Presentation

Plate presentations that include foods with an array of colors tend to provide a variety of vitamins and minerals. **See Figure 6-53.** For example, a salad that contains leafy greens, tomatoes, carrots, eggs, cucumbers, radishes, and cheddar cheese is more nutrient dense than a salad composed of leafy greens only. Retaining color through the proper cooking process also ensures the highest vitamin and mineral content of foods. Cooked vegetables that are still vibrant in color are not only attractive on the plate but are also more abundant in nutrients than overcooked vegetables with muted colors.

Vitamin and Mineral Presentation

National Turkey Federation

Figure 6-53. Plate presentations that include foods with an array of colors tend to provide a variety of vitamins and minerals.

Texture and Flavor

Using a wide variety of foods will increase the nutrient-density of menu items and showcase different textures and flavors. It is also important to vary the types of foods provided within the same food group. For example, a mixture of nuts, such as almonds, pecans, and walnuts, provides more vitamins and minerals than almonds only. A chef can also incorporate several grains such as quinoa, millet, and rice into an appealing dish. To add even more nutrients and flavor as well as add contrasting textures, the rice dish can be topped with toasted pine nuts, dried apricots, and chopped parsley. Using strategies that incorporate an array of textures and flavors from all food groups helps ensure that a variety of vitamins and minerals will adorn every plate on the menu.

Knowledge Check 6-10

1. Identify foods that help ensure that an abundance of vitamins and minerals are represented on the menu.
2. Explain how color can help indicate the nutrient content of a dish.
3. Summarize how texture and flavor relate to the nutrient content of a dish.

Chapter Summary

U.S. Apple® Association

Although water, vitamins, and minerals are non-energy-yielding nutrients, they are indispensable to overall health and well-being. Water is the nutrient required most by the body for survival. It is found in every cell and is necessary for digesting and absorbing nutrients; regulating body temperature; keeping tissues, organs, and joints healthy; and maintaining fluid balance. Water in its pure form is the best source of hydration for the body, but beverages and foods also supply the body with water.

Vitamins are classified as water-soluble vitamins or fat-soluble vitamins, and minerals are classified as major minerals or trace minerals. Vitamins and minerals play a significant role in regulating body processes and are found in a variety of foods from all food groups. Dietary reference intakes have been established for water, vitamins, and minerals to help ensure that the body has an adequate supply of nutrients and to prevent nutrient toxicities and deficiencies.

Some vitamins and minerals react adversely to air, light, heat, moisture, and pH. Understanding vitamin and mineral reactions gives chefs the opportunity to prepare dishes that retain nutrients while providing guests with visually appealing dishes full of flavor.

Chapter Review

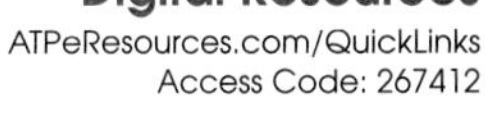
Digital Resources
ATPeResources.com/QuickLinks
Access Code: 267412

1. Describe the four main roles of water in the body.
2. Identify common sources of water.
3. Describe the effects of water toxicity and deficiency.
4. Summarize the way the body reacts when water output is high or water intake is low.
5. Describe antioxidants.
6. List the functions and sources of the nine water-soluble vitamins.
7. Describe the toxicities and deficiencies of the nine water-soluble vitamins.
8. List the functions and sources of the four fat-soluble vitamins.
9. Describe the toxicities and deficiencies of the four fat-soluble vitamins.
10. List the functions and sources of the six major minerals.
11. Describe the toxicities and deficiencies of the six major minerals.
12. List the functions and sources of the eight trace minerals.
13. Describe the toxicities and deficiencies of the eight trace minerals.
14. Identify the vitamins and minerals that react to air, light, and heat.
15. Describe the effect a hot cooking liquid has on vitamins and most minerals.
16. Describe how to incorporate vitamins and minerals into the menu while positively impacting presentation, texture, and flavor.

Nutritious Menu Planning

Foodservice Venues

Menu Styles

Chapter 7

Special Promotions

American Metalcraft, Inc.

Menus come in all shapes, sizes, colors, and types. This vital communication tool helps guests select the foods and beverages each time they enter a foodservice venue. Providing guests with options is key as an increasing number of people are looking for nutritious but flavorful dishes. Nutritious menu planning involves evaluating the composition of a dish and making modifications that transform recipes into healthy, flavorful dishes.

Chapter Objectives

1. Explain the functions of a menu.
2. Describe foodservice venues.
3. Differentiate between menu classifications and menu types.
4. Summarize regulations menus must legally follow.
5. Describe trends to analyze when evaluating menus.
6. Identify dietary considerations to assess when modifying menus.
7. Summarize the culinary nutrition recipe-modification process.
8. Describe effective ways to market menu items.

MENUS

The *menu* is a document that markets a foodservice operation to guests. Every purchasing decision, from ingredients and equipment to uniforms, is also driven by the menu. The menu determines the skill requirements of the staff and reflects an ambience that is unique to each foodservice venue. The menu is central to the success of an operation and an essential tool used to inform guests of the food and beverages offered as well as the price of each item. **See Figure 7-1.**

Menus

American Metalcraft, Inc.

Figure 7-1. The menu is the primary document that markets a foodservice operation to guests.

Foodservice Venues

Foodservice venues include a wide range of settings where food and beverages are available to customers or guests. Foodservice venues include restaurants, hotels, banquet halls, gaming and recreation complexes, as well as healthcare facilities, schools, colleges, and corporations. Each venue can be classified as a quick service, fast casual, casual dining, fine dining, institutional, beverage house and bar, or special event operation.

Quick Service. Quick service venues offer food and beverages that can be prepared and served quickly such as sandwiches and salads. Guests typically select items from a menu that remains fairly constant and pay for the order prior to eating. Orders are commonly placed and received at a counter or drive-through window. Quick service venues often include restaurant chains as well as food carts and food trucks.

Fast Casual. Fast casual venues offer limited table service where guests often place orders at a counter and then sit at a table to wait for a runner to deliver their food. Fast casual venues also include buffets where guests serve themselves and are then attended to by service staff. **See Figure 7-2.**

Fast Casual Venues

Barilla America, Inc.

Figure 7-2. Fast casual venues include buffets where guests serve themselves.

Casual Dining. Casual dining venues offer moderately priced food in an informal atmosphere. Guests are seated at tables where orders are taken and delivered by service staff. Casual dining venues include foodservice operations such as bistros and brasseries. Oftentimes casual dining venues revolve around a theme, such as sports, that influences the food, music, uniforms, and overall ambience of the operation. Many casual dining venues also include a bar.

Fine Dining. Fine dining venues provide seated guests with individualized service, refined presentations, and a sophisticated environment. **See Figure 7-3.** Meals typically last for an extended period of time, and menu prices are set higher to reflect the level of service and quality of food offered.

Fine Dining Venues

Daniel NYC/E. Laignel

Figure 7-3. Fine dining venues provide seated guests with individualized service, refined presentations, and a sophisticated environment.

Institutional. Institutional venues generally serve large numbers of people in an educational, health care, or correctional facility. Cafeteria-style menus are often found in institutional settings, such as universities and hospitals. A wide variety of healthy foods are becoming common offerings on these types of menus. This is especially true in health care facilities where a variety of dietary needs must be considered.

Beverage Houses and Bars. Beverage houses and bars include venues where the menu focus is often on drinks that feature items such as coffees, teas, or alcoholic beverages. **See Figure 7-4.** Although the emphasis is on beverages, many beverage house and bar menus also offer foods, such as appetizers, sandwiches, or baked goods. Specialty food bars, such as sushi bars and raw bars featuring raw and steamed seafood, are also popular venues where a specific type of food is showcased.

Beverage Houses and Bars

Daniel NYC/E. Kheraj

Figure 7-4. Beverage houses and bars include venues where the menu focus is often on drinks that feature items such as coffees, teas, or alcoholic beverages.

Special Events. Menus that are specifically planned for a special occasion are often featured at special event venues. Special event menus are often used for weddings and business conferences and can range from hors d'oeuvre receptions to formal dinners. Special event menus give managers and chefs the opportunity to showcase nutrient-dense dishes that are full of flavor.

Menu Classifications

Menu classifications reflect how frequently the menu changes. Some menus remain fairly constant while others change regularly based on factors such as product availability, seasonality, and sustainability concerns, as well as media influences and consumer trends. Common menu classifications include cycle menus, fixed menus, and market menus.

Cycle Menus. A *cycle menu* is a menu written for a specific period and is repeated once that period ends. **See Figure 7-5.** Institutional foodservice operations, such as retirement facilities, often use cycle menus. Cycle menus rotate on a regular basis depending on the institution. For example, a cycle menu with predetermined items might repeat every 11 days. Advantages of cycle menus include ease in scheduling staff, ordering products, and inventory control.

Fixed Menus. A *fixed menu* is a menu that remains the same or rarely changes. Fixed menus are common in chain restaurants and entertainment venues such as cinemas and sports arenas. Some foodservice operations enhance a fixed menu by attaching a list of daily specials. The predictability of a fixed menu can be appealing to guests because they know what to expect.

Market Menus. A *market menu* is a menu that changes frequently to coincide with changes in product availability. A market menu is also called a "du jour" menu or menu of the day. Market menus often revolve around seasonal items in order to take advantage of the freshest ingredients available and to highlight the creativity of the chef.

Eleven-Day Cycle Menus

Monday	Tuesday	Wednesday	Thursday	Friday
Butternut Squash Soup Broiled Cod Sliced Roasted Pork Sautéed Greens Mashed Potatoes Assorted Desserts	Vegetable Soup Chicken Tetrazzini Italian Meatloaf Grilled Zucchini Pesto Pasta Assorted Desserts	Creamy Mushroom Soup Roast Turkey Baked Ham Cranberry-Walnut Salad Corn Assorted Desserts	Chicken Noodle Soup Grilled Chicken Blackened Catfish Quinoa-Stuffed Peppers Braised Cabbage Assorted Desserts	Cream of Asparagus Soup Tofu Stir-Fry Teriyaki Chicken Fried Rice Grilled Pineapple Assorted Desserts
Minestrone Soup Meatballs Baked Scrod Mostaccioli Grilled Eggplant Assorted Desserts	Chicken and Rice Soup Pork Chop Balsamic-Glazed Salmon Caramelized Pearl Onions Green Beans Assorted Desserts	Lentil Soup Roast Beef Veggie Lasagna Roasted Potatoes Broccoli Slaw Assorted Desserts	Beef Barley Soup Oven-Fried Chicken Spaghetti & Meatballs Baked Acorn Squash Peas Assorted Desserts	Corn Chowder BBQ Pork Stuffed Peppers Roasted Herbed Potatoes Crinkle-Cut Carrots Assorted Desserts
Navy Bean Soup Chicken à la King Pasta Primavera Steamed Brown Rice Brussels Sprouts Assorted Desserts	Butternut Squash Soup Broiled Cod Sliced Roasted Pork Sautéed Greens Mashed Potatoes Assorted Desserts	Vegetable Soup Chicken Tetrazzini Italian Meatloaf Grilled Zucchini Pesto Pasta Assorted Desserts	Creamy Mushroom Soup Roast Turkey Baked Ham Cranberry-Walnut Salad Corn Assorted Desserts	Chicken Noodle Soup Grilled Chicken Blackened Catfish Quinoa-Stuffed Peppers Braised Cabbage Assorted Desserts
Cream of Asparagus Soup Tofu Stir-Fry Teriyaki Chicken Fried Rice Grilled Pineapple Assorted Desserts	Minestrone Soup Meatballs Baked Scrod Mostaccioli Grilled Eggplant Assorted Desserts	Chicken and Rice Soup Pork Chop Balsamic-Glazed Salmon Caramelized Pearl Onions Green Beans Assorted Desserts	Lentil Soup Roast Beef Veggie Lasagna Roasted Potatoes Broccoli Slaw Assorted Desserts	Beef Barley Soup Oven-Fried Chicken Spaghetti & Meatballs Baked Acorn Squash Peas Assorted Desserts
Corn Chowder BBQ Pork Stuffed Peppers Roasted Herbed Potatoes Crinkle-Cut Carrots Assorted Desserts	Navy Bean Soup Chicken à la King Pasta Primavera Steamed Brown Rice Brussels Sprouts Assorted Desserts	Butternut Squash Soup Broiled Cod Sliced Roasted Pork Sautéed Greens Mashed Potatoes Assorted Desserts	Vegetable Soup Chicken Tetrazzini Italian Meatloaf Grilled Zucchini Pesto Pasta Assorted Desserts	Creamy Mushroom Soup Roast Turkey Baked Ham Cranberry-Walnut Salad Corn Assorted Desserts

Figure 7-5. A cycle menu is a menu written for a specific period and is repeated once that period ends.

Menu Types

Planning menus involves choosing the menu type that best represents the food, service, and ambience of the operation. Menu types are influenced by the menu classification and can range from multiple courses offered for a set price to separate charges for individual items. A menu can be presented in various ways including on paper, menu boards, or verbally. Menus are also displayed electronically on restaurant websites, social media sites, and smartphone apps. Common menu types include à la carte, prix fixe, table d' hôte, and California.

À la Carte. An *à la carte menu* is a menu that offers separately priced food and beverages. For example, on an à la carte menu, a salad would be sold separately from an entrée. À la carte menus are used in a variety of venues ranging from casual to fine dining.

Prix Fixe. A *prix fixe menu* is a menu that offers limited choices within a collection of specific items for a multicourse meal at a set price. **See Figure 7-6.** A prix fixe menu typically includes an appetizer, soup, salad, entrée, and dessert. There are generally a few options within each category, so the guest is able to choose an item from each category. Prix fixe menus are often used in fine dining venues.

Prix Fixe Menus

Today's Menu

($65 per person)

First Course

Succulent shrimp and shaved carrot salad with honey coriander vinaigrette

– or –

Fresh and silky cream of butternut squash soup

Second Course

Roasted chicken with ricotta dumplings and a pearl onion and fresh pea ragout

– or –

Roasted strip loin of beef with a red wine sauce and a warm tomato and fennel purée

Third Course

Fresh fruit and cheese plate

– or –

Chocolate ganache cake

Figure 7-6. A prix fixe menu offers guests limited choices at a set price.

Table d'Hôte. A *table d'hôte menu* is a menu that identifies specific items that will be served for each course at a set price. Each dish is determined by the chef. Table d' hôte menus are commonly used at banquets and other special events.

California. A *California menu* is a menu that offers all food and beverage items for breakfast, lunch, and dinner throughout the entire day. California menus give guests the option to order breakfast items for dinner. Due to a large variety of choices, California menus are usually presented in a book format. Many restaurants that use California menus are open 24 hours.

National Honey Board

California menus give guests the option to order breakfast items for dinner.

Menu Regulations

When planning menus, culinary professionals must consider menu regulations and laws set forth by the federal government. Menu regulations are designed to protect guests from misleading or fraudulent food and beverage claims. Inaccurate menu labeling can jeopardize the health of guests, especially those with food allergies. Failure to comply with truth-in-menu guidelines, menu labeling laws, and nutrient and health claims can result in legal actions.

Truth-in-Menu Guidelines. Truth-in-menu guidelines were established by the federal government to ensure the accuracy of menu statements. **See Figure 7-7.** The guidelines are designed to protect consumers from misleading food and beverage claims such as falsified statements regarding product origins or the ingredients used to create a dish. For example, a menu description that lists an organic item when that item has been conventionally farmed would be in violation of truth-in-menu guidelines.

Truth-in-Menu Guidelines

Menu Label	Example of Misrepresentation
Portion size of item	Advertising 12 oz steak and serving 10 oz steak
Quality or grade of item	Listing USDA Prime beef and serving USDA Select beef
Preservation method	Advertising fish as fresh but serving fish that was previously frozen
Preparation method	Claiming food is house-made when it is prepackaged
Type of product served	Stating the use of extra-virgin olive oil when vegetable oil was used
Certified foods	Claiming food is organic that has not been certified as such
Point of origin	Stating "Florida" oranges when oranges came from different location
Nutrition information	Listing product as low-fat when it does not meet required criteria
Product brand	Serving different brand than listed

Figure 7-7. Truth-in-menu guidelines were established by the federal government to ensure the accuracy of menu statements.

In addition to written menu descriptions, the truth-in-menu guidelines govern menu photographs, illustrations, advertisements, and verbal descriptions. If there is a difference between what is shown, written, or stated and what is presented to the consumer, the operation can be held liable for fraud.

Menu Labeling Laws. Requirements for the nutrition labeling of standard menu items was established in the Patient Protection and Affordable Care Act of 2010. **See Figure 7-8.** This law requires foodservice operations with 20 or more locations to list the calories for standard menu items on restaurant menus and menu boards. Menu labeling laws were designed to provide a consistent means of disclosing nutritional information. However, the law recognizes that serving size and formula variations occur during food preparation and, therefore, require foodservice operations to show a "reasonable basis" for the nutrition data presented. By providing consumers with well-defined nutritional information, menu labeling laws can encourage consumers to make informed, healthy choices.

Nutrient and Health Claims. A *nutrient claim* is a statement that uses approved terminology to describe the nutrient content of a food. A *health claim* is a statement declaring a relationship between certain foods or nutrients and disease. **See Figure 7-9.** Nutrient and health claims that appear on menus and food labels must follow requirements and definitions established by the Food and Drug Administration (FDA). For example, if a nutrient claim on a menu refers to an item as "low sodium," then the item must contain no more than 140 mg of sodium per serving. If this is the case, the following health claim could be made: "Diets low in sodium may reduce the risk of high blood pressure."

Menu Labeling Laws

Types of Establishments

Restaurants with 20 or more locations nationally must add calorie counts to menus, menu boards, and drive-through menu boards for standard menu items; additional nutrition data must be available to customers on request. These types of establishments include the following:

- Restaurants that are part of chain that operates 20 or more locations under same trade name, regardless of ownership, that offer "substantially the same" menu items for sale
- "Retail food establishments" that are similar to foodservice operations such as convenience stores and mobile carts
- Vending machines owned by companies that operate 20 or more machines

Nutrition Labeling Requirements

- Calorie count for standard serving must appear next to menu item names
- Prominent clear and conspicuous statement about availability of additional nutrition information
- Succinct statement concerning suggested daily caloric intake posted prominently on menu to help guests understand calories in context of daily diet

Information Available on Request

- Calories
- Calories from fat
- Total fat
- Saturated fat
- Cholesterol
- Sodium
- Carbohydrates
- Sugars
- Dietary fiber
- Protein

Exempt Menu Items

- Temporary items appearing on menu for fewer than 60 days per calendar year
- Condiments and items placed on table or counter for general use that are not listed on menu
- Daily specials
- Custom orders
- Customary test market items appearing on menu for fewer than 90 days

Barilla America, Inc.

Figure 7-8. The Patient Protection and Affordable Care Act of 2010 established requirements for nutrition labeling of standard menu items.

Health Claims

Nutrient and Health	Claim
Calcium, vitamin D, and osteoporosis	Adequate calcium and vitamin D, as part of well-balanced diet, along with physical activity may reduce risk of osteoporosis
Dietary fat and cancer	Diets low in total fat may reduce risk of some cancers
Sodium and hypertension	Diets low in sodium may reduce risk of high blood pressure
Saturated fats, cholesterol, and heart disease	Diets low in saturated fats and cholesterol may reduce risk of heart disease
Dietary fiber and cancer	Low-fat diets rich in fiber-containing grain products, fruits, and vegetables may reduce risk of some types of cancer
Soy protein and heart disease	Diets low in saturated fats and cholesterol that include 25 g of soy protein daily may reduce risk of heart disease

Adapted from the Food and Drug Administration

Figure 7-9. A health claim is a statement declaring a relationship between certain foods or nutrients and disease.

If nutrient claims are made, the nutrition information need not appear on the menu but must be provided to guests upon request. Nutrient claims require that a dish be prepared consistently in order to uphold the validity of the claim. If the recipe is changed, the nutritional data must be reevaluated to ensure the nutrient claims pertaining to the dish are accurate.

Health claims are based on the strength of the scientific evidence supporting the claim. Health claims are ranked and the categories are described by the FDA as follows:

- **A:** Scientific evidence supports the claim.
- **B:** "...Although there is scientific evidence supporting the claim, the evidence is not conclusive."
- **C:** "Some scientific evidence suggests...however, FDA has determined that this evidence is limited and not conclusive."
- **D:** "Very limited and preliminary scientific research suggests...FDA concludes that there is little scientific evidence supporting this claim."

Knowledge Check 7-1

1. List four functions of a menu.
2. Differentiate between quick service, fast casual, and casual dining styles.
3. Explain why menu prices are set higher at fine dining operations.
4. Give three examples of an institutional venue.
5. Identify common specialty food bars.
6. Compare cycle, fixed, and market menus.
7. Differentiate between a prix fixe and a table d'hôte menu.
8. Explain why truth-in-menu guidelines were established.
9. Describe how menu labeling laws affect foodservice operations with 20 or more locations.
10. Give an example of a nutrient claim and a health claim.

EVALUATING MENUS

Evaluating menus is an ongoing process that is paramount to the success of every foodservice operation. Menus need to be evaluated to ensure that the overall goals of the operation are being met. Operational goals may be established based on a variety of factors such as flavor development, nutritional quality, costs, profitability, and ingredient sourcing. For example, print and electronic media have emphasized the "farm to table" movement, which focuses on increasing the transparency between where food is grown and where it is consumed. As a result, many foodservice operations have evaluated their menus in order to incorporate more locally produced items. Menu trends, menu mix, and purchasing options are important areas to consider when evaluating menus.

Menu Trends

Food and nutrition trends are ever changing as are guest preferences and the inspirations of chefs. For example, gluten-free dishes were virtually unknown in the past but are becoming top sellers on many menus. Current research indicates that significant menu trends also include locally sourced ingredients, healthy menu items for children, customized meals, and global flavor profiles.

Locally Sourced Ingredients. The trend of using locally sourced ingredients is reflected in the popularity of farmers' markets and on-site gardens where foodservice operations are growing produce on their land, in kitchen gardens, and even on rooftops. **See Figure 7-10.** Produce harvested at the height of freshness allow flavors to shine, and menus are highlighting this concept. In addition, there is a movement to double the amount of produce used within the foodservice industry over the next decade. Accomplishing this goal requires the cooperation of farmers, suppliers, distributors, retailers, and chefs. In general, locally produced food is transported less than 400 miles from its point of origin or the food comes from the state in which it was produced.

Farmers' Markets

Figure 7-10. The trend of using locally sourced ingredients is reflected in the popularity of farmers' markets.

Healthy Menu Items for Children. Menus that appeal to both children and parents are increasingly important in attracting and maintaining guests. Many children's menus have traded chicken nuggets and grilled cheese sandwiches in favor of meals reflecting complex flavors, whole grains, and fresh produce. For example, a mini black bean burger served on a whole grain bun with a side of strawberry-topped yogurt may replace a cheeseburger and French fry offering. In addition to parents, the medical community, commodity boards, and the federal government support this trend in an effort to help fight childhood obesity.

Customized Meals. As guests demand more control over their food choices, menus that encourage customization are becoming prevalent. Taking a cue from quick service restaurants where guests choose their favorite sandwich toppings, many foodservice operations are encouraging guests to customize everything from salads to burritos.

Global Flavor Profiles. Experimenting with flavor profiles that highlight seasonings and flavorings from around the world is gaining momentum. **See Figure 7-11.** Menus are reflecting the global flavors trend by incorporating creative dishes such as grilled pineapple topped with a hint of freshly cracked black pepper and mint leaves or zucchini drizzled with agave nectar and harissa (a spicy chili-based condiment from North Africa).

Global Spice Mixes

Spice Mix	Country of Origin	Common Ingredients
Baharat	Middle East	Black pepper, cinnamon, cloves, coriander, nutmeg, and paprika
Berbere	Ethiopia	Allspice, black pepper, cardamom, chiles, cloves, coriander, cumin, fenugreek, and ginger
Garam masala	India	Black peppercorns, cardamom, cinnamon, cloves, coriander, cumin seeds, mace, and nutmeg
Green curry	Thailand	Coriander, cumin, ginger, green chiles, lemongrass, turmeric, and white peppercorns
Harissa	North Africa	Caraway, chiles, coriander, cumin, and garlic
Quatre epices	France	Cinnamon, cloves, ginger, nutmeg, and pepper
Recado	Mexico	Achiote, allspice, black pepper, chiles, cinnamon, and cloves

Figure 7-11. Experimenting with flavor profiles that highlight seasonings and flavorings from around the world is gaining momentum.

Menu Mix

The success of a menu includes evaluating its menu mix. A *menu mix* is the assortment of items that may be ordered from a given menu. For example, a successfully balanced menu mix might include a variety of vegetarian options in addition to meat, poultry, and seafood dishes. A menu mix typically includes categories such as appetizers, soups, sandwiches, salads, entrées, sides, desserts, and beverages. With the exception of beverages, menu mix categories commonly appear on the menu in the order they are eaten.

Appetizers. An *appetizer* is a food that is larger than a single bite and is typically served as the first course of a meal. **See Figure 7-12.** Several appetizers can also be ordered and shared as a meal. An appetizer should provide a burst of flavor that awakens the taste buds and stimulates the appetite. Appetizers are an effective way to introduce nutritious options and can be promoted as smaller entrée portions.

Soups. Soups can vary from clear, stock-based creations to rich and creamy purées. Soups can be easily modified to create nutrient-dense versions without sacrificing flavor. For example, soups provide an ideal opportunity to incorporate vegetables and fruits, which increase nutrient density. Using a house-made stock instead of a ready-to-use stock can be an effective way to reduce sodium levels while enhancing flavor.

Sandwiches. One reason sandwiches are so popular is the wide variety of ingredients that can be used to create them. From lean animal-based proteins and whole grains to vegetables, fruits, legumes, nuts, and seeds, sandwiches can be filled with an assortment of healthy combinations that can be served hot or cold. Given their ingredient versatility, nutritious sandwich creations that amplify colors, textures, and flavors can be used to differentiate a foodservice operation. **See Figure 7-13.**

Figure 7-12. An appetizer is food that is larger than a single bite and is typically served as the first course of a meal.

Sandwiches

National Turkey Federation

Thai Turkey Burger

Courtesy of The National Pork Board

Italian Pork Pita

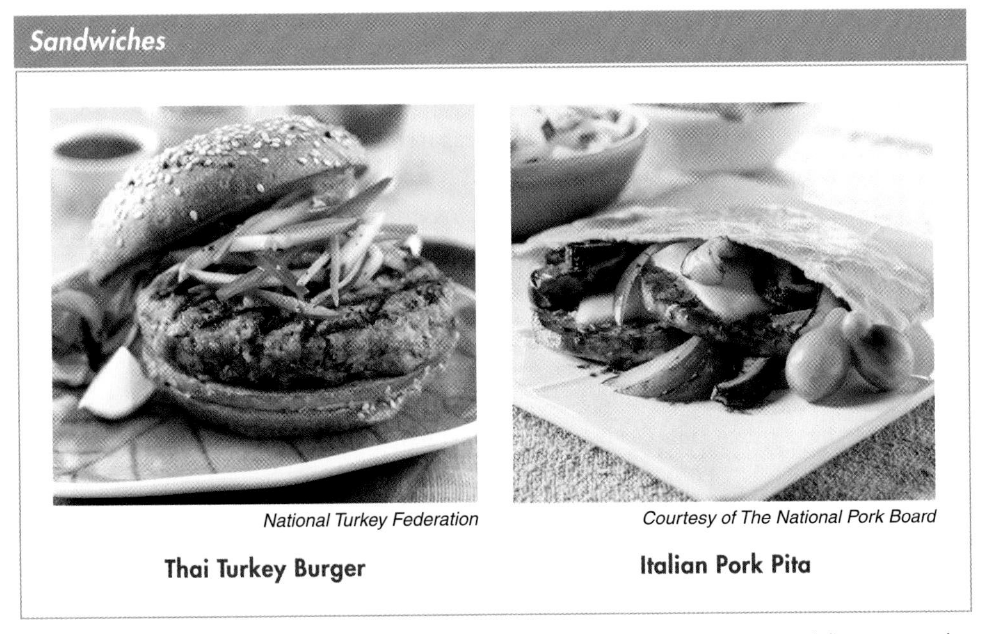

Figure 7-13. Nutritious sandwich creations that amplify colors, textures, and flavors can be used to differentiate a foodservice operation.

Salads. The first salads were greens sprinkled with or dipped in salt. Salads have since evolved into dishes that include a diverse range of nutrient-dense ingredients. Salads offer limitless opportunities to build and develop appealing flavors. Whether serving hot or cold as appetizers, entrées, sides, or even desserts, incorporating salads on the menu is an effective way to promote foods high in vitamins, minerals, and dietary fiber.

Chef's Tip

To create interest and increase nutrients consumed, use greens such as arugula, spinach, and endive instead of iceberg lettuce.

Salad dressings made from fruit or vegetable juices are an effective way to increase nutrients and provide a healthier alternative to cream-based salad dressings that are high in saturated fat. Substituting half of the cream with stock is another way to reduce fat and calories in cream-based dressings. Vinaigrette salad dressings made with unsaturated fats, such as olive oil, canola oil, or nut oils, also offer a flavorful and nutritious option to dressings high in saturated fat. **See Figure 7-14.**

Vinaigrette Salad Dressings

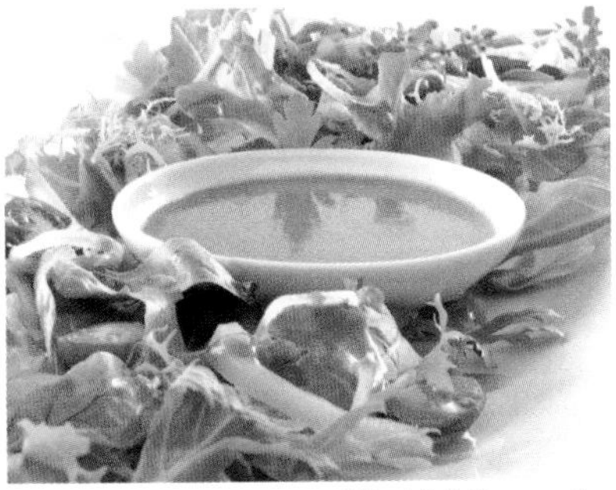

Vita-Mix® Corporation

Figure 7-14. Vinaigrette salad dressings made with unsaturated fats, such as olive oil, canola oil, or nut oils, offer a flavorful and nutritious option to dressings high in saturated fat.

Entrées. Entrées are commonly thought of as the focal point of the meal. Entrées typically include protein, starch, and vegetables. They can often be high in sugar, fat, sodium, and calories. Creating nutritious entrées is an opportunity to introduce guests to appropriate portions of nutrient-dense proteins, vegetables, fruits, and whole grains with textures and flavors that surpass their expectations.

Sides. Sides typically include a starch, vegetable, or fruit. A side can be plated as a single item, such as a baked potato or grilled asparagus, or a variety of ingredients can be combined to create side dishes such as rice pilaf or pasta salad. The use of sides is generally a cost-effective way to enhance an entrée with complementary colors, textures, and flavors.

Bread is a side that can be served before a meal, during the meal, or as part of the entrée. Healthy choices include breads made with whole grains, nuts, seeds, fruits, or vegetables. Butter, spreads, and dips commonly served with breads can add significant amounts of fat and calories. Nutritious alternatives to traditional spreads and dips may include Greek yogurt, hummus, and puréed vegetables such as avocado or eggplant. **See Figure 7-15.**

Wellness Concept

Reduce saturated fats by replacing traditional baked potato toppings, such as butter and sour cream, with salsa, pestos, or shaved, fresh parmesan cheese.

Breads and Nutritious Dips

Puréed eggplant

Vita-Mix® Corporation

Figure 7-15. Nutritious alternatives to traditional spreads and dips may include Greek yogurt, hummus, or puréed vegetables such as avocado or eggplant.

Desserts. Desserts are sweet finishes to meals and can include items such as baked goods, cheeses, fruits, nuts, and chocolates. Desserts can have flavor profiles ranging from incredibly rich and decadent to cool and refreshing. Desserts can be made more nutritious by incorporating whole grain flours, fruits, and nuts and by cutting back on processed sugars and high-fat ingredients such as butter and cream. **See Figure 7-16.** For example, grilled fruit served with a vanilla-infused yogurt and topped with almond granola is an effective way to reduce fat and calories. Offering smaller dessert portions or dessert samples on menus is an effective way of enticing guests to indulge in an appropriately sized sweet treat.

Desserts

Wisconsin Milk Marketing Board, Inc.

Figure 7-16. Desserts can be made more nutritious by incorporating ingredients such as fresh fruits and low-fat dairy products.

Beverages. Beverages are usually listed last on the menu or presented on a separate menu. A well-planned beverage menu often provides a significant source of revenue. For example, specialty drinks that incorporate nutrient-dense juices and fresh fruit purées can become signature items that not only appeal to any age but may also be an effective way to increase sales and profits.

Wellness Concept

It is important to consider the portion size of beverages as well as the number of beverages consumed, as they can be high in sugar, fat, and calories.

FOOD SCIENCE EXPERIMENT: Evaluating Entrées

7-1

Objective

- To evaluate entrées for nutrient density.

Materials and Equipment

- Menu

Procedure

1. Locate a menu that lists a breakfast, lunch, and dinner entrée.
2. Choose one breakfast, lunch, and dinner entrée from the menu to evaluate.
3. Access nutritional data online using the USDA National Nutrient Database as a standard reference for each of the three chosen entrées.
4. Record the amount of total calories, fat, sodium, and sugars for each entrée.
5. Record answers to the following questions:
   A. What are the total calories, fat, sodium, and sugars for each entrée?
   B. How does the nutritional data of the three entrées compare?
   C. Is the menu description effective? Why or why not?
   D. Is the menu item placement effective? Why or why not?
   E. What are the prices of the three menu entrées?
   F. How do the prices compare?
   G. What modifications could be made to each entrée to increase nutrient density while maintaining flavor?

Typical Results

In general, menu entrées can be modified to increase nutrient density while maintaining flavor.

Purchasing Options

As nutrition becomes even more important in menu planning, many chefs are taking a new approach to creating healthy menu items. This often entails choosing healthy starches and sides and then serving smaller portions of proteins. These are important considerations when purchasing.

Purchasing involves selecting and ordering the ingredients and products necessary to prepare and serve menu items. When purchasing ingredients, emphasis should be placed on selecting high-quality whole foods that have undergone the least amount of processing possible. Whole foods include meats, poultry, seafood, eggs, grains, and produce.

A *purchasing specification* is a written form listing the specific characteristics of a product that is to be purchased from a supplier. **See Figure 7-17.** Purchasing specifications may include product quality or grade, size, packaging, and food safety requirements. Additional factors to consider when purchasing include the availability and cost of items and whether items are locally produced.

Purchasing Specifications

Purchase Specification	
Item name	baking potatoes
Variety	Idaho Russet
Grade	US #1
Count per case	80
Net weight per case	50 lb
Packaging	heavy-duty cardboard box

Figure 7-17. A purchasing specification is a written form listing the specific characteristics of a product that is to be purchased from a supplier.

Availability. Quality and consistency are dependent on the availability of ingredients. Sources that change frequently or items that become limited can affect ingredient selection and pricing. For example, fresh cherries are generally at their peak and widely available during the summer months, making them an economical purchase. When cherries are at the height of freshness, their flavors can enhance dishes from appetizers to desserts. However, purchasing cherries in the winter months may not be cost effective and the quality may be substandard.

Cost. The cost of ingredients and the quantity ordered should be monitored closely. A par stock checklist can be used to help control costs. *Par stock* is the amount of a particular product that should be kept in inventory to ensure that an adequate supply is on hand for regular food production within a given operation. **See Figure 7-18.** Par stock values should be set high enough to ensure that the operation does not run out of a product. However, the values should be low enough to prevent wasting products because they were left in storage too long. Other factors, such as the production capability of staff and equipment needs, must also be considered to keep costs in line.

Par Stock

Item	Par Stock	Stock On Hand	Estimated Use Prior to Delivery	Quantity to Order
Tomatoes	5 cases	2 cases	1 case	4 cases
Lettuce	10 cases	5 cases	5 cases	10 cases
Cucumbers	2 cases	1 case	1 case	2 cases
Onions	1 case	15 cases	5 cases	0 cases

Figure 7-18. Par stock is the amount of a particular product that should be kept in inventory to ensure that an adequate supply is on hand for regular food production within a given operation.

Purveyors. Purveyors must be selected carefully in order to procure high-quality ingredients and supplies at the best possible price. Using more than one purveyor allows the opportunity for price comparisons and price breaks. Purveyors can also provide valuable information regarding the products that will best meet a foodservice operation's goals and objectives.

Locally Produced Foods. Buying locally allows chefs to have a greater understanding of growing conditions and product availability. It also promotes a positive relationship with local growers and suppliers. Seasonal menu items offer a chance to serve fresh local foods as well as highlight area farms, producers, and purveyors. Ingredients that are in season are generally more flavorful and higher in nutrients due to less time spent in transit.

U.S. Department of Agriculture

Fruits that are locally grown are in season and often more flavorful due to shorter transit times.

Some locally produced foods are grown using sustainable agriculture practices. Foods grown using a sustainable philosophy are produced in a manner that does not harm the environment, respects animals, provides a fair wage to the farmer, and supports and enhances rural communities.

Knowledge Check 7-2

1. Identify factors often used to establish operational goals.
2. Give four examples of current menu trends.
3. Describe how menu items for children are becoming healthier.
4. Define menu mix.
5. Give an example of how to modify a salad dressing high in saturated fat.
6. Describe ways to make desserts more nutritious.
7. Describe a purchasing specification.
8. Explain how availability affects purchasing.
9. Define par stock.
10. Explain the benefits of using more than one purveyor.
11. Identify the advantages of purchasing locally produced foods.

MENU MODIFICATIONS

Analyzing and evaluating a menu from several different perspectives can lead to successful menu modifications that not only retain the existing customer base but also attract new clientele. For example, after evaluating the menu, it may be decided that modifications need to be made to accommodate guests with dietary considerations such as food allergies. Whether changing existing recipes to make them healthier or adding nutritious menu items, it is essential to consider flavor development, cooking techniques, sensory qualities, and plate composition in order to create enticing meals that inspire repeat business.

Dietary Considerations

Individuals have different dietary considerations based on factors such as obesity, heart disease, diabetes, cancer, and celiac disease that may require menu item modifications. Considering the dietary needs of guests who follow a plant-based diet may also lead to menu modifications. Needs can be met by offering a well-balanced menu featuring a variety of flavorful dishes that are high in vitamins, minerals, and dietary fiber and low in sugar, fat, sodium, and calories. Offering a menu mix that incorporates plant-based, gluten-free, and lactose-free options is also an effective way to meet dietary concerns. **See Figure 7-19.**

Modifications for Special Dietary Considerations

Goal	Ingredients to Eliminate	Possible Ingredient Replacements
Plant-based dishes	Animal-based protein foods	Legumes, dried beans and peas, soy, and whole grains
	Animal-based stocks/bases	Vegetable stocks/bases
Gluten-free dishes	Wheat	Rice, quinoa, and amaranth
	Wheat flour	Potato, rice, and soy flours
	Processed spice mixes	Fresh herbs and/or toasted spices
	Soy sauce	Wheat-free soy sauce
Lactose-free dishes	Milk	Rice, soy, almond, coconut, or lactose-free milk
	Dairy products (cheeses, yogurts, and ice creams)	Soy-based or lactose-free dairy products

Figure 7-19. Offering a menu mix that incorporates plant-based, gluten-free, and lactose-free options is an effective way to meet dietary concerns.

Allergen-Free Menu Items

Menus often include allergy warnings for items containing milk, eggs, fish, shellfish, tree nuts, peanuts, wheat, and soybeans, which are the foods considered major allergens. In addition to menu warnings, safety is dependent on effective communication between guests and staff members.

> ***Wellness Concept***
>
> As an addition to the main menu, some foodservice operations offer allergen-free menus, such as gluten-free or peanut-free menus.

To accommodate guests with food allergies, many foodservice operations have implemented allergy awareness training programs. Allergy cards that list ingredients and preparation techniques can be provided to staff members in order to present clear and accurate information to guests. Some foodservice operations have established allergy-free work zones in the kitchen that prohibit the use of foods known to cause allergic reactions and cross-contamination. **See Figure 7-20.**

Nutritious Menu Items

Research suggests that flavor is the most important factor when selecting foods and, unfortunately, some individuals have the impression that "healthy" food lacks flavor. Chefs have the opportunity to change this misconception by using high-quality ingredients and healthy cooking techniques as well as by presenting foods attractively to create a memorable dining experience.

Flavor Development. Flavor development involves using the five primary tastes of sweet, sour, salty, bitter, and umami to create flavor combinations that are full of depth and complexity. The use of seasonings and flavorings, such as aromatics, herbs, and spices, is critical to flavor development. It is also important to add the right amount and type of ingredient before, during, and after the cooking process to enhance colors, create appealing textures, and layer flavors. **See Figure 7-21.**

Accommodating Guests with Food Allergies

- Establish allergy-free work zones
- Alert kitchen of food allergy
- Check all ingredients, including processed ingredients
- Review allergy-free preparation procedures
- Communicate thoroughly with guest, including all findings; and address any additional questions/concerns

Figure 7-20. To accommodate guests with food allergies, many foodservice operations have implemented allergy awareness training programs.

Flavor Development

Figure 7-21. It is important to add the right amount and type of ingredient before, during, and after the cooking process to enhance colors, create appealing textures, and layer flavors.

FOOD SCIENCE EXPERIMENT: Layering Flavors

7-2

Objective

- To compare methods of flavor development used to prepare stocks in terms of aroma, color, and flavor.

Materials and Equipment

- Marker
- 4 self-adhesive labels
- 2 small stock pots
- 2 sheet pans
- 4 tbsp canola oil, divided among 2 bowls
- 2 yellow onions
- 2 carrots
- 4 celery stalks
- 4 garlic cloves
- 2 cups white wine, divided among 2 bowls
- 2 bay leaves
- 2 sprigs of parsley
- 2 sprigs of thyme
- 8 black peppercorns
- Cutting board
- Knife
- Liquid measuring cup
- 10 cups water
- Wooden spoon
- Cheesecloth
- Chinois
- 2 stock buckets
- Tasting spoons

Procedure

1. Write "Straight Method" on two self-adhesive labels and "Layered Method" on two self-adhesive labels.
2. Attach a "Straight Method" label to one stock pot and a "Straight Method" label to one sheet pan. Attach a "Layered Method" label to the other stock pot and a "Layered Method" label to the other sheet pan.
3. Place the ingredients on the two sheet pans so that each sheet pan contains the following: bowl with 2 tbsp canola oil, 1 onion, 1 carrot, 2 celery stalks, 2 garlic cloves, bowl with 1 cup white wine, 1 bay leaf, 1 sprig of parsley, 1 spring of thyme, and 4 black peppercorns.
4. Use the cutting board and knife to small dice the onions, carrots, and celery, and mince the garlic from the "Straight Method" sheet pan.
5. Place all of the "Straight Method" ingredients into the "Straight Method" stock pot. Add 5 cups of water, bring to a boil, and then reduce to a simmer. Simmer and stir occasionally for 30 minutes.
6. Strain the "Straight Method" stock through a cheesecloth-lined chinois into a stock bucket.
7. Repeat step 4 for the ingredients on the "Layered Method" sheet pan.
8. Heat the empty "Layered Method" stock pot over medium heat. Add the canola oil and the onions. Cook the onions 5–6 minutes or until they begin to turn translucent.
9. Add the carrots and celery and cook for 5–6 more minutes or until the vegetables start to wilt.
10. Add the garlic and cook for 1 minute or until the garlic becomes fragrant.
11. Add the wine and herbs and cook for 1 minute.
12. Add 5 cups of water, bring to a boil, and then reduce to a simmer. Simmer and stir occasionally for 30 minutes.
13. Strain the "Layered Method" stock through a cheesecloth-lined chinois into a stock bucket.
14. Evaluate the aroma, color, and flavor of each stock.

Typical Results

There will be variances in the aroma, color, and flavor of each stock.

Cooking Techniques. Healthy cooking techniques should produce flavorful food while retaining the maximum amount of nutrients. A well-executed menu has dishes that highlight a variety of healthy cooking techniques such as grilling, broiling, roasting, sautéing, steaming, and poaching. Regardless of the cooking technique, the prepared menu items should highlight the natural flavors found in quality ingredients.

Sensory Qualities. Menu items should appeal to all five senses by using a variety of colors, textures, and flavors to provide guests with an exceptional dining experience. **See Figure 7-22.** For example, a fajita plays to the sense of hearing as it sizzles on the hot serving plate. As the fajita is set before the guest, the spicy aromas and colorful red, green, and yellow peppers appeal to the senses of smell and sight. The tortilla and meat paired with the contrasting crunch of sautéed vegetables creates a complementary texture and mouthfeel. When topped with a squeeze of lime, the fajita provides a depth of flavor that appeals to the taste buds.

Sensory Qualities

Photo Courtesy of Perdue Foodservice, Perdue Farms Incorporated

Figure 7-22. Menu items should appeal to all five senses by using a variety of colors, textures, and flavors to provide guests with an exceptional dining experience.

Plate Composition. The composition of food on the plate should create a visual impression that builds anticipation for the food about to be consumed. This can be achieved by using a variety of colors, shapes, textures, and heights to create a well-balanced, appealing presentation.

Plate composition also involves serving appropriate portion sizes to provide nutritionally balanced meals. A healthy plate consists of ½ of vegetables and fruits, ¼ of starch, and ¼ of protein. An effective strategy for ensuring that the plate composition is visually appealing and appropriately portioned is to map out the plated presentation.

Chef's Tip

Because garnishes enhance the presentation of a dish by contributing color, texture, and flavor they should be planned along with the menu item to achieve maximum impact.

Culinary Nutrition Recipe-Modification Process

The recipe-modification process involves changing an existing recipe to make it more nutritious while maintaining the integrity of the original dish. Recipes are commonly modified to reduce sugars, fats, sodium, and calories. **See Figure 7-23.** In some instances, recipes are modified based on a specific ingredient. For example, guests with celiac disease require that modifications be made to recipes containing gluten.

Modifying recipes to create healthier versions often involves ingredient substitutions and ingredient alternatives. An *ingredient substitution* is an ingredient that replaces an item of similar characteristics. Low-fat sour cream is an example of an ingredient substitution for full-fat sour cream. An *ingredient alternative* is an ingredient that replaces an item of different characteristics. Quinoa is an example of an ingredient alternative for rice. Changing ingredients may involve remaking a recipe several times in order to ensure that the integrity of the original recipe is upheld.

Modifications for Sugars, Fats, and Sodium

Goal	Ingredients to Evaluate	Possible Modifications
Reduce sugars	• Granulated sugar • Brown sugar • Molasses • Honey • Syrups	• Reduce total amount of sugars in recipe • Replace sugars with fruit, fruit purées, and fruit juice • Caramelize fruits and vegetables for natural sweetness • Replace sugars with sweet spices like cinnamon, allspice, nutmeg, and cardamom • Serve reduced sugar dishes at room temperature to enable taste buds to detect greater sweetness
Reduce fats	• Butter • Oil • Shortening • Lard • Hydrogenated oils • Animal-based foods	• Replace cream-based soups with stock-based soups • Replace fats with vegetable and fruit purées • Add small amounts of intensely flavored oils • Replace fats with vinegar, citrus zest/juice, or nectar • Use egg whites instead of whole eggs • Use lean cuts of meat and poultry • Remove skin from poultry • Trim fat from meats • Use healthy cooking techniques
Reduce sodium	• Salt and spice mixes • Purchased stocks, bases, salad dressings • Soy sauce • Worcestershire sauce • Canned goods • Capers and olives • Smoked/cured foods	• Replace salt and spice mixes with fresh herbs, spices, and acidic ingredients • Use reduced sodium stocks, bases, dressings, and sauces • Rinse heavily salted products • Use seasonal whole foods for optimal flavors

Figure 7-23. Recipes are commonly modified to reduce sugars, fats, sodium, and calories.

Cooking techniques may also change when recipes are modified. As a result, cooking times or temperatures may need adjusting.

Ingredient changes coupled with modified cooking techniques can make a dish more nutritious yet still appealing. For example, deep-fried fish coated in a thick flour batter is generally high in fat and calories. Instead, lightly coating fish in whole grain bread crumbs and toasted pecans and then baking the fish produces a nutrient-dense dish that is still crunchy and flavorful.

The culinary nutrition recipe-modification process helps ensure the best possible results when modifying recipes. The six steps of the culinary nutrition recipe-modification process are the following:

1. Evaluate the original recipe for sensory and nutritional qualities.
2. Establish goals for the recipe modification.
3. Identify modifications or substitutions.
4. Determine the functions of the identified modifications or substitutions.
5. Select appropriate modifications or substitutions.
6. Test the modified recipe to evaluate sensory and nutritional qualities.

Chefs are often inspired to create signature dishes after modifying a recipe. Signature dishes add a unique flavor profile to the original recipe and can still be prepared in a healthy manner.

CULINARY NUTRITION RECIPE-MODIFICATION PROCESS

Ranch Dressing

Ranch dressing features a delicious blend of herbs and spices combined with velvety mayonnaise, smooth sour cream, and tangy buttermilk. The result is a rich and creamy dressing noted for its versatility. For example, ranch dressing can be used as a traditional salad dressing, a sandwich spread, or a condiment. When full-fat mayonnaise, sour cream, and buttermilk are used, ranch dressing is high in both calories and total fat.

Yield: 12 servings, 1 fl oz each

Ingredients

¾ c	mayonnaise
⅓ c	sour cream
2 tbsp	parsley, dried
1 tsp	onion flakes
½ tsp	garlic powder
¼ tsp	kosher salt
¼ tsp	black pepper, ground
½ c	buttermilk

Preparation

1. Mix all ingredients except the buttermilk in a small bowl.
2. Slowly blend in the buttermilk.
3. Chill at least 2 hours before use.

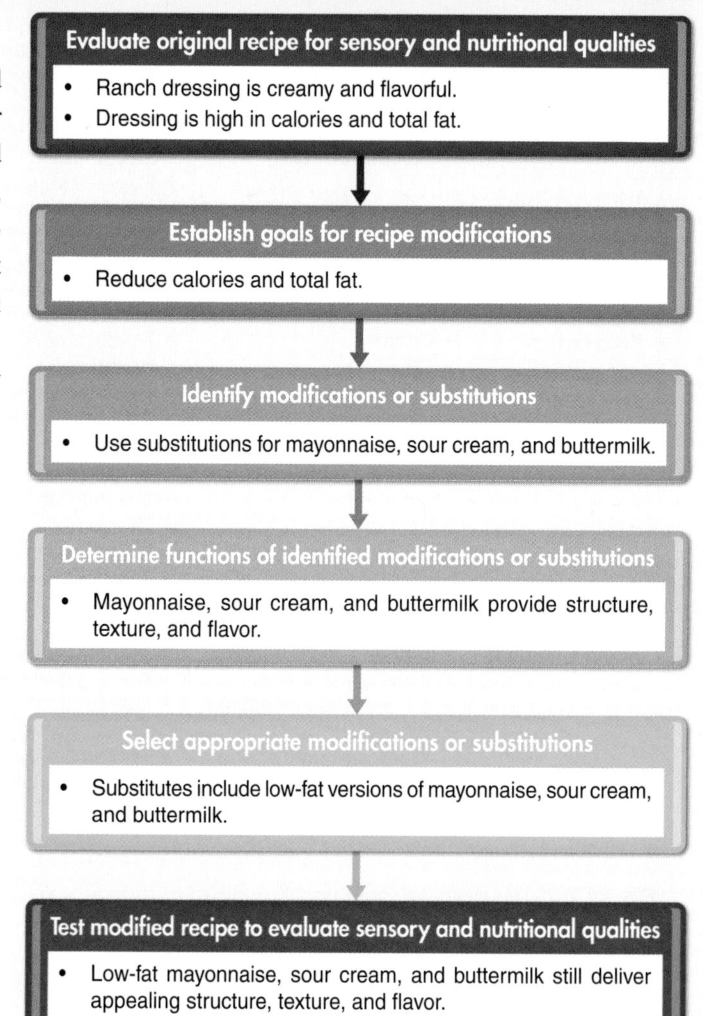

Modified Ranch Dressing

The original ranch dressing recipe has both an appealing mouthfeel and flavor but is high in calories and total fat content. To reduce calories and total fat, this modified ranch dressing recipe uses low-fat mayonnaise, sour cream, and buttermilk. The low-fat substitutions blend well with the herbs and spices and uphold the structure, texture, and flavor of the original recipe.

Yield: 12 servings, 1 fl oz each

Ingredients

¾ c	mayonnaise, low-fat
⅓ c	sour cream, low-fat
2 tbsp	parsley, dried
1 tsp	onion flakes
½ tsp	garlic powder
¼ tsp	kosher salt
¼ tsp	black pepper, ground
½ c	buttermilk, low-fat

Preparation

1. Mix all ingredients except the buttermilk in a small bowl.
2. Slowly blend in the buttermilk.
3. Chill at least 2 hours before use.

Ranch Dressing Nutritional Comparison

Nutrition Facts	Original	Modified
Calories	77.7	54.9
Total Fat	6.5 g	5.0 g

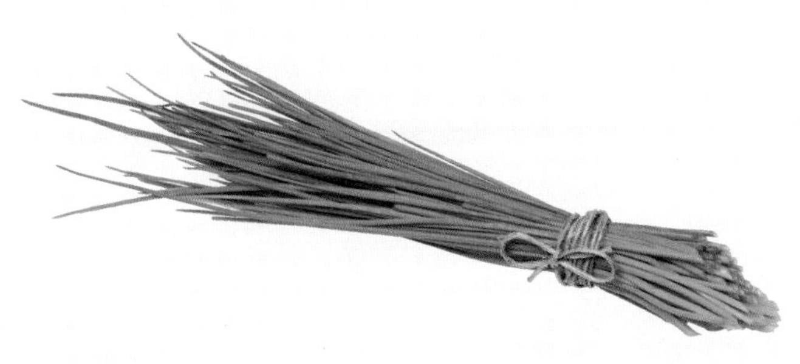

Barilla America, Inc.

Signature Zesty Ranch Dressing

This signature ranch dressing recipe uses smaller quantities of low-fat mayonnaise, sour cream, and buttermilk. By using less mayonnaise and sour cream and by eliminating the Parmesan cheese, the calories and total fat are reduced. To add a zesty flavor and enhance the aroma, fresh herbs are added. Cider vinegar adds complexity by creating a slightly sour taste that is balanced by the addition of a small amount of sugar. The overall result is a nutritious dressing with a blend of fresh flavors that maintains the appeal of traditional ranch dressing.

Yield: 8 servings, 1 fl oz each

Ingredients

¾ c	buttermilk, low-fat
2 tbsp	mayonnaise, low-fat
2 tbsp	sour cream, low-fat
2 tsp	cider vinegar
1 tbsp	basil, chopped
1 tbsp	chives, chopped
1 tsp	dry mustard
1 tsp	thyme, chopped
1 clove	garlic, minced
½ tsp	granulated sugar
¼ tsp	kosher salt
⅛ tsp	black pepper, ground

Preparation

1. Blend all ingredients in blender or food processor.
2. Chill at least 2 hours before using.

Ranch Dressing Nutritional Comparison

Nutrition Facts	Original	Modified	Signature
Calories	77.7	54.9	27.0
Total Fat	6.5 g	5.0 g	1.7 g

Knowledge Check 7-3

1. Identify common dietary considerations that can affect menus.
2. Explain ways to accommodate guests with food allergies.
3. Identify healthy cooking techniques that can be highlighted on menus.
4. Give an example of how menu items can appeal to all five senses.
5. Describe how to create an appealing plate composition.
6. Differentiate between an ingredient substitution and an ingredient alternative.
7. List the six steps of the culinary nutrition recipe-modification process.

MARKETING NUTRITIOUS MENU ITEMS

The menu is the most important marketing tool for a foodservice operation and should be planned carefully. The main functions of a menu are to communicate information to the customer and sell the menu items. An effective approach to marketing nutritious menu items is offering a variety of nutrient-dense dishes throughout each menu category.

Because flavor is the primary factor that influences menu selections, effective menu descriptions are of utmost importance. Successful menu descriptions are easy to read and informative and entice guests by the image that is created by words. Plate presentation is also a critical aspect in marketing items because food that looks appetizing is often perceived as better tasting. In addition, menu item placement, perceived value, and menu promotions are essential factors to consider when marketing nutritious menu items.

Menu Descriptions

Regardless of the menu type or classification, menu descriptions should be clear and pleasing. The menu description should communicate the main ingredients of a dish and the method of preparation. Effective menu descriptions can elevate a dish from sounding good to sounding exceptional with words that create appetizing images.

Some words have more selling power than others. For example, research indicates that the terms "low-fat," "low-carb," and "cholesterol-free" may not be effective menu descriptors because they send a message of deprivation. However, accentuating the positive aspects of an item with terms such as "whole-grains," "natural," or "organic" can help stimulate sales. Successful menu descriptions promote clarity, add appeal, increase the likelihood that an item will be ordered, and elevate the perceived value of menu items. **See Figure 7-24.**

Menu Item Placement

Menu items are typically grouped by course such as appetizers, entrées, and desserts. The placement of menu items affects the sale of nutrient-dense dishes. For example, if nutritious items are included along with traditional favorites, they have a greater likelihood of being ordered. However, when dishes are placed in special menu sections and emphasized as being "heart-healthy" or "gluten-free," generally only consumers seeking those types of foods will order them.

Courtesy of The National Pork Board

Menus should feature a variety of nutrient-dense dishes throughout each menu category.

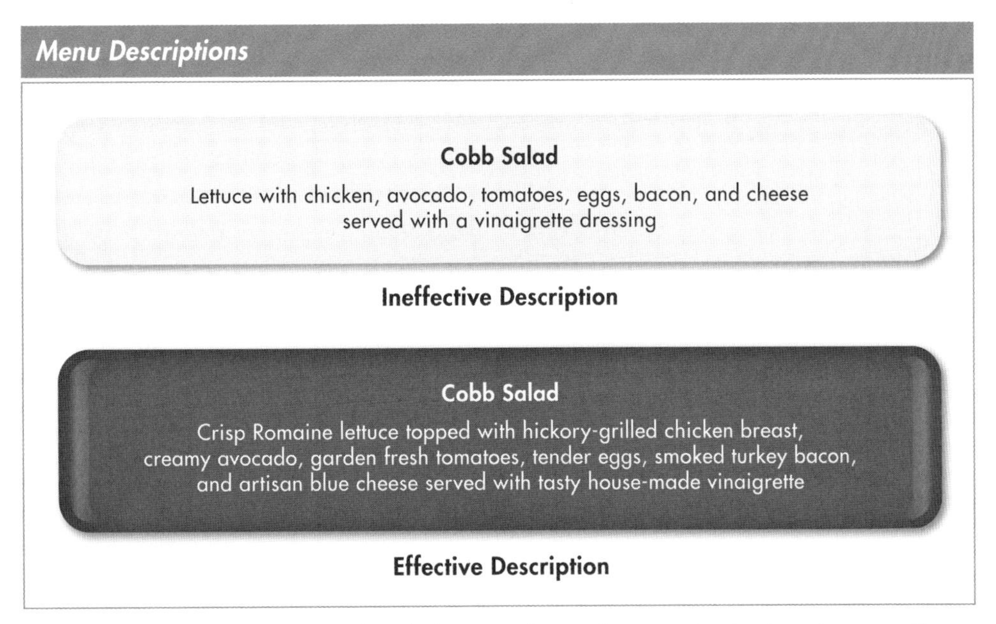

Figure 7-24. An effective menu description can elevate the perceived value of a menu item.

The eye follows a natural path when scanning a menu and has an influence over menu selections. For example, on a single-page menu, the eye naturally falls to the center of the page, which is an ideal location for marketing nutritious menu items.

Perceived Value

Perceived value represents the amount guests are willing to pay for a menu item, and menu prices should reflect guest expectations. For example, if most foodservice operations in a particular area sell a shrimp appetizer for $8.95 and one operation charges $15.95, the higher-priced appetizer will not align with guest expectations. Also, if a menu prices salads from $6.95 to $12.95 with the exception of one priced at $21.95, most guests are not likely to choose the highest-priced salad.

Once guests receive their food, plate presentation and portion size also influence perceived value. Skillful plate presentations can add value due to the artistic appearance. **See Figure 7-25.** Likewise, appropriate portions and the use of the correct plate size can influence the perceived value of the dish. For example, large amounts of food piled on a small plate can appear sloppy, while proper portions on an oversized plate give the perception there is less food than is actually there.

Perceived Value and Presentation

Alinea/Photo by Lara Kastner

Figure 7-25. Skillful plate presentations can add value due to the artistic appearance.

Chef's Tip

Perceived value is influenced by the overall dining experience. This makes it essential for foodservice operations to be clean and inviting and staffed with employees who demonstrate professionalism.

Other factors to consider with respect to perceived value include the labor involved in preparation and the cost of sourcing specialty ingredients. For example, cutting garnishes into elaborate shapes for plate presentation involves labor costs that are offset with higher menu prices. The expense of producing the garnish should be weighed against how much it affects the perceived value for the dish. Specialty ingredients, such as prestigious vintage wine, may be able to command a higher price because the guest perceives the high price as worth the indulgence.

Menu Promotions

Menu promotions are a form of marketing and advertising used to stimulate business. For example, offering free samples of a fruit smoothie is a promotional activity that has the potential to increase sales while introducing guests to a nutritious menu item. **See Figure 7-26.** Menu promotions can also include daily specials, coupons, and reward programs.

Menu promotions are commonly used to generate business during slow times, such as the beginning of the dinner hour. Discounted time periods, often referred to as "early-bird specials," are frequently used to increase business among specific demographic groups, such as families with children or senior citizens. The promotions used should build enough customer interest to increase sales and encourage repeat business.

Menu Promotions

California Strawberry Commission

Figure 7-26. Offering free samples of a fruit smoothie is a promotional activity that has the potential to increase sales while introducing guests to a nutritious menu item.

Knowledge Check 7-4

1. Identify the primary factor that influences menu selection.
2. Describe the characteristics of successful menu descriptions.
3. Identify words used in menu descriptions that may help stimulate sales.
4. Explain how menu placement can affect the sale of nutrient-dense dishes.
5. Identify factors that affect perceived value.
6. Explain the purpose of menu promotions.

Chapter Summary

As interest in nutrition continues to grow, the demand for healthy menu items is increasing. Nutrient-dense selections are now being offered in a variety of foodservice venues. Using healthy cooking techniques to prepare high-quality whole foods preserves nutrients and develops colors, textures, and flavors that appeal to the senses.

The culinary nutrition recipe-modification process can help create a greater awareness among chefs that foods do not have to be laden with excess sugars, fats, sodium, and calories to be flavorful. Marketing nutritious dishes with enticing menu descriptions, effective menu item placement, and appealing promotions can increase sales and encourage guests to eat nutrient-dense foods.

Service Ideas, Inc.

Chapter Review

1. Identify the main functions of a menu.
2. List the classifications of foodservice venues.
3. List and describe three common menu classifications.
4. List and describe four common menu types.
5. Identify common menu regulations.
6. Describe the four health claim rankings.
7. Describe four current menu trends.
8. Summarize the categories that can comprise a menu mix.
9. Identify factors to consider when modifying menus.
10. Explain how menu descriptions and menu item placement can be used to market nutritious menu items.
11. Describe what is meant by perceived value.

Beverages on the Menu

Chilled Beverages

Wisconsin Milk Marketing Board, Inc.

Hot Beverages

American Metalcraft, Inc

Chapter 8

Alcoholic Beverages

National Honey Board

Beverage menus can be as varied and exciting as food menus. Using influences such as seasonal and specialty ingredients, interesting flavor profiles, and unique interpretations adds interest and excitement to a beverage menu. Beverages not only quench thirst and provide nutrients, but also enhance the dining experience and generate significant revenue for foodservice operations. The beverage order is the first menu item addressed by a server, and since beverages are served throughout the entire dining experience, customers may order more than one beverage. This is especially true of dinner menus where customers may order a beverage before the meal, a different beverage to drink with the meal, and often a third beverage with dessert or following the meal.

Chapter Objectives

1. Identify factors that influence beverage choices.
2. Summarize the perceived value of beverages.
3. Describe the different beverage categories commonly represented on menus.
4. Identify nutritious beverage choices within each beverage category.
5. Describe beverages commonly served at breakfast, lunch, and dinner.
6. Summarize how foodservice venues promote beverages on the menu.

BEVERAGES

Beverages are ordered for many reasons, including thirst, flavor, social situations, and dietary considerations. The popularity of beverages from specialty water, coffees, and teas to craft beers and unique cocktails continues to grow. This popularity creates ideal opportunities to offer beverages that are not only flavorful and visually appealing, but also prepared and served to maximize nutrient density.

Perceived Value of Beverages

Beverages are an integral part of every menu and often profitable for foodservice operations. The cost of purchasing and preparing beverages is generally low in comparison to beverage menu prices. Because most guests order at least one beverage, the revenue generated from beverage sales can be significant.

The venue determines both the type of beverage served and the menu price. For example, a guest ordering coffee at a diner has different expectations in terms of quality and price than a guest ordering coffee at a fine dining restaurant. A guest at a diner might be served coffee in a mug along with sugar packets and individually packaged creamers. In contrast, a fine dining restaurant may offer a variety of coffees, prepare the coffee tableside with a French press, and serve it in china cups. The as-purchased cost of the coffee might not vary significantly between these two venues, but the menu price is considerably different due to the perceived value of the dining experience.

The perceived value of beverages is heavily influenced by the brand name of both alcoholic and nonalcoholic beverages. For example, top-shelf brands of wine, beer, and spirits demand a higher price than house brands because the quality and flavor are perceived as superior. Signature drinks and appealing presentations also add to the perceived value of beverages.

Carlisle FoodService Products

Beverages enhance the dining experience and can be a source of revenue for foodservice operations.

Basic categories of beverage menu items include water, dairy beverages, milklike beverages, coffees, teas, fruit and vegetable beverages, alcoholic beverages, and soft drinks. Some beverages are served alone, while others are combined to produce beverages such as café lattes, juice blends, and mixed drinks. Some beverages are calorie-free, while others contain macronutrients that supply calories. **See Figure 8-1.**

Water

Water is an essential nutrient that keeps the human body functioning properly. Water is obtained from food, beverages, and pure drinking water. Pure water is commonly referred to as tap water, or water that flows from a faucet.

Pure drinking water does not contain calories and is considered the best source of hydration for the body. Pure drinking water is often bottled and sold as still water. While guests are generally not charged for tap water, there is a charge associated with still water. Additional types of water found on menus can include carbonated water, flavored water, and enhanced water.

Nutritional Comparison of Beverages*

Type of Beverage	Calories	Fat	Carbohydrates	Protein
Water				
Still water	0	0 g	0 g	0 g
Carbonated water	0	0 g	0 g	0 g
Flavored water, sweetened, fruit flavored	52	0 g	13.0 g	0 g
Sports drink, low calorie	26	0 g	7.2 g	0 g
Dairy and Milklike				
Nonfat milk	86	0.4 g	11.9 g	8.4 g
2% milk	122	4.7 g	11.7 g	8.1 g
Whole milk	149	8.1 g	11.4 g	8.0 g
Chocolate milk	208	8.5 g	25.9 g	7.9 g
Soy milk	131	4.3 g	15.3 g	8.0 g
Coffee				
Regular and decaffeinated	5	0 g	0.9 g	0.2 g
Cappuccinos; 2% milk	60	2.0 g	6.0 g	4.5 g
Tea				
Brewed tea	2	0 g	0.7 g	0 g
Instant tea	5	0 g	1.2 g	0 g
Iced tea, presweetened	100	0 g	25.0 g	0 g
Fruit and Vegetable				
Orange	112	0.3 g	25.8 g	1.7 g
Tomato	41	0.1 g	10.3 g	1.8 g
Carrot	94	0 g	21.9 g	2.2 g
Alcohol				
Wine, white	161	0 g	1.9 g	0.2 g
Wine, red	170	0 g	4.0 g	0.5 g
Wine, rosé	168	0 g	3.3 g	0.5 g
Champagne	176	0 g	8.0 g	0 g
Beer, regular	97	0 g	8.8 g	0.7 g
Beer, light	66	0 g	3.1 g	0.5 g
Soft Drink				
Soft drink, regular	102	0 g	25.8 g	0 g
Soft drink, diet	0	0 g	0 g	0 g

* based on 8 fl oz serving size

Adapted from SPARKPEOPLE® and USDA National Nutrient Database for Standard Reference

Figure 8-1. Some beverages are calorie-free, while others contain macronutrients that supply calories.

Still Water. *Still water* is bottled water that does not contain carbonation. Still water has a long shelf life when stored at or below room temperature and away from sunlight. Still water originates from the same sources as tap water, including both surface water sources such as lakes and streams and groundwater sources. **See Figure 8-2.** Information on the source and processing of still water is often used to promote the product. The Food and Drug Administration (FDA) regulates the safety and quality of the following types of bottled still water and ensures that the label accurately represents the contents. Types of still water include purified, distilled, spring, artesian, and mineral water.

Water Sources

Canada Beef Inc.

Figure 8-2. Still water originates from the same sources as tap water, including surface water sources such as lakes and streams.

- *Purified water* is bottled still water that has been distilled or filtered to remove minerals and other impurities.
- *Distilled water* is purified water that has been boiled until it steams and then cooled until it condenses back to liquid form. Techniques such as reverse osmosis and deionization are used to eliminate impurities.
- *Spring water* is bottled still water obtained from a spring that flows naturally to the surface.
- *Artesian water* is bottled still water obtained by tapping an underground water source, which causes the water to rise into a well.
- *Mineral water* is bottled still water obtained from an underground water source that contains not less than 250 ppm (parts per million) of total dissolved solids such as calcium and magnesium.

Carbonated Water. *Carbonated water* is water that has been infused with carbon dioxide. **See Figure 8-3.** Carbonated water tastes similar to still water but has an effervescent texture. In its most basic form, carbonated water does not contain calories or caffeine and is low in sodium. Carbonated water is the main ingredient in soft drinks.

Carbonated Water

Figure 8-3. Carbonated water is water that contains carbon dioxide and has an effervescent texture.

Carbonated water is distributed as sparkling bottled water, seltzer water, and club soda. Sparkling bottled water is obtained from spring water that is naturally carbonated. In contrast, seltzer water and club soda are types of still water with added carbonation. Some varieties of carbonated water are flavored or sweetened with natural or artificial sweeteners and may or may not contain calories.

Flavored Water. Although flavored water is marketed as water, it also contains ingredients such as natural or artificial flavorings, sugar, or artificial sweeteners. Flavored water sweetened with sugar is higher in calories than flavored water made with artificial sweeteners.

Enhanced Water. Enhanced water contains added ingredients such as sugar, sugar replacers, artificial sweeteners, vitamins, minerals, and caffeine. Depending on the ingredients, enhanced water may or may not contain calories. Enhanced water is commonly marketed as vitamin-enhanced water, sports drinks, and energy drinks. They are often promoted as containing antioxidants or as able to restore electrolyte balance or provide energy.

Wellness Concept

Caffeine is the main stimulant in energy drinks. While caffeine has been shown to enhance mental alertness and physical performance, excessive amounts are associated with insomnia, irritability, and elevated heart rate.

FOOD SCIENCE EXPERIMENT: Analyzing Water 8-1

Objective

- To compare various types of water in terms of appearance and flavor.

Materials and Equipment

- Marker
- 6 tasting cups
- Masking tape
- 6 clear glasses
- Measuring cup
- 8 fl oz tap water
- 8 fl oz purified water
- 8 fl oz distilled water
- 8 fl oz spring water
- 8 fl oz artesian water
- 8 fl oz mineral water
- White paper

Procedure

1. Use the marker to label one tasting cup "Tap," the second cup "Purified," the third cup "Distilled," the fourth cup "Spring," the fifth cup "Artesian," and the sixth cup "Mineral."
2. Repeat step 1 using the masking tape to label the six glasses.
3. Pour ½ cup of each type of water into its corresponding glass.
4. Hold a piece of white paper behind each glass and assess the color and clarity of each type of water.
5. Fill each tasting cup with its corresponding water.
6. Taste each type of water.
7. Evaluate the color, clarity, and flavor of the six types of water.

Typical Results

There will be variances in the appearance and flavor of each type of water.

Vitamin-enhanced water contains vitamins as well as antioxidants and claims to boost metabolism or immunity. Sports drinks contain carbohydrates (sugar) and electrolytes (potassium, sodium, and chloride) and are meant to replenish water and other nutrients the body loses as a result of vigorous physical activity. Energy drinks contain stimulants such as caffeine, guarana, taurine, creatine, or ginseng to boost energy levels.

Dairy Beverages

Dairy beverages include products such as milk, buttermilk, and kefir. Dairy beverages provide nutrients including protein, carbohydrates, fat, calcium, and vitamin D. The amount of fat and calories vary among different varieties. **See Figure 8-4.** For example, whole milk is higher in fat and calories than low-fat and nonfat (skim) varieties. Buttermilk and kefir contain bacteria found to improve digestive health.

Milk Varieties

CROPP Cooperative

Figure 8-4. The amount of fat and calories vary among unflavored and flavored milk varieties.

Dairy beverages are commonly served alone, combined with ingredients to create specialty coffees and teas, or flavored with syrups. Dairy beverages can also be used as a base for smoothies, milkshakes and malts, and floats.

Smoothies. A smoothie is commonly made from milk or yogurt and blended with fruit and sometimes ice to form a thick consistency. **See Figure 8-5.** Frozen fruit can be used in smoothies to replace some of the ice and create a deeper fruit flavor. Frozen fruit also has a longer shelf life than fresh fruit, which can help control costs. In addition to fruit, ingredients such as vegetables, grains, nuts, seeds, herbs, and spices can be added to smoothies to create unique flavor profiles and increase nutrient density. While smoothies can be a healthy beverage choice, most are high in sugar.

Smoothies

National Honey Board

Figure 8-5. Smoothies are commonly made from milk or yogurt and are blended with fruit and sometimes ice to form a thick consistency.

Milkshakes and Malts. A milkshake is a sweet, cold beverage made by blending milk with ice cream or sweeteners and thickening agents. Milkshakes are often enhanced with ingredients such as fruits, nuts, candy, cookies, brownies, and syrups. Although milkshakes can be made with low-fat dairy products and nutrient-dense ingredients, most are high in fat and calories.

A malt is a type of milkshake made with malted milk powder. Malted milk powder is made from malted barley, dry milk, and wheat flour. Malts are not as sweet as milkshakes.

Floats. A float, such as a root beer float, is a carbonated beverage that is combined with ice cream. **See Figure 8-6.** Some venues offer milkshakes, malts, and floats flavored with liqueurs such as crème de menthe, vodka, or stout, a dark, rich beer with a robust flavor.

Ice Cream in Beverages

National Honey Board

Figure 8-6. Milkshakes, malts, and floats feature dairy products such as milk and ice cream.

Milklike Beverages

Milklike beverages are commonly derived from soybeans, nuts, seeds, or grains. Milklike beverages are not dairy products and do not contain lactose. They are most often gluten-free and vegan-friendly. **See Figure 8-7.** Milklike beverages are also popular for the flavor diversity they offer. Milklike beverages are available in low-fat and flavored varieties and are often available in shelf-stable packaging.

Nutritional Comparison of Milklike Beverages

	Calories*	Protein*	Lactose-Free	Gluten-Free
Soy milk, original	131	8 g	yes	yes
Almond milk, original	40	1 g	yes	yes
Hemp milk, original	100	2 g	yes	yes
Rice milk, original	120	1 g	yes	yes
Oat milk, original	130	4 g	yes	no

* based on 8 fl oz serving size

Figure 8-7. Milklike beverages are lactose-free, and most are gluten-free.

Soy-Based. Soy milk is made by extracting liquid from ground soybeans. Soy milk supplies approximately 8 g of protein per cup, is low in saturated fat, and contains vitamins and minerals such as calcium, magnesium, iron, selenium, copper, and manganese. Some soy milks contain added calcium, vitamin A, and vitamin D to increase their nutritional value. Soy milk is also sold in low-fat and flavored varieties.

Nut- and Seed-Based. Nut- and seed-based milklike beverages are made from ground nuts or seeds and water and typically contain emulsifiers to keep the fat in suspension. They are generally available in sweetened, unsweetened, and flavored varieties and range from 30 calories per cup to 170 calories per cup. Nut- and seed-based milklike beverages contain healthy unsaturated fats such as omega-3 fatty acids but are low in protein. Almond milk is a well-known nut-based beverage, but hazelnut and cashew milks are also becoming popular. Hemp milk is a type of milklike beverage derived from hemp seeds.

Although coconuts are a fruit, not a nut, coconut milk is often categorized with nut-based milklike beverages. Coconut milk is made from grated coconut meat that is mixed with water. Coconut milk is available fresh, frozen, and canned in regular and light varieties. Although coconut milk does not contain cholesterol, it is high in saturated fat and calories. For example, 1 cup of coconut milk contains approximately 51 g of saturated fat and 550 calories. Coconut milk is commonly added to mixed drinks, curried dishes, and soups. Some companies add water, flavorings, sweeteners, and nutrients to coconut milk to create a stand-alone beverage.

Grain-Based. Common grain-based milk-like beverages include rice milk and oat milk. Rice milk is made from partially milled rice that is mixed with water and is commonly enriched with B vitamins, vitamin A, vitamin D, and calcium. It is available in refrigerated and shelf-stable forms as well as in flavored and sweetened varieties. Rice milk is low in protein but high in carbohydrates. Rice milk is a hypoallergenic beverage choice, as it does not contain any of the eight major food allergens. Horchata is a Latin American beverage commonly made from chilled rice milk that is lightly sweetened and spiced with cinnamon.

Oat milk is made from oat groats and water and is sometimes blended with other grains and legumes, such as triticale, barley, brown rice, and soybeans. It has a mild, slightly nutty flavor and is low in saturated fat and cholesterol. Riboflavin, vitamin A, and vitamin D are commonly added to oat milk in order to increase its nutritional value. Oat milk is not gluten-free.

Coffees

Coffees are served both with meals and throughout the entire day. Coffee often draws attention because it is an aromatic beverage that stimulates the senses. **See Figure 8-8.**

Coffee beans are obtained from a small tree found in tropical and subtropical climates. The coffee tree produces small, red fruit called cherries that contain 1–2 coffee beans. Coffee beans are processed to remove their pulp, parchment, and silverskin. The light green to yellow beans that remain are then dried and roasted to produce coffee. The amount of roasting time produces different aromas, colors, and flavors and determines whether the coffee is classified as a light, medium, or dark roast. A longer roast produces a darker-colored bean with a more robust flavor.

Figure 8-8. Coffee is an aromatic beverage served both with meals and throughout the entire day.

Hundreds of coffee species are grown worldwide, but Arabica coffee beans and Robusta coffee beans are commonly used in commercially produced coffees. Arabica coffee beans have a smooth, complex flavor, while Robusta coffee beans are bitter. Coffee can be produced from a single type of coffee bean or blends of various species to produce different flavor profiles. Coffee may be purchased whole or ground. Coffee is also available in flavored, decaffeinated, and instant varieties.

Caffeine is present in all coffee beans, but most can be removed by the processor. Regular coffee has 80–200 mg of caffeine per 8 fl oz, while decaffeinated coffee averages 2–12 mg. According to the Mayo Clinic, caffeine consumption should be less than 500 mg per day. The American Academy of Pediatrics recommends adolescents consume no more than 100 mg per day.

Instant coffee is made by either heat-drying or freeze-drying brewed coffee to remove the moisture. Heat-drying produces powdered coffee. Freeze-drying produces coffee granules. Instant coffees are perceived as lower in quality than brewed coffees.

Teas

Tea is a beverage made by infusing the dried leaves of the tea plant, *Camellia sinensis,* in boiling water. **See Figure 8-9.** The sun, soil, precipitation, temperature, and elevation all contribute to the flavor and quality of tea leaves. Young, tender tea leaves are considered superior to older, coarser leaves. Tea plants also produce tiny white flowers once a year that are used for more expensive varieties of tea. The color, flavor, and caffeine content of tea are dependent upon the amount of processing and fermentation it undergoes. Tea generally contains between 14–61 mg of caffeine per 8 fl oz and is available in black, green, oolong, and white varieties.

Tea also contains tannins. A *tannin* is an astringent compound found in tea, the bark of some trees, and the skins, seeds, and stems of grapes. Tannins contain polyphenols, which are considered antioxidants. Research suggests that a subgroup of polyphenols, known as epigallocatechin gallate (EGCG), has powerful antioxidant qualities that can help reduce the risk of cardiovascular disease and some cancers. Because green tea and white tea go through minimal processing, they are considered to have the highest amount of antioxidants.

Figure 8-9. Tea is a beverage made by infusing the dried leaves of the tea plant, *Camellia sinensis,* in boiling water.

Black Teas. Black teas are fermented to produce flavors ranging from bold and smoky to light and fruity. Black teas have a long oxidation period that darkens the leaves. The leaves are allowed to wither and dry and are then rolled to break the membranes between the cells, which changes the aroma, color, and flavor of the leaves. A deep reddish-brown brew is produced by black tea leaves. Common varieties of black teas include Keemun, Lapsang Souchong, Darjeeling, Assam, and Ceylon.

Black tea is graded according to the size of its leaves. **See Figure 8-10.** For example, orange pekoe leaves are thin, wiry, and smaller than coarse, curly pekoe leaves. The grade of the tea only refers to the size of the leaves and not the quality. Larger leaves are generally packaged as loose tea, while smaller and broken leaves are packaged in bags. The smaller the tea leaf, the shorter the brewing time needed to release the flavor. The purpose of grading teas is to create teas that have a uniform leaf density and appearance so that the flavor will be extracted at the same rate when brewed.

Green Teas. Green teas are not fermented. They are dried and have a mild, slightly bitter flavor. Green teas lose their flavor more quickly than black teas and should not be stored for extended periods of time. Green tea leaves produce a greenish-yellow brew. Varieties include Sencha, Gunpowder, and Dragon Well.

Oolong Teas. Oolong teas are partially fermented, which creates an aroma, color, and flavor that is in between that of black and green teas. Oolong tea is often perceived as a superior tea. Formosa is the most common variety of oolong tea.

White Teas. White teas undergo the least amount of processing. Young tea leaves are typically steamed and dried to produce a subtle color and flavor. Common white tea varieties include White Silver Needle, White Peony, and White Darjeeling.

Herbal Beverages. Herbal beverages are not teas because they do not contain tea leaves, but they are brewed like teas. Herbal beverages are also known as tisanes. A *tisane* is an herbal beverage created by steeping herbs, spices, flowers, dried fruits, or roots in boiling water. The act of steeping creates an infusion. Rooibos is a tisane that is a deep red when brewed.

Common ingredients in herbal beverages include chamomile, ginseng, mint, ginger, cinnamon, rose hips, and lemon. Herbal beverages may be served for their flavor or function. For example, chamomile is said to have a calming effect, while mint aids digestion.

Common Tea Grades

Tea Grade	Grade Translation	Description
S	Souchong	Whole twisting leaf that is often light in color; China is the main producer
FOP	Flowery orange pekoe	Long whole leaf with a crushed flower appearance; India is the main producer
OP	Orange pekoe	Thin, wiry leaf with a tighter roll than FOP
T and G	Tippy and golden	Modifiers used to describe both whole leaf and broken grades; indicates colorful tips on the leaf
P	Pekoe	Curly, large broken leaf without a visible tip; Sri Lanka is the main producer
BOP	Broken orange pekoe	Small, squarish broken leaf with good body and strength; India is the main producer
F	Fannings	Smaller than BOP with less keeping quality; used for commercial tea bag
D	Dust	Smallest grade produced; quick liquorings

Figure 8-10. Black tea is graded according to the size of its leaves.

Fruit and Vegetable Beverages

Fruit and vegetable beverages are made by extracting juices from fruits and vegetables. **See Figure 8-11.** These beverages commonly feature one fruit or vegetable juice, such as orange or carrot juice. Juices also can be combined to create juice blends such as apple-cranberry juice or vegetable juice made from tomatoes and vegetables such as celery, spinach, and beets.

Fruits and Vegetables Commonly Used for Juices

- Apple
- Apricot
- Banana
- Beet
- Carrot
- Cherry
- Cranberry
- Grape
- Grapefruit
- Guava
- Kiwi
- Lemon
- Lime
- Mango
- Orange
- Papaya
- Passion fruit
- Peach
- Pear
- Pineapple
- Pomegranate
- Prune
- Pumpkin
- Raspberry
- Spinach
- Tangerine
- Tomato

Figure 8-11. Fruit and vegetable beverages are made by extracting juice from fruits and vegetables.

Juices made from 100% juice are labeled "juice" or "100% juice." A beverage labeled "fruit drink" or "juice cocktail" may only contain 10% pure juice, while the remainder is water, sugar, and flavorings. Juices can be purchased fresh, frozen, or canned. Some foodservice operations juice fruits and vegetables on the premises to take advantage of seasonal produce.

Alcoholic Beverages

Alcoholic beverages include wines, beers, and spirits. **See Figure 8-12.** The differences among these beverages lie in how they are produced. Wines are produced through the fermentation of fruits such as grapes and apples. Beers are produced through the fermentation of grains such as barley and rye. Spirits are distilled from grains, fruits, or vegetables.

Figure 8-12. Alcoholic beverages include wines, beers, and spirits.

In general, alcoholic beverages have a long shelf life. However, some wines and beers lose quality over time and can be sensitive to changes in temperature, humidity, light, and vibration.

There are many risks and liabilities associated with serving alcohol. Serving alcohol illegally can result in fines, loss of a liquor license, imprisonment, and/or the closing of a business. It is also necessary to protect guests when serving alcohol. Many states have mandatory or voluntary alcohol-service training programs that address alcohol laws, the checking of guest identification, and the prevention of intoxication. Serving alcohol responsibly can promote a pleasurable dining experience for all.

Wines. A *wine* is an alcoholic beverage commonly made from fermented fruit juice that has been aged for a period of time. Wines are often made from fruits such as grapes, plums, elderberries, and black currants. Wines can also be made from barley and rice. *Sake* is a Japanese wine made from rice and is not aged.

The degree of fermentation determines whether the wine is sweet or dry. Sweet wines contain higher concentrations of unfermented sugars than dry wines do. In dry wines, most of the sugars have been converted to alcohol. Some wines, especially those that are aged with the stems and skins of the fruits, are high in tannins. The tannins contained in wine produce a drying sensation in the mouth, which helps break up the richness of fatty foods and cleanses the palate.

Wines are commonly paired with food to enhance the dining experience. **See Figure 8-13.** The flavor of wine is greatly influenced by the region in which it is grown. Environmental factors, such as the climate and soil where the fruit or grain is grown, give the wine a unique aroma, color, and flavor. Wines are commonly named after the area of origin, such as Burgundy and Champagne, or the type of grape used to produce the wine, such as Chardonnay and Merlot. Wines are generally classified as still wines, sparkling wines, vintage wines, and fortified wines.

- Still wines do not contain carbonation and include white, red, and rosé (pink) wines. The color of wine depends mostly upon the type of grape used. Still wines are typically considered table wines and can be dry, semi-sweet, or sweet. Still wines contain 8%–15% alcohol by volume. **See Figure 8-14.**
- Sparkling wines are carbonated and have a bubbly, effervescent texture. Sparkling wine is often called Champagne, but only Champagne that originates from the Champagne region of France can be labeled as such. Sparkling wines are also produced in Italy, Spain, Germany, and Austria. Sparkling wines are rated by sweetness and categorized as brut, extra dry, sec, and demi-sec (the sweetest). Sparkling wines contain 8%–12% alcohol by volume.

Wine and Food Pairing Guidelines

Wine Characteristic	Ideal Menu Pairing	Menu Item Example
Light and crisp	Mild-flavored dishes	Grilled flounder
Bold and flavorful	Robust-flavored dishes	Pepper steak
Acidic	Acidic dishes	Pasta with marinara
Sweet	Spicy dishes	Green curry
High in tannins	Rich dishes	Cheese tray

Figure 8-13. Wines are commonly paired with food to enhance the dining experience.

Still Wines

L'Auberge Carmel
C.S. White, photographer

Figure 8-14. Still wines, such as white wines, are typically considered table wines.

- Vintage wines are made primarily from grapes that are harvested during a specific year. The label on the bottle indicates the year or vintage of the wine.
- Fortified wines have been supplemented with spirits such as brandy and may also contain aromatic herbs and spices. Popular fortified wines include Sherry, Vermouth, Marsala, Madeira, and Port. The addition of the spirit changes the flavor profile of the wine and raises the alcohol content to 16%–22% by volume. Fortified wines are commonly used in cooking to develop flavor.

Beers. A *beer* is an alcoholic beverage that is brewed from a mash composed of water and malted barley or other cereal grains, such as corn, rye, or wheat, flavored with hops, and fermented with yeast. Beer is 90% water, and the quality of the water greatly affects the finished product. The malting process requires soaking the cereal grain in water until it sprouts and then drying it. The malt contributes to the color, flavor, and sweetness of the beer, which is balanced by the bitterness of the hops. Hops are derived from the dried flowers of the hop plant. Most beers contain 3%–10% alcohol by volume, but some can contain as much as 32%. Light beers contain fewer calories than other beers.

INGREDIENT SPOTLIGHT: Red Wine

Carlisle FoodService Products

Unique Features

- Research suggests that moderate consumption promotes heart health by lowering LDL cholesterol levels and reducing the formation of harmful blood clots
- Adds depth of flavor to a variety of dishes
- Most of the alcohol evaporates during cooking
- Low in fat, cholesterol, and sodium

Menu Applications

- Serve as a beverage
- Use as a flavor enhancer
- Use as a tenderizer

Nutrition Facts
Serving Size 5 fl oz

Amount Per Serving	
Calories 125	Calories from Fat 0
	% Daily Value*
Total Fat 0g	0%
Saturated Fat 0g	0%
Trans Fat 0g	
Cholesterol 0mg	0%
Sodium 6mg	0%
Total Carbohydrate 4g	1%
Dietary Fiber 0g	0%
Sugars 1g	
Protein 0g	
Vitamin A 0% • Vitamin C	0%
Calcium 1% • Iron	4%

* Percent Daily Values (DV) are based on a 2,000 calorie diet.

A *craft beer* is a beer produced by a small, independently owned brewer. The prevalence of craft beers in beverage houses and bars continues to increase in response to the demand for unique flavors from local and regional sources.

Ale is a beer, such as a porter or a stout, that is fermented rapidly. *Lager* is a beer, such as a pilsner or a malt liquor, that is fermented slowly using cold temperatures. Like wines, different types of beer may be paired with different menu items to offer complementary flavor profiles. **See Figure 8-15.**

Beer and Food Interactions

Beer Characteristics	Interaction	Food Characteristics
• Hop bitterness • Roasted malt • Carbonation • Alcohol	→ Balances →	• Sweetness • Fattiness • Richness
• Sweetness • Maltiness	→ Balances →	• Spiciness • Acidity
• Hop bitterness	→ Emphasizes →	• Spiciness

Figure 8-15. Different types of beer may be paired with different menu items to offer complementary flavor profiles.

Spirits. A *spirit* is an alcoholic beverage made from distilled grains, fruits, or vegetables. Spirits are distilled by boiling the liquid from grains, fruits, or vegetables and condensing the resulting vapor. Spirits may be aged to develop aroma, color, and flavor. Unlike wines and beers, the alcohol content of spirits must be listed on the container. Most spirits are 35%–60% alcohol by volume. Alcohol content is often listed as "proof." Proof is twice the percentage of alcohol by volume. For example, 80 proof is 40% alcohol by volume.

Gin, whiskey, vodka, rum, tequila, and brandy are common spirits listed on beverage menus. Gin and most whiskeys are distilled from fermented grains, while vodka is distilled from cereal grains or potatoes. Rum is distilled from sugar cane, tequila is distilled from the fermented sap of the blue agave plant, and brandy is a distilled wine.

A *liqueur* is an alcoholic beverage made from distilled alcohol and flavored with fruits, spices, herbs, flowers, nuts, or cream, and bottled with added sugar. Liqueurs are quite sweet and have a syrupy consistency. Common liqueur flavors include orange, raspberry, almond, coffee, and crème de menthe.

Soft Drinks

The term "soft drink" originated to distinguish nonalcoholic beverages from "hard drinks" made with alcohol. A *soft drink* is a nonalcoholic beverage made from carbonated water, flavorings, sweeteners, colors, acids, and preservatives. **See Figure 8-16.** Diet soft drinks contain artificial sweeteners and have little to no calories. While cola is the most commonly purchased soft drink, other popular choices include lemon-lime, root beer, ginger ale, orange, grape, and cream soda. Despite its bitter flavor, tonic water contains sweeteners and is also classified as a soft drink.

Soft Drinks

Carlisle FoodService Products

Figure 8-16. A soft drink is a nonalcoholic beverage made from carbonated water, flavorings, sweeteners, colors, acids, and preservatives.

Soft drinks are found on almost every menu. Soft drink portion sizes and a lack of nutrients are of prime concern. Because soft drinks are mainly comprised of refined sugars, they lack nutrients and are high in empty calories. Studies have also linked soft drinks to dental caries and obesity.

Knowledge Check 8-1

1. Explain how the venue type and beverage brand influence perceived value.
2. List the basic beverage categories.
3. Compare five types of still water.
4. Define carbonated water.
5. List the ingredients commonly found in flavored and enhanced water.
6. Compare smoothies, milkshakes, malts, and floats.
7. Summarize the nutritional benefits of soy-based, nut- and seed-based, and grain-based beverages.
8. Explain how coffee beans are processed to create various flavor profiles.
9. Compare the four basic categories of tea.
10. Define tisane.
11. Explain how labeling of juices reflects nutrient density.
12. List the primary ingredients used to produce wines, beers, and spirits.
13. Compare the four classifications of wine.
14. Differentiate between craft beers, ales, and lagers.
15. Explain what is meant by "proof" on bottles of spirits.
16. Describe the health concerns associated with soft drinks.

PREPARING NUTRITIOUS BEVERAGES

In recent years, new beverages have entered the market based on functional ingredients. These "functional" beverages often claim to support health and well-being. For example, instead of stimulants such as caffeine, some energy drinks are being produced with natural energy-boosting ingredients such as herbs and complex carbohydrates. Likewise, functional dairy and milklike beverages may include probiotics that promote digestive health, and a variety of blended juices may include phytonutrients and antioxidants that enhance immunity.

Many beverages supply essential nutrients, but they can also be high in sugar, fat, calories, caffeine, and artificial ingredients, making portion control essential. For quenching thirst and hydrating the cells of the body, water is the best beverage.

Water

Pure water is calorie-free and drinking 8–10 cups per day typically keeps the body properly hydrated. Carbonated water is also calorie-free. Flavored and enhanced water often contains ingredients such as sugar, artificial flavors and sweeteners, and caffeine or other stimulants. Water with added sugar contains calories.

Water is often packaged in bottles that contain two or more servings. For example, a typical sports drink contains 2.5 servings per bottle. This can equate to approximately 51 g of sugar and 250 calories if the entire bottle is consumed.

Although the amount of caffeine found in soft drinks, such as cola, is limited by the FDA to 71 mg per 12 fl oz, there are no regulations on the amount of caffeine contained in energy drinks. Depending on the brand, the caffeine content can range from 72–250 mg. Because they contain several servings, containers of energy drinks can have as much as 500 mg of caffeine.

Dairy Beverages

Dairy beverages provide essential nutrients, including protein and calcium, but may also be high in fat. Low-fat and nonfat milk are nutrient-dense alternatives to whole milk.

Some dairy beverages have added sugars, such as chocolate milk or kefir with evaporated cane syrup, making them high in sugar and calories.

Smoothies. The nutritional content of smoothies varies greatly with the ingredients. Smoothies can be made from low-fat and nonfat milk, kefir, and milklike beverages. Fresh fruits add sweetness as well as nutrients to smoothies, but fruits packaged in syrup add empty calories.

Milkshakes and Malts. Milkshakes and malts are high in fat and cholesterol if made with ice cream and whole milk. However, combining low-fat milk with low-fat or low-sugar ice creams, ice milks, or frozen yogurts also makes a delicious milkshake or malt. Milkshakes and malts often contain empty-calorie ingredients such as candies, cookies, and flavored syrups. As an alternative, adding fruits such as strawberries, blueberries, peaches, bananas, and raspberries provides dietary fiber, vitamins, minerals, color, and flavor. To produce a lactose-free milkshake, milklike beverages can be blended with soy- or rice-based ice cream.

Floats. Floats made with ice cream and a carbonated beverage, such as root beer or cola, contain a large percentage of sugar, fat, and calories. Using a calorie-free or low-calorie carbonated beverage reduces sugar and calories. To reduce fat, low-fat ice creams, ice milks, and frozen yogurts can be effectively substituted for full-fat varieties.

Milklike Beverages

Milklike beverages, including soy milk, nut- and seed-based milks, and grain-based milks, supply essential nutrients and are flavorful, lactose-free alternatives to dairy beverages. With the exception of oat milk, milklike beverages are usually gluten-free. Using unsweetened varieties of milklike beverages allows greater control over the sugar content of a beverage, such as a smoothie.

Wellness Concept

Oats do not contain gluten on their own, but are usually processed alongside gluten-containing grains, which causes an increased risk of cross-contamination.

Coffees

Coffee flavor and quality vary according to where the coffee is grown and how it is processed. Light roasts are subtle, medium roasts are more pronounced, and dark roasts are the strongest in terms of aroma, color, and flavor. **See Figure 8-17.** *Espresso* is an intensely flavored coffee made from beans that have been roasted to the very dark, or espresso-roasted, stage.

Coffee should be brewed and presented to guests at an appropriate temperature and in the proper container. Small demitasse cups are used for strong coffees like espresso, while regular coffee is served in regular cups or mugs. Accompaniments such as cream and sugar, low-fat and nonfat milk, milk and sugar

Types of Coffee Roasts

Roast	Common Name	Description
Light-medium	City roast	Roasted to a medium-brown color, yielding a medium body and medium-flavored coffee
Medium-dark	Viennese roast	Roasted to a rich, dark-brown color, yielding a rich and intensely flavored coffee
Dark	French roast	Roasted to a very dark-brown color, yielding a very intensely flavored coffee with a strong aroma and a slightly bitter taste
Espresso	Italian roast	Roasted until beans are almost burnt, yielding a bitter, full-flavored, and intensely strong aromatic coffee

Figure 8-17. Light coffee roasts are subtle, medium roasts are more pronounced, and dark roasts are the strongest in terms of aroma, color, and flavor.

alternatives, and spices such as cinnamon and nutmeg may be offered. Specialty coffee drinks commonly include dairy or milklike beverages blended with flavored syrups such as vanilla, almond, hazelnut, and chocolate syrup. Coffee-like beverages offer a different flavor profile and include chicory coffee, dandelion coffee, and malted coffee.

When coffee is brewed, compounds that provide aroma and flavor are extracted. Medium grounds are commonly used for filtered or drip coffees. Finely ground coffee is ideal for brewing espresso, while coarse grounds are ideal for use in a French press. **See Figure 8-18.**

French Presses

Chef's Choice® by EdgeCraft Corporation

Figure 8-18. Coarsely ground coffee is ideal for use in a French press.

The strength of coffee is also influenced by the amount of coffee brewed. In general, 1 tbsp of ground coffee is used for every 6–7 fl oz of water. Coffee should be brewed with fresh, cold water for optimum flavor development. During brewing, the water is heated to 190°F–200°F in order to extract the flavor of the coffee.

Espresso can be served alone or used as the base for many specialty coffees, such as Americanos and macchiatos. An Americano is made using an equal ratio of espresso to very hot water. A macchioto is served with a small amount of frothy, steamed milk.

Some specialty coffees can be high in sugar, fat, and calories when made with sweeteners and whole milk. The amount of sugar, fat, and calories consumed in specialty coffees is often overlooked but can make a substantial contribution to total daily intake. Appropriate portion sizes of specialty coffees made with low-fat or nonfat dairy products can reduce fat and calories in the diet. **See Figure 8-19.**

Wellness Concept

Research suggests that coffee contains antioxidants, and moderate consumption may help reduce the risk of liver disease and type 2 diabetes.

Specialty Coffee Beverages

Coffee Beverage	Description	Whole Milk, 20 oz Variety			Nonfat Milk, 12 oz Variety		
		Calories	Total fat	Sugar	Calories	Total fat	Sugar
Cappuccino	Equal ratio of espresso to frothy steamed milk	190	9 g	14 g	60	0 g	8 g
Frappuccino®	Espresso blended with milk, ice, and sweeteners	350	5 g	69 g	160	0 g	36 g
Latte	Espresso with double the ratio of steamed milk to espresso	290	15 g	21 g	100	0 g	14 g
Latte, flavored*	Latte with added syrup	370	14 g	44 g	150	0 g	27 g
Mocha	Espresso mixed with chocolate sauce and served with whipped cream	450	23 g	45 g	230	8 g	28 g

* flavored with vanilla syrup

Figure 8-19. Appropriate portion sizes of specialty coffees made with low-fat or nonfat dairy products can reduce fat and calories in the diet.

CULINARY NUTRITION RECIPE-MODIFICATION PROCESS

Café Mocha

Café mocha is a sweet specialty coffee beverage flavored with chocolate. This recipe is high in calories, fat, and sugar due to the whole milk, granulated sugar, and chocolate sauce. It is also garnished with whipped cream and grated chocolate, which add more calories and fat.

Yield: 4 servings, 8 fl oz each

Ingredients

½ c	heavy cream
1 tsp	granulated sugar
3 c	whole milk
4 tbsp	granulated sugar
3 tbsp	unsweetened cocoa powder
1 c	strongly brewed coffee or espresso
¼ c	chocolate syrup
1 oz	dark chocolate, grated

Preparation

1. Whip heavy cream and 1 tsp sugar until soft peaks form. Set aside.
2. Pour whole milk into a saucepan over medium heat.
3. Stir in sugar and cocoa powder until well combined.
4. Add coffee and chocolate syrup, stir to incorporate, and bring to a simmer.
5. Pour hot mixture into mugs and garnish each mug with a fourth of the whipped cream and grated chocolate.

National Honey Board

Evaluate original recipe for sensory and nutritional qualities

- Café mocha is a warm, highly sweetened coffee beverage with a creamy mouthfeel and chocolate flavor.
- Garnish of whipped cream and shaved chocolate enhances texture, flavor, and presentation.
- Beverage is high in calories, fat, and sugar.

Establish goals for recipe modifications

- Reduce calories, fat, and sugar content.
- Maintain flavor and presentation.

Identify modifications or substitutions

- Use a substitution for the whole milk.
- Reduce the amount of granulated sugar and eliminate the chocolate syrup.
- Reduce the amount of whipped cream and grated chocolate for garnishing.

Determine functions of identified modifications or substitutions

- Whole milk provides a smooth, creamy base.
- Sugar and chocolate syrup provide sweetness.
- Whipped cream and grated chocolate enhance mouthfeel, flavor, and presentation.

Select appropriate modifications or substitutions

- Substitute nonfat milk for whole milk.
- Reduce granulated sugar by half and omit chocolate syrup to reduce empty calories.
- Reduce whipped cream and grated chocolate by half.

Test modified recipe to evaluate sensory and nutritional qualities

- Nonfat milk reduces calories and fat while providing an appealing mouthfeel.
- Reducing the granulated sugar and omitting the chocolate syrup reduces calories and sugar content but still yields a sweet, balanced flavor.
- Reducing the amount of whipped cream and chocolate garnish reduces calories and fat while maintaining presentation.

Modified Café Mocha

In this modified café mocha recipe, the calorie, fat, and sugar content is reduced by using nonfat milk, reducing the amount of sugar, and garnishing the beverage with less whipped cream and grated chocolate. The beverage still has a creamy texture, sweet yet balanced flavor, and appealing presentation.

Yield: 4 servings, 8 fl oz each

Ingredients

¼ c	heavy cream
½ tsp	granulated sugar
3 c	nonfat milk
2 tbsp	granulated sugar
3 tbsp	unsweetened cocoa powder
1 c	strongly brewed coffee or espresso
½ oz	dark chocolate, grated

Preparation

1. Whip heavy cream and ½ tsp sugar until soft peaks form. Set aside.
2. Pour nonfat milk into a saucepan over medium heat.
3. Stir in sugar and cocoa powder until well combined.
4. Add coffee, stir to incorporate, and bring to a simmer.
5. Pour hot mixture into mugs and garnish each mug with a fourth of the whipped cream and grated chocolate.

Café Mocha Nutritional Comparison

Nutrition Facts	Original	Modified
Calories	367.8	173.1
Total Fat	20.1 g	7.5 g
Saturated Fat	12.1 g	4.7 g
Sugar	36.1 g	18.5 g

Signature Soy Milk Café Mocha

In this signature café mocha, the whipped cream is eliminated and the milk is replaced with soy milk to create a lactose-free beverage. Eliminating the whipped cream significantly reduces calories and fat. In addition to grated chocolate, this beverage is garnished with cinnamon to elevate the flavor.

Yield: 4 servings, 8 fl oz each

Ingredients

3 c	nonfat soy milk
2 tbsp	granulated sugar
3 tbsp	unsweetened cocoa powder
1 c	strongly brewed coffee or espresso
2 tsp	cinnamon
½ oz	dark chocolate, grated

Preparation

1. Pour soy milk into a saucepan over medium heat.
2. Stir in sugar and cocoa powder until well combined.
3. Add coffee, stir to incorporate, and bring to a simmer.
4. Pour hot mixture into mugs and garnish each mug with a fourth of the cinnamon and grated chocolate.

Chef's Tip: *Add coffee liqueur to café mocha to create an alcoholic beverage with a deep coffee flavor.*

Café Mocha Nutritional Comparison

Nutrition Facts	Original	Modified	Signature
Calories	367.8	173.1	109.3
Total Fat	20.1 g	7.5 g	2.0 g
Saturated Fat	12.1 g	4.7 g	1.1 g
Sugar	36.1 g	18.5 g	14.7 g

Teas

Tea is available loose-leaf or in tea bags. Specialty tea blends and flavored teas offer a variety of unique flavors. Tea is calorie-free, contains a low to moderate amount of caffeine, and provides a variety of antioxidants.

Tea is commonly served hot or cold with accompaniments such as milk, sugar, and lemon. Appealing aromas, colors, and flavors that are neither too weak nor too strong are the result of properly brewed tea. Tea can also be used to create specialty beverages.

To properly brew tea, water is heated to just below the boiling point and the tea is steeped for 3–5 minutes. After the tea has steeped, the tea leaves or bag is removed. It is important not to steep the tea too long because this will make the tea bitter.

Various tea leaves can be combined to create flavorful tea blends. **See Figure 8-20.** Additional ingredients such as spices, citrus, sweeteners, and milk are often incorporated into specialty teas to add flavor, but may also add sugar, fat, and calories. Sweeteners and milk are used in tea lattes, which are often high in sugar, fat, and calories. For example, a 16 fl oz tea latte contains approximately 29 g of sugar, 7 g of fat, and 200 calories.

Fruit and Vegetable Beverages

Fruit and vegetable beverages are a healthy source of vitamins and minerals but also contain a lot of natural sugar. **See Figure 8-21.** Although 100% fruit and vegetable juices contain natural sugars, they are a healthier choice than juices with added sugars and artificial ingredients. Most foodservice operations offer guests 8–12 fl oz servings of juice yet a 4 fl oz serving is equivalent to one serving of fruit or vegetables.

Nutritional Comparison of Fruit and Vegetable Beverages*

Fruit or Vegetable Beverage	Calories	Sugar
Apple	114	24 g
Cranberry	116	31 g
Grape	152	36 g
Grapefruit	94	22 g
Orange	112	21 g
Pineapple	133	25 g
Tomato	41	9 g
Vegetable blend	53	9 g

* based on 8 fl oz serving size

Figure 8-21. Fruit and vegetable beverages are a healthy source of vitamins and minerals but also contain a lot of natural sugar.

Tea Blends

Type	Description
Caravan	Blend of Chinese, Indian, and Taiwan black teas; usually sweetened
Chai	Tea blended with warm, sweet spices, typically cardamon, cinnamon, cloves, ginger, and nutmeg
Earl Grey	Black tea blended with bergamot citrus oil
English black tea	Full-bodied blend of black teas; typically includes Keemun tea
Irish breakfast tea	Robust blend of black teas; typically includes Assam tea

Figure 8-20. Various tea leaves can be combined to create flavorful tea blends.

Smoothies made without dairy ingredients often highlight the flavors of fruits, vegetables, or juices. Green drinks, or green smoothies, contain green vegetables such as kale or spinach as a main ingredient. Green drinks are high in dietary fiber and antioxidants and are typically low in sugar and fat.

Alcoholic Beverages

Alcoholic beverages are common on menus, and many guests enjoy an alcoholic beverage that complements their meal. While alcoholic beverages can enhance the dining experience, they are generally high in both simple carbohydrates and calories and are often metabolized before other nutrients. Organs such as the liver, stomach, and kidneys are directly affected by alcohol consumption. **See Figure 8-22.**

Alcoholic beverages are commonly mixed with a variety of ingredients, including fruit and vegetable juices, simple syrups, soft drinks, and cream, all of which add to caloric intake. Reduced-calorie alcohols such as light beers, vodkas, and packaged mixed drinks are viable alternatives to high-calorie alcohols.

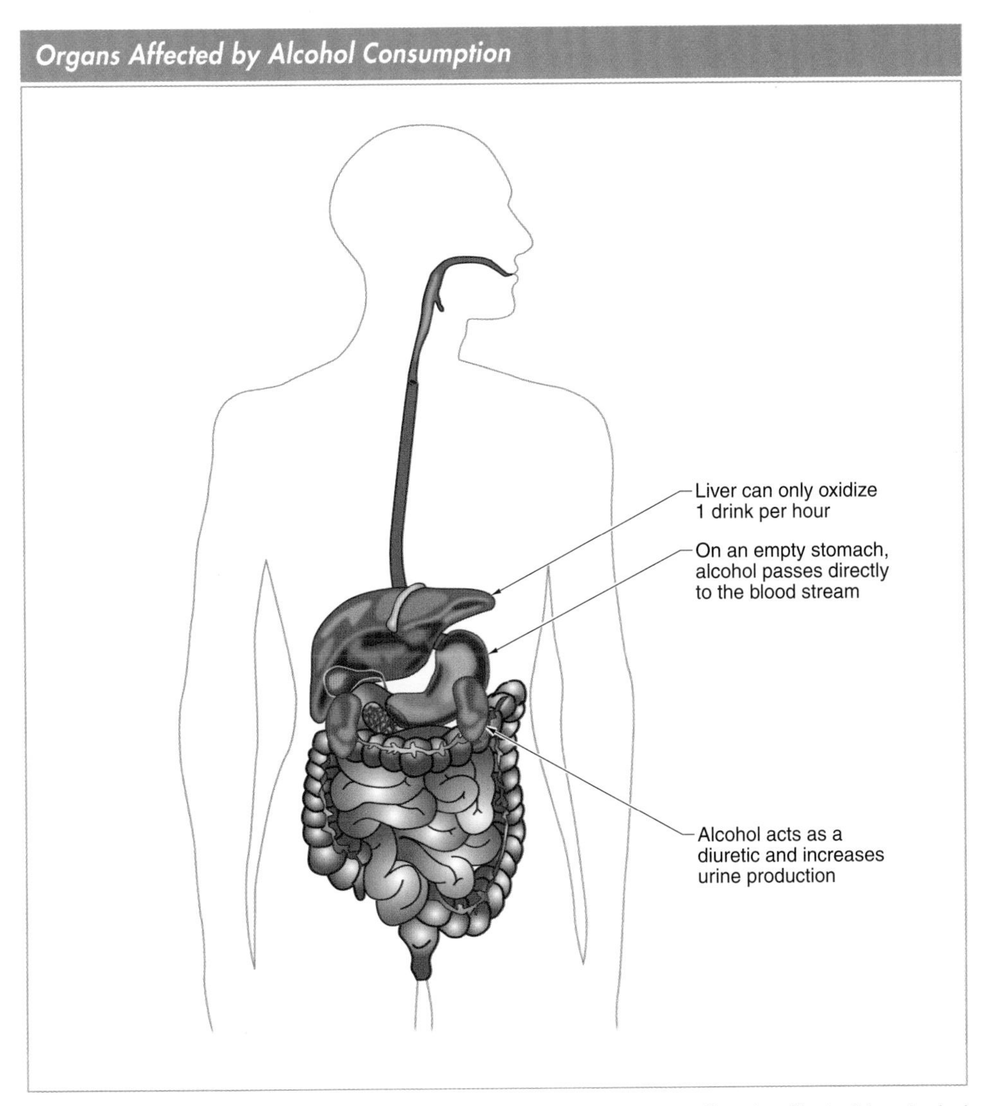

Figure 8-22. Organs such as the liver, stomach, and kidneys are directly affected by alcohol consumption.

Wines. Wine is served in glasses chosen for the type of wine being served. Wine glasses consist of a base, a stem, and a bowl. **See Figure 8-23.** All wine glasses have a bowl with a slightly narrow opening that widens toward the bottom. This allows air to mix with the wine. White wines are served in a narrow glass that is U-shaped to extract aroma. Red wine glasses have a wider bowl that allows the guest to swirl and aerate the wine in order to release the full aroma. Sparkling wines are served in tall, narrow flutes to retain carbonation.

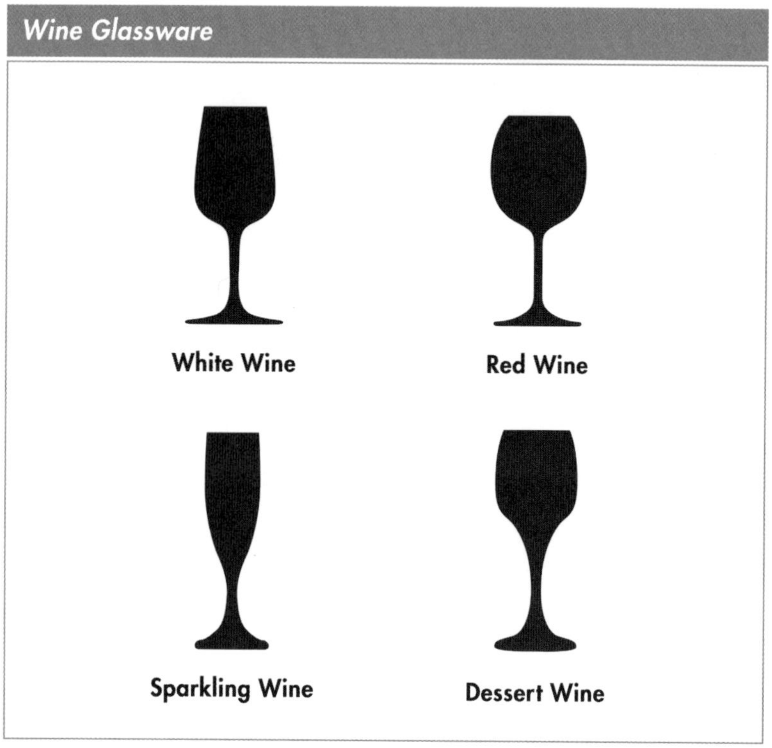

Wine Glassware

White Wine

Red Wine

Sparkling Wine

Dessert Wine

Figure 8-23. Wine glasses consist of a base, a stem, and a bowl and are chosen based on the type of wine being served.

The serving temperatures of wines vary. However, it is generally thought that the optimal temperatures to serve wines are as follows:

- white wines: between 45°F–55°F
- red wines: between 55°F–65°F
- sparkling wine: between 40°F–45°F

Beers. Beer is served in steins, pilsner glasses, goblets, weizen glasses, flutes, stemmed glasses, and pint glasses. **See Figure 8-24.** The width of the glass affects the color and head (foam) of the beer. For example, a dark beer served in a wide glass does not look as dark as one served in a narrow glass.

Figure 8-24. Beer is commonly served in steins, pilsner glasses, goblets, and weizen glasses.

Beer is produced worldwide, creating a range of global flavor profiles. Craft beers are brewed in small batches and often incorporate unique and regional ingredients such as chile peppers, ginger, honey, maple syrup, lemons, or raspberries. The resulting brews are innovative creations with complex flavors. Some restaurants even brew signature beers on-site.

Spirits. Although spirits are served in a variety of glassware, the most common glasses include shot glasses, rocks glasses, and snifters. For example, a rocks glass is used to serve spirits "on the rocks" (with ice), while a snifter is often used to serve brandy "neat" (without ice).

Spirits are commonly used as the base for many cocktails (mixed drinks) such as martinis, cosmopolitans, daiquiris, margaritas, and piña coladas. Cocktails are typically served in rocks glasses, tall highball glasses (8–12 fl oz), or Collins glasses (14 fl oz). Drinks that are shaken with ice and then strained are often served in stemmed glassware. **See Figure 8-25.**

Bartenders are adept at pouring traditional alcoholic beverages as well as those that are unique to the venue. Some bartenders are similar to chefs, using innovative combinations of ingredients such as seasonal fruits and exotic spice blends to create signature beverages.

Guests are seeking healthier menu items and foodservice operations are responding by offering lower calorie alcoholic beverages and incorporating fresh fruits, vegetables, herbs, spices, and sparkling water to create flavorful beverages that are lower in sugar, fat, and calories. For example, vodka mixed with seltzer water and a splash of cranberry juice and served with lime is a light, refreshing alternative to a calorie-dense cocktail. Cocktails such as piña coladas and white Russians can be made healthier by cutting the sugar in half and substituting low-fat dairy products for cream.

Chef's Tip

"Beertails" combine beer with teas, juices, spirits, or sugars to create inventive drinks that are gaining popularity.

Spirit and Cocktail Glassware

Daniel NYC/M. Hom
Rocks Glass

Browne Foodservice
Snifters

Daniel NYC
Stemmed Glass

Figure 8-25. Spirits and cocktails are commonly served in rocks glasses or snifters or strained into stemmed glassware.

CULINARY NUTRITION RECIPE-MODIFICATION PROCESS

White Sangria

Sangria is a refreshing alcoholic beverage that is traditionally made with red wine and seasonal fruits. White wines can also be used to make white sangrias. The fruits soak in a sugar and citrus mixture to soften and are served with the beverage to provide color, texture, and flavor. This recipe is made with white wine, orange liqueur, sugar, cantaloupe, peaches, and apricots to produce a golden-colored beverage that is sweet and flavorful but high in calories.

Yield: 4 servings, 8 fl oz each

Ingredients

1½ c	cantaloupe, rind and seeds removed, large dice
1 ea	peach, pit removed, ½ inch wedges
1 ea	apricot, pit removed, ½ inch wedges
¼ c	granulated sugar
2 tbsp	lime juice
2 tbsp	lemon juice
750 ml	white wine, chilled
¼ c	orange liqueur

Preparation

1. Place prepared cantaloupe, peaches, and apricots in a nonreactive pitcher.
2. Add sugar, lime juice, and lemon juice. Stir gently to combine.
3. Cover and refrigerate for 1 hour.
4. Add wine and orange liqueur. Stir to incorporate.
5. Serve chilled.

Evaluate original recipe for sensory and nutritional qualities

- White sangria is a refreshing alcoholic beverage with a sweet, fruity flavor.
- Beverage is light-gold in color with pieces of orange-colored fruit.
- Beverage is high in calories and sugar and low in dietary fiber.

Establish goals for recipe modifications

- Reduce calories and sugar content.
- Increase dietary fiber.
- Add more color, flavor, and nutrients.

Identify modifications or substitutions

- Reduce the amount of refined sugar.
- Use a substitution for orange liqueur.
- Add more colorful fruits.

Determine functions of identified modifications or substitutions

- Sugar provides sweetness.
- Orange liqueur provides a mild orange flavor.
- Adding more fruits will enhance color, flavor, and nutrient density.

Select appropriate modifications or substitutions

- Reduce sugar to 2 tbsp.
- Substitute orange juice for orange liqueur.
- Add oranges, limes, and lemons for color and nutrient density.

Test modified recipe to evaluate sensory and nutritional qualities

- Reducing the sugar lowers calories while producing a pleasing flavor that is not overly sweet.
- Substituting orange juice for orange liqueur adds color, a mild orange flavor, and vitamin C.
- Adding oranges, limes, and lemons adds color, flavor, natural sugar, dietary fiber, and vitamin C.

Modified White Sangria

In this modified white sangria, the sugar and alcohol contents are reduced and orange juice, oranges, limes, and lemons are added. These changes produce a colorful, refreshing beverage that is lower in calories and refined sugar and higher in dietary fiber and vitamin C.

Yield: 5 servings, 8 fl oz each

Ingredients

1½ c	cantaloupe, rind and seeds removed, large dice
1 ea	peach, pit removed, ½ inch wedges
1 ea	apricot, pit removed, ½ inch wedges
1 ea	orange, ¼ inch slices
1 ea	lime, ¼ inch slices
1 ea	lemon, ¼ inch slices
2 tbsp	granulated sugar
1 c	orange juice
2 tbsp	lime juice
2 tbsp	lemon juice
750 ml	white wine, chilled

Preparation

1. Place prepared cantaloupe, peaches, apricots, oranges, limes, and lemons in a nonreactive pitcher.
2. Add sugar, orange juice, lime juice, and lemon juice. Stir gently to combine.
3. Cover and refrigerate for 1 hour.
4. Add wine and stir to incorporate.
5. Serve chilled.

White Sangria Nutritional Comparison

Nutrition Facts	Original	Modified
Calories	288.9	212.6
Total Fat	0.2 g	0.4 g
Dietary Fiber	1.1 g	3.0 g
Sugar	20.3 g	18.1 g
Vitamin C	29.8 mg	83.1 mg

Signature White Sangria Fruit Medley

Additional fruit and club soda are added to create this signature white sangria. Incorporating an array of fruits adds contrasting colors and enhances both flavor and nutrient density. Club soda provides effervescence to elevate mouthfeel.

Yield: 8 servings, 8 fl oz each

Ingredients

1½ c	cantaloupe, rind and seeds removed, large dice
1 ea	peach, pit removed, ½ inch wedges
1 ea	apricot, pit removed, ½ inch wedges
1 ea	orange, ¼ inch slices
1 ea	lime, ¼ inch slices
1 ea	lemon, ¼ inch slices
1 qt	strawberries, stemmed, ¼ inch slices
1 c	raspberries
1 ea	mango, peeled, pit removed, ¼ inch slices
2 tbsp	granulated sugar
1 c	orange juice
2 tbsp	lime juice
2 tbsp	lemon juice
750 ml	white wine, chilled
2 c	club soda

Preparation

1. Place prepared cantaloupe, peaches, apricots, oranges, limes, lemons, strawberries, raspberries, and mangoes in a nonreactive pitcher.
2. Add sugar, orange juice, lime juice, and lemon juice. Stir gently to combine.
3. Cover and refrigerate for 1 hour.
4. Add wine and club soda. Stir to incorporate.
5. Serve chilled.

White Sangria Nutritional Comparison

Nutrition Facts	Original	Modified	Signature
Calories	288.9	212.6	184.2
Total Fat	0.2 g	0.4 g	0.7 g
Dietary Fiber	1.1 g	3.0 g	5.0 g
Sugar	20.3 g	18.1 g	19.7 g
Vitamin C	29.8 mg	83.1 mg	111.9 mg

Soft Drinks

Soft drinks are available at virtually all foodservice venues any time of the day. Many people consume soft drinks without realizing the nutritional impact of these beverages. Reducing both the serving size and the number of soft drinks consumed can make a significant impact on daily sugar and calorie intake. **See Figure 8-26.** Diet soft drinks are popular zero-calorie beverages sweetened with FDA-approved artificial sweeteners. Some soft drinks contain fruit juice, thereby adding nutrients. A healthy alternative to soft drinks is to mix carbonated water with fruit juice. This provides a similar flavor profile and texture as a soft drink, but increases nutrients while reducing sugars and calories.

> ***Chef's Tip***
>
> Sweet tea contains added sugar and has a similar sugar composition to soft drinks. A refreshing, healthier option is to mix equal parts of unsweetened tea with cranberry juice.

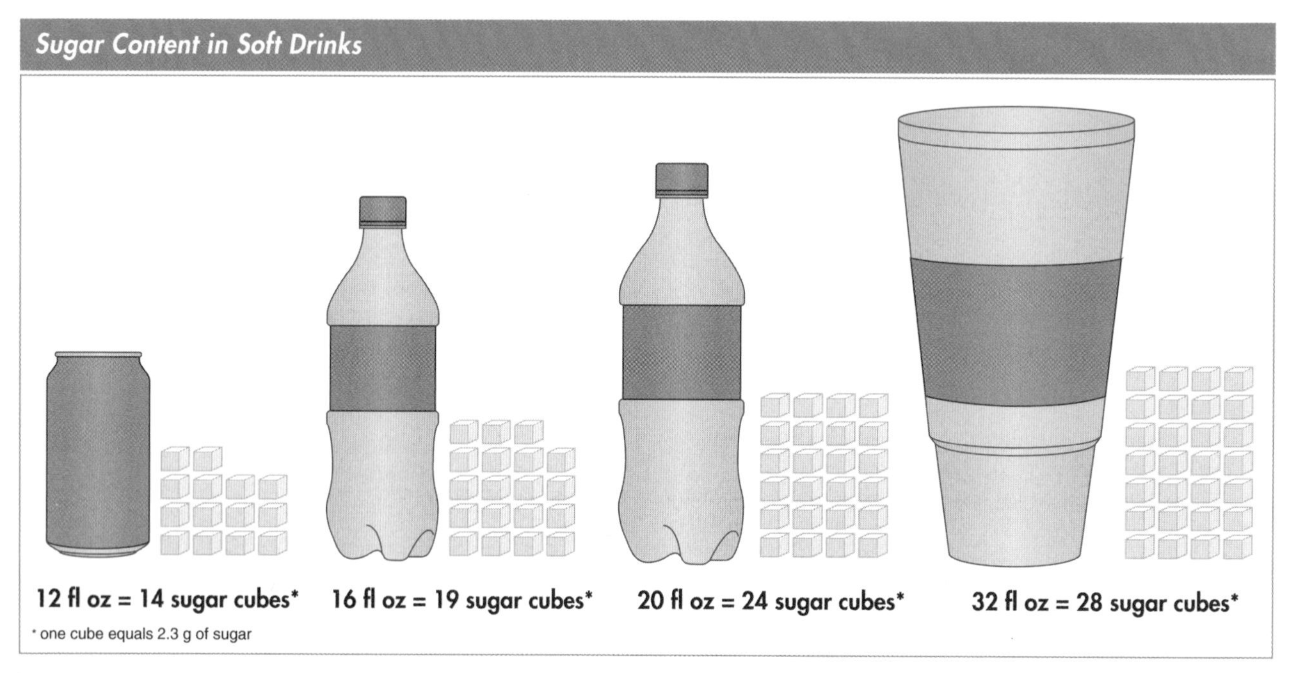

Figure 8-26. Reducing both the serving size and the number of soft drinks consumed can make a significant impact on daily sugar and calorie intake.

Knowledge Check 8-2

1. Identify the caffeine content of energy drinks.
2. Explain how smoothies and milkshakes can be prepared nutritiously.
3. Identify ways to make specialty coffees healthier.
4. Explain why specialty teas can be high in fat and calories.
5. Identify the serving size of juice that is equivalent to one serving of fruits or vegetables.
6. Explain why different-shaped wine glasses are used for different types of wine.
7. Describe how the type of glass used for beer affects presentation.
8. Explain how cocktails can be prepared more nutritiously.
9. Describe the impact of soft drinks on caloric intake.

BEVERAGE MENU MIX

Beverages generate revenue for every foodservice operation. Beverages are not only popular at meal times, but they are also regularly consumed to relax, recharge, and socialize throughout the day. Beverages may be listed on the main menu, promoted on special menus, or described verbally by a server. Offering a variety of nutritious beverage choices is appreciated by health-conscious guests and encourages others to make healthy beverage choices.

Beverage descriptions such as "lemonade" or "iced tea" are often straightforward. However, separate beverage menus are frequently used to promote alcoholic beverages and signature drinks. This allows room for effective menu descriptions that pique interest and encourage sales. **See Figure 8-27.**

Beverage Menu Descriptions

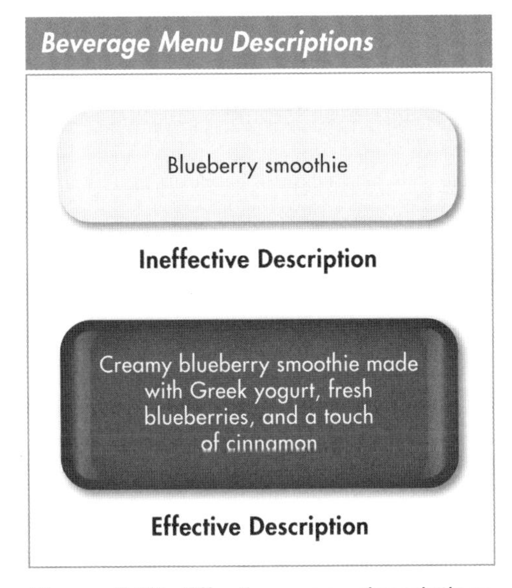

Figure 8-27. Effective menu descriptions pique interest and encourage sales.

Mealtime Beverages

Breakfast, lunch, and dinner menus feature a variety of beverages to complement the food choices offered. Specific beverages are often a draw for guests. For example, a restaurant may be known for having superior coffee or a superb wine selection. Offering guests a variety of beverage choices that are served at the appropriate temperature encourages beverage orders.

Breakfast. For most people, breakfast begins with coffee, tea, juice, or milk. **See Figure 8-28.** Specialty coffees, such as cappuccinos and mochas, and seasonally flavored coffees, such as pumpkin or cinnamon, are quite popular. Tea is also the foundation for many unique flavor combinations, such as tea lattes infused with spices.

Breakfast Beverages

Figure 8-28. Breakfast beverages often include coffee, tea, and juice.

Juices such as orange, grapefruit, apple, tomato, and vegetable juice are popular breakfast choices. Some foodservice operations feature freshly squeezed juices and may even offer guests the opportunity to create their own juice blends from a variety of seasonal fruits. For brunch, juices such as orange juice and peach nectar are often combined with sparkling wine to create Mimosas and Bellinis. Tomato juice may be combined with vodka and featured in Bloody Marys.

Milk and hot chocolate are also popular breakfast choices. Whole, low-fat, nonfat, or flavored milks may be offered. Milklike beverages such as soy, almond, and rice milks are often offered for their flavor and to meet the needs of lactose intolerant guests.

Lunch. Lunch menus often feature soft drinks, coffee, and tea along with alcoholic beverages such as beer if the venue is appropriate. Water, lemonades, iced teas, and coffees are commonly served with lunch. Infused water that features citrus, melons, berries, cucumbers, or fresh mint is also popular. **See Figure 8-29.**

Infused Water

Service Ideas, Inc.

Figure 8-29. Infused water that features citrus, melons, berries, cucumbers, or fresh mint is popular on lunch menus.

Dinner. Water, soft drinks, coffees, teas, and alcoholic beverages are commonly offered on dinner menus. Wines, beers, and spirits are served at most sit-down restaurants, where a beverage menu may consist of a wine list, craft beers, specialty coffees, and signature cocktails.

Beverages as Desserts

Beverages served as desserts include smoothies, milkshakes, malts, floats, hot chocolates, specialty coffees and teas, and liqueurs. Seasonality plays a role in dessert beverages. Hot apple cider and hot buttered rum are traditionally served in cooler months, while blended iced coffees and root beer floats are common in warmer months.

Coffee is served as a dessert beverage in a variety of ways, including cappuccinos and lattes. Coffee can be mixed with spirits to make specialty beverages, such as café brûlot, a flaming Creole coffee with brandy, citrus zest, and spices. Almond-, hazelnut-, coffee-, and orange-flavored liqueurs are commonly added to coffee to provide a sweet, warm finish to a meal. **See Figure 8-30.** Liqueurs and fortified wines are also served as dessert beverages.

Liqueurs Mixed with Coffee

Liqueur	Flavor Profile	Alcohol by Volume
Amaretto	Sweet almond and spice	28%
Chocolate	Cocoa and cream	12% – 24%
Coffee	Coffee and vanilla	20%
Frangelico	Hazelnut, cocoa, and vanilla	24%
Grand Marnier	Orange	42%
Irish cream	Sweet, rich, and nutty	17%

Adapted from The Gourmet's Guide to Cooking with Liquors and Spirits

Figure 8-30. Flavored liqueurs are commonly added to coffee to provide a sweet, warm finish to a meal.

Knowledge Check 8-3

1. List beverages commonly served during breakfast.
2. Compare the beverages served during lunch and dinner.
3. Explain how beverages can be served as desserts.

PROMOTING BEVERAGES ON THE MENU

An important first step in promoting beverages on the menu is to make sure the beverages complement the venue. Offering a range of beverage choices is also important. Promotions can be tailored to the venue and feature unique experiences. For example, many quick service and fast casual venues offer nonalcoholic beverages as self-serve and commonly offer free refills. **See Figure 8-31.** Other venues may offer pairings of wines, beers, or spirits for each course. Beverages can also be used to highlight local dairies, wineries, and breweries and to promote seasonal harvests. Beverage houses and bars as well as fine dining and special event venues should make the most of promoting beverages on the menu.

Self-Service Beverage Stations

Manitowoc Beverage Systems

Figure 8-31. Self-serve beverages are common at quick service and fast casual venues.

> **Chef's Tip**
>
> Truth-in-menu guidelines should always be adhered to when promoting beverages. If a particular brand is described on the menu, it must be used in the beverage that is served.

Beverage Houses and Bar Venues

Beverage houses and bars often feature specific types of coffees, teas, juices, wines, beers, and mixed drinks. For example, juice bars offer a menu of juices and juice blends that rely on fresh fruits and vegetables and often prepare beverages to guest specifications. In addition, guests are commonly given the option of adding ingredients such as wheat germ, flax seeds, and protein powder to increase the nutrient density of the beverage.

Coffee houses offer a variety of roasts, blends, and specialty beverages. In a similar fashion, tea houses specialize in unique and imported tea varieties. **See Figure 8-32.** Both the brewing process and tea service are commonly dictated by customs and traditions. Likewise, bars often specialize in brands from local wineries, craft breweries, or spirit distilleries. Beverage houses and bars frequently offer tastings to give guests the opportunity to sample featured beverages.

Tea Varieties

Rishi Tea

Figure 8-32. Tea houses specialize in unique and imported tea varieties.

Fine Dining and Special Event Venues

Rich specialty coffees, exceptional quality teas, and top-shelf alcoholic beverages are commonly offered at fine dining and special event venues. Attention to detail is a key focus at these venues and is often reflected in special services. For example, a sommelier or a mixologist is often available to advise guests on beverage choices. A *sommelier* is an individual formally trained in ordering, storing, and serving wines to complement a wide variety of menu items. **See Figure 8-33.** A *mixologist* is an individual formally trained in the preparation of mixed drinks, or cocktails. Signature drinks may also be featured to represent the fine dining operation or the theme of the special event.

Regardless of the venue, presenting beverages in the appropriate container is important. **See Figure 8-34.** Containers with a logo are often used to promote the foodservice operation or a special event. Appealing garnishes such as fresh fruits, vegetables, and herbs are used to elevate presentations and entice guests to have another beverage and to return to the venue.

Sommeliers

Daniel NYC/T. Schauer

Figure 8-33. A sommelier is an individual formally trained in ordering, storing, and serving wines to complement a wide variety of menu items.

Beverage Presentation

National Honey Board

Figure 8-34. Presenting beverages in the appropriate container can elevate their perceived value.

Knowledge Check 8-4

1. Describe the types of beverages featured at beverage houses and bar venues.
2. Differentiate between a sommelier and a mixologist.
3. Describe how beverage containers may be used to promote foodservice operations.

Chapter Summary

Beverages appear on every menu, are served across all meal periods, and are a significant source of revenue for foodservice operations. Beverages commonly represented on menus include water, dairy beverages, milklike beverages, coffees, teas, fruit and vegetable beverages, alcoholic beverages, and soft drinks. The venue type strongly influences the beverages that are served and the menu prices.

Water is necessary for life and pure drinking water is an optimal beverage choice for keeping the body hydrated. Other beverages also supply water and can be a source of key nutrients including protein, complex carbohydrates, essential fats, vitamins, and minerals. In contrast, some beverages are high in sugar, fat, calories, caffeine, or artificial ingredients and should be consumed in moderation. Offering guests a variety of nutrient-dense beverage options enhances the dining experience and meets today's demand for nutritious menu options.

National Honey Board

Chapter Review

1. Describe factors that influence the perceived value of beverages.
2. Compare four types of water beverages.
3. Compare the health benefits of dairy and milklike beverages.
4. Explain how processing affects coffee and tea.
5. Identify common market forms of fruit and vegetable beverages.
6. Compare the production of three types of alcoholic beverages.
7. Identify nutritional considerations for soft drinks.
8. Explain how to prepare each classification of beverages to maximize its nutrients.
9. List beverages commonly served at breakfast, lunch, dinner, and dessert.
10. Explain how beverages are promoted at various foodservice venues.

Egg, Soy & Dairy Products on the Menu

Daniel NYC/E. Kheraj

House Foo

Chapter 9

Dairy

Wisconsin Milk Marketing Board, Inc.

Egg, soy, and dairy products are often used in dishes to add texture and flavor, but they are also complete protein sources packed with essential fatty acids, vitamins, and minerals. Because of their versatility, chefs can use egg, soy, and dairy products in a variety of ways to create nutritious and satisfying menu options ranging from beverages to desserts. They can also be featured in some vegetarian dishes to meet the increasing demand for meatless menu items.

Chapter Objectives

1. Describe the four main parts of an egg.
2. Summarize the nutritional benefits of eggs.
3. Describe ways in which eggs are perceived.
4. Describe cooking techniques used to prepare eggs.
5. Identify ways eggs are used on the menu.
6. Describe ways to promote eggs on the menu.
7. Compare vegetable soybeans to field soybeans.
8. Summarize the perceived value of soy products.
9. Identify common market forms of soy products.
10. Describe how to prepare nutrient-dense and flavorful soy products.
11. Identify ways soy products are used on the menu.
12. Describe ways to promote soy products on the menu.
13. List calcium-rich dairy products.
14. Describe how dairy products can be perceived.
15. Identify liquid, semisolid, and solid dairy products.
16. Explain how to effectively prepare and plate dairy products.
17. Identify ways dairy products are used on the menu.
18. Describe ways to promote dairy products on the menu.

EGGS

Eggs have been a source of high-quality protein and nourishment for centuries and are one of the most versatile foods used in cuisines around the world. Although eggs from chickens are the most commonly used, eggs from quails, ducks, and geese are also used.

An egg consists of four main parts: the shell, shell membrane, yolk, and white (albumen). **See Figure 9-1.** An *eggshell* is the brittle, porous covering of an egg that protects the fragile yolk and white. The outside of the shell is covered by a protective layer called the bloom. Most egg producers wash off the bloom and spray the eggshell with mineral oil. The oil prevents air exchange and preserves the freshness of the egg.

Egg Composition

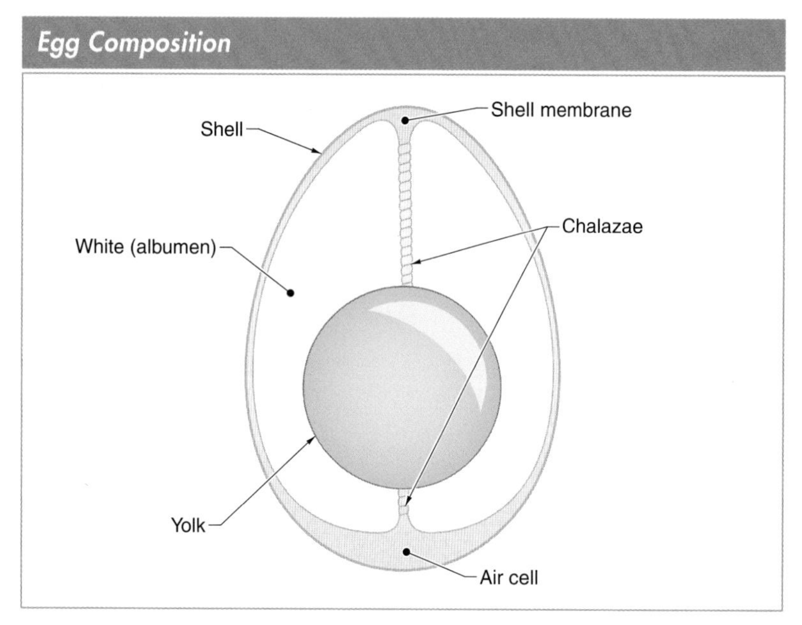

Figure 9-1. An egg consists of four main parts: the shell, shell membrane, yolk, and white (albumen).

The area just beneath the eggshell is the shell membrane. The *shell membrane* is the thin, skinlike material located directly under an eggshell. The *egg yolk* is the yellow portion of an egg. The *egg white,* also known as the albumen, is the clear portion of a raw egg, which makes up two-thirds of the egg and consists mostly of ovalbumin protein.

The internal structure of the egg is maintained by the chalazae, which are found in the albumin. The *chalazae* are twisted cordlike strands of egg white that anchor the yolk to the center of the egg white. A freshly laid egg contracts as it cools, which causes the membranes to separate and an air cell to form at the large end of the egg.

Food Science Note

For individuals with egg allergies, the immune system considers the proteins in eggs a threat and releases chemicals that cause an allergic reaction.

Egg Nutrients

Eggs are a complete protein. One large egg contains 6 g of protein. The majority of the protein is found in the egg white, which contains 4 g. The egg white also contains adequate amounts of all the essential amino acids, making the egg the standard by which other protein foods are compared.

In addition to protein, eggs also contain varying amounts of other essential nutrients. **See Figure 9-2.** For example, eggs are a good source of selenium, which is necessary for proper thyroid function. Eggs are also one of the few foods considered an excellent source of choline. Choline is a B vitamin found in egg yolk and is important for proper nerve and brain function. Egg yolk is also rich in vitamins A and D, phosphorus, iron, and the antioxidants lutein and zeaxanthin, which support eye health.

The yolk also contains fat and cholesterol. The yolk of one egg contains approximately 5 g of total fat, but less than one-third of this fat is saturated. Although egg yolks are high in cholesterol, research suggests that the saturated fat in foods raises a person's cholesterol more than the cholesterol contained in the food does.

Nutritional Data for a Large Egg (Raw)

Total calories	72
From fat (63%)	46.03
From protein (35%)	25.89
From carbohydrates (2%)	1.58
Total fat	4.75 g
Saturated fat	1.50 g
Monounsaturated fat	1.91 g
Polyunsaturated fat	0.68 g
Cholesterol	211.50 mg
Sodium	71 mg
Potassium	67 mg
Total carbohydrates	0.36 g
Dietary fiber	0 g
Sugar	0.18 g
Protein	6.28 g

Vitamins	Amount
Vitamin A	270 IU
Vitamin D	41 IU
Vitamin E	0.53 mg
Vitamin K	0.10 mcg
Vitamin C	0 IU
Thiamin (Vitamin B_1)	0.02 mg
Riboflavin (Vitamin B_2)	0.23 mg
Niacin (Vitamin B_3)	0.04 mg
Pantothenic acid (Vitamin B_5)	0.77 mg
Vitamin B_6	0.08 mg
Choline	125.50 mg
Folate (Vitamin B_9)	24 mcg
Vitamin B_{12}	0.45 mcg

Minerals	Amount
Calcium	28 mg
Copper	0.04 mg
Fluoride	0.6 mcg
Iodine	23.70 mcg
Iron	0.88 mg
Manganese	0.01 mg
Magnesium	6 mg
Phosphorus	99 mg
Potassium	69 mg
Selenium	15.30 mcg
Zinc	0.65 mg

Source: USDA National Nutrient Database, SR 23

Figure 9-2. Eggs contain many essential nutrients.

Perceived Value of Eggs

For foodservice operations, eggs are an economical source of high-quality protein. The common serving size on the menu is two eggs as a main dish. Eggs can help lower the total cost of a meal while still keeping the protein value high. For example, the amount of poultry or meat used in an entrée salad or sandwich can be reduced when an egg is added to the item. As a result, the dish is more cost effective, yet protein-rich and flavorful.

In addition to being used as a main ingredient, many popular menu items would not be possible without the functions that eggs provide. Eggs are used to bind, emulsify, thicken, coat, clarify, and leaven foods. They are also used to retard crystallization. **See Figure 9-3.**

Egg Functions

Function	Culinary Application
Binding agent	Dishes such as crab cakes, meatloaf, and lasagna
Emulsifying agent	Products such as mayonnaise, hollandaise sauce, and salad dressings
Thickening agent	Dishes such as puddings, custards, and sauces
Coating	Items to be breaded such as chicken and fish
Clarifying agent	Foods such as consommés
Leavening agent	Baked goods, omelets, and soufflés
Crystallization retardant	Whites are used in confections

Figure 9-3. Eggs are used to bind, emulsify, thicken, coat, clarify, and leaven foods and to retard crystallization.

Some people perceive eggs as a breakfast item. However, chefs often use eggs to turn common menu items into unique and elevated dishes, such as topping pizza with a poached egg. **See Figure 9-4.**

Perceived Value of Eggs

United States Potato Board

Figure 9-4. Eggs can be used to elevate menu items and increase the perceived value of a dish.

Although eggs are an excellent source of protein, the fat and cholesterol in eggs may pose a health concern to some individuals. Research suggests that individuals following a low-fat and/or low-cholesterol diet can consume 1–2 eggs per day without experiencing a rise in blood cholesterol levels. However, many guests prefer to order dishes made from egg whites so that they can have a high-protein meal that is also low in fat and cholesterol.

Market Forms of Eggs

Eggs are classified according to size based on the minimum weight per dozen eggs. Large eggs are the standard size used in foodservice operations. While whole shell eggs are the most common variety of egg used in foodservice, egg convenience products and egg substitutes are also used.

Whole Eggs. Whole eggs are available in many forms including eggs with brown shells, organic eggs, nutrient-enriched eggs, and pasteurized eggs. Eggs with brown shells are produced by a specific breed of hens and do not differ in terms of nutrition or flavor from eggs with white shells. Organic eggs have to follow strict FDA guidelines and come from hens that are not given hormones or antibiotics. Organic eggs also come from hens that are raised in a free-range environment with outdoor access. Some hens are fed special diets such as a vegetarian diet or a diet designed to produce eggs that are nutrient-enriched with omega-3 fatty acids.

Pasteurized eggs are often used in recipes that call for raw or undercooked eggs, such as hollandaise sauce. **See Figure 9-5.** *Pasteurization* is the process of destroying harmful microorganisms by heating and then quick-cooling a food product. Pasteurization makes foods such as eggs and milk safer to consume and improves their shelf life.

Eggs with Hollandaise Sauce

National Turkey Federation

Figure 9-5. Whole eggs are sometimes pasteurized for safety and used in recipes that call for raw or undercooked eggs, such as hollandaise sauce.

Egg Convenience Products. Frozen eggs, pasteurized liquid eggs, and powdered eggs are convenience products commonly used in the foodservice industry. Frozen eggs include shelled eggs that are usually pasteurized and available as whole eggs, egg whites, or egg yolks. Pasteurized liquid eggs are shelled and refrigerated. Like frozen eggs, liquid eggs are sold whole or separated. Powdered eggs are also available. Egg convenience products generally offer the same nutritional value as fresh eggs and are used most often in bakeries, volume kitchens, and manufacturing operations.

Chef's Tip

Egg whites and yolks can be frozen but should be separated before freezing for best results. Egg whites freeze well without any preparation. Yolks need to be mixed with salt, sugar, or acid to keep the protein molecules from grouping together and forming a paste.

Egg Substitutes. An *egg substitute* is an egg product made from pasteurized eggs that have had the yolks removed and vegetable gum, coloring, and flavoring added. Compared to whole eggs, egg substitutes are lower in calories and are generally fat- and cholesterol-free. **See Figure 9-6.** Egg substitutes can be used to replace whole eggs in a variety of items, including omelets, quiches, and some baked goods. They are usually frozen to provide a longer shelf life.

For individuals with egg allergies and some vegetarians, eggless egg substitutes are sometimes used. An *eggless egg substitute* is a yellow-colored liquid composed of soy, vegetable gums, and starches derived from corn or flour. Most eggless egg substitutes are sold in powdered form and are used most often in baked goods. Ground flaxseeds, puréed bananas, or silken tofu can also be used as a substitute for eggs in some recipes.

Nutritional Comparison of Whole Eggs and Egg Substitutes*

Egg Product	Calories	Total Fat	Cholesterol	Protein
Whole eggs	87	6 g	226 mg	8 g
Egg substitutes	29	0 g	0 g	6 g

* based on ¼ cup serving size

Figure 9-6. Compared to whole eggs, egg substitutes are lower in calories and are generally fat- and cholesterol-free.

Knowledge Check 9-1

1. Describe the four main parts of an egg.
2. Identify the nutrients found in the egg white and the egg yolk.
3. List common functions of eggs in culinary preparations.
4. Define pasteurization.
5. List common egg convenience products.

PREPARING NUTRITIOUS EGGS

In order to produce high-quality menu items, it is important to use eggs that have been stored and handled properly. Eggs should be refrigerated with large ends up to prevent movement of the air cell and preserve freshness.

Cooking eggs to an internal temperature of 160°F helps destroy potentially harmful bacteria like salmonella enteritidis. Salmonella enteritidis is found inside eggs that have become contaminated from infected laying hens. **See Figure 9-7.**

Salmonella Enteritidis

Agricultural Research Service, USDA

Figure 9-7. Eggs are tested for salmonella enteritidis, a bacteria found inside eggs that have become contaminated from infected laying hens.

Some recipes may call for only egg whites. Egg whites separate from yolks more easily when cold, but whip to a greater volume when at room temperature. Fat prevents egg whites from whipping, and therefore it is important to separate the eggs carefully and to use nonreactive bowls.

Cooking Eggs

For best results, eggs should be cooked at low temperatures for a short period of time. High temperatures and long cooking times toughen the protein in eggs, creating a rubbery finished product.

As eggs cook, the protein strands denature and unwind. The unwound protein strands form pockets that hold water. Denaturing changes the color of the egg white from translucent to opaque white and solidifies the protein. **See Figure 9-8.** When eggs are cooked at high temperatures and for long cooking times, proteins continue to denature, which toughens them. Overcooking eggs can also cause the yolk of an egg cooked in the shell to form a green outer ring due to the sulfur in the egg white reacting with the iron in the yolk.

Egg Protein Denaturing

Figure 9-8. Denaturing changes the color of the egg white from translucent to opaque white and solidifies the protein.

When adding eggs to a hot mixture, they must be tempered to prevent curdling. *Tempering* is a process in which a hot liquid is gradually added to eggs in order to slowly raise the temperature of the eggs, which stabilizes their texture and prevents curdling. Tempering is done by vigorously whisking a small amount of the hot mixture into the eggs to warm them. The warmed eggs are then added to the remaining hot mixture while being constantly stirred.

Chef's Tip

When beating whole eggs or egg whites, a bowl made of nonreactive material is most effective. Plastic bowls have pores that may absorb fat and prevent egg whites from being whipped.

Eggs can be cooked using both dry-heat and moist-heat cooking techniques. The most common egg preparations include eggs cooked in the shell, poached eggs, shirred eggs, fried eggs, scrambled eggs, and omelets. **See Figure 9-9.**

Sullivan University

Fried eggs are cooked using dry-heat cooking techniques.

Egg Preparations

Type of Preparation	Description
Soft-cooked in shell	Egg in its shell is simmered in water between 185°F and 205°F until it has a firm white and soft yolk
Hard-cooked in shell	Egg in its shell is simmered in water between 185°F and 205°F until it has a firm white and firm yolk
Poached	Egg is removed from shell and cooked in simmering water at 180°F
Shirred (baked)	Egg is cracked into a small bowl (ramekin) and baked
Sunny-side up	Lightly fried egg with an unbroken yolk that is cooked only on one side
Over-easy	Fried egg with an unbroken yolk that is cooked until the egg white has gone from translucent to white on the bottom and then flipped over and cooked on the other side until the white is no longer translucent
Over-medium	Fried egg with a completely cooked white and a yolk that is cooked nearly all the way through
Over-hard	Fried egg with a bright, firm white and a pale-yellow yolk that looks almost fluffy
Scrambled	Beaten eggs are cooked in a sauté pan or griddle while stirring (scrambling) until eggs reach the desired degree of doneness
Omelet	Beaten eggs that are typically cooked in a sauté pan or griddle and filled with ingredients such as vegetables, meats, and cheeses; frittata is an Italian omelet that is served open faced after being browned in a broiler or oven

Figure 9-9. The most common egg preparations include eggs cooked in the shell, poached eggs, shirred eggs, fried eggs, scrambled eggs, and omelets.

FOOD SCIENCE EXPERIMENT: Preparing Eggs Using Dry and Moist Heat 9-1

Objective

- To compare poached, shirred, fried, and scrambled eggs in terms of color, texture, and flavor.

Materials and Equipment

- 4 large eggs
- Water
- Saucepan
- Measuring spoons
- Salt
- Distilled white vinegar
- Nonreactive bowl(s)
- Slotted spoon
- 3 plates
- Tasting forks
- Brush
- Canola oil
- Ramekin
- Sauté pan
- High-heat spatula
- Whisk

Procedures

Allergen Alert

The materials used in this experiment may contain one or more food allergens.

To prepare the poached egg:

1. Fill a saucepan with 1 qt of water.
2. Add 1 tsp of salt and 1 tbsp of distilled white vinegar.
3. Bring the water to a boil and then lower the heat to approximately 180°F.
4. Break a large egg into a nonreactive bowl without breaking the yolk. Gently slide the egg into the simmering liquid.
5. Cook the egg for 3–5 minutes.
6. Use a slotted spoon to remove the egg and allow it to drain well before placing it on a plate.
7. Evaluate the poached egg in terms of color, texture, and flavor.

To prepare the shirred egg:

1. Brush a ramekin with 1 tsp canola oil.
2. Break a large egg into a nonreactive bowl without breaking the yolk. Gently slide the egg into the ramekin.
3. Place the ramekin in a 350°F oven and bake for approximately 12 minutes.
4. Remove the ramekin from the oven and evaluate the shirred egg in terms of color, texture, and flavor.

To prepare the fried egg:

1. Break a large egg into a nonreactive bowl without breaking the yolk.
2. Add 1 tsp of oil to a hot sauté pan.
3. Gently slide the egg into the pan.
4. Cook the egg over-medium and when the white is firm, use the spatula to flip the egg.
5. Cook the egg for 1 minute and then gently slide it onto a plate.
6. Evaluate the fried egg in terms of color, texture, and flavor.

To prepare the scrambled egg:

1. Break a large egg into a nonreactive bowl.
2. Use a whisk to beat the egg until it is fluffy.
3. Add 1 tsp of oil to a hot sauté pan.
4. Pour the beaten egg into the pan and slowly stir with a spatula until the egg is set but still moist.
5. Place the scrambled egg on a plate and evaluate its color, texture, and flavor.

Conclusions:

1. Compare the color, texture, and flavor of each egg.
2. Note the impact of dry versus moist heat on the color, texture, and flavor of each egg.

Typical Results

There will be variances in the color, texture, and flavor of each egg.

Flavor Development

In addition to their nutritive value, eggs provide color, texture, and a rich, yet mild flavor to menu items. Eggs can be paired with a wide range of ingredients to further develop flavors. For example, caramelized leeks, sautéed kale, and ripe tomatoes can be added to omelets and frittatas to layer flavors as well as increase color, texture, and nutrient density. Likewise, poached, fried, or scrambled eggs can be served atop toasted whole grain breads and garnished with fresh salsas or pestos for additional flavor and nutrients.

Food Science Note

Hens that are fed a diet rich in yellow-orange carotenoid pigments, such as those found in marigold petals or yellow corn, produce eggs with a darker yolk than those from hens that are fed wheat or barley.

Plating Eggs

Eggs can be a major component of both sweet and savory menu items, and egg presentations can be simple or elegant. **See Figure 9-10.** When served whole, eggs add vibrancy to the plate. For example, an elegant asparagus salad topped with a perfectly poached egg allows the guest to pierce the egg and experience the visual effect of the golden yolk dripping over the salad. The same asparagus salad can be garnished with a chopped hard-cooked egg, creating an entirely different visual and textural experience. In addition to being chopped, hard-cooked eggs also can be sliced, shaved, or cut into wedges and used to add shape, color, texture, and flavor to menu items.

Knowledge Check 9-2

1. Describe the proper storage of whole eggs.
2. Explain why eggs should not be cooked at high temperatures or overcooked.
3. Explain how to temper eggs.
4. List common egg preparations.
5. Describe how to develop flavor in egg dishes.
6. Describe various ways to effectively plate egg dishes.

Plating Eggs

Daniel NYC/N. Clutton

Asparagus Salad with a Soft-Cooked Egg

Courtesy of The National Pork Board

Shirred Egg with Canadian Bacon and Potatoes

Courtesy of The National Pork Board

Omelet with Sautéed Vegetables and Ham

Figure 9-10. Egg presentations can be simple or elegant.

CULINARY NUTRITION RECIPE-MODIFICATION PROCESS

Quiche Lorraine

Quiche is a versatile dish with ingredients that can be modified to change flavor profiles or increase nutrients. This version of quiche Lorraine has a rich crust and filling of bacon, Swiss cheese, egg, whole milk, and cream, making it high in calories and fat.

Yield: 6 servings, one 9-inch quiche

Ingredients

Crust

1 c	all-purpose flour
½ tsp	salt
⅓ c + 2 tbsp	vegetable shortening
4–5 tbsp	cold water

Filling

8 oz	bacon, medium dice
⅓ c	onion, small dice
4 oz	Swiss cheese, shredded
4 ea	whole eggs
1 pt	whole milk
½ c	heavy cream
¾ tsp	salt
¼ tsp	black pepper, ground
⅛ tsp	cayenne pepper

Courtesy of The National Pork Board

Preparation

Crust

1. Combine flour and salt in a mixing bowl.
2. Cut shortening into the flour with a pastry blender until the mixture is thoroughly blended and mealy in appearance.
3. Add cold water to the flour/shortening mixture 1 tbsp at a time until the mixture holds together and can be shaped into a ball of dough.
4. Place the dough on a lightly floured surface and roll into a circle that is 2 inches larger in diameter than the quiche pan.
5. Place dough in the pan and flute the edges.
6. Place pie weights on top of the crust and bake in a 350°F oven for 10 minutes.

Filling

7. Cook bacon in a sauté pan over medium heat until slightly crisp.
8. Add onions and cook until translucent (approximately 2–3 minutes).
9. Add bacon, onions, and cheese to the baked crust.
10. Break eggs into a nonreactive bowl and beat slightly with a whisk.
11. Add milk, cream, salt, black pepper, and cayenne pepper to the eggs. Whisk to combine.
12. Pour egg mixture over the other filling ingredients and bake at 350°F until the eggs are set and the crust is golden.

Evaluate original recipe for sensory and nutritional qualities

- Quiche Lorraine has a tender crust and a creamy filling with a savory, smoky flavor.
- Dish is high in calories, fat, cholesterol, and sodium.

Establish goals for recipe modifications

- Reduce calories, fat, cholesterol, and sodium.
- Maintain flavor profile.

Identify modifications or substitutions

- Use a substitution for the hydrogenated shortening in the crust.
- Use substitutions for the bacon, Swiss cheese, whole eggs, whole milk, and cream in the filling.
- Eliminate added salt in the filling.

Determine functions of identified modifications or substitutions

- Hydrogenated shortening produces a tender crust.
- Bacon provides texture and a smoky flavor.
- Swiss cheese adds creaminess and flavor.
- Eggs provide structure and a creamy texture.
- Whole milk and cream provide a rich mouthfeel.
- Salt enhances flavor.

Select appropriate modifications or substitutions

- Substitute canola oil for the hydrogenated shortening in the crust.
- Substitute lean, smoked ham for the bacon and use a smaller amount.
- Substitute low-fat Swiss cheese for the full-fat variety and use a smaller amount.
- Replace two of the whole eggs with four egg whites.
- Substitute nonfat milk for the whole milk and cream.
- Omit salt from the filling.

Test modified recipe to evaluate sensory and nutritional qualities

- Using canola oil in the crust reduces saturated fat yet the crust remains tender.
- Substituting ham for the bacon lowers calories, fat, and cholesterol while the quiche retains its texture and smoky flavor.
- Using low-fat Swiss cheese reduces calories, fat, and cholesterol yet maintains texture and flavor.
- Using egg whites reduces calories, fat, and cholesterol.
- Substituting nonfat milk for the whole milk and cream reduces calories, fat, and cholesterol yet still provides an appealing mouthfeel.
- Omitting salt from the filling reduces sodium, but smoked ham and Parmesan cheese maintain a salty flavor.

CULINARY NUTRITION RECIPE-MODIFICATION PROCESS

Modified Quiche Lorraine

In this modified recipe, canola oil replaces hydrogenated shortening in the crust to lower the amount of saturated fat. Lean, smoked ham is substituted for the bacon and low-fat Swiss cheese is used to further decrease calories and fat while maintaining the original flavor profile. Half of the whole eggs are replaced with egg whites, and nonfat milk is substituted for whole milk and cream to lower calories and fat as well as cholesterol. The smoked ham and Parmesan cheese provide saltiness, making it unnecessary to add salt to the filling.

Yield: 6 servings, one 9-inch quiche

Ingredients

Crust

1 c	all-purpose flour
½ tsp	salt
⅓ c	canola oil
3–4 tbsp	cold water

Filling

1 tsp	canola oil
⅓ c	onion, small dice
4 oz	lean, smoked ham, medium dice
2 oz	low-fat Swiss cheese, shredded
2 ea	whole eggs
4 ea	egg whites
1 pt	nonfat milk
¼ tsp	black pepper, ground
⅛ tsp	cayenne pepper

Preparation

Crust

1. Combine flour and salt in a mixing bowl.
2. Add oil and mix with a fork until pea-sized pieces of dough are formed.
3. Add cold water to the flour/oil mixture 1 tbsp at a time until the mixture holds together and can be shaped into a ball of dough.
4. Dampen the work surface and lay down a piece of parchment paper.
5. Place the dough in the center of the parchment paper and top with another sheet of paper.
6. Flatten and roll the dough until it is 2 inches larger in diameter than the quiche pan.
7. Remove the top layer of paper. Lift the bottom layer of paper and invert the dough over the pan.
8. Remove the paper and shape dough to the pan, fluting the edges.
9. Place pie weights on top of the crust and bake in a 350°F oven for 10 minutes.

Filling

10. Heat oil in a sauté pan over medium heat. Add onions and cook until translucent (approximately 2–3 minutes).
11. Add onions, ham, and cheese to the baked crust.
12. Break whole eggs into a nonreactive bowl. Add egg whites and beat mixture slightly with a whisk.
13. Add milk, black pepper, and cayenne pepper to the eggs. Whisk to combine.
14. Pour egg mixture over the other filling ingredients and bake at 350°F until the eggs are set and the crust is golden.

Quiche Lorraine Nutritional Comparison

Nutrition Facts	Original	Modified
Calories	629.4	331.3
Total Fat	51.4 g	17.4 g
Saturated Fat	20.1 g	2.6 g
Cholesterol	219.4 mg	95.9 mg
Sodium	923.95 mg	747.85 mg

Signature Roasted Vegetable, Pear, and Blue Cheese Frittata

A frittata can be prepared to have the same flavor profile as a quiche, but because frittatas do not have a crust, they are lower in calories and fat. This signature frittata recipe features an elevated flavor profile achieved by adding the caramelized flavors of roasted vegetables and pears. These ingredients, along with spinach, load the frittata with flavor, dietary fiber, vitamin A, and vitamin C. A small amount of blue cheese adds more flavor, and a mixture of whole eggs and egg whites keeps the dish low in calories, fat, and cholesterol.

Yield: 6 servings, one 10-inch frittata

Ingredients

to coat	nonstick cooking spray
¾ c	sweet potato, peeled, large dice
¾ c	Bosc pear, large dice
6 ea	Brussels sprouts, stemmed and sliced in half
¼ tsp	ground cinnamon
2 ea	whole eggs
4 ea	egg whites
¼ tsp	salt
¼ tsp	black pepper, ground
2 c	spinach, stems removed
2 oz	blue cheese, crumbled

Preparation

1. Spray three sheet pans with nonstick cooking spray. Place the sweet potatoes, pears, and Brussels sprouts on separate sheet pans. Sprinkle the pears with cinnamon.
2. In a 350°F oven, roast the sweet potatoes for 20 minutes and the pears and Brussels sprouts for 15 minutes, or until all items are tender.
3. Break whole eggs into a nonreactive bowl. Add egg whites, salt, and pepper. Whisk to combine.
4. Spray a 10-inch omelet pan with nonstick cooking spray and place over medium heat.
5. Add spinach to the pan. Cook spinach until it wilts (approximately 2 minutes).
6. Add roasted sweet potatoes, pears, and Brussels sprouts to the pan and sprinkle with blue cheese.
7. Pour egg mixture into the pan and bake in a 400°F oven until the eggs are set.

Quiche and Frittata Nutritional Comparison

Nutrition Facts	Original	Modified	Signature
Calories	629.4	331.3	106.6
Total Fat	51.4 g	17.4 g	4.6 g
Saturated Fat	20.1 g	2.6 g	2.3 g
Cholesterol	219.4 mg	95.9 mg	77.6 mg
Sodium	923.95 mg	747.85 mg	311.08 mg
Dietary Fiber	0.7 g	0.7 g	2.2 g
Vitamin A	724.2 IU	278.6 IU	3598.1 IU
Vitamin C	0.7 mg	0.6 mg	20.3 mg

EGG MENU MIX

Eggs have long been featured on menus as a main dish or a component of a dish. Although some egg dishes are high in fat and calories, recipe modifications can be made to produce lighter versions that are still full of flavor. **See Figure 9-11.** Eggs are also used throughout the menu for their functional properties such as emulsification and leavening.

When highlighting eggs in menu descriptions, it is important to list the preparation techniques. Menu descriptions should also encourage sales by emphasizing the unique qualities of the eggs, as well as flavorful combinations served with the eggs. **See Figure 9-12.**

Eggs on the Menu

Menu Item	Description	Options for a Lighter Version
Eggs Sardou	Poached eggs topped with artichoke hearts, Canadian bacon, anchovies, truffles, and hollandaise sauce	• Replace the hollandaise sauce with a dollop of Greek yogurt • Replace the hollandaise sauce with puréed roasted red peppers
Roasted vegetable quiche	Buttery crust filled with roasted tomatoes, asparagus, squash, and fontina cheese	• Eliminate the crust and serve as a frittata • Reduce the amount of cheese
Egg foo yung	Savory pancake-shaped omelet blended with crisp bean sprouts, bok choy, scallions, and ham; served with soy sauce	• Use low-sodium ham and soy sauce • Replace the ham with shrimp • Eliminate the ham and serve as a vegetarian menu item

Figure 9-11. Some egg dishes are high in fat and calories, but recipe modifications can produce lighter versions of the dishes that are still full of flavor.

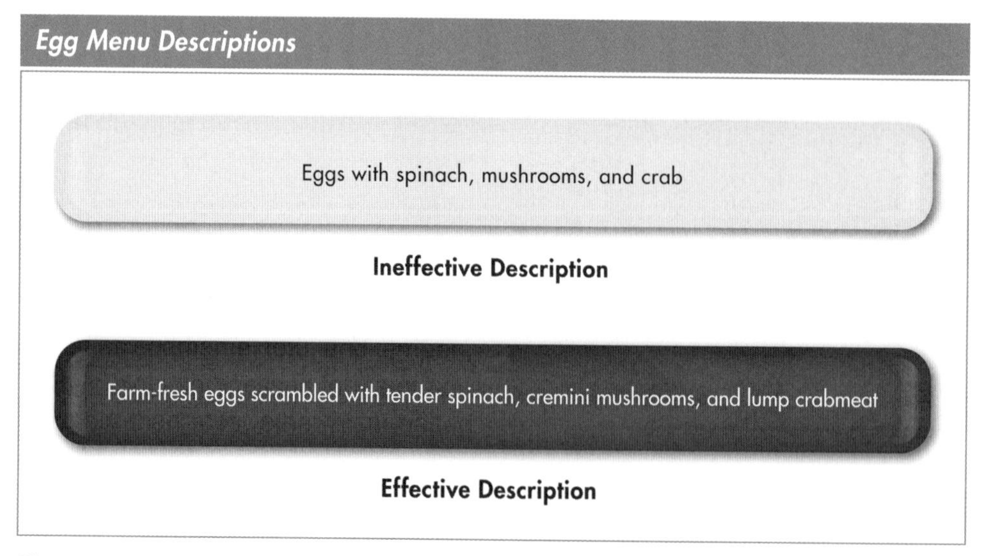

Figure 9-12. Effective menu descriptions should highlight preparation techniques, unique qualities of the eggs, and flavorful ingredients served with the eggs.

Eggs in Beverages

Eggs are used in both nonalcoholic and alcoholic beverages. While eggs may add flavor and nutrients to beverages, they are typically used to provide structure, enhance texture, or emulsify ingredients. For example, egg yolks are often used to thicken beverages, while egg whites can be whipped and folded into a beverage to create a lighter texture. This technique is often used when making eggnog. Eggnog is high in fat and calories but can be lightened by substituting low-fat milk or low-fat vanilla ice cream for the cream.

Eggs in Appetizers

Whole eggs, egg yolks, and egg whites can add color, texture, and flavor to appetizer menu items. Deviled or pickled eggs are served chilled. As a hot appetizer, eggs can be used in mini quiches. **See Figure 9-13.** Chopped hard-cooked eggs are a traditional garnish for appetizers featuring smoked salmon.

Eggs in Appetizers

Courtesy of The National Pork Board

Figure 9-13. As a hot appetizer, eggs can be used in mini quiches.

Eggs in Soups

Eggs are used in soups in different ways. For example, egg yolks add richness to soups such as Greek avgolemono, which also includes chicken broth, lemons, and rice. Egg whites replace noodles when they are poured directly into a hot broth to form streams of ribbons in egg-drop soup. Chopped hard-cooked eggs are a traditional garnish for black bean soup, providing a contrast of color.

Eggs in Sandwiches

Eggs are used in sandwiches alone and in combination with other foods. Egg sandwiches with ham or sausage as well as egg-based wraps are popular items. **See Figure 9-14.** Egg salad sandwiches are prevalent on lunch menus. Both lunch and dinner menus may offer sandwiches such as hamburgers topped with a fried egg. With the increasing number of individuals following a flexitarian diet, eggs can be featured in sandwiches filled with grilled or sautéed vegetables for a meatless meal that is high in protein, dietary fiber, vitamins, and minerals. Egg yolks are a main ingredient in mayonnaise, a popular sandwich condiment and binding agent for sandwich fillings such as chicken salad.

Eggs in Sandwiches

Courtesy of The National Pork Board

Figure 9-14. Egg-based wraps are an effective way to use eggs in sandwiches.

Eggs in Salads

While eggs can be the main ingredient in a salad, they are more commonly used as a salad component. Hard-cooked eggs are used in potato salads, a variety of tossed salads, and composed salads such as Cobb and chef's salads. Poached eggs can be added to salads to elevate the presentation as well as to add color, texture, flavor, and nutrient density. Egg yolks are commonly used to emulsify salad dressings such as Caesar dressing.

Eggs in Entrées

Eggs play a significant role in entrées for all meal periods. As a breakfast entrée, eggs are commonly served scrambled, poached, or fried. Eggs are also the primary ingredient in omelets, frittatas, quiches, and soufflés. The use of eggs on the breakfast menu extends to their use in sauces, such as hollandaise, and in menu items such as French toast, pancakes, waffles, and muffins.

Eggs are also used in a variety of lunch and dinner entrées. For example, fried rice and many pasta dishes commonly contain eggs. Egg yolks are used in sauces such as carbonara and allemande. Whole eggs are also used to bind fillings and are a key ingredient in breading. For example, fried chicken is often coated in flour, eggs, and then bread crumbs. The eggs allow the bread crumbs to adhere to the food.

Eggs in Sides

Eggs are common side dish options on à la carte breakfast menus. It is less common to see eggs as a side dish during other meal periods.

Eggs in Desserts

Eggs are one of the most widely used ingredients in desserts because of the different functions they perform. Eggs provide structure and help bind ingredients in baked goods, promote leavening in soufflés and cakes, and add texture to custards. Eggs are also used in confections in order to reduce crystallization.

Daniel NYC/T. Schauer

Eggs are used in desserts, such as soufflés, to promote leavening.

Knowledge Check 9-3

1. Identify the functions of eggs in beverages.
2. Explain how eggs can be used in appetizers and soups.
3. Describe the ways in which eggs are used in sandwiches and salads.
4. List common preparations of eggs in entrées.
5. Identify when eggs are commonly served as side dishes.
6. Describe the uses of eggs in desserts.

PROMOTING EGGS ON THE MENU

Chefs are finding new and interesting ways to promote eggs across the menu. For example, some foodservice operations are promoting eggs on "Meatless Mondays" to accommodate guests following this growing trend. Because of their nutrient density and versatility, eggs are an ideal food to include in complete-protein vegetarian entrées. Featuring eggs in place of other animal proteins can also be cost effective for both the foodservice operation and the guest. In addition to being a nutrient-dense food, eggs are gluten-free and can be promoted in dishes ranging from comfort foods to elegant menu items.

Most foodservice venues serve eggs in salads, on some sandwiches, and in a variety of breakfast and brunch menu items. Egg promotions vary according to the hours of operation and menu type. For example, a foodservice operation that serves breakfast all day promotes eggs more heavily than an operation open only for dinner.

Some bar venues include pickled eggs as an accompaniment to alcoholic beverages. Eggs are also a traditional garnish for sushi. Fine dining and special event venues often promote local farm-raised eggs, cage-free eggs, or varieties such as duck or quail eggs to create interest and elevate the menu. **See Figure 9-15.** These venues also often feature eggs in a variety of sauces and custard-based desserts.

Roasted Scallops with Quail Eggs

Daniel NYC/F. Miterre

Figure 9-15. Fine dining and special event venues often promote local farm-raised eggs, cage-free eggs, or egg varieties such as duck or quail eggs.

INGREDIENT SPOTLIGHT: Quail Eggs

Photo Courtesy of D'Artagnan, Photography by Doug Adams Studio

Nutrition Facts

Serving Size 3 eggs (27 g)

Amount Per Serving	
Calories 43	Calories from Fat 27
	% Daily Value*
Total Fat 3g	5%
Saturated Fat 1g	5%
Trans Fat	
Cholesterol 228mg	76%
Sodium 38mg	2%
Total Carbohydrate 0g	0%
Dietary Fiber 0g	0%
Sugars 0g	
Protein 4g	
Vitamin A 3% • Vitamin C	0%
Calcium 2% • Iron	6%

* Percent Daily Values (DV) are based on a 2,000 calorie diet.

Unique Features

- Tiny with a speckled brown shell
- Rich flavor
- Three weigh approximately 1 oz
- Low in saturated fat and sodium
- Source of protein, riboflavin, folate, pantothenic acid, vitamin B_{12}, phosphorus, iron, and selenium

Menu Applications

- Soft-cook, hard-cook, poach, or pickle to use in appetizers, salads, and garnishes
- Use as an accompaniment with sushi and tapas

Wellness Concept

Two egg whites can be substituted for one whole egg. This substitution saves 5 g of fat and 38 calories while contributing 7 g of protein.

Knowledge Check 9-4

1. Explain how hours of operation and menu type affect the way in which eggs are promoted on the menu.
2. Identify the unique eggs that may be promoted at fine dining and special event venues.

SOY PRODUCTS

Soy product is a general term that refers to items made from soybeans. A *soybean* is the oil-rich seed from the soybean plant and is a complete protein. Soybeans are often categorized as vegetable soybeans and field soybeans.

Vegetable soybeans are generally bred for their large seed size, appealing texture and flavor, and short cooking time. Vegetable soybeans are commonly served fresh or processed into soy products such as tofu, meat substitutions, and condiments. In contrast, field soybeans have smaller seeds and are generally bred for their higher oil content and yield. Field soybeans are used to produce soybean oil. Soybean oil is inexpensive and used in many culinary applications from baking to frying. Because field soybeans are grown specifically for their oil, they are sometimes considered oilseeds. An *oilseed* is an oil-rich seed grown primarily for its oil.

Soybeans have been used as a protein source for hundreds of years. They are native to Asia and play an important role in Asian cuisine. Soybeans are higher in total fat than other plant-based foods, but most of the fat is unsaturated and includes omega-3 fatty acids. Because soybeans are a plant-based food, they are naturally cholesterol-free. Soybeans are rich in dietary fiber, riboflavin, vitamin K, phosphorus, magnesium, iron, copper, and manganese. **See Figure 9-16.**

Perceived Value of Soy Products

Soybeans are generally an inexpensive yet high-quality protein. This makes soy products cost-effective ingredients that can be used as substitutes for or supplements to animal-based proteins. For example, crumbled tofu can be added to a pork dumpling filling, lowering the cost of the meat filling while still providing protein.

Key Nutrients in Soybeans*

Nutrition Facts	Amount	Percent Daily Value
Calories	173	—
Protein	17 g	—
Total fat	9 g	14%
Omega-3 fatty acids	4466 mg	—
Dietary fiber	6 g	24%
Riboflavin	0.3 mg	17%
Vitamin K	19.2 mcg	24%
Phosphorus	245 mg	24%
Magnesium	86 mg	21%
Iron	5.1 mg	29%
Copper	0.4 mg	20%
Manganese	0.8 mg	41%

* based on 100 g of mature seeds, cooked

Figure 9-16. Soybeans are loaded with dietary fiber, riboflavin, vitamin K, phosphorus, magnesium, iron, copper, and manganese.

When considering the value of soy products, chefs recognize their versatility and contribution to a nutrient-dense menu. Generally, soy products have a longer shelf life than other protein products and can be purchased in bulk at a reduced price. Some soy products also come in shelf-stable packaging so that they can be placed in dry storage and do not use valuable refrigerator space. Once reconstituted, these products must be refrigerated.

The positive perception of soy is increasing as consumers become educated about its health benefits. Research suggests that when soybeans are consumed closer to their whole food state, whether cooked, sprouted, or fermented, they contain more isoflavones than the more processed forms of soy. An *isoflavone* is an organic compound found in legumes and is considered a phytonutrient. *Phytonutrients,* also known as phytochemicals, are naturally occurring substances found in plants that have been found to protect against disease.

Market Forms of Soy Products

Market forms of soy products include whole forms of soy such as fresh, frozen, and dried soybeans and a variety of processed soy products. **See Figure 9-17.** Because soy is easily converted into products, it is used as the base for a variety of menu items. Soy may be shaped to simulate an animal protein product such as hot dogs or ground into a meal for use in soup bases. Soy products, such as soy milk, yogurts, and cheeses, do not contain lactose and make an ideal dairy substitute for those with lactose intolerance.

Soy is also used to produce many Asian sauces and flavorings, including soy sauce, miso, kecap, and natto. Asian sauces and flavorings are often used as condiments and dipping sauces and to develop flavor in soups, sauces, and marinades.

Common Market Forms of Soy Products

- Asian condiments
- Meat analogs
- Miso
- Soy cheeses
- Soy flour
- Soy ice creams
- Soy margarines
- Soy mayonnaises
- Soy milks
- Soy nuts
- Soy protein bars
- Soy sour creams
- Soy yogurts
- Tempeh
- Tofu

Figure 9-17. Soy products are available in a wide range of market forms.

- *Soy sauce* is a type of Asian sauce made from fermented soybeans, wheat, salt, and water. Tamari soy sauce is made with little or no wheat and is darker, thicker, and milder in taste than regular soy sauce.
- *Miso* is a type of Asian flavoring made from fermented soybeans, starch such as barley or rice, salt, and water. Miso has a paste-like consistency similar to peanut butter.
- *Kecap* is a type of Asian sauce made from fermented soybeans, palm sugar, and usually star anise and garlic. Kecap has a thick consistency and a sweet, complex flavor.
- *Natto* is a type of Asian sauce made from fermented soybeans and has a distinct aroma and cheese-like flavor. Natto is typically served as a condiment for rice dishes.

A variety of soy products can be used to produce healthy and flavorful menu items. Common soy products used on menus include edamame, tofu, tempeh, soy milk, textured soy protein, and meat analogs.

Soy sauce is used as a condiment, in dipping sauces, and to develop flavor in soups, sauces, and marinades.

Edamame. *Edamame* are green soybeans. Edamame is commonly available fresh or frozen and served in or out of the pod. Edamame can be served as a nutrient-dense side dish and adds protein to soups, salads, sandwich spreads, and dips. Soy nuts are ripe soybeans that may be served raw or roasted. Soy nuts add crunch as well as nutrients to a variety of menu items.

Tofu. *Tofu,* also known as soybean curd or bean curd, is the curd of ground, cooked soybeans that has been pressed into a cake. Tofu has a mild, slightly nutty flavor that easily takes on the flavor of other ingredients. It has a smooth, creamy texture that ranges in firmness from silken (soft) to extra firm. Tofu can be grilled or sautéed and makes an effective addition to salads and stir-fries. Tofu can also be scrambled as an egg substitute.

Tempeh. *Tempeh* is a flat, dense cake made from fermented soybeans. Most of the flavor in tempeh comes from the mold produced during fermentation. Rice or millet is often added to tempeh to produce a yeasty, nutty flavor and chewy texture. Tempeh is available fresh or frozen.

Soy Milk. *Soy milk* is the liquid expressed from soybeans that have been soaked and finely ground. Soy milk is available in regular, low-fat, and flavored varieties. **See Figure 9-18.** Liquid soy milk is sold fresh and in aseptic, shelf-stable packaging. Powdered soy milk is also available. Soy milk is an ideal milk substitute for lactose intolerant individuals and can replace milk in many recipes.

Yuba and okara are soy milk by-products used in culinary preparations. *Yuba* is the protein skin that forms when soy milk is heated. It is removed from heated soy milk and dried in sheets or sticks. Yuba is used to make wraps and as a meat substitute. When yuba is cut and added to stir-fries, it has the texture and consistency of scrambled eggs. *Okara* is the ground soybean pulp left over from the production of soy milk. Okara can be added to baked goods and fillings for extra protein and dietary fiber.

Soy Milks

CROPP Cooperative

Figure 9-18. Soy milk is the liquid expressed from soybeans that have been soaked and finely ground and is available in regular, low-fat, and flavored varieties.

Textured Soy Protein. *Textured soy protein (TSP)* is a granular, meatless protein product made from soy flour. TSP is rehydrated before it is cooked or during the cooking process. TSP is low in fat and has a mild flavor with a slightly chewy texture. TSP is an effective substitute for ground or chopped animal-based proteins and can be used as a meat extender to lower the cost of a recipe while maintaining the protein content.

Meat Analogs. A *meat analog* is a meatless product made from soybean by-products that is formed into the familiar shape of a processed meat product. Meat analogs are shaped into popular meat items such as hot dogs, hamburgers, bacon, sausage, deli meats, pork chops, chicken patties, chicken tenders, and chicken nuggets. Meat analogs can be flavorful, nutrient-dense additions to healthy menus. However, some meat analogs include ingredients that also make them high in fat and sodium.

FOOD SCIENCE EXPERIMENT: Preparing Soy Milk

9-2

Objective

- To compare house-made soy milk to commercially prepared soy milk in terms of color, texture, and flavor.

Materials and Equipment

- 2 cups soybeans, soaked overnight, drained and rinsed
- Measuring cup
- Blender
- 12 cups cold water, divided
- 8 qt pot
- Spoon
- Cheesecloth
- Sieve
- Medium bowl
- 1½ qt commercially prepared soy milk, plain
- Tasting cups

Procedure

1. Place half of the soybeans in a blender with 3 cups of cold water.
2. Blend the soybeans and water on high until the soybeans are finely ground.
3. Place the soybean mixture in a pot.
4. Repeat steps 1–3 for the remaining soybeans.
5. Add 6 cups of water to the pot and place over medium-high heat.
6. Bring the mixture to a boil while stirring frequently, then reduce the heat and simmer for 7 minutes.
7. Remove the pot from the heat to stop the simmer.
8. Repeat steps 6–7 two more times.
9. Line a sieve with three layers of cheesecloth and set the sieve over a bowl.
10. Pour the mixture over the cheesecloth.
11. Discard the cheesecloth and reserve the liquid in the bowl. This is the soy milk.
12. Let the soy milk cool and then refrigerate until cold.
13. Pour the house-made soy milk and commercially prepared soy milk into separate tasting cups.
14. Evaluate the color, texture, and flavor of each cup of soy milk.

Allergen Alert

The materials used in this experiment may contain one or more food allergens.

Typical Results

There will be variances in the color, texture, and flavor of each soy milk.

Knowledge Check 9-5

1. Compare the uses of vegetable and field soybeans.
2. Give examples of nutrients found in soybeans.
3. Summarize the perceived value of soy products.
4. List four common soy-based Asian sauces and flavorings.
5. Describe six common soy products.

PREPARING NUTRITIOUS SOY PRODUCTS

Soy products can offer health benefits over animal-based proteins in terms of dietary fiber, cholesterol content, and phytonutrients. These health benefits are more predominant in whole soy products such as edamame and tofu.

Cooking Soy Products

Many soy products, such as soy nuts, soy milks, yogurts, cheeses, and frozen desserts, can be eaten as-purchased. Other soy products, such as edamame, dried soybeans, tofu, tempeh, and TSP, may be cooked. Cooked soy products can be served alone or added to a variety of different dishes.

Edamame. Edamame can be served raw or steamed to preserve its color, texture, and mild flavor. Sautéing caramelizes edamame and intensifies the flavor. Edamame can also be seasoned and roasted to produce a crisp product that makes a healthy side in place of potato chips or French fries.

Dried Soybeans. Dried soybeans are often soaked in water to rehydrate them and reduce cooking time. They are then simmered in water until they reach the desired texture. Adding an acid, sugar, or calcium-rich food affects the cell structure of legumes and slows the softening process. Therefore, these ingredients should be added after the soybeans have been cooked. Canned soybeans are a convenience product that can be kept in dry storage for use in dishes such as soups and bean salads.

Tofu. Silken tofu has a creamy, custard-like appearance and texture. **See Figure 9-19.** Silken tofu does not hold its shape well and is better puréed or blended into soups, sauces, dips, salad dressings, and puddings. Silken tofu can be blended into smoothies, used as a pasta filling, added to risottos, frittatas, and macaroni and cheese, and used as a substitute for high-fat dairy products in desserts like cheesecake.

Firm tofu can be drained to remove its packaging liquid and compressed to make the product even firmer. Firm tofu is solid and dense, enabling it to withstand dry-heat cooking techniques such as grilling, pan-searing, stir-frying, and sautéing. Preparing tofu with dry-heat cooking techniques produces a crisp exterior while the interior texture remains soft. Firm tofu is commonly added to pasta dishes, stir-fries, and curries.

INGREDIENT SPOTLIGHT: Edamame

Frieda's Specialty Produce

Unique Features

- Green, crescent-shaped pods of soybeans approximately 2–3 inches long
- Mild, slightly nutty flavor
- Low in saturated fat, cholesterol, and sodium
- Source of dietary fiber, protein, thiamin, folate, vitamin K, phosphorus, iron, magnesium, copper, and manganese

Menu Applications

- Steam, sauté, or roast until crisp and serve as an appetizer with dipping sauces
- Use in soups, salads, stir-fries, and rice and pasta dishes

Nutrition Facts
Serving Size 1 cup, prepared (155 g)

Amount Per Serving	
Calories 189	Calories from Fat 67
	% Daily Value*
Total Fat 8g	12%
Saturated Fat 1g	5%
Trans Fat 0g	
Cholesterol	0%
Sodium 9mg	0%
Total Carbohydrate 15g	5%
Dietary Fiber 8g	32%
Sugars 3g	
Protein 17g	
Vitamin A 0% • Vitamin C	16%
Calcium 10% • Iron	20%

* Percent Daily Values (DV) are based on a 2,000 calorie diet.

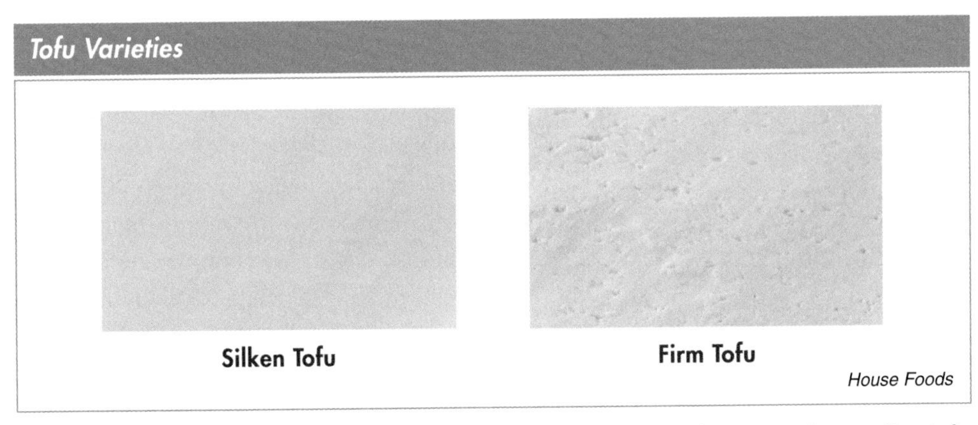

Figure 9-19. Silken tofu has a creamy, custard-like appearance and texture, whereas firm tofu is more solid and dense.

Tempeh. Tempeh has a meaty firmness and is often sliced and then baked, grilled, or sautéed. It can be marinated and used in place of meat on kebabs or added to sandwiches, salads, soups, and stews. Tempeh is often marinated with ingredients that feature Asian or Mediterranean flavor profiles.

TSP. TSP takes approximately 5–10 minutes to reconstitute after it is covered with boiling water. Using stock as the reconstituting liquid provides additional flavor. Dry TSP can also be added to dishes such as chili, but additional moisture may need to be added into the dish. TSP can also be substituted for any ground or chopped meat.

Meat Analogs. Many chefs make their own meat analogs by mixing a combination of tofu, tempeh, and TSP with vegetables and flavorings. This allows the chef to control the ingredients and produce a more flavorful item. Meat analogs are cooked like the meat product they are replacing, but the cooking time is generally shorter.

Flavor Development

The subtle flavor of many soy products and their ability to take on both sweet and savory flavors allows the chef to be creative. Soy products such as tofu naturally absorb the flavor of the ingredients they are paired with. **See Figure 9-20.** For example, tofu is often prepared with ginger, scallions, and soy sauce to take on an Asian flavor profile. Tofu is equally delicious when prepared in Indian curries. To create an Italian flavor profile, tofu can be used in ravioli fillings.

Figure 9-20. Dressings used in salads are absorbed by soy products, helping develop flavor.

One of the greatest challenges in cooking with soy products is compensating for the loss of savory aromas and flavors when it is substituted for meats or poultry. Incorporating mushrooms, Parmesan cheese, or tomatoes can add a umami component and complexity of flavor to soy menu items. Using fresh herbs or spice mixtures with complex flavors and adding ingredients such as chiles, caramelized onions, and roasted garlic can enhance both aroma and flavor.

Plating Soy Products

The different shapes, colors, and textures of soy products offer plating versatility. Edamame can enhance the plate with its bright green color and fresh flavor. Silken tofu can be scrambled and used as a substitute for eggs or blended into a nutrient-dense smoothie. Tofu, tempeh, TSP, and meat analogs are effectively plated in the same manner as animal-based proteins. Soy products such as milk, yogurts, and cheeses can also be plated with the same visual appeal as the dairy products they replace.

Knowledge Check 9-6

1. Describe how edamame and dried soybeans are commonly prepared.
2. Compare silken tofu to firm tofu.
3. Explain how tempeh is commonly prepared.
4. Describe the methods used to cook TSP and meat analogs.
5. Explain how to effectively develop flavor when using a soy product as a substitute for meat.
6. Describe how soy products can enhance plated presentations.

SOY PRODUCT MENU MIX

Soy products can be served across all meal periods. While soy products have traditionally been limited to vegetarian menu items, they are being offered more frequently in nonvegetarian menu items. Many existing dishes on the menu can be modified to provide a soy option. For instance, a stir-fry or pasta dish that is offered with a choice of beef, poultry, or seafood can easily be offered with tofu or tempeh as options. Soy products are easily used in menu items due to their ability to take on different flavor profiles. **See Figure 9-21.**

Menu item descriptions featuring soy products should be enticing as well as informative. A soy product can be an item that guests are not familiar with and may be hesitant to try without a tempting description. Highlighting cooking techniques, textures, flavors, and accompaniments can create inviting menu descriptions that result in sales. **See Figure 9-22.**

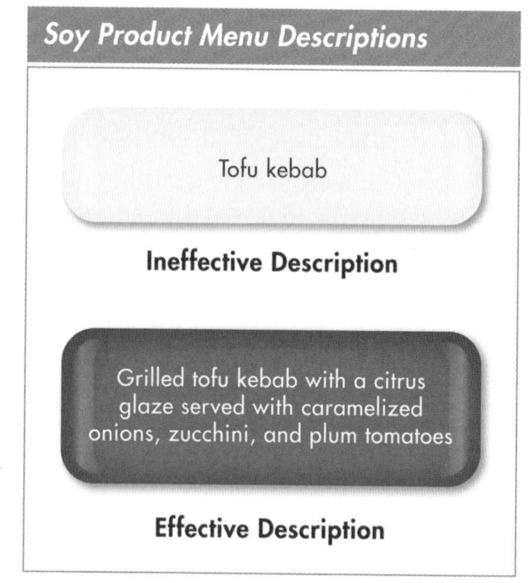

Figure 9-22. Highlighting cooking techniques, textures, flavors, and accompaniments can create effective menu descriptions of soy products.

Food Safety

Because soy is one of the eight most common food allergens, soy should be included in menu descriptions to alert people to the presence of soy in that menu item.

Soy Products on the Menu

Menu Item	Description
Marinated tempeh steak	Tempeh marinated in a blend of ginger, garlic, lemongrass, and soy sauce, grilled, and served with quinoa
Thai tofu stir-fry	Stir-fried tofu, onions, bell peppers, and snow peas in a spicy Thai green curry sauce
Tropical soy smoothie	Soy yogurt blended with pineapple and mango and garnished with toasted coconut

Figure 9-21. Soy products are easily used in menu items due to their ability to take on different flavor profiles.

Soy Products in Beverages

Soy products are used in a variety of hot and cold beverages. Soy milk is often used to replace cream or milk in hot coffee, tea, and cocoa or blended with these beverages to create specialty drinks. Soy milk, soy yogurt, soy powder, and silken tofu are common substitutions for dairy in smoothies. **See Figure 9-23.** Soy products can be mixed with flavorings, fruits, vegetables, and juices to create flavorful, nutrient-dense beverages.

Soy Products in Beverages

Frieda's Specialty Produce

Figure 9-23. Soy milk, soy yogurt, soy powder, and silken tofu are common substitutions for dairy in smoothies.

Soy Products in Appetizers

Soy products are used in appetizers for color, texture, and nutritive value. Edamame is often steamed and served plain or seasoned with sea salt. Chefs are developing specialty seasoning mixes and dipping sauces to add more appeal and originality to edamame appetizers. Marinated tofu or tempeh that has been baked or grilled on skewers has gained popularity when served alone or in lettuce wraps. Tofu is also commonly puréed and combined with ingredients to create flavorful dips and spreads. **See Figure 9-24.** Soy cheese is being offered more frequently in appetizers to meet the needs of vegetarians and lactose-intolerant guests.

Soy Products in Appetizers

House Foods

Figure 9-24. Tofu is commonly puréed and combined with ingredients to create flavorful appetizers, such as tofu mushroom spread.

Soy Products in Soups

One of the most common soups containing a soy product is miso soup. Miso soup often features seaweed and small pieces of tofu mixed into a flavorful broth. Hot and sour soup is a popular soup that differs throughout Asia but commonly contains tofu. Puréed silken tofu is often used in soups to add thickness and texture. It is also used as a substitute for cream to lower the saturated fat content in cream-based soups.

Soy Products in Salads

Soy products are often added to salads to provide protein. Edamame can be mixed into potato, pasta, and grain salads for a burst of color, while tofu is popular grilled and served warm on top of salad greens. Marinated, chilled tofu is also a popular addition to salads. **See Figure 9-25.** Garnishing salads with soy nuts adds a crunchy texture, and seasoned soy nuts can impart even more flavor.

Soy Products in Salads

House Foods

Figure 9-25. Marinated, chilled tofu is a popular addition to salads.

Soy Products in Entrées

Soy products are commonly used as the main source of protein in plant-based entrées. While tofu and tempeh are used most often, menu items made from TSP and meat analogs such as chicken tenders and burgers are also available. Silken tofu is a common replacement for scrambled eggs and can be used to make an "egg" salad sandwich. Firm tofu and tempeh is often used to mimic meats in stir-fries, pasta dishes, curries, and sandwiches. **See Figure 9-26.**

Soy Products in Entrées

House Foods

Figure 9-26. Soy products can be used to replace meats in entrées such as enchiladas.

Soy Products in Desserts

Silken tofu, soy yogurt, soy milk, and soy cream cheese are used to replace dairy in desserts. Cheesecakes, mousses, puddings, custards, cream fillings, and ice creams can be made from soy "dairy" products.

Planet Hollywood International, Inc.

Soy products can be substituted for dairy products in desserts such as cheesecake.

Knowledge Check 9-7

1. Identify soy products commonly used in beverages.
2. Explain how soy products can be used in appetizers.
3. Give examples of how soy products are used in soups and salads.
4. Describe the ways in which soy products are commonly incorporated into entrées.
5. List the types of soy products commonly used in desserts.

PROMOTING SOY PRODUCTS ON THE MENU

Because soy is an inexpensive source of cholesterol-free protein that is rich in phytonutrients, it is an ideal food to promote as a nutritious menu item. Soy products can help meet the growing demand for plant-based dishes and dairy-free options. Some soy products can also be promoted as gluten-free menu items.

Soy products are versatile and can be featured alone, in combination with animal proteins, or paired with seasonal fruits and vegetables to enhance color, texture, and flavor. Soy products can be promoted throughout the menu in hot or cold dishes. **See Figure 9-27.**

Tofu Hummus Wrap

House Foods

Figure 9-27. Soy products can be promoted throughout the menu in hot or cold dishes, such as wraps.

Salads with tofu as the main protein are popular, as are soy yogurts, soy milk, and soy-based smoothies. Soy nuts are commonly used to add crunch and protein to salads. Scrambled tofu as well as soy analogs in the form of sausages and bacon are popular breakfast items, while stir-fries with tofu, soy-based burgers, and tempeh sandwiches are popular for lunch and dinner. Soy products are also commonly used in soups such as chili.

Beverage houses and bars may offer soy products in a variety of hot and cold beverages. Fine dining and special event venues often feature soy products in less processed forms such as tofu, tempeh, or sprouts, and are more likely to highlight house-made soy analogs. **See Figure 9-28.**

Tofu-Stuffed Mushrooms

House Foods

Figure 9-28. Fine dining and special event venues may offer soy products in every menu category, from appetizers, such as stuffed mushrooms, through desserts.

Knowledge Check 9-8

1. Identify ways in which soy products are promoted in breakfast menu items.
2. Give examples of lunch and dinner menu items that promote soy products.

DAIRY PRODUCTS

Dairy products are derived from the milk of cows as well as other hoofed animals such as goats and sheep. Dairy products such as milk, cheese, and yogurt are complete-protein foods and rich in calcium. **See Figure 9-29.** Dairy products also supply essential vitamins and minerals, such as riboflavin, vitamin B_{12}, and phosphorus.

Calcium-Rich Dairy Products

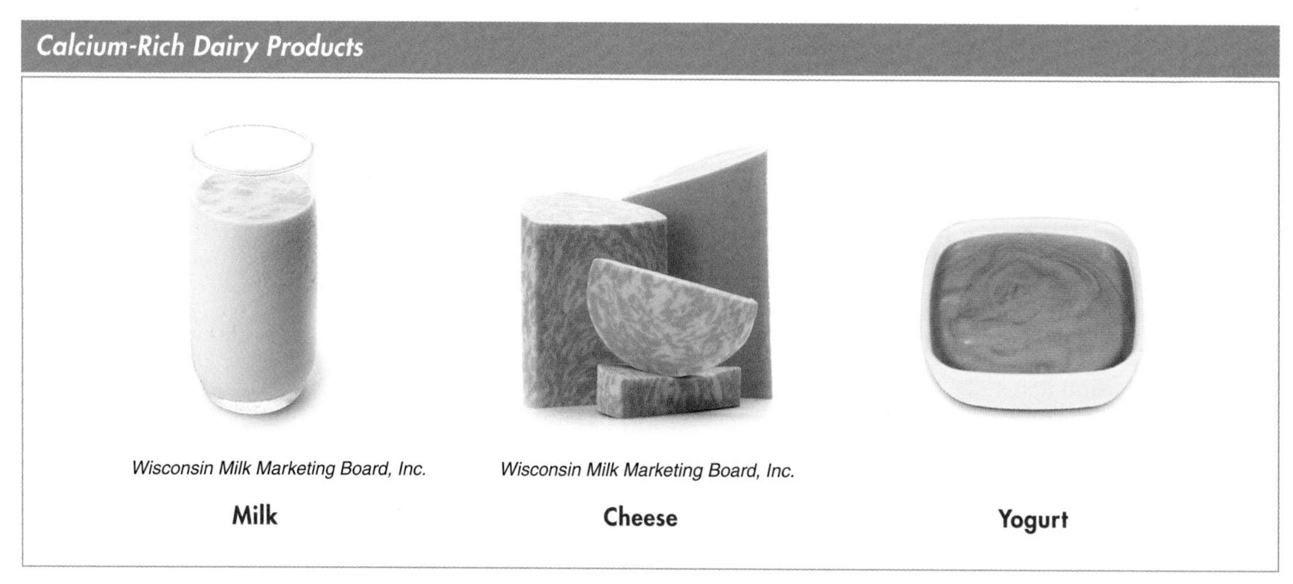

Wisconsin Milk Marketing Board, Inc.

Milk

Wisconsin Milk Marketing Board, Inc.

Cheese

Yogurt

Figure 9-29. Dairy products such as milk, cheese, and yogurt are complete-protein foods and rich in calcium.

Items made from milk that have little to no calcium, such as butter, cream, and sour cream, are often grouped with dairy products because they are made from milk. These items are not as nutrient dense as other dairy products because they do not contain calcium.

Many dairy products are high in fat. However, a variety of dairy products are available in low-fat and nonfat varieties. Calcium-rich, low-fat dairy products have less fat and fewer calories than the full-fat versions but still provide essential nutrients. **See Figure 9-30.**

Nutritional Comparison of Whole and Low-Fat Milk*

Type of Milk	Calories	Total Fat	Saturated Fat	Calcium
Whole milk	149	8 g	5 g	276 mg
Low-fat milk	102	2 g	2 g	305 mg

* based on 1 cup serving size

Figure 9-30. Low-fat dairy products have less fat and fewer calories than the full fat versions but still provide essential nutrients.

Most market forms of dairy products are pasteurized to destroy potentially harmful microorganisms. Milk, yogurt, some cheeses, and some forms of cream are homogenized. *Homogenization* is the process of emulsifying the fat particles in milk in order to provide a uniform consistency and prevent fat separation. The result of homogenization is a product with a smooth, even texture.

Most dairy products contain lactose, the primary sugar in milk. Individuals who are lactose intolerant are unable to properly digest and absorb lactose. This intolerance produces gas, bloating, and diarrhea. Some milks, yogurts, cheeses, and ice creams have had the lactose removed in order to make them easier for people with lactose intolerance to digest. Other dairy product alternatives such as almond milk, soy milk, rice milk, coconut milk, and hemp milk are naturally lactose-free. These plant-based dairy product alternatives are also appropriate for vegetarian diets.

Perceived Value of Dairy Products

For foodservice operations, dairy products offer a range of choices that add diverse textures and flavors to menu items while increasing nutrients. Dairy products can be stand-alone items, as well as ingredients in appetizers, soups, sandwiches, salads and dressings, sauces, entrées, sides, and desserts. Food costs for dairy products vary, but the cost of higher-priced items can be offset by the function, flavor, and value that they add to the menu.

Individuals perceive dairy products in many different ways. **See Figure 9-31.** For example, yogurt is typically perceived as healthy. Dairy products are perceived as comfort foods when used in dishes such as macaroni and cheese or a cold glass of milk served with cookies. The perception of comfort foods can be elevated when dairy products are used in unique ways. For example, topping a cheeseburger with an unexpected type of cheese, such as Gorgonzola, enhances the perceived value of the menu item.

Dairy products can also be perceived as sophisticated foods. For example, a cheese tray elegantly presented with handcrafted artisan cheeses has an elevated perceived value. Artisan cheeses are made in small batches with milk from a limited number of farms in order to offer a selection of cheeses not found in major marketplaces.

Liquid Dairy Products

Liquid dairy products include fluid forms of milk products such as milk and cream. *Milk* is a liquid dairy product produced by the mammary glands of cows, goats, sheep, or water buffalo. Milk is comprised mainly of water but also contains protein, carbohydrates, fat, vitamins, and minerals. *Casein* is the main group of proteins found in milk. Casein is a complete protein, containing all essential amino acids necessary for proper growth and cell maintenance. Lactose is the primary carbohydrate in milk and provides sweetness. *Butterfat* is the natural fat found in milk. Butterfat provides flavor and a rich mouthfeel. Milk also contains many vitamins and minerals, including vitamins A and D, riboflavin, calcium, phosphorus, magnesium, potassium, and selenium.

Figure 9-31. Dairy products are perceived in many ways, such as a healthy yogurt smoothie, a comforting macaroni and cheese entrée, or a sophisticated cheese tray.

Cream is a liquid dairy product made from the butterfat that separates from non-homogenized milk. Milk and cream classifications are generally determined by butterfat content. **See Figure 9-32.** Some milks have flavorings added or are fermented due to the addition of bacterial cultures. Flavored milks contain flavoring ingredients such as chocolate or strawberry syrup plus added sugars or sweeteners. As a result, flavored milk is typically higher in sugar and calories than plain milk.

Butterfat Percentage in Liquid Dairy Products

Liquid Dairy Product	Butterfat Percentage
Whole milk	3.5%
Low-fat milk	0.5%–2%
Nonfat (skim) milk	< 0.5%
Heavy whipping cream	≥ 36%
Light whipping cream	30%–36%
Light cream	18%–30%
Half-and-half	10%–18%

Figure 9-32. Milk and cream classifications are generally determined by butterfat content.

Buttermilk is a fermented liquid dairy product made by adding bacterial cultures to low-fat or skim milk to create a thickened texture and tangy flavor. Buttermilk contains little or no butterfat and has a longer shelf life than milk because the cultures inhibit harmful bacterial growth.

Acidophilus milk is a fermented liquid dairy product made by adding lactobacillus acidophilus bacteria to milk. The addition of the harmless bacteria has been found to promote digestive health. Sweet acidophilus milk has the bacteria added but is processed differently so that fermentation does not occur. Sweet acidophilus milk is similar in appearance and flavor to milk.

Food Science Note

Milk contains tryptophan, an amino acid that the body converts into niacin. Niacin, a B vitamin, has been found to raise HDL (good cholesterol) levels.

Semisolid Dairy Products

Semisolid dairy products are fermented with cultures such as bacteria or yeast to create a thick texture and tangy flavor. Semisolid dairy products such as yogurt, kefir, and sour cream often contain probiotics. A *probiotic* is a live microorganism that has been found to be beneficial to the digestive tract. Cultured dairy products that contain probiotics often feature labels that read, "contains active cultures." Semisolid dairy products are commonly made from whole milk or cream, but low-fat and nonfat varieties are also available.

Yogurt is a semisolid dairy product that has been fermented by adding bacterial cultures to milk. Bacterial cultures commonly added to yogurt include lactobacillus bulgaricus or streptococcus thermophilus. Yogurt is sold in many forms, including whole, low-fat, and nonfat varieties that are available plain or flavored. Yogurt is also sold frozen and in beverage form. In addition to natural sugars, flavored yogurts often contain refined sugars or artificial sweeteners. **See Figure 9-33.**

In addition to regular yogurt, Greek yogurt is commonly available. When regular yogurt is produced, it is strained two times to remove part of the whey. In contrast, Greek yogurt is strained three times, producing a thicker consistency than that of regular yogurt. Regular and Greek yogurt also differ in terms of nutrients. Greek yogurt has more protein, less sodium, fewer carbohydrates, and less calcium than regular yogurt has. **See Figure 9-34.**

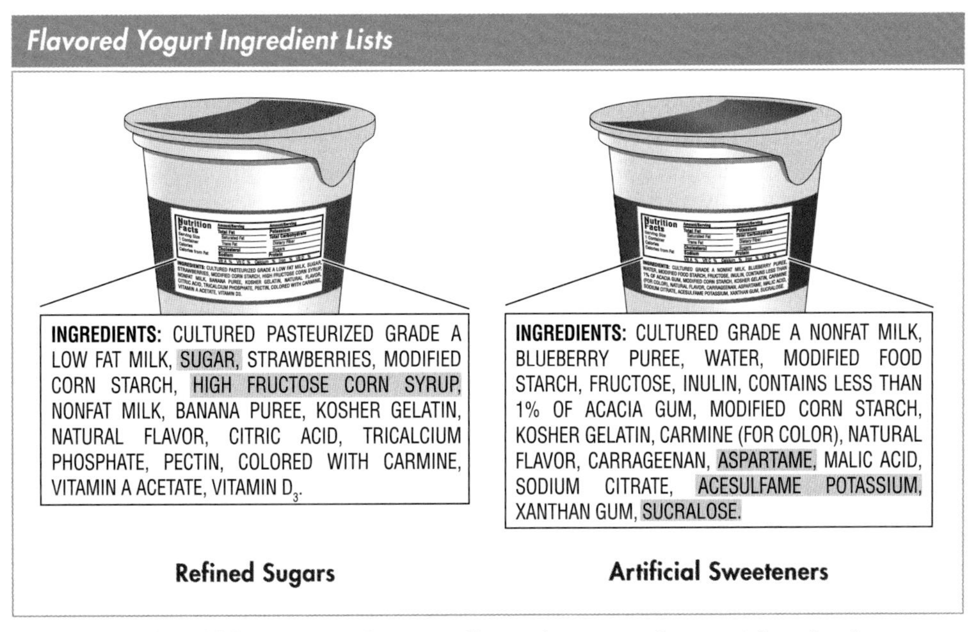

Figure 9-33. In addition to natural sugars, flavored yogurts often contain refined sugars or artificial sweeteners.

*Nutritional Comparison of Regular and Greek Nonfat Yogurts**

Type of Yogurt	Calories	Sodium	Carbohydrates	Protein	Calcium
Regular yogurt	127	175 mg	17 g	13 g	452 mg
Greek yogurt	134	82 mg	8 g	23 g	250 mg

* based on 8 oz serving size

Figure 9-34. Greek yogurt has more protein, less sodium, fewer carbohydrates, and less calcium than regular yogurt.

INGREDIENT SPOTLIGHT: Greek Yogurt

Unique Features

- Strained to remove the whey
- Has a thicker texture and tangier flavor than regular (non-strained) yogurt
- Withstands heat better than regular (nonstrained) yogurt
- Low in total fat and cholesterol when made from nonfat milk
- Source of protein, riboflavin, pantothenic acid, vitamin B_{12}, calcium, potassium, phosphorus, and zinc

Menu Applications

- Substitute for mayonnaise or sour cream
- Use to replace some of the oil and/or eggs in baked goods
- Use in savory and sweet dishes as well as dips

Nutrition Facts
Serving Size 1 container; 170 g (plain, nonfat)

Amount Per Serving	
Calories 100	Calories from Fat 6
	% Daily Value*
Total Fat 0g	0%
Saturated Fat 0g	0%
Trans Fat 0g	
Cholesterol 8mg	3%
Sodium 61mg	3%
Total Carbohydrate 6g	2%
Dietary Fiber 0g	0%
Sugars 6g	
Protein 17g	
Vitamin A 0% • Vitamin C	0%
Calcium 19% • Iron	1%

* Percent Daily Values (DV) are based on a 2,000 calorie diet.

Kefir is a semisolid dairy product that has been fermented by adding strains of active bacteria and yeast to milk. This process creates a beverage with a yogurt-like consistency that has a tangy, rich flavor. Kefir is available plain or flavored.

Sour cream is a semisolid dairy product that has been fermented by adding strains of active bacteria to cream. The bacteria are derived from lactic acid and mixed with cream to produce a thick consistency and tangy flavor. Sour cream contains approximately 18%–20% butterfat. Low-fat varieties contain approximately 40% less fat. *Crème fraîche* is a soured cream containing approximately 28% butterfat. It is less tangy than sour cream.

Solid Dairy Products

Solid dairy products have a wide range of textures and degrees of firmness. Solid dairy products include butters and cheeses. Both butters and cheeses contain fat and should always be used in appropriate portions.

Butters. *Butter* is a solid dairy product made by churning butterfat until it reaches a solid state. Butter contains a minimum of 80% butterfat. Butter is available in unsalted, salted, and whipped varieties. Salted butters are sometimes labeled as "sweet cream butter." Whipped butter has a softer consistency and is more spreadable when cold than other butter varieties. Whipped butter is available both salted and unsalted.

Butter is considered a solid fat and is therefore abundant in both total fat and saturated fat. For example, one tablespoon of butter supplies 12 g of total fat, 7 g of saturated fat, and 102 calories. Nutritionists recommend limiting the amount of saturated fat in the diet, so it is important to use butter in moderation when developing healthy menu items.

Cheeses. *Cheese* is a solid dairy product made from milk curds. Most cheese is produced by adding enzymes to milk, causing the milk to thicken and separate into curds and whey. The *curd* is the thick, casein-rich part of coagulated milk. The *whey* is the watery part of milk. After the curds and whey separate, the whey is drained off and the curds are typically salted. The curds are then processed to produce different types of cheeses.

Cheese is a concentrated source of many nutrients because it takes approximately 10 lb of whole milk to produce just 1 lb of cheese. Cheese is an animal-based product and, therefore, contains saturated fat and cholesterol. In general, a 1 oz serving of cheese made from whole milk contains 2–6 g of saturated fat and 14–32 mg of cholesterol. In contrast, low-fat cheeses contain 3 g or less of saturated fat and less than 20 mg of cholesterol per 1 oz serving. Cheeses also contain salt, so consideration should be given to the amount of sodium in the cheese when preparing menu items. **See Figure 9-35.**

Sodium Content of Common Cheeses*

Type of Cheese	Sodium Content	% Daily Value
Cheddar	174 mg	7%
Mozzarella	176 mg	7%
Provolone	245 mg	10%
Blue cheese	391 mg	16%
Parmesan	428 mg	18%

* based on 1 oz serving size

Figure 9-35. Consideration should be given to the amount of sodium in cheeses when preparing menu items.

Cheese is available in a wide range of textures and flavors as well as many low-fat and nonfat varieties. In general, cheeses are classified as fresh, soft, semisoft, blue-veined, hard, or grating cheeses. Soft cheeses tend to be lower in fat than hard cheeses because they have a higher water content.

Food Science Note

Whey is often dried and used in commercially prepared salad dressings, protein shakes, sauces, cheese products, canned fruits and vegetables, and meat, pasta, and milk products. Whey is also used as a foaming agent, a gelling agent, and an emulsifier.

- A *fresh cheese* is a cheese that is not aged or allowed to ripen. Baker's cheese, chèvre, cottage cheese, cream cheese, feta, mascarpone, ricotta, Neufchâtel, and mozzarella are fresh cheeses. In general, fresh cheeses have a creamy texture, are mild in flavor, and spoil easily. Most fresh cheeses are also available in low-fat varieties. Neufchâtel is often used in place of cream cheese because it has less fat and fewer calories than cream cheese. **See Figure 9-36.**
- A *soft cheese,* also known as a rind-ripened cheese, is a cheese that has been sprayed with a harmless live mold to produce a thin skin or rind. Brie and Camembert are soft cheeses. The mold that ripens soft cheeses reacts with the rind to produce a soft, suede-like outer coating and a soft interior. Soft cheeses typically have a full-bodied flavor. **See Figure 9-37.**
- Semisoft cheeses have either a dry rind or a washed rind. A *dry-rind cheese* is a semisoft cheese that is allowed to ripen through exposure to air. Although the rind is hard and dry, the interior of the cheese remains smooth and the flavor is generally mild to slightly sharp. Common examples of dry-rind semisoft cheeses are bel paese, Havarti, and Monterey Jack.

Nutritional Comparison of Cream Cheese and Neufchâtel*

Type of Cheese	Calories	Total Fat	Saturated Fat
Cream cheese	96	10 g	5 g
Neufchâtel	71	6 g	4 g

* based on 1 oz serving size

Figure 9-36. Neufchâtel is often used in place of cream cheese because it has less fat and fewer calories.

Soft Cheeses

Figure 9-37. Soft cheeses such as Brie and Camembert have been sprayed with a harmless live mold to produce a thin skin or rind.

A *washed-rind cheese* is a semisoft cheese with an exterior rind that is washed with a brine, wine, olive oil, nut oil, or fruit juice. Washing the rind generates the growth of harmless bacteria that penetrate the cheese to ripen the interior. Common examples of washed-rind semisoft cheeses are brick, Limburger, Muenster, and Port Salut. Washed-rind cheeses are somewhat firm and range in flavor from mild to robust. **See Figure 9-38.**

- A *blue-veined cheese* is a cheese produced by inserting live mold spores into the center of the ripening cheese with a needle. Common blue-veined cheeses include blue cheese, Gorgonzola, Roquefort, and Stilton. Blue-veined cheeses have a creamy to crumbly texture with a sharp, tangy flavor. **See Figure 9-39.**

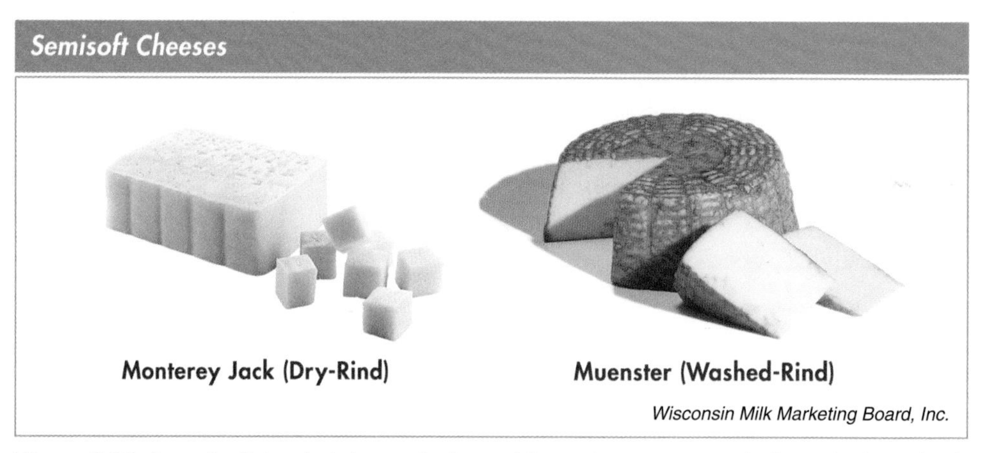

Wisconsin Milk Marketing Board, Inc.

Figure 9-38. A semisoft dry-rind cheese is ripened through exposure to air. A semisoft washed-rind cheese is ripened when bacteria that grow as a result of an exterior wash of a brine, wine, olive oil, nut oil, or fruit juice penetrate the cheese.

Wisconsin Milk Marketing Board, Inc.

Figure 9-39. A blue-veined cheese, such as blue cheese, is produced by inserting live mold spores into the center of the ripening cheese with a needle.

Wisconsin Milk Marketing Board, Inc.

Figure 9-40. A hard cheese, such as Cheddar, is a firm, somewhat pliable and supple cheese with a slightly dry texture and buttery flavor.

- A *hard cheese* is a firm, somewhat pliable, and supple cheese with a slightly dry texture and buttery flavor. Cheddar, Cheshire, Gruyère, Manchego, provolone, and Swiss are hard cheeses. Because hard cheeses have a firm texture, they slice and grate well, making them well suited for sandwiches and salads. **See Figure 9-40.**
- A *grating cheese* is a hard, crumbly, dry cheese that is commonly grated or shaved and placed on top of food. The most common grating cheeses are Asiago, Parmesan, and Romano. The crumbly texture of grating cheeses makes them difficult to slice. **See Figure 9-41.**

Barilla America, Inc.

Figure 9-41. A grating cheese, such as Parmesan, is a hard, crumbly, dry cheese that is commonly grated or shaved and placed on top of food.

FOOD SCIENCE EXPERIMENT: Preparing Ricotta Cheese

9-3

Objective

- To compare house-made and commercially prepared ricotta cheese in terms of texture and flavor.

Materials and Equipment

- Scissors
- Cheesecloth
- Sieve
- Water
- Stainless steel bowl
- Measuring cup
- Measuring spoons
- 3 cups whole milk
- 3 cups nonfat milk
- 1 tsp kosher salt
- Saucepan
- Spoon
- 3 tbsp white wine vinegar
- 2 plates
- 16 oz commercially prepared part-skim ricotta cheese
- Tasting spoons

Procedure

Allergen Alert

The materials used in this experiment may contain one or more food allergens.

1. Cut two layers of cheesecloth large enough to cover the inside of a sieve.
2. Dampen the cheesecloth with water and line the sieve with the damp cheesecloth.
3. Place the sieve over a stainless steel bowl and set aside.
4. Add the whole milk, nonfat milk, and salt to a saucepan and stir.
5. Place the saucepan over medium heat and bring to a boil, stirring occasionally.
6. Turn off the heat and stir in the vinegar.
7. Allow the mixture to stand for 7–10 minutes until it curdles and separates into curds and whey.
8. Pour the mixture into the cheesecloth-lined sieve and allow the whey to drain into the bowl for approximately 20–30 minutes.
9. Remove the ricotta cheese from the sieve and place it on a plate.
10. Place the commercially prepared ricotta cheese on a second plate.
11. Evaluate the texture and flavor of each plate of ricotta cheese.

Typical Results

There will be variances in the texture and flavor of each ricotta cheese.

Canned and Dried Dairy Products

Canned and dried dairy products offer convenience, a long shelf life, and are typically less expensive than liquid dairy products. Evaporated milk, sweetened condensed milk, and dry milk are the most common canned and dried dairy products used in professional kitchens.

Evaporated milk is produced by removing 60% of the water from milk and is available in whole, low-fat, and nonfat varieties. During the canning process, the Maillard reaction occurs. Therefore, evaporated milk may be slightly darker in color than milk.

Sweetened condensed milk is produced by removing 60% of the water from milk and adding sugar. The final product is 40%–45% sugar and very thick. It is available in whole, low-fat, and nonfat varieties and is similar in color to evaporated milk.

Dry milk is produced by removing all of the moisture from liquid milk to form a powder. **See Figure 9-42.** It is available in whole and nonfat varieties. Dry milk can be used in its powdered form or reconstituted with water.

Dry Milk

Figure 9-42. Dry milk is made by removing all of the moisture from liquid milk to form a powder.

Knowledge Check 9-9

1. List essential nutrients supplied by dairy products.
2. Define homogenization.
3. Describe the perceived values of dairy products.
4. Describe the composition of milk.
5. Explain how milks and creams are classified.
6. List three semisolid dairy products.
7. Define probiotic.
8. Identify the amount of butterfat found in butters.
9. Compare the characteristics of the six cheese classifications.
10. Describe three common canned and dried dairy products.

PREPARING NUTRITIOUS DAIRY PRODUCTS

Dairy products can be served as purchased or used to enhance the textures and flavors of a variety of menu items. Using milks, yogurts, and cheeses in menu items increases the protein and calcium content of the dish. However, the fat content of many dairy products should also be considered. Low-fat dairy products and smaller servings of high-fat dairy products provide guests with nutritious, flavorful, and appealing menu choices.

Cooking Dairy Products

Dairy products are commonly used to add moisture, color, and flavor to recipes. They can also help tenderize items such as cakes and breads. When cooking with dairy products, it is important to consider the type of product and the desired outcome. For example, cooking high-protein dairy products on low heat will prevent the proteins from toughening. Similarly, cooking low-fat dairy products on low heat will prevent them from curdling or burning easily.

Cooking with Milks and Creams. Milk scorches when cooked at high temperatures. **See Figure 9-43.** In contrast, cream tolerates heat better than milk due to its high fat content. When milk scorches, the whey proteins become insoluble and form a film on the bottom and sides of the pan. A skin can also form on the surface of the milk due to water evaporation and an increased concentration of casein, fat, and sodium. Cooking milk at a low temperature in a double boiler and stirring constantly helps prevent milk from scorching.

When milk is heated along with acids, enzymes, salts, or polyphenols, the casein precipitates from the milk to form curds. A *polyphenol* is an antioxidant that can cause curding. The casein particles or curds that form may or may not be desirable. For example, curds are useful as the base of products such as cottage cheese and cream cheese, but they are not desired in a sauce. To prevent curds from forming, the acids, enzymes, salts, or polyphenols should be added to the milk instead of the milk being added to any of these ingredients.

Scorched Milk

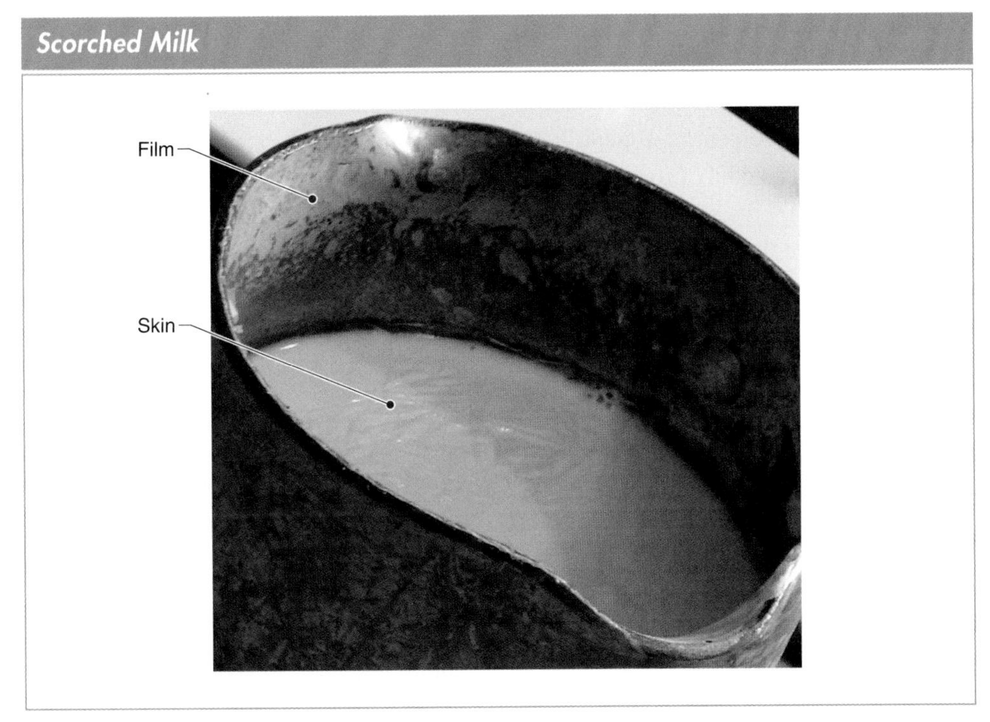

Figure 9-43. Milk scorches at high temperatures, producing a film on the bottom and sides of a pan and a skin on the surface of the milk.

Cooking with Semisolid Dairy Products. Semisolid dairy products contain bacteria, making them more acidic. For this reason, dairy products such as yogurt are commonly used in marinades to help tenderize proteins. Likewise, many bakeshop preparations rely on sour cream to help promote tenderness as well as retain moisture. Semisolid dairy products such as yogurt, sour cream, and crème fraîche are commonly used to make dips, toppings, soups, salad dressings, and sauces.

Yogurt is one of the most nutrient-dense dairy products and can often be substituted for higher-fat dairy products such as whole milk, cream, sour cream, and crème fraîche. For example, yogurt can be used in place of sour cream as a garnish for soups, salads, and baked potatoes. **See Figure 9-44.** Yogurt can also be substituted in many baked goods. Yogurt makes an ideal low-fat substitute in creamy soups, dips, dressings, and mayonnaise-based dishes such as chicken or tuna salad.

Sour Cream Substitutions

National Turkey Federation

Figure 9-44. A dollop of yogurt can be used in place of sour cream on chili to lower the fat and calories in the dish.

Cooking with Butters. Butter adds flavor and mouthfeel to foods. Unsalted butter allows the chef to better control the amount of salt added to a recipe. Most foodservice operations use clarified butter, which has had the milk solids and water removed, because it has a higher smoke point.

Butter is sometimes browned to produce a more complex flavor. Because brown butter is so flavorful, less is needed, which can help lower the fat content of a menu item. Likewise, a small amount of compound butter enhanced with herbs, spices, or garlic can be used to elevate flavor while maintaining a moderate fat content. A *compound butter* is a flavorful butter made by mixing cold, softened butter with flavoring ingredients such as fresh herbs, garlic, vegetable purées, dried fruits, preserves, or wine reductions.

Cooking with Cheeses. The type of cheese selected for a recipe depends on the flavor desired in the finished product and how the cheese will react during the cooking process. For example, when making a sauce, cheese should be incorporated toward the end of the cooking process using low heat to create a smooth, even texture. In contrast, items topped with cheese, such as pizza, can tolerate intense heat for a short time.

Shredding cheese allows less to be used while still covering the item. This maintains flavor and reduces fat and calories. Shredding cheese also allows it to melt faster and more evenly. **See Figure 9-45.**

Cooking with Cheeses

Courtesy of The National Pork Board

Figure 9-45. Shredding cheese allows it to melt faster and more evenly.

Some high-fat cheeses will "oil-off" during cooking and release a visible layer of fat on the surface of the cheese. Blisters sometimes develop when cooking aged cheeses, which can negatively affect the appearance of a dish. The elasticity of cheese is important for menu items such as pizza and lasagna, and chefs often mix different cheeses to get the desired result.

Cooking with Canned and Dried Dairy Products. Evaporated milk is commonly used in custards. It has a slightly thick consistency that can be substituted for cream-based sauces, reducing fat and calories. When evaporated milk is mixed with water, it can also be used as a substitute for fresh milk. Sweetened condensed milk is most commonly used in baked goods and confections such as fudge. It can be cooked to produce a caramel-like product. Dry milk is used in baked goods. It can be reconstituted but does not have the flavor of liquid milk. Once reconstituted, dry milk can be whipped.

Flavor Development

Dairy products offer a range of flavors, from sweet and mild to strong and robust. For example, the flavor of cream is subtle, while the flavor of aged cheese can be quite intense. When developing flavor, it is important to consider the type and amount of dairy product used to ensure that the recipe is both flavorful and healthy. For example, a small amount of a strongly flavored cheese can be used in place of a large amount of a mild cheese in order to produce a dish that is lower in fat and calories.

Changing the types of dairy products in dishes and layering nutrient-dense ingredients helps create different flavor profiles and produce unique menu items. For example, a traditional grilled cheese sandwich offers a mild flavor, but changing the cheese to smoked mozzarella and adding tomatoes, roasted garlic, and fresh basil provides a completely different flavor profile. Likewise, a sandwich of Brie, grilled pears, and toasted slivered almonds adds layers of textures and flavors not found in a traditional grilled cheese sandwich. When nutrient-dense, flavorful ingredients are added, less cheese is required to impact flavor, which can lead to healthier menu items.

Plating Dairy Products

Dairy products should be served at appropriate temperatures. For example, milk should be served cold. Butter spreads easily at room temperature. Ice cream and frozen desserts should be served frozen. When cheeses are not cooked, they should be presented at room temperature to allow the taste buds to detect the distinct flavor of each cheese.

Because many dairy products are high in fat, portion size should be considered. **See Figure 9-46.** For example, a dollop of whipped cream or sour cream as a garnish still provides flavor and is a healthier option than an excessive amount of dairy products.

Pairing dairy products with colorful fruits and vegetables is an effective way to elevate presentation and increase nutrient density. A fruit or vegetable tray adorned with cheeses of various shapes, colors, textures, and flavors elevates visual appeal, nutrient value, and flavor. Whether it is a bowl of creamy yogurt topped with dried fruit, a marinara pasta dish topped with ribbons of Parmesan, or fresh berries topped with a dollop of whipped cream, nutritious presentations can be enhanced by dairy products.

Wisconsin Milk Marketing Board, Inc.

When cheeses are not cooked, they should be presented at room temperature for optimum flavor.

Knowledge Check 9-10

1. Describe the proper methods for cooking with milks.
2. Identify common uses of semisolid dairy products.
3. Compare clarified, brown, and compound butters.
4. Explain how a cheese is chosen for a recipe.
5. List common uses of canned and dried dairy products.
6. Describe techniques for developing flavors in dishes that contain dairy products.
7. Explain how dairy products can be used to enhance plated presentations.

Nutritional Comparison of Sour Cream Portion Sizes

Portion	Calories	Total Fat	Saturated Fat
3 tbsp	69	7.11 g	4.14 g
1 tbsp	23	2.37 g	1.38 g

Figure 9-46. An appropriate portion of high-fat dairy products, such as sour cream, is a healthier option than excessive amounts of any dairy product.

CULINARY NUTRITION RECIPE-MODIFICATION PROCESS

Chocolate Mousse

This chocolate mousse recipe is produced by folding whipped egg whites and whipped cream into a base of melted chocolate, butter, and egg yolks. The resulting dish has a light, airy texture and rich chocolate flavor. However, even a ½ cup serving size is very high in both calories and fat.

Yield: 8 servings, ½ cup each

Ingredients

8 oz	bittersweet chocolate, finely chopped
5 oz	unsalted butter
2 oz	pasteurized egg yolks
5 oz	pasteurized egg whites
1½ oz	granulated sugar
½ c	heavy whipping cream

Preparation

1. Melt chocolate and butter in a double boiler over low heat while stirring constantly.
2. Remove chocolate mixture from the heat and let stand 2–3 minutes to cool slightly.
3. Whisk each egg yolk, one at a time, into the chocolate mixture. Set the mixture aside to cool.
4. Place egg whites in a nonreactive bowl and beat until soft peaks form.
5. Add sugar and continue beating until stiff peaks form.
6. Fold whipped egg whites into the chocolate mixture. Set aside.
7. Whip heavy whipping cream in a nonreactive bowl until soft peaks form.
8. Fold whipped cream into the cooled chocolate mixture.
9. Spoon into serving dishes, refrigerate, and serve when chilled.

Wisconsin Milk Marketing Board, Inc.

Evaluate original recipe for sensory and nutritional qualities

- Chocolate mousse has a velvety texture and rich chocolate flavor.
- Dish is high in calories and fat.

Establish goals for recipe modifications

- Reduce calories and fat.
- Maintain light texture and chocolate flavor.

Identify modifications or substitutions

- Reduce the amount of chocolate.
- Eliminate the butter.
- Use a substitution for the egg yolks and whipped cream.
- Add more sugar.

Determine functions of identified modifications or substitutions

- Chocolate provides flavor.
- Butter adds flavor and a rich mouthfeel.
- Egg yolks and whipped cream provide structure, texture, and flavor.
- Sugar provides sweetness.

Select appropriate modifications or substitutions

- Reduce the amount of bittersweet chocolate to 2 oz and add cocoa powder.
- Omit the butter.
- Substitute Greek yogurt for the egg yolks and whipped cream.
- Increase the amount of sugar from 1½ oz to 3½ oz.

Test modified recipe to evaluate sensory and nutritional qualities

- Using less bittersweet chocolate with the addition of cocoa powder maintains color and flavor while reducing calories and fat.
- Omitting the butter decreases calories and fat.
- Using Greek yogurt with the whipped egg whites provides structure and a velvety texture.
- Increasing the amount of sugar balances the tanginess of the Greek yogurt.

Modified Chocolate Mousse

To reduce the calories and fat of the original chocolate mousse recipe, Greek yogurt is substituted for the whole eggs and heavy cream. Whipped egg whites are folded into the Greek yogurt to achieve a light texture. Sugar is increased to compensate for the reduction in chocolate and to balance the tanginess of the Greek yogurt. The resulting dish is a rich, creamy chocolate mousse with 25 less grams of fat and less than half the calories of the original recipe.

Yield: 8 servings, ½ cup each

Ingredients

2 oz	bittersweet chocolate, finely chopped
1 lb	Greek yogurt
1½ oz	unsweetened cocoa powder
7 oz	pasteurized egg whites
3½ oz	granulated sugar

Preparation

1. Melt chocolate in a double boiler over low heat while stirring constantly.
2. Remove chocolate from the heat and let stand 2–3 minutes to cool slightly.
3. Add Greek yogurt and stir to combine.
4. Fold cocoa powder into the yogurt mixture. Set aside to cool.
5. Place egg whites in a nonreactive bowl and beat until soft peaks form.
6. Add sugar and continue beating until stiff peaks form.
7. Fold whipped egg whites into the cooled yogurt mixture.
8. Spoon into serving dishes, refrigerate, and serve when chilled.

Chocolate Mousse Nutritional Comparison

Nutrition Facts	Original	Modified
Calories	366.3	142.6
Total Fat	30.2 g	3.8 g
Saturated Fat	18.2 g	2.2 g

Signature Tofu Chocolate Mousse

This signature chocolate mousse recipe is a lactose-free dessert that gets its structure from silken tofu instead of egg yolks and whipped cream. Silken tofu is low in calories and fat and provides a rich mouthfeel. Espresso powder is added to bring forth a more intense chocolate flavor. Because tofu has a mild flavor, sugar is added to provide sweetness. A garnish of chocolate-covered espresso beans enhances the presentation.

Yield: 8 servings, ½ cup each

Ingredients

2 oz	bittersweet chocolate, finely chopped
1 lb	silken tofu, drained overnight in cheesecloth-lined sieve
1½ oz	unsweetened cocoa powder
1 tsp	espresso powder
7 oz	pasteurized egg whites
3½ oz	granulated sugar
8 ea	chocolate-covered espresso beans

Preparation

1. Melt chocolate in a double boiler over low heat while stirring constantly.
2. Remove chocolate from the heat and let stand 2–3 minutes to cool slightly.
3. Place tofu in a food processor. Add melted chocolate, unsweetened cocoa powder, and espresso powder. Purée until smooth.
4. Pour puréed tofu mixture into a bowl and set aside to cool.
5. Place egg whites in a nonreactive bowl and beat until soft peaks form.
6. Add sugar and continue beating until stiff peaks form.
7. Fold whipped egg whites into the cooled tofu mixture.
8. Spoon into serving dishes and garnish each dish with a chocolate-covered espresso bean. Refrigerate and serve when chilled.

Chocolate Mousse Nutritional Comparison

Nutrition Facts	Original	Modified	Signature
Calories	366.3	142.6	149.5
Total Fat	30.2 g	3.8 g	5.4 g
Saturated Fat	18.2 g	2.2 g	2.2 g

DAIRY PRODUCT MENU MIX

Dairy products provide structure to many menu items and serve as essential components in sauces, fillings, and garnishes. Dairy products are served alone or as part of menu items across all menu categories. **See Figure 9-47.** Milks, coffees and teas enriched with dairy products, and dairy-based signature drinks are offered throughout the day. Breakfast menus feature dairy products in their original form such as milk, cream, and yogurt, as well as dairy products used in the preparation of items such as omelets. Lunch and dinner menus commonly include dairy products in a variety of items ranging from beverages to desserts.

Dairy Products on the Menu

Menu Item	Description
Bruschetta	Toasted bread rubbed with garlic and topped with pesto, heirloom tomatoes, and grated Parmigiano-Reggiano cheese
Ravioli	Whole wheat ravioli filled with chèvre and spinach and served with a sage brown butter
Yogurt parfait	Greek yogurt layered with grilled nectarines and toasted pine nuts and drizzled with maple syrup

Figure 9-47. Dairy products are served alone or as part of menu items across all menu categories.

Wisconsin Milk Marketing Board, Inc.

Dairy-based signature drinks may be offered on breakfast, lunch, and dinner menus.

Menu descriptions of dairy products are often overlooked because of their role as a menu staple. However, featuring the main dairy product along with its complementary ingredients creates an effective menu description and can increase revenue. **See Figure 9-48.**

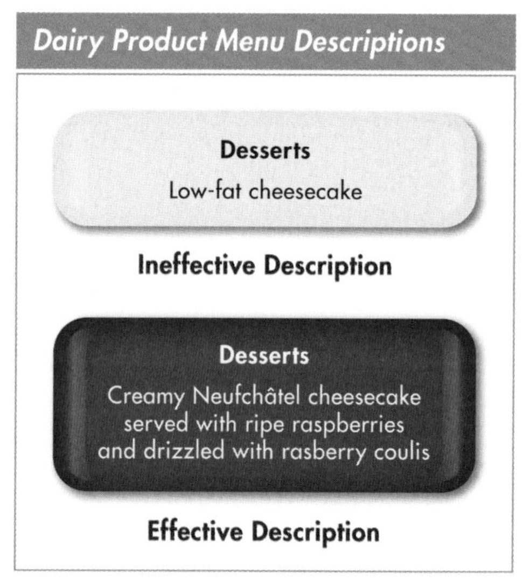

Figure 9-48. Featuring the main dairy product along with its complementary ingredients creates an effective menu description and can increase revenue.

Dairy Products in Beverages

Milk is a nutritious stand-alone beverage that is also added to other beverages. The acidity of coffee and tea is slightly neutralized when milk or cream is added. Also, milk, cream, and ice cream are commonly featured in milkshakes, floats, and alcoholic beverages. Milk and yogurt are common ingredients in smoothies. Low-fat varieties of dairy products help promote nutritionally balanced beverages.

Dairy Products in Appetizers

Dairy products are featured in both hot and cold appetizers. Yogurt, sour cream, cottage cheese, and cream cheese are often used as base ingredients for dips and spreads. Milk, cream, and cheese play key roles in sauces that are served with appetizers. Chefs often rely on local dairies for artisan cheeses as well as global cheese varieties to use as appetizers.

Dairy Products in Soups

Dairy products are used in soups as a main ingredient and as an accompaniment. Creamed vegetable soups such as broccoli, potato, and asparagus soups are quite popular. Milk or cream also forms the base for many chowders. **See Figure 9-49.** Chilled fruit-based soups often include yogurt to balance the sweetness of the fruit, add body, and enhance mouthfeel. While dairy products can garnish many soups, they are an expected component of some soups, such as the melted Swiss cheese on French onion soup or the sour cream used to garnish Mexican tortilla soup.

Dairy Products in Soups

Courtesy of The National Pork Board

Figure 9-49. Milk or cream forms the base for soups such as chowders.

Dairy Products in Sandwiches

Cheese is often used in its natural state or melted in sandwiches. It is commonly sliced, grated, or featured in sauces. Sandwiches are sometimes layered or topped with cheese and toasted or served open face. Wraps, burgers, and deli meat sandwiches with cheese are common choices. Cheese is also a featured ingredient in pizzas and tacos.

Yogurt and low-fat sour cream can be served as a condiment to add flavor and tanginess to sandwiches. As a condiment, yogurt offers a lower fat alternative to spreads such as butter or mayonnaise. For example, a savory peach yogurt spread pairs well with a grilled chicken breast sandwich.

Dairy Products in Salads

Dairy products are added to salads for color, texture, and flavor or can be one of the main ingredients in a salad. For example, a Caprese salad features buffalo mozzarella layered with fresh tomatoes and basil. Diced, crumbled, and sliced cheeses are often used to top salads. Cottage cheese adds a contrasting texture to crisp salad greens and provides protein. **See Figure 9-50.**

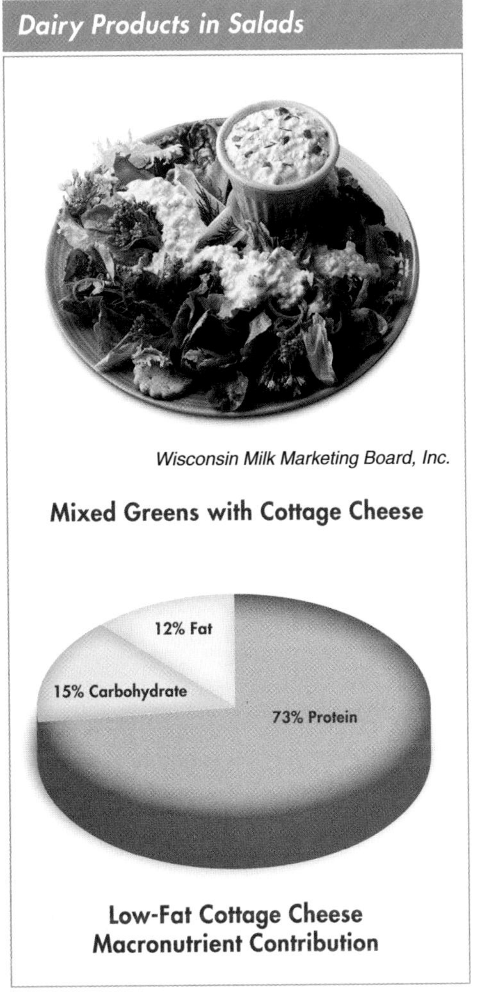

Figure 9-50. Cottage cheese adds a contrasting texture to crisp salad greens and provides protein.

Dairy products can be used two ways in salad dressings. For example, with blue cheese dressing, buttermilk or sour cream serves as the base of the dressing, and then the blue cheese is folded in to develop more texture and flavor. Yogurt plays an important role in the modification of dairy-based dressings because it can replace high-fat dairy products such as cream and sour cream. Yogurt can be used in creamy dressings for both savory salads and fruit salads.

Dairy Products in Entrées

Dairy products add texture, flavor, and nutrients to a variety of entrées. For breakfast, yogurt is commonly served plain or garnished with fresh fruits, nuts, granola, or a drizzle of honey. Cheese plays an important role in omelets and soufflés, and both cheese and milk are used in casseroles served throughout the day. Sauces served with entrées incorporate different forms of dairy products. For example, milk plays an important role as the base in a béchamel sauce, and cream, yogurt, sour cream, butter, and cheese are commonly used to add richness and enhance the consistency of various sauces. A variety of cheeses are commonly featured in pasta dishes as well as in cheese fondues. **See Figure 9-51.**

Dairy Products in Sides

Dairy products add color, a creamy texture, and flavor to side dishes. Cheese is sometimes used as a topping for vegetables such as broccoli and asparagus. Potatoes au gratin is a traditional side dish made with cream or milk and garnished with cheese. Pasta dishes such as gnocchi may be served in a cream sauce or sprinkled with cheese and served as a side.

Stuffed vegetables such as peppers or tomatoes often use béchamel sauce to bind ingredients and grated or shaved cheese as a garnish. Risotto can be finished with milk or cheeses such as Parmesan or fontina. Butter is widely used in side dishes to finish sauces, garnish vegetables, and serve as a condiment for breads.

Dairy Products in Entrées

Barilla America, Inc.

Figure 9-51. Pasta entrées often feature sauces made from dairy products such as milk, cream, and cheese.

Dairy Products in Desserts

Dairy products such as milk, cream, and butter are integral components of many desserts. Milk and cream are used to make ice creams, custards, and mousses. Butter is often a key fat in baked goods such as pies, cookies, cakes, and breads. Low-fat dairy products such as milk, sour cream, yogurt, and cream cheese can be substituted for higher-fat dairy products in desserts. Pungent cheeses are used less frequently in desserts, but chefs can incorporate them in unique ways. For example, sharp Cheddar cheese can be incorporated into pie crusts, and goat cheese can be used to make a tangy ice cream that is topped with fresh fruit.

Chef's Tip

In desserts such as key lime tarts, sweetened condensed milk is combined with lime juice, lime zest, and eggs to make a flavorful filling. Using nonfat sweetened condensed milk reduces fat and calories while maintaining texture and flavor.

Knowledge Check 9-11

1. List uses of dairy products in beverages and appetizers.
2. Explain how dairy products are commonly used in soups and sandwiches.
3. Identify the functions of dairy products in salads.
4. Describe how dairy products are incorporated into entrées.
5. Explain the functions of dairy products in side dishes.
6. List uses of dairy products in desserts.

PROMOTING DAIRY PRODUCTS ON THE MENU

Dairy products such as milk, yogurt, and ice cream are popular stand-alone menu items. Nutrient-dense dairy products can be promoted to meet the demand for nutritious menu items. For example, Greek yogurt topped with seasonal fruits can be promoted as a nutritious protein-rich breakfast, a side dish, or a light dessert. **See Figure 9-52.** Likewise, offering tasting menus with artisan and house-made cheeses of various textures, flavors, and origins is an effective promotional strategy. Interesting combinations of fruits, herbs, spices, peppers, and teas can be used to promote signature ice creams or frozen yogurts. In addition, promoting dairy products from local dairies can help attract guests.

Blueberry Yogurt Parfait

Wisconsin Milk Marketing Board, Inc.

Figure 9-52. Greek yogurt topped with seasonal fruits can be promoted as a nutritious protein-rich breakfast, a side dish, or a light dessert.

Quick service and fast casual venues offer dairy products in menu items that are easy to eat and transport, such as sandwiches, pizzas, salads, and soups. Dairy products such as milk and cream are emphasized at beverage houses and bar venues in smoothies and specialty drinks that feature coffee, tea, and alcohol. Artisan and imported cheeses are likely to be showcased in appetizers, salads, sauces, entrées, and elegant desserts at fine dining and special event venues. **See Figure 9-53.**

Baked Alaska

Daniel NYC/B. Milne

Figure 9-53. Elegant dairy-based desserts, such as baked Alaska, are often served at fine dining and special event venues.

Knowledge Check 9-12

1. List popular items featuring dairy products at quick service and fast casual venues.
2. Describe the ways in which dairy products are featured at beverage houses and bar venues.
3. Explain how dairy products are often used at fine dining and special event venues.

Chapter Summary

Eggs, soy, and calcium-rich dairy products are excellent sources of high-quality protein. They supply essential vitamins and minerals and are abundant in antioxidants. Egg, soy, and dairy products can be used across all menu categories to create flavorful meals that meet the demand for healthy menu items. Many forms of these products can also be used to produce gluten-free and vegetarian dishes.

Whole eggs, egg convenience products, and egg substitutes are used in professional kitchens. Eggs are commonly cooked in the shell, poached, shirred, fried, scrambled, and prepared as omelets. In addition to providing color, texture, and flavor, eggs are used to bind, emulsify, thicken, coat, clarify, and leaven foods.

A variety of soy products are available, making soy a versatile ingredient that can be used across all meal periods. Less processed soy products, such as edamame, tofu, and tempeh, have more phytonutrients than highly processed forms of soy. Soybeans contain more protein than most other plant-based foods and are plentiful in dietary fiber.

Courtesy of The National Pork Board

Dairy products, including milk, yogurt, and cheese, are highly nutritious but may be high in fat. A variety of low-fat and nonfat dairy products are available. High-fat dairy products such as butter and cream can be part of a healthy dish when used in moderation.

Chapter Review

Digital Resources
ATPeResources.com/QuickLinks
Access Code: 267412

1. Describe the structure of eggs.
2. List the nutrients supplied by eggs.
3. Describe the factors that influence the perceived value of eggs.
4. Compare the three market forms of eggs.
5. Explain how to cook ten popular egg preparations.
6. Identify methods for developing flavors in and enhancing the presentation of egg dishes.
7. Explain how eggs are incorporated throughout the menu mix.
8. Describe ways in which eggs are promoted on the menu.
9. Identify the nutritional benefits of soy products.
10. Explain how soy products are perceived on the menu.
11. Describe the characteristics of common market forms of soy products.
12. Compare the methods of cooking various soy products.
13. Explain how to develop flavors in and enhance the presentation of dishes that feature soy products.
14. Describe the functions of soy products throughout the menu mix.
15. Explain how soy products are featured at various foodservice venues.
16. Explain the nutritional benefits of dairy products.
17. Identify factors that influence the perceived value of dairy products.
18. Compare the main characteristics of liquid, semisolid, solid, canned, and dried dairy products.
19. Explain how to effectively cook with dairy products.
20. Describe how to develop flavors in and enhance the presentation of dishes featuring dairy products.
21. Identify the functions of dairy products throughout the menu mix.
22. Explain how dairy products can be incorporated into the menus at various foodservice venues.

Poultry & Meats on the Menu

Turkey

National Turkey Federation

Beef

Photo Courtesy of the Beef Checkoff

Pork

Courtesy of The National Pork Board

Poultry and meats are excellent sources of high-quality protein that can be prepared to create healthy and tender menu items. The flavor of poultry and meats can be elevated by using rubs, marinades, and other flavor-building techniques. Serving appropriate portion sizes of poultry and meats with colorful vegetables and whole grains creates nutritious dishes with colors, textures, and flavors that are appealing to guests.

Chapter Objectives

1. Describe factors that influence the perceived value of poultry.
2. Describe the six poultry classifications and how each kind is further classified.
3. Identify the terminology used to identify fabricated cuts of poultry.
4. Describe the cooking techniques commonly used for poultry.
5. Explain how rubs, marinades, and bastings can help develop the flavor of poultry.
6. Describe ways in which poultry can be plated.
7. Describe healthy ways to use poultry in the menu mix.
8. Summarize how foodservice venues promote poultry on the menu.
9. Describe factors that influence the perceived value of meats.
10. Describe the characteristics of the meats most commonly used in the professional kitchen.
11. Identify the leanest cuts of meat.
12. Describe the cooking techniques commonly used for meats.
13. Explain ways to develop the flavor of meats.
14. Describe healthy ways to use meats in the menu mix.
15. Summarize how foodservice venues promote meats on the menu.

POULTRY

Poultry is the collective term for various kinds of birds that are raised for human consumption. Poultry provides essential dietary protein and typically contains less fat than other animal proteins because the muscle tissue does not contain marbling, which is the fat found within the muscle. Poultry is also plentiful in vitamins and minerals, such as riboflavin, niacin, vitamin B_{12}, magnesium, iron, and zinc.

Perceived Value of Poultry

Poultry is a fairly inexpensive form of protein and can help control menu costs. Because the cost savings can be passed onto guests and because it is a versatile ingredient, poultry is a popular menu item for any venue. Nutritionally, a skinless, boneless grilled chicken breast is perceived as a healthy menu item because of its low caloric and fat content. In contrast, crispy deep-fried chicken may be perceived as an indulgence by those seeking to eat healthy. **See Figure 10-1.**

Uncommon poultry items such as duck, Cornish hen, goose, and squab are often perceived by guests as sophisticated menu items. Therefore, these forms of poultry often command a higher price.

The sides chosen for a poultry entrée also influence its perceived value. For example, fried chicken served with French fries and coleslaw is a common meal with a relatively low food cost. In contrast, fried chicken served with a baked sweet potato and sautéed chard offers a more nutrient-dense meal. The chicken is the same, yet the flavor profile, nutrient density, and presentation are enhanced by the sides offered.

Poultry Classifications

The kinds of poultry recognized by the United States Department of Agriculture (USDA) that are commonly served in foodservice operations include chickens, turkeys, ducks, and geese. Each bird is classified based on age and/or gender. As a bird ages, its breastbone becomes less flexible and the flesh and skin toughen, intensifying the overall flavor. Therefore, younger birds are often desired for their tenderness and mild flavor. Many operations also serve ratites such as ostrich and emu and game birds such as pheasant and quail.

Figure 10-1. A skinless, boneless grilled chicken breast is perceived as a healthy menu item in contrast to crispy deep-fried chicken.

Chef's Tip

If a bird is labeled as free-range poultry, it means that the bird was given access to the outdoors but was not necessarily raised outdoors. A bird labeled as organic poultry is fed a diet of organically grown feed and not treated with antibiotics or hormones.

Chickens. Chicken is a relatively inexpensive form of poultry that is simple to prepare. Chickens are classified by age and sometimes gender. Depending on the classification, chickens can range in weight from 1½ pounds up to 8 pounds. Factors such as age, gender, and weight affect the flavor, tenderness, and techniques used to prepare and cook chicken. Chicken classifications include Cornish game hens, broilers/fryers, roasters, capons, and stewers. **See Figure 10-2.**

Turkeys. Like chicken, turkey is a relatively inexpensive protein found on many menus. Turkeys are classified by age and range from 4–30 pounds. Most turkeys are fairly tender but toughen with age. Tougher cuts require moist-heat cooking techniques such as braising or stewing. Classifications of turkeys include fryer/roaster turkeys, young turkeys, yearling turkeys, and mature turkeys. **See Figure 10-3.**

National Honey Board

Some forms of poultry can be roasted whole.

Chicken Classifications

Classification	Characteristics and Uses
Cornish game hens	• Male or female • Less than 5 weeks old • Weigh up to 1½ pounds • Tender with mild flavor • Commonly stuffed and roasted whole
Broilers/fryers	• Male or female • Less than 5 weeks old • Weigh approximately 1½–3½ pounds • Tender flesh with smooth skin • Slightly higher fat content than Cornish hens • Prepared using any cooking technique
Roasters	• Male or female • 2–3 months old • Weigh 5 pounds or more • Tender flesh with smooth skin • Prepared using any cooking technique
Capons	• Castrated male • Less than 4 months old • Weigh approximately 4–7 pounds • Produce large, well-formed breasts • More tender than broilers/fryers • Commonly roasted
Stewers	• Female • More than 10 months old • Weigh approximately 3–8 pounds • Tough but flavorful flesh and skin • Commonly braised or stewed

Figure 10-2. Chicken classifications include Cornish game hens, broilers/fryers, roasters, capons, and stewers.

Turkey Classifications

Classification	Characteristics and Uses
Fryer/roaster turkeys	• Female • Less than 3 months old • Weigh approximately 4–9 pounds • Tender with soft, flexible skin • Males referred to as "toms"; females as "hens" • Commonly roasted, sautéed, or pan-fried
Young turkeys	• Female • Less than 8 months old • Weigh approximately 8–22 pounds • Tender flesh, smooth skin, and firm breastbone • Commonly roasted or stewed
Yearling turkeys	• Male or female • Less than 15 months old • Weigh approximately 10–30 pounds • Tender flesh • Commonly roasted or stewed
Mature turkeys	• Male or female • More than 15 months old • Weigh approximately 10–30 pounds • Tough but flavorful flesh and skin • Commonly roasted or stewed

Figure 10-3. Classifications of turkeys include fryer/roaster turkeys, young turkeys, yearling turkeys, and mature turkeys.

Ducks. Ducks are slightly larger than chickens and consist of all dark flesh. Ducks have less flesh in proportion to bone and fat than most other kinds of poultry. For example, a duck yields half as much flesh as a chicken of the same size. The majority of fat in a duck is located in and just beneath the skin. With the fatty skin removed, duck is lean in comparison to other kinds of poultry. Ducks are classified by age and include broiler/fryer ducklings, roaster ducklings, and mature ducks. **See Figure 10-4.**

Foie gras is the fattened liver of a duck or goose. Duck and goose liver is smooth and yellow in color, unlike the grainy texture and reddish-brown color of chicken liver. A duck or goose that is bred to produce foie gras is fed a rich diet until the liver becomes almost solid fat. Foie gras is typically seared in a very hot sauté pan and served immediately. Foie gras may also be poached, cooled, and then puréed to make a liver pâté. Foie gras should never be overcooked since it is very high in fat and will melt, reducing the size of the liver and altering its creamy texture. Foie gras is high in saturated fat and cholesterol and should be consumed in small portions. **See Figure 10-5.**

Duck Classifications

Classification	Characteristics and Uses
Broiler/fryer ducklings	• Less than 2 months old • Weigh approximately 3–6½ pounds • Tender flesh and soft windpipe • Commonly roasted or stewed
Roaster ducklings	• Less than 4 months old • Weigh approximately 4–7½ pounds • Tender flesh with windpipe starting to harden • Commonly roasted
Mature ducks	• More than 6 months old • Tough flesh with hardened windpipe • Commonly braised

Figure 10-4. Ducks are classified by age and include broiler/fryer ducklings, roaster ducklings, and mature ducks.

Foie Gras

Photo Courtesy of D'Artagnan, Photography by Doug Adams Studio

Figure 10-5. Foie gras is high in saturated fat and cholesterol and should be consumed in small portions.

Geese. Geese are nearly as large as turkeys with dark flesh and a large amount of fat in both the skin and the flesh. Only young geese are used in foodservice operations. A *young goose* is a goose that is usually less than six months of age and weighs approximately 4–10 pounds. Young geese have tender flesh that has a rich flavor due to the high fat content. A young goose is commonly roasted at very high temperatures to aid in rendering some of the fat from the skin and the flesh. Because goose flesh is high in fat, it is often served with a tart sauce to balance the fatty flavor.

Ratites. A *ratite* is a flightless bird that has a flat breastbone and small wings in relation to its body size. Unlike other forms of poultry, ratites have almost no breast flesh. Most of the flesh on a ratite is located on the back, thigh, and forequarter of the bird. The fan, taken from a muscle in the thigh, is the most tender cut. The top loin is also a very tender cut.

Ratite flesh is dark red and has a similar appearance to beef after it is cooked. The flavor is also similar to beef but sweeter, and the flesh contains less fat. Even though ratites are classified as poultry, the flesh should be cooked no higher than 155°F since it will develop a metallic flavor at higher temperatures. The loin can be roasted whole or cut into steaks, fillets, or medallions and served with vegetables and wild rice mixed with dried fruits. Cuts from the loin can also be stuffed with a variety of fillings and rolled. Less tender cuts from the thigh and leg are often cut into smaller pieces or ground and then braised or stewed for use with grains, pastas, or vegetables. Ostriches and emus are examples of ratites. **See Figure 10-6.**

Game Birds. A *farm-raised game bird* is a game bird that is raised for legal sale. Farm-raised game birds include pheasants, quails, grouses, partridges, and wild turkeys. **See Figure 10-7.** These birds have dark flesh and a fairly strong flavor. Farm-raised game birds are generally leaner than poultry, so care must be taken not to overcook these birds.

Ratite Classifications

Classification	Characteristics and Uses
Ostriches	• Weigh approximately 300–400 pounds • Up to 8 feet in height • Tender cuts in the fan or thigh are commonly grilled, broiled, or sautéed • Tougher cuts from the leg are commonly marinated then roasted, braised, or stewed
Emus	• Weigh approximately 125–140 pounds • Up to 6 feet in height • Tender cuts are commonly grilled or broiled • Tougher cuts are commonly braised or stewed

Figure 10-6. Ostriches and emus are examples of ratites.

INGREDIENT SPOTLIGHT: Emu

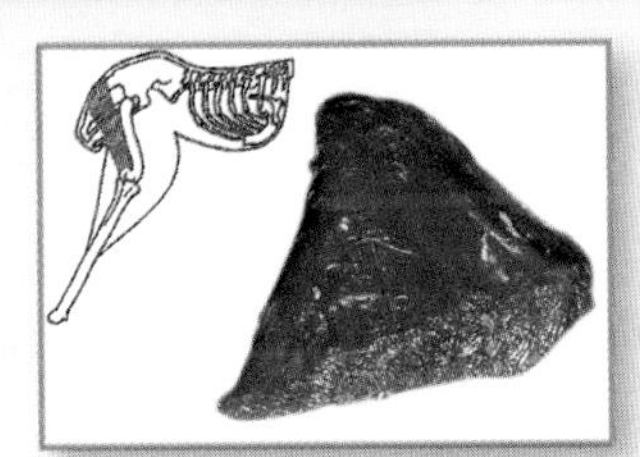

Emu Today and Tomorrow

Unique Features

- Flightless bird comprised of red meat
- Considered a sustainable product
- Low in total fat and sodium
- Source of protein, riboflavin, niacin, pantothenic acid, vitamin B_6, vitamin B_{12}, iron, and selenium

Menu Applications

- Add to the menu as a unique poultry item with red meat
- Use in a variety of items including kebabs, burgers, and sausages

Nutrition Facts

Serving Size 3 oz (fan fillet, broiled)

Amount Per Serving	
Calories 131	Calories from Fat 18
	% Daily Value*
Total Fat 2g	3%
Saturated Fat 0g	2%
Trans Fat	
Cholesterol 70mg	23%
Sodium 45mg	2%
Total Carbohydrate 0g	0%
Dietary Fiber 0g	0%
Sugars 0g	
Protein 27g	
Vitamin A 0% • Vitamin C	0%
Calcium 1% • Iron	22%

* Percent Daily Values (DV) are based on a 2,000 calorie diet.

Game Bird Classifications

Classification	Characteristics and Uses
Pheasants	• Weigh approximately 1½–2½ pounds • Light colored flesh with mild flavor • Commonly roasted or braised
Quails	• Smallest game bird • Breast weighs only 1–2 ounces • Commonly skewered to be grilled or broiled, or boned out and filled with stuffing or rice and roasted whole
Grouses	• Resemble chickens, but with thicker legs and dark flesh • Small grouses are commonly sautéed; large grouses are commonly roasted
Partridges	• Weigh approximately 1 pound • White flesh • Yield an edible portion for 2 people • Slightly chewy with gamey flavor • Commonly roasted or broiled
Wild turkeys (farm-raised)	• Dark flesh that is tougher, leaner, less moist, and less meaty than domesticated turkey flesh • Strong, gamey flavor • Commonly braised, roasted, or smoked

Figure 10-7. Farm-raised game birds include pheasants, quails, grouses, partridges, and wild turkeys.

Market Forms of Poultry

Poultry can be purchased fresh, frozen, canned, and dried. Fresh and frozen poultry are available in many convenience forms, including boneless and skinless breasts, whole bone-in pieces, and cooked shredded poultry. Canned poultry is convenient for making salads, sandwiches, and other dishes that call for cooked poultry. Due to processing, canned poultry is high in sodium and is generally more expensive than fresh or frozen poultry. Dried poultry is used in food production to make reconstituted products, such as dry soups.

Poultry is also commonly available based on the type of flesh preferred. Some poultry has both white and dark flesh. **See Figure 10-8.** Other poultry has only dark flesh. Leg and thigh flesh is always dark, but breasts and wings can be either white or dark. White flesh is found in muscles that are used less often, such as a chicken breast. In contrast, chickens spend a majority of time on their feet and therefore have dark flesh in their legs and thighs due to developed muscles. Because ducks and geese fly long distances, they have dark flesh throughout their entire bodies. The more a muscle is used, the darker and more flavorful the muscle becomes.

Pheasant Flesh

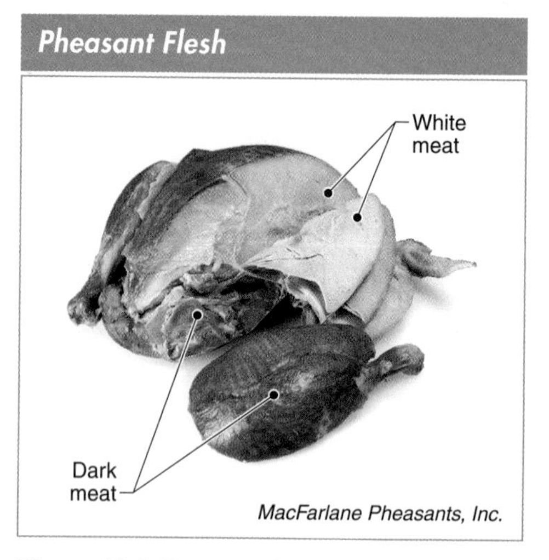

Figure 10-8. Some poultry, such as pheasant, has both white and dark flesh.

Whole and Fabricated Poultry. Poultry is sold whole or cut-up, bone-in or boneless, ground, or processed into prepared forms such as chicken tenders. Specific terminology is used to describe fabricated pieces of poultry. **See Figure 10-9.** Purchasing whole birds is more economical than purchasing fabricated pieces.

Whole poultry is commonly cut into halves, quarters, and eighths. To halve a bird, the bird is split from top to bottom between the breasts and along the backbone to the tail into two equal portions. Poultry can then be cut into quarters for grilling, broiling, or roasting. To do this, the halved bird is divided into two leg and thigh sections and two wing and breast sections, yielding four quarters. To cut poultry into eighths for grilling, broiling, roasting, and frying, the quarters are cut into two breasts, two wings, two thighs, and two legs. **See Figure 10-10.**

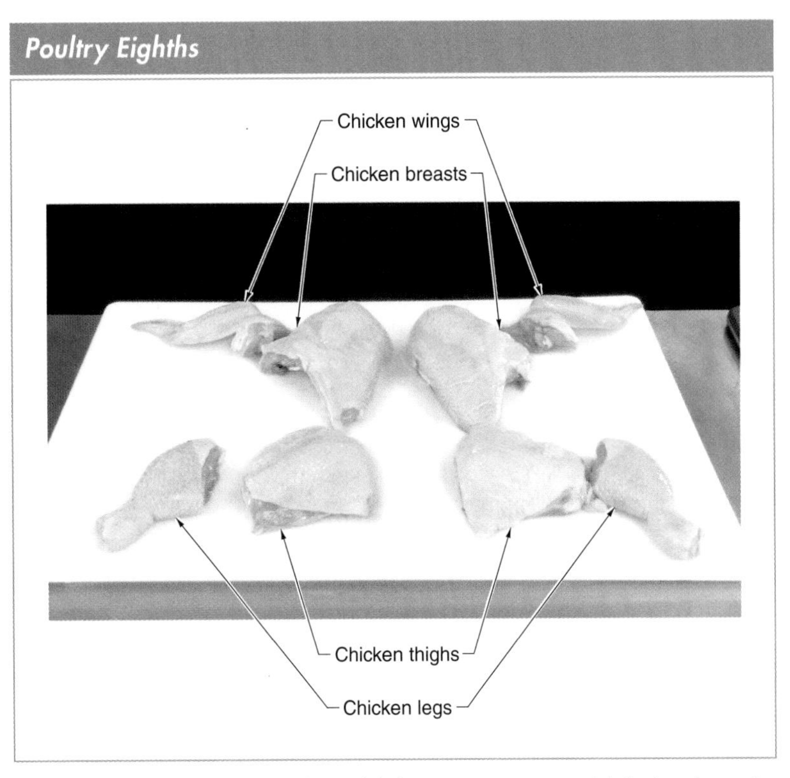

Figure 10-10. To cut a bird into eighths, quarters are cut into two breasts, two wings, two thighs, and two legs.

Fabricated Poultry Terminology

Poultry Cut	Description
Poultry half	Full half-length of a bird split down the breast and spine
Breast quarter	Half of a breast, a wing, and a portion of the back
Breast	Top front portion of the flesh above the rib cage
Airline breast	Boneless skin-on chicken breast with the first wing section attached
Poultry tenderloin	Inner muscle that runs alongside the breastbone of a bird
Tender	Small strip of a breast
Wing	Tip, paddle, and drummette
Tip	Outermost section of a wing
Paddle	Second section of a wing located between the two wing joints
Drummette	Innermost section of a wing located between the first wing joint and the shoulder
Leg quarter	Thigh, drumstick, and a portion of the back
Thigh	Upper section of the leg located below the hip and above the knee joint
Drumstick	Lower portion of the leg located below the hip and above the knee joint
Poultry leg	Drumstick and thigh

Figure 10-9. Specific terminology is used to describe fabricated pieces of poultry.

Food Safety

All knives, cutting boards, equipment, and surfaces, as well as hands, must be washed in hot soapy water and sanitized after coming into contact with raw poultry. Also, poultry should be stored on the bottom shelf of a refrigeration unit so raw juices do not drip onto other products.

In addition to fabricated pieces of poultry, giblets are also commonly used in the professional kitchen. *Giblets* is the name for the grouping of the neck, gizzard, liver, and heart of a bird. **See Figure 10-11.** Giblets are often finely chopped and used to make gravies. Chicken gizzards are often breaded or battered and then fried and served as entrées. Chicken livers are commonly breaded and then sautéed or fried and served topped with caramelized onions. Chicken livers are also used to make liver pâtés. Most giblet preparations contain significant amounts of fat and calories. However, giblets are a key source of protein, vitamins A and C, B vitamins, iron, and selenium.

Giblets

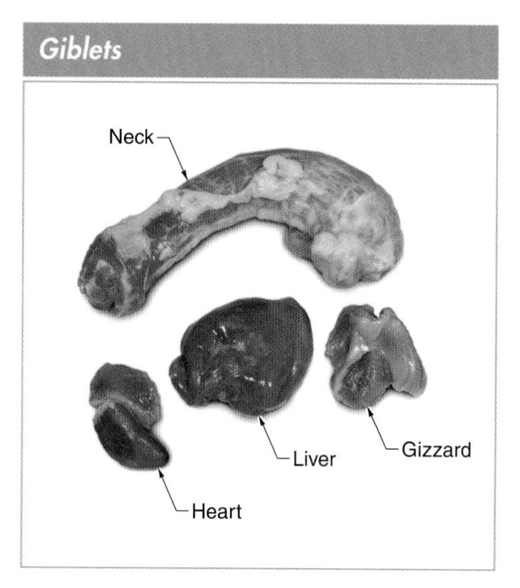

Figure 10-11. Giblets is the name for the grouping of the neck, gizzard, liver, and heart of a bird.

Poultry Convenience Products. Poultry convenience products include fully cooked and vacuum-wrapped poultry as well as boned and canned poultry. Precooked and portioned convenience poultry products include chicken wings and tenders, hot dogs, lunchmeat, sausages, and other convenience items such as frozen dinners and soups. **See Figure 10-12.** Many convenience poultry products have been marinated, injected, or stuffed with fillers and are generally high in sodium. While some items are 100% poultry, others may contain skin, bones, and cartilage as fillers. Processed poultry products are mechanically deboned, and binders are commonly added before the product is stamped to produce consistent shapes and sizes.

Poultry Convenience Products

Breaded Chicken

Boneless Chicken Breasts

Photo Courtesy of Perdue Foodservice, Perdue Farms Incorporated

Figure 10-12. Common poultry convenience products include precooked chicken tenders and deboned products.

Knowledge Check 10-1

1. Identify the nutritional benefits of poultry.
2. Contrast the perceived value of a grilled chicken breast and of fried chicken.
3. Identify the six poultry classifications.
4. Summarize the classifications of chickens, turkeys, and ducks.
5. Describe the type of geese most often used in foodservice operations.
6. Describe the flesh and flavor of ratites.
7. Identify five farm-raised game birds.
8. Identify the cuts that result from fabricating a bird into eighths.
9. Name key nutrients found in giblets.
10. List poultry convenience products commonly used in foodservice.

PREPARING NUTRITIOUS POULTRY

Poultry is a low-fat, high-quality protein that is easy to cook. The majority of fat in poultry is located just under the skin, near the tail, and on the abdomen. Preparing poultry with the skin and bones helps preserve moisture and keeps the poultry from drying out. However, consuming the skin adds saturated fat and calories. **See Figure 10-13.** To lower both fat and calories, the skin can be removed. White and dark flesh also contain varying amounts of fat, so cooking temperatures need to be monitored closely so that leaner pieces, such as boneless, skinless cuts, do not overcook and become dry. To promote tenderness, it is important to let cooked poultry rest so juices redistribute and keep the bird moist.

The cooking technique used to prepare poultry depends on the age and tenderness of the bird. **See Figure 10-14.** Young and tender birds can be cooked using any cooking technique, but older birds are best cooked with moist heat. For example, a chef may choose a young bird such as a broiler/fryer for broiling or grilling, but use a mature bird, such as a stewer, to flavor a stock.

Nutritional Comparison of Poultry With and Without Skin*

Type of Poultry	Calories	Total Fat	Saturated Fat
Boneless chicken breasts roasted with skin	197	8 g	2 g
Boneless chicken breasts roasted without skin	165	4 g	1 g

* based on 100 g serving size

Figure 10-13. Chicken with skin contains more fat and calories than chicken without skin.

Cooking Poultry, Ratites, and Game Birds

Bird	Class	Cooking Methods
Chicken	Cornish game hen	Broiling, grilling, or roasting
	Broiler/fryer	Any
	Roaster	Any
	Capon	Roasting
	Stewer	Braising or stewing
Turkey	Fryer/roaster turkey	Roasting, sautéing, or pan-frying
	Young turkey	Roasting or stewing
	Yearling turkey	Roasting or stewing
	Mature turkey	Roasting or stewing
Duck	Broiler/fryer duckling	Roasting, sautéing, or stewing
	Roaster duckling	Roasting
	Mature duck	Braising
Goose	Young goose	Roasting
Ratite	Ostrich	Grilling, broiling, sautéing, braising, or stewing
	Emu	Grilling, broiling, sautéing, braising, or stewing
Game bird	Pheasant	Roasting or braising
	Quail	Grilling, broiling, roasting, or sautéing
	Grouse	Roasting or sautéing
	Partridge	Broiling, roasting, or sautéing
	Wild turkey	Roasting, smoking, or braising

Figure 10-14. The cooking technique used to prepare poultry depends on the age and tenderness of the bird.

Cooking Poultry

Although poultry can be prepared using any cooking technique, dry-heat techniques such as grilling, broiling, smoking, barbequing, roasting, sautéing, and frying are among the most popular. Combination techniques, such as stewing, are also often used to cook poultry.

Grilling. Because only a small amount of oil is added to the grill to prevent sticking, grilling is a healthy cooking technique for poultry. Grilling also enhances presentation by creating char lines and imparts an appealing smoky flavor. **See Figure 10-15.** However, care should be taken when grilling because it produces intense heat that can quickly dry poultry out. Small birds, such as quails, are typically skewered prior to being grilled so that they lie flat on the grates. Cornish hens are often split in half or butterflied before being grilled. Birds larger than Cornish hens are typically broken down into quarters or eighths before being grilled to promote even cooking.

Grilled Chicken Breasts

Photo Courtesy of Perdue Foodservice, Perdue Farms Incorporated

Figure 10-15. Grilling poultry enhances presentation by creating char lines and imparts an appealing smoky flavor.

Smoked Duck Breast

Photo Courtesy of D'Artagnan, Photography by Doug Adams Studio

Figure 10-16. Smoking is a dry-heat cooking technique in which food such as duck has been cooked over smoldering hardwoods in a vented enclosure.

Broiling. Like grilling, broiling does not require a lot of oil. Also like grilling, broiling requires intense heat, so care needs to be taken not to overcook the poultry. Most poultry can be broiled in the same manner as it is grilled.

Smoking. *Smoking* is a dry-heat cooking technique in which food is slowly cooked over smoldering hardwoods in a vented enclosure. **See Figure 10-16.** Mesquite, hickory, maple, apple, and cherry woods are used to impart different smoked flavors. Turkey, duck, and pheasant are commonly smoked.

Barbequing. *Barbequing* is a dry-heat cooking technique in which food is slowly cooked over hot coals or smoldering hardwoods. Barbequing can be described as a combination of grilling and roasting, depending on the source of the heat. Most poultry that is barbequed is brushed with a sauce, which adds layers of flavor but may also add refined sugars.

Roasting. Properly roasted poultry results in a beautifully browned exterior that is tender and juicy inside. Whole poultry is often placed on a rotisserie and spit-roasted. *Spit-roasting* is the process of cooking food by skewering it and suspending and rotating it above or next to a heat source. **See Figure 10-17.**

Spit-Roasting Poultry

Henny Penny Corporation

Figure 10-17. Spit-roasting is the process of cooking food by skewering it and suspending and rotating it above or next to a heat source.

When roasting a whole bird, basting helps keep the bird moist and crisps the skin. Small birds, such as squabs, pheasants, and Cornish hens, are roasted at lower temperatures to produce a golden exterior without drying out the flesh. Poultry that is high in fat, such as ducks and geese, is typically roasted at high temperatures to melt the fat out and away from the skin. As the fat seeps from the skin, it provides a barrier to keep the poultry moist and crisps the skin, which can be removed after cooking to reduce fat and calories.

Sautéing. Boneless poultry and poultry pieces are often sautéed. Sautéing uses very little additional fat, which can help keep the poultry low in fat and calories. **See Figure 10-18.** Sautéed poultry develops its aroma and flavor as the exterior begins to brown. It is important not to overcrowd the pan when sautéing because the poultry will not brown properly. To elevate the flavor of sautéed poultry, the pan can be deglazed with a little stock, wine, or juice to form a flavorful sauce.

Sautéed Turkey Medallions

National Turkey Federation

Figure 10-18. Sautéing is a low-fat cooking technique often used on poultry.

Frying. Poultry is often stir-fried, pan-fried, or deep-fried. Like sautéing, stir-frying requires little additional fat, keeping the poultry lower in fat and calories than pan- or deep-fried poultry. To prevent poultry from absorbing large amounts of fat when frying, it is essential to heat the cooking oil to the proper temperature. Whether breaded or battered, fried poultry should be crispy and golden brown but not oily.

Simmering. Poultry pieces, such as chicken breasts, are typically simmered in a flavorful liquid to enhance moisture and flavor. If simmered poultry is to be served cold, it is best to cool the poultry in the simmering liquid to absorb additional moisture and flavor. Menu items featuring simmered poultry are often served with a sauce that has been enriched with the cooking liquid.

Stewing. In stewed poultry items, the cooking liquid permeates throughout the dish to develop flavor. Sautéing poultry and then adding aromatics and root vegetables at the beginning of the stewing process maximizes the flavors absorbed by the poultry as it cooks. Chicken cacciatore and coq au vin are examples of stewed chicken dishes. **See Figure 10-19.** Duck is also commonly stewed. To decrease the fat content of stewed dishes, they can be chilled to let the fat rise to the top of the pot where it can easily be removed.

Coq au Vin

Daniel NYC/E. Laignel

Figure 10-19. In stewed dishes such as coq au vin, the cooking liquid permeates throughout the dish to develop flavors.

FOOD SCIENCE EXPERIMENT: Effect of Heat and Brine on Poultry 10-1

Objective

- To compare the effect of heat on brined and nonbrined chicken breasts in terms of weight (portion size), texture, and flavor.

Materials and Equipment

- Measuring cup
- 4 cups water, divided
- 2 bowls
- ½ cup salt, divided
- Spoon
- 4 boneless, skinless chicken breasts, approximately 6 oz each
- 4 boneless chicken breasts with skin, approximately 6 oz each
- Kitchen scale
- 2 sheet pans
- Parchment paper
- Marker
- Instant-read thermometer
- Tongs

Procedure

1. Make a brine by adding 2 cups of water and ¼ cup of salt to each of the two bowls and stir.
2. Place two skinless chicken breasts in one of the brines and refrigerate for 2 hours.
3. Place two chicken breasts with skin in the other brine and refrigerate for 2 hours.
4. Weigh each of the eight chicken breasts. Record the weight and type of chicken breast.
5. Line each sheet pan with parchment paper and use the marker to label a space for each type of chicken breast as follows:
   - "Chicken Breast with Skin"
   - "Skinless Chicken Breast"
   - "Brined Chicken Breast with Skin"
   - "Brined Skinless Chicken Breast"
6. Place one chicken breast of each type on each sheet pan. *Note:* There should be four chicken breasts that correspond to their respective labels on each pan.
7. Place one sheet pan in a 325°F oven and cook until an instant read thermometer placed in the thickest part of each breast reaches 165°F.
8. Place the second sheet pan in a 425°F oven and cook until an instant read thermometer placed in the thickest part of each breast reaches 165°F.
9. Use tongs to place one of the cooked chicken breasts on the scale. Record the weight, oven temperature, and type of chicken breast.
10. Repeat step 9 for each chicken breast, recording the weight, oven temperature, and type of chicken breast.
11. Calculate and compare the difference in weight between each type of chicken breast before and after cooking.
12. Evaluate the texture and flavor of each chicken breast.

Typical Results

There will be variances in the weight of the raw and cooked chicken breasts as well as the texture and flavor of each cooked chicken breast.

Flavor Development

Poultry has a mild taste that allows it to blend well with other ingredients to build and layer flavors. Effective ways to enhance and develop flavors in poultry menu items include using rubs and marinades as well as basting.

Rubs. There are endless combinations of dry and wet rubs that can be applied to poultry before cooking to elevate flavors. Dry rubs are made by grinding herbs and spices together into a fine powder and rubbing the mixture into the meat prior to cooking. Dry rubs can be as simple as salt and pepper mixed with dried oregano, or a more complex blend of herbs and spices, such as jerk or curry seasonings. Often dry rubs feature ingredients such as coffee, tea, or cocoa powder to create unique flavor combinations. **See Figure 10-20.** A wet rub is made by incorporating wet ingredients, such as Dijon mustard, flavored oils, puréed garlic, or honey, into a dry rub mixture and then applying it to the meat prior to cooking.

Tea Powder

Figure 10-20. Dry rubs can use ingredients such as tea powder to create unique flavor combinations.

Chef's Tip

When cooking whole poultry at high temperatures, fresh herb rubs applied to the exterior of the skin have a tendency to burn. However, fresh herbs can be rubbed under the skin or stuffed into the cavity of the bird to promote flavor development.

Marinades. Marinades help develop flavor, provide moisture, and tenderize poultry. Marinades typically contain oil and an acidic ingredient, such as wine, citrus juices, or yogurt, which acts as a tenderizer in addition to providing flavor. The flavor imparted by marinades can be mild to robust depending on the ingredients. For example, a marinade of oil and lemon juice lightly flavors poultry, while some sauces in global cuisines can be used as marinades to intensify flavors. **See Figure 10-21.**

Global Sauces Used as Marinades

Sauce	Typical Ingredients
Caribbean mojo	Olive oil, garlic, cumin, vinegar, and lemon, lime, or grapefruit juice
Italian pesto	Olive oil, basil, pine nuts, Parmesan cheese, and lemon juice
African piri-piri	Olive oil, chile peppers, garlic, and vinegar or lemon juice

Figure 10-21. Some sauces in global cuisines can be used as marinades to intensify flavors.

Basting. *Basting* is the process of using a brush or a ladle to place pan drippings or sauces over an item during the cooking process to help retain moisture and enhance flavor. For example, basting a cherry sauce on a duck as it roasts helps keep the poultry moist and provides a layer of flavor. Likewise, basting a chicken kebab with Japanese ponzu (citrus-based sauce) will produce a flavorful, juicy menu item.

Food Safety

Marinades can be cooked and reduced for use as a sauce since cooking kills bacteria. However, uncooked marinades should be disposed of immediately.

Plating Poultry

Poultry is a versatile entrée that can be plated in a variety of ways. **See Figure 10-22.** For example, chicken may be presented in quarters or individual pieces such as a breast. Chicken may also be presented skewered or used in combination dishes such as stir-fries. Turkey is generally cut into slices and fanned for serving. Ducks are often presented as a half or quarter, and Cornish hens are typically served whole. If the bird is very small, such as a quail, two may be presented to a guest. Because poultry is neutral in color, the addition of fruits and vegetables along with healthy grains elevates the presentation with heightened colors and textures and adds flavor and nutrients.

Seasonal menus may alter the way poultry is plated and presented. For example, grilled and barbequed poultry are popular in the warmer months, while roasted poultry is often featured in the cooler months. Sides of vegetables and grains also differ depending upon the season. Sweet corn, zucchini, or heirloom tomatoes harvested at the peak of summer ripeness may accompany grilled poultry, while a healthy whole grain stuffing and root vegetables pair well with roasted poultry served in autumn.

Plating Poultry

Daniel NYC/T. Schauer

Roasted Duck

National Honey Board

Chicken Kebabs

Photo Courtesy of Perdue Foodservice, Perdue Farms Incorporated

Whole Cornish Hen

National Turkey Federation

Turkey Dim Sum

Figure 10-22. Poultry is a versatile entrée that can be plated in a variety of ways.

CULINARY NUTRITION RECIPE-MODIFICATION PROCESS

Fried Chicken

Fried chicken is a comfort food with flavor profiles that vary regionally. In this recipe, bone-in chicken quarters with skin are marinated in whole buttermilk and pan-fried, causing the dish to be high in calories and fat.

Yield: 4 servings, 1 quarter each

Ingredients

1 c	whole buttermilk
1 ea	garlic clove, minced
¾ tsp	salt
½ tsp	marjoram, dried
½ tsp	sage, dried
½ tsp	thyme, dried
⅛ tsp	cayenne pepper
4 ea	chicken quarters
6 oz	all-purpose flour
16 fl oz	vegetable oil

Preparation

1. Combine buttermilk, garlic, salt, marjoram, sage, thyme, and cayenne pepper in a mixing bowl and set aside.
2. Rinse chicken and pat dry.
3. Place chicken in buttermilk mixture, cover, and refrigerate 1–2 hours.
4. Dredge chicken in flour and place on a sheet pan. Cover chicken and return to refrigerator for 30 minutes.
5. Heat oil in a saucepan to 350°F.
6. Place chicken in saucepan, skin-side down, and pan-fry until light golden brown (approximately 4 minutes per side).
7. Transfer chicken to a sheet pan and finish cooking in a 400°F oven until an internal temperature of 165°F is reached (approximately 25–30 minutes).

Evaluate original recipe for sensory and nutritional qualities

- Fried chicken has a crisp exterior, moist interior, and appealing flavor.
- Dish is high in calories and fat.

Establish goals for recipe modifications

- Reduce calories and fat.
- Enhance flavor.

Identify modifications or substitutions

- Modify cooking technique.
- Use a substitution for flour.
- Use a substitution for bone-in, skin-on chicken.

Determine functions of identified modifications or substitutions

- Pan-frying cooks the chicken and seals in moisture.
- Flour provides a crisp crust.
- Chicken skin adds flavor.

Select appropriate modifications or substitutions

- Change cooking technique from pan-frying to baking.
- Substitute bread crumbs, Parmesan cheese, and parsley for the flour.
- Substitute boneless, skinless chicken breasts for bone-in, skin-on chicken quarters.

Test modified recipe to evaluate sensory and nutritional qualities

- Baking reduces calories and fat by eliminating oil as the cooking medium.
- Replacing flour with bread crumbs, Parmesan cheese, and parsley produces a crisp exterior with enhanced flavors.
- Using boneless, skinless chicken breasts reduces calories and fat.

Modified Baked Fried Chicken

In this modified fried chicken recipe, baking boneless, skinless chicken breasts instead of frying skin-on chicken quarters lowers calories and decreases the total fat content by more than half. A crispy texture is maintained, and flavor is elevated with a crust of seasoned bread crumbs, Parmesan cheese, and parsley.

Yield: 4 servings, 1 breast each

Ingredients

1 c	low-fat buttermilk
1 ea	garlic clove, minced
½ tsp	marjoram, dried
½ tsp	sage, dried
½ tsp	thyme, dried
⅛ tsp	cayenne pepper
4 ea	boneless, skinless chicken breasts
¾ c	bread crumbs, dried
1 oz	Parmesan cheese, grated
2 tbsp	fresh parsley, chopped

Carlisle FoodService Products

Preparation

1. Combine buttermilk, garlic, marjoram, sage, thyme, and cayenne pepper in a mixing bowl and set aside.
2. Rinse chicken and pat dry.
3. Place chicken in buttermilk mixture, cover, and refrigerate 1–2 hours.
4. Combine bread crumbs, Parmesan cheese, and parsley in a shallow dish.
5. Dredge chicken in bread crumb mixture and place on a sheet pan.
6. Bake chicken in a 375°F oven until an internal temperature of 165°F is reached (approximately 30–40 minutes).

Chicken Nutritional Comparison

Nutrition Facts	Original	Modified
Calories	449.7	385.3
Total Fat	34.7 g	6.4 g
Saturated Fat	5.5 g	2.4 g

CULINARY NUTRITION RECIPE-MODIFICATION PROCESS

Signature Crispy Baked Chicken Cutlets

In this signature recipe, boneless, skinless chicken breasts are pounded thin to enhance presentation. The chicken is baked, making it lower in calories and fat than the original recipe. A thin crust of panko bread crumbs seasoned with lemon zest, smoked paprika, and pepper amplifies the texture and flavor while adding few calories and fat.

Yield: 4 servings, 1 breast each

Ingredients

4 ea	boneless, skinless chicken breasts
½ c	panko bread crumbs
¼ c	fresh parsley, chopped
1 tbsp	lemon zest
2 tsp	smoked paprika
¼ tsp	black pepper, ground
½ c	low-fat milk

Preparation

1. Pound chicken breasts between 2 pieces of plastic wrap to a uniform ⅓ inch thickness.
2. Combine panko bread crumbs, parsley, lemon zest, smoked paprika, and black pepper in a shallow dish.
3. Dip chicken breasts in milk and then dredge in panko mixture.
4. Place chicken on a sheet pan and bake in a 375°F oven until an internal temperature of 165°F is reached (approximately 20 minutes).

Chicken Nutritional Comparison

Nutrition Facts	Original	Modified	Signature
Calories	449.7	385.3	307.0
Total Fat	34.7 g	6.4 g	3.7 g
Saturated Fat	5.5 g	2.4 g	1.0 g

Knowledge Check 10-2

1. Explain how poultry skin affects preparation and nutrition.
2. Describe the techniques commonly used to cook poultry.
3. Differentiate between roasting small birds and large birds.
4. Name the types of poultry cuts commonly sautéed.
5. Explain how to reduce the fat content of stewed poultry dishes.
6. Explain the purpose of marinades.
7. Describe the process of basting.
8. Describe effective ways to plate poultry.

POULTRY MENU MIX

Poultry can be featured throughout the menu in appetizers, soups, sandwiches, salads, and entrées. Because of its naturally mild flavor, poultry of all types can be paired with additional ingredients such as fruits, vegetables, or whole grains to create enticing nutrient-dense menu items. **See Figure 10-23.** The mild flavor of poultry also adds to its appeal among all age groups. For example, roasted chicken often represents comfort food and may have increased appeal among a mature demographic, while drumsticks can easily be picked up, making them ideal for children.

When describing poultry on the menu, it is important to inform guests of the type and cut of poultry used in the menu item as well as the cooking technique used. **See Figure 10-24.** This information provides guests with flavor profiles as well as nutritional values. For example, individuals seeking a more nutrient-dense meal may tend to choose a poultry dish that features a grilled chicken breast over a dish of fried chicken wings. Mentioning the type of poultry, such as emu, duck, or quail, is also tied to perceived value and menu price.

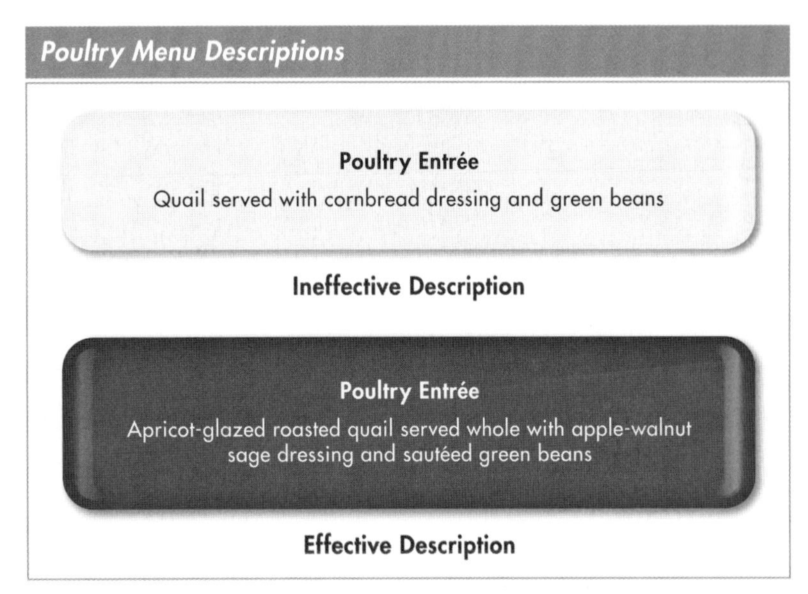

Figure 10-24. Effective menu descriptions should inform guests of the type and cut of poultry used in the menu item as well as the cooking technique.

Poultry on the Menu

Poultry Type	Menu Item Description
Chicken	Grilled chicken tenders accented with Maytag blue cheese, pecans, and crisp romaine lettuce wrapped in a whole-wheat tortilla
Turkey	Spicy turkey chili verde with hominy and butternut squash
Duck	Pan-seared and roasted duck breast with grilled peaches and walnut-parsley brown rice
Cornish hen	Roasted whole, boneless Cornish hen stuffed with spinach, goat cheese, and sun-dried tomatoes

Figure 10-23. Poultry of all types can be paired with additional ingredients such as fruits, vegetables, or whole grains to create enticing nutrient-dense menu items.

Poultry in Appetizers

Poultry can be used to create nutritious appetizers that can be served either hot or cold. For example, poultry can be sliced into strips or chunks, skewered, grilled, and then served warm with various dipping sauces such as honey mustard or Greek yogurt enhanced with garlic and dill. Poultry served chilled is a filling popular for lettuce wraps, which is also ideal for those seeking gluten-free foods.

Other popular poultry appetizers include chicken wings and spring rolls. Chicken wings are a sweet-to-spicy appetizer that is often high in fat and calories because it is fried. Wings can be grilled instead of fried to reduce fat and calories while still maintaining flavor. **See Figure 10-25.** Rice paper spring rolls filled with poultry and an array of colorful vegetables are rich in protein, vitamins, and minerals, including antioxidants. Poultry can also be shredded and used to make tacos and empanadas.

Poultry in Appetizers

Photo Courtesy of Perdue Foodservice, Perdue Farms Incorporated

Figure 10-25. Chicken wings are a popular fried appetizer that can be grilled to reduce fat and calories.

Poultry in Soups

Poultry has long been used in soups, from broths to hearty cream-based soups. Chicken stock is a staple base for many popular soups. For example, clear Asian soups are often flavored with broth made from chicken stock. Other poultry menu favorites include chicken tortilla soup, Thai chicken-coconut soup, and white chicken chili. **See Figure 10-26.**

Poultry in Soups

Irinox USA

Figure 10-26. Chicken tortilla soup is a popular soup featuring poultry.

Poultry in Sandwiches

Poultry is commonly used to make both hot and cold sandwiches. Sandwiches featuring poultry include closed and open-faced sandwiches, wraps, and panini. Chicken salad is a common cold sandwich filling made from cooked, chopped chicken usually mixed with mayonnaise. Substituting low-fat mayonnaise or Greek yogurt for mayonnaise can provide an appealing flavor while decreasing fat and calories. Turkey is often served as a hot open-faced sandwich. Whole breasts, shredded poultry, and ground poultry patties are also used to make both hot and cold sandwiches.

To improve the nutrient density of sandwiches, the poultry skin should be removed and the sandwich can be topped with fruits or vegetables, lower fat cheeses, nuts, and seeds and served with whole grain breads. Healthier sandwich spreads such as flavored mustards also provide flavor while being low in fat and calories.

Poultry in Salads

Poultry is commonly served either warm or chilled in salads. In addition to leafy green salads, poultry may be added to grain, potato, and pasta salads. Common examples of poultry salads include a Caesar salad with a grilled and sliced chicken breast, a spinach salad with roast duck, and a taco salad with crumbled turkey sausage. Adding poultry to salads helps build textures and flavors and can turn salads into nutritious entrées. **See Figure 10-27.**

Poultry in Salads

National Turkey Federation

Figure 10-27. Adding poultry to salads helps build textures and flavors and can turn salads into nutritious entrées.

Poultry in Entrées

Poultry is a universally appealing entrée across all types of cuisines. From fried chicken to chicken and dumplings, from Asian stir-fries to Italian cacciatore, and from Mexican tacos to French coq au vin, poultry can be prepared in multiple ways and featured in nutritious, flavorful meals. Poultry is served grilled, broiled, smoked, barbequed, roasted, baked, sautéed, fried, simmered, braised, and stewed. Whole poultry is often roasted with stuffing and vegetables. Poultry also combines well with grains or pastas. Ground poultry can be substituted where other types of ground meats are typically used, such as in sausages, chili, and meatloaf, to lower saturated fats.

Ratite steaks, fillets, medallions, whole roasts, and ground meat can be showcased in entrées and prepared in the same manner as veal. Most cuts are best prepared medium-rare by grilling, broiling, or sautéing. If a ratite is to be cooked well-done, it is best to braise or stew it. Mushrooms, garlic, red wine, and compotes complement the flavor of ratites.

Wild game bird dishes, such as quail stuffed with wild rice and mushrooms, may be served with roasted root vegetables or grilled marinated vegetables. Game birds may also be grilled and served with fresh asparagus. Pheasant is often roasted, smoked, or braised and flavored with Cognac and apples. With its mild flavor, poultry is a flexible ingredient that can be used to create both familiar and unique entrées that are healthy and appealing to guests. **See Figure 10-28.**

Food Safety

Salmonella and Campylobacter jejuni are bacterial concerns when handling, processing, and cooking raw poultry. Never use the same cutting boards or tools that were used with raw poultry for other products until they have been washed and sanitized.

Poultry in Entrées

Photo Courtesy of D'Artagnan, Photography by Doug Adams Studio

Teriyaki Duck Breast

Cape Cod Cranberry Growers' Association

Chicken Filled with Wild Rice

Figure 10-28. Poultry is a flexible ingredient that can be used to create entrées that are healthy and appealing to guests.

Knowledge Check 10-3

1. Identify nutrient-dense ways to use poultry in appetizers.
2. Explain the importance of chicken stock in soups.
3. Describe ways to create healthy poultry-based sandwiches.
4. List common uses of poultry in salads.
5. Give three examples of healthy entrées featuring poultry.

PROMOTING POULTRY ON THE MENU

Promoting poultry on the menu can be as simple as featuring seasonal fare. In the summer, grilled items are popular, so adding chicken kebabs to the menu is usually well received. During the winter months, hearty soups and stews are popular poultry menu items. Poultry is typically featured on lunch and dinner menus but may also appear on breakfast menus.

With the increased demand for nutritious meals, healthy poultry cooking techniques such as grilling, roasting, sautéing, or simmering can easily be promoted on the menu. In addition, promoting unique types of poultry, such as ostrich, emu, or quail, can pique guests' interest.

Quick Service and Fast Casual Venues

Most quick service and fast casual venues often limit poultry menu items to chicken and turkey. Due to the expectation that meals will be delivered quickly, poultry is commonly fried or grilled and often incorporated into sandwiches and salads. Wraps made with poultry are gaining popularity, as are gluten-free lettuce wraps. **See Figure 10-29.**

Turkey Wraps

National Turkey Federation

Figure 10-29. Wraps made with poultry are gaining popularity at quick service and fast casual venues.

Casual Dining and Institutional Venues

Weekly specials or cycle menus at casual dining and institutional venues may feature items such as chicken pot pie and barbeque chicken. In these venues, one type of poultry can usually be substituted for another. For example, a turkey tostada can easily become a chicken tostada. Poultry can also be substituted for higher-fat proteins, such as red meat, in a variety of sandwiches and salads.

Poultry is often repurposed at these venues. For example, if a dinner menu features roasted chicken, the lunch special for the following day could feature a roasted chicken and wild rice soup made from the leftover chicken.

Beverage Houses and Bar Venues

Beverage houses and bar venues often feature poultry in items such as soups, sandwiches, and salads. Sweet as well as hot and spicy chicken wings are popular on bar menus. Some beverage and bar venues also offer appetizers such as chicken satay (skewered and grilled poultry) with dipping sauces.

Fine Dining and Special Event Venues

Fine dining and special event venues often attract guests seeking unique flavors and elegant presentations. These venues are more apt to feature poultry menu items such as organic chicken, duck, goose, ratites, and game birds. **See Figure 10-30.** For example, a fine dining restaurant may serve pan-roasted quail in a port wine sauce paired with grilled asparagus and red heirloom rice. Fine dining and special event venues will garner a higher price for poultry items.

Rosewater Duck

National Honey Board

Figure 10-30. Fine dining and special event venues are more apt to feature poultry menu items such as rosewater duck.

Knowledge Check 10-4

1. Name the varieties of poultry commonly served in quick service and fast casual venues.
2. Explain how poultry can be repurposed at casual dining and institutional venues.
3. List common poultry preparations served at beverage houses and bar venues.
4. Describe the types of poultry dishes that are served at fine dining and special event venues.

MEATS

An *herbivore* is an animal that feeds on grass and other plant-based foods. Many types of herbivores are raised for human consumption. The flesh of these herbivores is commonly referred to as meat. The most common herbivores raised for their meat include cattle, hogs, and sheep. Meat is a source of high-quality protein, B vitamins, iron, and zinc.

Perceived Value of Meats

Meat is typically the most expensive item on the plate and is often the largest portion of the food budget. The cost of meat is usually reflected in a higher menu price and customers often expect to pay more for higher quality cuts and grades of meat. For example, menu items such as steaks, center-cut pork chops, or racks of lamb are often perceived as sophisticated menu items and have higher prices. In contrast, less expensive cuts of meat are often used in dishes perceived as "comfort foods," such as beef stew, chili, and meatloaf. Therefore, guests expect to see lower menu prices for these types of menu items. **See Figure 10-31.**

Accompaniments served with meats can elevate perceived value. Pork tenderloin medallions served with wild rice, dried cherries, and roasted pecans will demand a higher price than a pulled pork sandwich with sweet potato fries. Highlighting meats from local farms is also an effective way to heighten perceived value and encourage sales. Guests often attach value to locally raised foods and order menu items featuring such foods.

Some people perceive all meats as high in fat. However, in addition to being an excellent source of protein, lean meats prepared using healthy cooking techniques are relatively low in fat. **See Figure 10-32.** When properly prepared, lean meats, such as pork tenderloin, sirloin tips, veal stew meat, and lamb chops, are not only delicious, but also provide guests with a nutritious and satisfying dining experience.

Entourage

Meat is typically the most expensive item on the plate, and customers expect to pay more for higher quality cuts and grades of meat.

Meat Classifications

Cattle, hogs, and sheep provide most of the meat used in the professional kitchen. Beef and veal come from cattle, pork from hogs, and lamb from sheep. Bison and game meats, such as venison and rabbit, are also featured on some menus. Most game animals are herbivores. However, some game animals, such as wild boar, also eat meat.

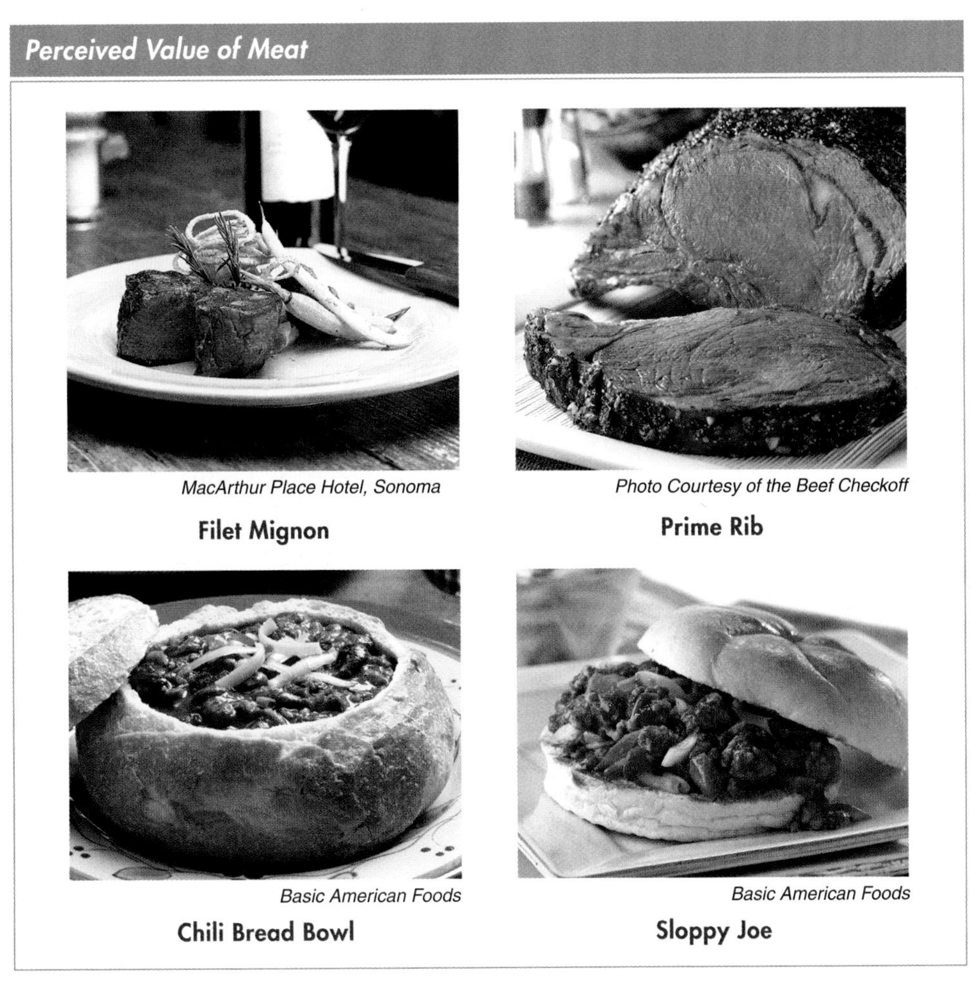

Figure 10-31. Menu items such as steak have a higher perceived value than dishes such as chili and sandwiches.

Lean Meats

Type of Meat*	Calories	Total Fat	Saturated Fat
Beef, eye of round (roasted)	162	4 g	1 g
Beef, top loin steak (broiled)	182	6 g	2 g
Beef, top round steak (braised)	199	5 g	2 g
Veal, leg (roasted)	150	3 g	1 g
Pork, boneless top loin chop (broiled)	131	4 g	1 g
Pork, boneless top loin roast (roasted)	173	6 g	2 g
Pork, tenderloin (broiled)	187	6 g	2 g
Lamb, leg (roasted)	182	7 g	3 g
Lamb, sirloin chop (broiled)	188	8 g	3 g
Lamb, loin (broiled)	192	9 g	4 g

* based on 100 g serving size

Figure 10-32. Lean meats prepared using healthy cooking techniques are relatively low in fat.

Beef and Bison. *Beef* is meat from domesticated cattle. The age and gender of domesticated cattle affect the quality and flavor of the meat. *Grain-fed beef* is meat from cattle that were grain-fed in confined feeding operations for 90 days to 1 year. *Grass-fed beef* is meat from cattle that were raised on grass with little or no special feed. Grass-fed beef contains less marbling than grain-fed beef, making it lower in saturated fat. The difference in marbling is often attributed to the different nutrients found in the cattle's respective diets. **See Figure 10-33.**

Grass-Fed Beef

Photo Courtesy of D'Artagnan, Photography by Doug Adams Studio

Figure 10-33. Grass-fed beef is generally lower in saturated fat than grain-fed beef.

A bison is a large animal that is over 6 feet in height and 10 feet in length and can weigh over 2000 lb. Most bison are raised free-range and grass-fed, but some are grain-fed. Grass-fed bison meat contains more unsaturated fat and less saturated fat than beef. Bison is extremely lean with very little marbling. **See Figure 10-34.** Bison is higher in protein and iron than beef and lower in cholesterol than skinless chicken. Bison is sold in market forms similar to those of beef and is prepared using the same cooking techniques. Bison meat is comparable in flavor to beef, yet richer and sweeter. Because it has a similar texture and flavor to beef, bison can be used in any recipe that calls for beef.

Bison

Photo Courtesy of D'Artagnan, Photography by Doug Adams Studio

Figure 10-34. Most bison meat is lean and contains little marbling.

Wellness Concept

A 100 g serving of ground meat from grass-fed bison contains approximately 146 calories, 20 g of protein, and 7 g of fat. In contrast, 100 g of ground meat from grain-fed bison contains approximately 223 calories, 19 g of protein, and 16 g of fat.

Photo Courtesy of D'Artagnan, Photography by Doug Adams Studio

Ground bison meat can be formed into patties and offered on menus as bison burgers.

INGREDIENT SPOTLIGHT: Bison

Photo Courtesy of D'Artagnan, Photography by Doug Adams Studio

Unique Features

- Similar texture to beef with a slightly sweeter flavor
- Requires a lower cooking temperature than beef
- Higher in protein and iron than beef
- Low in total fat and sodium
- Source of protein, riboflavin, niacin, vitamin B_6, vitamin B_{12}, phosphorus, iron, zinc, and selenium

Menu Applications

- Roast, broil, grill, or stir-fry for use in appetizers, salads, and entrées
- Use as a substitute for beef and poultry

Nutrition Facts

Serving Size 3 oz (separable lean, roasts)

Amount Per Serving	
Calories 122	Calories from Fat 19
	% Daily Value*
Total Fat 2g	3%
Saturated Fat 1g	4%
Trans Fat	
Cholesterol 70mg	23%
Sodium 48mg	2%
Total Carbohydrate 0g	0%
Dietary Fiber 0g	0%
Sugars 0g	
Protein 24g	
Vitamin A 0% • Vitamin C	0%
Calcium 1% • Iron	16%

* Percent Daily Values (DV) are based on a 2,000 calorie diet.

Veal. *Veal* is meat from calves, which are young cattle. Veal is lighter in color than beef and has a firm texture and little fat. **See Figure 10-35.** The meat is light pink in color, delicate in flavor, and very tender. Free-range veal is allowed some exercise and is slightly less tender. A calf that is over nine months of age is sold as baby beef. Baby beef has a stronger flavor and brighter color than meat from younger calves.

Pork. *Pork* is meat from slaughtered hogs that are less than a year old. Selective breeding and careful diets are used to produce an animal that is lean and tender. Pork is light pink in color and has a very tender texture and delicate flavor. **See Figure 10-36.** Approximately one third of the meat is sold fresh and two thirds is cured and smoked to produce ham, sausage, and bacon.

Chef's Tip

Heritage breeds of pork such as Kurobuta (Berkshire) and Duroc produce more well-marbled and full-flavored meat that many guests prefer.

Veal

Photo Courtesy of D'Artagnan, Photography by Doug Adams Studio

Figure 10-35. Veal is lighter in color than beef and has a firm texture and little fat.

Pork

Courtesy of The National Pork Board

Figure 10-36. Pork is light pink in color and has a very tender texture and delicate flavor.

Lamb. *Lamb* is meat from slaughtered sheep that are less than a year old. Quality lamb is pinkish to deep red in color, firm, and has some marbling. **See Figure 10-37.** Young lamb has a tender and delicate flavor, while mature sheep (mutton) is less tender and has a strong, distinct flavor.

Game Meats. Game meat comes from the meat of wild animals. Game is typically raised on farms or ranches where its diet is carefully controlled. The tenderness and flavor of game depends on the age and type of animal as well as its diet. According to the USDA, common game meats include venison, rabbit, and goat. **See Figure 10-38.**

Lamb

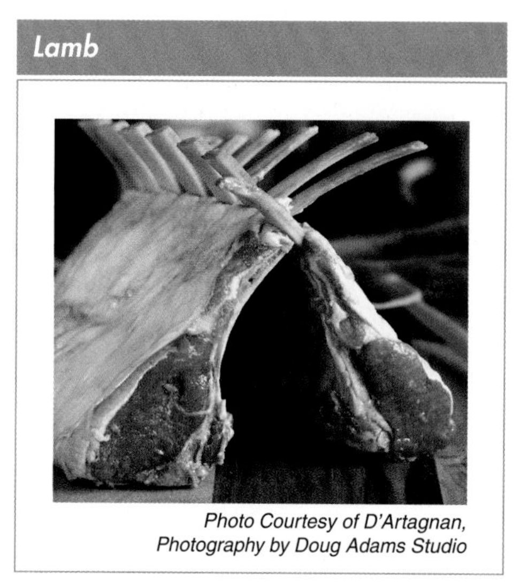

Photo Courtesy of D'Artagnan, Photography by Doug Adams Studio

Figure 10-37. Lamb is pinkish to deep red in color, firm, and has some marbling.

Photo Courtesy of D'Artagnan, Photography by Doug Adams Studio

Game meat comes from the meat of wild animals, such as venison.

Game Meats

Type of Game	Description	Common Culinary Applications	Key Nutrients*
Venison	• Includes meat from deer, elk, antelope, moose, or pronghorn • Dark, tender meat with little marbling	• Loin cooked using dry-heat cooking techniques • Often cooked using combination cooking techniques • Ground and used for sausages and pâtés	• Loin steak, broiled: Calories: 150 Total fat: 2 g • Protein, thiamin, riboflavin, niacin, and iron
Rabbit	• Similar appearance and texture to skinless chicken meat • Very lean and tender	• Can be substituted for chicken • Effectively grilled, roasted, sautéed, braised, or stewed	• Composite of cuts, roasted: Calories: 197 Total fat: 8 g • Protein, niacin, vitamin B_6, vitamin B_{12}, phosphorus, and selenium
Goat	• Flavor is similar to a combination of beef and lamb • Young goats (kids) have a sweet, tender meat	• Various cuts can be grilled, broiled, roasted, sautéed, fried, or braised	• Composite of cuts, roasted: Calories: 143 Total fat: 3 g • Protein, riboflavin, niacin, vitamin B_{12}, phosphorus, zinc, selenium, copper

* based on 100 g serving size

Figure 10-38. Common game meats include venison, rabbit, and goat.

Composition of Meats

Meat from cattle, hogs, and sheep is composed of bundles of muscle fibers held together by two types of connective tissues called collagen and silverskin. *Collagen* is a soft, white connective tissue that breaks down into gelatin when heated. *Silverskin* is a tough, rubbery, silver-white connective tissue that does not break down when heated. Silverskin is trimmed from meat prior to cooking because it is inedible and removing it helps prevent meat from curling as it cooks. The amount of connective tissue increases as the animal ages.

Collagen and fat provide flavor and help to thicken sauces made from meat drippings. The amount of marbling affects the flavor, tenderness, and quality of the meat, but it does not guarantee flavor or tenderness. Fat cap is the fat that surrounds a muscle. **See Figure 10-39.** The fat cap is left on some cuts of meat during cooking to keep the meat moist and add flavor, but it is often removed prior to service. Most visible fat can be trimmed prior to or after the cooking process, thus lowering the amount of fat and calories in the finished product.

Market Forms of Meats

Fresh and frozen meats are used in food-service operations. Fresh meats have an appropriate color and pleasant smell and should be firm and elastic. Frozen meats should be wrapped tightly and handled properly to preserve their quality.

Whole and Fabricated Meats. Common market forms of cattle, hogs, and sheep include whole and partial carcasses, primal cuts, and fabricated cuts. Hogs and sheep can be purchased as whole carcasses. However, most meats are broken down into primal or fabricated cuts. A *primal cut* is a large cut of meat from a whole or a partial carcass. Primal cuts are often turned into fabricated cuts in-house. A *fabricated cut* is a ready-to-cook cut of meat that is packaged to certain size and weight specifications. Fabricated cuts typically cost more per pound, but they are often chosen for their convenience.

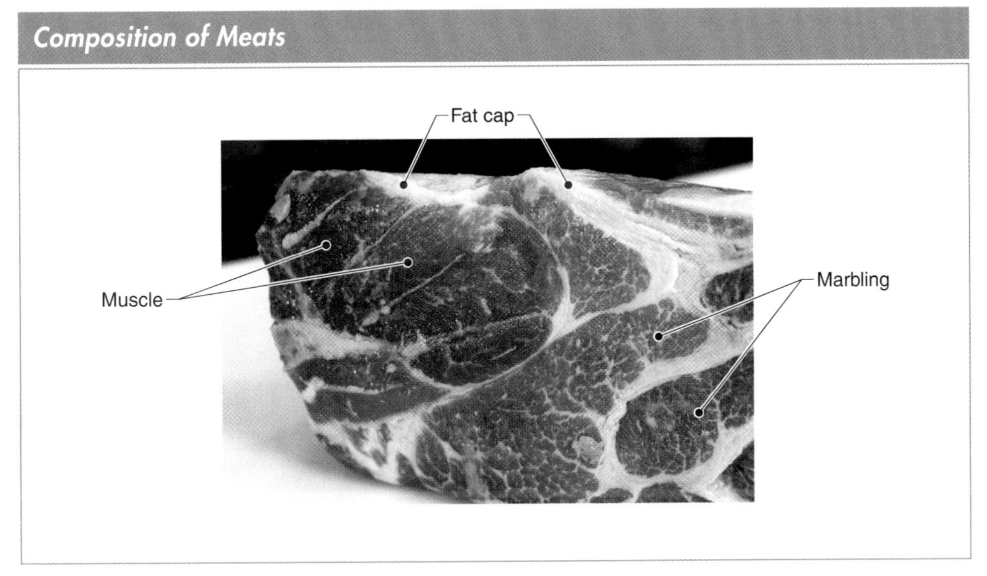

Figure 10-39. Meat is composed of muscle held together by connective tissues called collagen and silverskin. Marbling is found within the muscle, and fat cap surrounds the muscle.

Nutritionally, the tenderloin is a lean and tender cut of meat. In general, a 3 oz serving of meat from the tenderloin has less than 6 g of fat. The tenderloin is commonly fabricated into cuts such as Chateaubriand, filet mignon, tornedos, and tenderloin tips. **See Figure 10-40.**

In addition to primal cuts, offals have surged in popularity as more chefs feature menu items that emphasize the use of the entire animal. An *offal* is the edible part of an animal that is not part of a primal cut. Offals can include the heart, brains, kidney, liver, sweetbreads (thymus gland), tongue, tripe (stomach lining), oxtail (tail from cattle), jowls, and feet. Offals are a very good source of protein but vary significantly in terms of fat and calories. For example, 4 oz of kidney contains approximately 3 g of fat and 116 calories, whereas 4 oz of sweetbreads contains 23 g of fat and 267 calories.

Meat Convenience Products. Meat convenience products are available in many different forms, including canned meats, luncheon meats, and hot dogs. Convenience products generally use additives. An *additive* is a substance added to a food to maintain or improve nutritional quality, maintain quality and freshness, aid in processing or preparation, or make food more appealing. For example, sodium nitrates are used in products such as ham, bologna, and hot dogs to keep the meat from turning brown and to prevent bacterial growth. Although the use of additives is strictly regulated by the FDA, some food manufacturers limit or avoid the use of them whenever possible.

Tenderloin Cuts

Cut	Nutritious Culinary Applications	Positive Impacts of Applications
Chateaubriand	• Pan-sear and then roast • Serve with a mushroom ragout made with olive oil, Zinfandel, onions, thyme, and low-fat sour cream	• Pan-searing creates a brown crust • Roasting maintains tenderness • Mushroom ragout is low in fat, sodium, and calories
Filet mignon	• Broil • Serve with a chimichurri sauce made with olive oil, parsley, cilantro, garlic, vinegar, lemon juice, and crushed red pepper	• Broiling is a low-fat cooking technique that quickly browns the exterior while maintaining a moist interior • Olive oil supplies unsaturated fat • Sauce is low in sodium with a fresh, zesty flavor and a vibrant green color that enhances presentation
Tournedos	• Pan-sear in canola oil • Serve with a deglazed pan sauce made with shallots, garlic, brown stock, cognac, Dijon mustard, and peppercorns	• Pan-searing creates a flavorful, brown crust while maintaining a moist interior • Canola oil supplies unsaturated fat • Pan sauce is low in fat, sodium, and calories
Tips	• Skewer and grill to make a satay • Serve with a dipping sauce made with peanut butter, low-sodium soy sauce, chile paste, brown sugar, and lime juice	• Grilling creates a smoky flavor, and char lines enhance the presentation • Peanut butter supplies unsaturated fat • Low-sodium soy sauce, chile paste, and lime juice provide saltiness while keeping the sodium content down • Dipping sauce has a balance of salty, spicy, sweet, sour, and umami flavors

Figure 10-40. The tenderloin is commonly fabricated into cuts such as Chateaubriand, filet mignon, tournedos, and tenderloin tips.

Knowledge Check 10-5

1. List the key nutrients provided by meats.
2. Explain how the cut of meat affects the menu price.
3. Contrast grain-fed and grass-fed beef.
4. Compare the flavor and nutritional characteristics of beef to those of bison and veal.
5. Describe the appearance, texture, and flavor of pork and lamb.
6. Differentiate between collagen and silverskin.
7. Explain how marbling affects meat.
8. Differentiate between primal and fabricated cuts.
9. Define offal.
10. Explain why additives are used in meat convenience products.

PREPARING NUTRITIOUS MEATS

The age and composition of an animal, as well as the type of cut, determines the best method of preparation. Muscles are made up of long, stringy fibers held together by connective tissue. Young animals have very fine muscle fibers, but as the animal ages, the fibers become thicker from exercise. This is why younger animals have tender muscles and can be cooked using dry-heat cooking techniques, while older animals have less tender muscles that require slow, combination cooking techniques.

Muscle fibers make up what is called the "grain" of the meat. These muscle fibers are easily seen in less tender cuts of meat, including beef brisket and flank steak. Slicing meats across the grain maintains tenderness, while slicing along the grain produces a stringy product. Pounding or cutting meat tenderizes the muscle and connective tissue. **See Figure 10-41.** The following methods are used to tenderize meats:

- pounding meat with a mallet to break up the protein structure and muscle fibers
- using a hand tenderizer with needlelike knives that pierce and gently cut the connective tissues and muscle fibers
- grinding the meat to completely break apart strands of connective tissue and muscle fibers

Tenderizing Meat

Slicing Across the Grain

Pounding with a Mallet

Using a Hand Tenderizer

Figure 10-41. Slicing meat across the grain maintains tenderness and pounding or cutting meat tenderizes the muscle and connective tissue.

Cooking Meats

During the cooking process, proteins and sugars in meat react to create appealing aromas, colors, and flavors. Meat changes in texture as its proteins coagulate, causing it to become firmer as it loses water. *Shrinkage* is the loss of volume and weight of food as it cooks. Shrinkage is the reason a 20 lb roast may only be 18.5 lb when fully cooked. **See Figure 10-42.**

Cooking meat at too high a temperature can toughen the protein. However, grilling and stir-frying use very high temperatures for short periods and result in only the exterior of an item receiving high amounts of heat. High heat quickly cooks the exterior of the meat to a crispy texture while slowly cooking the interior of the meat. This is the reason that a grilled steak is crispy and somewhat dry on the outside yet remains tender and juicy on the inside.

Determining Doneness. When cooking meats to the desired degree of doneness, the type of meat, the thickness of the meat, the temperature of the meat when it begins to cook, and the intensity of the heat are factors that must be considered. With the exception of braised and stewed meats, the most accurate way to determine doneness is by measuring the internal temperature of the meat with an instant-read thermometer inserted into the thickest part of the meat. The following are temperatures that meats should be cooked to in order to be considered safe for consumption:

- Beef and veal should be cooked to an internal temperature of 145°F for at least 15 seconds and rested for 3 minutes.
- Pork should be cooked to an internal temperature of 145°F for at least 15 seconds and rested for 3 minutes.
- Game steaks and chops should reach an internal temperature of 145°F for 15 seconds. Game roasts should reach 145°F for 4 minutes. Stuffed game meat should reach 165°F for 15 seconds.
- All ground meats must be cooked to an internal temperature of 160°F.

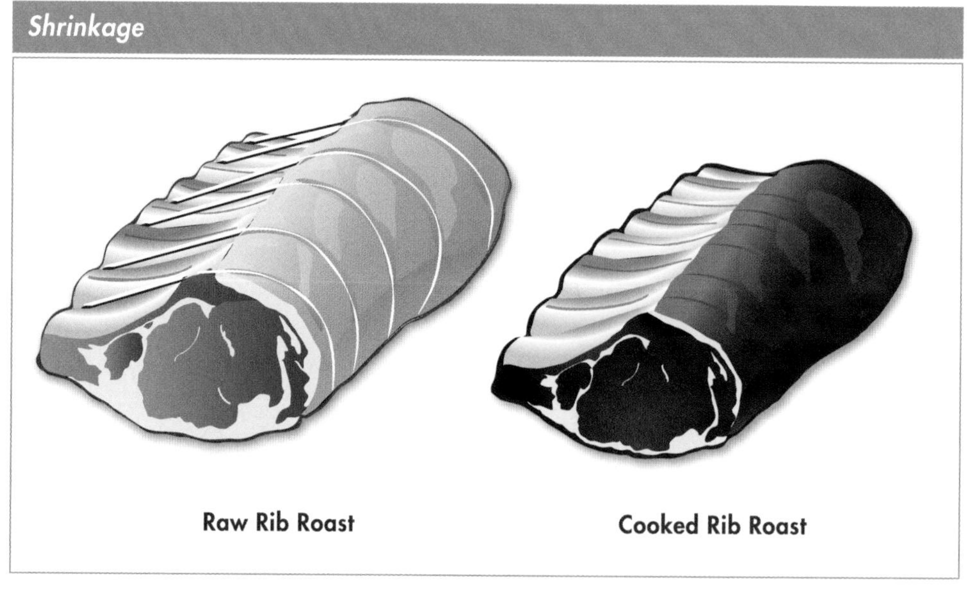

Figure 10-42. Shrinkage is the reason a 20 lb roast may only be 18.5 lb when fully cooked.

Beef, veal, and lamb can be served to varying degrees of doneness, including rare, medium, medium-well, and well-done. **See Figure 10-43.** Braised or stewed meats are cooked until they are fork tender, or until the meat can be easily separated with a fork. When cooking meats, healthy cooking techniques that preserve the integrity of the meat and add flavor result in a nutritious menu item. Choosing the best cooking technique involves consideration of the animal source, the cut, and customer preference. Meats are commonly grilled, broiled, smoked, barbequed, roasted, sautéed, fried, braised, and stewed.

Courtesy of The National Pork Board

Ground meats are used for menu items such as meatballs and must be cooked to an internal temperature of 160°F.

Chef's Tip

It is important to allow meats to rest so that the juices are allowed to redistribute within the meat in order to keep the meat tender and juicy.

Determining Degrees of Doneness

Beef and Veal

Degree of Doneness	Internal Temperature
Very rare	130°F
Rare	140°F
Medium-rare	145°F
Medium	160°F
Medium-well	165°F
Well-done	170°F

Photo Courtesy of the Beef Checkoff

Lamb

Degree of Doneness	Internal Temperature
Medium-rare	145°F
Medium	160°F
Medium-well	165°F
Well-done	170°F

Photo Courtesy of D'Artagnan, Photography by Doug Adams Studio

Figure 10-43. Beef, veal, and lamb can be served to varying degrees of doneness, including rare, medium, medium-well, and well-done.

CULINARY NUTRITION RECIPE-MODIFICATION PROCESS

Beef Bourguignon

Beef bourguignon is a traditional French stew consisting of beef cooked in red wine, mushrooms, and onions. Most of the calories and fat in this recipe come from the beef, bacon, and butter used to make this flavorful dish.

Yield: 6 servings (8 oz beef and vegetables with ½ cup stewing liquid each)

Ingredients

Marinade and beef

2 ea	garlic cloves, crushed
5 oz	onions, ¼ inch slices
3 oz	carrots, ¼ inch rondelles
2 tbsp	parsley, chopped
5 ea	peppercorns, crushed
1½ tsp	salt
1½ c	dry red wine
2 lb	beef, choice stew meat

Bouquet garni

½ ea	carrot
½ ea	leek
2 sprigs	thyme
1 ea	bay leaf

Braise

6 oz	bacon
2 tbsp	all-purpose flour
1½ tsp	tomato paste
8 oz	tomatoes, quartered and seeded
1 c	beef stock
2 oz	butter
8 oz	button mushrooms, sliced
8 ea	pearl onions, boiled and peeled
1 tsp	salt
½ tsp	black pepper, ground

Preparation

Marinade and bouquet garni

1. Combine marinade ingredients in a mixing bowl.
2. Tie bouquet garni ingredients together and add to marinade.
3. Place beef in the marinade, cover, and refrigerate for 4 hours.
4. Remove beef from marinade and pat dry. Discard bouquet garni and reserve marinade.

Braise

5. Cook bacon in a sauté pan over medium heat until slightly golden and the fat is rendered.
6. Add beef and cook until browned on both sides.
7. Add flour and cook approximately 1 minute.
8. Add tomato paste and stir to incorporate.
9. Add reserved marinade, tomatoes, and beef stock. Stir to combine. Bring to a boil and then reduce to a simmer.
10. Cover and place in a 350°F oven until beef is tender (approximately 2–3 hours).
11. Remove beef and strain the cooking liquid through a china cap or chinois. Reserve beef and cooking liquid.
12. Melt butter in a sauté pan over medium heat. Add mushrooms and cook until tender (approximately 2–3 minutes).
13. Add pearl onions, reserved beef and cooking liquid, salt, and pepper.
14. Simmer for 15 minutes.

Carlisle FoodService Products

Evaluate original recipe for sensory and nutritional qualities

- Beef bourguignon has a meaty texture and full-bodied flavor.
- Dish is high in calories, fat, and sodium.

Establish goals for recipe modifications

- Reduce amount of calories, fat, and sodium.
- Maintain flavor.

Identify modifications or substitutions

- Increase marinating time.
- Use less beef and a leaner cut.
- Eliminate bacon.
- Use a substitute for the bacon fat and butter.
- Eliminate the added salt.

Determine functions of identified modifications or substitutions

- Marinating adds flavor.
- Beef adds flavor, a meaty texture, and appealing mouthfeel.
- Bacon provides flavor and the fat used to brown the beef.
- Butter provides flavor and is used to sauté the mushrooms.
- Salt enhances flavor.

Select appropriate modifications or substitutions

- Increase marinating time from 4 hours to 8 hours.
- Use 1½ lb of trimmed bottom round beef instead of 2 lb of beef chuck, and add more mushrooms to compensate for the reduction in beef.
- Remove bacon from the recipe.
- Substitute olive oil for the bacon fat used to brown the beef and for the butter used to sauté the mushrooms.
- Omit the salt.

Test modified recipe to evaluate sensory and nutritional qualities

- Increasing the marinating time maintains flavor.
- Using less beef and a leaner cut reduces calories and fat, and adding more mushrooms provides texture and flavor.
- Eliminating the bacon reduces the amount of calories, fat, and sodium.
- Substituting olive oil for both the bacon fat and butter reduces saturated fat.
- Omitting the salt reduces the sodium content.

CULINARY NUTRITION RECIPE-MODIFICATION PROCESS

Modified Beef Bourguignon

To reduce calories and fat, this modified version of beef bourguignon uses a smaller portion of beef and a leaner cut. To compensate for less meat, more mushrooms are added to maintain texture and mouthfeel. To further reduce calories and fat, bacon is eliminated, which also lowers the sodium content. Allowing the meat to marinate longer produces a full-bodied, flavorful bourguignon.

Yield: 6 servings (8 oz beef and vegetables with ½ cup stewing liquid each)

Ingredients

Marinade and beef

2 ea	garlic cloves, crushed
5 oz	onions, ¼ inch slices
3 oz	carrots, ¼ inch rondelles
2 tbsp	parsley, chopped
5 ea	peppercorns, crushed
½ tsp	salt
1½ c	dry red wine
1½ lb	beef, trimmed bottom round, cubed

Bouquet garni

½ ea	carrot
½ ea	leek
2 sprigs	thyme
1 ea	bay leaf

Braise

2 tbsp	olive oil
2 tbsp	all-purpose flour
1½ tsp	tomato paste
8 oz	tomatoes, quartered and seeded
1 c	beef stock
12 oz	button mushrooms, sliced
8 ea	pearl onions, boiled and peeled
½ tsp	black pepper, ground

Preparation

Marinade and bouquet garni

1. Combine marinade ingredients in a mixing bowl.
2. Tie bouquet garni ingredients together and add to marinade.
3. Place beef in the marinade, cover, and refrigerate for 8 hours.
4. Remove beef from marinade and pat dry. Discard bouquet garni and reserve marinade.

Braise

5. Heat 1 tbsp olive oil in a sauté pan over medium heat. Add beef and cook until browned on both sides.
6. Add flour and cook approximately 1 minute.
7. Add tomato paste and stir to incorporate.
8. Add reserved marinade, tomatoes, and beef stock. Stir to combine. Bring to a boil and then reduce to a simmer.
9. Cover and place in a 350°F oven until beef is tender (approximately 2–3 hours).
10. Remove beef and strain the cooking liquid through a china cap or chinois. Reserve beef and cooking liquid.
11. Heat 1 tbsp of olive oil in a sauté pan over medium heat. Add mushrooms and cook until tender (approximately 2–3 minutes).
12. Add pearl onions, reserved beef and cooking liquid, and pepper.
13. Simmer for 15 minutes.

Beef Bourguignon Nutritional Comparison

Nutrition Facts	Original	Modified
Calories	450	290.2
Total Fat	26.3 g	9.8 g
Saturated Fat	11.5 g	2.4 g
Sodium	1457.6 mg	421.8 mg

Signature Stewed Bison with Root Vegetables

This signature recipe uses bison instead of beef, incorporates wild mushrooms, and adds a variety of vegetables. Replacing the beef with bison reduces fat, and the wild mushrooms add a more robust flavor. Beef stock is replaced with low-sodium vegetable stock to reduce sodium and complement the addition of carrots, parsnips, and turnips. Incorporating more vegetables also adds a variety of nutrients, with a significant increase in dietary fiber and vitamin A.

Yield: 6 servings (8 oz bison and vegetables with ½ cup stewing liquid each)

Ingredients

Marinade and bison

2 ea	garlic cloves, crushed
5 oz	onions, ¼ inch slices
3 oz	carrots, ¼ inch rondelles
2 tbsp	parsley, chopped
5 ea	peppercorns, crushed
½ tsp	salt
1½ c	dry red wine
1½ lb	bison, trimmed of fat, cubed

Bouquet garni

½ ea	carrot
½ ea	leek
2 sprigs	thyme
1 ea	bay leaf

Braise

2 tbsp	olive oil
2 tbsp	all-purpose flour
1½ tsp	tomato paste
8 oz	tomatoes, quartered and seeded
1 c	vegetable stock, low-sodium
12 oz	wild mushrooms, sliced
6 oz	carrots, peeled, 1 inch slices
5 oz	parsnips, peeled, 1 inch slices
2 oz	turnips, peeled, medium dice
8 ea	pearl onions, boiled and peeled
½ tsp	black pepper, ground

Preparation

Marinade and bouquet garni

1. Combine marinade ingredients in a mixing bowl.
2. Tie bouquet garni ingredients together and add to marinade.
3. Place bison in the marinade, cover, and refrigerate for 8 hours.
4. Remove bison from marinade and pat dry. Discard bouquet garni and reserve marinade.

Braise

5. Heat 1 tbsp olive oil in a sauté pan over medium heat. Add bison and cook until browned on both sides.
6. Add flour and cook approximately 1 minute.
7. Add tomato paste and stir to incorporate.
8. Add reserved marinade, tomatoes, and vegetable stock. Stir to combine. Bring to a boil and then reduce to a simmer.
9. Cover and place in a 350°F oven until bison is tender (approximately 2–3 hours).
10. Remove bison and strain the cooking liquid through a china cap or chinois. Reserve bison and cooking liquid.
11. Heat 1 tbsp olive oil in a sauté pan over medium heat. Add mushrooms and cook until tender (approximately 2–3 minutes).
12. Add carrots, parsnips, turnips, pearl onions, reserved bison and cooking liquid, and pepper.
13. Simmer for 15 minutes or until the vegetables are tender.

Stewed Beef and Bison Nutritional Comparison

Nutrition Facts	Original	Modified	Signature
Calories	450	290.2	300.4
Total Fat	26.3 g	9.8 g	7.8 g
Saturated Fat	11.5 g	2.4 g	1.7 g
Sodium	1457.6 mg	421.8 mg	366.6 mg
Dietary Fiber	2.6 g	2.6 g	4.7 g
Vitamin A	3343.9 IU	2890.9 IU	6306.5 IU

Grilling. Grilling uses a hot flame to sear and cook foods quickly. **See Figure 10-44.** This healthy cooking technique is effective for tender cuts of meat that are fabricated from the rib and the loin. Grilling to the well-done stage is not recommended because most of the moisture cooks out and causes the meat to become dry. Although meats should generally be trimmed of excess fat, keeping a thin layer of fat on the meat while it cooks helps the meat retain flavor and moisture. Removing this layer of fat after cooking will help reduce the amount of saturated fat.

Grilling

Figure 10-44. Grilling uses a hot flame to sear and cook foods quickly.

Broiling. Broiling is another healthy way to cook meats cut from the rib or loin. Broiling creates intense heat, so care should be taken to not overcook and dry out the meat.

Smoking. With smoking, an appealing smoky flavor is achieved based on the type of wood used to create the smoke. Meats that are smoked usually contain higher amounts of fat and collagen, so little additional fat is needed. Pork is smoked more often than any other meat. Some beef, such as beef brisket, is also commonly smoked. **See Figure 10-45.**

Smoked Beef Brisket

Photo Courtesy of the Beef Checkoff

Figure 10-45. Beef brisket is commonly smoked.

Barbequing. Various cuts of pork, beef ribs, and veal chops are commonly barbequed. Depending on the cut of meat and the procedures used to develop flavors, barbequing can be a healthy way to prepare meats. For example, a blend of herbs and spices rubbed into the meat imparts flavor with little or no extra calories. Sauces made from fruit purées, vinegars, mustards, or natural sugars such as honey or maple syrup can be added to barbequed meats to increase nutrients, provide flavor, and help retain moisture. **See Figure 10-46.**

Barbequed Pork Sandwich

Courtesy of The National Pork Board

Figure 10-46. Sauces made from fruit purées, vinegars, mustards, or natural sugars such as honey or maple syrup can be added to barbequed meats to increase nutrients, provide flavor, and help retain moisture.

Roasting. Meat from the loin and leg are often roasted. Meat from a rib of beef or veal as well as meat from a rack of lamb is also commonly roasted. Smaller cuts of meat should be roasted at higher temperatures for shorter cooking times to allow the meat to caramelize on the exterior without overcooking the interior. Larger cuts require longer cooking times and should be roasted at lower temperatures to prevent excessive shrinkage. Lean cuts of meat have a tendency to become dry when roasted, so some form of fat is typically added to help keep the meat moist.

Sautéing. Tender cuts of meat are typically sautéed. Sautéing requires only a small amount of hot fat to sear and cook the meat. Using an unsaturated fat such as canola oil to cook the meat provides healthy mono- and polyunsaturated fats.

Frying. Stir-frying is similar to sautéing and only requires a small amount of fat, typically an unsaturated oil, to cook tender cuts of meat. Vegetables are often stir-fried with the meat, adding nutrients to the menu item. **See Figure 10-47.** Tender cuts of meat are also commonly breaded and pan- or deep-fried. Oil used for pan- and deep-frying should be the correct temperature so that the meat does not absorb excess oil or burn.

Pork Stir-Fry

Courtesy of The National Pork Board

Figure 10-47. Vegetables are often stir-fried with meat, adding nutrients to the menu item.

Care should be taken when broiling meats as the intense heat from a broiler can cause foods to overcook and become dry.

CULINARY NUTRITION RECIPE-MODIFICATION PROCESS

Veal Scallopini

Scallopini is an Italian term for a small, thin slice of meat. In this veal scallopini recipe, veal is dredged in flour and sautéed in butter and oil, and a Marsala wine sauce is finished with butter to produce a smooth texture and rich flavor. This recipe relies on a large amount of butter and oil, which yields a flavorful dish that is high in fat.

Yield: 4 servings (4 oz veal with 1 oz sauce each)

Ingredients

4 ea	4 oz veal scallopinis, pounded ⅛ inch thick
½ tsp	salt
⅛ tsp	black pepper, ground
2 oz	all-purpose flour
4 tbsp	butter
2 tbsp	canola oil
3 oz	button mushrooms, sliced
½ c	beef stock
½ c	Marsala wine

Preparation

1. Season both sides of veal with salt and pepper.
2. Dredge veal in flour, shaking off the excess.
3. Heat 2 tbsp of butter and canola oil in a sauté pan over medium heat.
4. Add veal and cook 3–4 minutes.
5. Turn veal over and add mushrooms. Cook 3–4 minutes or until veal is browned, while occasionally stirring the mushrooms. Remove veal and mushrooms and set aside.
6. Deglaze pan by whisking in ¼ cup beef stock.
7. Add wine and stir to combine. Simmer until liquid is reduced by half.
8. Add remaining beef stock and bring to a boil.
9. Reduce heat to low and add 2 tbsp of butter. Stir to incorporate.
10. Add veal and mushrooms. Coat with sauce and serve warm.

Chef's Tip: *Refrigerating veal after it has been dredged in flour helps the coating adhere.*

Evaluate original recipe for sensory and nutritional qualities

- Veal scallopini has an appealing texture and a rich, flavorful sauce.
- Dish is moderate in calories yet has a significant amount of fat and sodium.

Establish goals for recipe modifications

- Reduce calories, fat, and sodium.

Identify modifications or substitutions

- Eliminate added salt.
- Reduce butter and oil.
- Eliminate flour.

Determine functions of identified modifications or substitutions

- Salt provides flavor.
- Butter and oil prevent veal from sticking to the sauté pan and provide flavor.
- Flour combines with the butter and oil to slightly thicken the sauce.

Select appropriate modifications or substitutions

- Omit the added salt.
- Reduce the amount of butter and oil to 1 tbsp each.
- Omit the flour used for dredging.

Test modified recipe to evaluate sensory and nutritional qualities

- Omitting the salt reduces sodium.
- Reducing the amount of butter and oil reduces calories and fat while still preventing the veal from sticking to the sauté pan.
- Omitting the flour reduces calories but does not alter the texture and rich flavor of the sauce.

Modified Veal Scallopini

In this modified veal scallopini recipe, flour is eliminated to reduce calories, and the salt used for seasoning is omitted to decrease sodium. The amount of butter and oil are also reduced to lower both calories and fat. These changes create a healthier menu item that still has a full-bodied flavor.

Yield: 4 servings (4 oz veal with 1 oz sauce each)

Ingredients

4 ea	4 oz veal scallopinis, pounded ⅛ inch thick
¼ tsp	black pepper, ground
1 tbsp	butter
1 tbsp	canola oil
3 oz	button mushrooms, sliced
½ c	beef stock
½ c	Marsala wine

Preparation

1. Season both sides of veal with pepper.
2. Heat butter and canola oil in a sauté pan over medium heat.
3. Add veal and cook 3–4 minutes.
4. Turn veal over and add mushrooms. Cook 3–4 minutes or until veal is browned, while occasionally stirring the mushrooms. Remove veal and mushrooms and set aside.
5. Deglaze pan by whisking in ¼ cup beef stock.
6. Add wine and stir to combine. Simmer until liquid is reduced by half.
7. Add remaining beef stock and bring to a boil.
8. Reduce heat to low and add veal and mushrooms. Coat with sauce and serve warm.

Veal Scallopini Nutritional Comparison

Nutrition Facts	Original	Modified
Calories	352.2	219.1
Total Fat	22.2 g	10.0 g
Saturated Fat	9.2 g	3.4 g
Sodium	425.6 mg	133.6 mg

Signature Veal Scallopini with Grilled Picante

In this signature recipe, the veal is grilled, which eliminates the butter and oil used for sautéing and reduces calories and fat. A nutrient-dense grilled picante keeps the dish low in sodium and elevates the veal with color and a unique smoky flavor.

Yield: 4 servings (4 oz veal with 2 oz picante each)

Ingredients

¾ lb	plum tomatoes, halved and seeded
½ ea	red onion
1 ea	jalapeno pepper
½ tsp	garlic, minced
¼ c	cilantro, chopped
1 tbsp	lime juice
4 ea	4 oz veal scallopinis, pounded ⅛ inch thick

Preparation

1. Grill plum tomatoes, red onion, and jalapeno pepper until charred on both sides and then place in a blender.
2. Add garlic, cilantro, and lime juice and purée until smooth. Set aside.
3. Place veal on an oiled grill for 2–3 minutes per side, or until done.
4. Plate veal and top with picante.

Veal Scallopini Nutritional Comparison

Nutrition Facts	Original	Modified	Signature
Calories	352.2	219.1	156.4
Total Fat	22.2 g	10.0 g	3.7 g
Saturated Fat	9.2 g	3.4 g	1.4 g
Sodium	425.6 mg	133.6 mg	77.8 mg

Braising. Braising uses slower, combination cooking techniques that are effective on either tender or tougher cuts of meat. Larger cuts of beef, such as the chuck and shank, are commonly braised, as are larger cuts of pork. Tender cuts of veal and lamb, such as chops, can also be braised. Braised meats are first seared to add a layer of flavor. Aromatics are then added, followed by enough liquid, such as stock or wine, to come halfway up the side of the meat. Depth of flavor is created as the meat is cooked until fork tender. Root vegetables can also be added to braised meats to enhance the color, texture, flavor, and nutrients of the dish.

Stewing. Stewing is similar to braising, except the meat is cut into small pieces, seared, and covered completely with liquid. **See Figure 10-48.** Both braised and stewed items can be chilled to allow the fat to rise to the surface, where it can be easily removed to reduce the amount of fat and calories in the dish.

Stewing Meats

Figure 10-48. Stewing is similar to braising, except the meat is cut into small pieces, seared, and covered completely with liquid.

Flavor Development

Meat quality has changed over the last several years due to selective breeding and customer demand for leaner products. Therefore, meats that were once well marbled and full of flavor are now lean and sometimes lacking in flavor. Although the process of cooking meats develops aromas, textures, and flavors, some meats benefit from flavor enhancers prior to cooking. Effective ways to develop flavors prior to cooking include aging, curing, rubs, and marinades. Sauces may also be added at the end of the cooking process or served with meats to add moisture and further develop flavor.

Wellness Concept

Studies suggest that six common cuts of pork are approximately 16% lower in total fat and 27% lower in saturated fat than 15 years ago.

Aging. *Aging* is the period of rest that occurs for a length of time after an animal has been slaughtered. Immediately after slaughter, rigor mortis sets in as the tissues in the meat seize and become stiff. Meat going through rigor mortis is extremely tough and virtually flavorless. However, two to three days after slaughter, enzymes begin to break down the muscles, tenderizing the tissues. The meat is then ready to be fabricated, frozen, or cooked.

Veal is typically not aged longer than the initial few days after slaughter. However, beef can be wet-aged or dry-aged to make it more tender and flavorful.

Wet aging is the process of aging meat in vacuum-sealed plastic. **See Figure 10-49.** The beef carcass is typically broken down into smaller cuts that are vacuum-sealed and aged under refrigeration for one to six weeks. During this time, enzymes further break the meat down and tenderize it, intensifying its natural flavor.

Wet Aging

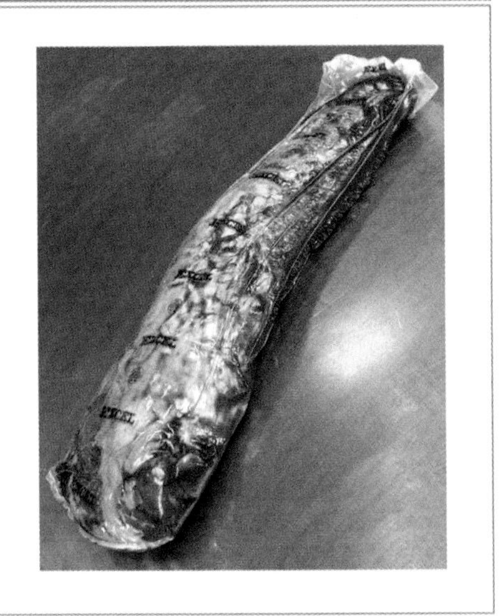

Figure 10-49. Wet aging is the process of aging meat in vacuum-sealed plastic.

Dry aging is the process of aging larger cuts of meat by hanging them in a well-controlled environment. During dry aging, temperature, humidity, and airflow are strictly monitored for up to six weeks. Over this time period, enzymes break down tissues and water evaporates, concentrating flavor.

Curing. *Curing* is the process of using salt and sodium nitrite alone or with flavorings or sugar to preserve a food item. Beef and pork are often cured. Summer sausage, beef jerky, and bresaola are cured beef products. **See Figure 10-50.** More than two-thirds of all pork is cured. For example, ham, bacon, sausage, and prosciutto are cured pork products. Although cured meats are extremely flavorful, they are high in sodium and should be used in moderation when preparing nutrient-dense menu items.

Rubs. Both dry rubs and wet rubs can be used to develop flavor in all types of meats. **See Figure 10-51.** When a meat is cooked, the dry rub creates a crust that enhances flavor and preserves tenderness. Wet rubs adhere to meat more thoroughly than dry rubs, often resulting in deeper flavors.

Cured Beef Products

Figure 10-50. Summer sausage and beef jerky are cured beef products.

Dry Rubs

Figure 10-51. Dry rubs can be used to develop flavor in all types of meats.

FOOD SCIENCE EXPERIMENT: Preparing Beef Jerky Sliced Along and Across the Grain 10-2

Objective

- To compare beef jerky sliced along the grain and across the grain in terms of appearance and mouthfeel.

Materials and Equipment

- 2 lb flank steak
- Cutting board
- Chef's knife
- 3 plastic bags
- Marker
- Measuring cup
- Measuring spoons
- ⅓ cup low-sodium Worcestershire sauce
- ⅓ cup low-sodium soy sauce
- ½ tbsp honey
- 1 tsp freshly ground black pepper
- 1 tsp onion powder
- ½ tsp liquid smoke
- ½ tsp red pepper flakes
- Bowl
- Spoon
- Paper towels
- Parchment paper
- 2 sheet pans
- 2 wire racks

Procedure

1. Place the flank steak on a cutting board and use the knife to trim away visible fat.
2. Place the flank steak in a plastic bag, seal, and freeze for approximately 15 minutes.
3. Use the marker to label one of the empty plastic bags "Along the Grain" and the remaining bag "Across the Grain."
4. Add Worcestershire sauce, soy sauce, honey, pepper, onion powder, liquid smoke, and red pepper flakes to a bowl and stir.
5. Remove the flank steak from the freezer and thinly slice half of steak along the grain into ⅜ inch strips.
6. Place these strips in the plastic bag labeled "Along the Grain" and add half of the marinade. Seal the bag.
7. Thinly slice the remaining half of the steak across the grain into ⅜ inch strips.
8. Place these strips in the plastic bag labeled "Across the Grain" and add the remaining marinade. Seal the bag.
9. Refrigerate both bags for 3–6 hours, allowing the marinade to penetrate the meat.
10. Remove strips from the bag and pat dry with paper towels.
11. Line each sheet pan with parchment paper and place a wire rack on each pan.
12. Arrange the "Along the Grain" strips on one sheet pan and the "Across the Grain" strips on the other sheet pan.
13. Place both pans in a 250°F oven and bake for 3½–4 hours or until the jerky dries.
14. Evaluate the appearance and mouthfeel of each pan of jerky.

Typical Results

There will be variances in the appearance and mouthfeel of the jerky sliced along the grain compared to the jerky sliced across the grain.

Marinades. Like rubs, marinades can be used with all types of meats. A marinade develops flavor and tenderizes at the same time. The acidic base of a marinade tenderizes meat by breaking down its protein structure. Any herb, spice, or condiment can be added to a marinade to help develop aromas and flavors. The length of time that meat should be left in the marinade is determined by the size of the piece of meat and the flavor that is desired.

Any herb, spice, or condiment can be added to a marinade to help develop aromas and flavors.

Plating Meats

Properly preparing meats and side dishes is not only essential to developing nutritious and flavorful menu items, but it also enhances plated presentations. For example, a beef stir-fry with vibrant, colorful vegetables presents a more attractive and nutrient-dense plate than a stir-fry with overcooked vegetables that are dull in color. A pork chop served with mashed potatoes and sauerkraut appears bland with its neutral color palate, whereas a pork chop served with roasted sweet potatoes and sautéed spinach presents a visually enticing plate and increases nutrients.

With its rich aroma and flavor, meat has strong menu appeal. However, the saturated fat content of certain meats is concerning to some guests. To meet the demand for more nutritious menu items, it is effective to start with high-quality meat, cut away visible fat, and plate smaller portions. Serving smaller, flavorful portions that are thoughtfully presented along with sides that are full of color and texture can provide a satisfying meal that is both visually appealing and nutrient dense. In addition, fanning or stacking meat on the plate gives the impression of a larger portion. **See Figure 10-52.**

Plating Meats

National Honey Board

Figure 10-52. Fanning or stacking meat on the plate gives the impression of a larger portion.

Knowledge Check 10-6

1. Describe three methods of tenderizing meats.
2. Define shrinkage.
3. Identify the time and temperature guidelines for safely cooking meats.
4. Describe nine common cooking techniques used to prepare meats.
5. List four ways to develop the flavor of meats prior to cooking.
6. Compare wet and dry aging.
7. Give examples of cured beef and pork products.
8. Explain how to offer nutrient-dense meat dishes to guests.

CULINARY NUTRITION RECIPE-MODIFICATION PROCESS

Stuffed Pork Chops

Stuffed pork chops offer high-quality protein in an appealing presentation. However, the portion size, along with the butter in the stuffing and the shortening used to sear the pork, contributes a large amount of calories and fat.

Yield: 4 servings (8 oz chop with ¼ cup stuffing and ¼ cup sauce each)

Ingredients

4 oz	butter
¼ c	onion, small dice
¼ c	celery, small dice
1½ c	fresh white bread, ½ inch cubes
½ tsp	sage, dried
½ tsp	salt
¼ tsp	black pepper, ground
¼ c	chicken stock
4 ea	8 oz bone-in center-cut pork chops
2 oz	shortening
2 c	chicken stock
2 tbsp	cornstarch
2 tbsp	water

Preparation

1. Melt butter in a sauté pan over medium heat. Add onions and celery and cook until tender.
2. Add bread cubes, sage, salt, pepper, and chicken stock. Stir to combine and then remove from heat.
3. Cut a horizontal pocket in each chop.
4. Fill each pocket with a fourth of the stuffing mixture.
5. Melt shortening in a sauté pan over medium-high heat. Add the stuffed chops and sear on both sides.
6. Add chicken stock and bring to a simmer. Cover and simmer until the chops reach an internal temperature of 160°F.
7. Remove chops and continue simmering until the stock is reduced by half.
8. Mix cornstarch with water to make a slurry.
9. Add slurry to the reduced stock and stir until thick.
10. Top each chop with one-fourth of the sauce.

Courtesy of The National Pork Board

Evaluate original recipe for sensory and nutritional qualities

- Pork chop is flavorful and moist with white bread and sage stuffing and a flavorful sauce.
- The portion size is large.
- Dish is low in dietary fiber.
- Dish is high in calories, fat, sodium, and protein.

Establish goals for recipe modifications

- Reduce portion size.
- Increase dietary fiber.
- Reduce calories, fat, and sodium.

Identify modifications or substitutions

- Reduce serving size of the pork chop and replace stuffing with dressing.
- Replace refined simple carbohydrates with unrefined complex carbohydrates.
- Eliminate added salt.
- Reduce total fats and replace saturated fats (butter and shortening) with unsaturated fats.

Determine functions of identified modifications or substitutions

- Pork chops provide high-quality protein, and stuffing provides a contrasting texture and complementary flavor.
- White bread provides structure to the stuffing.
- Salt seasons the stuffing.
- Butter provides moisture and a rich flavor to the stuffing, and shortening prevents chops from sticking to the pan and promotes browning.

Select appropriate modifications or substitutions

- Reduce pork chop portion to a 4 oz serving size, and modify stuffing to serve as a dressing.
- Substitute whole wheat bread for white bread in the dressing.
- Omit added salt in the dressing.
- Add apples and walnuts to the dressing.
- Substitute olive oil for the butter and nonstick cooking spray for the shortening.

Test modified recipe to evaluate sensory and nutritional qualities

- Reducing portion size lowers calories, fat, sodium, and protein content, and serving a side of dressing instead of stuffing maintains presentation.
- Using whole wheat bread in the dressing increases dietary fiber.
- Omitting added salt from the dressing reduces sodium.
- Adding apples and walnuts to the dressing increases nutrient density and elevates texture and flavor.
- Using olive oil instead of butter replaces saturated fat with unsaturated fat.
- Using nonstick cooking spray instead of shortening reduces calories and fat but still prevents the pork chops from sticking.

CULINARY NUTRITION RECIPE-MODIFICATION PROCESS

Modified Pork Chops with Dressing

In this modified recipe, the serving size of the pork is reduced by half to represent an appropriate portion and reduce calories and fat. By replacing stuffing with a side of dressing, the plate still looks full. Using whole wheat bread and adding an apple and walnuts to the dressing elevates texture, flavor, and nutrient density. Replacing the shortening with nonstick cooking spray decreases calories and saturated fat.

Yield: 4 servings (4 oz chop with ½ cup dressing and ½ cup sauce each)

Ingredients

Dressing

2 tsp	olive oil
¼ c	onion, small dice
¼ c	celery, small dice
1 ea	Granny Smith apple, medium, peeled, small dice
1 oz	walnuts, chopped
2 c	whole wheat bread, ½ inch cubes
1 tbsp	sage, fresh, chopped
¼ tsp	black pepper, ground
¼ c	chicken stock

Pork chops

to cover	nonstick cooking spray
4 ea	4 oz boneless center-cut pork chops, trimmed
2 c	chicken stock
2 tbsp	cornstarch
2 tbsp	water

Preparation

Dressing

1. Heat olive oil in a sauté pan over medium heat. Add onions, celery, and apple and cook until tender.
2. Add walnuts, bread cubes, sage, pepper, and chicken stock. Stir to combine.
3. Cover and bake in a 350°F oven for 30 minutes.
4. Remove cover and bake an additional 10 minutes.

Pork chops

5. Heat a sauté pan sprayed with nonstick cooking spray over medium heat. Sear chops on both sides.
6. Add chicken stock and bring to a simmer. Cover and simmer until the chops reach an internal temperature of 145°F.
7. Remove chops and continue simmering until the stock is reduced by half.
8. Mix cornstarch with water to make a slurry.
9. Add slurry to the reduced stock and stir until thick.
10. Top each chop with one-fourth of the sauce and serve with a side of dressing.

Pork Chops Nutritional Comparison

Nutrition Facts	Original	Modified
Calories	769.2	327.6
Total Fat	52.6 g	13.8 g
Saturated Fat	25.4 g	2.9 g
Sodium	691.7 mg	325.9 mg
Dietary Fiber	0.8 g	3.2 g
Protein	54.3 g	29.9 g

Signature Grilled Pork Chops with Pineapple Salsa

This signature recipe changes the cooking technique for the pork chops from searing and simmering to grilling. Grilling builds a layer of flavor without adding calories and fat to the cooking process. A pineapple salsa provides vibrant color, adds vitamin C, and creates a balance of flavors.

Yield: 4 servings (4 oz chop with ¼ cup salsa each)

Ingredients

Salsa

4 ea	pineapple rings, ½ inch thick
4 oz	red onion, ½ inch slices
1 ea	jalapeno pepper, seeded and minced
2 tsp	lime juice
⅛ tsp	salt

Marinade and chops

1 tbsp	olive oil
1 tbsp	lime juice
¼ tsp	salt
¼ tsp	black pepper, ground
4 ea	4 oz boneless center-cut pork chops

Preparation

Salsa

1. Place pineapple and onion on an oiled grill for 2–4 minutes per side or until tender.
2. Chop pineapple and onion and combine with jalapeno, lime juice, and salt.

Marinade and chops

3. Mix olive oil, lime juice, salt, and pepper until well combined.
4. Brush marinade on each side of the chops and allow to sit for 10 minutes.
5. Place chops on an oiled grill for approximately 4 minutes per side or until an internal temperature of 145°F is reached.
6. Top each chop with one-fourth of the pineapple salsa.

Pork Chops Nutritional Comparison

Nutrition Facts	Original	Modified	Signature
Calories	769.2	327.6	225.1
Total Fat	52.6 g	13.8 g	8.6 g
Saturated Fat	25.4 g	2.9 g	2.7 g
Sodium	691.7 mg	325.9 mg	270.9 mg
Dietary Fiber	0.8 g	3.2 g	1.5 g
Protein	54.3 g	29.9 g	25.5 g
Vitamin C	2.1 mg	3.5 mg	28.0 mg

MEAT MENU MIX

Meats make up a major component of the menu in almost every foodservice operation. They are typically the first ingredient chosen for a menu item. Meats are also an integral part of global cuisines. For example, many different cuisines have versions of meat pies where various types of meats are mixed with an array of vegetables and served in a crust. Meats can be used in appetizers, such as charcuterie plates, Asian pot stickers, and grilled skewered satays. In salads, steak may be featured with blue cheese and lettuce in one restaurant and gingered sesame noodles in another. Entrées cover all categories of meats and may include a simple roast beef sandwich, a burger made from ground ostrich meat, or an elegant dish of veal piccata or rack of lamb.

When writing menu descriptions, it is important to identify and describe both the cut of meat and the preparation technique. Effective descriptions allow guests to make informed decisions. They also help create a sense of excitement in anticipation of a delicious meal. **See Figure 10-53.**

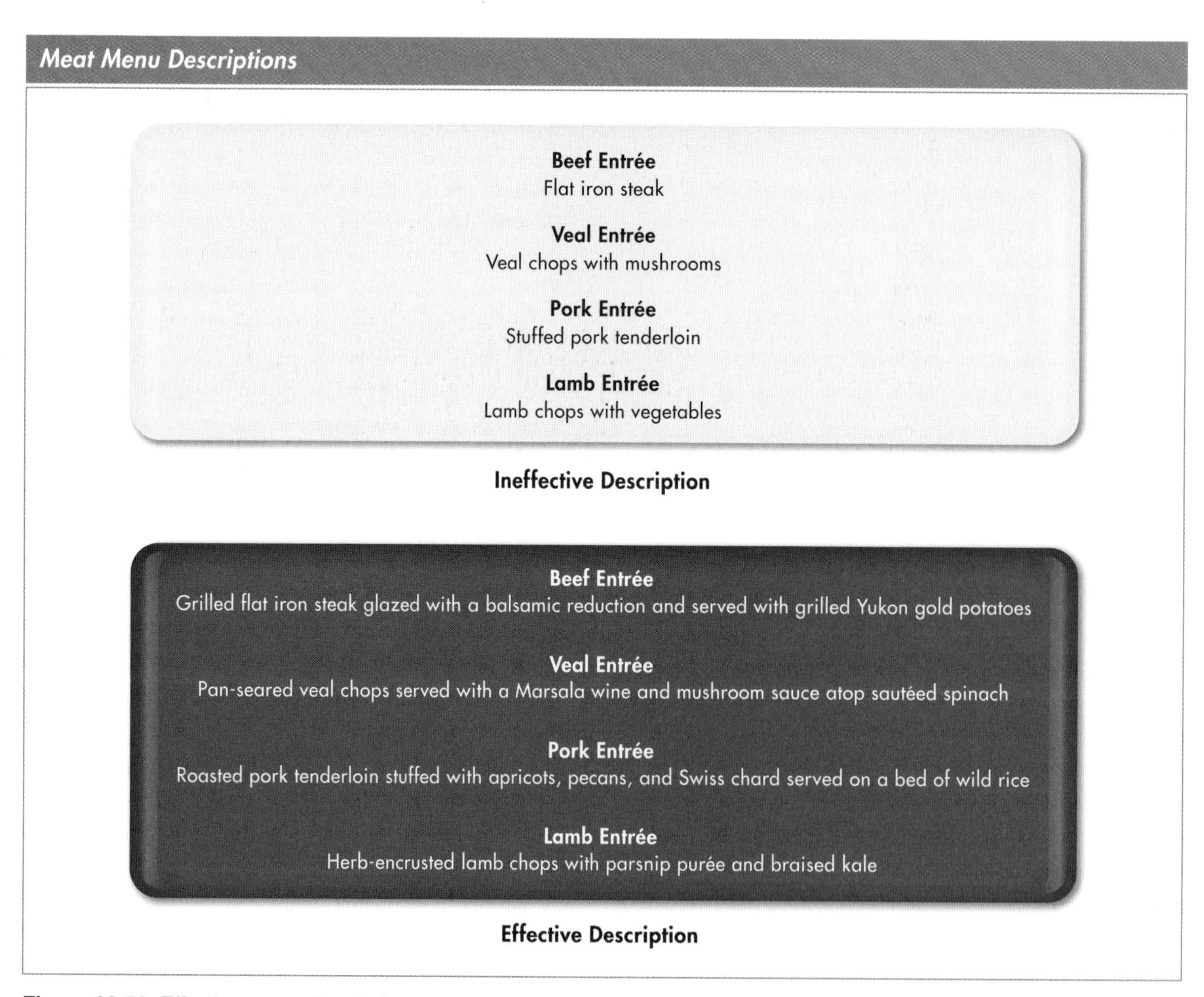

Figure 10-53. Effective menu descriptions create a sense of excitement and anticipation of a delicious meal.

Meats in Appetizers

Based on the sustainability philosophy of using an animal from head to tail, many foodservice operations are using meats to prepare charcuterie selections in-house, producing flavorful sausages, pâtés, terrines, and galantines. Skewered meats, such as kebabs, are also popular. These items are generally prepared using healthy cooking techniques such as grilling or broiling. Flavorful and nutritious dipping sauces for meats include pestos, salsas, vegetable or fruit purées, and yogurt blended with fresh herbs. Ground meats are often made into meatballs, meat pies, or empanadas, in which the meat is combined with vegetables, fruits, or whole grains.

Meat is also served raw in appetizers such as carpaccio and tartare. *Carpaccio* is thin slices of meat or seafood that are served raw. **See Figure 10-54.** The thinly sliced meat is attractively arranged on a serving plate, seasoned with salt and pepper, and often drizzled with extra-virgin olive oil or lemon or lime juice. *Tartare* is freshly ground or chopped raw meat or seafood. Tartare is usually seasoned and served with toast points.

Carpaccio

Courtesy of Chef Gui Alinat

Figure 10-54. Carpaccio is thin slices of meat or seafood that are served raw.

Meats in Soups

Popular soups featuring meat include beef vegetable, beef barley, ham and pea, and chili. The meat in soups not only adds flavor, but also provides contrasting textures. Beef and veal are commonly used to make stock, which provides a rich flavor base for soups. Stocks can also be reduced to create flavorful sauces and hearty grain dishes, such as risottos and pilafs.

The meat in soups adds flavor and contrasting textures.

Meats in Sandwiches

Meat sandwiches featuring sausage, bacon, or ham are often found on breakfast menus, while lunch and dinner menus may feature heartier sandwiches made from grilled, roasted, or braised meats. Sandwiches can range from a simple hamburger or hot dog to more complex specialty sandwiches. Specialty sandwiches often layer flavors by adding ingredients such as sun dried tomatoes, a crisp broccoli slaw, or roasted garlic and are served on artisanal bread. Meats are also used in sandwiches in many global cuisines, such as Cuban sandwiches, gyros, tacos, pizzas, and barbequed meat sandwiches. **See Figure 10-55.**

To keep sandwiches nutrient dense, consideration needs to be given to portion size and added ingredients. For example, reducing the amount of meat in a taco and adding more garnishes, such as beans, lettuce, avocadoes, radishes, and tomatoes, can create a flavorful, satisfying, and healthy menu item.

Meat Sandwiches in Global Cuisines

Sandwich	Origin	Key Ingredients
Arepa	Venezuela	A corn cake filled with chicken, avocado, cheese, shredded beef, eggs, or other ingredients; often toasted on a griddle
Cubano	Cuba	Cuban bread filled with ham and roast pork marinated in mojo sauce, dill pickles, and Swiss cheese
Gyro	Greece	A pita filled with roasted lamb or beef cut from a vertical spit; garnishes include lettuce, tomatoes, onions, cucumbers, and a yogurt-based tzatziki sauce
Muffuletta	New Orleans	A Sicilian round bread filled with a marinated olive salad, mortadella or smoked ham, salami, and provolone cheese; olive salad contains olives, assorted pickled vegetables, celery, pimentos, garlic, and cocktail onions
Po' boy	Louisiana	A baguette typically filled with fried shrimp, oysters, or catfish; ham, turkey, or roast beef are considered alternate fillings; may also include lettuce, tomatoes, and remoulade or tartar sauce
Roti	India	An Indian flatbread filled with curried meats, potatoes, and vegetables
Shawarma	Middle East	A pita filled with lamb or chicken flavored with vinegar and spices such as cardamom, cinnamon, and nutmeg; garnishes include onions, tomatoes, lettuce, pickled turnips, and cucumbers; often topped with hummus, tahini, or hot sauce
Torta	Mexico	A crusty or a soft roll filled with marinated steak, fried pork, or marinated pork; garnishes include avocados, lettuce, jalapenos, tomatoes, and cheese; often grilled in a sandwich press or on a griddle

Figure 10-55. Meats are used in sandwiches in many global cuisines.

Meats in Salads

Meats can be the main ingredient in a salad such as ham salad. **See Figure 10-56.** Meats can also be used as a component of a leafy green salad such as a chef's salad or Cobb salad. When meats are served warm atop chilled salad greens, the contrasting temperatures and textures can enhance the dining experience. For example, a simple bed of salad greens topped with pan-seared lamb or grilled beef featuring Thai flavors such as lime, soy, garlic, curry, and cilantro makes an appealing, nutrient-dense menu option. Salads featuring meats are often hearty and satisfying enough to be served as an entrée.

Meats in Entrées

Meats have long been regarded as the focal point of entrées. For example, meats at the center of the plate have traditionally included items such as a slab of ribs, a 16 oz porterhouse steak, or two thick-cut pork chops. However, these large portions have been criticized for contributing to the rising level of obesity and related health issues. Many menus now offer smaller portions of meat and increase the amount of vegetables and whole grains in order to fill the plate. Offering smaller meat options, such as a petite filet or a half portion, is often appreciated by guests seeking to reduce, but not eliminate, their meat consumption.

Curried Pork and Apple Salad

Courtesy of The National Pork Board

Figure 10-56. Meats can be featured ingredients in salads.

With the wide variety of meats available as well as different cuts, meats can be featured on the menu in numerous ways. **See Figure 10-57.** Less tender and large cuts of meat can be braised or stewed with an array of vegetables and served with a whole grain or pasta. Some larger tender cuts of meat such as a prime rib roast or ham may be roasted whole and then carved to order to allow guests control over portion size. Tender cuts of meat are often portioned into smaller pieces for use in kebabs or stir-fries. Ground meats are commonly used for meatloaves and casseroles and as fillings for items such as pasta shells and bell peppers.

Meats in Entrées

Photo Courtesy of the Beef Checkoff

Stuffed Veal Pinwheels

Venison World

Grilled Venison

Figure 10-57. Meats can be featured as entrées on the menu in numerous ways.

Knowledge Check 10-7

1. Give examples of healthy dipping sauces to serve with meat appetizers.
2. Explain how meats can be used in soups and salads.
3. Identify ways to keep sandwiches featuring meats nutrient dense.
4. Explain why smaller portions of meat are being used in entrées.

PROMOTING MEATS ON THE MENU

People often select a restaurant based on the entrées offered on the menu. For example, if a person desires a steak, hamburger, or barbequed ribs, this often influences or even determines the chosen destination. Promotions such as "daily specials" can encourage guests to try a new dish.

Many foodservice operations offer meat as entrée specials to simplify their food purchases and as an outlet for repurposing leftovers. For example, a restaurant may offer a prime rib special on Sunday evening and a prime rib sandwich for Monday lunch. These daily or weekly specials often attract repeat business.

Sometimes promotions are planned to create unique dishes that inspire guests to try new menu items. Special promotions can be centered around meat, such as bison or venison, as the showcased ingredient. Seasonality also affects the promotion of meats on the menu. For example, grilled meats are frequently promoted during the summer months, and warm comfort foods, such as beef bourguignon or a lamb stew, are often promoted during the winter months. **See Figure 10-58.**

Chef's Tip

Offering half portions of meat menu items helps meet the growing demand for healthier menu options.

Promoting Meats on the Menu

Photo Courtesy of the Beef Checkoff

Grilled Beef

Courtesy of The National Pork Board

Pork Stew

Figure 10-58. Grilled meats are commonly promoted during the summer, whereas stews are often promoted during the winter.

Quick Service and Fast Casual Venues

Quick service and fast casual venues usually feature beef and pork items on the menu in the form of sandwiches. Grilled burgers, sausages, sandwiches, and wraps can be prepared quickly and are commonly offered on carry-out menus. Tortilla or gluten-free lettuce wraps filled with beef or pork are popular menu items at these venues.

Casual Dining and Institutional Venues

Casual dining and institutional venues often feature beef and pork dishes that can be cooked in quantity and held without loss of quality. Grilled to order meats are also popular, as are soups, braised dishes, pastas, and global cuisine menu items such as burritos.

Beverage Houses and Bar Venues

Beverage houses and bars may limit the number of meat items on their menus. Meat served at these venues is often featured in soups, salads, and sandwiches.

Fine Dining and Special Event Venues

At fine dining and special event venues, special care is typically given to planning, preparing, and presenting visually appealing meat dishes. Therefore, in addition to beef and pork entrées, veal, lamb, and game meats are often featured at these venues. **See Figure 10-59.** Fine dining and special event venues may also promote organic and free-range meats and dry aged beef.

Venison Entrée

Charlie Trotter's

Figure 10-59. Meats such as venison are commonly featured at fine dining venues.

Knowledge Check 10-8

1. List common meat preparations served at quick service and fast casual venues.
2. Explain how meats can be served at casual dining and institutional venues.
3. Identify uses of meats at beverage houses and bar venues.
4. Name specialty meats that may be featured at fine dining and special event venues.

Chapter Summary

Poultry and meats supply a variety of nutrients to the body, especially protein, which builds, repairs, and maintains cells and tissues. Poultry comes from birds raised for human consumption and includes chickens, turkeys, ducks, geese, ratites, and game birds. Meats include beef, pork, lamb, veal, bison, and game meats such as venison and rabbit. Poultry and meats are typically purchased fresh or frozen and can be fabricated in-house.

Preparing poultry and meats using healthy cooking techniques keeps the menu item lower in fat and calories. Keeping portion sizes appropriate and incorporating nutrient-dense ingredients helps meet the growing demand for healthy menu items across all types of foodservice venues.

Chef Eric LeVine

Chapter Review

Digital Resources
ATPeResources.com/QuickLinks
Access Code: 267412

1. List factors that affect the perceived value of poultry.
2. Compare the characteristics of chickens, turkeys, ducks, geese, and ratites.
3. List common market forms of poultry.
4. Describe healthy cooking techniques used to prepare poultry.
5. Compare three methods of developing flavors in poultry.
6. Explain how to effectively plate poultry dishes.
7. Give examples of how to incorporate poultry into the menu mix.
8. Identify ways in which poultry is promoted at various foodservice venues.
9. List factors that affect the perceived value of meats.
10. Compare the characteristics of beef, bison, veal, pork, and lamb.
11. Describe the composition of meats.
12. List common market forms of meats.
13. Explain how to determine the doneness of meats.
14. Describe healthy cooking techniques commonly used to prepare meats.
15. Compare four methods of developing flavors in meats.
16. Explain how to effectively plate meat dishes.
17. Give examples of how to incorporate meats into the menu mix.
18. Identify ways in which meats are promoted at various foodservice venues.

Fish & Shellfish on the Menu

Lean Fish

Daniel NYC/B. Milne

Fatty Fish

Daniel NYC/M. Pric

Shellfish

New Zealand Greenshell™ Mussels

Fish and shellfish are found in both freshwater and saltwater and are collectively known as seafood. Seafood is a high-quality protein that is low in saturated fat and often rich in essential fatty acids. A wide variety of sustainable seafood is available because of aquaculture, or fish farming, and because of sustainable fishing practices. With the large variety of nutrient-dense seafood available, chefs can prepare fish and shellfish in a variety of ways to meet the growing demand for nutritious menu items.

Chapter Objectives

1. Describe factors that influence the perceived value of fish.
2. Compare lean and fatty fish.
3. Describe roundfish, flatfish, and cartilaginous fish.
4. Identify terminology used to identify fabricated cuts of fish.
5. Describe cooking techniques commonly used to cook fish.
6. Identity ways to develop the flavor of fish.
7. Describe ways in which fish can be plated.
8. Describe healthy ways to use fish in the menu mix.
9. Summarize how foodservice venues promote fish on the menu.
10. Describe factors that influence the perceived value of shellfish.
11. Describe crustaceans and mollusks.
12. Identify market forms of shellfish.
13. Describe cooking techniques commonly used to cook shellfish.
14. Explain ways to develop the flavor of shellfish.
15. Describe healthy ways to use shellfish in the menu mix.
16. Summarize how foodservice venues promote shellfish on the menu.

FISH

Fish is the classification of animals that have fins, gills, and an internal skeleton made of bones or cartilage. Nutritionally, fish supply protein, fat, water, vitamins, and minerals. Most fish are low in saturated fat, cholesterol, and sodium and are a key source of niacin, vitamins E and K, potassium, and iodine. Some fish are also rich in essential omega-3 fatty acids.

Perceived Value of Fish

The perceived value of fish as a nutritious menu item has increased due to mounting evidence proclaiming the health benefits of consuming fish at least twice a week. Sustainability also plays a role in the perceived value of fish. For example, chefs often inform guests that only sustainable fish are used on their menus. Sustainable fish are not endangered or overharvested. **See Figure 11-1.** Using sustainable fish on menus communicates to guests that a foodservice operation is ecologically responsible and also increases the perceived value of the fish on the menu.

The availability of fish can be unpredictable, causing menu prices to change. Fish are often listed as "market price" because the price fluctuates based on availability. Guests often view market price menu items as part of a refined dining experience.

Chef's Tip

In addition to providing consumers with lists of the best fish choices and fish to avoid, the Monterey Bay Aquarium Seafood Watch® program publishes a list of fish that can be substituted for endangered species.

Fish Classifications

Fish are often grouped by their freshwater or saltwater habitat and by whether they are lean or fatty. **See Figure 11-2.** Lean fish contain very little fat and typically have light-colored flesh and a mild flavor. Northern pike, tilapia, cod, perch, haddock, and halibut are considered lean fish. In contrast, fatty fish contain more fat, are rich in omega-3 fatty acids, and typically have dark-colored flesh that is more flavorful. Catfish, trout, herring, salmon, mackerel, and tuna are considered fatty fish. Fish are commonly classified as roundfish, flatfish, or cartilaginous fish based on their shape and structure.

Some varieties of fish are raised on aquafarms. These farm-raised fish are fattier than wild-caught fish of the same varieties. For example, farm-raised catfish can be as much as five times fattier than their wild counterparts. These farm-raised fish also contain more omega-6 fatty acids and fewer omega-3 fatty acids.

Roundfish. A *roundfish* is a fish with a cylindrical body, an eye located on each side of the head, and a backbone that runs from head to tail in the center of the body. Roundfish are obtained from freshwater as well as saltwater habitats. Cod, perch, trout, grouper, and salmon are roundfish.

Sustainable Fish

- Albacore tuna (Canada and U.S. Pacific, troll/pole)
- Arctic char (farmed)
- Barramundi (U.S. farmed)
- Catfish (U.S. farmed)
- Cod, Pacific (U.S. non-trawled)
- Halibut, Pacific (U.S.)
- Rainbow trout (U.S. farmed)
- Sablefish (Alaska and Canada)
- Salmon (Alaska wild)
- Sardines, Pacific (U.S.)
- Striped bass (farmed and wild)
- Tilapia (U.S. farmed)
- Yellowfin tuna (U.S. troll/pole)

Adapted from the Monterey Bay Aquarium Seafood Watch®

Figure 11-1. Sustainable fish are not endangered or overharvested.

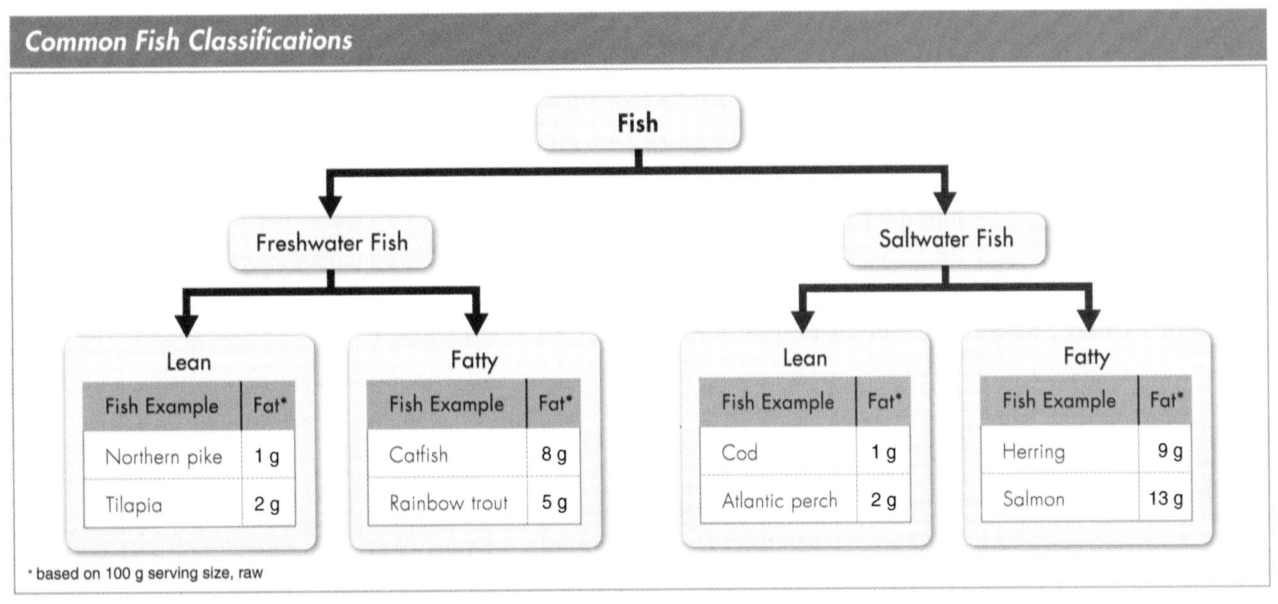

Figure 11-2. Fish are often grouped by their freshwater or saltwater habitat and by whether they are lean or fatty.

INGREDIENT SPOTLIGHT: Rainbow Trout

U.S. Fish & Wildlife Service

Unique Features

- Fatty, freshwater roundfish native to North America
- Sustainable fish that is aquafarmed
- Skin has shades of green, blue, and yellow with a band of pink extending from head to tail
- Firm flesh with a rich, delicate flavor
- Low in sodium
- Source of omega-3 fatty acids, protein, niacin, and vitamin B_{12}

Nutrition Facts

Serving Size 3 oz cooked, dry-heat (85g)

Amount Per Serving	
Calories 144	Calories from Fat 55
	% Daily Value*
Total Fat 6g	9%
Saturated Fat 2g	9%
Trans Fat	
Cholesterol 58mg	19%
Sodium 36mg	1%
Total Carbohydrate 0g	0%
Dietary Fiber 0g	0%
Sugars 0g	
Protein 21g	
Vitamin A 5% • Vitamin C	5%
Calcium 7% • Iron	2%

* Percent Daily Values (DV) are based on a 2,000 calorie diet.

Menu Applications

- Grill, broil, sauté, pan-fry, smoke, or cook on a plank for use in appetizers, salads, sandwiches, and entrées
- Pair with fruit salsas, vegetable purées, fresh herbs, roasted nuts, capers, and citrus

Flatfish. A *flatfish* is a thin, wide fish with both eyes located on one side of the head and a backbone that runs from head to tail through the lateral line of the body. Flatfish have a thin, wide body and swim parallel to the surface of the water with both eyes facing the surface. The skin is darker on the top side of a flatfish and lighter on the bottom side. Flounder, halibut, and sole are flatfish.

Cartilaginous Fish. A *cartilaginous fish* is a fish that has a skeleton composed of cartilage instead of bones. Cartilaginous fish have tough, smooth skin instead of scales. Sharks, skates, and stingrays are cartilaginous fish.

Market Forms of Fish

Fish can be purchased fresh, frozen, or processed. The best indicators of a fresh fish are its smell, appearance, and texture. **See Figure 11-3.** Fresh fish should pass through rigor mortis before fabrication or the muscle fibers will be tough. A fish that has a stiff body is an indication that the fish is still in rigor mortis and therefore fresh.

Fresh fish is best stored between 30°F and 34°F and used within 1–2 days. Frozen fish should be stored at a temperature below 0°F or the fish may become dry, tough, and off in flavor. Fish are commonly sold whole or fabricated. The more a fish is processed, the higher the cost per pound.

Whole and Fabricated Fish. A *whole fish* is the market form of a fish that has not been treated or had any parts removed. Although whole fish cost less per pound, they yield more waste than fabricated fish. Fabricated fish are sold drawn, pan-dressed, and as steaks, fillets, butterflied fillets, wheels, and loins. **See Figure 11-4.** The following terminology is used to describe fabricated fish:

- A *drawn fish* is a fish that has had only the viscera (internal organs) removed.
- A *dressed fish* is a fish that has been scaled and has had the viscera, gills, and fins removed.
- A *pan-dressed fish,* also known as a headed and gutted fish, is a dressed fish that has had its head removed.
- A *fish steak* is a cross section of a dressed fish.
- A *fish fillet* is a lengthwise piece of flesh cut away from the backbone of a fish.
- A *butterflied fillet* is two single fillets from a dressed fish that are held together by the uncut back or belly of the fish.
- A *wheel* is the round center cut of a large fish from which steaks are cut.
- A *fish loin* is a lengthwise cut from either side of the backbone of a large roundfish.

Purchasing Fresh Fish

	Acceptable	Unacceptable
Smell	Slight smell of seaweed or the ocean	Strong fishy smell
	Slight smell of fish	Ammonia odor
Appearance	Round, clear, bulging eyes	Cloudy or sunken eyes
	Bright-red or reddish-pink gills	Brown gills or missing gills
	Firmly attached scales	Loose scales
	Properly covered flesh	Bruised, discolored, or damaged flesh
	Moist and solid fillets	Flesh of fillets separates when slightly bent
Texture	Wet, slightly slippery exterior surface	Slimy internal cavity
	Smooth scales lying flat against body	Rough scales
	Moist, intact fins	Dry fins
	Firm flesh	Mushy flesh

Figure 11-3. The best indicators of a fresh fish are its smell, appearance, and texture.

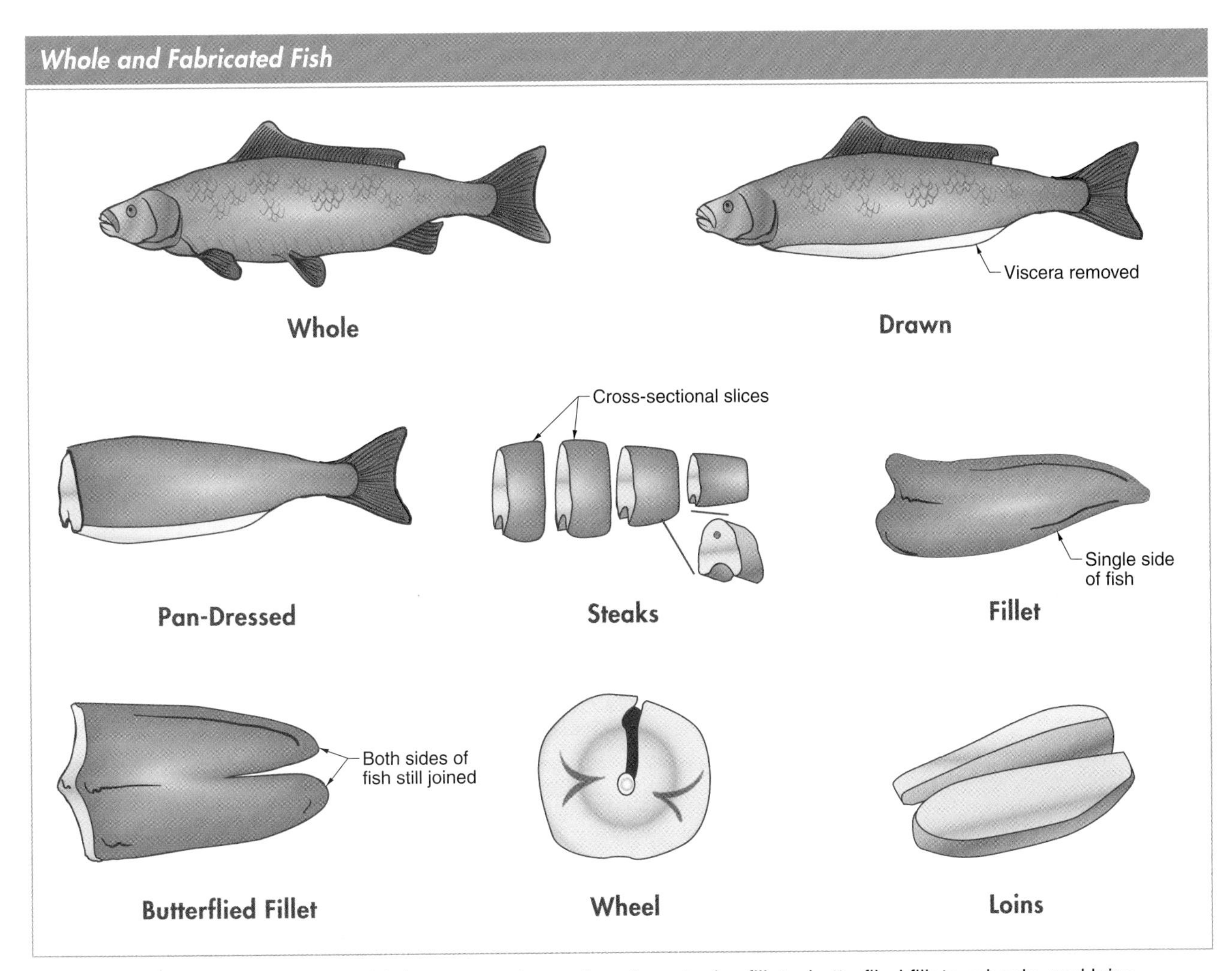

Figure 11-4. Fabricated fish are sold drawn, pan-dressed, and as steaks, fillets, butterflied fillets, wheels, and loins.

Fish Convenience Products. Fish convenience products are products that have been processed and include items such as canned, smoked, cured, salted, pickled, and dried fish. Tuna, salmon, sardines, anchovies, and herring are often sold canned. Canned fish are commonly packed in water or oil. Fish packed in water instead of oil contains fewer calories and less fat. **See Figure 11-5.** Canned fish such as salmon, sardines, and anchovies also contain bones, which provide calcium. Popular smoked and cured fish include herring, cod, haddock, salmon, and sturgeon. Cod is often salted and herring is often pickled. Dried fish is commonly found in Asian markets and can be used to add texture and flavor to soups and sauces.

Nutritional Comparison of Light Tuna Canned in Oil and Water*

Type of Tuna	Calories	Total Fat
Canned in oil (1 can, 171 g)	339	14 g
Canned in water (1 can, 165 g)	142	2 g

* based on 1 can, drained

Figure 11-5. Canned fish packed in water instead of oil contains fewer calories and less fat.

Preformed fish products, such as surimi and breaded fish sticks, are also fish convenience products. *Surimi* is imitation shellfish made from fish that is colored, flavored, and shaped to resemble a particular type of shellfish. **See Figure 11-6.** Surimi is often used as a substitute for crab because it is less expensive. This type of seafood product must be labeled "imitation" and is purchased fully cooked in fresh or frozen form. Surimi is high in protein and low in fat.

Another fish convenience product is caviar. Although roe, or eggs, are harvested from other varieties of fish, such as salmon and whitefish, only sturgeon roe can be labeled as caviar. **See Figure 11-7.** Caviar supplies many nutrients, including protein, calcium, and iron. The salting and drying process used to produce caviar adds flavor but also increases the sodium content.

Food Science Note

A 1 oz serving of caviar contains 75 calories, 5 g of fat, 7 g of protein, 78 mg of calcium, 3 mg of iron, and 425 mg of sodium.

Surimi

Harbor Seafood, Inc.

Figure 11-6. Surimi is imitation shellfish made from fish that is colored, flavored, and shaped to resemble a particular type of shellfish.

Caviar

Fortune Fish Company

Figure 11-7. Although roe, or eggs, are harvested from other varieties of fish, such as salmon and whitefish, only sturgeon roe can be labeled as caviar.

Knowledge Check 11-1

1. List the nutritional benefits of fish.
2. Explain why fish are often listed on menus as "market price."
3. Differentiate between lean and fatty fish.
4. Contrast roundfish, flatfish, and cartilaginous fish.
5. Identify the storage requirements for fresh and frozen fish.
6. Describe eight common forms of fabricated fish.
7. Describe common fish convenience products.

PREPARING NUTRITIOUS FISH

The composition of fish is quite different from land animals. Most fish have 3% collagen as compared to 15% in meat or poultry. Fish also have short muscle fibers and less connective tissue, which makes them more tender. The muscle tissues of fish are arranged in sheets or layers that flake when the connective tissue gelatinizes during the cooking process. Dark muscle fibers are thinner and have more connective tissue than white muscle fibers, giving dark-fleshed fish a more gelatinous texture than white-fleshed fish. **See Figure 11-8.**

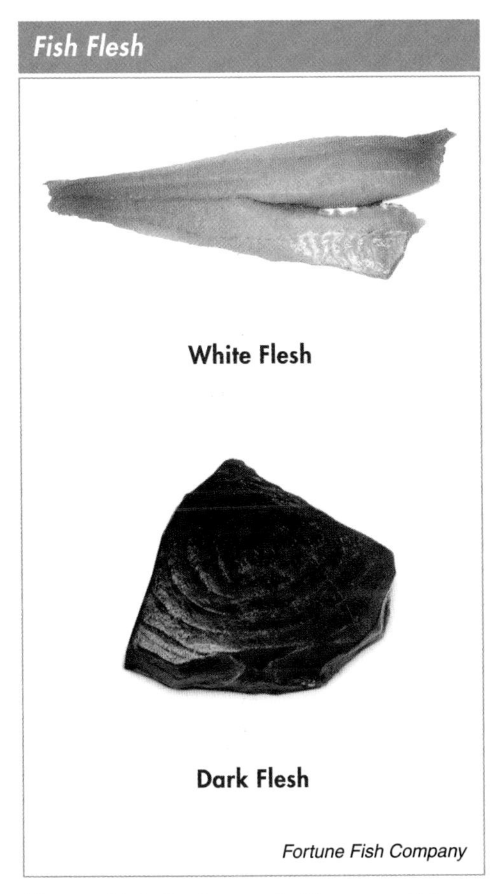

Figure 11-8. Dark muscle fibers are thinner and have more connective tissue than white muscle fibers giving dark-fleshed fish a more gelatinous texture than white-fleshed fish.

Fish recipes are generally simple and rely on the freshness of the fish. The quality of fish is a critical factor in creating an appealing finished dish. Fish must also be cooked carefully to preserve its delicate texture.

Raw Fish

Because fish have tender muscles and a minimal amount of connective tissue, they can be served raw. *Sashimi* is very thin slices of raw fish served with garnishes and sauces. Sushi may also feature raw fish. *Sushi* is a vinegar-seasoned rice dish garnished with raw fish, cooked seafood, eggs, or vegetables. **See Figure 11-9.** *Ceviche* is extremely thin slices of raw fish that have been marinated in lemon or lime juice. Heat is not applied; instead, the acid denatures the protein in the fish, resulting in an opaque flesh with a firm texture.

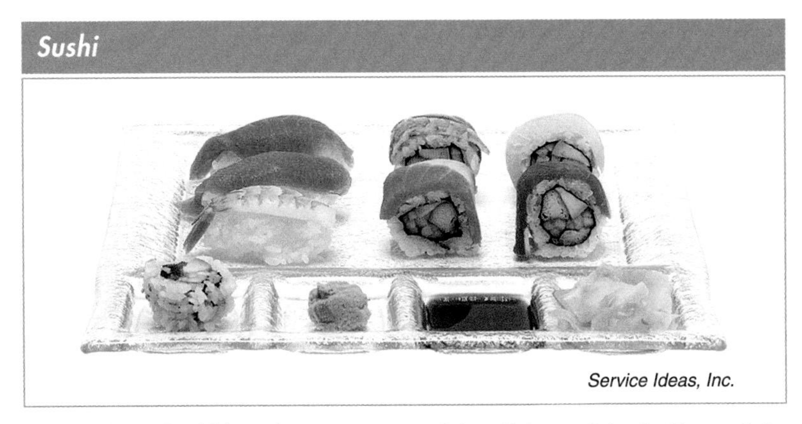

Figure 11-9. Sushi is a vinegar-seasoned rice dish garnished with raw fish, cooked seafood, eggs, or vegetables.

Cooking Fish

Fish are naturally tender and require very little cooking. It is essential to begin the cooking process with the highest quality fish available and stop the cooking process when the flesh of the fish becomes opaque and begins to flake. **See Figure 11-10.** Overcooking fish causes the proteins in the fish to shrink and squeeze out water, which results in a dry finished product.

Appearance of Cooked Fish

Figure 11-10. It is essential to begin the cooking process with the highest quality fish available and stop the cooking process when the flesh of the fish becomes opaque and begins to flake.

Fish can be cooked with dry-heat and moist-heat cooking techniques. The fat content and thickness of the flesh play a role in determining the best cooking technique to use. For example, a fatty, thick-fleshed fish like salmon or tuna is ideal for dry-heat cooking techniques such as grilling, broiling, smoking, and baking. However, lean fish have a tendency to fall apart using dry-heat cooking techniques. Moist-heat techniques such as poaching or steaming are good choices when preparing lean fish. Both lean and fatty fish can be sautéed or fried.

FOOD SCIENCE EXPERIMENT: Effect of Dry Heat and Moist Heat on Lean Fish 11-1

Objective

- To compare lean fish cooked with dry-heat and moist-heat cooking techniques in terms of appearance, texture, and flavor.

Materials and Equipment

- 2 skinless lean fish fillets, approximately 6 oz each (e.g., tilapia, cod, or sole)
- Vegetable oil
- Broiler pan
- Broiler
- Spatula
- 2 plates
- Parchment paper
- Measuring spoons
- Water
- Sheet pan

Procedure

Allergen Alert

The materials used in this experiment may contain one or more food allergens.

1. Place one fish fillet on a lightly oiled broiler pan.
2. Broil the fillet 3–4 inches from the heat source until the fillet is cooked through (approximately 5–6 minutes).
3. Use the spatula to remove the fillet from the broiler and set it aside on a plate.
4. Fold the parchment paper in half lengthwise.
5. Place the second fillet parallel to and approximately 2 inches away from the fold.
6. Pour 2 tbsp of water over the fillet.
7. Fold the parchment paper over the fillet and then carefully fold all the edges over to seal the packet.
8. Place the packet on a sheet pan and bake in a 375°F oven until the fillet is cooked through (approximately 12 minutes).
9. Remove the packet from the oven and place it on the second plate.
10. Evaluate the appearance, texture, and flavor of each fillet.

Typical Results

There will be variances in the appearance, texture, and flavor of each lean fish fillet.

Grilling. Grilling is recommended for fatty, firm-fleshed fish because lean, delicate fish may stick to the grill or fall apart easily. **See Figure 11-11.** The fish should be seasoned and then cooked on both sides if it is a thick cut or cooked on one side if it is a thin cut. To prevent drying out the fish, it should be grilled at a low temperature.

Grilled Fish

Eloma Combi Ovens

Figure 11-11. Grilling is recommended for fatty, firm-fleshed fish because lean, delicate fish may stick to the grill or fall apart easily.

Broiling. Broiling works well with fatty fish, but lean fish can be successfully broiled if they are brushed with oil and basted during the cooking process to help retain moisture. Broiled fish pair well with salsas and chutneys, which also add nutritive value.

Smoking. Both lean and fatty fish can be smoked. Smoked fish have an intense aroma and flavor that corresponds to the type of wood used during the smoking process. Smoked fish are commonly served as appetizers or in salads.

Baking. Fatty fish are often baked because their naturally high fat content helps keep them moist during the cooking process. Leaving the head, tail, and skin on the fish as it bakes seals in moisture and adds flavor. The cavity of the fish can be filled with herbs and spices or lemon to impart additional flavor.

Sautéing. Lean and fatty fish are suitable for sautéing in a minimal amount of oil. To ensure sautéed fish cooks evenly, uniform cuts should be used. Topping sautéed fish with slivered almonds or wilted spinach enhances both flavor and presentation while adding nutrients.

Frying. Frying is a popular cooking technique used for both lean and fatty fish. Pan-fried fish should be cooked over medium heat and turned over approximately three-fourths of the way through the cooking process. If done correctly, the fish will not be oily.

Deep-frying is a popular method for cooking battered or breaded lean fish. Whole small fish, fish fillets, and fish steaks are often fried. Although deep-fried fish is higher in fat and calories, smaller portions served with vegetables and whole grains can create a nutritious dish.

Poaching. Fish is a delicate food that is ideal for poaching. Care should be taken to keep the poaching liquid at the proper temperature, between 160°F and 180°F, to prevent the fish from becoming tough and losing flavor. A well-seasoned poaching liquid, such as a court bouillon or fish stock, can enhance the flavor of poached fish. **See Figure 11-12.** The poaching liquid can be reduced and used as a sauce to add flavor and preserve the water-soluble vitamins that leach from the fish into the cooking liquid.

Simmering. Simmering is similar to poaching, but with a water temperature between 185°F and 205°F. Sole is commonly used to produce paupiettes, which are thin fish fillets that have been rolled and simmered in a court bouillon.

Poaching Fish

Figure 11-12. A well-seasoned poaching liquid, such as a court bouillon or fish stock, can enhance the flavor of poached fish.

Steaming. Steaming is an effective low-fat cooking technique used with both lean and fatty fish. Steaming heats fish in their own juices and locks in aroma and flavor. Steaming also preserves more nutrients than poaching or simmering because water-soluble vitamins are not lost to the cooking liquid.

A fish is commonly wrapped in parchment paper, foil, or grape leaves before it is steamed. Fish wrapped in parchment paper is "en papillote." Aromatic vegetables and seasonings are usually added before the wrap is sealed to infuse the fish with more flavor. **See Figure 11-13.** The steamed fish is typically plated still in the parchment paper for the guest to open and watch the steam escape.

En Papillote

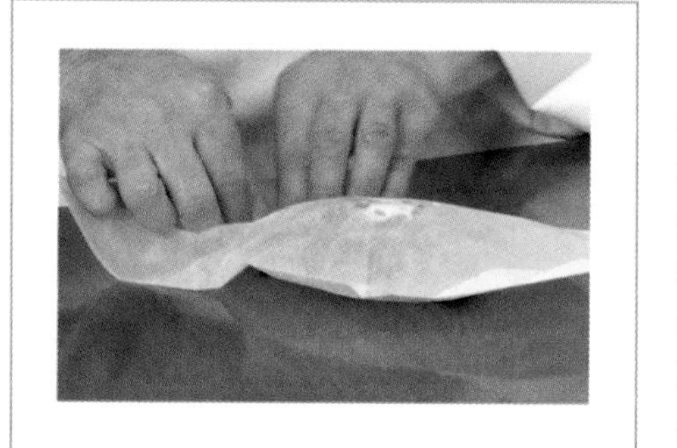

Figure 11-13. Aromatic vegetables and seasonings are usually added to fish cooked en papillote to infuse the fish with more flavor.

Food Safety

Fish such as shark, tilefish, swordfish, and mackerel may be high in mercury, which affects brain function and development. Therefore, it is recommended that pregnant women and children do not eat these varieties.

Flavor Development

The flavor of a fish depends upon its type and its habitat. Saltwater fish typically have more flavor because they have a higher amino acid content in their cells, which is partly responsible for their umami flavor. Fatty fish also have more flavor than lean fish, and wild fish generally have a more pronounced flavor than farm-raised fish.

When developing flavors in fish menu items, it is important to use the freshest fish possible, cook the fish properly, and use flavor enhancers that complement the fish. Flavor can be enhanced by the way in which the fish is prepared as well as the cooking technique. For example, fish can be marinated to start developing flavor before the cooking process begins. The fish can then be grilled to add another layer of flavor. However, it is important to only marinate the fish for a short period of time because it absorbs flavors quickly. Also, acidic ingredients in marinades rapidly denature the proteins, which can result in an unappealing texture if the fish is marinated too long.

Herbs and spices, as well as acidic ingredients such as citrus, mangoes, pineapples, and tomatoes, complement the flavor of most fish. **See Figure 11-14.** Aromatics such as onions and garlic also add flavor to fish dishes. For additional flavor development, fish can be wrapped in banana leaves and grilled to impart a smoky flavor or baked in a salt crust to hold in moisture and flavor. The preparation and cooking technique used with fish not only enhance flavor, but can also elevate the presentation.

Salmon with Fruit Salsa

Figure 11-14. Herbs and spices, as well as acidic ingredients such as citrus, mangoes, pineapples, and tomatoes, complement the flavor of most fish.

Plating Fish

Daniel NYC/T. Schauer

Figure 11-15. A beautifully prepared and garnished fish adds value for both the foodservice operation and the guest.

Plating Fish

A beautifully prepared and garnished fish adds value to the foodservice operation and for the guest. **See Figure 11-15.** For example, fish can be served whole or prepared en papillote for an elegant presentation. Sometimes fish is deboned tableside, which elevates the presentation.

Different preparation techniques can also elevate plated fish dishes by incorporating a variety of shapes, colors, and textures. For example, fresh herbs, roasted nuts, or a fresh fruit salsa can be used to accentuate the appearance and flavor of grilled fish. Serving fish with nutrient-dense sauces and side dishes is an effective way to add flavors that complement the fish, increase the nutritive value of the dish, and create a stunning presentation.

Knowledge Check 11-2

1. Explain how the composition of fish differs from that of land animals.
2. Compare sashimi, sushi, and ceviche.
3. Identify the best types of fish for grilling, broiling, smoking, and baking.
4. Explain how to sauté and fry fish.
5. Differentiate between poaching, simmering, and steaming fish.
6. Explain how to effectively develop flavors in a fish dish.
7. Give two examples of how to elevate the presentation of a plated fish dish.

CULINARY NUTRITION RECIPE-MODIFICATION PROCESS

Fried Fish Tacos

Fried fish tacos consist of battered fish wrapped in corn or flour tortillas. The tacos are often garnished with a sauce and shredded cabbage. Frying the fish adds calories and fat. This fish taco recipe features a creamy sauce that also adds calories and fat due to its base of sour cream and mayonnaise.

Yield: 8 servings (1 taco, 1¼ oz sauce, and ½ oz cabbage each)

Ingredients

Sauce

½ c	sour cream
½ c	mayonnaise
1 tbsp	lime juice
1 ea	jalapeno pepper, minced
1 tsp	capers, minced
½ tsp	oregano, dried
½ tsp	cumin, ground
½ tsp	dill, dried
½ tsp	cayenne pepper, ground

Tacos

1 c	flour
2 tbsp	cornstarch
1 tsp	baking powder
1 ea	whole egg
1 c	beer
16 fl oz	vegetable oil
1 lb	cod fillets
¼ c	flour (for dredging)
8 ea	corn tortillas
4 oz	red cabbage, finely shredded

Preparation

Sauce

1. Combine all sauce ingredients. Cover and refrigerate.

Tacos

2. Combine 1 cup flour, cornstarch, and baking powder in a mixing bowl.
3. Blend egg and beer in a separate bowl and then add to the flour mixture. Stir to combine.
4. Heat oil in a saucepan to 375°F.
5. Dredge fillets in flour and dip in beer batter before carefully placing them in the heated oil.
6. Cook until crisp and golden (approximately 5–7 minutes).
7. Place tortillas, one at a time, in the hot oil for approximately 30 seconds to heat them.
8. Divide fillets evenly and place on the tortillas.
9. Top each taco with sauce and shredded cabbage.

Evaluate original recipe for sensory and nutritional qualities

- Flavorful tortilla is filled with crispy fish.
- Sauce is creamy, well-balanced, and flavorful.
- Cabbage adds color, texture, and flavor.
- Dish is high in calories and fat.

Establish goals for recipe modifications

- Reduce calories and fat.
- Maintain flavor profile.

Identify modifications or substitutions

- Modify cooking technique.
- Eliminate the flour, cornstarch, baking powder, egg, and beer.
- Eliminate oil for frying.
- Use substitutions for the sour cream and mayonnaise in the sauce.

Determine functions of identified modifications or substitutions

- Fish are fried to produce a crispy texture.
- Flour, cornstarch, baking soda, egg, and beer provide the batter for frying, and additional flour is used for dredging.
- Oil is the cooking medium for the fish and tortilla.
- Sour cream and mayonnaise form the base of the sauce.

Select appropriate modifications or substitutions

- Sauté fish instead of frying.
- Omit the oil for frying, the flour for dredging, and the batter.
- Heat tortillas on a griddle or grill instead of hot oil.
- Add lime for flavor.
- Substitute Greek yogurt and low-fat mayonnaise for the sour cream and mayonnaise in the sauce.

Test modified recipe to evaluate sensory and nutritional qualities

- Sautéing the fish reduces calories and fat.
- Heating tortillas on a griddle or grill reduces calories and fat.
- Lime enhances flavor.
- Greek yogurt and low-fat sour cream reduce calories and fat, yet the sauce remains creamy and flavorful.

Modified Sautéed Fish Tacos

This fish taco recipe modifies the cooking technique from frying to sautéing. As a result, the fish are lower in calories and fat. The tortillas are heated on a griddle or grill to further reduce calories and fat, and lime juice is added to increase flavor. Greek yogurt and low-fat mayonnaise maintain the texture and flavor of the sauce but are lower in calories and fat than the sour cream and mayonnaise used in the original recipe.

Yield: 8 servings (1 taco, 1¼ oz sauce, and ½ oz cabbage each)

Ingredients

Sauce

½ c	Greek yogurt
½ c	low-fat mayonnaise
1 tbsp	lime juice
1 ea	jalapeno pepper, minced
1 tsp	capers, minced
½ tsp	oregano, dried
½ tsp	cumin, ground
½ tsp	dill, dried
½ tsp	cayenne pepper, ground

Tacos

1 lb	cod fillets
¼ tsp	black pepper, ground
2 tbsp	olive oil
3 tbsp	lime juice
8 ea	corn tortillas
4 oz	red cabbage, finely shredded

Preparation

Sauce

1. Combine all sauce ingredients. Cover and refrigerate.

Tacos

2. Season fillets with pepper.
3. Heat olive oil in a sauté pan over medium heat. Carefully lay the fillets in the hot pan and cook until golden brown on both sides.
4. Season fillets with lime juice.
5. Place tortillas on a griddle or grill and warm them on both sides.
6. Divide fillets evenly and place on the tortillas.
7. Top each taco with sauce and shredded cabbage.

Fish Taco Nutritional Comparison

Nutrition Facts	Original	Modified
Calories	375.5	190.2
Total Fat	20.3 g	7.5 g
Saturated Fat	3.7 g	1.2 g

CULINARY NUTRITION RECIPE-MODIFICATION PROCESS

Signature Marinated Fish Tacos

The fish in this recipe are marinated and grilled to elevate flavor. The sauce is made with low-fat ingredients but remains rich and creamy. This recipe features a nutrient-dense salsa verde that also further enhances flavor. The dish keeps calories and fat in moderation and is plentiful in vitamin C.

Yield: 8 servings (1 taco, 1¼ oz sauce, and ½ oz cabbage each)

Ingredients

Marinade

1 c	white onions, medium dice
¼ c	cilantro, chopped
¼ c	olive oil
3 tbsp	lime juice
3 tbsp	orange juice
2 ea	garlic cloves, minced
1 tsp	oregano, dried

Sauce

½ c	Greek yogurt
½ c	low-fat mayonnaise
2 tbsp	lime juice
1 tbsp	nonfat milk

Salsa verde

½ lb	tomatillos, husk removed and sliced in half
1 ea	garlic clove
1 tbsp	serrano chile, chopped with seeds
2 tbsp	cilantro, chopped
2 tsp	water

Tacos

1 lb	tilapia fillets
8 ea	corn tortillas
4 oz	red cabbage, finely shredded

Preparation

Marinade

1. Combine all marinade ingredients.
2. Place fillets in the marinade, cover, and refrigerate.
3. After 30 minutes, turn the fillets, recover, and refrigerate for an additional 30 minutes.

Sauce

4. Combine all sauce ingredients. Cover and refrigerate until needed.

Salsa verde

5. Place salsa ingredients in a blender. Purée until well blended.

Tacos

6. Place marinated fillets on an oiled grill for 3–5 minutes per side or until cooked through.
7. Divide fillets evenly and place on the tortillas.
8. Top each taco with sauce, salsa verde, and shredded cabbage.

Recipe Title Nutritional Comparison

Nutrition Facts	Original	Modified	Signature
Calories	375.5	190.2	224.0 g
Total Fat	20.3 g	7.5 g	9.3 g
Saturated Fat	3.7 g	1.2 g	1.6 g
Vitamin C	11.4 mg	13.1 mg	20.5 mg

Vita-Mix® Corporation

FISH MENU MIX

There is an incredible variety of fish to choose from depending upon location and season, and almost every menu features fish in some way. Fish are used across the menu in dishes that range from comfort foods such as soups, stews, and casseroles to more unique dishes such as fish cooked on a wooden plank. Because of the delicate flavor of fish, it can be showcased in appetizers, soups, salads, sandwiches, and entrées.

Effective menu descriptions can encourage guests to order a fish dish. An effective menu description will identify the type of fish, key ingredients, and cooking techniques. An effective menu description also emphasizes the uniqueness of the fish and the flavorful components of the dish. **See Figure 11-16.**

Fish Menu Descriptions

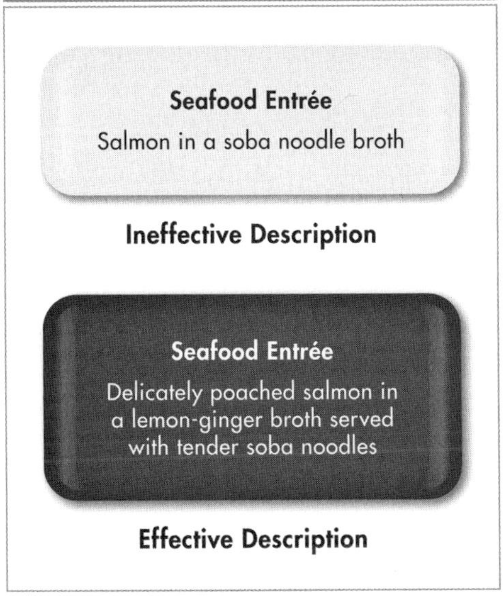

Seafood Entrée
Salmon in a soba noodle broth

Ineffective Description

Seafood Entrée
Delicately poached salmon in a lemon-ginger broth served with tender soba noodles

Effective Description

Figure 11-16. An effective menu description emphasizes the uniqueness of the fish and the flavorful components of the dish.

Fish in Appetizers

Fish can be used in appetizers to provide guests with a variety of flavorful and nutritious menu items. For example, fish used in appetizers can be raw, lightly seared, grilled, or broiled and served with complementary garnishes or sauces. **See Figure 11-17.** Fish can be chopped, minced, or ground and mixed with various ingredients to become a flavorful spread to serve on toast points. Appetizers featuring smoked fish are commonly accompanied by crackers or crudités. Caviar and other fish roe can also be used as the focal point of an appetizer or serve as a colorful accent that adds unique texture and flavor. It is important to remember that any raw fish product, such as caviar, needs to be handled properly to prevent the possibility of contamination.

Fish in Appetizers

Irinox USA

Figure 11-17. Nutritious fish appetizers can include raw, lightly seared, grilled, or broiled fish and can be served with complementary garnishes or sauces.

Transforming a fried fish entrée into an appetizer portion is an effective way to promote moderate consumption of fried foods. For example, serving a duo of miniature fried fish cakes accompanied by a cucumber-dill relish provides guests with a more nutrient-dense option than an entrée of two large, fried fish cakes served with tartar sauce.

Fish in Soups

There are many soups from around the world that feature fish. **See Figure 11-18.** Fish soups, chowders, and stews are often made with local fish and are prepared using regional flavor profiles. Fish are versatile ingredients in soups because their flesh not only provides texture, flavor, and nutrients such as protein, but their bones can be used to make flavorful stocks. Fish soups range from clear, flavorful broths to creamy soups with multiple layers of flavor.

Clear soups featuring fish provide a lighter option and are often made from fish bones that have been simmered with wine, citrus, and herbs. In contrast, fish chowders are heavier soups, often built on a flavorful base of sautéed mirepoix that has been thickened with a roux and enriched with milk and cream. Fish chowders with a clear base offer a more nutrient-dense option than cream-based chowders. When preparing fish soups, using a variety of colorful vegetables adds structure to the soup and increases dietary fiber, vitamins, and minerals.

Fish in Sandwiches

Fish is served both hot and cold in a variety of sandwiches. Many foodservice operations have signature fish sandwiches that are either simple or sophisticated, depending on the venue. Grilled red snapper accompanied by lettuce and tomato on a Kaiser roll may be popular at a casual venue, while spice-rubbed blackened grouper with mango-papaya salsa on a Hawaiian roll may be offered at a more upscale venue.

Fried fish sandwiches are also popular, but they are high in fat and calories. Instead of deep-frying the fish, it can be breaded and baked to provide guests a healthier option that is still crispy and flavorful. **See Figure 11-19.** As a breakfast sandwich, cured salmon (lox) on a bagel is a popular menu item. Fish tacos are popular on both lunch and dinner menus.

Fish in Sandwiches

Florida Department of Agriculture and Consumer Services, Bureau of Seafood and Aquaculture Marketing

Figure 11-19. Instead of being deep-fried, fish can be breaded and baked to provide guests a healthier option that is still crispy and flavorful.

Fish in Soups Around the World

Soup	Origin	Key Ingredients
Bouillabaisse	France	Fish stock, cooked fish and shellfish, garlic, orange peel, basil, bay leaf, fennel, and saffron
Fanesca	Ecuador	Figleaf gourd, pumpkin, twelve grains, and salt cod
Fufu and Egusi	Ghana	Vegetables, meat, fish, and balls of wheat gluten
Miso	Japan	Fish broth, fermented soy, and dashi
Tom yam	Thailand	Stock, lemon grass, kaffir lime leaves, galangal, shallots, lime juice, fish sauce, tamarind, and crushed chiles

Figure 11-18. There are many soups from around the world that feature fish.

Fish in Salads

Fish can be the main ingredient in a salad or accompany other salad ingredients. Popular cold salads featuring fish as the main ingredient include tuna and salmon salads. In these salads, the fish is cooked, flaked apart, and bound with a dressing and other flavorful ingredients, such as herbs and vegetables. Leafy green salads featuring fish are often served as entrées. For example, mixed greens and garden fresh vegetables tossed in a vinaigrette and topped with grilled mahi-mahi makes a flavorful and nutrient-dense entrée.

Charlie Trotter's

Fish, such as salmon, can be the main ingredient in a salad or accompany other salad ingredients.

Fish in Entrées

Because of the health benefits associated with eating fish, it is often the protein of choice for guests seeking nutritious meals. This gives chefs greater opportunities to feature different types of fish entrées. By choosing a variety of healthy cooking techniques and serving fish entrées with fresh vegetables, fruits, and whole grains, guests are provided with nutritious menu options. **See Figure 11-20.** For example, flavorful, nutrient-dense fish entrées may include grilled tilapia with a roasted corn relish atop quinoa, steamed bass with black bean salsa and sautéed vegetables, or baked halibut with a tomato, fennel, and olive salad.

Fish in Entrées

U.S. Apple® Association

Figure 11-20. By choosing a variety of healthy cooking techniques and serving fish entrées with fresh fruits, vegetables, and whole grains, guests are provided with nutritious menu options.

Knowledge Check 11-3

1. Identify the components of effective fish menu descriptions.
2. Give two examples of appetizers featuring fish.
3. Explain why fish are versatile ingredients in soups.
4. Describe how fish are used in sandwiches and salads.
5. Give an example of a nutrient-dense fish entrée.

PROMOTING FISH ON THE MENU

Guests order fish for many reasons, including flavor, nutrition, and to experience the value of a daily special. Menu items featuring flavorful fish prepared using healthy techniques can encourage sales. Educating guests on fish varieties as well as sustainable seafood practices are also effective ways to promote fish on the menu.

Fish nomenclature can be confusing because the same fish can be named differently in various regions of a country. Using a familiar name for a fish makes the fish more marketable. The Regulatory Fish Encyclopedia (RFE) attempts to standardize fish nomenclature and provides the market and common names for both imported and domestic fish. **See Figure 11-21.** It is a useful reference for promoting fish on the menu in any venue.

Market and Common Names of Fish

Market Name	Common Name
Alaska cod	Pacific cod
Alaska pollock	Walleye pollock
Cod	Atlantic cod
Grouper	Coney
Marlin	Blue or striped marlin
Medium red or silver salmon	Coho salmon
Ocean perch	Golden redfish
Sole	Dover sole
Tuna	Albacore, kawakawa, skipjack, or yellowfin

Adapted from FDA Regulatory Fish Encyclopedia

Figure 11-21. The Regulatory Fish Encyclopedia (RFE) attempts to standardize fish nomenclature and provides the market and common names for both imported and domestic fish.

Elegant Fish Presentations

Daniel NYC/T. Schauer

Figure 11-22. Fine dining and special event venues often prepare fish simply, yet in sophisticated presentations.

Quick service and fast casual venues may serve fish sandwiches, fried fish and chips, or more nutrient-dense menu options, including grilled or broiled fish dishes. Casual dining and institutional venues may offer fish specials that correspond with seasonality or availability.

Beverage houses and bars may offer a selection of fish dishes such as sandwiches, salads, and spreads made with fish. Sushi bars and tapas bars offer small portions that enable guests to sample a variety of fish. Fine dining and special event venues often prepare and present different varieties of fish simply, yet in sophisticated presentations. **See Figure 11-22.**

Florida Department of Agriculture and Consumer Services, Bureau of Seafood and Aquaculture Marketing

Nutrient-dense menu options often include grilled fish.

Knowledge Check 11-4

1. Name the reference that helps standardized fish nomenclature.
2. Identify common fish dishes served at quick service and fast casual venues.
3. Describe factors that may influence fish menu items at casual dining and institutional venues.
4. Explain how fish can be served at beverage houses and bar venues.
5. Describe the use of fish at fine dining and special event venues.

SHELLFISH

Shellfish is the classification of aquatic invertebrates that may or may not have a hard, external shell. Nutritionally, shellfish are similar to fish in that they provide protein, fat, water, vitamins, and minerals to the diet. Like fish, shellfish are also low in saturated fat. However, some shellfish may be higher in cholesterol or sodium than fish. Crustaceans, such as shrimp and lobster, and cephalopods, such as squid, are higher in cholesterol than other seafood, but they still have very low levels of saturated fat. Shellfish are also a key source of nutrients such as vitamin B_{12}, iron, zinc, selenium, and manganese.

Perceived Value of Shellfish

Some shellfish are perceived as luxurious menu items. For example, guests expect whole lobster with drawn butter or mussels simmered in white wine to carry a higher menu price. In contrast, some shellfish menu items are perceived by guests as comfort foods, such as fried shrimp platters and lobster roll sandwiches. **See Figure 11-23.**

Many people also perceive shellfish such as lobster, crab legs, and oysters as a food they would rather eat at a restaurant because these foods are labor intensive to prepare. Certain shellfish are labor intensive. For example, shucking oysters can take a significant amount of time, which needs to be factored into labor and menu costs. Because fresh shellfish have a short shelf life, chefs must also carefully plan so that the product is not wasted. Planning for the use of leftovers helps alleviate problems with storage time.

Shellfish Classifications

Shellfish are generally classified by the type of shell they have. There are two classifications of shellfish: crustaceans and mollusks. **See Figure 11-24.** While crustaceans always have an external shell, or exoskeleton, some mollusks, known as cephalopods, do not. Crustaceans and mollusks can be found in both freshwater and saltwater habitats.

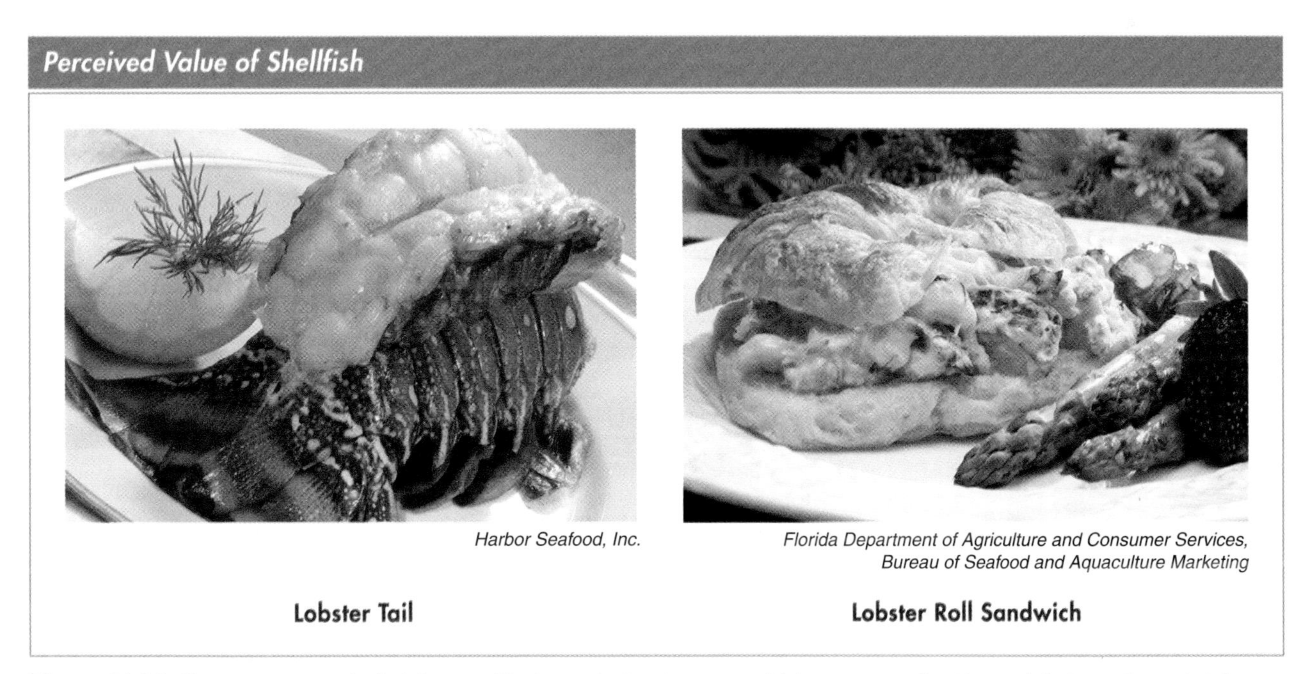

Figure 11-23. Guests expect whole lobster with drawn butter to carry a higher menu price than a lobster roll sandwich.

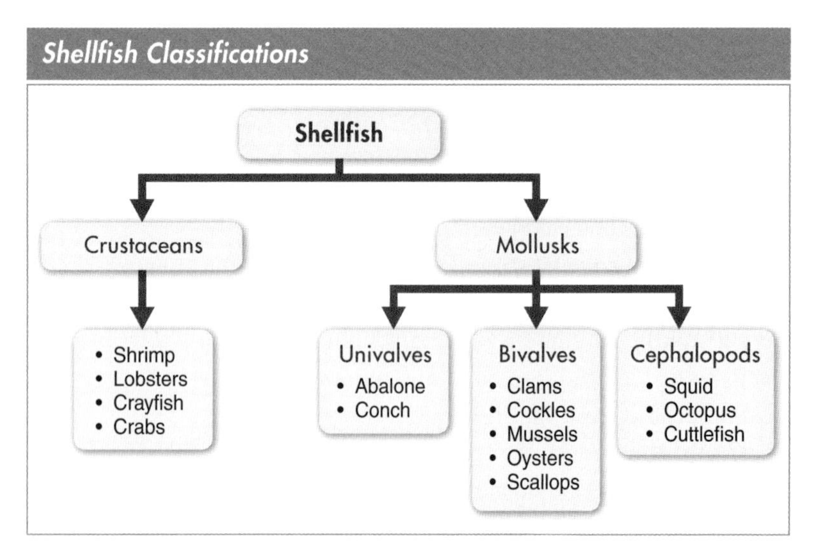

Figure 11-24. There are two classifications of shellfish: crustaceans and mollusks.

Crustaceans. A *crustacean* is a shellfish that has a hard, segmented shell protecting its soft flesh and lacks an internal bone structure. **See Figure 11-25.** Shrimp, prawns, lobsters, crayfish, and crabs are crustaceans. Prawns resemble large shrimp and are often used in place of shrimp. Unlike fish, crustaceans can live outside of the water for several days if they are kept moist.

> **Food Science Note**
>
> Shellfish that supply omega-3 fatty acids include oysters, shrimp, squid, and mussels. Oysters supply 629 mg, shrimp supply 459 mg, squid supply 422 mg, and mussels supply 411 mg of omega-3 fatty acids per 3 oz serving (raw).

Mollusks. A *mollusk* is a shellfish with a soft, nonsegmented body. Mollusks are grouped into univalves, bivalves, and cephalopods.

- A *univalve,* also known as a gastropod, is a mollusk that has a single solid shell and a single foot. **See Figure 11-26.** Univalves use their foot to secure themselves to objects and to move. Univalves include abalone and conch.

Crustaceans

Type of Crustacean	Description and Characteristics	Key Nutrients*
Shrimp *National Cancer Institute, Renee Comet (photographer)*	• Tender, sweet, white flesh • Commonly sold headless with or without shell • Sold fresh or frozen • Shell turns from gray/blue to pink/red when cooked • Often grilled, broiled, sautéed, or fried and served hot or cold	• Calories: 71 • Total fat: 1 g • Protein: 14 g • Niacin, vitamin B_{12}, vitamin D, phosphorus, copper, and selenium
Lobsters *Fortune Fish Company*	• Cold water lobsters include large Atlantic lobsters • Warm water lobsters include smaller spiny or rock lobsters • Majority of flesh is in tail and claws • Lobsters sold live must be kept alive until cooked • Often steamed, simmered, or poached	• Calories: 77 • Total fat: 1 g • Protein: 16 g • Vitamin B_{12}, pantothenic acid, phosphorus, zinc, selenium, and copper

* based on 100 g serving size, raw

Figure 11-25. (continued on next page)

Crustaceans

Type of Crustacean	Description and Characteristics	Key Nutrients*
Crayfish  *Eloma Combi Ovens*	• Freshwater crustacean that resembles a tiny lobster • Sometimes called crawfish or crawdad • Flavor similar to shrimp with a slightly tougher texture • Commonly used in Creole and Cajun cuisines	• Calories: 72 • Total fat: 1 g • Protein: 15 g • Niacin, folate, phosphorus, magnesium, potassium, selenium, and copper
Crabs  *Harbor Seafood, Inc.*	• Varieties include king, blue, Dungeness, snow, and stone crabs • Hard shell with sweet-tasting flesh • Blue crabs harvested within 6 hours of molting are sold as soft-shell crabs • Available fresh, frozen, and canned • Commonly steamed and served whole or used in soups, sandwiches, and salads	• Calories: 84 • Total fat: 1 g • Protein: 18 g • Vitamin C, folate, vitamin B_{12}, phosphorus, magnesium, zinc, selenium, and copper

* based on 100 g serving size, raw

Figure 11-25. A crustacean is a shellfish that has a hard, segmented shell protecting its soft flesh and lacks an internal bone structure.

Mollusks: Univalves

Type of Univalve	Description and Characteristics	Key Nutrients*†
Abalone  *National Oceanic and Atmospheric Administration/Department of Commerce*	• Brown, bowl-shaped shell with iridescent, multicolored interior • Sweet, slightly salty flavor and tender texture • Commonly imported from Japan, New Zealand, and Mexico and sold canned or frozen	• Calories: 105 • Total fat: 1 g • Protein: 17 g • Thiamin, pantothenic acid, vitamin B_{12}, vitamin E, vitamin K, phosphorus, magnesium, and selenium
Conch  *U.S. Fish & Wildlife Service*	• Pink-orange shell with interior resembling a large snail • Flesh is rubbery and typically sliced thin and tenderized before cooking • Native to Florida Keys and Caribbean	• Calories: 130 • Total fat: 1 g • Protein: 26 g • Folate, vitamin B_{12}, vitamin E, magnesium, and selenium

* abalone based on 100 g serving size, raw
† conch based on 100 g serving size, baked

Figure 11-26. A univalve, also known as a gastropod, is a mollusk that has a single solid shell and a single foot.

- A *bivalve* is a mollusk that has a top shell and a bottom shell connected by a central hinge. **See Figure 11-27.** Mollusks contain a siphon, which is a tubular organ that is used to draw in or eject fluids. Bivalves include clams, cockles, mussels, oysters, and scallops. Live bivalves should close tightly when gently tapped. Bivalves that are broken or do not close when tapped are dead and should be discarded.

Mollusks: Bivalves

Type of Bivalve	Description and Characteristics	Key Nutrients*
Clams *United States Department of Agriculture*	• Varieties include soft-shell and hard-shell clams • Soft-shell clams have tender, sweet flesh but are susceptible to drying out because their shell does not fully close • Hard-shell clams are encased in a blue-gray shell, have a chewy texture, and are commonly sold as littleneck, cherrystone, topneck, and chowder clams, depending on size • Commonly steamed or fried and a key ingredient in clam chowders	• Calories: 86 • Total fat: 1 g • Protein: 15 g • Vitamin C, riboflavin, niacin, phosphorus, potassium, selenium, copper, and manganese
Cockles *Fortune Fish Company*	• Deep-ridged shell, approximately 1 inch wide • Commonly used to make paella and Southeast Asian and Thai dishes	• Calories: 79 • Total fat: 1 g • Protein: 14 g • Riboflavin, niacin, and iron
Mussels *New Zealand Greenshell™ Mussels*	• Varieties include blue (most common) and greenlip mussels • Tender, sweet flesh • Whisker-like threads (beard) extending outside the shell are commonly removed for presentation • Commonly steamed and served hot or cold in appetizers	• Calories: 86 • Total fat: 2 g • Protein: 12 g • Vitamin C, thiamin, riboflavin, folate, phosphorus, selenium, and manganese
Oysters *Florida Department of Agriculture and Consumer Services, Bureau of Seafood and Aquaculture Marketing*	• Varieties include Atlantic (most common), Pacific, and European oysters • Tender, salty-sweet flesh • Commonly served raw, roasted, baked, breaded and fried, or poached	• Calories: 51 • Total fat: 2 g • Protein: 6 g • Vitamin C, riboflavin, niacin, vitamin B_{12}, phosphorus, iron, zinc, selenium, copper, and manganese
Scallops	• Varieties include small bay scallops and larger-sized sea scallops • Creamy white flesh with sweet, delicate flavor • Sold as "wet" (soaked in preservative to whiten and prevent spoilage), "dry" (untreated), or quick-frozen • Commonly sautéed, broiled, or marinated in lemon or lime juice for ceviche	• Calories: 69 • Total fat: 0.5 g • Protein: 12 g • Vitamin B_{12}, phosphorus, magnesium, potassium, and selenium

* based on 100 g serving size, raw

Figure 11-27. A bivalve is a mollusk that has a top shell and a bottom shell connected by a central hinge.

- A *cephalopod* is a mollusk that does not have an external shell. **See Figure 11-28.** Cephalopods have arms that extend from a prominent head and may have an internal bone called a cuttlebone. They are sometimes called "ink fish" because of their ability to expel a dark cloud of ink to confuse prey. Cephalopods include squids, octopuses, and cuttlefish. Cephalopods become tough if overcooked.

Mollusks: Cephalopods

Type of Cephalopod	Description and Characteristics	Key Nutrients*
Squid	• Translucent with two tentacles, eight sucker-equipped arms, two lateral fins, and a flat, internal cuttlebone • Often called by its Italian name, calamari • Ink is used to color and flavor grain and pasta dishes • Flesh is white, firm, and sweet • Large squid are cut into steaks and sold frozen • Commonly sautéed, breaded and fried, or used to make stews	• Calories: 92 • Total fat: 1 g • Protein: 16 g • Riboflavin, niacin, vitamin B_{12}, phosphorus, selenium, and copper
Octopus	• Gray with eight sucker-equipped arms, a birdlike beak, and no internal or external shell • White, firm, and sweet flesh • Typically sold whole and available fresh or frozen • Commonly cooked and chilled to use in salads	• Calories: 82 • Total fat: 1 g • Protein: 15 g • Niacin, vitamin B_6, vitamin B_{12}, phosphorus, potassium, iron, selenium, and copper
Cuttlefish	• Translucent with two tentacles, eight sucker-equipped arms, a hard internal cuttlebone, and large eyes at base of the head • Ink is used to color and flavor grain and pasta dishes • Flesh is white, firm, and sweet • Tentacles and arms commonly cut into rings and sautéed, breaded and fried, or cooked in soups or stews	• Calories: 79 • Total fat: 1 g • Protein: 16 g • Vitamin C, riboflavin, vitamin B_{12}, calcium, phosphorus, potassium, zinc, selenium, and copper

* based on 100 g serving size, raw

Figure 11-28. A cephalopod is a mollusk that does not have an external shell.

INGREDIENT SPOTLIGHT: Blue Mussels

Fortune Fish Company

Nutrition Facts
Serving Size 3 oz cooked, moist-heat (85 g)

Amount Per Serving	
Calories 146	Calories from Fat 34
	% Daily Value*
Total Fat 4g	6%
Saturated Fat 1g	4%
Trans Fat	
Cholesterol 48mg	16%
Sodium 314mg	13%
Total Carbohydrate 6g	2%
Dietary Fiber 0g	0%
Sugars	
Protein 20g	
Vitamin A 5% • Vitamin C	19%
Calcium 3% • Iron	32%

* Percent Daily Values (DV) are based on a 2,000 calorie diet.

Unique Features

- Bivalve mollusk found in both the Atlantic and Pacific Oceans
- Aquafarmed mussels have a thinner, blue-black shell, while wild-caught mussels have a thicker, silver-blue shell
- Tender, sweet flesh with a yellow-orange color
- Low in saturated fat
- Source of protein, thiamin, riboflavin, folate, vitamin B_{12}, phosphorus, iron, zinc, selenium, and manganese

Menu Applications

- Steam, sauté, simmer, or bake for use in appetizers, soups, stews, salads, and entrées
- Pair with tomatoes, herbs, garlic, lemon, and white wine
- Serve hot or cold

Market Forms of Shellfish

Shellfish can be found in various market forms, including fresh, frozen, and processed convenience products. Shellfish may also be sold in the shell or shucked.

Live Shellfish. When shellfish are sold fresh, they are often still alive. Lobsters as well as hard-shell and soft-shell crabs are commonly sold live. Live clams, oysters, and scallops are also available for purchase in some regions.

Shucked Shellfish. Clams, oysters, mussels, scallops, and some types of crab are often sold with their shells removed and are available fresh or frozen. Shrimp can be sold with the head removed or with both the head and shell removed.

Frozen Shellfish. Some shellfish, such as lobsters, shrimp, and prawns, can be frozen in the shell. Typically the lobster tail is frozen and the rest of the lobster flesh is used in convenience products. With a shrimp or prawn, the head and thorax (body) are typically removed, and the tail is frozen while still at sea to ensure maximum freshness.

Shellfish Convenience Products. Shellfish convenience products include canned, smoked, dried, and breaded items. Lobster meat, crabmeat, shrimp, and oysters are available canned and may be used in a variety of menu items, such as soups and salads. Smoked shellfish products, such as clams, oysters, shrimp, scallops, and mussels, are often used in appetizers. Dried shellfish is typically found in Asian cuisine and often incorporated into appetizers, soups, and entrées. Breaded shellfish products include shrimp, clams, oysters, and calamari (squid).

Knowledge Check 11-5

1. Identify nutrients provided by shellfish.
2. Describe the factors that contribute to the perceived value of shellfish.
3. Name the two general classifications of shellfish.
4. List common varieties of crustaceans.
5. Describe the three classifications of mollusks.
6. List common shellfish convenience products.

PREPARING NUTRITIOUS SHELLFISH

Purchasing shellfish live, shucked, or further processed influences how the shellfish is prepared. Shellfish generally require short cooking times and overcook quickly. Some shellfish may need tenderizing prior to cooking. Shellfish are naturally low in fat, and care should be taken to prepare and serve them in a manner that highlights their flavor.

Plating shellfish with nutrient-dense sides and serving light, flavorful sauces, such as vegetable purées or fruit salsas, instead of butter can elevate flavor while preserving the nutritional integrity of the menu item.

Raw Shellfish

Mollusks are commonly served raw. **See Figure 11-29.** However, mollusks are prone to contaminants and consuming them raw is a potential health risk. Oysters can be pasteurized for safety and are listed as such on some menus. Octopus is also served raw as sashimi and in sushi.

Raw Oysters

Florida Department of Agriculture and Consumer Services, Bureau of Seafood and Aquaculture Marketing

Figure 11-29. Mollusks, such as oysters, are commonly served raw.

Cooking Shellfish

Shellfish can be cooked using dry-heat or moist-heat cooking techniques. Shellfish can be grilled, broiled, smoked, baked, sautéed, fried, poached, simmered, or steamed. Color is a good indicator of the doneness of shellfish. For example, a lobster shell turns red when it is done, and shrimp curl and turn slightly pink when cooked. **See Figure 11-30.** Crayfish turn a reddish-brown when cooked and the center of a scallop turns opaque. Like fish, shellfish can quickly overcook. Overcooking shellfish toughens the proteins, causing dryness and flavor loss.

Cooking Lobster

Figure 11-30. Color is a good indicator of the doneness of shellfish. A lobster shell turns red when it is done.

Grilling. Lobsters, shrimp, prawns, clams, oysters, scallops, and squid can be successfully grilled. **See Figure 11-31.** Shrimp and prawns can be skewered and grilled as kebabs. Clams and oysters can be set on a grill until their shells open and then served as appetizers. Although grilling is a healthy cooking technique that can enhance flavor and presentation, it is important to properly coat the grill with oil so that the shellfish do not stick.

Grilled Shellfish

Florida Department of Agriculture and Consumer Services, Bureau of Seafood and Aquaculture Marketing

Figure 11-31. Many types of shellfish, including clams, can be successfully grilled.

Baked Clams

Florida Department of Agriculture and Consumer Services, Bureau of Seafood and Aquaculture Marketing

Figure 11-32. Baking is a nutritious way to cook shellfish and is typically used to cook mollusks and squid.

Broiling. Most shellfish can be broiled. The high heat of a broiler makes it essential to closely monitor shellfish so that they stay tender and moist. Lobster flesh can best tolerate the intense heat of a broiler.

Smoking. Shellfish such as clams and oysters are commonly smoked. Smoking can be a healthy way to prepare shellfish and add flavor to menu items.

Baking. Baking is a nutritious way to cook shellfish and is typically used to cook mollusks and squid. **See Figure 11-32.** Mollusks are commonly shucked and stuffed with a filling or topped with ingredients, such as spinach and bread crumbs, and baked. Squid is commonly stuffed and baked. Other shellfish, such as lobsters and soft-shell crabs, can also be baked.

Sautéing. Sautéing is a low-fat cooking technique that can be effectively used to enhance the flavor of shellfish such as shrimp, prawns, crayfish, soft-shell crabs, mussels, scallops, squid, and octopuses. Tenderizing shellfish before sautéing leads to a moist finished product.

Frying. Vegetable stir-fry dishes commonly include shellfish such as shrimp or scallops for added protein, texture, and flavor. Many types of shellfish are often pan- or deep-fried, including shrimp, prawns, soft-shell crabs, oysters, and squid. Frying shellfish in oil that has been heated to the correct temperature prevents the shellfish from absorbing excess fat.

All-Clad Metalcrafters

Frying shellfish in oil heated to the correct temperature prevents the shellfish from absorbing excess fat.

Poaching. Poaching produces tender and flavorful shellfish that can be incorporated into other preparations. For example, lobster is commonly poached in a court bouillon and mixed with various ingredients to form a lobster roll. Poached octopus can be chilled, tossed with a vinaigrette, and served as a cold salad.

Simmering. Although "boiled shrimp" may be a common menu entry, shellfish such as crustaceans should be simmered rather than boiled. Boiling shellfish creates an unappealing texture. Simmering crustaceans in their shell adds flavor. Mussels are commonly simmered in their shell until the shell opens. **See Figure 11-33.** Cuttlefish may be simmered in cooking liquid when preparing seafood risotto.

Simmered Mussels

Daniel NYC/E. Kheraj

Figure 11-33. Mussels are commonly simmered in their shell until the shell opens.

FOOD SCIENCE EXPERIMENT: Cooking Shellfish

11-2

Objective

- To compare properly cooked shrimp and overcooked shrimp in terms of appearance, texture, and flavor.

Materials and Equipment

- 8 qt pot
- Water
- 1 lb shrimp, peeled and deveined
- Slotted spoon
- 2 plates

Procedure

1. Fill the pot halfway full with water and bring to a simmer.
2. Place the shrimp in the simmering water and cook until they turn pink.
3. Use the slotted spoon to remove half of the shrimp and set aside on a plate.
4. Continue simmering the remaining shrimp for an additional 5 minutes before placing them on a second plate.
5. Evaluate the appearance, texture, and flavor of each plate of shrimp.

Allergen Alert

The materials used in this experiment may contain one or more food allergens.

Typical Results

There will be variances in the appearance, texture, and flavor of each plate of shellfish.

Steaming. Lobster tails, shrimp, prawns, crayfish, hard-shell crabs, clams, and mussels are commonly steamed and may be served hot or cold. **See Figure 11-34.** Clambakes involve the use of underground steamers to steam shellfish, not to bake them. Seaweed may be added to steamed shellfish to keep it moist and to add flavor.

Steamed Shellfish

Florida Department of Agriculture and Consumer Services, Bureau of Seafood and Aquaculture Marketing

Figure 11-34. Shellfish, such as stone crabs, are commonly steamed and may be served hot or cold.

Flavor Development

The flavor of shellfish is delicate and best appreciated when it is not overpowered by other ingredients. Dry-heat and moist-heat cooking techniques affect flavor differently. Dry heat will brown or caramelize ingredients, while moist heat does not. For example, pan-searing, grilling, and sautéing add depth of flavor, while steaming emphasizes the natural flavor of shellfish. Shellfish are typically added to the plate at the last second. However, flavorful ingredients like onions, garlic, spinach, and grains can be cooked before adding the shellfish to develop depth of flavor in dishes such as soups and stews.

Developing flavor in shellfish is also achieved by using rubs and marinades to create different flavor profiles. Cooking shellfish in their shells enhances flavor, and pan sauces can add complexity to shellfish dishes. The flavor of shellfish is often enhanced with an acidic ingredient, such as lemon or lime juice. The acidic flavor of citrus balances the sweetness of the shellfish and adds another layer of flavor.

The temperature at which shellfish is served also affects flavor development. For example, steamed shrimp may be served hot or chilled. Shellfish should always be served at the intended temperature to uphold the texture, flavor, and overall quality of the dish.

Plating Shellfish

Shellfish have varied colors, shapes, and sizes that can enhance plated presentations in a myriad of ways. **See Figure 11-35.** For example, the vibrant red and pink hues of crustaceans add visual appeal to the plate by contrasting with yellow and green vegetables or by adding a burst of color to neutral grains and pastas. Blue crabs and black mussels also provide unique colors to the plate.

The dark ink from cephalopods, such as cuttlefish, can add color to pastas, sauces, and risottos and result in intriguing presentations. Lobster contains tomalley, a soft green substance that can be used to accentuate the plate. Lobster roe, also known as coral, may be used in this manner as well. Serving shellfish in their shells can also elevate a presentation and prompt a higher menu price.

Food Safety

Cooked shellfish cannot be placed on plates that previously held raw seafood because the bacteria from the raw food can contaminate the cooked food.

Plating Shellfish

New Zealand Greenshell™ Mussels

Mussels on the Half Shell

Daniel NYC/B. Milne

Crawfish with Leeks

Irinox USA

Shrimp with Piperade Sauce

Daniel NYC/M. Price

Shrimp with Couscous

Figure 11-35. Shellfish have varied colors, shapes, and sizes that can enhance plated presentations.

Knowledge Check 11-6

1. Explain how color can be used to determine the doneness of shellfish.
2. Identify the types of shellfish best prepared by grilling, broiling, smoking, and baking.
3. Describe how to prepare shellfish by sautéing, frying, poaching, simmering, and steaming.
4. Explain how to effectively develop flavors in a shellfish dish.
5. Describe how to elevate the presentation of a plated shellfish dish.

CULINARY NUTRITION RECIPE-MODIFICATION PROCESS

Crab Louie Salad

Crab Louie salad is an arranged salad consisting of lump crabmeat, eggs, and tomatoes on a bed of lettuce. It is traditionally served with a dressing comprised of mayonnaise, cream, chili sauce, lemon juice, green peppers, scallions, and hard-cooked eggs. There are many versions of Crab Louie salad. This recipe includes avocados, asparagus, and black olives. Crab Louie salad provides high-quality protein and nutrient-dense ingredients, but it is also high in calories, fat, cholesterol, and sodium.

Yield: 6 servings (3 cups salad and ¼ cup dressing each)

Ingredients

Dressing

1 c	mayonnaise
2 tbsp	cream
3 tbsp	chili sauce
1 tsp	lemon juice
1½ tsp	green pepper, minced
1½ tbsp	scallions, minced
1 ea	hard-cooked egg, minced

Salad

9 c	iceberg lettuce, chopped
1 lb	Dungeness crab meat, cooked and torn into 2 inch pieces
3 ea	avocados, large dice
1 ea	tomato, medium dice
18 ea	asparagus spears, blanched and cut into 2 inch pieces
6 ea	hard-cooked eggs, quartered
½ c	black olives, sliced

Preparation

Dressing

1. Combine all dressing ingredients. Set aside until needed.

Salad

2. Place lettuce on plates.
3. Carefully arrange crab, avocados, tomatoes, asparagus, eggs, and black olives on the lettuce.
4. Drizzle dressing over the salad.

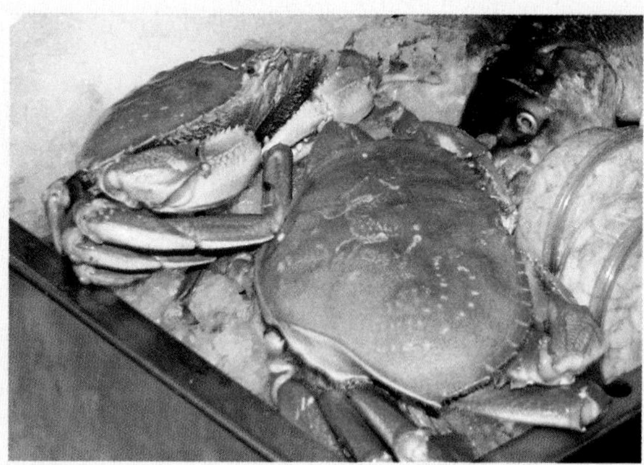

Fortune Fish Company

Evaluate original recipe for sensory and nutritional qualities

- Crab Louie is a colorful, arranged salad with a creamy dressing.
- Serving size is an entrée portion that provides protein and dietary fiber.
- Dish is high in calories, fat, cholesterol, and sodium.

Establish goals for recipe modifications

- Reduce calories, fat, cholesterol, and sodium.
- Maintain flavor and presentation.

Identify modifications or substitutions

- Use a substitution for the mayonnaise in the dressing.
- Eliminate the cream and egg in the dressing.
- Reduce the number of avocados and eggs in the salad.
- Increase the number of tomatoes in the salad.
- Eliminate black olives in the salad.

Determine functions of identified modifications or substitutions

- Mayonnaise is the base for the dressing and provides a creamy texture and rich flavor.
- Cream thins the dressing to an appealing consistency, and egg adds texture and flavor to the dressing.
- Avocados provide a creamy texture and mild flavor to the salad, and eggs provide contrasting color, texture, and richness to the salad.
- Tomatoes add color, texture, and sweetness to the salad.
- Black olives provide color, texture, and saltiness to the salad.

Select appropriate modifications or substitutions

- Substitute low-fat mayonnaise and plain, nonfat yogurt for the mayonnaise in the dressing.
- Omit the cream and egg from the dressing.
- Reduce the number of avocados to one and eggs to three in the salad.
- Increase the number of tomatoes to three in the salad.
- Omit black olives from the recipe.

Test modified recipe to evaluate sensory and nutritional qualities

- Low-fat mayonnaise and nonfat yogurt provide a creamy, flavorful base for the dressing and reduce calories, fat, and cholesterol.
- Dressing has an appealing consistency and does not need to be thinned with cream, and omitting the egg lowers fat and cholesterol, yet the texture and flavor of the dressing are maintained.
- Reducing the number of avocados and eggs lowers calories and fat, and using fewer eggs also lowers cholesterol.
- Increasing the number of tomatoes adds color and nutrients.
- Omitting black olives reduces calories, fat, and sodium.

CULINARY NUTRITION RECIPE-MODIFICATION PROCESS

Modified Crab Louie Salad

In this modified version of Crab Louie salad, low-fat mayonnaise and nonfat yogurt are used as the base of the dressing. This change, along with omitting the egg, substantially reduces the calories, fat, and cholesterol in the dressing. To continue the reduction of calories, fat, and cholesterol, the number of avocados and eggs are decreased. The number of tomatoes is increased to enhance presentation. The black olives are omitted to lower sodium.

Yield: 6 servings (3 cups salad and ¼ cup dressing each)

Ingredients

Dressing

½ c	low-fat mayonnaise
½ c	plain, nonfat yogurt
3 tbsp	chili sauce
1 tsp	lemon juice
1½ tsp	green pepper, minced
1½ tbsp	scallions, minced

Salad

9 c	iceberg lettuce, chopped
1 lb	Dungeness crab meat, cooked and torn into 2 inch pieces
1 ea	avocado, large dice
3 ea	tomatoes, medium dice
18 ea	asparagus spears, blanched and cut into 2 inch pieces
3 ea	hard-cooked eggs, quartered

Preparation

Dressing

1. Combine all dressing ingredients. Set aside until needed.

Salad

2. Place lettuce on plates.
3. Carefully arrange crab, avocados, tomatoes, asparagus, and eggs on the lettuce.
4. Drizzle dressing over the salad.

Crab Louie Salad Nutritional Comparison

Nutrition Facts	Original	Modified
Calories	539.1	273.5
Total Fat	37.6 g	12.7 g
Saturated Fat	7.4 g	2.5 g
Cholesterol	355.4 mg	183.9 mg
Sodium	823.6 mg	668.5 mg
Dietary Fiber	8.6 g	4.9 g
Protein	29.4 g	24.6 g

Signature Crab Louie Spinach Salad

A variety of ingredients including herbs, spices, Dijon mustard, and lemon juice add a burst of flavor to the dressing in this signature Crab Louie salad. The dressing remains low-fat due to a base of nonfat yogurt and low-fat sour cream. Nutrient-dense spinach replaces the iceberg lettuce, and red peppers, apples, and carrots are added for their nutritive value and to elevate color, texture, and flavor. The addition of these ingredients allows the amount of crab to be reduced, while portion size remains plentiful.

Yield: 6 servings (3 cups salad, ¼ cup dressing each)

Ingredients

Dressing

½ c	plain, nonfat yogurt
¼ c	low-fat sour cream
⅛ tsp	paprika
4 tsp	Dijon mustard
2 tsp	lemon juice
¼ tsp	black pepper, ground
1 tsp	Old Bay® seasoning
3 tbsp	parsley, chopped

Salad

9 c	baby spinach
¾ lb	Dungeness crab meat, cooked and torn into 2 inch pieces
1 ea	avocado, large dice
2 ea	tomatoes, medium dice
16 ea	spears asparagus, blanched and sliced into 2 inch pieces
2 ea	hard cooked eggs, chopped
1 ea	Fugi apple, cut into 12 slices
½ c	red pepper, medium dice
½ c	carrots, julienned

Preparation

Dressing

1. Combine yogurt, sour cream, paprika, Dijon mustard, lemon juice, pepper, Old Bay®, and parsley in a mixing bowl. Set aside.

Salad

1. Place spinach on plates.
2. Carefully arrange crab, avocado, tomatoes, asparagus, eggs, apple, red pepper, and carrots on the lettuce.
3. Drizzle dressing on top.

Chef's Tip: *Substitute the vegetables used in Crab Louie salad based upon seasonality.*

Crab Louie Salad Nutritional Comparison

Nutrition Facts	Original	Modified	Signature
Calories	539.1	273.5	216.7
Total Fat	37.6 g	12.7 g	9.3 g
Saturated Fat	7.4 g	2.5 g	2.4 g
Cholesterol	355.4 mg	183.9 mg	128.6 mg
Sodium	823.6 mg	668.5 mg	358.2 mg
Dietary Fiber	8.6 g	4.9 g	5.9 g
Protein	29.4 g	24.6 g	20.1 g

SHELLFISH MENU MIX

Shellfish are used across the menu. For example, smoked oysters can be added to an omelet for a luxurious breakfast item. Lunch menus often incorporate shellfish into a variety of soups, sandwiches, and salads. Dinner menus commonly highlight shellfish as appetizers and entrées.

Shellfish are indigenous to the cuisines of countries that border large bodies of water and reflect the flavor profiles of that region. Mentioning the point of origin in a menu description can be a powerful selling tool. For example, lobsters harvested from the waters of the Atlantic Ocean should be marketed as Atlantic lobsters since they are prized for their sweet flavor. Also, effective menu descriptions should be written using words that highlight both the shellfish and flavorful components of the dish. **See Figure 11-36.**

Shellfish Menu Descriptions

Seafood Entrée
Moules Marinière

Ineffective Description

Seafood Entrée
Moules Marinière—Mussels simmered in a flavorful white wine broth served with French bread

Effective Description

Figure 11-36. Effective menu descriptions should be written using words that highlight both the shellfish and the flavorful components of the dish.

Shellfish in Appetizers

One of the most popular chilled shellfish appetizers is a shrimp cocktail. Clams casino, oysters Rockefeller, and raw oysters on the half shell served with hot sauce, lemon wedges, horseradish, and crackers are also classic shellfish appetizers.

Although the classics remain popular, shellfish appetizers have evolved to include more contemporary offerings. **See Figure 11-37.** For example, a chef may choose to serve a tequila lime cocktail sauce or a ginger mignonette instead of tartar sauce to accompany a grilled shrimp or scallop kebab. Likewise, a flavorful broth of coconut milk and lemongrass can be served with steamed mussels, or an array of salsas may be presented with a crab cake appetizer.

Figure 11-37. Although the classic shrimp cocktail remains popular, shellfish appetizers have evolved to include more contemporary offerings.

Shellfish in Soups

Shellfish soups can range from light, clear broths to rich, thick bisques. Shellfish such as lobster and shrimp are often used to make flavorful stocks that serve as the base for soups, stews, and gumbos. **See Figure 11-38.** Shellfish such as mussels are left in their shells to enhance both the flavor and presentation of soups. Shellfish are typically added to soups near the end of the cooking process to preserve their delicate texture. When serving rich shellfish soups, such as lobster bisque or clam chowder, it is important to serve smaller portion sizes.

Florida Department of Agriculture and Consumer Services, Bureau of Seafood and Aquaculture Marketing

Figure 11-38. Shellfish such as lobster and shrimp are often used to make flavorful stocks that serve as the base for soups, stews, and gumbos.

Shellfish in Sandwiches

Shellfish sandwiches are often associated with a particular geographical region where locally sourced shellfish are featured in signature sandwiches. For example, New Orleans highlights local oysters, shrimp, and catfish in po' boy sandwiches. New England showcases Atlantic lobsters in lobster rolls, and San Francisco features Dungeness crabs in specialty crab sandwiches.

Although shellfish are a lean source of protein, care should be taken to use healthy cooking techniques when preparing sandwiches. **See Figure 11-39.** For example, instead of deep-frying oysters for po' boys, the oysters can be grilled or breaded and baked. In order to increase nutrient density, puréed avocado or yogurt mixed with Dijon mustard can be used instead of mayonnaise to bind ingredients. Shellfish can also be chopped up and formed into patties. Crab or shrimp "burgers" are popular menu items, particularly as sliders.

Alpha Baking Co., Inc.

Figure 11-39. Nutrient-dense shellfish sandwiches can be prepared with healthy cooking techniques and nutritious ingredients.

Shellfish in Salads

Shellfish are popular salad ingredients because of their textures, flavors, and versatility. For example, shellfish can be added to virtually any leafy green salad to enhance texture, flavor, and protein content. Shellfish can be used in hot or cold salads, such as a warm grilled garlic prawn and radish salad or a chilled crab Louie salad.

To use shellfish as a cost-effective ingredient and to promote menu prices that add value for the guest, a smaller amount of shellfish is used in salads as compared to other protein sources. For example, instead of a 4 oz portion of beef in an entrée salad, a 3 oz portion of shellfish can be used.

Shellfish in Entrées

Entrées featuring shellfish offer chefs flexibility in both cooking techniques and presentation styles. **See Figure 11-40.** For example, shellfish is often grilled, baked, sautéed, fried, simmered, or steamed and can be presented alone or with sides. A grilled or broiled shellfish entrée might feature spicy Thai shrimp or grilled soft-shell crabs. Oysters baked with flavorful ingredients are often served in their shell to elevate the presentation. Shellfish can also be baked into pasta dishes such as lasagna for a unique twist on a traditional dish.

A wide variety of entrées feature sautéed shellfish, such as shrimp, clams, mussels, scallops, and squid. Sautéed scallops make an appealing presentation, with the caramelized exterior contrasting the tender white interior. Sautéed squid blended with sun-dried tomatoes, olives, and scallions makes a colorful, nutrient-dense entrée.

Fried shellfish entrées can be kept lower in saturated fat content when unsaturated oils are used for frying instead of hydrogenated oils. Simmering or steaming shellfish can provide guests with a healthy entrée that is both tender and flavorful.

Figure 11-40. Entrées featuring shellfish offer chefs flexibility in both cooking techniques and presentation styles.

Knowledge Check 11-7

1. Name three common appetizers that feature shellfish.
2. Explain how shellfish can be incorporated into soups and salads.
3. Describe how to prepare healthy sandwiches featuring shellfish.
4. Give two examples of nutrient-dense entrées featuring shellfish.

PROMOTING SHELLFISH ON THE MENU

Offering a diverse selection of shellfish helps educate guests on the wide variety of shellfish available. The sweet, rich flavor and tender texture of shellfish can be highlighted in menu descriptions of both daily specials and signature dishes. Including shellfish on the menu can appeal to health-conscious guests by providing a lean source of protein that contains essential fatty acids, vitamins, and minerals. **See Figure 11-41.**

Quick service and fast casual venues typically serve grilled and fried shellfish, which is often accompanied by French fries or hush puppies. Crabmeat or surimi is frequently used to make bound salads for sandwiches served at these venues. Combination plates of several different types of shellfish are popular at casual dining venues.

Beverage Houses and Bar Venues

Many beverage houses and bars offer shellfish menu items such as peel-and-eat shrimp or sushi. Shellfish is also featured in salads and sandwiches or as appetizers, such as calamari or steamed clams.

Shellfish are a major component of raw bars. A *raw bar* is a presentation of a variety of raw and steamed seafood presented and served on a bed of ice. **See Figure 11-42.** Common items found on raw bars include shrimp, crab legs, clams, mussels, and oysters. Shellfish shooters are also popular. Shellfish shooters consist of raw or cooked seafood presented in a shot glass or on the half shell with a flavorful sauce or liquid. Common shellfish shooters include oysters in a horseradish cocktail sauce, mussels on the half shell with a creamy vinaigrette, and shrimp in a roasted pepper and spicy tomato juice cocktail.

Raw Bars

Daniel NYC/B. Milne

Figure 11-42. A raw bar is a presentation of a variety of raw and steamed seafood presented and served on a bed of ice.

Key Health Benefits of Shellfish

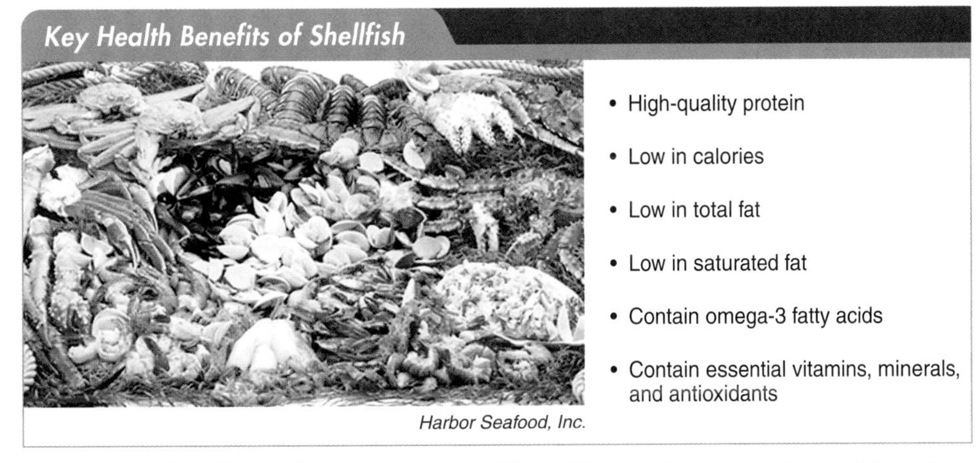

Harbor Seafood, Inc.

- High-quality protein
- Low in calories
- Low in total fat
- Low in saturated fat
- Contain omega-3 fatty acids
- Contain essential vitamins, minerals, and antioxidants

Figure 11-41. Shellfish on the menu can appeal to health-conscious guests by providing a lean source of protein that contains essential fatty acids, vitamins, and minerals.

Fine Dining and Special Event Venues

Fine dining venues often offer a catch of the day or creative surf-and-turf entrées, such as pan-seared scallops with veal or pork kebabs. Special event venues commonly serve fish and shellfish as hors d'oeuvres and entrées.

L. Isaacson and Stein Fish Company

Shellfish such as clams are often served as hors d'oeuvres.

Knowledge Check 11-8

1. Explain how shellfish can be served at quick service and fast casual venues.
2. List items commonly served on raw bars.
3. Describe the use of shellfish at fine dining and special event venues.

Chapter Summary

Freshwater and saltwater fish are commonly classified as roundfish, flatfish, or cartilaginous fish. Fish can also be lean or fatty. Fatty fish contain essential fatty acids and are plentiful in omega-3 fatty acids, which are part of a heart-healthy diet. Shellfish are classified as crustaceans or mollusks. Mollusks are further classified as univalves, bivalves, and cephalopods.

Both fish and shellfish supply protein and a wide variety of vitamins and minerals, including antioxidants. The preparation and cooking techniques used with seafood vary depending on the seafood type and fat content. Using healthy cooking techniques, serving appropriate portion sizes, and incorporating fruits, vegetables, and whole grains all provide the opportunity to serve nutrient-dense, flavorful menu items.

Chapter Review

1. Compare the three classifications of fish.
2. Describe common market forms of fish.
3. Explain how fish can be served raw.
4. Describe common cooking techniques used to prepare healthy fish dishes.
5. Explain how to elevate the flavor and presentation of fish dishes.
6. Explain how fish can be incorporated into the menu mix.
7. Summarize the use of fish at various foodservice venues.
8. Compare the two classifications of shellfish.
9. Identify common market forms of shellfish.
10. Describe common cooking techniques used to prepare healthy shellfish dishes.
11. Explain how to elevate the flavor and presentation of shellfish dishes.
12. Identify how shellfish can be incorporated throughout the menu mix.
13. Describe the use of shellfish at a variety of foodservice venues.

Digital Resources
ATPeResources.com/QuickLinks
Access Code: 267412

Vegetables & Legumes on the Menu

Vegetables

Tanimura & Antle®

Legumes

Courtesy of The National Pork Board

Combination

Courtesy of The National Pork Board

Vegetables and legumes make significant contributions to a healthy diet. Both provide complex carbohydrates, dietary fiber, and a variety of vitamins and minerals. Legumes are also a cholesterol-free source of protein. Vegetables and legumes can both be served as the main item in a dish or combined with other ingredients, such as whole grains or animal proteins, to create a nutritious, well-balanced menu item. The many varieties of vegetables and legumes can add unique colors, varied textures, and appealing flavors across all menu categories.

Chapter Objectives

1. Describe factors that influence the perceived value of vegetables.
2. Identify the classifications of vegetables.
3. Explain health benefits linked to the pigments found in vegetables.
4. Describe the cooking techniques commonly used with vegetables.
5. Identify ways to develop flavors when preparing vegetables.
6. Explain how vegetables can enhance plated presentations.
7. Explain how to use vegetables throughout the menu mix.
8. Describe ways to promote vegetables on the menu.
9. Describe factors that influence the perceived value of legumes.
10. Identify the classifications of legumes.
11. Explain how to prepare fresh, frozen, canned, and dried legumes.
12. Describe how to use legumes as a starch and a protein in plated presentations.
13. Explain how to use legumes throughout the menu mix.
14. Describe ways to promote legumes on the menu.

VEGETABLES

A *vegetable* is an edible root, tuber, bulb, stem, leaf, flower, or seed of a nonwoody plant. Vegetables are receiving heightened attention as chefs use more seasonal ingredients and as health-conscious consumers replace meats as the focus of the plate. Vegetables supply the body with nutrients such as dietary fiber and an abundant amount of vitamins and minerals, including antioxidants and phytonutrients. In addition to providing health benefits, vegetables are a cost-effective way to enhance the menu with a profusion of colors, textures, and flavors. **See Figure 12-1.**

Seasonal Dishes

Irinox USA

Figure 12-1. Seasonal vegetables are a cost-effective way to enhance the menu with a profusion of colors, textures, and flavors.

Perceived Value of Vegetables

When considering the perceived value of vegetables on the menu, it is important to account for the cost of raw ingredients and the price guests are willing to pay for different menu items. For example, jumbo carrots are much less expensive than heirloom carrots. If the carrots are going to be puréed into a soup, it is more economical to use jumbo carrots. However, a side dish consisting of glazed multicolored carrots may be more expensive but can also increase the perceived value and command a higher menu price.

Vegetables are also perceived as highly nutritious, and the demand for healthy meals has increased their value on the menu. For example, rich side dishes are being replaced with lighter vegetable dishes, such as seasonal tomatoes tossed with olive oil, balsamic vinegar, and basil. This trend has also carried over into children's menus, where potato chips and French fries are commonly being replaced with raw vegetables, such as celery sticks.

The heightened interest in the "farm to table" movement influences the perceived value of vegetables as well. For example, incorporating locally grown vegetables into the menu can positively impact perceived value as guests anticipate the fresh flavors of vegetables harvested at the peak of ripeness. In addition, local farms often grow unique varieties of vegetables such as white or purple asparagus that add intrigue to the menu and inspire guests to select menu items based on the desire to try something new.

An increasing number of individuals are choosing to follow a vegetarian or flexitarian diet. Vegetarian diets vary and can be classified by the types of foods consumed. **See Figure 12-2.** With the vast array of colorful, texturally diverse, and flavorful vegetables, chefs can create plant-based menu items that increase the perceived value of vegetables for both vegetarian and nonvegetarian guests. To create nutrient-dense, flavorful items using vegetables, it is essential to understand the various vegetable classifications.

Vegetable Classifications

Vegetables are a large class of foods because of their many differences. For example, some vegetables are large, fibrous, and dense, while others are small, tender, and delicate. The common classifications for vegetables include roots, tubers, bulbs, stems, leaves, flowers, and seeds as well as fruit-vegetables and sea vegetables. **See Figure 12-3.** Although not a vegetable, fungi such as mushrooms are generally prepared in the same manner as vegetables.

Vegetarian Classifications

Diet Type	Foods Consumed	Foods Not Consumed
Vegan	• Plant-based foods	• Animal-based foods (meats, poultry, seafood, eggs, milk and dairy products, honey)
Lacto-vegetarian	• Plant-based foods • Milk and dairy products	• Meats • Poultry • Seafood • Eggs
Lacto-ovo vegetarian	• Plant-based foods • Milk and dairy products • Eggs	• Meats • Poultry • Seafood
Flexitarian	• Plant-based foods • Animal-based foods in limited quantity and/or frequency	

Figure 12-2. Vegetarian diets vary and can be classified by the types of foods consumed.

Vegetable Classifications

Type	Common Examples			Common Characteristics
Root vegetables	• Carrots • Parsnips • Salsify • Radishes	• Turnips • Rutabagas • Beets • Celeriac	• Jicamas • Lotus roots • Bamboo shoots	• Grows underground • Leaves extend above ground • Hard, dense texture
Tubers	• Potatoes • Sweet potatoes	• Yams • Ocas	• Sunchokes • Water chestnuts	• Grows underground • Buds produce new plants
Bulb vegetables	• Garlic • Shallots	• Onions • Scallions	• Leeks	• Grows underground • Buds enclosed in overlapping membranes/leaves • Strong flavor and aroma
Stem vegetables	• Asparagus • Celery	• Fennel • Rhubarb	• Kohlrabi • Hearts of palm	• Edible bulbs and shoots • Harvested when young and tender
Leaf vegetables	• Cabbage • Bok choy • Brussels sprouts • Lettuces • Chicory • Watercress	• Spinach • Sorrel • Chard • Kale • Collards • Mustard greens	• Turnip greens • Beet greens • Dandelion greens • Nopales • Tatsoi • Fiddlehead ferns	• Edible leaves • Leafstalks and shoots often edible • Mild flavor becomes bitter with age
Flower vegetables	• Squash blossoms • Broccoli	• Cauliflower • Artichokes		• Edible flowers of nonwoody plants
Seed vegetables	• Beans • Peas	• Pulses • Lentils		• Edible seeds of nonwoody plants • High in protein and starch
Fruit-vegetables	• Tomatoes • Tomatillos • Cucumbers • Sweet peppers	• Hot peppers • Eggplants • Avocados • Okra	• Sweet corn • Summer squashes • Winter squashes • Pumpkins	• Botanical fruits prepared as vegetables
Sea vegetables	• Nori • Kombu	• Arame • Wakame	• Dulse • Hijiki	• Edible saltwater plants • Naturally salty taste

Figure 12-3. The common classifications for vegetables include roots, tubers, bulbs, stems, leaves, flowers, and seeds as well as fruit-vegetables and sea vegetables.

Root Vegetables. A *root vegetable* is an earthy-flavored vegetable that grows underground and has leaves that extend above ground. Root vegetables include carrots, parsnips, salsify, radishes, turnips, rutabagas, beets, celeriac, jicamas, lotus roots, and bamboo shoots. There are also some edible roots that are not classified as root vegetables, such as ginger and horseradish.

Tubers. A *tuber* is a short, fleshy vegetable that grows underground and bears buds capable of producing new plants. Examples of tubers include potatoes, sweet potatoes, yams, ocas, sunchokes, and water chestnuts. While some tubers such as ocas, sunchokes, and water chestnuts can be served raw, others such as potatoes, sweet potatoes, and yams require cooking. Both roots and tubers are commonly used in global cuisines. **See Figure 12-4.**

Roots and Tubers in Global Cuisines

Vegetable	Common Dishes	Origin of Dish
Carrots	• Carrots Vichy • Shredded Carrot Salad • Carrot Soup • Glazed Carrots	• France • Morocco • Turkey • Southern United States
Sweet potatoes	• Sweet Potato Stew • Boiled Sweet Potatoes • Coconut-Curry Sweet Potatoes • Sweet Potato Pie	• Caribbean • Africa • India • New England
Celeriac	• Celeriac with Horseradish • Steamed Celeriac • Celeriac Salad • Roasted Celeriac	• Great Britain • Spain • France • Pacific Northwest

Figure 12-4. Roots and tubers are frequently used in global cuisines.

Bulb Vegetables. A *bulb vegetable* is a strongly flavored vegetable that grows underground and consists of a short stem base with one or more buds that are enclosed in overlapping membranes or leaves. Bulb vegetables are members of the lily family and include garlic, shallots, onions, scallions, and leeks. **See Figure 12-5.** Bulb vegetables are often sautéed or roasted and used as building blocks of flavor as well as for their aromatic qualities.

Bulb Vegetables

Garlic

National Onion Association

Yellow Onions

Figure 12-5. Bulb vegetables are members of the lily family and include garlic and yellow onions.

Stem Vegetables. A *stem vegetable* is the main trunk of a plant that develops edible buds and shoots instead of roots. Stem vegetables include asparagus, celery, fennel, rhubarb, kohlrabi, and hearts of palm. Stem vegetables are most often harvested when they are young and tender and require only a minimal amount of cooking.

Leaf Vegetables. *Leaf vegetables,* also known as greens, are edible plant leaves that are often accompanied by edible stems and shoots. Cabbages, bok choy, Brussels sprouts, lettuces, chicory, watercress, spinach, chard, kale, and mustard greens are classified as leaf vegetables. Plentiful in dietary fiber, vitamins, minerals, and antioxidants, leaf vegetables are regarded as a highly nutrient-dense food. Leaf vegetables have a mild flavor when they are young, but their flavor becomes bitter with age. Leaf vegetables are commonly sautéed, roasted, steamed, or braised. Some leaf vegetables, such as Romaine lettuce and bok choy, can also be grilled to elevate flavors and presentation. **See Figure 12-6.**

INGREDIENT SPOTLIGHT: Rainbow Chard

National Garden Bureau Inc.

Nutrition Facts
Serving Size 1 cup, simmered (175 g)

Amount Per Serving	
Calories 35	Calories from Fat 1
	% Daily Value*
Total Fat 0g	0%
Saturated Fat	0%
Trans Fat	
Cholesterol 0mg	0%
Sodium 313mg	13%
Total Carbohydrate 7g	2%
Dietary Fiber 4g	15%
Sugars 2g	
Protein 3g	
Vitamin A 214% • Vitamin C	53%
Calcium 10% • Iron	22%

* Percent Daily Values (DV) are based on a 2,000 calorie diet.

Unique Features

- Deep green leaves with stems and veins in vibrant shades of white, pink, red, orange, and yellow
- Contains a variety of phytonutrients believed to have anti-inflammatory and antioxidant properties
- Low in calories and total fat
- Source of dietary fiber, riboflavin, vitamin B_6, vitamin A, vitamin C, vitamin E, vitamin K, calcium, magnesium, potassium, copper, and manganese

Menu Applications

- Cook to reduce bitterness and create a sweeter flavor
- Add stalks and leaves to a variety of menu items, such as omelets, stir-fries, pastas, and grain dishes
- Use as a substitute for dark leafy greens

Grilled Romaine Lettuce

Tanimura & Antle®

Figure 12-6. Some leaf vegetables, such as Romaine lettuce, can be grilled to elevate flavors and presentation.

Flower Vegetables. A *flower vegetable* is the edible flower of a nonwoody plant that is prepared as a vegetable. Examples of flower vegetables include cauliflower, broccoli, squash blossoms, and artichokes. Flower vegetables are available in hues ranging from white and green to orange and purple, often making plated presentations more vibrant. **See Figure 12-7.**

Seed Vegetables. A *seed vegetable* is the edible seed of a nonwoody plant. Legumes such as beans, peas, pulses, and lentils are also classified as seed vegetables. Seed vegetables have certain characteristics that set them apart from other vegetables. For example, legumes have a higher protein content than other vegetables and therefore are often considered a protein source as well as a starch in menu items.

Figure 12-7. Flower vegetables are available in hues ranging from white and green to orange and purple, often making plated presentations more vibrant.

Fruit-Vegetables. A *fruit-vegetable* is a botanical fruit that is sold, prepared, and served as a vegetable. Included in this classification are tomatoes, tomatillos, cucumbers, sweet peppers, hot peppers, eggplants, okra, sweet corn, summer squashes, winter squashes, and pumpkins. **See Figure 12-8.** Fruit-vegetables come in a multitude of varieties and can be prepared in many different ways. For example, summer squash such as zucchini can be served raw on a crudité platter, grilled, sautéed, roasted, or steamed.

Figure 12-8. A fruit-vegetable, such as a winter squash, is a botanical fruit that is sold, prepared, and served as a vegetable.

Sea Vegetables. A *sea vegetable* is an edible saltwater plant that contains high amounts of dietary fiber, vitamins, and minerals. Sea vegetables have a naturally salty taste because of the minerals they absorb from the ocean. Therefore, salt is typically not added to dishes containing sea vegetables. Common varieties of sea vegetables include nori, kombu, arame, wakame, dulse, and hijiki. **See Figure 12-9.** Sea vegetables are available in both fresh and dried forms and are typically served raw, pickled, sautéed, or steamed. Agar agar is a sea vegetable containing alginic acid, which is used as a thickening agent in ice creams, puddings, and pie fillings and as a vegetarian substitute for gelatin.

Figure 12-9. Common varieties of sea vegetables include nori, kombu, and wakame.

Fungi. Mushrooms belong to a classification known as fungi. While mushrooms are not vegetables, they are used in the same manner. A *mushroom* is the fleshy, spore-bearing body of an edible fungus that grows above the ground. **See Figure 12-10.** Mushrooms commonly used in foodservice operations include button mushrooms, criminis, portobellos, enokitakes, wood ears, shiitakes, oyster mushrooms, chanterelles, morels, porcinis, blue foot mushrooms, and truffles. The flavors of mushrooms range from very mild to intense and earthy.

Fresh Mushrooms

Mushroom Council

Figure 12-10. A mushroom is the fleshy, spore-bearing body of an edible fungus that grows above the ground.

Market Forms of Vegetables

The selection, storage, and preparation of vegetables vary depending on the form in which they are purchased. Factors such as seasonality, cost, skill level of the staff, and time available for tasks also play a role in how vegetables are purchased and used on the menu. Vegetables can be purchased fresh, frozen, canned, or dried.

Fresh Vegetables. Most fresh vegetables should be refrigerated between 34°F and 40°F. The length of storage varies according to the vegetable, but most fresh vegetables soften, wilt, and discolor with prolonged storage. There is also a loss of vitamins and minerals with prolonged storage as well as a change in flavor as the sugars in the vegetables turn to starch. Certain vegetables, such as potatoes, onions, winter squashes, and tomatoes, require dry storage in a cool, dark place at a temperature between 55°F and 60°F.

Frozen Vegetables. Frozen vegetables retain nutrients because they are picked at the peak of ripeness, flash frozen, and packaged. A large variety of frozen vegetables are available. They offer convenience because many are precut before being frozen and packaged. **See Figure 12-11.** Some frozen vegetables are also precooked. However, the texture of frozen vegetables is often compromised due to the freezing process, which breaks down cell walls and causes the vegetables to become softer. Frozen vegetables are generally prepared by sautéing, baking, or steaming.

Frozen Vegetables

Figure 12-11. Vegetables are often precut before they are frozen and packaged.

Canned Vegetables. *Canning* is a process of preserving foods in an airtight container to prevent spoilage. Canned vegetables offer an advantage over other market forms because they have an extensive shelf life. **See Figure 12-12.** However, canned vegetables lack more nutrients than other forms of vegetables because they are packaged in liquid, which facilitates the loss of water-soluble nutrients. The packaging liquid usually contains added sodium as well, but low-sodium varieties are available.

Canned Vegetables

Sullivan University

Figure 12-12. Canned vegetables offer an advantage over other market forms because they have an extensive shelf life.

Dried Vegetables. Drying preserves foods by removing enough moisture from the food to prevent decay and spoilage. **See Figure 12-13.** Vegetables are dried as quickly as possible in order to minimally affect the color, texture, and flavor. However, drying can result in the loss of water-soluble vitamins. Depending on the recipe and use, dried vegetables such as mushrooms must be reconstituted in liquid and then squeezed to remove the excess liquid before they are used. Dried vegetables should be stored in a cool, dark, dry place and used within a year.

Dried Porcini Mushrooms

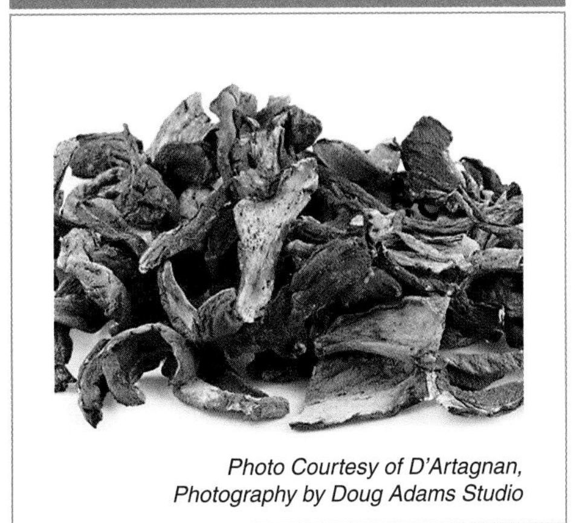

Photo Courtesy of D'Artagnan, Photography by Doug Adams Studio

Figure 12-13. Drying preserves foods by removing enough moisture from the food to prevent decay and spoilage.

Knowledge Check 12-1

1. Explain how the "farm to table" movement influences the perceived value of vegetables.
2. List the nine classifications of vegetables.
3. Differentiate between root vegetables, tubers, and bulb vegetables.
4. Describe stem vegetables.
5. Explain why leaf vegetables are considered nutrient dense.
6. Identify a characteristic of seed vegetables that sets them apart from other vegetables.
7. Give six examples of fruit-vegetables.
8. Explain why salt is not typically added to dishes containing sea vegetables.
9. Define mushroom.
10. Identify the storage requirements for fresh vegetables.
11. Differentiate between frozen, canned, and dried vegetables.

FOOD SCIENCE EXPERIMENT: Evaluating Different Market Forms of Vegetables 12-1

Objective

- To compare dried, fresh, frozen, and canned mushrooms in terms of color, texture, and flavor.

Materials and Equipment

- Masking tape
- Marker
- 4 small bowls
- Dried white button mushrooms, rehydrated
- Fresh white button mushrooms, washed
- Frozen white button mushrooms, thawed
- Canned white button mushrooms, drained
- Tasting spoons

Procedure

1. Use the masking tape and marker to label one bowl "Dried Mushrooms," the second bowl "Fresh Mushrooms," the third bowl "Frozen Mushrooms," and the fourth bowl "Canned Mushrooms."
2. Place the appropriate forms of mushrooms in their corresponding bowls.
3. Evaluate the color, texture, and flavor of each bowl of mushrooms.

Typical Results

There will be variances in the color, texture, and flavor of each bowl of mushrooms.

PREPARING NUTRITIOUS VEGETABLES

Most vegetables benefit from minimal preparation and can be served raw or cooked until just tender so that their nutrients are retained and their natural colors, textures, and flavors can shine. When vegetables are overcooked, have prolonged contact with water, or are held improperly, their cellular structure breaks down. This results in a loss of nutrients and color and a soft or mushy texture. Also, health benefits have been linked to specific pigments within vegetables.

- Red vegetables, such as beets, red cabbage, red potatoes, radishes, rhubarb, and tomatoes, obtain their color from lycopene or anthocyanins. Tomatoes contain a high amount of lycopene, which has been found to reduce several types of cancer, especially prostate cancer. Unlike most nutrients, lycopene is better absorbed by the body when foods are cooked. Anthocyanins contain antioxidants that protect cells from damage caused by free radicals.
- Orange and yellow vegetables, including butternut squash, carrots, pumpkin, rutabagas, sweet corn, sweet potatoes, yellow peppers, and yellow summer and winter squashes, are colored by beta-carotene, which converts to vitamin A in the body. Scientists have reported that foods rich in beta-carotene help reduce the risk of cancer and heart disease and can improve the immune system.
- Green vegetables contain the pigment chlorophyll. Vegetables rich in chlorophyll, such as broccoli, leafy greens, and asparagus, are also rich in phytonutrients and B vitamins that may help protect the body against some forms of cancer.

- Blue and purple vegetables include eggplants, purple asparagus, purple cabbage, purple carrots, and purple potatoes. Bluish-purple vegetables contain anthocyanins, which act as powerful antioxidants.
- White vegetables such as garlic, cauliflower, onions, and mushrooms contain anthoxanthin. Allicin is a chemical found in anthoxanthin that may help lower cholesterol and blood pressure and may also reduce the risk of stomach cancer and heart disease.

Raw Vegetables

Raw vegetables typically have a crisp texture. Raw vegetables are commonly served as crudités with dips or incorporated into salads. **See Figure 12-14.** The texture and flavor of raw vegetables can also enliven cooked dishes. For example, a blend of arugula and julienned jicama can be served with braised meats to provide an appealing contrast of colors, textures, and flavors. A wide variety of raw vegetables, including lettuces, tomatoes, onions, cucumbers, carrots, and peppers, are also used effectively in sandwiches to help layer textures and flavors.

Frieda's Specialty Produce

There are many eggplant varieties, with colors ranging from white to deep purple.

Cooking Vegetables

Fresh vegetables are often regarded as more flavorful and nutrient dense than cooked vegetables. In order to maintain their flavor and health benefits, it is important to follow proper cooking techniques. Techniques used to cook fresh vegetables can be broken down into two main categories: fast cooking techniques and slow cooking techniques. **See Figure 12-15.** Fast cooking techniques include blanching, steaming, sautéing, stir-frying, and grilling. Slow cooking techniques include simmering, roasting, braising, and stewing. Regardless of the cooking technique, tender vegetables such as spinach and sprouts will cook more quickly than more fibrous vegetables such as collards and artichokes.

Crudités

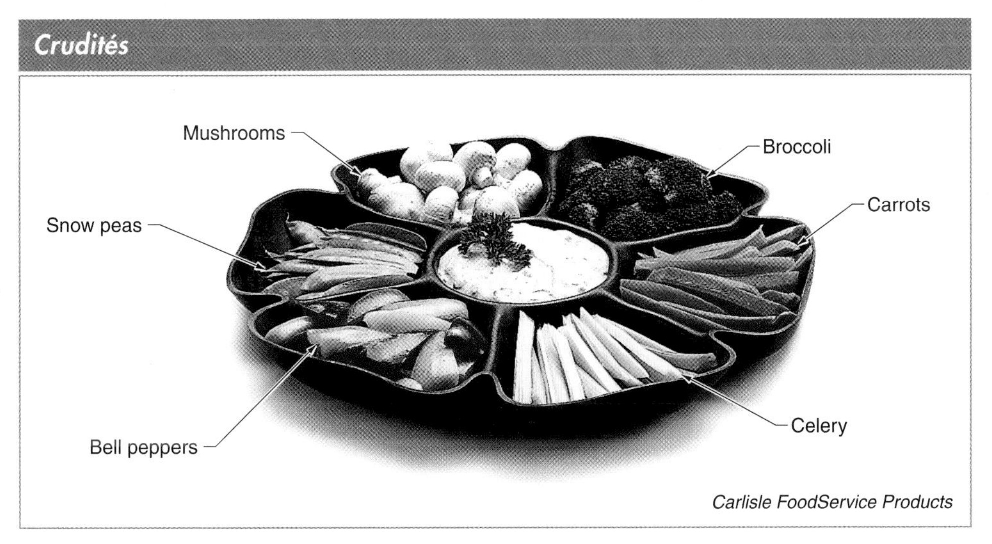

Carlisle FoodService Products

Figure 12-14. Raw vegetables are commonly served as crudités with dips.

Cooking Vegetables

Fast Cooking Techniques	Vegetables Commonly Used
Blanching	Asparagus, broccoli, carrots, and cauliflower
Steaming	Broccoli, Brussels sprouts, carrots, and cauliflower
Sautéing	Carrots, celery, dark leafy greens, garlic, onions, summer squash, and tomatoes
Stir-frying	Bell peppers, bok choy, broccoli, mushrooms, and snap peas
Slow Cooking Techniques	**Vegetables Commonly Used**
Simmering	Cabbage, potatoes, and sweet potatoes
Roasting	Asparagus, beets, Brussels sprouts, carrots, cauliflower, garlic, mushrooms, onions, potatoes, and winter squash
Braising/ Stewing	Carrots, celery, eggplant, onions, tomatoes, and winter squash

Figure 12-15. Techniques used to cook fresh vegetables can be broken down into two main categories: fast cooking techniques and slow cooking techniques.

When blanching vegetables, it is important to start with rapidly boiling water to quickly cook the vegetables, and then immediately shock them in ice water to stop the cooking process. Blanching, steaming, sautéing, stir-frying, and grilling are effective techniques for cooking vegetables until they reach the crisp-tender stage, thus preserving nutrients.

Although most vegetables can be cooked using fast cooking techniques, some root vegetables, tubers, and winter squashes are commonly simmered or roasted. When simmering these types of vegetables, starting them in cold water and slowly heating the water to a simmer allows the fibers in the vegetables to soften more evenly and create a consistent texture in the final product.

Roasting is a cooking technique that can be used successfully with a wide variety of vegetables, including carrots, beets, parsnips, sweet potatoes, cauliflower, tomatoes, eggplant, and all varieties of squash. Not only does roasting soften the fibers in vegetables gradually and evenly, it caramelizes their natural sugars and intensifies their colors and flavors.

A variety of leafy vegetables, such as cabbage, turnips, and collards, can be successfully braised or stewed. To retain nutrients, it is important to serve braised and stewed vegetables with their cooking liquid.

Irinox USA

Ratatouille is a flavorful vegetable stew consisting of bell peppers, eggplants, onions, tomatoes, zucchini, garlic, and herbs.

Flavor Development

With the assortment of vegetables available, texture and flavor development can be used in limitless ways to enliven dishes. For example, raw radishes add crunch and a fresh, spicy flavor to dishes, while roasted and puréed butternut squash adds a creamy and sweet, earthy flavor. A raw onion can be used to add a crunchy texture and a pungent burst of flavor to Mexican tortilla soup, whereas French onion soup highlights the brown color, soft texture, and sweet flavor of caramelized onions.

One of the most effective cooking techniques used to caramelize vegetables and develop flavors is roasting. Roasting vegetables slowly draws out the natural sugars and results in a sweeter vegetable with a softened texture. **See Figure 12-16.**

Roasted Potatoes

United States Potato Board

Figure 12-16. Roasting vegetables slowly draws out the natural sugars and results in a sweeter vegetable with a softened texture.

Blanching vegetables is an effective way to prevent overcooking and retain nutrients, but it yields limited flavor. To elevate the flavor of blanched vegetables, they can be reheated for service by sautéing, grilling, or roasting. The reheating process enhances the flavors of the vegetables.

The classic combination of carrots, celery, and onions known as mirepoix is often used to develop flavors in items such as stocks, broths, soups, and stews. Mirepoix can be sweated or roasted depending on the level of color and flavor development desired.

Plating Vegetables

Vegetables can be used creatively in plated presentations to highlight their many colors, shapes, textures, and flavors. **See Figure 12-17.** When plating vegetables, it is important to think about the final presentation of a dish before the preparation begins. Vegetables can be prepared in numerous ways to add visual contrast to the overall presentation. For example, potatoes can be diced, sautéed, and seasoned, or they can be tournéed, simmered, and dressed with herbs.

Plating Vegetables

Daniel NYC/E. Laignel

Figure 12-17. Vegetables can be used creatively in plated presentations to highlight their many colors, shapes, textures, and flavors.

Knowledge Check 12-2

1. Describe the health benefits linked to the pigments found in vegetables.
2. Give two examples of dishes that use raw vegetables.
3. List the fast and slow cooking techniques used to prepare vegetables.
4. Describe the procedure for blanching vegetables.
5. Explain why roasting is an effective way to develop the flavors of vegetables.

CULINARY NUTRITION RECIPE-MODIFICATION PROCESS

Braised Swiss Chard

Braised Swiss chard is a classic side dish that provides dietary fiber, vitamins A and C, calcium, and iron. Both the stems and leaf portions of the chard are chopped and cooked with smoked pork, bacon, or smoked ham hocks until very tender. In this recipe, bacon adds an appealing smoky flavor but also contributes fat and sodium to the dish.

Yield: 4 servings, ½ cup each

Ingredients

2 lb	Swiss chard
1 tbsp	canola oil
⅓ c	bacon, diced
½ c	yellow onion, small dice
1 tsp	garlic, minced
3 c	water
2 tsp	kosher salt
1 tsp	red pepper flakes
TT	white vinegar

Preparation

1. Wash Swiss chard under running water.
2. Separate the leaves from the stems. Tear the leaves into large pieces and dice the stems.
3. Heat oil in a saucepan and add bacon. Cook until crispy.
4. Add onion and cook until translucent (approximately 5 minutes).
5. Add the leaves, stems, and garlic, stirring frequently until chard begins to wilt.
6. Add water to the saucepan and bring to a boil.
7. Reduce heat and cover the saucepan. Simmer until chard is tender (approximately 25 minutes).
8. Season with salt, red pepper, and white vinegar.

Chef's Tip: *A variety of different greens, such as mustard greens, turnip greens, kale, and collard greens, can be used for braising. The cooking times will vary depending on the type of green.*

Evaluate original recipe for sensory and nutritional qualities

- Braised Swiss chard has a smoky flavor and a slightly mushy texture.
- The dish has a moderate fat content and is high in sodium.

Establish goals for recipe modifications

- Reduce total fat and sodium content.
- Increase flavor and nutrients.

Identify modifications or substitutions

- Use a substitution for the bacon.
- Use a substitution for the white vinegar and add nutrient-dense ingredients.
- Reduce the amount of water and cooking time.

Determine functions of identified modifications or substitutions

- Bacon provides a smoky flavor.
- White vinegar provides flavor.
- Water is the cooking medium.

Select appropriate modifications or substitutions

- Substitute a smaller amount of smoked turkey sausage for the bacon.
- Substitute balsamic vinegar for the white vinegar and add golden raisins.
- Reduce water from 3 cups to 1½ cups and cook until chard is tender.

Test modified recipe to evaluate sensory and nutritional qualities

- Substituting smoked turkey for the bacon reduces calories, fat, and sodium yet maintains the smoky flavor.
- Using balsamic vinegar adds depth of flavor, and using golden raisins provides color, texture, flavor, and additional nutrients.
- Reducing water and cooking time preserves nutrients and texture.

Modified Braised Swiss Chard

To keep the appealing smoky flavor of the original recipe, smoked turkey sausage is used instead of bacon. This substitution also reduces fat and, in conjunction with using less salt, helps lower the sodium content. The addition of sweet golden raisins increases nutrient density while providing a contrasting color, texture, and flavor. Reducing the water and cooking time helps preserve the color, texture, and nutrients.

Yield: 4 servings, ½ cup each

Ingredients

2 lb	Swiss chard
1 tbsp	canola oil
½ c	yellow onion, small dice
1 tsp	garlic, minced
¼ c	smoked turkey sausage, diced
¼ c	golden raisins
1½ c	water
1 tsp	kosher salt
1 tsp	red pepper flakes
TT	balsamic vinegar

Frieda's Specialty Produce

Preparation

1. Wash Swiss chard under running water.
2. Separate the leaves from the stems.
3. Tear the leaves into large pieces and dice the stems.
4. Heat oil in a saucepan and add onions. Cook until translucent (approximately 5 minutes).
5. Add garlic, sausage, and stems. Cook 2 minutes, stirring occasionally.
6. Add raisins and three-fourths of the water to the saucepan. Bring to a simmer and cover.
7. Simmer until stems are soft (approximately 3 minutes).
8. Add the leaves, salt, and pepper flakes.
9. Add the rest of the water and simmer, partially covered, until leaves are tender (approximately 3 minutes). Stir occasionally.
10. Season with balsamic vinegar.

Chef's Tip: *For different flavor profiles, experiment with various types of sausages, such as fennel and olive chicken sausage.*

Swiss Chard Nutritional Comparison

Nutrition Facts	Original	Modified
Calories	127.2	125.8
Total Fat	8.3 g	4.9 g
Sodium	1506.5 mg	808.6 mg
Dietary Fiber	3.9 g	4.3 g
Vitamin A	13,875.0 IU	13,871.5 IU
Vitamin C	69.5 mg	69.8 mg
Calcium	125.5 mg	130.4 mg
Iron	4.2 mg	4.4 mg

CULINARY NUTRITION RECIPE-MODIFICATION PROCESS

Signature Marinated Swiss Chard Salad

A common belief is that dark leafy greens like Swiss chard require long cooking times to become tender and palatable. However, marinating Swiss chard in a slightly acidic vinaigrette softens the greens and enhances flavor. In the original and modified recipes, water-soluble vitamin C leached into the cooking liquid, but it is retained when Swiss chard is served raw. The lemon juice in the vinaigrette used in this recipe allows less salt to be used, reducing sodium. The addition of fennel, currants, and walnuts provides more nutrients, colors, textures, and layers of flavor.

Yield: 4 servings, ½ cup each

Ingredients

2 lb	Swiss chard
½ c	fennel, thinly shaved
1 tbsp	currants
1 tbsp	olive oil
1 tbsp	lemon juice
½ tsp	kosher salt
½ tsp	black pepper, ground
1 tbsp	walnut pieces

Preparation

1. Wash Swiss chard under running water.
2. Separate the leaves from the stems. Tear the leaves into large pieces and dice the stems.
3. Combine the leaves, stems, fennel, and currants in a large bowl.
4. Whisk together the olive oil, lemon juice, salt, and pepper in a separate bowl.
5. Pour the vinaigrette over the salad ingredients and toss to combine.
6. Marinate for 20 minutes.
7. Garnish with walnuts before service.

Chef's Tip: *To add even more lemon flavor, roast lemons at 350°F for 30 minutes. Press the solids through a fine mesh strainer and use the roasted lemon juice for the vinaigrette.*

Swiss Chard Nutritional Comparison

Nutrition Facts	Original	Modified	Signature
Calories	127.2	125.8	96.6
Total Fat	8.3 g	4.9 g	5.1 g
Sodium	1506.5 mg	808.6 mg	724.2 mg
Dietary Fiber	3.9 g	4.3 g	4.3 g
Vitamin A	13,875.0 IU	13,871.5 IU	13,889.2 IU
Vitamin C	69.5 mg	69.8 mg	71.3 mg
Calcium	125.5 mg	130.4 mg	126.2 mg
Iron	4.2 mg	4.4 mg	4.4 mg

VEGETABLE MENU MIX

Vegetables are an integral component of the menu mix because they can serve as the main item of a dish or accent other menu items. Vegetables also provide essential nutrients and enhance the dining experience with contrasting colors, textures, and flavors. For example, the flavors of meats and grains are fairly consistent, but the flavors provided by vegetables vary due to the different preparation methods and combinations. Global flavor profiles can provide inspiration for creating interesting vegetable dishes. **See Figure 12-18.**

Vegetables can be served in every course. For example, vegetables are often used in breakfast items such as breakfast breads and omelets. During lunch and dinner, there are limitless uses of vegetables in salads, soups, sandwiches, and entrées. This provides an opportunity to create enticing menu descriptions that highlight vegetables and healthy preparation techniques. **See Figure 12-19.** It is important to choose words carefully when describing vegetables because some words have more appeal than others. For example, while simmering may be the technique used to prepare a vegetable, there are better ways to describe the flavor. For example, "simmered potatoes" could be effectively described as "tender new potatoes seasoned with rosemary."

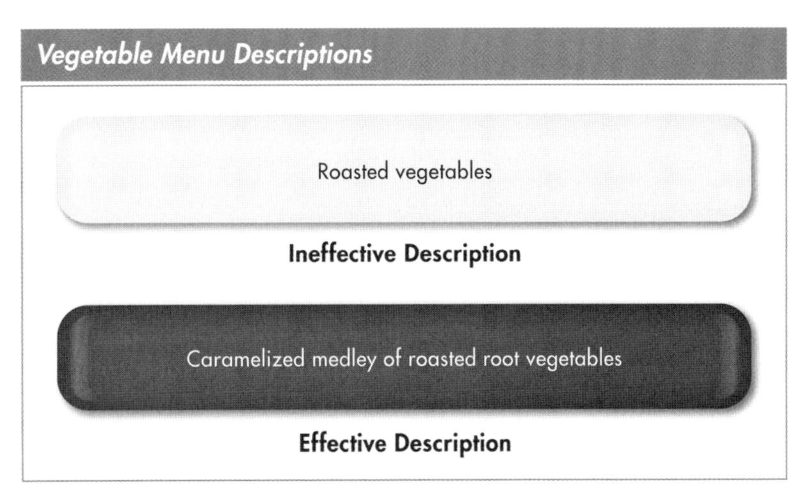

Figure 12-19. Effective menu descriptions sound enticing and emphasize healthy preparation techniques.

Vegetables in Beverages

Vegetables can be served in nutritious and appealing drinks such as juices and smoothies. For example, leafy greens such as spinach add a pleasing taste and vibrant color to fruit-based smoothies. Tomatoes, cucumbers, garlic, and seasonings can be blended to create a gazpacho-inspired beverage. Sea vegetable powders, such as spirulina, can be added to beverages to boost nutrient content. Beverages can also be garnished with vegetables. For example, asparagus spears, ribbons of cucumber, stalks of celery, and sun-dried tomatoes can elevate the presentation and increase the flavor of beverages.

Vegetables on the Menu

Menu Item	Description
Coleslaw	Eastern European salad of shredded cabbage and carrots dressed with vinaigrette
Kinpira vegetables	Japanese preparation that combines sautéed and simmered vegetables with ingredients such as mirin, sake, and soy sauce
Moussaka	Mediterranean casserole of eggplant, potatoes, and ground lamb
Ratatouille	French preparation of stewed garden vegetables

Figure 12-18. Global flavor profiles can provide inspiration for creating interesting vegetable dishes.

Vegetables in Appetizers

Vegetables can make excellent additions to appetizers or serve as the main ingredient. They can also be prepared in infinite ways to create nutritious appetizers. For example, grilled bruschetta can highlight the colorful and succulent flavors of red and yellow tomatoes, a tartlet can be filled with the earthy flavors of sautéed mushrooms, and a zesty dip for a crudité platter can contain roasted red peppers. Vegetables such as onions, artichokes, zucchini, and squash blossoms can be stuffed with various fillings to create enticing appetizers. **See Figure 12-20.**

Vegetables in Appetizers

Tanimura & Antle®

Figure 12-20. Vegetables such as onions can be stuffed with various fillings to create enticing appetizers.

Vegetables in Soups

Vegetables provide the foundation for many soups and can be used to effectively develop flavors. For example, tomatoes provide the basis for traditional tomato soup, but fire-roasting the tomatoes and topping the soup with caramelized shallots can elevate the flavor. In addition to providing color, texture, and flavor to soups, vegetables also provide guests with nutrient-dense options and flavor profiles from all regions of the globe. **See Figure 12-21.** Soups featuring vegetables can be served hot or cold, smooth or chunky, and cooked or raw.

Courtesy of The National Pork Board

Vegetables add color, texture, flavor, and nutrients to soups.

Vegetables in Sandwiches

Many sandwiches include vegetables such as lettuce, tomatoes, and onions as a garnish. However, vegetables can also be featured as the main ingredient in meatless sandwiches, such as a whole wheat tortilla filled with grilled or roasted vegetables. To create sandwiches with vegetables as the main ingredient, it is important to consider flavor profiles that work well in other vegetable dishes and apply those profiles to the sandwiches. For example, tomatoes, mozzarella, basil, and balsamic vinaigrette can be served as a panini, and a grilled portobello mushroom stuffed with sautéed spinach and feta cheese can be served on a whole grain bun. Sandwiches featuring lettuce wraps are an ideal way to reduce calories and serve gluten-free menu options. **See Figure 12-22.**

Vegetables Used in Soups Around the World

Soup	Origin	Key Ingredients
Borscht	Ukraine	Beets, onions, beef stock or water, red-wine vinegar, dill, sugar, sour cream, and optional vegetables such as cabbage, beans, and celery root
Caldo verde	Portugal	Mashed potatoes, minced collards, savoy cabbage, kale, onions, and chorizo
Callaloo	Trinidad and Tobago	Callaloo leaves or spinach, okra, crab meat, chicken stock, onions, thyme, chile pepper, and salt beef
Fanesca	Ecuador	Figleaf gourd, pumpkin, twelve grains, and salt cod
Gazpacho	Spain	Tomatoes, peppers, cucumbers, garlic, oil, and vinegar
Ginataan	Philippines	Coconut milk, rice flour, jackfruit, yams, taro root, saba, sugar, and water
Minestrone	Italy	Beans, onions, celery, carrots, stock, tomatoes, and optional pasta
Pozole	Colombia	Hominy, pork, chiles, cabbage, oregano, cilantro, avocados, radishes, and lime juice
Shchi	Russia	Cabbage, beef brisket, and a variety of root vegetables
Tarator	Bulgaria	Yogurt, cucumbers, garlic, nuts, dill, oil, and water
Trahana	Turkey	Cracked wheat, yogurt, and dried fermented vegetables

Figure 12-21. In addition to providing color, texture, and flavor to soups, vegetables also provide guests with nutrient-dense options and flavor profiles from all regions of the globe.

Vegetables in Sandwiches

Courtesy of The National Pork Board

Figure 12-22. Lettuce wraps are an ideal way to reduce calories and serve gluten-free menu options.

Vegetables in Salads

Many salads feature leafy greens paired with countless combinations of other vegetables, fruits, cheeses, nuts, seeds, meats, poultry, and seafood. Radicchio and Bibb lettuce are often used to hold salad ingredients and increase visual appeal. Vegetables can be served raw in salads to showcase their natural textures and flavors, or they can be cooked using techniques such as grilling and sautéing to create an appealing contrast in temperature and more developed flavors. The opportunities to use vegetables in salads are vast, allowing chefs to showcase their creativity and imagination in innovative ways. **See Figure 12-23.**

Chef's Tip

Caramelizing vegetables such as beets creates a warm, creamy texture and sweet flavor that complement crisp, bitter salad greens.

Vegetables in Salads

Daniel NYC/M. Price

Figure 12-23. Vegetables can be showcased in salads in creative and innovative ways.

Vegetables in Entrées

Whether vegetables are featured as the center of the plate or used to accent the main item, they increase flavor and nutrient density as well as elevate the presentation. For example, pasta dishes can incorporate an array of vegetables, such as grilled eggplant, Swiss chard, or pumpkin, to create nutrient-dense menu items. In addition, vegetables can be featured in sauces such as salsas and relishes, adding both visual interest and a layer of flavor to other dishes.

Vegetables in Sides

Although vegetables are becoming more popular as entrées, side dishes are where vegetables have traditionally been emphasized. Vegetable sides can include medleys of carrots, cauliflower, and broccoli as well as elevated presentations such as grilled eggplant with roasted tomato purée and toasted pine nuts.

Modifying vegetable side dishes high in fats and calories creates an opportunity to incorporate healthy cooking techniques. For example, baked French fries provide guests with a satisfying, nutrient-dense alternative to traditional deep-fried varieties, which are higher in fat and calories. Baked sweet potato fries offer even more nutrients, color, and flavor without added fat and calories. **See Figure 12-24.**

Vegetables as Sides

Photo Courtesy of McCain Foods USA

Figure 12-24. Baked sweet potato fries are a nutrient-dense alternative to traditional French fries.

Vegetables in Desserts

Vegetables are often featured in dessert items such as sweet potato pie, pumpkin pie, and carrot cake. **See Figure 12-25.** Vegetables such as roasted squash, beets, parsnips, and legumes can also be puréed and added to dessert items to provide moisture, replace fat, and enhance flavor. For example, corn, sweet potatoes, and pumpkins can produce appetizing custards and ice creams. Cucumber or tomato sorbets provide guests with a refreshing dessert that is not overly sweet. Although using vegetables in desserts adds nutritive value, it is still important to serve appropriate portion sizes.

Knowledge Check 12-3

1. Explain how vegetables can enhance beverages and appetizers.
2. Describe different ways to use vegetables in soups and salads.
3. Explain how to develop sandwiches that use vegetables as the main ingredient.
4. Identify three ways to effectively use vegetables in entrées.
5. Give an example of how to modify a vegetable side dish so it is more nutrient dense.
6. Describe different ways to use vegetables in desserts.

Vegetables in Desserts

National Honey Board

Figure 12-25. Carrot cake is a traditional dessert that highlights a vegetable.

PROMOTING VEGETABLES ON THE MENU

Promoting vegetables on all menus is important due to the increased demand for healthier options. Menu items can showcase seasonal vegetables, which are at their peak in terms of flavor and nutrition. It is also important to offer vegetarian options that will appeal to both vegetarian and nonvegetarian guests, such as a meatless chili.

Packaged salads and raw vegetables with dips are popular "grab-and-go" items. **See Figure 12-26.** Vegetables are also commonly used in soups and sandwiches. Side dishes featuring fresh vegetables offer a nutrient-dense alternative to items such as French fries.

Melissa's Produce

Vegetables such as beets can be cooked and puréed and added to desserts to provide moisture, replace fat, and enhance flavor.

Prepackaged Salads

Sullivan University

Figure 12-26. Prepackaged salads are often served at quick service and fast casual venues.

Incorporating vegetables in casseroles, savory pies, and pasta dishes is another way vegetables can be promoted. Seasonal vegetables are often elegantly presented in fine dining and special event venues. **See Figure 12-27.**

Beet Salad

Daniel NYC/P. Medilek

Figure 12-27. Seasonal vegetables are often elegantly presented at fine dining and special event venues.

Knowledge Check 12-4

1. Explain why seasonal vegetables can be promoted on menus.
2. Give three examples of menu items that promote vegetables.

LEGUMES

A *legume* is the edible seed of a nonwoody plant and grows in multiples within a pod. Sometimes the pod is eaten along with the seeds. Legumes are extremely nutrient dense and serve as a rich source of protein and complex carbohydrates, including dietary fiber. Legumes are also rich in vitamins and minerals. Legumes have become increasingly popular due to their strong nutritional profile, pleasing flavors, and low cost as compared to other forms of proteins.

Perceived Value of Legumes

Regarded as nutrient-dense and cost-effective ingredients, legumes are highly valued by chefs for their versatility and potential for increasing profits. For example, legumes can add body to dishes that feature animal proteins, requiring less animal protein in those dishes and making them more cost effective. Although most individuals are familiar with legumes such as black beans and chickpeas, offering exotic and heirloom varieties can pique the interest of guests, increase perceived value, and boost sales. **See Figure 12-28.**

Heirloom Legumes

National Garden Bureau Inc.

Figure 12-28. Offering exotic and heirloom varieties of legumes can pique the interest of guests and increase perceived value.

An effective way to use legumes is to incorporate them into a familiar dish, such as beef stew. This allows guests to feel comfortable with the menu item but also perceive the dish as a more sophisticated option. Because legumes are high in protein, they are ideal for vegetarian dishes, and their high dietary fiber content promotes satiety.

Legume Classifications

There are a large variety of legumes, some of which are among the oldest recorded sources of food. Legumes such as beans, peas, pulses, lentils, and sprouts are commonly used to create a wide variety of menu items. Although all legumes can be cooked, some can also be eaten raw.

Beans. A *bean* is the oval or kidney-shaped edible seed of various plants in the legume family. Beans generally grow in a row inside a pod. Some bean varieties, such as green beans, are called edible pods because both the pod and the beans are edible. Other varieties, such as black beans, have tough, stringy pods, so only the beans are edible. Beans are usually kidney-shaped or round and are available fresh, frozen, canned, or dried. Popular varieties of beans include limas, cannellinis, anasazis, peruanos, calypsos, flageolets, pintos, kidney beans, great northern beans, and black beans. **See Figure 12-29.**

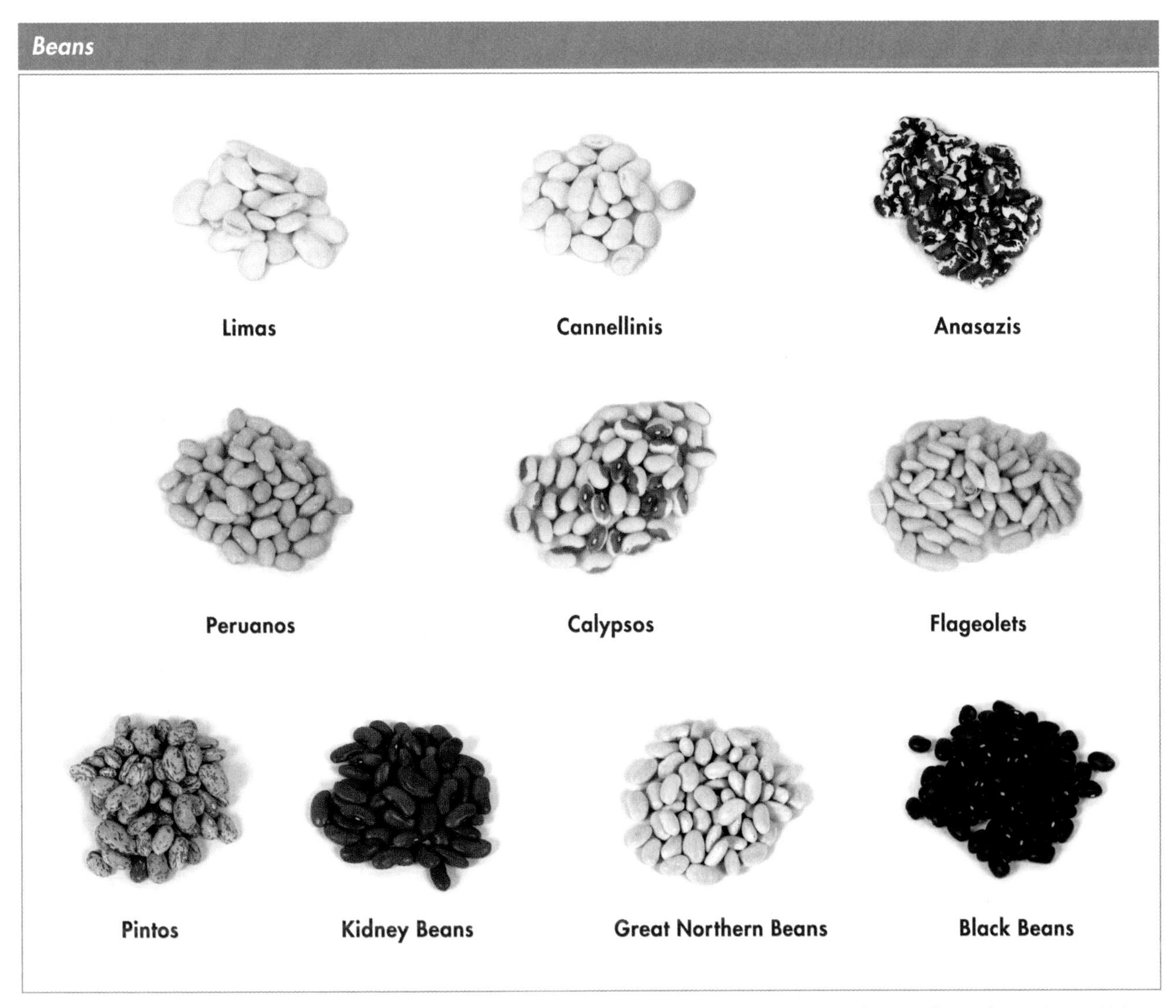

Figure 12-29. Popular varieties of beans include limas, cannellinis, anasazis, peruanos, calypsos, flageolets, pintos, kidney beans, great northern beans, and black beans.

INGREDIENT SPOTLIGHT: Chickpeas

Unique Features

- Also referred to as garbanzo or ceci beans
- Round, slightly irregular in shape with a light beige color
- Firm texture and mild, nutty flavor
- Commonly used in Indian, Greek, and Spanish cuisines
- Low in saturated fat, cholesterol, and sodium
- Source of dietary fiber, protein, folate, copper, and manganese

Nutrition Facts
Serving Size 1 cup, simmered (164 g)

Amount Per Serving	
Calories 269	Calories from Fat 36
	% Daily Value*
Total Fat 4g	7%
Saturated Fat 0g	2%
Trans Fat	
Cholesterol 0mg	0%
Sodium 11mg	0%
Total Carbohydrate 45g	15%
Dietary Fiber 12g	50%
Sugars 8g	
Protein 15g	
Vitamin A 1% • Vitamin C	4%
Calcium 8% • Iron	26%

* Percent Daily Values (DV) are based on a 2,000 calorie diet.

Menu Applications

- Use to make the Mediterranean dip hummus
- Incorporate into soups, stews, and salads
- Purée and add to dishes to provide a creamy consistency
- Roast and use as a garnish to provide a crunchy texture

Figure 12-30. The pods of green peas are not edible, but the pods and seeds of snow peas and sugar snap peas are edible.

Peas. A *pea* is the round edible seed of various plants in the legume family. Green peas, snow peas, and sugar snap peas are popular varieties of peas. The pods of green peas are not edible, but both the pods and peas of snow peas and sugar snap peas are edible. **See Figure 12-30.** Split peas are harvested fully mature, left to dry, and then split. Split peas are commonly used to make soups.

Pulses. A *pulse* is the dried seed of a legume. Dried beans and peas, such as cannellini beans and black-eyed peas, are shelled and then left to dry until they become rock hard. Most pulses must be rehydrated before they can be cooked.

Lentils. A *lentil* is a very small pulse that has been split in half. Lentils come in a variety of colors, including shades ranging from white to green. **See Figure 12-31.** Lentils do not have to be rehydrated like other dried legumes because they are smaller and already split in half so they cook more evenly. Split peas are not lentils, but, like lentils, they do not have to be soaked before cooking.

Figure 12-31. Lentils come in a variety of colors ranging from white to green.

Sprouts. A *sprout* is an edible strand of a germinated bean or seed. Common varieties of sprouts include mung bean, soybean, alfalfa, and radish sprouts. Some chefs grow their own sprouts from legumes such as garbanzo beans, pinto beans, lima beans, and lentils as well as various grains, nuts, and seeds. Depending on the plant of origin, sprouts range in taste from mild to spicy. Sprouts are considered a nutrient-dense food with protein, dietary fiber, vitamins, minerals, and phytonutrients. Sprouts make a healthy addition to menu items such as soups, salads, sandwiches, and entrées. **See Figure 12-32.**

Sprouts

Courtesy of The National Pork Board

Figure 12-32. Sprouts are rarely served alone but make a healthy addition to menu items such as salads.

Knowledge Check 12-5

1. Define legume.
2. List the primary macronutrients found in legumes.
3. Identify factors that influence the perceived value of legumes.
4. List five classifications of legumes.
5. Compare beans and peas.
6. Explain what is meant by “edible pods.”
7. Differentiate between pulses and lentils.
8. Define sprout.

PREPARING NUTRITIOUS LEGUMES

Legumes have a unique nutritional profile since they are considered both a starch and a protein. With the exception of soybeans and peanuts, most legumes are low in fat. In addition, legumes provide key vitamins and minerals such as folate, calcium, magnesium, potassium, and iron. **See Figure 12-33.**

Legumes can be purchased frozen, canned, or dried. Some legumes, such as green peas and green beans, can also be purchased fresh. The way in which legumes are to be prepared determines the form that will be purchased. For example, canned legumes are already cooked and therefore require minimal advance preparation. In contrast, dried legumes must be soaked in water and then simmered. This helps the legumes maintain their shape and a consistent texture.

When preparing legumes, care should be taken to not overcook them. Legumes that have been overcooked will become soft and mushy. Once legumes are fully cooked, they should be seasoned. Legumes can be used in both hot and cold dishes across the menu.

Cooking Legumes

The various purchased forms of legumes are prepared by different cooking techniques. For example, fresh legumes such as sugar snap peas and snow peas only require blanching to produce an appealing, crisp, and tender texture. Frozen legumes are fully cooked and only need to be thawed before incorporating them into menu items. Canned legumes are also fully cooked but are typically preserved in a salt solution and should be rinsed off before proceeding with preparation.

Nutrient Density of Common Legumes*

Legumes	Calories	Total Fat	Dietary Fiber	Protein	Additional Nutrients
Lima beans	150	0 g	9 g	10 g	Folate, phosphorus, magnesium, potassium, iron, copper, and manganese
Pinto beans	167	1 g	8 g	10 g	Thiamin, phosphorus, magnesium, potassium, selenium, copper, and manganese
Kidney beans	153	0 g	12 g	11 g	Thiamin, phosphorous, magnesium, potassium, iron, copper, manganese
Great northern beans	155	1 g	9 g	10 g	Thiamin, folate, phosphorus, magnesium, potassium, copper, and manganese
Black beans	165	1 g	8 g	10 g	Thiamin, folate, phosphorus, magnesium, potassium, copper, and manganese
Chickpeas	182	3 g	9 g	10 g	Folate, copper, and manganese
Green split peas	168	1 g	13 g	12 g	Thiamin, folate, phosphorus, copper, and manganese
Lentils	169	1 g	15 g	12 g	Thiamin, folate, phosphorus, iron, and manganese

* based on ¼ cup serving size, raw

Figure 12-33. Legumes are a nutrient-dense food rich in dietary fiber, protein, vitamins, and minerals, and are typically low in fat.

Food Science Note

Soybeans contain 207 calories and 9 g of fat, and peanuts contain 207 calories and 18 g of fat, based on a ¼ cup serving size (raw).

The preparation method of dried legumes depends on the type being used. **See Figure 12-34.** For example, lentils are very small and can be cooked without soaking. To cook lentils, they should be covered with cold water, brought to a boil, reduced to a simmer, and then cooked until tender. Larger varieties of dried legumes, such as great northern beans, need to be soaked before cooking. The purpose of soaking dried legumes is to soften them so that when they are cooked, the exterior of the legume does not fall apart yet the interior remains tender. After dried legumes have been soaked, they should be rinsed with fresh, cold water. This step helps with digestion by removing some of the gaseous properties associated with legumes. Legumes are typically cooked in water. However, beans are sometimes cooked in stock to impart additional flavor.

Courtesy of The National Pork Board

Properly cooked legumes have a tender interior and an exterior that remains intact.

Cooking Dried Legumes

Dried Legume	Soak Time	Cooking Time
Anasazis	At least 4 hours	45 to 60 minutes
Black beans	At least 4 hours	1 to 1½ hours
Black-eyed peas	None required	45 minutes to 1 hour
Cannellini beans	At least 4 hours	45 minutes to 1 hour
Chickpeas	8 to 12 hours	1½ to 2½ hours
Fava beans	8 to 12 hours	45 to 60 minutes
Great northern beans	At least 4 hours	1 to 1½ hours
Green split peas	None required	45 minutes
Yellow split peas	None required	1 to 1½ hours
Kidney beans	At least 4 hours	1 hour
Green lentils	None required	30 to 45 minutes
Mung beans	At least 4 hours	45 to 60 minutes
Pinto beans	At least 4 hours	1 to 1½ hours
Soybeans	8 to 12 hours	3 to 3½ hours

Figure 12-34. The preparation and cooking methods of dried legumes depend on the type of legumes being used.

FOOD SCIENCE EXPERIMENT: Cooking Legumes in Water, Acid, Alkali, and Salt 12-2

Objective

- To compare legumes cooked in water, acid, alkali, and salt in terms of texture and flavor.

Materials and Equipment

- Masking tape
- Marker
- 4 large saucepans
- Measuring cup
- 4 cups dried black-eyed peas
- Water
- ½ cup tomato juice
- Tablespoon measure
- 1 tbsp baking soda
- 3 tbsp kosher salt
- Tasting spoons

Procedure

1. Use the masking tape and marker to label one saucepan "Water," the second saucepan "Acid," the third saucepan "Alkali," and the fourth saucepan "Salt."
2. Add 1 cup of black-eyed peas to each of the four saucepans.
3. Add 6 cups of water to each saucepan.
4. Add ½ cup tomato juice to the saucepan labeled "Acid."
5. Add 1 tbsp baking soda to the saucepan labeled "Alkali."
6. Add 3 tbsp kosher salt to the saucepan labeled "Salt."
7. Heat all of the saucepans until the liquid comes to a boil.
8. Reduce the heat and simmer the black-eyed peas for 45 minutes.
9. Remove each saucepan from the heat and allow the black-eyed peas to cool slightly.
10. Evaluate the texture and flavor of each saucepan of black-eyed peas.

Typical Results

There will be variances in the texture and flavor of each saucepan of legumes.

Flavor Development

The flavor development of legumes is strongly influenced by the way in which they are used in dishes. For example, the flavor of legumes is enhanced when they are incorporated into soups, stews, and ragouts. Likewise, when legumes are used in a salad, the dressing will add flavor. Aromatic ingredients like onions and garlic as well as seasonings and flavorings such as herbs and spices can also be used to effectively develop and enhance the flavor of legumes.

Chef's Tip

Spices and herbs that are commonly paired with legumes include basil, cayenne, chili, chives, cumin, parsley, sage, savory, and thyme.

Plating Legumes

Due to the nutritional profile of legumes, they can serve as the main component of a dish or can be used as an accompaniment. **See Figure 12-35.** When legumes are the focal point of the plate, it is usually to highlight their protein content. Regardless of their role, legumes can enhance plated presentations with their shapes, colors, patterns, textures, and flavors. For example, black beans served with an egg dish provide a pleasing contrast of colors. Likewise, a soup or salad featuring different types of beans can be visually appealing.

National Turkey Federation

Snow peas are a colorful and flavorful addition to salads.

Plating Legumes

White Chili

Pork Chops with Cannellini Salad

Courtesy of The National Pork Board

Figure 12-35. Due to the nutritional profile of legumes, they can serve as the main component of a dish or can be used as an accompaniment.

Knowledge Check 12-6

1. Contrast the preparation techniques for fresh, frozen, and canned legumes.
2. Explain why dried legumes are soaked before cooking.
3. Give an example of how to develop flavor when preparing legumes.
4. Identify the reason legumes can be the focal point of a dish.

CULINARY NUTRITION RECIPE-MODIFICATION PROCESS

Red Beans and Rice

Red beans and rice is a well-known New Orleans dish and is also popular in Caribbean cuisine. The flavor base comes from onions, celery, and green bell peppers that have been sautéed in bacon fat. The creamy beans provide dietary fiber, which is complemented by the contrasting textures of smoked sausage, ham hocks, and rice. Despite an appealing smoky flavor, the bacon fat, smoked sausage, and ham hocks contribute significant calories and fat, especially saturated fat. Also, white rice is refined and therefore low in dietary fiber.

Yield: 4 servings, 1 cup each

Ingredients

½ lb	dried kidney beans
2 tbsp	bacon fat
1 c	smoked sausage, chopped
½ c	yellow onion, small dice
⅓ c	celery, small dice
⅓ c	green bell pepper, small dice
1 tbsp	garlic, minced
1 tsp	kosher salt
½ tsp	cayenne pepper
½ lb	smoked ham hocks
5 c	chicken stock
2 c	white rice, cooked
¼ c	spring onions, thinly sliced

Preparation

1. Soak dried beans overnight.
2. Heat bacon fat in a saucepan over medium-high heat.
3. Add sausage and stir until browned.
4. Add onions, celery, bell peppers, and garlic.
5. Season with salt and cayenne pepper and cook, stirring occasionally, until the vegetables are soft (approximately 4 minutes).
6. Drain and rinse the soaked beans.
7. Add ham hocks, beans, and chicken stock. Stir well and bring to boil.
8. Reduce heat and simmer uncovered, stirring occasionally, until the beans are tender and the liquid starts to thicken (approximately 2 hours).
9. Remove ham hocks and separate the meat from the bones.
10. Dice the meat and return it to the saucepan. Reserve bones for making stock or soup.
11. Continue cooking the beans until they are creamy (approximately 15–20 minutes).
12. Serve over white rice and garnish with spring onions.

Evaluate original recipe for sensory and nutritional qualities

- Smoky-flavored beans in a rich sauce are served over white rice.
- The dish is high in calories, total fat, and saturated fat with a moderate amount of dietary fiber.

Establish goals for recipe modifications

- Reduce calories and fats while maintaining flavor.
- Increase dietary fiber.

Identify modifications or substitutions

- Use substitutions for the bacon fat, smoked sausage, and white rice.
- Eliminate the ham hocks.

Determine functions of identified modifications or substitutions

- Bacon fat is used to cook the vegetables and provides flavor.
- Smoked sausage and ham hocks provide a smoky flavor to the sauce.
- White rice provides refined carbohydrates, structure, and texture.

Select appropriate modifications or substitutions

- Substitute olive oil for bacon fat.
- Substitute Canadian bacon for smoked sausage and omit ham hocks.
- Substitute brown rice for white rice.

Test modified recipe to evaluate sensory and nutritional qualities

- Substituting olive oil for the bacon fat reduces saturated fat while providing an effective cooking medium.
- Substituting Canadian bacon for smoked sausage and omitting ham hocks reduces calories and fat while maintaining a smoky flavor.
- Substituting brown rice for white rice provides complex carbohydrates and increases dietary fiber while providing structure and texture.

Modified Red Beans and Rice

This heart-healthy version of red beans and rice is significantly lower in saturated fat while twice as high in dietary fiber as compared to the original recipe. To reduce saturated fat, olive oil (an unsaturated fat) is substituted for the bacon fat. To further reduce saturated fat while maintaining a smoky flavor, Canadian bacon is used in place of the smoked sausage and ham hocks. Canned beans are used for convenience, but because they are a high-sodium item, additional salt for seasoning is omitted. Brown rice is a complex carbohydrate and makes a substantial contribution of dietary fiber.

Yield: 4 servings, 1 cup each

Ingredients

2 tbsp	olive oil
½ c	yellow onion, small dice
⅓ c	celery, small dice
⅓ c	green bell pepper, small dice
1 tbsp	garlic, minced
½ tsp	cayenne pepper
2 c	canned kidney beans
½ c	thick-cut Canadian bacon, diced
2½ c	vegetable stock, low sodium
2 tsp	fresh thyme, minced
2 c	brown rice, cooked
¼ c	spring onions, thinly sliced

Courtesy of The National Pork Board

Preparation

1. Heat olive oil in a saucepan over medium-high heat.
2. Add onions and stir occasionally until they begin to caramelize (approximately 8 minutes).
3. Add celery, bell peppers, and garlic to the saucepan.
4. Season with cayenne pepper and cook, stirring occasionally, until the vegetables are soft (approximately 4 minutes).
5. Drain and rinse the canned beans.
6. Add Canadian bacon and beans to the pan, stirring to incorporate.
7. Add vegetable stock and stir well. Bring to a boil.
8. Reduce heat and simmer uncovered, stirring occasionally, until the liquid begins to thicken (approximately 5 minutes).
9. Stir in thyme and check seasoning.
10. Serve over cooked brown rice and garnish with spring onions.

Beans and Rice Nutritional Comparison

Nutrition Facts	Original	Modified
Calories	550.3	441.5
Total Fat	23.5 g	12.5 g
Saturated Fat	8.0 g	2.4 g
Dietary Fiber	5.5 g	11.7 g
Protein	31.2 g	19.3 g

CULINARY NUTRITION RECIPE-MODIFICATION PROCESS

Signature Chestnut Lima Beans and Rice

This signature beans and rice recipe incorporates chestnut lima beans and black rice to create a dish with unique colors and appealing flavors. Heart-healthy olive oil contributes flavor and keeps the dish low in saturated fat. Eliminating meat makes this an ideal vegan recipe that is also a complete protein.

Yield: 4 servings, 1 cup each

Ingredients

½ lb	dried chestnut lima beans
2 tbsp	olive oil
½ c	yellow onion, small dice
⅓ c	celery, small dice
⅓ c	green bell pepper, small dice
1 tbsp	garlic, minced
2 tsp	kosher salt
½ tsp	cayenne pepper
2½ c	vegetable stock, low-sodium
2 tsp	fresh thyme, minced
2 c	black rice, cooked
¼ c	spring onions, thinly sliced

Preparation

1. Soak the chestnut lima beans overnight.
2. Heat olive oil in a saucepan over medium-high heat.
3. Add onions and stir occasionally until they begin to caramelize (approximately 8 minutes).
4. Add celery, bell peppers, and garlic to the saucepan.
5. Season with salt and cayenne pepper. Cook until the vegetables are soft (approximately 4 minutes), stirring occasionally.
6. Drain and rinse the beans before adding them to the pan.
7. Add vegetable stock and stir well. Bring to a boil.
8. Reduce heat and simmer uncovered, stirring occasionally, until the beans are tender and the liquid starts to thicken (approximately 1½ hours).
9. Continue cooking the beans until they are creamy (approximately 15–20 minutes).
10. Stir in thyme and check seasoning.
11. Serve over cooked black rice and garnish with spring onions.

Chef's Tip: *Chestnut limas are just one of a very large selection of heirloom beans that add color and flavor to a menu.*

Beans and Rice Nutritional Comparison

Nutrition Facts	Original	Modified	Signature
Calories	550.3	441.5	354.7
Total Fat	23.5 g	12.5 g	10.3 g
Saturated Fat	8.0 g	2.4 g	1.7 g
Dietary Fiber	5.5 g	11.7 g	8.8 g
Protein	31.2 g	19.3 g	11.1 g

LEGUME MENU MIX

With the vast assortment of legumes available, there are countless ways they can be incorporated into nutritious and delicious menu items. Legumes are being used more frequently because they are an inexpensive source of protein, are abundant in complex carbohydrates, and have a mild flavor that makes them highly versatile.

Legumes can be used effectively in breakfast items such as omelets and breakfast burritos. Stewed beans also make a flavorful breakfast accompaniment. On the lunch menu, legumes can be used as sandwich spreads, as part of a salad, or as the main ingredient in a salad, such as a multicolored bean salad. Legumes used in soups and stews are popular on both lunch and dinner menus. Legumes can also be used as fillings in savory pies or added to pasta dishes.

When incorporating legumes into the menu mix, their textures and flavors should be highlighted in menu descriptions. Describing legumes with words such as "creamy" and "rich" and listing the cooking technique on the menu will entice guests to try dishes featuring legumes. **See Figure 12-36.**

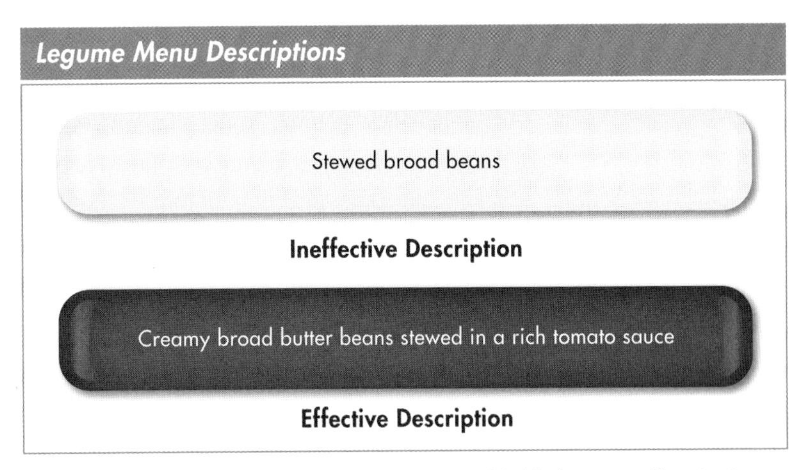

Figure 12-36. Effective menu descriptions highlight appealing textures, flavors, and cooking techniques.

Legumes in Appetizers

Some of the most popular and nutrient-dense appetizers often feature puréed legumes, such as hummus or other bean spreads and dips. These spreads and dips often include olive oil, making them rich in unsaturated fats in addition to the protein, complex carbohydrates, vitamins, and minerals found in legumes. Adding legumes to savory pancakes and vegetable patties is another effective way to incorporate legumes into appetizers.

Wellness Concept

Studies suggest that legumes contain phytonutrients such as flavonoids, lignans, and phytosterols that may help prevent disease.

Legumes in Soups

Legumes are used extensively in soups and stews across many cuisines. Legumes can be left whole, puréed, or used as a mixture of both to create different texture profiles. In some soups, legumes such as lentils serve as the main ingredient. Legumes are also used to enhance the color, texture, and flavor of soups that feature meats, poultry, or seafood as the main ingredient. **See Figure 12-37.** For example, white chili commonly showcases chicken as the main ingredient, with cannellini or great northern beans providing a rich, creamy texture, extra protein, and dietary fiber.

Legumes in Soups

Courtesy of The National Pork Board

Figure 12-37. Legumes can be used to enhance the color, texture, and flavor of soups that feature meat.

Legumes in Sandwiches

Legumes are commonly used as spreads in sandwiches. Legume spreads can be added to whole grain breads to help layer flavors and to wraps and pitas to bind sandwich ingredients. When legumes are formed into patties, they can be the featured component of a sandwich. For example, falafel, a Middle Eastern sandwich made with chickpeas or fava beans, and black bean burgers are two sandwiches highlighting legumes as the main ingredient. Sprouted legumes, such as mung beans or alfalfa sprouts, are also used in sandwiches to provide a pleasing crisp texture. **See Figure 12-38.**

Legumes in Sandwiches

Courtesy of The National Pork Board

Figure 12-38. Sprouted legumes, such as alfalfa sprouts, are used in sandwiches to provide a pleasing crisp texture.

Legumes in Salads

When legumes are used in lettuce-based salads, they are typically added to enhance color and texture and to provide additional nutrients. Kidney beans and chickpeas are often used in this way on salad buffets or as add-on ingredients to a salad menu. A variety of legumes can also be used in salads that do not contain greens. For example, legumes can be combined with other ingredients, such as onions and peppers, and tossed with salad dressing to create a bean salad.

Legumes in Entrées

Legumes can be formed into patties and served as veggie burgers, mixed into pasta dishes and stir-fries, or incorporated into casseroles to create nutrient-dense entrées that are flavorful and satisfying. **See Figure 12-39.** Legume-based soups and stews can also be showcased as entrées. When legumes are the featured item in vegetarian entrées, they should be served with whole grains, such as red beans with brown rice, to provide a complete protein. In nonvegetarian entrées, legumes provide contrasting colors and textures as well as flavor. For example, as a starch component of an entrée, legumes can be stewed with root vegetables and served as an accompaniment to meats.

Legumes in Entrées

Courtesy of The National Pork Board

Figure 12-39. Legumes are commonly used to create nutrient-dense entrées that are flavorful and satisfying.

Legumes in Sides

Legumes play an important role in side dishes because they often serve as the starch component of the plate. For example, green peas with fresh mint and a pinch of sea salt is a pleasing starch-based side dish that pairs well with lamb and pork. Ragouts featuring legumes such as white beans are also popular side dishes. Beans, lentils, and sprouted legumes can also be added to dried fruits or vegetables, mixed with a vinaigrette, chilled, and served cold as a side dish with contrasting colors, textures, and flavors.

Legumes in Desserts

Although legumes are a unique ingredient in desserts, they are being used more frequently in pastries and brownies. For example, adzuki beans and mung beans are often mashed, lightly sweetened, and added to pastries. Puréed black beans can be added to brownies as a fat replacement. As the popularity of gluten-free desserts rises, bean flours can be used effectively to create gluten-free baked items.

National Honey Board

Puréed black beans can be added to brownies as a fat replacement.

Knowledge Check 12-7

1. Identify three reasons legumes are being used more frequently on the menu.
2. Give two examples of appetizers that feature legumes.
3. Describe the ways in which legumes are incorporated into soups, sandwiches, and salads.
4. Explain how legumes are commonly used in entrées and sides.
5. Give two examples of desserts that feature legumes.

PROMOTING LEGUMES ON THE MENU

Promoting legumes on the menu can showcase the nutrient density of menu items. Promoting heirloom varieties of legumes is also an effective way to gain the attention of guests. For example, using a unique variety of a more common legume, such as the green garbanzo bean, can be an effective way to pique the interest of guests and promote legumes on the menu. Some legumes such as peas and fresh beans can be promoted on seasonal menus.

Because of their mild flavor and ability to blend well with other ingredients, legumes can be featured on breakfast, lunch, dinner, and even dessert menus. It is especially effective to promote legumes in vegetarian dishes. **See Figure 12-40.**

Legume Dishes

Dish	Description
Dahl and bulgur	Dahl is a thick lentil stew and pairs well with bulgur
Adzuki bean and brown rice salad	Small red beans native to Asia that pair well with the nutty flavor of brown rice
Red beans and rice	Often contains pork products, but pork can easily be omitted to make the dish vegetarian
Black bean and teff burger	Mashed black beans combined with cooked teff and aromatic vegetables formed into a patty

Figure 12-40. Legumes are used in a variety of vegetarian dishes.

Latin-inspired egg dishes often feature beans in some form. Beans and lentils are also popular in soups and salads because they are nutrient dense and travel well when offered as take-out items. Legumes are common buffet items because they maintain their quality for an extended period of time. **See Figure 12-41.**

A variety of legumes are used to make soups, stews, salads, and stir-fries. Legumes are essential menu items for vegetarians because they contain essential amino acids. Legumes are also used across venues due to their versatility as a featured item as well as an accompaniment to a variety of entrées. Legume preparations at fine dining and special event venues tend to be more refined in presentation. **See Figure 12-42.**

Legumes on Buffets

Bean salad

Sullivan University

Figure 12-41. Legumes are commom buffet items because they maintain their quality for an extended period of time.

Elegant Pea Purée

Courtesy of Chef Eric LeVine

Figure 12-42. Legume preparations at fine dining and special event venues tend to be more refined in presentation.

Knowledge Check 12-8

1. Explain why legumes are commonly offered on buffets.
2. Explain why legumes are often featured in vegetarian menu items.

Chapter Summary

Vegetables are classified as roots, tubers, bulbs, stems, leaves, flowers, and seeds as well as fruit-vegetables and sea vegetables. Mushrooms are a fungi prepared in the same manner as vegetables. Legumes include beans, peas, pulses, lentils, and sprouts.

Daniel NYC/T. Schauer

Vegetables and legumes are essential components of nutritious menus, providing nutrients such as protein, complex carbohydrates, dietary fiber, vitamins, minerals, antioxidants, and phytonutrients. They also add color, texture, and flavor to menu items across all meal periods. Vegetables and legumes are often served as accompaniments, accents, and side dishes, but they can also be the focal point of the plate. Legumes are unique because they can be the starch component on the plate or incorporated with whole grains to provide cholesterol-free complete proteins. Vegetables and legumes featured in plant-based meals or used in combination with lean animal proteins provide nutrient-dense menu items that can be found in many different types of foodservice venues.

Chapter Review

Digital Resources
ATPeResources.com/QuickLinks
Access Code: 267412

1. Compare the nine vegetable classifications and fungi.
2. Summarize the advantages and disadvantages of the four market forms of vegetables.
3. Describe the health benefits of eating a wide assortment of colorful vegetables.
4. Describe the cooking techniques commonly used to prepare vegetables.
5. Identify ways to enhance the flavor and presentation of vegetables.
6. Describe how vegetables can be used throughout the menu mix.
7. Summarize the use of vegetables at various foodservice venues.
8. Compare the five legume classifications.
9. Describe the cooking techniques commonly used to prepare legumes.
10. Identify ways to enhance the flavor and presentation of legumes.
11. Describe how legumes can be used throughout the menu mix.
12. Summarize the use of legumes at various foodservice venues.

Fruits, Nuts & Seeds on the Menu

California Strawberry Commission

National Turkey Federation

Chapter

13

Seeds

National Turkey Federation

Fruits, nuts, and seeds are nutrient-dense components of a healthy diet. While all are rich in vitamins, minerals, and dietary fiber, fruits supply carbohydrates to the body, and nuts and seeds supply proteins and essential fats. The plentiful colors, textures, flavors, and shapes of fruits, nuts, and seeds can enliven menus with nutritious dishes that add value for both guests and the foodservice operation.

Chapter Objectives

1. List the classifications of fruits.
2. Compare the market forms of fruits.
3. Describe the health benefits associated with fruit pigments.
4. Explain how to prepare fruits to enhance colors, textures, and flavors of dishes.
5. Identify ways fruits can be used throughout the menu mix.
6. Describe how fruits can be promoted at various foodservice venues.
7. List the classifications of nuts and seeds.
8. Explain how to prepare nuts and seeds to enhance colors, textures, and flavors of dishes.
9. Identify the ways in which nuts and seeds can be used throughout the menu mix.
10. Describe how nuts and seeds can be promoted at various foodservice venues.

FRUITS

A *fruit* is the edible, ripened ovary of a flowering plant that usually contains one or more seeds. The large variety of fruits, with their array of colors, textures, flavors, and shapes, give chefs the opportunity to enhance a variety of dishes throughout the menu. Choosing the freshest fruits of the season promotes the use of the highest quality ingredients on the menu. Using fruits on the menu also helps meet the demand of health-conscious consumers seeking lighter menu items that feature both seasonal and local ingredients. When fruits are harvested at the peak of ripeness, they are nutrient dense and provide superior flavor profiles. **See Figure 13-1.** Nutritionally, fruits provide a wide assortment of vitamins and minerals, including antioxidants and dietary fiber.

Ripe Strawberries

California Strawberry Commission

Figure 13-1. When fruits are harvested at the peak of ripeness, they are nutrient dense and provide superior flavor profiles.

Perceived Value of Fruits

For chefs to maximize the use of fruits on the menu, guests must be enticed to order dishes where fruits are not only featured but also used to enhance flavor. **See Figure 13-2.** For example, the likelihood of a duck entrée being ordered may increase when the guest can envision the duck paired with the sweet, tart flavors of a complementary cherry or orange sauce. Likewise, side dishes such as wild rice have heightened value when dried cranberries are incorporated. Also, fruits can be used to garnish plated presentations to enhance visual appeal.

Grilled Pears

Pear Bureau Northwest

Figure 13-2. To maximize the use of fruits on the menu, guests must be enticed to order dishes where fruits are featured.

Research indicates that the demand for healthy menu items has increased. This creates an ideal opportunity to raise the perceived value of dishes by replacing highly caloric items such as mashed potatoes or French fries with items such as fresh fruit salads. Incorporating unique and varied seasonal fruits into the menu can also influence perceived value and profits. **See Figure 13-3.** For example, fruits that are available for a limited time often inspire guests to select dishes featuring that fruit in order to experience the exceptional flavors of the season. These dishes often command a higher price due to the limited supply of seasonal fruits. To further strengthen perceived value, elevate flavors, and increase nutrient density, it is important for chefs to understand how fruits are classified and the various forms in which fruits are marketed.

Dragon Fruit

Frieda's Specialty Produce

Figure 13-3. Incorporating unique and varied seasonal fruits into the menu can influence perceived value and profits.

Fruit Classifications

A vast assortment of fruits is available due to agricultural practices that have led to the creation of many varieties and hybrids. A *variety fruit* is a fruit that is the result of breeding two or more fruits of the same species that have different characteristics. For example, a Jonagold is a variety of apple created by breeding a Jonathan apple with a golden delicious apple. A *hybrid fruit* is a fruit that is the result of crossbreeding two or more fruits of different species to obtain a completely new fruit. For example, raspberries and blackberries were crossbred to produce loganberries. The major classifications of fruits are berries, grapes, pomes, drupes, melons, citrus fruits, and tropical and exotic fruits. **See Figure 13-4.**

Fruit Classifications

Type	Common Examples		Common Characteristics
Berries	• Strawberries • Blueberries • Cranberries • Gooseberries • Currants	• Blackberries • Raspberries • Loganberries • Boysenberries	• Small and soft • Tiny, edible seeds
Grapes	Table grapes • Red flame • Thompson • Concord • Raisins (when dried)	Wine grapes • Chardonnay • Pinot Grigio • Merlot	• Oval • Smooth skin • Grown on vines
Pomes	• Apples • Pears • Asian pears		• Edible skin • Fleshy fruit • Core with seeds
Drupes (stone fruits)	• Peaches • Nectarines • Apricots • Avocados	• Dates • Plums • Cherries • Olives	• Stone-like seed or pit • Most have edible skins
Melons	• Cantaloupes • Muskmelons • Honeydews • Canary melons	• Watermelons • Casabas • Crenshaws	• Hard, outer rind • Soft, inner flesh • Most contain many seeds
Citrus fruits	• Oranges • Mandarins • Tangerines • Lemons	• Limes • Grapefruits • Ugli fruits	• Thick, colorful rind • Segmented, pulpy flesh
Tropical fruits	• Bananas • Plantains • Coconuts • Kiwifruit • Pineapples • Papayas	• Figs • Pomegranates • Prickly pears • Persimmons • Mangoes	• Grown in humid tropics • Widely cultivated
Exotic fruits	• Star fruit • Dragon fruit • Mangosteens • Lychees • Rambutans • Durians	• Passion fruit • Breadfruit • Kiwanos • Kumquats • Jackfruit	• Grown in moderately humid subtropics • Not widely cultivated

Figure 13-4. Fruits are commonly classified as berries, grapes, pomes, drupes, melons, citrus fruits, tropical fruits, and exotic fruits.

Berries. A *berry* is a type of fruit that is small and has many tiny, edible seeds. Common types of berries include strawberries, blueberries, cranberries, blackberries, raspberries, and loganberries. Nutritionally, berries are a key source of antioxidants as well as vitamin C and potassium. Berries peak during the late summer months and should be harvested when ripe because they do not continue to ripen after harvesting. Berries can be eaten raw but are also commonly used in beverages, sauces, jams, preserves, vinegars, salads, and pie fillings.

Grapes. A *grape* is an oval fruit that has a smooth skin and grows on woody vines in large clusters. Table grapes such as red flame, Thompson, and Concord are suitable for eating, whereas Chardonnay, Pinot Grigio, and Merlot grapes are used for making wines. **See Figure 13-5.** High in vitamins C and K, grapes are often served on fruit and cheese trays, used in sauces and salads, and made into jams, preserves, and vinegars. Grapes that have been dried are referred to as raisins. Raisins have a unique textural quality that complements both sweet and savory dishes.

Grapes

Red flame
Concord
Thompson

Figure 13-5. Table grapes such as red flame, Thompson, and Concord are suitable for eating.

Pomes. A *pome* is a fleshy fruit that contains a core of seeds and an edible skin. The most well-known examples of pomes are apples and pears, both of which are available in hundreds of varieties. **See Figure 13-6.** Some lesser known pomes include quinces and loquats. Quinces have a tart flavor and are grown in warm climates. Loquats are also referred to as Japanese plums and are known for having a sour flesh and thin skin. Pomes are considered a good source of vitamin C and dietary fiber. Pomes are often poached in various liquids and presented as a dessert.

Common Apple Varieties

Name	Flavor Profile
 Cortland	Sweet, hint of tartness, and juicy
 Crispin	Sweet, juicy, and crisp
 Empire	Sweet, tart, and juicy
 Fuji	Spicy, sweet, and juicy
 Gala	Sweet, juicy, and crisp
 Granny Smith	Tart, crisp, and juicy
 Idared	Sweet, tart, juicy, and firm
 Jonagold	Tangy and sweet
 McIntosh	Sweet, tangy, and juicy
 Rome	Mildly tart and firm

U.S. Apple® Association

Figure 13-6. Apples are classified as pomes and are available in many varieties.

Drupes. A *drupe,* also known as a stone fruit, is a type of fruit that contains one hard seed or pit. Drupes include fruits such as peaches, nectarines, apricots, avocados, dates, plums, cherries, and olives. The nutrients supplied by drupes vary but typically include vitamins A, C, and K, plus niacin, potassium, copper, manganese, folate, and dietary fiber. Drupes continue to ripen after harvesting and are often eaten raw. They are also popular additions to fruit salads, chutneys, compotes, and pie fillings.

Melons. A *melon* is a type of fruit with a hard outer rind (skin) and a soft inner flesh that typically contains many seeds. Melons such as cantaloupes, muskmelons, honeydews, canary melons, watermelons, casabas, and Crenshaws are commonly used in the professional kitchen. The flesh of melons varies from pale green to deep orange and red in color with rinds that are smooth to ridged. Melons are an excellent source of vitamin C and potassium. Some melons supply additional nutrients including niacin, vitamin B_6, folate, vitamin K, and copper. Melons are harvested just before ripeness. Melons are commonly used raw in fruit salads, but they can be incorporated throughout the whole menu. **See Figure 13-7.** Melons can also be cooked to produce unique sauces, breads, and cakes.

INGREDIENT SPOTLIGHT: Asian Pears

Frieda's Specialty Produce

Nutrition Facts

Serving Size 1 medium (approximately 122g)

Amount Per Serving	
Calories 51	Calories from Fat 2
	% Daily Value*
Total Fat 0g	0%
Saturated Fat 0g	0%
Trans Fat	
Cholesterol 0mg	0%
Sodium 0mg	0%
Total Carbohydrate 13g	4%
Dietary Fiber 4g	18%
Sugars 9g	
Protein 1g	

Vitamin A	0%	Vitamin C	8%
Calcium	0%	Iron	0%

* Percent Daily Values (DV) are based on a 2,000 calorie diet.

Unique Features

- Indigenous to Japan, China, and Korea, with new plantings occurring in New Zealand, Australia, Chile, France, and the United States
- Round in shape, with colors ranging from slightly green to golden yellow to bronze
- Recognized for its firm flesh, crunchy texture, and sweet flavor
- Sometimes referred to as an apple pear because of its appearance and texture
- Harvested when fully ripe with a long shelf life when held properly in cold storage
- Low in fat, cholesterol, and sodium
- Source of dietary fiber and vitamin C

Menu Applications

- Add to fruit salads, granola, and salad greens to enhance texture and provide sweetness
- Use grated in marinades, especially with cuts of beef

California Fresh Apricot Council

Drupes, such as apricots, continue to ripen after being harvested.

Melons

Service Ideas, Inc.

Figure 13-7. Melons are commonly used raw in fruit salads but can also be incorporated throughout the menu.

Citrus Fruits. A *citrus fruit* is a type of fruit with a brightly colored, thick rind and pulpy, segmented flesh that grows on trees in warm climates. Citrus fruits, including oranges, mandarins, tangerines, lemons, limes, grapefruits, and ugli fruits, are characterized by their high acid content and strong aroma. Citrus fruits are a key source of vitamin C, potassium, folate, and dietary fiber. All parts of a citrus fruit are edible, but the rind, which includes the peel and white pith, is bitter. **See Figure 13-8.** Citrus rinds can be cooked in sugar syrup for use as a garnish or as an ingredient in baked goods. Citrus fruits are highly prized for their flavorful zest and juice, which are often used to enhance or balance flavors. Citrus fruits also have the unique ability to naturally enhance the saltiness of a dish without having to add salt.

Tropical and Exotic Fruits. Although hundreds of tropical and exotic fruits exist, only about 50 are well known throughout most of the world. A *tropical fruit* is a type of fruit grown in the humid tropics where temperatures average 80°F with little temperature variation between winter and summer. Common tropical fruits include bananas, plantains, coconuts, kiwifruit, pineapples, papayas, pomegranates, and mangoes. **See Figure 13-9.** An *exotic fruit* is any fruit grown in the moderately humid subtropics where the coldest months average 50°F and summers are hotter than winters. Examples of exotic fruits include mangosteens, lychees, rambutans, durians, passion fruit, breadfruit, jackfruit, and guavas. In general, growth is virtually nonexistent for both tropical and exotic fruits when temperatures fall below 50°F.

Figure 13-8. All parts of a citrus fruit are edible, but the rind, which includes the peel and white pith, is bitter.

Frieda's Specialty Produce

Tropical and exotic fruits can be prepared in a variety of ways and used in sweet and savory menu items.

Countries Producing Tropical Fruits

Fruit	Countries
Bananas	Burundi, Nigeria, Costa Rica, Mexico, Colombia, Ecuador, Brazil, India, Indonesia, Philippines, Papua New Guinea, and Spain
Plantains	Colombia, Ecuador, Peru, Venezuela, Ivory Coast, Cameroon, Sri Lanka, and Myanmar
Coconuts	Indonesia, Philippines, India, Sri Lanka, Brazil, Thailand, Mexico, Vietnam, Malaysia, and Papua New Guinea
Papayas	Nigeria, Mexico, Brazil, China, India, Indonesia, Thailand, and Sri Lanka
Mangoes	India, Indonesia, Philippines, Thailand, Mexico, Haiti, Brazil, and Nigeria

Figure 13-9. Tropical fruits tend to be more recognizable than exotic fruits because they are more widely cultivated.

Tropical fruits tend to be more recognizable than exotic fruits because they are more widely cultivated. In contrast, the consumption and trade of exotic fruits is more limited because they are not widely cultivated. With the exception of breadfruit, which is commonly baked, boiled, or fried, most tropical and exotic fruits are consumed raw. However, tropical and exotic fruits can be cooked and prepared in a variety of ways. For example, they can be used to make breads and pies, jams, chutneys, and compotes for use in both sweet and savory applications.

Fresh Blackberries

Oregon Raspberry & Blackberry Commission

Figure 13-10. Fresh fruits at the height of ripeness require minimal preparation and are ideal for serving raw to highlight their colors, textures, and flavors.

Market Forms of Fruits

Fruits are available fresh, frozen, canned, or dried. The selection of fruits will vary based on factors such as availability and ripeness. Fruits that have been processed may be convenient and save time in the kitchen but may contain added ingredients, such as sugar or sodium.

Fresh Fruits. Fresh fruits at the height of ripeness require minimal preparation and are ideal for serving raw to highlight their colors, textures, and flavors. **See Figure 13-10.** Minimal processing also allows for maximum nutrient retention. Fresh fruits are best used within three days and should be washed just before use. Storing washed fruit will result in a shorter shelf life.

Frozen Fruits. Frozen fruits are closest in form to fresh because they have been harvested at the peak of ripeness and quickly frozen to preserve nutrients and quality. However, frozen fruits are softer than fresh fruits because the cell membranes of the fruit rupture and release water during the freezing process. Most frozen fruits are free of added sugars, and their extended shelf lives compared to fresh fruits make them convenient. Frozen fruits are used effectively in smoothies, blended cocktails, baked goods, pie and pastry fillings, ice creams, sorbets, and syrups. **See Figure 13-11.** Stewed frozen fruits are used to make sauces such as chutneys.

Peach Sorbet

Vita-Mix® Corporation

Figure 13-11. Frozen fruits are used effectively in smoothies, blended cocktails, baked goods, pie and pastry fillings, ice creams, sorbets, and syrups.

Canned Fruits. Canned fruits are convenient, consistent, and available in a variety of forms, such as whole, halved, sliced, chunked, and puréed. However, canned fruits packed in fruit juice, light syrup, or heavy syrup can significantly alter the sugar and caloric contents of a dish. **See Figure 13-12.** Because canned fruits are softer than fresh fruits, they should be minimally heated if shape and texture are to be retained. Canned fruits can be used effectively to make chutneys, coulis, and pie and pastry fillings.

Nutritional Comparison of Canned Peaches*

Type	Calories	Sugar
Canned in water	59	12 g
Canned in juice	110	26 g
Canned in light syrup	136	33 g
Canned in heavy syrup	194	49 g

* based on 1 cup serving size

Figure 13-12. Canned fruits packed in water, fruit juice, light syrup, or heavy syrup can significantly alter the sugar and caloric contents of a dish.

Dried Fruits. Dried fruits are unique because they can be used in more culinary applications than other types of fruits. For example, dried fruits can be used whole or chopped to provide a chewy texture in dishes ranging from soups and salads to side dishes and desserts. They can also be puréed and used as a binding agent. When dried fruits are rehydrated in a warm liquid, they produce a soft texture and can be blended into sauces or chopped and added to dishes such as rice pilaf. Dried fruits commonly used to make baked goods, such as breads, muffins, scones, cakes, and cookies, provide added texture and a more developed flavor. **See Figure 13-13.**

Dried Fruits

Frieda's Specialty Produce

Dried Cranberries

Cranberry Bread

Figure 13-13. Dried fruits commonly used to make baked goods such as breads provide added texture and a more developed flavor.

Knowledge Check 13-1

1. Give an example of how fruits can be used as flavor enhancers.
2. Differentiate between fruit varieties and fruit hybrids.
3. List the seven fruit classifications.
4. Identify the nutrients provided by berries.
5. Identify common culinary uses for grapes.
6. Define pome.
7. Name fruits classified as drupes.
8. Name fruits classified as melons.
9. Identify the edible parts of citrus fruits.
10. Differentiate between tropical fruits and exotic fruits.
11. Describe four market forms of fruits.

PREPARING NUTRITIOUS FRUITS

The nutritional qualities of fruits are closely linked to their pigments. Therefore, preparing and incorporating fruits containing all colors of the rainbow not only adds visual appeal but also increases the nutrient density of menu items. **See Figure 13-14.** There are specific health benefits associated with red, orange and yellow, green, blue and purple, and white fruits.

- Red fruits such as strawberries, raspberries, red grapes, red apples, cherries, watermelons, pink grapefruits, and pomegranates supply lycopene, which may reduce the risk of some cancers, most notably prostate cancer.
- Orange and yellow fruits such as pears, peaches, nectarines, apricots, cantaloupes, yellow watermelons, oranges, lemons, pineapples, persimmons, and mangoes supply beta-carotene. Beta-carotene is converted into vitamin A, which promotes healthy vision, proper growth and development, and a healthy immune system.
- Green fruits such as green grapes, green apples, avocados, honeydew melons, limes, and kiwifruits supply chlorophyll. Chlorophyll contains chemicals that promote bone health and may reduce the risk of eye diseases.
- Blue and purple fruits, such as blueberries, blackberries, purple grapes, raisins, plums, and prunes, are rich in the pigment anthocyanin. Anthocyanin works as an antioxidant to promote cellular health.
- White fruits such as bananas and plantains supply anthoxanthins, which some researchers have found to contain health-promoting chemicals that lower blood pressure and cholesterol.

Incorporating Colorful Fruits on the Menu

Original Menu Items	Menu Items with Additional Fruits
Breakfast • Oatmeal	Breakfast • Oatmeal with bananas, raspberries, and blueberries
Lunch • Chicken salad pita with celery • Salad greens	Lunch • Chicken salad pita with celery and green grapes • Salad greens with dried cranberries and pears
Dinner • Pan-seared salmon • Brown rice with almonds • Glazed carrots	Dinner • Pan-seared salmon with mango-lime salsa • Brown rice with almonds and grilled plantains • Glazed carrots with raisins

Figure 13-14. Incorporating fruits containing all colors of the rainbow into the menu not only adds visual appeal, but also increases the nutrient density of the menu items.

Knowing how to prepare fruits is important because fruits increase the nutrient density of dishes, provide a contrast of color and texture, and can be paired with other foods to balance and enhance flavors. For example, adding blackberries to vanilla yogurt adds a burst of color and elevates the texture and flavor. The sweetness of fruits can also be used to contrast with robust, earthy, or spicy flavors such as a sweet and spicy pineapple-jalapeno preserve. The acidic nature of many fruits also helps reduce the reliance on salt to enhance flavor. Fruits can be prepared and served in their raw state or cooked using various methods.

Fruits increase the nutrient density of dishes, provide a contrast of color and texture, and can be paired with other foods to balance and enhance flavors.

Raw Fruits

Since fruits picked at the peak of ripeness are among the most delicious, raw fruits can be showcased to enhance their colors, textures, flavors, and nutrient densities. Using fresh seasonal fruits is also generally the most cost effective approach. Incorporating raw fruits into the menu can be achieved in a variety of ways. For example, raw bananas or berries can enhance smoothies, salads, cereals, and puddings. Fresh berries can also be used to garnish a variety of menu items, from appetizers and entrées to desserts.

Although raw fruits such as apples or pears can be served whole, sliced apples and pears make pleasing additions to salads. When cut into small dices, apples and pears can elevate grain dishes and effectively garnish cold soups. Diced honeydew, cantaloupe, and pineapple can be made into a raw salsa for a nice addition to dishes featuring grilled fish.

It is important to note that some raw fruits, such as apples, pears, bananas, and avocados, are prone to enzymatic browning. **See Figure 13-15.** *Enzymatic browning* is the oxidation of fruits and vegetables, resulting in a brown surface color. The browning of the fruit is caused by a color pigment called melanin, which turns the flesh of the fruit brown when it is exposed to air. Enzymatic browning can be slowed by tossing the fruit with citric acid, ascorbic acid powder, or an acidic juice such as lemon or lime juice. Fruits can also be wrapped tightly in plastic wrap to prevent browning.

Enzymatic Browning

Figure 13-15. Some fresh fruits, such as bananas, are prone to enzymatic browning.

FOOD SCIENCE EXPERIMENT: Fruit Oxidation

13-1

Objective

- To test the rate of oxidation.

Materials and Equipment

- Masking tape
- Marker
- 4 small bowls
- 2 avocados
- Cutting board
- Chef's knife
- Spoon
- Medium bowl
- Fork
- Juice from fresh lime
- Plastic wrap

Procedure

1. Use the masking tape and marker to label the bowls 1, 2, 3, and 4.
2. Place the avocados on the cutting board and cut each one in half lengthwise.
3. Remove and reserve the pits from both avocados.
4. Scoop the flesh from the two avocados into the medium bowl and mash with a fork.
5. Divide mashed avocado evenly between the four small bowls.
6. Mix lime juice with the avocado in bowl 1 and cover the bowl tightly with plastic wrap.
7. Press the two pits into the avocado in bowl 2 and cover the bowl tightly with plastic wrap.
8. Cover bowl 3 tightly with plastic wrap.
9. Press the plastic wrap into the avocado in bowl 4, smoothing out any air bubbles.
10. Place all four bowls in the refrigerator.
11. After 24 hours, check the avocados in each bowl for oxidation and record the results.
12. Repeat step 11 for the next 5 days.
13. On the fifth day, compare the rate of oxidation for the four bowls of avocado.

Typical Results

Each avocado will display variances in the rate of oxidation based on the addition of acidic ingredients and the exposure to air.

Cooking Fruits

Cooking fruits causes the natural sugars to concentrate, resulting in a sweeter taste and softer texture. **See Figure 13-16.** The colors and shapes of cooked fruits typically change. When fruits are cooked to the point where their colors become muted and textures becomes mushy, most of the nutrients have been lost. However, nutrient loss can be minimized if fruits are served in their cooking liquid. While cooking fruits until just soft is generally ideal for nutrient retention, extended cooking times may be appropriate when making items such as jams and sauces. As fruits cook, pectin is released, which naturally thickens the mixture.

Cooked Apples

Figure 13-16. Cooking fruits causes their natural sugars to concentrate, resulting in a sweeter taste and softer texture.

Common cooking techniques used with fruits include grilling, roasting, sautéing, poaching, and stewing. **See Figure 13-17.** Fruits that are typically grilled include pears, drupes such as peaches and nectarines, and firmer fruits like watermelons and pineapples. This technique produces appealing grill marks while caramelizing the fruit sugars to increase sweetness. Roasting also caramelizes the sugars and is most effective with firmer fruits such as apples or underripe bananas.

Cooking Fruits

Cooking Technique	Fruits Commonly Used
Grilling	Pears, peaches, nectarines, watermelons, and pineapples
Roasting	Apples, pears, peaches, bananas, and figs
Sautéing	Berries (e.g., strawberries, blueberries), apples, pears, and bananas
Poaching	Pears, apples, and drupes (e.g., peaches, apricots)
Stewing	Apples, apricots, plums, cherries, bananas, pineapples, mangoes, and dried fruits

Figure 13-17. Common cooking techniques used with fruits include grilling, roasting, sautéing, poaching, and stewing.

Virtually any fruit can be sautéed to create unique textures and flavors that lend themselves to both sweet and savory applications.

Although many fruits can be poached, firmer fruits such as pears are often used, often in a simple syrup flavored with wine and aromatics. A wide variety of fruits can be stewed, including apples, apricots, plums, cherries, bananas, pineapples, mangoes, and dried fruits such as prunes and dates, with excellent results.

Flavor Development

The flavors of fruits develop as they ripen. Therefore, it is important to consider the ripeness of fruits as well as appropriate preparation techniques to fully develop flavors in a dish. When a fruit is perfectly ripe, it is best to preserve the integrity of the fruit by highlighting its natural flavor. For example, berries at the height of sweetness need minimal preparation and can be added in their natural state to dishes such as salads and desserts to develop colors, textures, and flavors.

The ripeness of a fruit also determines its texture. For example, a ripe mango is soft whereas an underripe mango is firm. Puréeing a ripe mango creates a silky sweet purée that can add a sweet layer of flavor to dishes such as grilled poultry and fish. In contrast, a firm mango can be stewed for use in a chutney. **See Figure 13-18.**

Flavor Development

Pear Bureau Northwest

Figure 13-18. It is important to consider the ripeness of fruits as well as appropriate preparation techniques to fully develop flavors in a dish.

Plating Fruits

Because fruits come in a wide assortment of colors, textures, and shapes, they can be used to enliven plated presentations. **See Figure 13-19.** For example, placing chunks of juicy red watermelon on a bed of salad greens creates an attractive contrast in color, texture, and flavor. To elevate the presentation, crumbled feta cheese can be added for another burst of color. The creamy texture and salty flavor of the feta cheese also pairs nicely with the sweet watermelon and crisp salad greens.

The vibrant colors and sweet flavors of fruits also accentuate the neutral colors and savory flavors of animal-based proteins. In addition to fruits, vegetable side dishes also add color, texture, and flavor to the plate. However, vegetables can have even greater appeal when paired effectively with fruits. For example, the rich color of chopped dried apricots can accent the deep green color of braised Swiss chard to create a signature dish of contrasting textures and balanced flavors.

Fruits can be cut into simple or elaborate shapes as well as left whole for use as garnishes. **See Figure 13-20.** Fruit-based sauces also make effective garnishes and can be used to "paint" the plate with designs that create interest. There are virtually endless ways in which chefs can use fruits to create dishes that not only increase nutrient density, but also enhance flavor and presentation.

Plating Fruits

National Honey Board

Grapefruit-Blueberry Salad

Courtesy of The National Pork Board

Pork with Plum Sauce

Figure 13-19. Because fruits come in a wide assortment of colors, textures, and shapes, they can easily enliven plated presentations.

Fruit Garnishes

National Honey Board

Figure 13-20. Fruits can be cut into simple or elaborate shapes as well as left whole for use as garnishes.

CULINARY NUTRITION RECIPE-MODIFICATION PROCESS

Peach Pie

This peach pie recipe is appropriate for any season because it uses canned peaches for the filling. However, the peaches are canned in syrup, which contributes a significant amount of sugar. This recipe also includes a rich butter crust that is delicious and flaky but high in calories and fat.

Yield: 8 servings, one 9-inch pie

Ingredients

Crust

12½ oz	all-purpose flour
2 tbsp	granulated sugar
1 tsp	kosher salt
12 oz	unsalted butter, cubed
6 tbsp	cold water

Filling

32 oz	sliced peaches, canned in syrup
½ c	granulated sugar
2 tbsp	all-purpose flour
¼ tsp	ground nutmeg
⅛ tsp	kosher salt
2 tbsp	unsalted butter
1 tbsp	fresh lemon juice

Preparation

Crust

1. Combine two-thirds of the flour with sugar and salt in food processor bowl. Pulse twice to incorporate.
2. Spread butter cubes evenly over flour mixture. Pulse until no dry flour remains and dough begins to collect in clumps.
3. Spread dough evenly around food processor bowl with rubber spatula. Sprinkle with remaining flour, and pulse until dough is just barely broken up. Transfer dough to a large bowl.
4. Sprinkle dough with water. Fold and press dough until it comes together into a ball. Divide ball in half.
5. Form each half of the dough into a 4-inch disk. Wrap tightly in plastic and refrigerate for at least 2 hours.

Filling

6. Drain peaches, reserving ⅓ cup syrup.
7. In saucepan, mix the sugar, flour, nutmeg, and salt. Add reserved syrup and cook over medium heat, stirring constantly, until mixture bubbles and thickens.
8. Stir in butter, lemon juice, and peaches. Remove from heat and cool.

To assemble and bake

9. Roll dough on a lightly floured surface to form two disks large enough to cover a 9-inch pie pan.
10. Line pie pan with one disk of dough and add cooled peach filling.
11. Lay remaining disk of dough over top of filling and crimp edges.
12. Place pie in 400°F oven and bake 40–45 minutes until golden brown.
13. Remove pie from oven and allow to cool.

Evaluate original recipe for sensory and nutritional qualities

- Pie has a light, flaky crust and sweet peach filling.
- Pie is high in calories, fat, and sugar.

Establish goals for recipe modifications

- Reduce calories, fat, and sugar.
- Enhance flavor.

Identify modifications or substitutions

- Modify crust.
- Use a substitution for the peaches canned in syrup.
- Reduce the amount of granulated sugar and eliminate the butter in the filling.

Determine functions of identified modifications or substitutions

- Crust provides texture and flavor.
- Peaches canned in syrup provide a soft texture and sweet flavor.
- Granulated sugar adds sweetness, and butter provides richness to the filling.

Select appropriate modifications or substitutions

- Remove top layer of crust.
- Substitute fresh, in-season peaches for canned peach filling.
- Reduce the amount of granulated sugar from ½ cup to ⅓ cup, and omit the butter in the filling.
- Add cinnamon and cardamom to filling.

Test modified recipe to evaluate sensory and nutritional qualities

- Removing top layer of crust reduces calories and fat.
- Substituting fresh peaches for canned peaches reduces calories and sugar while adding an appealing texture and fresh fruit flavor.
- Reducing the amount of granulated sugar lowers calories and sugar, and eliminating the butter lowers calories and fat.
- Additional spices enhance flavor.

Modified Peach Pie

This modified peach pie recipe showcases fresh peaches and is most appropriate in the summer when ripe peaches are brimming with natural sweetness. Because summer peaches are at the height of flavor, less sugar is needed in the filling. Cinnamon, cardamom, and nutmeg are added to further accentuate the sweetness of the peaches and develop depth of flavor. Calories and fat are reduced by removing the top crust and eliminating butter from the filling.

Yield: 8 servings, one 9-inch pie

Ingredients

Crust

6 oz	all-purpose flour
1 tbsp	granulated sugar
½ tsp	kosher salt
6 oz	unsalted butter, cubed
3 tbsp	cold water

Filling

4 lb, as purchased	fresh peaches, peeled and sliced
⅓ c	granulated sugar
¼ c	cornstarch
¼ tsp	ground cinnamon
⅛ tsp	ground cardamom
⅛ tsp	ground nutmeg

Preparation

Crust

1. Combine two-thirds of flour with sugar and salt in food processor bowl. Pulse twice to incorporate.
2. Spread butter cubes evenly over flour mixture. Pulse until no dry flour remains and dough begins to collect in clumps.
3. Spread dough evenly around food processor bowl with rubber spatula. Sprinkle with remaining flour, and pulse until dough is just barely broken up. Transfer dough to large bowl.
4. Sprinkle dough with water. Fold and press dough until it comes together into a ball.
5. Form dough into a 4-inch disk. Wrap tightly in plastic and refrigerate for at least 2 hours.

Filling

6. Combine peaches, sugar, cornstarch, and spices in large bowl.
7. Toss to mix well.

To assemble and bake

8. Roll out dough on lightly floured surface to form disk large enough to cover a 9-inch pie pan.
9. Line pie pan with disk of dough and add peach filling.
10. Place pie in 350°F oven and bake 1 hour or until filling is set.
11. Remove pie from oven and allow to cool.

Chef's Tip: *For additional flavor development, add 2 tsp of finely minced candied ginger and 1 tsp of lemon zest to the pie filling before baking.*

Peach Pie Nutritional Comparison

Nutrition Facts	Original	Modified
Calories	626.4	350.2
Total Fat	38.0 g	17.9 g
Saturated Fat	23.8 g	11.0 g
Sugar	27.8 g	24.2 g

CULINARY NUTRITION RECIPE-MODIFICATION PROCESS

Signature Peach Tartlet

In this signature gluten-free recipe, almonds in the crust replace flour to remove gluten and to provide protein, healthy unsaturated fats, and flavor. An almond-vanilla filling adds complexity and a rich mouthfeel. The filling is also made with milk and almond paste to supply protein and calcium. Sweet caramelized peaches are enhanced with agave nectar and tossed with lemon zest to balance flavors. A topping of toasted almonds adds more protein, a contrasting texture, and depth of flavor.

Yield: 8 servings, 4½-inch tartlets

Ingredients

Crust

4 c	slivered almonds
4 oz	unsalted butter, melted
4 tbsp	granulated sugar
¼ tsp	kosher salt

Pudding filling

12 oz	milk, 2%
½ c	almond paste
4 tbsp	granulated sugar
1 tbsp + 1½ tsp	cornstarch
⅛ tsp	salt
1 each	egg yolk
¼ tsp	unsalted butter
¾ tsp	vanilla extract
½ tsp	almond extract

Peach filling

1 lb, as purchased	fresh peaches, peeled and halved
2 tsp	agave nectar
½ tsp	lemon zest, minced
2 tbsp	almonds, sliced and toasted

Preparation

Crust

1. Place almonds in food processor and pulse until finely ground.
2. Transfer almonds to bowl and stir in melted butter, sugar, and salt until almonds soak up the butter.
3. Form almond dough mixture into a ball. Cover tightly in plastic wrap and refrigerate for at least 1 hour.
4. Roll out dough onto lightly floured surface to ⅛ inch thick. Cut dough to fit tart pans.
5. Line tart pans with dough.
6. Place tart pans on baking sheet and bake in 350°F oven for 15–20 minutes until golden brown.
7. Remove baking sheet from oven and allow tarts to cool before filling.

Pudding filling

8. Combine milk and almond paste in blender and blend until smooth.
9. In saucepan, combine sugar, cornstarch, and salt. Gradually stir in milk mixture.
10. Place saucepan over medium heat. Stir until mixture bubbles and starts to thicken.
11. Reduce heat to low and stir 2 minutes longer. Remove saucepan from heat.
12. Whisk egg yolk in bowl. Add small amount of warm mixture to yolk while stirring constantly.
13. Add yolk mixture to saucepan, stirring constantly to incorporate.
14. Place saucepan over heat and bring to a simmer. Stir 1 minute and remove from heat.
15. Gently stir in butter as well as vanilla and almond extracts.
16. Transfer pudding mixture to bowl, cover with plastic wrap, and refrigerate 2 hours or until chilled.

Peach filling

17. Preheat grill over medium-high heat.
18. Place peaches cut side down on lightly oiled grill.
19. Grill peaches for 5 minutes or until caramelized. Turn peaches over and grill 2–3 minutes.
20. Place peaches in bowl and add agave nectar and lemon zest. Mix well.
21. After the peaches cool, slice each peach half into three wedges.

To assemble

22. Divide almond pudding among tart shells.
23. Spread pudding to form an even layer over crust.
24. Arrange three peach slices in each tart shell.
25. Garnish with toasted almonds.

Chef's Tip: *Agave nectar, a natural sweetener derived from the agave plant, often makes an effective substitute for granulated sugar.*

Peach Pie Nutritional Comparison

Nutrition Facts	Original	Modified	Signature
Calories	626.4	350.2	593.4
Total Fat	38.0 g	17.9 g	44.9 g
Saturated Fat	23.8 g	11.0 g	10.6 g
Sugar	27.8 g	24.2 g	26.0 g
Protein	5.8 g	3.9 g	15.5 g
Calcium	22.2 mg	19.8 mg	238.6 mg

Knowledge Check 13-2

1. Identify the health benefits associated with fruits of different colors.
2. Describe the benefits of using fruits in a variety of dishes.
3. Explain how to minimize nutrient loss in cooked fruits.
4. Identify the cooking techniques used with fruits.
5. Describe how ripeness affects the flavor and texture of fruit dishes.
6. Describe how fruits can be used to enhance plated presentations.

FRUIT MENU MIX

Breakfast items such as oatmeal with dried cherries, pancakes topped with mixed berry compote, and scrambled egg burritos with sliced avocado offer nutrient-dense dishes with added layers of texture and flavor. Likewise, fruit pies and pastries are a tasty way to enjoy a variety of fruits. While fruits are easily recognizable in many breakfast and desserts menu items, their unique flavors and natural sweetness are also used to enhance savory items. For example, fruit-based salsas, chutneys, and coulis can be used to elevate entrées featuring items such as pork, poultry, and fish. Creamy fruit-based smoothies, platters of local and exotic fruits, and cheese trays showcasing both fresh and preserved fruits are additional ways fruits can be folded into the menu mix.

Descriptions of fruits on the menu should highlight the names of the fruits and the preparation techniques in ways that clearly communicate an appealing dish. Extremely simple descriptions such as "watermelon sorbet" do little to elevate the dish. A brief description that features the fruit along with other unique components of the dish is most effective at encouraging sales. **See Figure 13-21.**

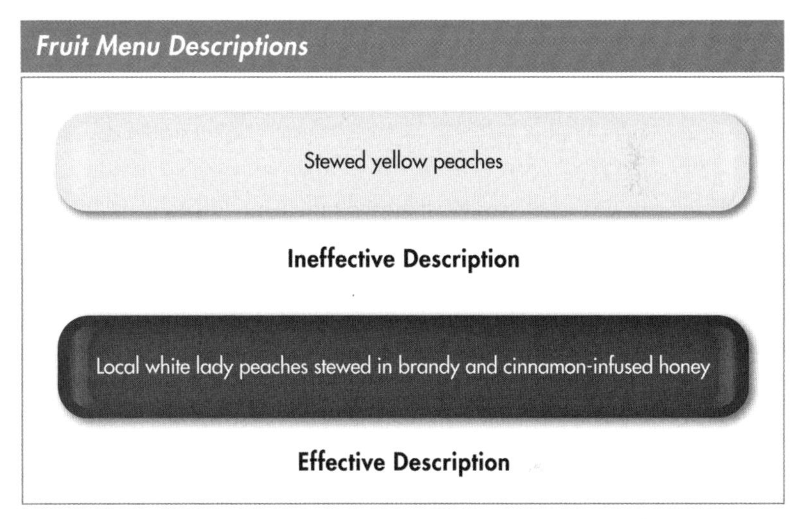

Figure 13-21. Featuring fruit with other unique dish components creates an effective menu description.

Fruits in Beverages

Fruits in beverages are available on virtually all menu types. For example, breakfast menus commonly contain a section that features orange, apple, grapefruit, and cranberry juices. Fruit smoothies are often showcased in beverage houses and bars and feature fruits ranging from limes, bananas, cherries, and coconuts to strawberries, pineapples, and apricots. **See Figure 13-22.** Some foodservice operations promote fruit-based signature drinks as an alternative to beers, wines, and spirits. Lemon wedges, orange slices, or cherries are effective beverage garnishes.

Fruits in Beverages

Fruit Smoothie	Key Ingredients	Key Nutrients
Variation 1	Frozen cherries, grapes, blueberries, bananas, Greek yogurt, and agave nectar	Protein, dietary fiber, vitamin C, riboflavin, vitamin B_6, vitamin B_{12}, vitamin K, calcium, phosphorus, potassium, and manganese
Variation 2	Papayas, kiwifruit, bananas, lemon juice, tofu, honey, ginger, and ice cubes	Protein, dietary fiber, vitamin C, thiamin, vitamin B_6, folate, vitamin A, potassium, phosphorus, magnesium, copper, and manganese
Variation 3	Pears, dates, bananas, almond milk, maple syrup, and cinnamon	Dietary fiber, vitamin C, vitamin B_6, vitamin K, vitamin D, vitamin E, calcium, potassium, iron, zinc, and manganese

Figure 13-22. Fruit smoothies are often showcased in beverage houses and bars and feature a wide variety of fruits.

Fruits in Appetizers

Fruits often play a supporting role in a variety of hot and cold appetizers. For example, a warm fig jam may be used to garnish baked brie, a fresh mango relish could be served with grilled chicken skewers, or chilled melon wrapped with prosciutto may accompany a charcuterie selection. Because fruits lend themselves to many preparation techniques, chefs can also use fruits to bring intrigue to appetizers in ways that may surprise or amuse guests. For example, marinated and roasted peaches have been presented on appetizer selections as "faux gras," a creative nod to the goose liver dish "foie gras." Using raw fruits, such as a sliced star fruit, is an effective way to garnish appetizers while adding color and texture to the plate.

Fruits in Soups

Although fruits are more commonly featured as the main ingredient in soups during the warmer months, they can be used effectively year round. **See Figure 13-23.** For example, a puréed honeydew melon soup finished with a squeeze of lime and garnished with Greek yogurt and chopped mint leaves is a refreshing chilled soup to offer during the summer. As a garnish, pomegranate seeds can add vibrant color and flavor to soups. Fruits can also be used to create a balance of flavors. For example, adding dried fruits like raisins to spicier soups is an effective way to balance the flavors between spicy and sweet.

Figure 13-23. Fruits are more commonly featured as the main ingredient in soups during the warmer months but can be used effectively year round.

Fruits in Salads

The sweetness of fruits is naturally compatible with the bitter flavors of many salad greens. For example, fruits ranging from strawberries, grapes, apples, and mandarins to Asian pears, papayas, lychees, and dried cherries harmonize well with the peppery and bitter flavors of greens such as arugula and chicory. Many chefs use vinegars made from fermented fruit juices, such as balsamic and apple cider vinegars, to make salad dressings as well as sauces. Fruit juices, concentrates, and purées are also used to make salad dressings that are generally lower in fat and calories than those made with saturated fats.

Fruits in Entrées

While fruits are not served as entrées, they are often used to enhance the flavor of a dish by adding natural sweetness or acidity. **See Figure 13-24.** This is why fruit-based salsas pair well with the richness of salmon and why caramelized Granny Smith apples complement savory pork chops. Acidic fruits also play an essential role in marinades by imparting flavor and acting as a tenderizer.

Fruits in Sides

Fresh and dried fruits play a vital role in side dishes by adding vibrant colors, unique textures, and complementary flavors. Side dishes often feature mild-flavored grains that pair nicely with many different fruits. This allows the development of limitless flavor combinations, from quinoa with fresh lemon zest to rice with grilled pineapple. The flavors of legumes and vegetables can also be elevated by adding fruits to create side dishes such as chickpeas braised with dried dates or carrot salad with grapes. Fruits are commonly added to stuffings and fillings as well, creating combinations such as cornbread stuffing with currants. Fruits used in preserves, jams, jellies, marmalades, and butters are often served with breads, which is yet another way to incorporate fruits into the menu mix. **See Figure 13-25.**

Fruits in Sides

Figure 13-25. Fruits used in preserves, jams, jellies, marmalades, and butters are often served with breads.

Fruits in Entrées

Entrée	Description
Normandy-style pork filet	Traditional French preparation featuring pork cutlets served in a sauce of apples, onions, and brandy
Roast turkey breast with fig-cranberry compote	Roasted turkey served with an enhanced version of cranberry sauce featuring dried figs, cinnamon, and toasted pecans
Duck à l'orange	Crispy whole roasted duck with a pan sauce made from the roasted duck and oranges
Grilled mahi-mahi with tropical fruit salsa	Grilled mahi-mahi topped with a sweet and spicy salsa of pineapple, mango, and chiles
Cannelloni stuffed with Swiss chard and raisins	Cannelloni filled with braised Swiss chard, golden raisins, and goat cheese, topped with a roasted pepper coulis

Figure 13-24. While fruits are not served as entrées, they are often used to enhance the flavor of a dish by adding natural sweetness or acidity.

Fruits in Desserts

Fruits are a predominant ingredient in many desserts and are considered by many chefs to be a necessary component of a dessert menu. As a featured component, fruits can be used in their natural form, puréed to flavor crèmes, gelées, gelatos, and ice creams, juiced to make sorbets, or candied and cooked into sauces and compotes. Fruits can also be the showcase ingredient in pies and tarts. **See Figure 13-26.** Even when fruits are not the focus of a dessert, they often serve as a beautiful garnish.

Fruits in Desserts

Sullivan University

Figure 13-26. Fruits can be the showcase ingredient in pies and tarts.

Knowledge Check 13-3

1. Identify ways to incorporate fruits into beverages.
2. Describe different uses for fruits in appetizers.
3. Identify ways to use fruits in soups and salads.
4. Explain how fruits are used in entrées.
5. Describe the benefits of using fruits in sides.
6. Describe the use of fruits in desserts.

PROMOTING FRUITS ON THE MENU

Research suggests that flavor is the most influential factor for guests when deciding what to order from a menu. Therefore, menu descriptions that promote the flavors fruits bring to a dish can be a key selling point. Words such as succulent, juicy, sweet, tart, ripe, crisp, and refreshing are adjectives that promote the attributes fruits bring to menu items.

Fruits can be promoted on breakfast, lunch, and dinner menus. Breakfast menus commonly offer fruits with cereals, yogurts, pancakes, waffles, and French toast. Lunch menus can effectively promote fruits in salads and on sandwiches, such as applewood-smoked turkey topped with sweet apple slices. **See Figure 13-27.** Dinner menus often promote fruit as an accompaniment to savory items, such as pan-roasted fish served with an orange and fennel salad.

Fruits on Sandwiches

U.S. Apple® Association

Figure 13-27. Lunch menus can effectively emphasize the appeal of fruits on sandwiches.

Quick service and fast casual venues may offer guests whole fruits, cut fruits, or prepackaged fruits in items such as fruit salads, yogurts, and hot cereals. Fresh fruits are also incorporated into smoothies, salads, and muffins. Casual dining and institutional venues commonly feature fruits such as berries, pears, peaches, melons, or grapes on menus as side dishes or garnishes.

Fruit-based smoothies, blended drinks, and fruit juices are often featured items on beverage house menus. Fruits are a staple ingredient for bars since both alcoholic and nonalcoholic drinks often use cut fruits as garnishes and fruit juices as drink ingredients. In fine dining and special event venues, fruits are showcased in appetizers, soups, salads, side dishes, entrées, and desserts.

Vita-Mix® Corporation

Fruits are a staple ingredient in both alcoholic and nonalcoholic drinks.

Knowledge Check 13-4

1. List words that can be used to effectively promote fruits on the menu.
2. Explain how fruits can be promoted on breakfast, lunch, and dinner menus.
3. Identify the ways in which fruits are commonly offered at different foodservice venues.

NUTS AND SEEDS

In the culinary world, the terms nuts and seeds are often used together. A *nut* is a dry, hard-shelled fruit or seed that contains an inner kernel. A *seed* is a fruit or unripened ovule of a nonwoody plant. Both nuts and seeds have historically been an important source of nutrients throughout the world. Nutritionally, nuts and seeds provide protein, dietary fiber, essential fatty acids, vitamins, minerals, and antioxidants.

Perceived Value of Nuts and Seeds

Nuts and seeds are much like herbs and spices in the way they add value to menu items. While they may not be the main ingredients, nuts and seeds emphasize the flavor and detail a chef adds to a dish. For example, a stuffing may still be appetizing without incorporating walnuts, but adding walnuts can elevate its appeal. The same principle applies to adding pine nuts to a marinara pasta dish or garnishing a pumpkin pie with roasted pumpkin seeds. Although these dishes may be flavorful without nuts or seeds, adding these ingredients builds flavors and provides both a textural and visual element that can increase the perceived value of the dish. **See Figure 13-28.**

Perceived Value of Nuts and Seeds

Chef Eric LeVine

Marinated Beef Skewers with Pecan Chèvre

Courtesy of The National Pork Board

Pork Tenderloin with Roasted Almonds

National Turkey Federation

Marinated Shrimp with Sesame Seeds

Figure 13-28. Adding nuts and seeds builds flavors and provides both a textural and visual element that can increase the perceived value of a dish.

In addition to providing contrasting textures to both sweet and savory menu items, the health benefits associated with nuts and seeds can also increase the perceived value of a dish. Not only are nuts and seeds cholesterol-free, they are recognized as a source of heart-healthy mono- and polyunsaturated fats. Likewise, studies show that more people are following a flexitarian approach to eating by incorporating meatless dishes into their diets. This creates an opportunity for chefs to increase the perceived value of plant-based dishes by combining nuts and seeds with grains or legumes to form dishes that provide complete proteins.

Research also indicates that gluten-free menu items have increased in popularity. Therefore, using nut flours such as almond flour instead of wheat flours in baked goods can increase the perceived value of nuts and seeds and increase profits.

Nut and Seed Classifications

Whole nuts and seeds are available both in and out of the shell. Nuts and seeds also come in more processed forms such as halved, chopped, sliced, blanched, dry-roasted, oil-roasted, salted, smoked, candied, and flavored. Some nuts and seeds are milled into flours, while others, such as cashews, hazelnuts, peanuts, and sunflower seeds, are commonly ground into butters.

Nuts commonly used in the professional kitchen include almonds, Brazil nuts, cashews, chestnuts, hazelnuts, macadamia nuts, peanuts, pecans, pine nuts, pistachios, soy nuts, and walnuts. **See Figure 13-29.** Seeds commonly used in the professional kitchen include chia seeds, flaxseeds, pumpkin seeds, sesame seeds, poppy seeds, and sunflower seeds.

Almonds. An *almond* is a teardrop-shaped fruit that grows on a small almond tree. Almonds come in two distinct types. One type of almond provides a sweet flavor and is the primary type used in food preparation. The second type of almond is bitter and used to produce oils and extracts. Almond flavor pairs well with fruits like cherries and raspberries. For example, toasted almonds and dried cherries mixed into a quinoa pilaf creates an appealing dish. Almond milk is a popular dairy alternative made from water and ground almonds to produce a rich, creamy product that can be used as a lactose-free substitution for milk.

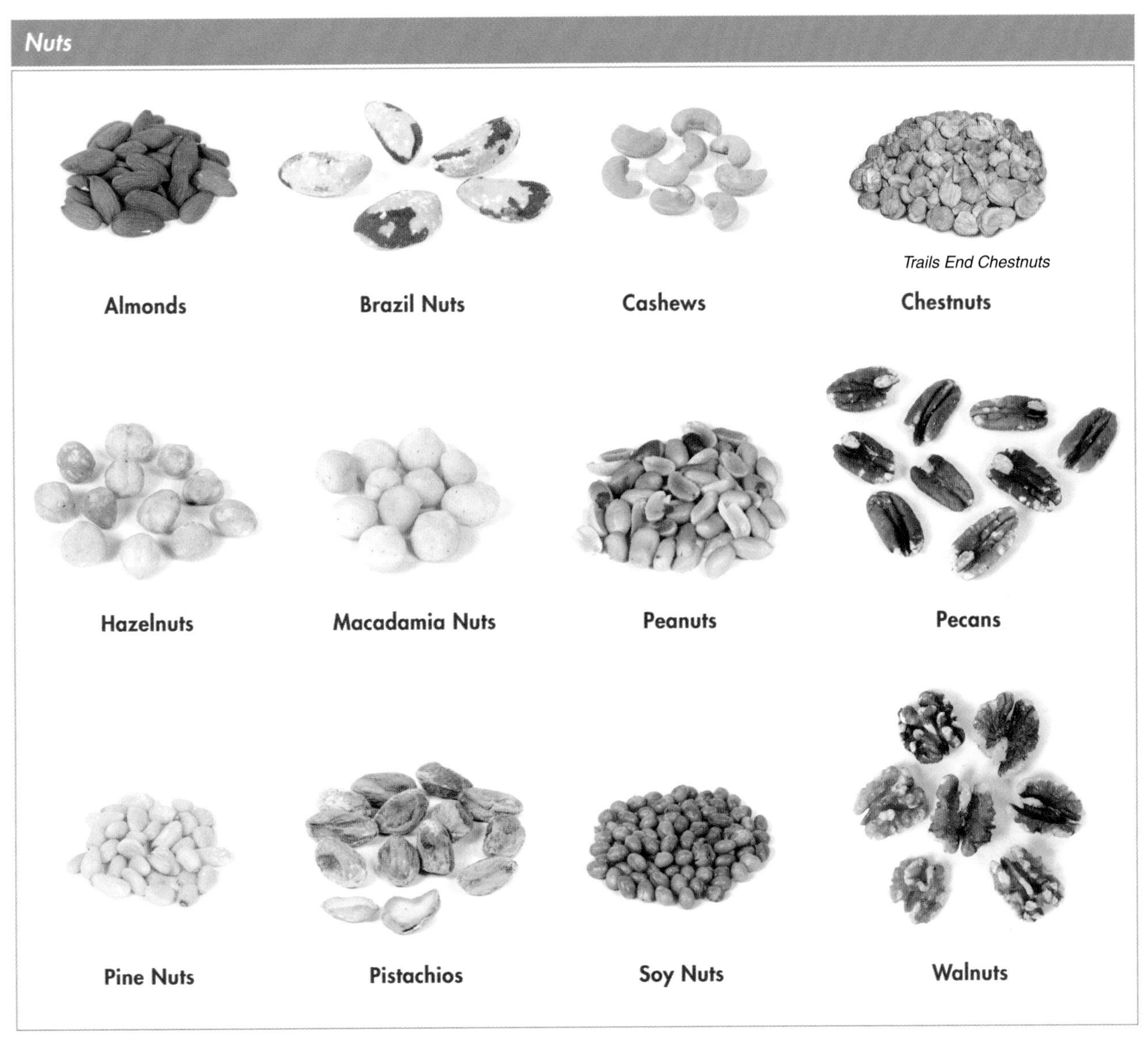

Figure 13-29. Nuts commonly used in the professional kitchen include almonds, Brazil nuts, cashews, chestnuts, hazelnuts, macadamia nuts, peanuts, pecans, pine nuts, pistachios, soy nuts, and walnuts.

Brazil Nuts. A *Brazil nut* is a 1–2 inch long, richly flavored, white seed of the very large fruit of a Brazil nut tree. Brazil nuts have a very creamy texture and a pleasant, nutty flavor that complements tropical fruits such as pineapples and mangoes. Care should be taken to monitor the shelf life of Brazil nuts because their high fat content causes them to become rancid quickly.

Cashews. A *cashew* is a nut extracted from the kidney-shaped kernel inside the fruit of the cashew tree and has a buttery flavor. Cashews are slightly softer than other nuts. The flavor of cashews complements dark chocolate as well as different berries. For example, mixing roasted, unsalted cashews into a blend of dried fruits, toasted coconut, and dark chocolate chips can make an appealing and healthy topping for yogurt and cereals. Likewise, adding cashews to chicken with peapods and water chestnuts makes a tasty entrée.

INGREDIENT SPOTLIGHT: Cashews

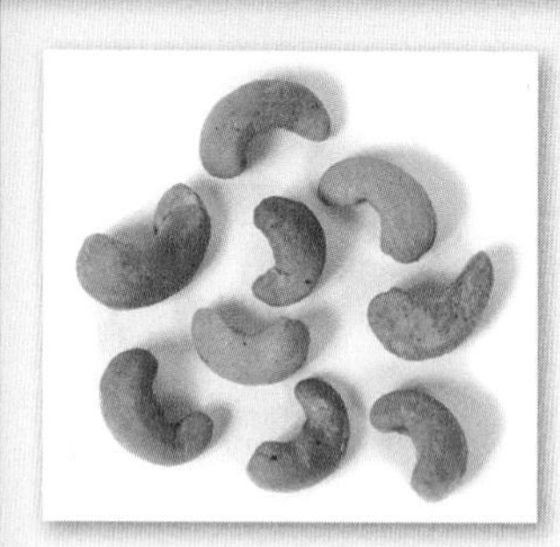

Nutrition Facts
Serving Size 1 oz (28 g)

Amount Per Serving	
Calories 155	Calories from Fat 103
	% Daily Value*
Total Fat 12g	19%
Saturated Fat 2g	11%
Trans Fat	
Cholesterol 0mg	0%
Sodium 3mg	0%
Total Carbohydrate 9g	3%
Dietary Fiber 1g	4%
Sugars 2g	
Protein 5g	
Vitamin A 0% • Vitamin C	0%
Calcium 1% • Iron	10%

* Percent Daily Values (DV) are based on a 2,000 calorie diet.

Unique Features

- Indigenous to Central and South America, but grown worldwide
- Grows on the outside of a tropical fruit known as a cashew apple
- Shaped like a large, round kidney bean and has a buttery flavor
- Only available commercially shelled because the shells contain irritants similar to poison ivy
- Rich in monounsaturated fat
- Source of protein, magnesium, copper, and manganese

Menu Applications

- Add roasted cashews to cereals, salads, vegetables, grain dishes, and desserts for enhanced texture and flavor
- Add cashew butter to smoothies, salad dressings, soups, and sauces for a creamy texture and nutty flavor

Chestnuts. A *chestnut* is a nut from a rounded-off, triangular-shaped kernel found inside the burrlike fruit of the chestnut tree. Chestnuts have a higher starch and lower fat content than other nuts and are generally served cooked. The most common preparation methods include roasting, boiling, and steaming. Chestnuts have a slightly chewy texture and a very mild flavor that works well in savory items such as soups and stuffings.

Hazelnuts. A *hazelnut,* also known as a filbert, is a grape-sized nut found in the fuzzy outer husks of the hazel tree. Hazelnut oil is used for its aromatic scent and rich flavor. Hazelnuts can be used whole or ground in sweet and savory dishes. Hazelnuts are also commonly roasted to enhance their aromatic qualities and provide a crunchy texture to dishes. Hazelnuts make a nice addition to salads. For example, roasted hazelnuts, strawberries, and blueberries make an appealing addition to a fresh spinach salad.

Macadamia Nuts. A *macadamia nut* is a tan, marble-sized nut with a hard shell grown on the macadamia tree. Macadamia nuts have the highest ratio of heart-healthy monounsaturated fats compared to any other nut. The creamy flavor and crunchy texture of macadamia nuts complement chocolate and tropical fruits, such as a grilled pineapple and banana filled crêpe topped with chopped macadamia nuts and shaved dark chocolate.

Peanuts. A *peanut,* also known as a groundnut, is a legume found within a thin, netted, tan-colored pod that grows underground. Although peanuts are legumes, culinarians classify them with nuts and seeds because they are used similarly. Peanuts are high in protein and unsaturated fats, with a distinct flavor and a somewhat softer texture than other nuts. Peanuts are used successfully in both sweet and savory dishes. For example, chicken satay is a popular dish served with a sauce featuring a blend of peanut butter, soy sauce, ginger, garlic, and cilantro. **See Figure 13-30.** Peanut oil has a high smoke point that makes it ideal for cooking.

Satays

National Honey Board

Figure 13-30. Chicken satay is a popular dish served with a sauce featuring a blend of peanut butter, soy sauce, ginger, garlic, and cilantro.

Pecans. A *pecan* is a light-brown nut from the fruit of the pecan tree. Pecans are soft when untoasted and have a mild, sweet flavor that is versatile enough for use in dishes ranging from appetizers to desserts. Roasting pecans brings out a deep, rich aroma and flavor that can enhance a variety of dishes, such as a sautéed apple stuffing.

Pine Nuts. A *pine nut,* also known as a pignoli, is an ivory-colored, torpedo-shaped nut from the pine cone of various types of pine trees. Pine nuts are smaller and softer than most nuts with a mild, slightly sweet flavor. They are commonly used in Asian, Mediterranean, and Middle Eastern cuisines to create items such as basil pesto sauce and pine nut cookies. **See Figure 13-31.** Due to their high unsaturated fat content, pine nuts turn rancid easily and should always be kept refrigerated or frozen.

Pine Nut Pesto Ingredients

Pine nuts

Barilla America, Inc.

Figure 13-31. Pine nuts, along with basil, olive oil, garlic, and Parmesan cheese, are commonly used to make pesto sauce.

Pistachios. A *pistachio* is a green, bean-shaped nut from the pistachio tree. Pistachios are known for their distinctive color, which can vary from deep, bright green to a pale, yellow green. Pistachios have a crunchy texture and mild, sweet flavor that matches well with dried fruits. For example, a warm compote of pistachios, dried apricots, and currants adds an appealing contrast of colors, textures, and temperatures when served with angel food cake.

Soy Nuts. A *soy nut* is a soybean that has been soaked in water and roasted or baked. Although a legume, soy nuts have a crunchy, nutlike texture that can be used similarly to nuts. For example, wilted spinach with garlic, olive oil, and roasted soy nuts makes an appetizing dish. Soy nuts are also an excellent source of protein.

Walnuts. A *walnut* is a nut from the fruit of the walnut tree. Walnuts are slightly harder than most nuts and range in size from small to jumbo. English walnuts have a distinct, dry flavor that is used more often in recipes than black walnuts. English walnuts are popular in both sweet and savory dishes and pair well with most types of meat. Black walnuts have a more intense flavor than English walnuts and are mostly used in baked products. Walnut oil can be used as an effective garnish.

Chia Seeds. A *chia seed* is a small brown seed from a plant in the mint family. While chia seeds were consumed regularly in ancient times, they are only recently making a comeback. Chia seeds are a rich source of essential fatty acids. An effective way to use chia seeds is to add them to breads and muffins.

Flaxseeds. A *flaxseed* is a dark-amber-colored seed of the flowering flax plant. Flaxseeds are one of the richest sources of omega-3 fatty acids. Unlike chia seeds, flaxseeds should be ground to facilitate the absorption of their essential fatty acids. **See Figure 13-32.** Ground flaxseeds can be incorporated into a variety of doughs to make items such as muffins, scones, and bread more nutrient dense. The omega-3 fatty acids in flaxseeds are highly sensitive to both light and heat. Therefore, flaxseeds should be purchased in lightproof containers and kept frozen.

Figure 13-32. Flaxseeds should be ground to facilitate the absorption of their essential fatty acids.

Vita-Mix® Corporation

Figure 13-33. Sesame seeds are available in a paste known as tahini, which is a primary ingredient in hummus.

Pumpkin Seeds. A *pumpkin seed,* also known as a pepita, is a green seed that comes from a pumpkin plant. Pumpkin seeds are a good source of protein, vitamin K, and iron. The slightly chewy texture and subtle sweet flavor of raw or toasted pumpkin seeds can enhance an array of dishes. For example, serving warm toasted pumpkin seeds with a dash of soy sauce on a bed of cool salad greens creates a variance in texture, flavor, and temperature that elevates the dish.

Sesame Seeds. A *sesame seed* is a small teardrop-shaped seed of the sesame plant. Sesame seeds have a slightly sweet, nutty flavor and come in shades of ivory, brown, black, and red. Sesame seeds are commonly used whole in breads, cakes, cookies, and candies as well as to garnish soups, salads, and entrées. The sesame seed is also valued for its oil because it adds flavor and resists rancidity better than other oils. Toasted sesame oil is often used in Asian dishes and is darker in color with a much stronger flavor than untoasted sesame oil. Sesame seeds are also available in a paste known as tahini, which is a primary ingredient in hummus. **See Figure 13-33.**

Poppy Seeds. A *poppy seed* is a tiny, blue-gray seed from the poppy plant. Poppy seeds have a mild, nutty flavor and supply a variety of nutrients including dietary fiber, calcium, and iron. Poppy seeds are best known for garnishing breads but are also used in salad dressings and as a garnish for a variety of menu items.

Sunflower Seeds. A *sunflower seed* is a tan seed that comes from the sunflower plant. Sunflower seeds can be served raw or cooked and are commonly added to breads and cakes or sprinkled over salads and breakfast cereals. The oil extracted from sunflower seeds is used as a cooking oil. In addition, for individuals with peanut allergies, sunflower butter is often an effective substitute for peanut butter. For example, a sunflower butter sauce provides an effective substitute for a peanut sauce when preparing Thai chicken and noodles.

Knowledge Check 13-5

1. Differentiate between nuts and seeds.
2. Identify the key nutrients found in nuts and seeds.
3. Summarize ways that nuts and seeds can impact perceived value.
4. Describe the characteristics of twelve nuts commonly used in professional kitchens.
5. Describe the characteristics of five seeds commonly used in professional kitchens.

PREPARING NUTRITIOUS NUTS AND SEEDS

When preparing menu items with nuts and seeds, using them in different forms, such as whole, chopped, or ground into butters, can add both texture and flavor to a dish. Because nuts and seeds contain a high proportion of unsaturated fat, they are also used to produce cooking and flavoring oils. Nut and seed oils not only make appealing and nutrient-dense dips and vinaigrettes, but also provide unique flavors that can enhance and elevate cooked dishes. **See Figure 13-34.**

To produce an appealing finished dish, it is essential to use nuts and seeds that are high quality. Because nuts and seeds are already ripe when they are harvested, there is minimal fluctuation in flavor throughout the year. However, flavor variances may be detectable based on the geographical region where nuts and seeds are grown.

The high fat content in nuts and seeds makes them susceptible to rancidity if not properly stored in a cool, dry location. Therefore, it is important to strictly monitor the shelf life of nuts and seeds as well as nut and seed oils to ensure freshness. Their high fat content also means nuts and seeds are calorie-dense, so it is important to monitor portion sizes when preparing menu items containing nuts and seeds.

Food Science Note

Research has shown that people who regularly eat nuts and seeds have a lower risk of heart disease.

Nut and Seed Oils

Oil Type	Unsaturated Fats*	Usage Options	Smoke Point
Almond	• Monounsaturated: 9.5 g • Polyunsaturated: 2.4 g	• Dip for bread • Salad dressings • Vegetable dishes • Chicken, duck, and fish dishes	420°F
Hazelnut	• Monounsaturated: 10.6 g • Polyunsaturated: 1.4 g	• Dip for bread • Salad dressings • Potato and legume dishes • Fish dishes	430°F
Peanut	• Monounsaturated: 6.2 g • Polyunsaturated: 4.3 g	• Salad dressings • Pasta dishes • Stir-fry dishes • Pastries	470°F
Walnut	• Monounsaturated: 3.1 g • Polyunsaturated: 8.6 g	• Dip for bread • Fish dishes • Meat dishes • Baked fruits • Cakes	375°F
Pumpkin seed	• Monounsaturated: 3.5 g • Polyunsaturated: 8.0 g	• Dips • Salad dressings • Pasta dishes • Breads • Pastries • Marinades	375°F
Sesame seed	• Monounsaturated: 5.4 g • Polyunsaturated: 5.7 g	• Salad dressings • Stir-fry dishes • Asian noodle dishes • Marinades	410°F
Sunflower seed	• Monounsaturated: 11.7 g • Polyunsaturated: 0.5 g	• Salad dressings • Vegetable dishes • Poultry, fish, and meat dishes • Quick breads	465°F

* based on 1 tbsp serving size

Figure 13-34. Nut and seed oils not only make appealing and nutrient-dense dips and vinaigrettes, but also provide unique flavors that can enhance and elevate cooked dishes.

Cooking with Nuts and Seeds

Nuts and seeds are typically added in the final stages of preparing a dish to add texture, but they also can be incorporated earlier in the cooking process. For example, when making soups and stews, nut butters are often integrated at the beginning stages to develop and layer flavors throughout the cooking liquid. Whole nuts, such as cashews and peanuts, are also commonly added during the cooking process so that their textures can soften and their flavors can permeate the dish. In addition to supplying texture and flavor, ground almonds, walnuts, and pistachios are used to thicken both sweet and savory dishes.

Flavor Development

The flavors of nuts and seeds deepen and intensify when they are toasted or roasted. For example, topping a puréed butternut squash soup with toasted pumpkin seeds adds a crunchy element that contrasts with the creamy mouthfeel of the soup and builds another layer of flavor. Nuts and seeds can also be seasoned or candied to add a unique flavor element to a variety of menu items.

Toasting and Roasting Nuts and Seeds. Toasting and roasting causes the proteins and sugars in nuts and seeds to form a brown exterior surface. As a result, toasted and roasted nuts and seeds develop a deep, rich flavor. **See Figure 13-35.** Toasting can be accomplished in a dry sauté pan over high heat. This is the preferred method for toasting small nuts and seeds. In contrast, roasting is better for larger nuts because they tend to cook more evenly in the oven.

Spiced and Candied Nuts and Seeds. Nuts and seeds are often flavored with spices or candied with sugar to enhance their colors, textures, and flavors. Nuts and seeds can be flavored with a myriad of spices to create flavors that range from spicy to sweet. Spiced nuts and seeds can serve as garnishes for dishes throughout the menu. In contrast, the sweetness of candied nuts and seeds make them an ideal garnish for salads and desserts. Although candied nuts and seeds are higher in sugar than other forms of nuts and seeds, only a small amount is needed to make a flavorful impact on a menu item.

Developing Flavors with Roasted Nuts

Figure 13-35. Roasting enlivens the oils contained within nuts and causes their proteins and sugars to form a brown exterior surface with a deep, rich flavor.

Plating Nuts and Seeds

Nuts and seeds come in a variety of unique shapes, colors, and forms that can enhance the presentation of all menu items. **See Figure 13-36.** For example, grilled salmon garnished with chopped green pistachios adds a pop of color to the plate and allows the guest to anticipate the textural difference between the crisp pistachios and the tender salmon. Nut butters can be made into a smooth sauce that can add both flavor and a decorative element to plated presentations, such as roasted chicken lettuce wraps. A drizzle of a nut or seed oil over poached eggs or a creamy puréed soup can also enhance a plated presentation while layering flavors and supplying essential fatty acids.

FOOD SCIENCE EXPERIMENT: Comparing Raw, Toasted, and Roasted Nuts

13-2

Objective

- To compare raw, toasted, and roasted nuts in terms of color, texture, and flavor.

Materials and Equipment

- Masking tape
- Marker
- 3 plates
- Measuring cup
- 1 cup raw nuts (e.g., pine nuts, pecans, cashews, walnuts)
- 2 sauté pans

Procedure

1. Use the masking tape and marker to label one plate "Raw Nuts," the second plate "Toasted Nuts," and the third plate "Roasted Nuts."
2. Place ⅓ cup of the raw nuts in each of the two sauté pans.
3. Place remaining ⅓ cup of raw nuts on the plate labeled "Raw Nuts."
4. Set one sauté pan over medium-high heat until the nuts are toasted.
5. Place toasted nuts on the plate labeled "Toasted Nuts."
6. Place remaining sauté pan in a 350°F oven and roast the nuts for 5 minutes.
7. Remove nuts from oven and place them on the plate labeled "Roasted Nuts."
8. Evaluate the color, texture, and flavor of each plate of nuts.

Allergen Alert

The materials used in this experiment may contain one or more food allergens.

Typical Results

There will be variances in the color, texture, and flavor of each plate of nuts.

Plating Nuts and Seeds

National Honey Board

Gorgonzola with Honey and Walnuts

Chef Eric LeVine

Ceviche with Black Sesame Seeds

Figure 13-36. Nuts and seeds come in a variety of unique shapes, colors, and forms that can enhance plated presentations.

CULINARY NUTRITION RECIPE-MODIFICATION PROCESS

Candied Pecans

This candied nut recipe produces pecans with a rich, sweet flavor and a crisp candy coating. However, the pecans are deep-fried in oil, which adds calories and fat to a food that is already high in calories and unsaturated fats.

Yield: 8 servings, 1 oz each

Ingredients

for frying	canola oil
1 c	pecans
4 c	water
1 c	confectioners' sugar
1 tsp	kosher salt

Preparation

1. Heat canola oil in a deep fryer to 350°F.
2. Place pecans in a large saucepan, cover with water, and bring to a boil. Boil nuts for 10 minutes.
3. Strain nuts from the water and place in a large bowl.
4. Sprinkle confectioners' sugar over the nuts and stir to coat.
5. Rest nuts in the confectioners' sugar for five minutes.
6. Place nuts in fryer and cook until golden brown (approximately 1–2 minutes).
7. Remove nuts and drain on paper towels.
8. Season with salt and allow nuts to cool and coating to harden before serving.

Evaluate original recipe for sensory and nutritional qualities

- Candied pecans are crunchy with a hard, sweet coating.
- Deep-frying adds calories and fat to the candied pecans.

Establish goal for recipe modifications

- Reduce calories and total fat.
- Increase flavor and nutrients.

Identify modifications or substitutions

- Change cooking technique from deep-frying.
- Use a substitution for confectioners' sugar.

Determine functions of identified modifications or substitutions

- Deep-frying creates a hard candy coating.
- Confectioners' sugar adds sweetness.

Select appropriate modifications or substitutions

- Change cooking technique from deep-frying nuts in oil to boiling nuts in maple syrup.
- Substitute maple syrup for the confectioners' sugar.

Test modified recipe to evaluate sensory and nutritional qualities

- Changing the cooking technique reduces calories and fat and the nuts maintain a hard, sweet candy coating.
- Substituting maple syrup for confectioners' sugar increases nutrients and flavor.

Modified Candied Pecans

In this modified recipe, the pecans are added to boiling maple syrup instead of being deep-fried. This modification reduces the calories and fat, while still producing a hard candy coating. The maple syrup replaces the confectioners' sugar and adds a warm sweet flavor that complements and enhances the richness of the pecans. Maple syrup also provides essential nutrients such as zinc and manganese that are not found in refined sugars.

Yield: 8 servings, 1 oz each

Ingredients

⅔ c	maple syrup
½ tsp	kosher salt
1 c	pecans

Preparation

1. Place maple syrup in a medium saucepan and bring to a boil.
2. Add salt and pecans and stir.
3. Continue stirring until the syrup thickens and crystallizes.
4. Pour nuts onto a parchment-lined sheet pan to cool and harden.

Candied Nuts Nutritional Comparison

Nutrition Facts	Original	Modified
Calories	154.5	136.7
Total Fat	12.3 g	9.0 g
Saturated Fat	1.1 g	0.8 g
Sugar	10.3 g	12.4 g
Zinc	0.6 mg	0.9 mg
Manganese	0.6 mg	1.1 mg

Chef's Tip: *Candied nuts contribute sugar, but because they are so flavorful, only a small amount needs to be used to elevate texture and flavor in a dish.*

Signature Candied Spiced Walnuts

The cooking technique for this signature recipe focuses on roasting the nuts instead of deep-frying or boiling them in a sugar. Roasting intensifies the natural flavor of the nuts and since additional oil is not needed, the nuts are lower in calories and fat than in the original recipe. The heat of the spices is balanced by the sweetness of the brown sugar. To increase the essential polyunsaturated fats, walnuts are used instead of pecans. This recipe also works well with almonds, cashews, pistachios, and pumpkin seeds.

Yield: 8 servings, 1 oz each

Ingredients

1 c	walnuts
4 c	water
⅔ c	light brown sugar
2 tsp	ground cinnamon
1 tsp	ground cumin
½ tsp	ground cayenne pepper
½ tsp	ground mustard
½ tsp	kosher salt

Preparation

1. Place walnuts in a large saucepan, cover with water, and bring to a boil. Boil nuts for 10 minutes and then strain.
2. Combine sugar, spices, and salt in a large bowl and mix well.
3. Add nuts to the sugar mixture and stir to coat.
4. Rest nuts in the sugar mixture for 5 minutes.
5. Place nuts on a sheet pan and bake in a 250°F oven for approximately 45 minutes, stirring the nuts every 7–8 minutes.
6. Remove nuts from the oven and allow to cool before serving.

Chef's Tip: *Spiced nuts can be used in a variety of menu items, from oatmeal to salads to pastas.*

Candied Nuts Nutritional Comparison

Nutrition Facts	Original	Modified	Signature
Calories	154.5	136.7	133.2
Total Fat	12.3 g	9.0 g	8.2 g
Saturated Fat	1.1 g	0.8 g	0.8 g
Polyunsaturated Fat	3.2 g	2.7 g	5.9 g
Sugar	10.3 g	12.4 g	12.9 g
Zinc	0.6 mg	0.9 mg	0.4 mg
Manganese	0.6 mg	1.1 mg	0.5 mg

Knowledge Check 13-6

1. Summarize the storage requirements for nuts and seeds.
2. Identify the point at which nuts and seeds are added to dishes during the cooking process.
3. Describe two methods for developing the flavor of nuts and seeds.
4. Describe how to plate nuts and seeds to enhance presentation.

NUT AND SEED MENU MIX

Nuts and seeds play an important role in adding nutritive value to both sweet and savory menu items. Their variety of colors, textures, and flavors make them a valuable addition when incorporated into a menu item or when used as a garnish. Because nuts and seeds complement many foods, they can be used effectively in breakfast, lunch, and dinner menus. **See Figure 13-37.** For example, whole nuts and seeds can be added to breakfast dishes such as pancakes, fresh fruit, yogurt parfaits, or hot and cold cereals to increase nutrient density. During lunch, nuts and seeds can elevate salads, while nut butters can be used as dips or sandwich spreads. For dinner, nuts and seeds can be used to accent dishes in all menu categories by adding richness as well as a crunchy textural element.

It is important to emphasize the use of nuts and seeds in menu descriptions because it shows care has been taken to maximize texture and flavor and enhance the presentation. **See Figure 13-38.** Including nuts and seeds in menu descriptions can also guide guests towards dishes that are nutrient dense, such as a vegetarian dish that features nuts to form a complete protein. Additionally, highlighting the use of nuts and seeds on the menu is helpful to guests with food allergies.

Food Safety

Allergies to nuts can be life-threatening. Therefore, items containing nuts must be clearly identified on the menu, and staff must be trained to accommodate guests with food allergies.

Nuts and Seeds in Beverages

Nut-based milks such as soy milk, almond milk, hazelnut milk, and cashew milk are becoming popular beverage choices. Nut-based milks have a delicate nutty flavor and are lactose-free. When choosing nut-based milks, it is important to consider the levels of added sugar and whether the product has been fortified with additional vitamins and minerals, such as vitamins A and D and calcium.

Nuts and Seeds on the Menu

Menu Item	Description
Maple sweet potato pancakes with pumpkin seeds	Nutrient-dense sweet potato pancake accented with candied pumpkin seeds and maple syrup
Marinated chicken with macadamia nut fruit salsa	Teriyaki-marinated grilled chicken breast with fresh pineapple-macadamia nut salsa
Almond-crusted halibut	Pan-seared halibut filet encrusted with herbed almond bread crumbs
Zesty rice pilaf with pine nuts	Rice pilaf flavored with lemon, garlic, and oregano and topped with toasted pine nuts
Chocolate hazelnut tofu cheesecake	Decadent chocolate tofu cheesecake baked in a hazelnut crust

Figure 13-37. Nuts and seeds complement many foods and can be used effectively in a variety of menus items.

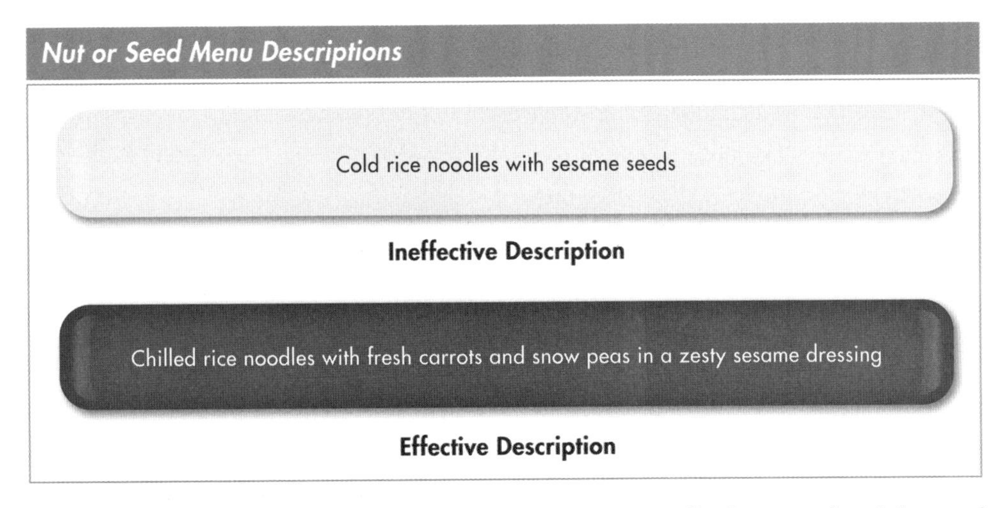

Figure 13-38. An emphasis on the use of nuts and seeds creates effective menu descriptions and demonstrates that care has been taken to maximize the texture and flavor of the presentation.

Smoothies are also a nutrient-dense meal and snack option. Using nut butters in smoothies imparts flavor as well as protein, dietary fiber, vitamins, and minerals. Peanut butter and banana is a common smoothie combination. Nut butters such as almond and cashew butter also combine nicely with various fruits to make appealing smoothies.

Nuts and Seeds in Appetizers

Because nuts and seeds easily absorb flavors, they can be seasoned, toasted, or candied to create unique flavor blends to serve as appetizers. For example, an appetizer of candied cashews, peanuts, and macadamia nuts flavored with orange rind and chiles offers guests an appealing combination of shapes, colors, textures, and flavors to begin their dining experience. Nuts and seeds prepared in a similar fashion, or even left raw, can elevate a cheese plate and also pair well with dried fruits.

Using chopped nuts to bread and then bake foods is an effective way to incorporate nuts into appetizers. For example, baked almond-crusted chicken tenders or cashew-crusted shrimp is a nutrient-dense alternative to deep-fried appetizers. In all menu categories, nuts and seeds make excellent garnishes that add color, texture, and flavor.

Nuts and Seeds in Soups

In addition to increasing nutritive value and flavor, nuts and seeds can be used to enhance the texture of soups in other ways. **See Figure 13-39.** For example, peanut butter can be whisked into a roasted tomato soup to add a creamy texture and rich mouthfeel, while roasted peanuts can be used to garnish the tomato soup with a contrasting crunch. There are unlimited ways in which nuts and seeds can be used in soups to create a nutrient-dense blend of ingredients with appealing textures and flavors.

Nuts in Soups

Figure 13-39. In addition to increasing nutrients and flavor, nuts are commonly used to enhance the texture of soups.

Nuts and Seeds in Salads

Similar to the way nuts and seeds can create dual textures in soups, nuts and seeds can create a silky smooth salad dressing or add an element of crunch to the salads they garnish. For example, a salad dressing can be made by whisking in a nut butter and an acidic ingredient plus seasonings to create a thick and creamy salad dressing that is rich in unsaturated fats. Also, nut oils can be used to enhance the flavor of salad dressings. Nuts can then be added as a salad garnish to build texture as well as accentuate the nutty flavor of the dressing. Garnishing salads with nuts and seeds also offers a healthy alternative to croutons, which are high in saturated fats and simple carbohydrates.

Daniel NYC/B. Milne

Nuts can be added as a salad garnish to build texture.

Nuts and Seeds in Entrées

Nuts and seeds are primarily used as visual and textural garnishes for entrées but can impart inviting flavors as well. For example, using a combination of black and white sesame seeds to coat a piece of pan-seared tuna elevates the presentation while adding a unique texture and appealing flavor to the entrée. Nuts and seeds can also be used to make a more nutrient-dense and flavorful breading. For example, sesame seeds or ground nuts such as cashews or peanuts can be added to a mixture of spices and bread crumbs and used to coat a protein such as a turkey cutlet or a pork tenderloin. **See Figure 13-40.**

Nuts in Entrées

Courtesy of The National Pork Board

Figure 13-40. Ground nuts or seeds can be added to a mixture of spices and bread crumbs and used to coat a protein such as pork.

Nuts and seeds can be added to plant-based ingredients to create dishes that are complete proteins. For example, whole-grain pasta topped with a pine nut pesto or a vegetarian burger featuring black beans and toasted walnuts on a whole-grain bun offers guests nutrient-dense protein alternatives to meat-based meals.

Nuts and Seeds in Sides

Nuts and seeds can be used to heighten the appeal of side dishes ranging from vegetables to starches. They can be sprinkled directly over an item just prior to service or cooked into a dish in order to infuse the dish with flavor. For example, pecans can be used to garnish braised mustard greens or mixed into cornbread stuffing before baking. In both cases, the nuts add texture, flavor, and a variety of nutrients.

Nuts and seeds are commonly mixed into bread dough or added to the top of yeast breads or quick breads. For example, oatmeal muffins may have almonds or sunflower seeds blended into the dough before baking and dinner rolls are often topped with sesame seeds. **See Figure 13-41.** Bread is a side that is often served with olive oil for dipping. Instead of olive oil, serving bread with almond, hazelnut, pistachio, or walnut oil can be an unexpected and unique way to incorporate nuts into the menu mix.

Dinner Rolls with Seeds

Figure 13-41. Dinner rolls are often topped with seeds such as sesame seeds.

Nuts and Seeds in Desserts

Not only are nuts and seeds commonly featured in desserts, they are viewed by some chefs as essential elements when building a dessert menu. Nuts and seeds can increase the appeal of items ranging from cakes, chocolates, custards, and creams to cookies, pastries, crêpes, and frozen desserts. Nuts such as pecans and walnuts are often the featured ingredient in pies and tarts, while the green hues of pistachios make them an effective garnish in desserts such as cannoli. Although these desserts can be abundant in fat and calories, offering sample-size portions can provide guests with a small indulgence that fits appropriately into a balanced diet. In addition, nut and seed flours have risen in popularity as more gluten-free desserts are incorporated into the menu.

National Honey Board

Nut butters, such as peanut butter, are commonly featured in desserts.

Chef's Tip

Adding ingredients such as cinnamon, ginger, honey, maple syrup, or lemon zest to nut butters adds complexity and depth of flavor.

Knowledge Check 13-7

1. Identify two beverages that incorporate nuts and seeds.
2. Describe how nuts and seeds can be used as an alternative to fried appetizers.
3. Explain how nuts and seeds can enhance the texture of soups and salads.
4. Identify the benefits of using nuts and seeds in entrées.
5. Describe the use of nuts and seeds in sides.
6. Explain why nut and seed flours are being used more often to prepare desserts.

PROMOTING NUTS AND SEEDS ON THE MENU

One of the best ways to promote nuts and seeds on a menu is to emphasize their nutrient density. **See Figure 13-42.** Nuts and seeds increase the protein content of a dish while providing dietary fiber, heart-healthy fats, and an abundance of micronutrients. Nuts and seeds also offer a variety of textures and flavors. The fact that nuts and seeds meld with both savory and sweet flavors increases their use in menu items across all foodservice venues. For example, baseball parks may sell bags of peanuts in the shell, while a fine dining venue may use chia seeds to accentuate seared scallops or cocoa-dusted hazelnuts to garnish a signature dessert.

Food Science Note

Studies suggest that an amino acid found in nuts and seeds called arginine can increase the flexibility of artery walls, which helps reduce the risk of blood clots.

Nutrient Density of Nuts and Seeds*

Common Nuts						
Nut	**Calories**	**Monounsaturated Fat**	**Polyunsaturated Fat**	**Dietary Fiber**	**Protein**	**Additional Nutrients**
Almonds	161	8.6 g	3.4 g	3 g	6 g	Riboflavin, vitamin E, magnesium, and manganese
Brazil nuts	184	6.9 g	5.8 g	2 g	4 g	Phosphorus, magnesium, selenium, and copper
Cashews	155	6.7 g	2.2 g	1 g	5 g	Phosphorus, magnesium, copper, and manganese
Chestnuts	60	0.2 g	0.3 g	2 g	1 g	Vitamin C, copper, and manganese
Hazelnuts	176	12.8 g	2.2 g	3 g	4 g	Vitamin E, copper, and manganese
Macadamia nuts	201	16.5 g	0.4 g	2 g	2 g	Thiamin and manganese
Peanuts	159	6.8 g	4.4 g	2 g	7 g	Niacin, folate, and manganese
Pecans	193	11.4 g	6.1 g	3 g	3 g	Manganese
Pine nuts	188	5.3 g	9.5 g	1 g	4 g	Manganese
Pistachios	156	6.5 g	3.8 g	3 g	6 g	Thiamin, vitamin B_6, copper, and manganese
Soy nuts	126	1.3 g	3.4 g	2 g	11 g	Folate, vitamin K, phosphorus, magnesium, copper, and manganese
Walnuts	183	2.5 g	13.2 g	2 g	4 g	Copper and manganese

* based on 1 oz serving size

Figure 13-42. One of the best ways to promote nuts and seeds on any menu is to emphasize their nutrient density. (continued on next page)

Nutrient Density of Nuts and Seeds*

Common Seeds						
Seed	**Calories**	**Monounsaturated Fat**	**Polyunsaturated Fat**	**Dietary Fiber**	**Protein**	**Additional Nutrients**
Chia seeds	137	0.6 g	6.5 g	11 g	4 g	Calcium, phosphorus, and manganese
Flaxseeds	150	2.1 g	8.0 g	8 g	5 g	Thiamin, phosphorus, magnesium, and manganese
Pumpkin seeds	151	4.0 g	5.9 g	1 g	7 g	Vitamin K, phosphorus, magnesium, iron, copper, and manganese
Sesame seeds	160	5.3 g	6.1 g	3 g	5 g	Calcium, phosphorus, magnesium, iron, copper, and manganese
Poppy seeds	147	1.7 g	8.0 g	5 g	5 g	Thiamin, calcium, phosphorus, magnesium, iron, copper, and manganese
Sunflower seeds	164	5.2 g	6.5 g	2 g	6 g	Thiamin, vitamin B_6, vitamin E, phosphorus, magnesium, selenium, copper, and manganese

* based on 1 oz serving size

Figure 13-42. (continued from previous page)

Nuts and seeds are often used to garnish items such as breads and muffins, yogurts, cereals, soups, salads, sauces, and desserts. **See Figure 13-43.** Nut butters such as almond and hazelnut butters are also becoming popular sandwich spreads. Nuts and seeds often garnish a variety of menu items, such as soups, salads, and sauces.

Beverage Houses and Bar Venues

Beverage houses and bar venues incorporate nuts and seeds in a variety of ways. Nut butters are often used to add flavor to smoothies, and nut-based milks are increasingly being used to create specialty drinks that are lactose-free or more nutrient dense than beverages made from traditional dairy products. In addition, beverage houses often feature baked goods that incorporate nuts and/or seeds.

Garnishing with Nuts or Seeds

Figure 13-43. Nuts and seeds often garnish menu items such as salads.

Chef's Tip

Nut butters are made by grinding nuts in a food processor until a thick paste is formed. In general, there is a two to one ratio of nuts to nut butter (2 cups of nuts yields 1 cup of nut butter).

Fine Dining and Special Event Venues

Fine dining and special event venues are more likely to feature dishes with toasted, spiced, or candied nuts and seeds. These venues may also feature nuts such as chestnuts or red sesame seeds that are not commonly offered in more casual venues.

Knowledge Check 13-8

1. Identify menu items that are commonly garnished with nuts and seeds.
2. Summarize ways nuts and seeds are used at beverage houses and bar venues.
3. Describe how nuts and seeds may be featured at fine dining and special event venues.

hapter Summary

Fruits from across the color spectrum not only make a positive nutritional impact, but their unique textures and flavors can create exceptional-tasting dishes and elevate plated presentations. The array of fruits available from the various fruit classifications allows chefs to develop flavors in ways that can also enhance each category of the menu. While fruits may be the featured ingredient in items such as beverages and desserts, they can also be used to complement, balance, and heighten the flavor of items such as appetizers and entrées.

Whether nuts and seeds are raw, roasted, spiced, candied, or ground into butters, they are outstanding sources of heart-healthy fats that also supply protein, dietary fiber, and a multitude of vitamins and minerals. Nuts and seeds provide chefs with an opportunity to impart extraordinary textures and flavors into dishes that can be promoted on breakfast, lunch, and dinner menus across all types of foodservice venues.

Courtesy of The National Pork Board

Chapter Review

Digital Resources
ATPeResources.com/QuickLinks
Access Code: 267412

1. List the seven fruit classifications.
2. Identify the nutrients provided by each fruit classification.
3. Describe the advantages and disadvantages of the four market forms of fruit.
4. Summarize the health benefits associated with the various fruit color pigments.
5. Describe common ways to prepare fruits.
6. Describe the ways in which fruits can be used to enhance presentation.
7. Summarize how fruits can be showcased throughout the menu.
8. Identify ways to promote fruits on the menu.
9. Summarize the health benefits of nuts and seeds.
10. Contrast common nuts and seeds.
11. Describe how to effectively develop the flavors of nuts and seeds.
12. Explain how nuts and seeds can be used to enhance presentation.
13. Summarize how nuts and seeds can be showcased throughout the menu.
14. Identify the ways in which nuts and seeds are promoted on the menu.

Pastas, Grains & Breads on the Menu

Courtesy of The National Pork Board

Indian Harvest Specialtifoods, Inc./Rob Yuretich

Breads

National Cancer Institute/Daniel Sone (photographer)

Pastas, grains, and breads are the primary sources of carbohydrates in a balanced diet. They are popular around the globe and come in a variety of forms. Pastas and breads can be prepared from whole grains, which can also be served alone. Whole grains and whole grain flours are good sources of dietary fiber, vitamins, and minerals. Whole grain menu items pair well with a variety of fruits, vegetables, nuts, seeds, and lean proteins to create nutrient-dense dishes full of color, texture, and flavor.

Chapter Objectives

1. Summarize the perceived value of pasta.
2. Describe common pasta classifications and how they are prepared.
3. Identify flours used to make gluten-free pastas.
4. Explain how to effectively cook and plate pasta dishes.
5. Summarize how to promote nutrient-dense pasta dishes throughout the menu mix.
6. Describe how foodservice venues promote pastas.
7. Differentiate between whole grains and refined grains.
8. Describe cereal grains and pseudocereals.
9. Identify gluten-free and ancient grains.
10. Explain how grains are cooked.
11. Summarize how to promote nutrient-dense grains throughout the menu mix.
12. Describe how foodservice venues promote grains.
13. Compare and contrast yeast breads and quick breads.
14. Identify ways to increase the nutrient density and flavor of breads.
15. Summarize how to promote nutrient-dense breads throughout the menu mix.
16. Describe how foodservice venues promote breads.

PASTAS

Pasta is a rolled or extruded dough product composed of flour, water, salt, oil, and sometimes eggs. The flour used to make pasta can come from any grain, but semolina flour is most common. Semolina comes from durum wheat and is high in gluten. *Gluten* is a type of protein found in grains such as wheat, rye, barley, and some varieties of oats. Gluten enables pasta dough to maintain its shape, form, and texture.

Pasta dough can be fashioned into a variety of shapes and sizes. It can be enhanced with vegetable purées, spices, or squid ink to give the dough a specific color. The mild flavor of pasta pairs well with a variety of ingredients and allows it to take on different flavor profiles. Pasta is a staple in Mediterranean, North Eastern European, and Asian cuisines. Pasta dishes range from simple, fresh ingredients tossed with a little oil to layered dishes containing proteins, nutrient-dense vegetables, and flavorful sauces. Many pasta dishes, such as macaroni, fettuccine, spaghetti, rigatoni, lasagna, and tortellini, are named for the type of pasta used in the dish. **See Figure 14-1.**

Perceived Value of Pastas

Pasta has a low food cost and offers incredible versatility due to its mild flavor and variety of shapes and sizes. Depending on the venue, chefs may choose to prepare classic pasta dishes or create signature dishes. House-made pastas, baked pasta dishes, and stuffed pastas mixed with high-quality ingredients such as seasonal vegetables can garner a higher price. Through the use of flavorful sauces and unique ingredient combinations, pastas can become a refined and sophisticated menu item. **See Figure 14-2.**

Figure 14-1. Many pasta dishes, such as macaroni, fettuccine, spaghetti, rigatoni, lasagna, and tortellini, are named for the type of pasta used in the dish.

Perceived Value of Pasta Dishes

Irinox USA

Figure 14-2. Through the use of flavorful sauces and unique ingredient combinations, pasta dishes such as linguine and clams can become a refined and sophisticated menu item.

While some foodservice operations make fresh pastas in-house, a large selection of dried and frozen pastas are also used. These pasta selections range from dried linguini with squid ink to frozen shells stuffed with seafood. The chef can use these products to create a variety of pasta dishes. For example, seafood-stuffed pasta dressed with lemon juice and olive oil makes a simple, flavorful appetizer. In contrast, the same seafood-stuffed pasta combined with a white wine sauce and tossed with lobster meat can have a higher price point as an entrée.

Appropriately sized portions of whole grain pastas supply complex carbohydrates that deliver sustained energy. When paired with an array of vegetables and lean proteins, pasta menu items provide a flavorful, well-balanced meal.

Pasta Classifications

Pasta can be made fresh or purchased in dried or frozen form. The common classifications of pastas include ribbon, shaped, tube, and stuffed pastas. There are also a variety of Asian noodles and gluten-free pastas available.

Ribbon Pastas. *Ribbon pasta* is pasta that is cut into thin, round strands or flat, ribbonlike strands. **See Figure 14-3.** Ribbon pastas are made from pasta dough that has been rolled very thin by hand or with a pasta machine and is cut into ribbons of various widths. Pasta dough for ribbon pasta is typically made using whole eggs and commonly available fresh, frozen, or dried. Common types of ribbon pastas include capellini, egg noodles, fettuccine, lasagna, linguine, and spaghetti.

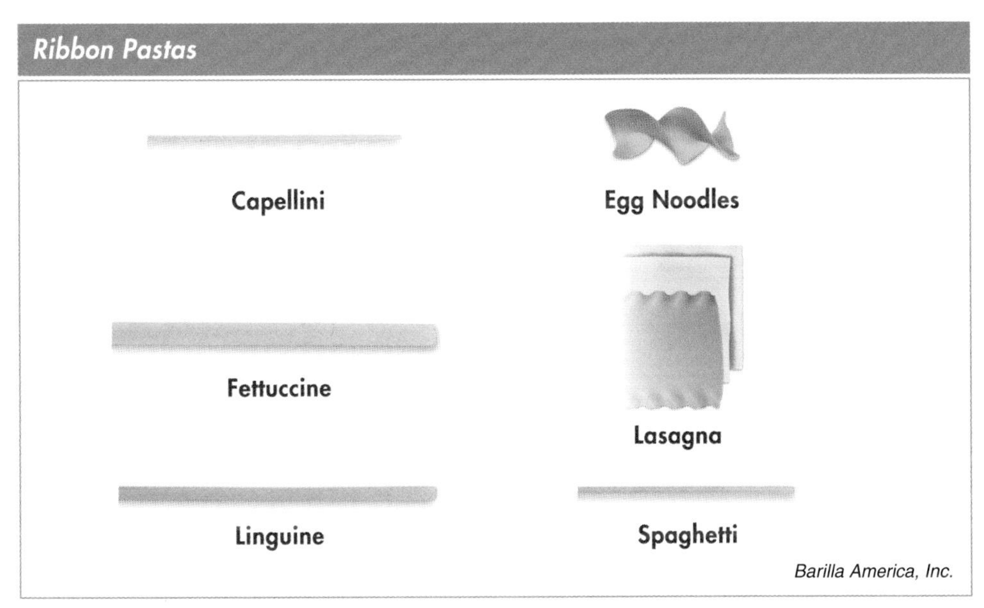

Figure 14-3. A ribbon pasta is a thin, round strand or flat, ribbonlike strand of pasta.

Shaped Pastas. *Shaped pasta* is pasta that has been extruded into complex shapes such as corkscrews, bowties, shells, flowers, or stars. **See Figure 14-4.** Shaped pastas can be smooth or ridged. The crevices in ridged pastas effectively hold sauce or other ingredients. Shaped pastas are primarily available dried, but some varieties are sold fresh or frozen. Campanelle, conchiglie, couscous, farfalle, fiori, gemelli, jumbo shells, orecchiette, orzo, radiatori, rotini, and stelline are common types of shaped pastas.

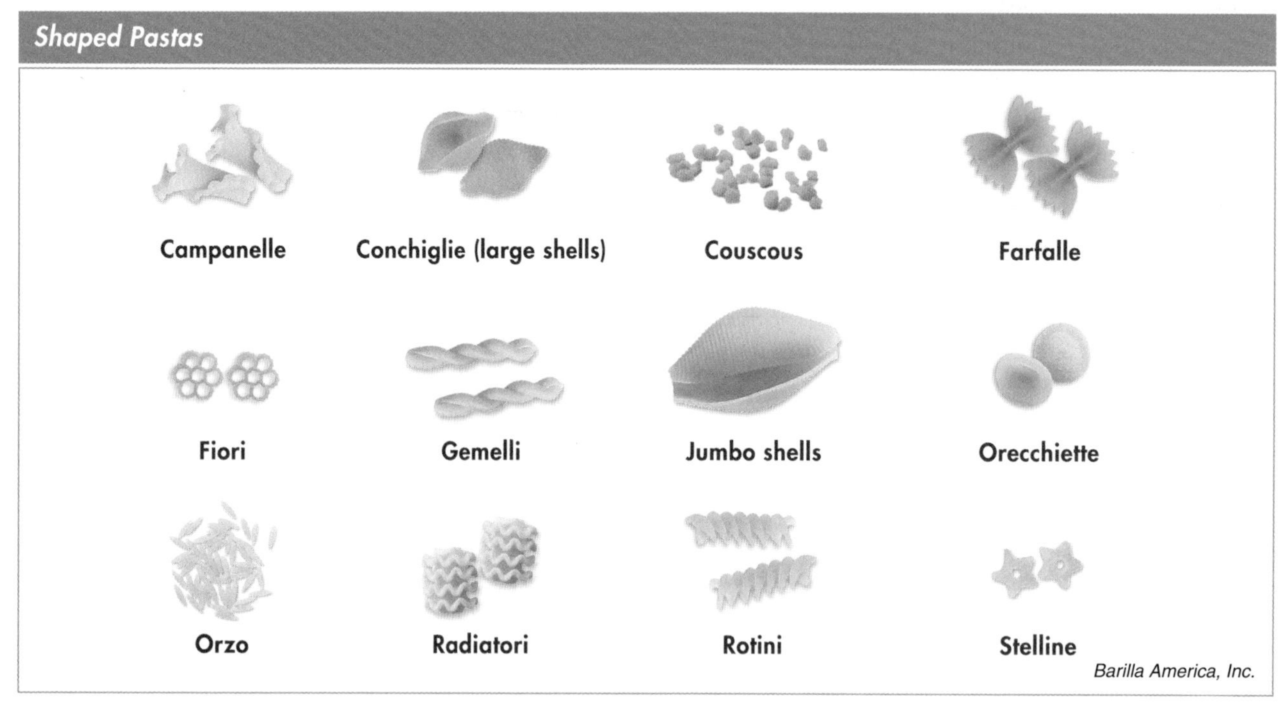

Figure 14-4. A shaped pasta is a pasta that has been extruded into a complex shape such as a corkscrew, bowtie, shell, flower, or star.

INGREDIENT SPOTLIGHT: Couscous

Indian Harvest Specialtifoods, Inc./ Rob Yuretich

Unique Features

- Granular pasta typically made from semolina flour; also available in whole grain options
- Moroccan couscous comparable in size to coarse kosher salt grains
- Israeli couscous comparable in size to whole peppercorns
- Low in fat, cholesterol, and sodium
- Source of selenium

Menu Applications

- Add to soups, stews, and salads
- Mix with fruits, nuts, sweet spices, and natural sugars to create a hot breakfast item or dessert

Nutrition Facts
Serving Size 1 cup, cooked (157 g)

Amount Per Serving	
Calories 176	Calories from Fat 2
	% Daily Value*
Total Fat 0g	0%
Saturated Fat 0g	0%
Trans Fat	
Cholesterol 0mg	0%
Sodium 8mg	0%
Total Carbohydrate 36g	12%
Dietary Fiber 2g	9%
Sugars 0g	
Protein 6g	

Vitamin A	0%	Vitamin C	0%
Calcium	1%	Iron	3%

* Percent Daily Values (DV) are based on a 2,000 calorie diet.

Tube Pastas. *Tube pasta* is pasta that has been pushed through an extruder to form a tube shape and then fed through a cutter that cuts the tube to the desired length. **See Figure 14-5.** Tube pastas can be straight, curved, smooth, or ridged. Some large varieties of tubes are stuffed with ingredients such as cheese, seafood, or vegetables. Most tube pastas are only available in dried form. Common varieties of tube pastas include cellentani, ditalini, macaroni, manicotti, penne, pipettes, rigatoni, and ziti.

Stuffed Pastas. *Stuffed pasta* is pasta that has been formed by hand or machine to hold a filling. **See Figure 14-6.** Stuffed pastas start out as sheets of rolled pasta dough. Mounds of filling are evenly spaced on the sheets of dough and a wash is brushed around each filling. A second sheet of dough is laid on top and the dough is cut and formed into individual pieces. Stuffed pastas are commonly filled with meats, poultry, seafood, cheeses, vegetables, or a combination of ingredients. Examples of stuffed pastas include ravioli, tortellini, and tortelloni.

Asian Noodles. Asian noodles come in long, ribbonlike strands and can be classified as vegetable-based, egg-based, or grain-based noodles. **See Figure 14-7.** Asian noodles are typically not cooked in boiling water. Instead, hot water is poured over the noodles until they soften.

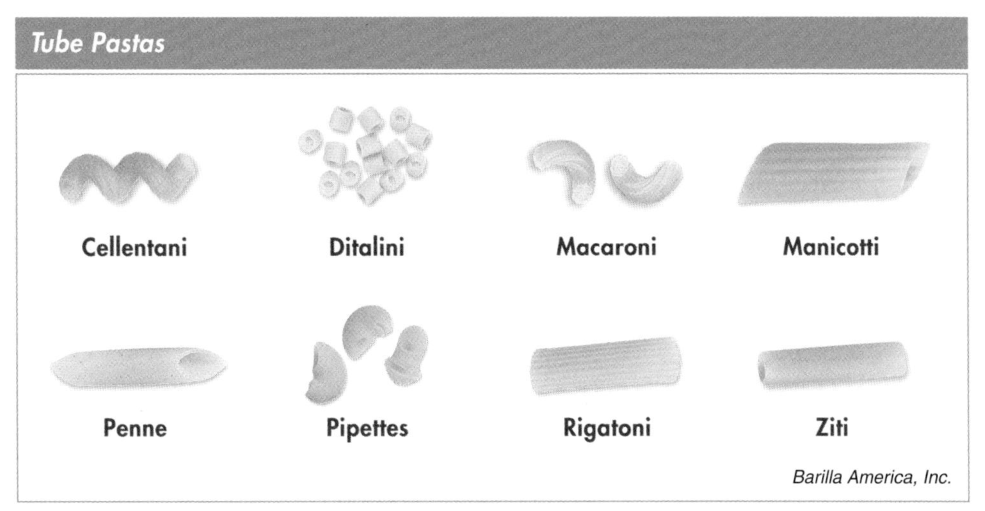

Figure 14-5. A tube pasta is a pasta that has been pushed through an extruder to form a tube and then fed through a cutter that cuts the tube to the desired length.

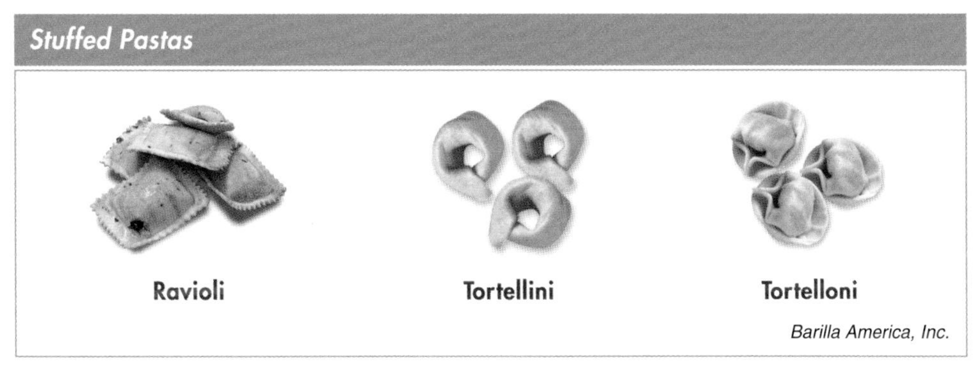

Figure 14-6. A stuffed pasta is a pasta that has been formed by hand or machine to hold a filling.

Figure 14-7. Asian noodles come in long, ribbonlike strands and can be classified as vegetable-based, egg-based, or grain-based noodles.

- Vegetable-based Asian noodles are made from bean or vegetable starches instead of wheat flour. Vegetable-based Asian noodles include cellophane and shirataki noodles.

  Cellophane noodles, also known as glass noodles, are made from mung bean starch and are fairly transparent. Cellophane noodles can be used either hot or cold and are often used in soups or stir-fries or as fillings, especially in vegetarian dishes. Similar noodles made from sweet potato starch are available. Shirataki noodles are long, slender noodles made from Japanese yams and do not contain gluten, fat, or calories. Some varieties of shirataki noodles include tofu, which adds a small amount of calories.

- Egg-based Asian noodles are made from wheat flour, eggs, and water. This style of noodle often has a yellow hue and includes hokkien, lo mein, and somen noodles.

  Hokkien noodles resemble spaghetti and are sold fresh, dried, or vacuum-sealed. Hokkien noodles work well in soups or stir-fries. Lo mein noodles, also called Cantonese noodles, come in varying widths and can be purchased fresh, dried, or frozen. Lo mein is a popular stir-fry dished named for the noodle itself. Somen noodles are long, thin wheat noodles sold as bundled rods. They are used in soups and stir-fries and are served cold in noodle dishes.

- Grain-based Asian noodles are typically made from wheat, rice, or buckwheat flours but do not contain eggs. Common types of grain-based Asian noodles include la mian, rice, soba, and udon noodles.

La mian noodles are Chinese wheat noodles available fresh or dried in round strands or ribbons of various widths. They are used in a variety of dishes including soups, stir-fries, and chow mein. Rice noodles are made from ground rice flour and are available dried in various shapes. They should be soaked in hot water before cooking and rinsed after cooking to remove excess starch. Soba noodles are made from buckwheat flour and have a brown-gray color. They can be used hot or cold in soups, entrées, and side dishes. Udon noodles are thick, white wheat noodles and are available fresh or dried. They have a chewy, slippery texture and are typically served hot in soups, stews, and stir-fries. Ramen noodles are similar to udon noodles, only thinner and longer. They are commonly served in a broth with ingredients such as meat and vegetables.

Gluten-Free Pastas. Most pastas are made from flour that contains gluten. Gluten causes an adverse reaction in individuals with gluten intolerance or celiac disease. To meet the growing demand for gluten-free menu items, gluten-free pastas made from rice, wild rice, corn, quinoa, buckwheat, or potato flours can be used. Gluten-free pastas are available in varieties of ribbon, shaped, and tube pastas. While some gluten-free pastas are available fresh or frozen, most are available dried.

PREPARING NUTRITIOUS PASTAS

The large variety of pastas available can be used to produce flavorful, nutrient-dense menu items. Pastas can be added to soups, salads, stews, and casseroles to develop colors, textures, and flavors. By using whole grain pastas enhanced with seasonal vegetables and aromatic herbs and spices, as well as lean meats, poultry, or seafood, chefs can offer dishes rich in protein, complex carbohydrates, vitamins, and minerals. **See Figure 14-8.** It is also important to plate appropriately sized portions of pasta.

Preparing Nutritious Pastas

Fresh parsley
Fire-roasted tomatoes
Whole grain pasta
Lean pork meatballs
Zucchini
Courtesy of The National Pork Board

Figure 14-8. By using whole grain pastas enhanced with seasonal vegetables and aromatic herbs and spices, as well as lean meats, poultry, or seafood, chefs can offer dishes rich in protein, complex carbohydrates, vitamins, and minerals.

Knowledge Check 14-1

1. Name the type of flour most often used to make pastas.
2. Explain the purpose of gluten in pasta doughs.
3. Give an example of how a chef can enhance the perceived value of a pasta dish.
4. Explain the benefits of consuming appropriate portions of whole grain pastas.
5. Differentiate between ribbon and tube pastas.
6. List four varieties of shaped pastas.
7. Explain how to prepare stuffed pastas.
8. Explain how Asian noodles are typically cooked.
9. List six flours used to make gluten-free pastas.

Cooking Pastas

Pasta should be cooked in boiling water. The water is salted to enhance flavor. Four to six quarts of water should be used for every pound of pasta. The boiling water dilutes the starch and allows the pasta to move around, preventing it from sticking together. Most pasta doubles in volume when cooked. For example, 1 lb of uncooked pasta typically yields 2–2½ lb of cooked pasta.

Fresh pasta generally cooks within 3–5 minutes once it is added to boiling water. Cooking times for dried pastas range from 3–15 minutes. **See Figure 14-9.** Frozen pastas are often precooked before being frozen and typically take 6–8 minutes to cook.

Pasta is cooked until it becomes al dente, meaning "to the tooth." Al dente pasta should have a slight resistance in the center when it is chewed. Undercooked pasta remains somewhat brittle, while overcooked pasta is mushy.

Pasta should be drained, not rinsed. Rinsing pasta washes away starch and flavor and prevents sauces from adhering.

Approximate Cooking Times for Dried Pasta

Pasta	Minutes	Pasta	Minutes
Capellini	3–5	Spaghetti	9–10
Egg noodles	7–8	Penne and mostaccioli	9–11
Macaroni	7–8	Tortellini	10–11
Conchiglie	8–9	Farfalle	11–12
Lasagna	8–9	Fettuccine	12–13
Manicotti and connelloni	8–9	Fusilli	12–13
Linguine	9–10	Jumbo shells	12–13
Orzo	9–10	Vermicelli	12–13

Figure 14-9. Cooking times for dried pastas range from 3–15 minutes.

Flavor Development

Pasta is versatile and pairs well with other ingredients to create a variety of flavorful dishes. For example, sauces and flavor enhancers such as aromatic vegetables, herbs, and spices help develop flavors in pasta menu items. The mildness of pasta allows the flavors of other ingredients to shine, making it essential to use high-quality, flavorful ingredients. **See Figure 14-10.** By using high-quality ingredients such as olive oils and cheeses, and by incorporating just enough sauce or dressing to coat the pasta, the amount of fat and calories in the dish can be reduced while raising the flavor profile.

The type of pasta used in a dish also affects flavor development. Egg-based pastas are slightly richer than pastas made without eggs, and pasta doughs that incorporate vegetable purées, cheeses, and herbs increase the flavor of the pasta.

Plating Pastas

For more nutrient-dense pasta dishes, chefs can replace refined pastas with whole grain varieties and serve appropriate portion sizes. Pastas are typically neutral in color, so enriching the dish with colorful, nutrient-dense ingredients not only makes the dish healthier, but also elevates the presentation. For example, using basil pesto sauce gives pasta a vibrant green hue. The dish can then be garnished with bright red tomatoes and shavings of fresh Parmesan cheese to present a nutrient-dense and visually appealing menu item. Pasta doughs made with purées of vegetables such as beets and spinach also add color, flavor, and interest to the plate. **See Figure 14-11.**

Pastas come in shapes and sizes that can be plated to enhance presentation. For example, ravioli can be plated in a unique pattern, fanned, or used as a base to provide height to other ingredients. Star-shaped pastas are often used in soups for visual effect and because they stay on a spoon easily. Pastas need to be sauced appropriately because using too much sauce not only adds excess calories, but often creates a puddle on the plate that detracts from the presentation.

FOOD SCIENCE EXPERIMENT: Textural Differences in Cooked Pasta 14-1

Objective

- To compare undercooked, al dente, and overcooked pasta in terms of texture.

Materials and Equipment

- Masking tape
- Marker
- 3 medium bowls
- Pot
- Water
- 1 lb dried elbow macaroni
- Spoon
- Timer
- Slotted spoon
- Colander
- Tasting spoons

Procedure

Allergen Alert

The materials used in this experiment may contain one or more food allergens.

1. Use the masking tape and marker to label one bowl "Undercooked Pasta," the second bowl "Al Dente Pasta," and the third bowl "Overcooked Pasta."
2. Fill the pot three-fourths full with water and bring to a boil.
3. Read the directions on the package of pasta to determine cooking times.
4. Add the pasta to the boiling water and stir.
5. Set the timer for five minutes less than package directions.
6. When the timer goes off, use the slotted spoon to place one-third of the pasta in the bowl labeled "Undercooked Pasta." Quickly reset the timer for five minutes.
7. When the timer goes off, use the slotted spoon to place one-third of the pasta in the bowl labeled "Al Dente Pasta." Quickly reset the timer for five minutes.
8. When the timer goes off, use the colander to strain the remaining pasta and then place the pasta in the bowl labeled "Overcooked Pasta."
9. Evaluate the texture of each bowl of pasta.

Typical Results

There will be distinct variances in the texture of each bowl of pasta.

Pasta Flavor Development

Barilla America, Inc.

Figure 14-10. The mildness of pasta allows the flavors of ingredients such as fresh tomatoes, herbs, and spices to shine, making it essential to use high-quality ingredients.

Colorful Pastas

Figure 14-11. Pasta doughs made with purées of vegetables such as beets and spinach add color, flavor, and interest to the plate.

CULINARY NUTRITION RECIPE-MODIFICATION PROCESS

Fettuccine Alfredo

Fettuccine Alfredo features fettuccine noodles tossed in a sauce comprised of butter, heavy cream, Parmesan cheese, and freshly ground black pepper. The creamy sauce coats each noodle, adding rich flavor. However, the sauce is made with dairy products that cause the dish to be high in calories, total fat, and saturated fat. The fettuccine noodles are made from refined semolina flour, making the dish low in dietary fiber.

Yield: 4 servings, 2 cups each

Ingredients

1 gal.	water
2 tsp	salt
1 lb	fettuccine
6 tbsp	unsalted butter
1 ea	shallot, minced
1 c	heavy cream
1 c	Parmesan cheese, grated
½ tsp	salt
¼ tsp	black pepper, ground
3 tbsp	parsley, chopped

Preparation

1. Bring water to a boil in a stockpot and season with salt.
2. Add pasta to boiling water and cook until al dente.
3. Reserve ¼ cup of the pasta water and then drain the pasta. Set aside.
4. Melt butter in a large saucepan over medium heat. Add shallot and cook until tender.
5. Add heavy cream and bring to a boil.
6. Cook until the sauce has thickened and reduced slightly (approximately 5 minutes). Reduce heat to low.
7. Return pasta and reserved pasta water to the stockpot on low heat.
8. Add sauce and half the Parmesan cheese to the pasta.
9. Season with salt and pepper. Toss to combine.
10. Sprinkle each dish with remaining Parmesan cheese and garnish with parsley.

Evaluate original recipe for sensory and nutritional qualities

- Fettuccine Alfredo has a rich, creamy texture and savory flavor.
- Dish is high in calories, total fat, and saturated fat.
- Fettuccine is low in dietary fiber.

Establish goals for recipe modifications

- Reduce calories, total fat, and saturated fat.
- Maintain creamy texture and savory flavor.
- Increase the amount of dietary fiber.

Identify modifications or substitutions

- Use a substitution for the butter and heavy cream.
- Reduce the amount of Parmesan cheese.
- Use a substitution for the fettuccine.

Determine functions of identified modifications or substitutions

- Butter and heavy cream provide a smooth mouthfeel and rich flavor.
- Parmesan cheese adds flavor.
- Fettuccine provides structure and texture.

Select appropriate modifications or substitutions

- Substitute olive oil and evaporated milk for the butter and heavy cream.
- Reduce the amount of Parmesan cheese to 1 tbsp.
- Add aromatics and flavorings, including roasted garlic, wine, and herbs, to elevate flavor.
- Substitute whole wheat fettuccine for the fettuccine made with refined semolina flour.

Test modified recipe to evaluate sensory and nutritional qualities

- Substituting olive oil and evaporated milk for the butter and heavy cream reduces calories and fats while still providing a creamy texture.
- Reducing the amount of Parmesan cheese lowers calories and fats.
- Adding aromatics and flavorings creates depth of flavor.
- Using whole wheat fettuccine adds dietary fiber.

Modified Fettuccine Alfredo

To create a healthier version of fettuccine Alfredo, the calories, total fat, and saturated fat need to be reduced. To accomplish these goals while maintaining a rich, creamy texture and flavor, the butter and heavy cream are replaced with olive oil and evaporated milk, and the quantity of Parmesan cheese is reduced. Roasted garlic, wine, and thyme are added to create depth of flavor, and whole wheat fettuccine is used to increase dietary fiber.

Yield: 4 servings, 2 cups each

Ingredients

1 gal.	water
2 tsp	salt
1 lb	fettuccine, whole wheat
2 tsp	olive oil
½ c	yellow onion, julienned
1 tbsp	garlic, roasted
⅓ c	white wine
14 oz	evaporated milk
1 tsp	kosher salt
1 tsp	thyme, minced
1 tbsp	Parmesan cheese, grated
¼ tsp	black pepper, ground
3 tbsp	parsley, chopped

Preparation

1. Bring water to a boil in a stockpot and season with salt.
2. Add pasta to boiling water and cook until al dente. Drain and set aside.
3. Heat oil in a saucepan over low heat. Add onions and cook until caramelized (approximately 10 minutes).
4. Add garlic and wine and reduce until almost dry.
5. Add evaporated milk and bring to a simmer, but do not boil. Allow the liquid to simmer until it is reduced by a third.
6. Transfer the mixture to a blender and purée until smooth.
7. Stir in salt, thyme, Parmesan cheese, and pepper.
8. Return pasta to the stockpot and apply low heat.
9. Add sauce to the pasta and toss to combine.
10. Garnish with parsley.

Fettuccine Alfredo Nutritional Comparison

Nutrition Facts	Original	Modified
Calories	958.8	603.1
Total Fat	48.4 g	12.2 g
Saturated Fat	29.2 g	5.1 g
Dietary Fiber	2.8 g	12.4 g

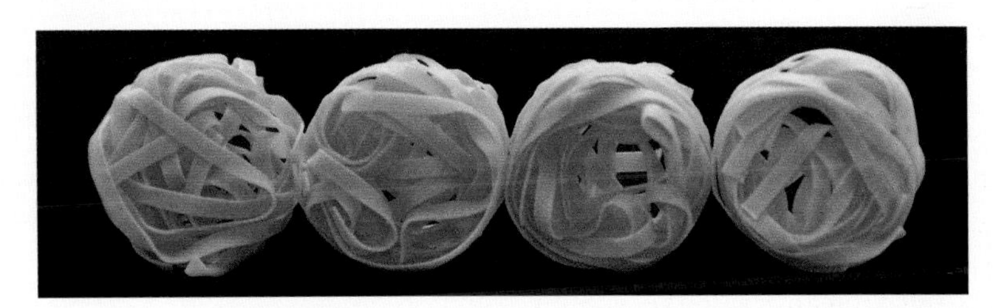

CULINARY NUTRITION RECIPE-MODIFICATION PROCESS

Signature Lemony Fettuccine Alfredo with Peas

This signature recipe features a lower calorie fettuccine Alfredo with enhanced color and flavor. Less pasta is used, the amount of butter is reduced, and low-fat milk is used. Adding a small amount of low-fat cream cheese to the sauce enhances its creamy texture. To elevate flavor, garlic and lemon zest are added. Incorporating fresh peas adds nutrients such as dietary fiber and folate as well as color, texture, and sweetness.

Yield: 4 servings, 1½ cups each

Ingredients

1 gal.	water
2 tsp	salt
12 oz	fettuccine
1 tbsp	unsalted butter
1 clove	garlic, minced
1 tsp	lemon zest
1 c	2% milk
2 tbsp	low-fat cream cheese
¾ c	Parmigiano-Reggiano cheese, grated
¼ tsp	black pepper, ground
2 c	peas, blanched
3 tbsp	parsley, chopped

Preparation

1. Bring water to a boil in a stockpot and season with salt.
2. Add pasta to boiling water and cook until al dente.
3. Reserve ¼ cup of the pasta water and then drain the pasta. Set aside.
4. Melt butter in a large saucepan over medium heat. Add garlic and lemon zest and cook until garlic is soft (approximately 1 minute).
5. Add milk and cream cheese and stir until smooth.
6. Stir in Parmigiano-Reggiano cheese and pepper.
7. Return pasta and reserved pasta water to the stockpot on low heat.
8. Add sauce and peas to the pasta. Toss to combine.
9. Garnish with parsley.

Fettuccine Alfredo Nutritional Comparison

Nutrition Facts	Original	Modified	Signature
Calories	958.8	603.1	468.6
Total Fat	48.4 g	12.2 g	12.9 g
Saturated Fat	29.2 g	5.1 g	7.0 g
Dietary Fiber	2.8 g	12.4 g	4.6 g
Folate	63.9 mcg	71.3 mcg	210.9 mcg

Knowledge Check 14-2

1. Identify the volume of water needed to cook 1 lb of pasta.
2. Explain the meaning of "al dente."
3. Explain why cooked pasta should not be rinsed.
4. List the reasons why it is important to use high-quality ingredients in pasta dishes.
5. Give an example of how to enhance the color of a plated pasta dish.

PASTA MENU MIX

Pasta commonly serves as the starch component of many dishes on lunch and dinner menus. This provides the opportunity to use a variety of pastas in dishes across the menu and to highlight pastas made from various types of flours. Pastas can also be used to add nutrients to appetizers, soups, salads, entrées, and sides. Because pasta has a mild flavor, it can be used as the base for highlighting different flavor profiles. **See Figure 14-12.**

Menu item descriptions that call attention to different pasta varieties and interesting flavor combinations often encourage sales. Many pasta dishes have a signature feature that can be emphasized. For example, the menu description of a signature lasagna can emphasize its fire-roasted tomato sauce and layers of grilled zucchini. Menu descriptions that showcase fresh house-made pastas also can be an effective part of the menu. **See Figure 14-13.**

Pastas in Appetizers

Pastas served as appetizers include gnocchi and stuffed pastas, such as ravioli, tortellini, and pot stickers. Gnocchi is typically made using flour, water, potatoes, and sometimes cheese. To enrich the color and nutrient density of gnocchi, puréed vegetables, such as beets, carrots, winter squash, or spinach, can be incorporated into the dough. Ravioli and tortellini can be filled with cooked meats, cheeses, or vegetables such as spinach. Pot stickers are filled with similar ingredients and served with a dipping sauce. Serving these pastas with healthy sauces such as vegetable purées offers guests a healthy menu option.

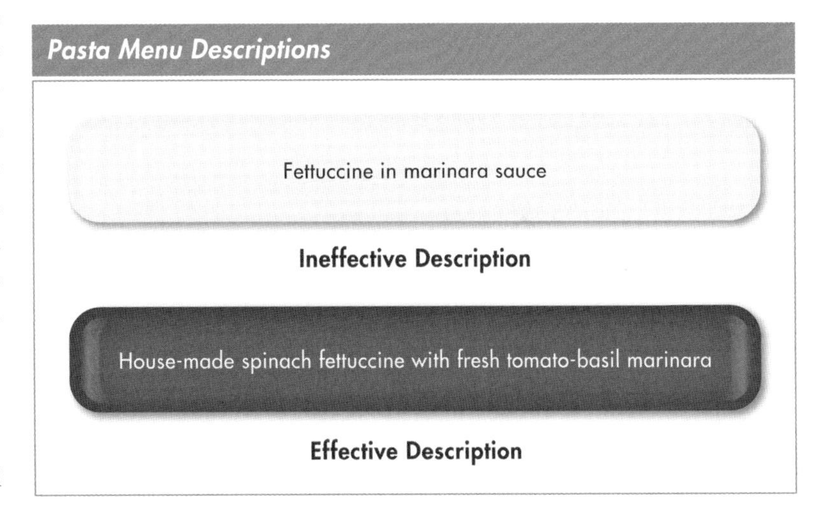

Figure 14-13. Menu descriptions that showcase fresh house-made pastas can be an effective part of the menu.

Pastas on the Menu

Menu Item	Description
Chicken and couscous soup	Flavorful chicken, tender Israeli couscous, and seasonal vegetables in a lemon-thyme infused chicken broth
Root vegetable gnocchi with pork and arugula	Colorful root vegetable dumplings accented with roasted pork shoulder and a baby arugula salad
Beef tenderloin with orzo and broccolini	Grilled beef tenderloin with a rich demi-glace accompanied by savory wild mushroom orzo and sautéed broccolini

Figure 14-12. Pastas can be used to highlight different flavor profiles and add nutrients to a variety of menu items.

Pastas in Soups

Some soups, such as chicken noodle and minestrone, traditionally showcase pastas. **See Figure 14-14.** However, pastas can be added to almost any soup. Adding different types of pastas to soups is an effective way to add complex carbohydrates to the menu mix. Smaller pasta varieties, such as couscous, orzo, and ditalini, work well in soups because they easily fit on a spoon. Ribbon pastas such as spaghetti, fettuccine, and capellini can also be used but need to be broken down into smaller pieces before being incorporated into a soup. Pasta is usually cooked separately and then added to a soup, but it can be cooked in the soup to help thicken it.

Pastas in Soups

Figure 14-14. Some soups, such as chicken noodle, traditionally showcase pastas.

Pastas in Salads

Salads that include pasta as a main ingredient are very popular. **See Figure 14-15.** Pasta salads are commonly made with shaped pastas, such as farfalle, rotini, and orzo. Once the pasta has been cooked and cooled, it is mixed with ingredients that range from cooked meats, poultry, or seafood to raw or cooked fruits and vegetables. Pasta salads are often bound by a vinaigrette- or mayonnaise-based dressing and served cold.

Pastas in Salads

Courtesy of The National Pork Board

Figure 14-15. Salads that include pasta as a main ingredient are very popular.

Combining a whole grain pasta with an array of vegetables, a moderate portion of lean meats, and a small amount of oil creates a healthy menu item that is rich in protein, complex carbohydrates, vitamins, minerals, and unsaturated fat. In contrast, pasta salads made with refined grains and dressed in mayonnaise-based dressings lack essential nutrients and are high in saturated fat.

Tanimura & Antle®

Pastas, such as small shells, may be added to leafy green salads to enhance texture.

Pastas in Entrées

There are virtually endless applications of pastas in entrées. Common entrées that feature pastas include raviolis, stuffed shells and tube pastas, spaghetti with tomato sauce, fettuccine Alfredo, and lasagna. Noodles are also popular in Asian entrées as a foundation for stir-fry dishes. Pasta can be the main item on a plate or part of an overall dish. Using whole grain pasta provides greater nutrition and satiety. Pastas paired with an assortment of vegetables and lean proteins, including tofu or beans, create nutritious, flavorful, and well-balanced menu entrées.

It is also important to serve proper portion sizes of pastas. Large servings of pasta are high in fat and calories. Replacing rich sauces that are high in saturated fat with flavorful sauces made from garden-ripe ingredients and unsaturated oils is another way to offer pasta entrées that are both appealing and nutritious. **See Figure 14-16.**

Pastas in Sides

Pasta salad is a popular side dish. Macaroni and cheese is also a classic side dish, but it is typically high in fat and calories. Macaroni and cheese can be more nutritious by incorporating vegetables and low-fat dairy products.

Pasta side dishes can be found on à la carte menus in which the main item and pasta side are ordered separately. Offering a whole wheat linguini is an effective way to provide guests with a healthy à la carte option that can complement a variety of menu items. Adding an array of fruits and vegetables to pasta sides provides appealing colors, textures, and flavors and enhances nutrition.

Chef's Tip

Pasta that has been parcooked and stored for later service can be brought back to temperature by rinsing it in warm water or dropping it in simmering water.

Nutritional Comparison of Pasta Entrées*

Entrée	Calories	Total Fat	Saturated Fat
Pasta carbonara	830	56 g	30 g
Linguini with scallops, capers, and sun-dried tomatoes	376	15 g	5 g

* based on an 8 oz serving size

Figure 14-16. Sauces made from garden-ripe ingredients and unsaturated oils are more nutrient dense than cream-based sauces, which are high in saturated fat.

Knowledge Check 14-3

1. List popular pastas served as appetizers.
2. Describe how pastas can be used in soups and salads.
3. Explain how to create nutrient-dense pasta entrées.
4. Give an example of a nutrient-dense pasta side dish.

PROMOTING PASTAS ON THE MENU

Many pasta dishes are regarded as comfort foods and are easily promoted. However, when promoting pasta dishes, it is important to maintain a nutritionally balanced menu. Serving whole grain pastas with lean proteins and seasonal vegetables meets the demand for healthy menu items. Also, promoting pastas made from gluten-free grains meets the growing demand for gluten-free menu items.

Pasta salads, such as spaghetti, cucumbers, and tomatoes dressed in a vinaigrette, are a nutritious carry-out item at quick service venues. Baked pastas, such as lasagna and stuffed shells, are popular buffet items. **See Figure 14-17.** Nutrient-dense pasta items include spaghetti with pesto sauce, fettuccine with grilled vegetables, eggplant lasagna, and ziti with roasted chicken.

Fine dining venues may prepare pastas in-house and incorporate seasonal produce, artisan cheeses, and a variety of meats, poultry, and seafood into their pasta menu items. **See Figure 14-18.** Special event venues commonly top pasta with grilled or roasted meats and vegetables.

Baked Lasagna

Barilla America, Inc.

Figure 14-17. Baked pastas such as lasagna are popular buffet items.

Pastas at Fine Dining Venues

Daniel NYC/P. Medilek

Spring Pea Ravioli

Daniel NYC/E. Kheraj

Butternut Squash Ravioli

Figure 14-18. Fine dining venues may prepare pastas in-house and incorporate seasonal ingredients.

Knowledge Check 14-4

1. Give an example of how pasta can be used to create a nutritionally balanced menu item.
2. Describe the ways in which pastas are used to showcase seasonal foods at fine dining venues.

GRAINS

A *grain* is the edible fruit, in the form of a kernel or seed, of a grass. Grains contain an abundance of complex carbohydrates, which serve as the main source of energy for the body. Many grains are also rich in dietary fiber, B vitamins, vitamin E, magnesium, potassium, iron, zinc, copper, and selenium. There are four components to a kernel of grain: the husk, bran, germ, and endosperm. **See Figure 14-19.** Most grains require processing to make it easier for the body to digest and absorb their nutrients. However, the amount of processing has a direct effect on nutritional value, as highly processed grains are less nutrient dense. The different forms of grains include whole, cracked, refined, milled, pearled, and flaked.

A *whole grain* is a grain that has only had the husk removed. A *cracked grain* is a whole kernel of grain that has been cracked by being placed between rollers. Because the bran, germ, and endosperm are preserved in whole grains, such as brown rice, wheat berries, and oat groats, these grains are rich in dietary fiber and B vitamins. **See Figure 14-20.** Whole grains also contain antioxidants and phytonutrients. Consuming whole grains has been shown to improve cardiovascular and digestive health, reduce cholesterol, and promote stable blood glucose levels.

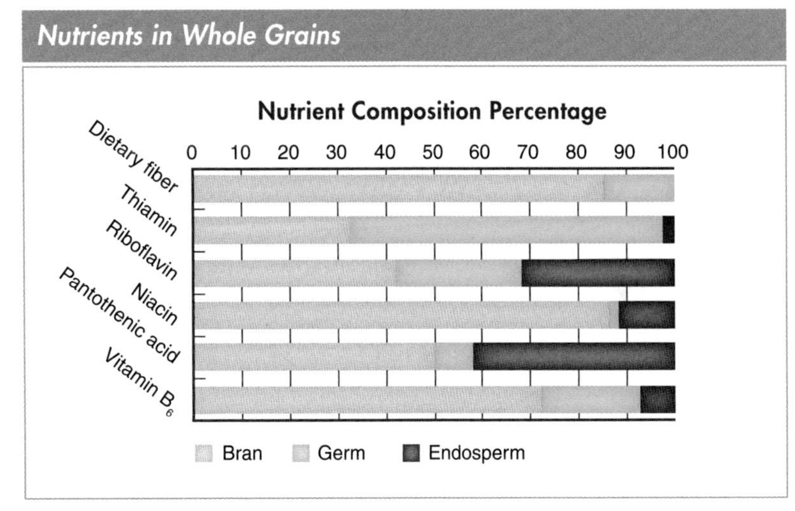

Figure 14-20. Whole grains are rich in dietary fiber and B vitamins.

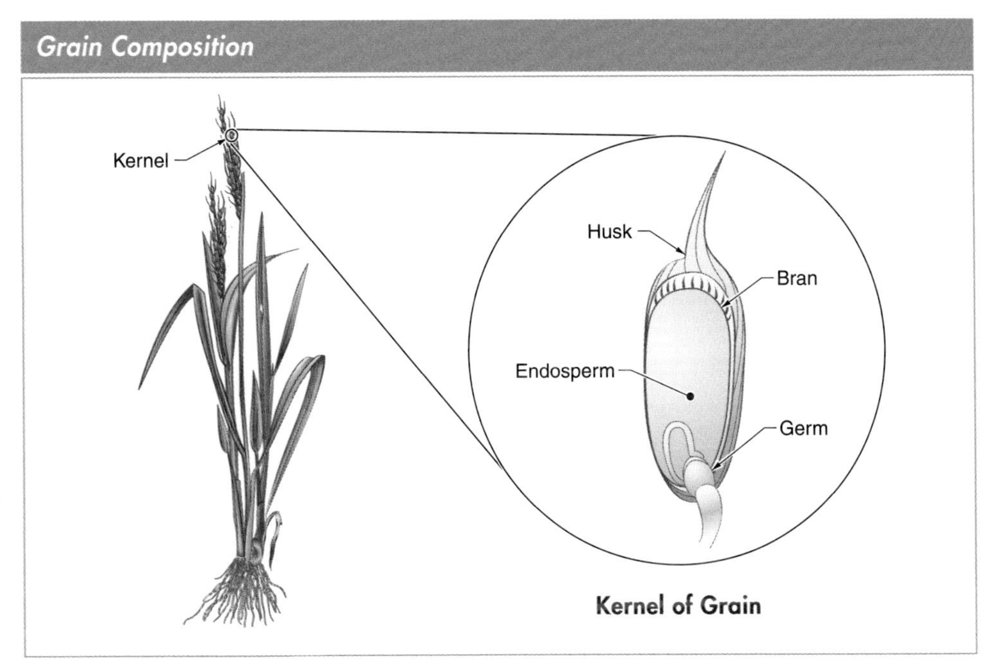

Figure 14-19. There are four components to a kernel of grain: the husk, bran, germ, and endosperm.

A *refined grain* is a grain that has been processed to remove the bran, germ, or both. When a grain kernel is stripped of both the bran and germ, dietary fiber is lost. The remaining endosperm is mainly comprised of starch. Refined grains are easier to digest and take less time to cook than whole grains.

Refined grains can be milled, pearled, or flaked. A *milled grain* is a refined grain that has been ground into a fine meal or powder. Cornmeal and all flours are milled grains. A *pearled grain* is a refined grain with a pearl-like appearance that results from having been scrubbed and tumbled to remove the bran. Barley is often pearled. A *flaked grain,* also known as a rolled grain, is a refined grain that has been rolled to produce a flake. Oatmeal is a flaked grain.

Perceived Value of Grains

Whole grains and ancient grains are perceived as healthier menu options than refined grains. These grains add value to the menu because of their nutritional benefits. Increasing the portion of whole or ancient grains and vegetables on the plate and serving an appropriate amount of meat, poultry, or seafood is often cost effective and still perceived by guests as a plentiful portion. **See Figure 14-21.**

In order to further increase the perceived value of grains on the menu, chefs can choose more exotic grain varieties such as red rice, whole wheat berries, black barley, or quinoa. These grains attract the attention of guests and can be offered at higher prices.

Perceived Value of Grains

Photo Courtesy of the Beef Checkoff

Figure 14-21. Increasing the portion of whole or ancient grains and vegetables and serving an appropriate amount of meat is often cost effective and still perceived by guests as a plentiful portion.

Cereal Grains

A *cereal grain* is a grain that is derived from plants in the grass family. Examples of cereal grains include rice, wheat, oats, barley, corn, rye, triticale, millet, teff, farro, and spelt. Cereal grains are available in whole form as well as processed into flours for use in products such as packaged cereals, pastas, and baked goods.

Millet, teff, farro, and spelt are considered ancient grains and have been consumed in their natural form for thousands of years. In contrast, grains such as rice, wheat, and corn have been selectively bred or genetically modified to make them easier to grow, harvest, and refine.

Rice. Rice is grown around the globe and serves as a staple food for half of the world's population. Rice is gluten-free and many varieties are available. Rice can be white, brown, purple, black, red, or green. Brown rice has a chewier texture and nuttier flavor and is more nutrient dense than white rice because it is less processed. **See Figure 14-22.** While brown rice has had only the husk removed, white rice has had the husk, bran, and germ removed. As a result of milling, white rice loses a large percentage of B vitamins, phosphorus, magnesium, iron, dietary fiber, and fatty acids. White rice that has been completely milled and polished is enriched with thiamin, niacin, and iron. Regardless of its color and flavor, rice is classified by the size of the grain. Grain sizes of rice include short-grain, medium-grain, and long-grain. **See Figure 14-23.**

Nutritional Comparison of Brown and White Rice*

Type of Rice	Dietary Fiber	Omega-6 Fatty Acids	Manganese
Brown	4 g	603 mg	1.8 mg
White	1 g	98 mg	0.7 mg

* based on 1 cup serving size, cooked

Figure 14-22. Brown rice is more nutrient dense than white rice because it is less processed.

Rice

Figure 14-23. Grain sizes of rice include short-grain, medium-grain, and long-grain.

Wild rice is not a true rice, but comes from an aquatic grass and has a rich, nutty flavor. Wild rice is low in fat, cholesterol, and sodium. Wild rice is also a good source of phosphorus, magnesium, zinc, and manganese.

Food Science Note

Purple, black, and red rice varieties are significantly higher in anthocyanin, an antioxidant that promotes cellular health.

Wheat. *Wheat* is a light-yellow cereal grain cultivated from an annual grass that yields the flour used in many pastas and baked goods. The most common forms of wheat used in cooking include durum wheat, wheat berries, and bulgur wheat.

- Durum wheat is the hardest type of wheat. Durum is commonly milled into semolina. *Semolina* is the granular product that results from milling the endosperm of durum wheat. Semolina is high in protein and, therefore, high in gluten. Due to its high gluten strength, semolina is used to make most pastas. **See Figure 14-24.**

Protein in Wheats*

Durum	Wheat Berries	Bulgur
7 g	6 g	4 g

* based on ¼ cup serving size

Figure 14-24. Durum wheat is commonly milled into semolina flour and used to make pastas because of its high gluten (protein) strength.

- Wheat berries have had only the husk removed and provide a distinct chewy texture. Wheat berries can be used in salads, pilafs, and stuffings.
- Bulgur wheat kernels are a golden-brown color and nutty in flavor. The husks and bran of bulgur wheat are removed before the wheat is steamed and dried. Bulgur is commonly simmered and used in salads.

Wellness Concept

Freekeh wheat is an ancient grain derived from young green wheat that has been toasted. Freekeh is rich in dietary fiber, protein, selenium, potassium, and magnesium.

Oats. *Oats* are a cereal grain derived from the berries of oat grass. **See Figure 14-25.** Oats come in many different forms. An *oat groat* is an oat grain from which only the husk has been removed. *Steel-cut oats* are oat groats that have been toasted and cut into small pieces. *Rolled oats,* also known as old-fashioned oats, are oats that have been steamed and flattened into small flakes. Rolled oats require less cooking time than steel-cut oats. Oats are commonly served as a hot cereal or incorporated into baked items. Oats can also be served in a fashion similar to rice pilaf or risotto.

Barley. *Barley* is a cereal grain that resembles brown rice in shape, yet takes longer to cook. **See Figure 14-26.** Barley in its whole grain form is either brown or black. Pearled barley is polished barley with the bran removed. Cooked barley has a very chewy texture and a robust, earthy flavor. Barley is often used as a natural thickener and flavor enhancer in soups and stews. It is also a common ingredient in the production of beers.

Grains are commonly incorporated into baked menu items.

Oats

Key Nutrients*	% Daily Value
Manganese	96%
Phosphorus	21%
Thiamin	20%
Magnesium	17%
Dietary fiber	17%

- Calories: 152
- Protein: 7 g

* based on ¼ cup serving size, raw

Figure 14-25. Oats are a cereal grain derived from the berries of oat grass.

Barley

Key Nutrients*	% Daily Value
Manganese	45%
Dietary fiber	32%
Selenium	25%
Thiamin	20%
Magnesium	15%

Canada Beef Inc.

- Calories: 163
- Protein: 6 g

* based on ¼ cup serving size, raw

Figure 14-26. Barley is a cereal grain that resembles brown rice in shape, yet takes longer to cook.

Corn. *Corn,* also known as maize, is a gluten-free cereal grain cultivated from an annual grass that bears kernels on large woody cobs called ears. Corn is prepared both as a grain and as a vegetable. When corn is dried and ground, it can be made into cornmeal. **See Figure 14-27.** Cornmeal is commonly used to make polenta, grits, and tortillas. When cornmeal is cooked with a liquid such as milk or stock, its gritty raw texture becomes creamy as the starches swell and then gelatinize.

Cornmeal

Key Nutrients*	% Daily Value
Magnesium	10%
Dietary fiber	9%
Thiamin	8%
Manganese	8%
Phosphorus	7%

- Calories: 111
- Protein: 3 g

* based on ¼ cup serving size, raw

Figure 14-27. Corn is a gluten-free cereal grain that bears kernels on large woody cobs called ears and can be dried and ground into cornmeal.

Rye. *Rye* is a hearty cereal grain with dark-brown berries that are long and thin. **See Figure 14-28.** Rye has a distinct, heavy flavor and is darker in color than wheat. Rye is commonly used to make flour, breads, crackers, and whiskey. Soaked and cooked rye berries are used in salads, pilafs, and hot breakfast cereals because they remain chewy and do not solidify as they cool.

Triticale. *Triticale* is a hybrid cereal grain made by crossing rye and wheat. It has a sweet, nutty flavor and contains more protein and less gluten than wheat. Like rye, triticale makes heavy, hearty loaves of bread.

Rye

Key Nutrients*	% Daily Value
Manganese	56%
Dietary fiber	25%
Selenium	21%
Phosphorus	16%
Magnesium	13%

- Calories: 142
- Protein: 6 g

* based on ¼ cup serving size, raw

Figure 14-28. Rye is a hearty cereal grain with dark-brown berries that are long and thin.

Millet. *Millet* is a small, round, butter-colored ancient cereal grain that is gluten-free. **See Figure 14-29.** Millet has a mild flavor and can be toasted to bring forth a slightly crunchy texture, nutty flavor, and deep-yellow color. Millet pairs well with aromatics such as chives, green onions, and garlic. It is often used in salads, casseroles, and stuffings.

Teff. *Teff* is a gluten-free ancient cereal grain and is the smallest grain in the world. **See Figure 14-30.** Teff has a distinct nutty flavor and comes in brown, white, red, and purple varieties. Teff is native to Africa and a primary ingredient in injera, a type of fermented bread served in Ethiopia. The flavor of teff pairs well with both savory and sweet ingredients.

Farro. *Farro* is a hearty ancient cereal grain that tastes similar to wheat yet resembles brown rice. **See Figure 14-31.** Farro is low in gluten and has a chewy texture and nutty flavor. It is cultivated across the Mediterranean and commonly used in Italy to enhance the texture and flavor of soups. Farro can be purchased as a whole grain or flour. Farro is often pearled and can be used to make risottos.

Millet

Key Nutrients*	% Daily Value
Manganese	41%
Copper	19%
Dietary fiber	17%
Magnesium	14%
Phosphorus	14%

- Calories: 189
- Protein: 6 g

* based on ¼ cup serving size, raw

Figure 14-29. Millet is a small, round, butter-colored ancient cereal grain that is gluten-free.

Teff

Key Nutrients*	% Daily Value
Manganese	223%
Magnesium	22%
Phosphorus	21%
Iron	21%
Copper	20%

- Calories: 177
- Protein: 7 g

* based on ¼ cup serving size, raw

Figure 14-30. Teff is a gluten-free ancient cereal grain and is the smallest grain in the world.

Farro

Key Nutrients*	% Daily Value
Dietary fiber	20%
Niacin	20%
Magnesium	15%
Zinc	15%
Iron	10%

Indian Harvest Specialtifoods, Inc./ Rob Yuretich

- Calories: 170
- Protein: 6 g

* based on ¼ cup serving size, raw

Figure 14-31. Farro is a hearty ancient cereal grain that tastes similar to wheat, yet resembles brown rice.

Spelt. *Spelt* is an ancient cereal grain with a nutty flavor and high protein content. **See Figure 14-32.** Spelt is commonly mistaken for farro because of their similar appearance. Spelt can be used as a hot breakfast cereal or ground and used as a substitute for wheat flour in pastas and baked goods.

Pseudocereals

A *pseudocereal* is a seed that is classified as a grain but is derived from broadleaf plants instead of grasses. Pseudocereals include ancient grains such as quinoa, amaranth, and buckwheat. Although pseudocereals are seeds, they are commonly referred to as grains because they are used in a similar manner. Like cereal grains, pseudocereals are often processed into flours to be used in products such as packaged cereals, pastas, and baked goods.

Spelt

Key Nutrients*	% Daily Value
Manganese	65%
Dietary fiber	19%
Phosphorus	18%
Niacin	15%
Magnesium	15%

- Calories: 147
- Protein: 6 g

* based on ¼ cup serving size, raw

Figure 14-32. Spelt is an ancient cereal grain with a nutty flavor and high protein content.

Quinoa. *Quinoa* is a gluten-free pseudocereal with a small, round seed and is classified as a complete protein. **See Figure 14-33.** As a complete protein, quinoa contains adequate amounts of all the essential amino acids, including lysine and isoleucine, which are inadequate in most other grains. This ancient grain also contains more heart-healthy monounsaturated fats than other cereal grains. Quinoa has a somewhat crunchy texture and mild, nutty flavor. It is available in ivory, red, pink, brown, and black varieties. Quinoa must be rinsed before cooking to remove its bitter coating. As it cooks, the germ partially detaches, appearing as a tail-like extension from the seed. Quinoa is often used in salads, stuffings, and quick breads and served as a side dish.

Quinoa

Indian Harvest Specialtifoods, Inc./Rob Yuretich

- Calories: 157
- Protein: 6 g

Key Nutrients*	% Daily Value
Manganese	43%
Magnesium	21%
Folate	20%
Phosphorus	20%
Copper	13%

* based on ¼ cup serving size, raw

Figure 14-33. Quinoa is a gluten-free pseudocereal with a small, round seed and is classified as a complete protein.

INGREDIENT SPOTLIGHT: Quinoa

Indian Harvest Specialtifoods, Inc./Rob Yuretich

Unique Features

- Hearty ancient grain dating back to 3000 BC
- Contains all essential amino acids, making it a complete protein
- Studies suggest it contains anti-inflammatory phytonutrients
- Low in cholesterol and sodium
- Source of protein, dietary fiber, magnesium, phosphorus, and manganese

Menu Applications

- Use to turn vegetarian dishes into complete protein menu items
- Use as a substitute for rice in hot cereals, soups, salads, sides, entrées, and desserts
- Toast in a dry skillet before cooking to enhance its nutty flavor

Nutrition Facts

Serving Size 1 cup cooked (185 g)

Amount Per Serving	
Calories 222	Calories from Fat 32
	% Daily Value*
Total Fat 4g	5%
Saturated Fat	0%
Trans Fat	
Cholesterol 0mg	0%
Sodium 13mg	1%
Total Carbohydrate 39g	13%
Dietary Fiber 5g	21%
Sugars	
Protein 8g	

Vitamin A	0%	Vitamin C	0%
Calcium	3%	Iron	15%

* Percent Daily Values (DV) are based on a 2,000 calorie diet.

Amaranth. *Amaranth* is a gluten-free pseudocereal with a small, yellow, black-flecked seed and is high in protein. **See Figure 14-34.** This ancient grain has a mild, nutty, malt-like flavor. When amaranth is cooked in its raw form, it has a consistency similar to porridge. In contrast, when amaranth is toasted before it is cooked, the resulting dish is more like a pilaf. Amaranth flour is used to increase the protein content in some breads.

Buckwheat. *Buckwheat* is a gluten-free pseudocereal with a dark, three-cornered seed. **See Figure 14-35.** Buckwheat has a nutty, earthy flavor and is commonly ground into a coarse flour. This ancient grain is used to make a variety of foods, including pancakes and soba noodles. Buckwheat is also popular in salads, pilafs, and stuffings. *Kasha* is roasted buckwheat.

Buckwheat flour is used to make a variety of foods, including pancakes.

Amaranth

Indian Harvest Specialtifoods, Inc./Rob Yuretich

- Calories: 179
- Protein: 7 g

Key Nutrients*	% Daily Value
Manganese	81%
Magnesium	30%
Phosphorus	27%
Vitamin B_6	14%
Selenium	13%

* based on ¼ cup serving size, raw

Figure 14-34. Amaranth is a gluten-free pseudocereal with a small, yellow, black-flecked seed and is high in protein.

Buckwheat

- Calories: 146
- Protein: 6 g

Key Nutrients*	% Daily Value
Magnesium	98%
Manganese	28%
Copper	23%
Dietary fiber	17%
Niacin	15%

* based on ¼ cup serving size, raw

Figure 14-35. Buckwheat is a gluten-free pseudocereal with a dark, three-cornered seed.

Gluten-Free Grains

Gluten is best known for giving elasticity and structure to baked products, but it has an adverse effect on digestion and absorption of nutrients for some individuals. Gluten-free grains can provide safe and healthy alternatives for guests with gluten intolerance or celiac disease. Rice, wild rice, corn, millet, teff, quinoa, amaranth, and buckwheat are all gluten-free grains.

FOOD SCIENCE EXPERIMENT: Baking with Various Grain Flours 14-2

Objective

- To compare muffins prepared with different grain flours in terms of color, texture, mouthfeel, and flavor.

Materials and Equipment

- 2 muffin pans
- 24 baking cup liners
- 2 whisks
- 2 measuring cups
- 2 medium stainless steel bowls
- Sifter
- 2 sets of measuring spoons
- 2 mixing bowls
- 2 portion scoops
- 2 toothpicks
- 2 cooling racks
- Muffin ingredients

1 c	canola oil
2 c	granulated sugar
4 ea	whole eggs
2 c	all-purpose flour
2 c	oat, quinoa, or rice flour
2 tsp	baking powder
1 tsp	baking soda

Procedure

Allergen Alert
The materials used in this experiment may contain one or more food allergens.

1. Preheat oven to 350°F.
2. Line each muffin pan with 12 baking cup liners.
3. Whisk ½ cup oil, 1 cup sugar, and 2 eggs in a medium bowl until blended.
4. Sift 2 cups all-purpose flour, 1 tsp baking powder, and ½ tsp baking soda in a mixing bowl.
5. Add the egg mixture to the flour mixture and stir to incorporate.
6. Scoop batter into one of the prepared muffin pans.
7. Bake for 25 minutes or until a toothpick inserted in the middle of a muffin comes out clean.
8. Allow the muffins to cool on a wire rack before removing them from the pan.
9. Repeat steps 3–8 using oat, quinoa, or rice flour instead of all-purpose flour.
10. Evaluate the color, texture, mouthfeel, and flavor of each type of muffin.

Typical Results

There will be variances in the color, texture, mouthfeel, and flavor of each type of muffin.

Knowledge Check 14-5

1. List the key nutrients found in grains.
2. Define whole grain.
3. Differentiate between cracked, milled, pearled, and flaked grains.
4. Identify two ways to increase the perceived value of grains on the menu.
5. Define cereal grain.
6. Describe how rice is classified.
7. List three common forms of wheat and their uses in the kitchen.
8. Identify common uses of oats, barley, cornmeal, and rye.
9. Describe four ancient cereal grains.
10. List three pseudocereals.
11. Name eight gluten-free grains.

PREPARING NUTRITIOUS GRAINS

When preparing grains, it is important to cook them properly. Grains are cooked with liquid, and correct proportions are needed to ensure that properly cooked grains have an appealing chewy texture. Through the cooking process, there are a variety of ways grains can be flavored. For example, grains can be toasted before cooking, simmered in a flavorful stock, or sautéed with aromatic vegetables.

Cooking Grains

The cooking process for grains begins by combining liquid and grains in the appropriate ratio. Most grains are simmered in a hot liquid until the liquid is absorbed. As grains absorb the cooking liquid, they typically expand in volume. **See Figure 14-36.** Cooking times vary depending on the type and amount of grain being cooked. Once the grain has simmered for the desired amount of time, it is removed from the heat and allowed to rest for approximately 5 minutes. Resting grains allows any remaining cooking liquid to be absorbed.

Grains should be checked for doneness after the recommended cooking time has elapsed. Properly simmered grains should have a slight resistance to them. However, the overall consistency varies according to the grain being cooked. For example, grains such as quinoa and wild rice separate when cooked properly, while grains such as amaranth and rolled oats have a porridge-like consistency. Grains that separate can be fluffed with a fork after cooking.

The cooking process for grains begins by combining liquid and grains in the appropriate ratio.

Cooking Grains

Dry Grain (1 cup)	Liquid	Yield*	Dry Grain (1 cup)	Liquid	Yield*
Arborio rice	2½ c	2½ c	Hominy	5 c	3 c
Barley, pearled	3 c	3½–4 c	Jasmine rice	1½ c	2 c
Basmati rice, brown	2 c	3½ c	Millet	3 c	5 c
Basmati rice, white	1¾ c	3½ c	Oats, steel-cut	4 c	2 c
Brown rice, long-grain	2 c	3½ c	Quinoa	2 c	4 c
Brown rice, short-grain	2 c	3¾ c	Rye, berries	3 c	3 c
Buckwheat groats, unroasted	2 c	3½ c	Rye, flakes	3 c	2½ c
Bulgur wheat	2 c	2½–3 c	Spelt, soaked overnight	3½ c	2½ c
Cornmeal polenta	2½ c	3½ c	Sweet rice	2 c	2 c
Couscous	1¼ c	2¼ c	Wheat berries	2½ c	3 c
Forbidden (black) rice	1¾ c	2¾ c	Wheat flakes	4 c	2 c
Grits	3 c	3½–4 c	White rice	2 c	2½ c

* yields are approximate

Figure 14-36. As grains absorb cooking liquid, they typically expand in volume.

While most grains are cooked solely by simmering, grains used to make risottos and pilafs are sautéed before they are simmered. **See Figure 14-37.** Risotto is a classic Italian dish traditionally made by sautéing and then simmering Arborio rice. Risotto is cooked slowly to release the starches from the grain, resulting in a creamy finished product. In a pilaf, the flavoring ingredients and grains are sautéed in fat, which prevents clumping, before adding the liquid.

Figure 14-37. While most grains are cooked solely by simmering, grains used to make risotto and pilafs are sautéed before they are simmered.

Flavor Development

The subtle flavor of grains blends well with both savory and sweet ingredients to produce dishes with exceptional flavors. For example, cooking grains in stocks, bouillons, consommés, or juices infuses the grains with flavor. Building a base of flavor by sautéing aromatics like onions and garlic also promotes flavor development in grain dishes. When making grain pilafs, the flavor of vegetables such as carrots, celery, and fennel complements grains. (Vegetables in pilafs also increase the amount of nutrients in the dish.) The mild flavor of grains also pairs well with robust earthy flavors such as sautéed mushrooms. Rehydrating dried mushrooms and using that liquid to cook a grain is an effective way to impart flavor.

Dried fruits and nuts are often added to grain dishes, resulting in enhanced color, texture, and flavor. A small amount of cheese can impart richness, while fresh herbs can add a floral finish to grain dishes. Grains can also be toasted prior to cooking to impart an extra layer of flavor to a finished dish.

Plating Grains

Grains can provide shape, color, texture, and height to plated presentations. Grains are often placed in molds to create shapes that form the foundation for other ingredients. For example, shellfish can be arranged on top of a molded grain to add height and artistry to the plate. Colorful grain varieties can brighten the plate, while neutral grains provide a visual contrast for more vibrant ingredients. **See Figure 14-38.**

Plating Grains

Farro-Stuffed Tomatoes **Brown Rice and Barley Salad**

Indian Harvest Specialtifoods, Inc./Rob Yuretich

Figure 14-38. Neutral grains provide a visual contrast to more vibrant ingredients.

Knowledge Check 14-6

1. Describe the most common method of cooking grains.
2. Compare the risotto and pilaf preparation methods.
3. Explain how to effectively develop flavor in grain dishes.

CULINARY NUTRITION RECIPE-MODIFICATION PROCESS

Risotto

While the creamy texture and rich flavor of this risotto is appealing, the dish is also very decadent. The butter, heavy cream, and Parmesan cheese used to finish the dish are the primary sources of both calories and saturated fat. This dish is also deficient in dietary fiber because it uses Arborio rice, which is a refined grain.

Yield: 6 servings, ⅔ cup each

Ingredients

2 tbsp	vegetable oil
½ c	yellow onion, medium dice
1⅔ c	Arborio rice
½ c	white wine
5 c	chicken stock, hot
3 tbsp	unsalted butter
3 tbsp	heavy cream
½ c	Parmesan cheese, grated
1 tbsp	Italian parsley, chopped
TT	kosher salt
TT	black pepper, ground

Preparation

1. Heat oil in a large saucepan over medium heat. Add onions and cook until translucent.
2. Add rice and stir to coat with oil.
3. Cook until a lightly toasted aroma can be detected. Do not allow the rice to brown.
4. Add white wine and stir continuously until almost dry.
5. Add approximately ¾ cup of the hot stock and stir continuously until all the moisture has been absorbed.
6. Repeat step 5 until all the stock has been absorbed.
7. Add butter, heavy cream, Parmesan cheese, and parsley. Stir to incorporate.
8. Season with salt and pepper to taste and serve immediately.

Chef's Tip: *Cook risotto close to service because it can become pasty if held. A properly cooked risotto should form a soft, creamy mound when plated.*

Evaluate original recipe for sensory and nutritional qualities

- Risotto has a rich, creamy texture and full-bodied flavor.
- Dish is high in fat.
- Dish lacks color and is low in dietary fiber.

Establish goals for recipe modifications

- Reduce both total fat and saturated fat.
- Maintain creamy texture and full-bodied flavor.
- Increase color and nutrient density.

Identify modifications or substitutions

- Reduce the amount of butter, heavy cream, and Parmesan cheese.
- Add a variety of colorful vegetables.

Determine functions of identified modifications or substitutions

- Butter, heavy cream, and Parmesan cheese provide a creamy texture and rich flavor.
- Vegetables add color and nutrients.

Select appropriate modifications or substitutions

- Reduce the amount of butter to 2 tbsp, heavy cream to 1 tbsp, and Parmesan cheese to ¼ cup.
- Add carrots, red peppers, peas, corn, and zucchini.

Test modified recipe to evaluate sensory and nutritional qualities

- Reducing the amount of butter, heavy cream, and Parmesan cheese lowers total fat and saturated fat without compromising texture and flavor.
- Incorporating a variety of vegetables adds color, texture, and flavor while increasing dietary fiber, vitamin A, and vitamin C.

Modified Risotto

In this modified risotto recipe, the amounts of butter, heavy cream, and Parmesan cheese are reduced to lower the fat content. A variety of garden-fresh vegetables are added to increase dietary fiber, vitamin A, and vitamin C. The addition of vegetables also builds color, texture, and flavor into the dish. Vegetable stock is used in place of chicken stock to complement the flavor of the vegetables.

Yield: 6 servings, ⅔ cup each

Ingredients

2 tbsp	vegetable oil
½ c	yellow onion, medium dice
½ c	carrots, medium dice
½ c	red bell peppers, medium dice
1½ c	Arborio rice
½ c	white wine
1 qt	vegetable stock, hot
⅓ c	green peas
⅓ c	corn kernels
⅓ c	zucchini, medium dice
2 tbsp	unsalted butter
1 tbsp	heavy cream
¼ c	Parmesan cheese, grated
1 tbsp	Italian parsley, chopped
TT	kosher salt
TT	black pepper, ground

Preparation

1. Heat oil in a large saucepan over medium heat. Add onions and cook until translucent.
2. Add carrots and red peppers and cook until they begin to soften (approximately 3–4 minutes).
3. Add rice and stir to coat with oil.
4. Cook until a lightly toasted aroma can be detected. Do not allow the rice to brown.
5. Add white wine and stir continuously until almost dry.
6. Add approximately ¾ cup of the hot stock and stir continuously until all the moisture has been absorbed.
7. Repeat step 6 two more times.
8. Add peas, corn, and zucchini to the saucepan.
9. Continue adding the stock in ¾ cup increments, while stirring, until all the stock has been absorbed.
10. Add butter, heavy cream, Parmesan cheese, and parsley. Stir to incorporate.
11. Season with salt and pepper to taste and serve immediately.

Chef's Tip: *Any combination of vegetables can be incorporated into risotto for a colorful and nutrient-dense dish.*

Risotto Nutritional Comparison

Nutrition Facts	Original	Modified
Calories	432.7	396.6
Total Fat	18.0 g	10.8 g
Saturated Fat	7.8 g	4.1 g
Dietary Fiber	0.2 g	1.6 g
Vitamin A	384.5 IU	2003.1 IU
Vitamin C	2.2 mg	28.7 mg

CULINARY NUTRITION RECIPE-MODIFICATION PROCESS

Signature Steel-Cut Oat Risotto

In this signature version of risotto, the Arborio rice is replaced with steel-cut oats. Replacing the refined grain with a whole grain adds more dietary fiber to the dish while maintaining a creamy texture. The earthy flavor of the added mushrooms complements the mildness of the oats. The addition of garlic and thyme helps layer and develop flavor. Both the steel-cut oats and mushrooms add protein and manganese to the dish.

Yield: 8 servings, ⅔ cup each

Ingredients

2 tbsp	vegetable oil
¾ c	yellow onion, medium dice
1 tsp	garlic, minced
4 c	assorted wild mushrooms, chopped
2 c	steel-cut oats
⅔ c	white wine
6 c	vegetable stock, hot
2 tsp	thyme, minced
1 tbsp	butter
1 tbsp	heavy cream
⅓ c	Parmesan cheese, grated
TT	kosher salt
TT	black pepper, ground

Preparation

1. Heat oil in a large saucepan over medium heat. Add onions and cook until translucent.
2. Add garlic and mushrooms and cook until the mushrooms begin to release water (approximately 5–6 minutes).
3. Add oats and cook until oats become fragrant and begin to brown (approximately 3 minutes).
4. Add white wine and stir continuously until almost dry.
5. Add approximately ¾ cup of the hot stock and stir continuously until all the moisture has been absorbed.
6. Repeat step 5 until all the stock has been absorbed.
7. Add thyme, butter, heavy cream, and Parmesan cheese. Stir to incorporate.
8. Season with salt and pepper to taste and serve immediately.

Chef's Tip: *When substituting steel-cut oats for rice in risotto, additional liquid is required because steel-cut oats take longer to cook than rice.*

Risotto Nutritional Comparison

Nutrition Facts	Original	Modified	Signature
Calories	432.7	396.6	369.9
Total Fat	18.0 g	10.8 g	12.5 g
Saturated Fat	7.8 g	4.1 g	3.5 g
Dietary Fiber	0.2 g	1.6 g	7.2 g
Protein	12.4 g	14.6 g	14.0 g
Vitamin A	384.5 IU	2003.1 IU	3249.8 IU
Vitamin C	2.2 mg	28.7 mg	5.8 mg
Manganese	0.1 mg	0.2 mg	2.1 mg

GRAIN MENU MIX

Grains can be used across the menu in both sweet and savory applications. **See Figure 14-39.** Grains such as oats or spelt might be featured on breakfast menus as hot cereals paired with fruits, honey, and spices such as cinnamon, nutmeg, or cardamom. As the starch component on the plate, grains serve an important role as an accompaniment to meats, poultry, and seafood. Grains such as wheat berries, brown rice, millet, farro, and quinoa can be used to create flavorful appetizers, soups, salads, risottos, and pilafs.

Menu descriptions that highlight the unique characteristics of grains encourage sales. For example, a chicken curry dish sounds more enticing when it is served "over jasmine-scented rice" as opposed to "with steamed white rice." Likewise, "stone-ground cornmeal" indicates a higher quality ingredient than "cornmeal." Chefs can also denote flavor profiles that add value to a dish. For example, "millet pilaf enhanced with caramelized beets and fresh herbs" will draw more attention than "millet pilaf." **See Figure 14-40.**

Chef's Tip

Seitan or "wheat meat" can be used on the menu in vegetarian dishes. This high-protein food made from wheat gluten has a chewy, meatlike texture that makes an effective meat substitute.

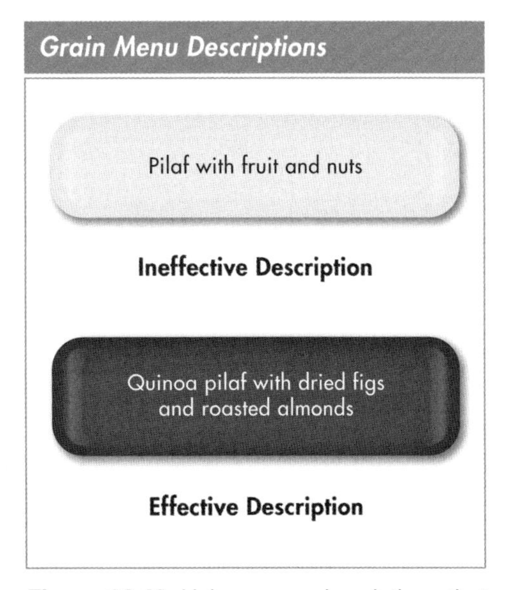

Figure 14-40. Using menu descriptions that denote flavor profiles is an effective way to add value to a dish.

Grains in Beverages

Grain-based beverages, such as rice milk and oat milk, are lactose-free and suitable for guests who are lactose intolerant. They also make ideal milk alternatives for those following a vegan diet because they are not made with animal products but still supply essential nutrients like calcium and iron. **See Figure 14-41.** The bran and germ of grains such as oats, wheat, or quinoa can be added to beverages such as smoothies to increase dietary fiber, vitamins, and minerals. Grains are also used to make beers and spirits.

Grains on the Menu

Menu Item	Description
Roasted beef tenderloin with polenta	Locally raised beef encrusted with fine herbs served atop creamy stone-ground polenta accented with goat cheese
Stuffed poblano pepper	Vegetarian entrée featuring fire-roasted poblanos stuffed with spelt and lentils on a bed of arugula dressed with a light tomatillo sauce
Millet pudding	Vanilla-infused millet pudding sweetened with agave nectar and garnished with fresh blueberries and roasted pecans

Figure 14-39. Grains can be used across the menu in both sweet and savory applications.

Nutritional Comparison of Milks*

Types of Milk	Calories	Total Fat	Calcium†	Iron†
Cow milk (2%)	137	5 g	35%	1%
Oat milk	130	3 g	35%	10%
Rice milk	130	2 g	30%	6%

* based on 1 cup serving size
† based on percent daily value

Figure 14-41. Oat and rice milks are lactose-free and supply essential nutrients like calcium and iron.

Grains in Appetizers

Grains can be used to add nutrient density to appetizers. For example, incorporating grains into appetizer fillings adds structure, increases nutrients, and enhances visual appeal. **See Figure 14-42.** Grains can also be molded into a shape and used as a base to elevate other items on the plate.

Grains in Soups

Grains are commonly used to add body, texture, and flavor to soups. Adding whole grains such as brown rice or quinoa to soups also provides protein, complex carbohydrates, and an array of vitamins and minerals. Grains can be used in soups ranging from beef barley to chicken noodle to gumbo.

Grains in Appetizers

Figure 14-42. Grains can be used in appetizers to add structure, increase nutrients, and enhance visual appeal.

Grains in Salads

Salads made from whole grains are often served as side dishes. Tabbouleh, a Middle Eastern salad consisting of cooked and chilled bulgur wheat mixed with olive oil, lemon, mint, and parsley, is a classic whole grain salad. To increase the nutrient density of tabbouleh, a chef may choose to replace the bulgur with quinoa, making the vegetarian dish a complete protein. Rice is another popular grain in cold salads. However, virtually any grain can be mixed with a dressing and ingredients such as lean meats, fruits, vegetables, nuts, or seeds to create a flavorful, nutritious salad. **See Figure 14-43.**

Grain-Based Salads

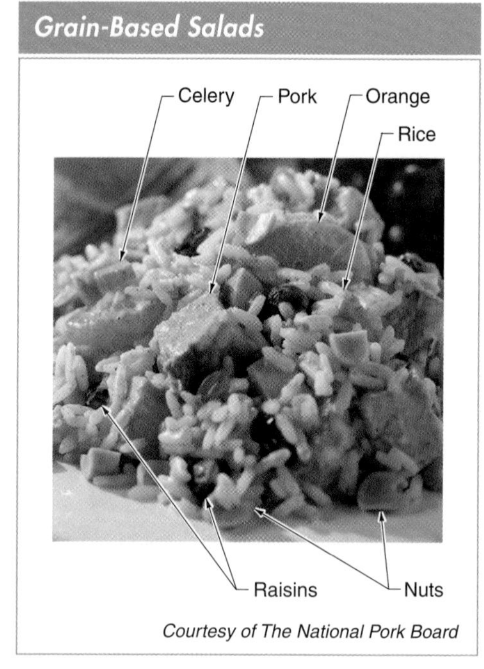

Figure 14-43. Grains can be mixed with ingredients such as lean meats, fruits, vegetables, nuts, or seeds to create flavorful, nutritious salads.

Grains in Entrées

Most entrées include a main protein accompanied by a grain or a vegetable. With the increased demand for more nutritious meals, chefs can reduce the portion size of the animal-based protein by increasing the amount of whole grains and vegetables on the plate. This makes grains a vital component of a well-balanced, nutrient-dense entrée.

Grains also play an important role in vegetarian entrées because they provide protein and promote satiety. Although quinoa is a complete protein, most grains, including rice, wheat, and corn, must be paired with legumes, nuts, or seeds to make a complete-protein menu item.

There are many ways to incorporate whole grains into entrées to create flavorful, satisfying meals, including stir-fries, pilafs, risottos, and polentas. **See Figure 14-44.** While cornmeal is the traditional grain used to make polenta, amaranth can be an appealing substitute. Likewise, Arborio rice is the standard grain used to make risotto, but the same cooking technique can be applied to steel-cut oats to create an oat risotto. Breakfast entrées can feature grains such as oats or teff as hot cereals, and buckwheat flour can be used to increase the nutrient density of pancakes and waffles.

Grains in Sides

Grains can be used in both hot and cold side dishes. Hot grain dishes, such as pilafs, often accompany entrées, while cold grain salads are popular accompaniments to sandwiches. Some menus offer grain side dishes such as grilled polenta or wheat berries with dried fruits and nuts as à la carte items. Side dishes should emphasize whole grain varieties because they are more nutrient dense than refined grains.

Grains in Desserts

Grains are most often associated with savory applications, but their unique texture and light flavor also make them ideal for dessert applications. Rice pudding can be made by combining cooked rice with cow's milk, nut milk, coconut milk, or juice and cooking the mixture until it becomes thick and creamy. The rice pudding can be sweetened with an unrefined sugar such as maple syrup or honey and mixed with fruits and nuts to provide a delicious and nutrient-dense dessert. **See Figure 14-45.** Cooked grains can also be incorporated into pies and tarts.

Grains in Entrées

Courtesy of The National Pork Board

Figure 14-44. There are many ways to incorporate whole grains into entrées, including flavorful stir-fries.

Grains in Desserts

Daniel NYC

Figure 14-45. The unique texture and light flavor of grains make them ideal for desserts such as rice pudding.

Knowledge Check 14-7

1. Provide two reasons why grains are used in beverages.
2. Describe how grains can be used in appetizers.
3. Identify the functions of grains in soups and salads.
4. Describe how whole grains can be incorporated into entrées.
5. Explain how grains are used in side dishes.
6. Give an example of a grain-based dessert.

PROMOTING GRAINS ON THE MENU

Promoting whole grains on the menu helps meet the growing demand for nutritious menu items and can also provide guests with a variety of inspired flavors. **See Figure 14-46.** For example, an enchilada featuring a blend of quinoa and sautéed vegetables wrapped in a blue corn tortilla is a unique and nutrient-dense menu item that can encourage sales. Grains are served at a variety of venues and can be incorporated into menu items across all meal periods.

Whole Grains to Promote on the Menu

- Amaranth
- Brown/black barley
- Brown rice
- Buckwheat
- Bulgur (cracked wheat)
- Cornmeal
- Farro
- Millet
- Oats
- Popcorn
- Quinoa
- Rye
- Sorghum
- Spelt
- Teff
- Triticale
- Wheat berries
- Whole grain breads
- Whole grain cereal flakes
- Whole wheat pasta
- Wild rice

Figure 14-46. Promoting whole grains on the menu helps meet the demand for nutritious menu items.

Grain menu items such as hot and cold breakfast cereals and grain-based salads are commonly promoted as "grab-and-go" offerings at quick service venues. Oatmeal and grits are popular breakfast menu items. Buffets often feature grain dishes made with rice, barley, or quinoa. Grains such as wild rice, bulgur, and millet may be used in salads, soups, and stews. **See Figure 14-47.** Grains are often paired with other nutritious foods such as seasonal vegetables and lean proteins to create a variety of savory and sweet menu items.

Turkey and Wild Rice Soup

National Turkey Federation

Figure 14-47. Grains, such as wild rice, are commonly used in soups.

Indian Harvest Specialtifoods, Inc./Rob Yuretich

Grains are often paired with seasonal vegetables to create savory menu items.

Grains may be the main salad ingredient or added to leafy green salads for enhanced nutrition. Grains are also commonly added to soups and stews to make these menu items heartier. At sushi bars, rice is a staple ingredient. **See Figure 14-48.** Beverage houses and bars often serve popcorn, which is a whole grain, as a snack.

Grains at Bar Venues

House Foods

Figure 14-48. At sushi bars, rice is a staple ingredient.

At fine dining and special event venues, unique varieties of grains, such as black rice or red quinoa, may be showcased in appetizers and entrées. Labor intensive grain preparations, such as risottos, are often featured menu items and help meet guest expectations for an elevated dining experience.

BREADS

Bread is one of the oldest staple foods. Made by combining flour and water to form a dough, breads are flavored with ingredients such as milk, eggs, sugar, butter, fruits, vegetables, nuts, seeds, herbs, and spices. When made with whole grains, breads can be an excellent source of complex carbohydrates and dietary fiber. Some types of bread, however, rely on ingredients such as sugar and butter to create a rich product and should be served in smaller portions.

Perceived Value of Breads

For a foodservice operation, bread is an attractive menu item because of its low food cost and strong profit margin. The aroma and flavor of freshly baked bread evokes positive memories and encourages sales. For example, guests tend to pay more for garden salads served with warm breadsticks, sandwiches made from artisan breads, and fresh-baked dinner rolls enhanced with ingredients such as sun-dried tomatoes, basil, and Asiago cheese. From quick service to fine dining venues, the appealing aroma and flavor of freshly baked breads is used as a marketing tool to attract guests.

Chef's Tip

Research suggests that consumer demand is growing for freshly baked bread. Advances in parbaked breads are offering foodservice operations more options to meet this growing demand.

Knowledge Check 14-8

1. Identify two grain dishes commonly served at quick service venues.
2. Give an example of how grains can be served at beverage houses and bar venues.
3. Explain how grains are featured at fine dining and special event venues.

With the increased awareness of farm-to-table cooking, more artisan breads are being featured on menus. **See Figure 14-49.** Artisan breads are crafted by bakers on a smaller scale, rather than mass produced. Artisan breads are made using quality ingredients and a minimal amount of preservatives. Chefs often procure artisan breads from local bakeries and farmers' markets rather than from a purveyor. Offering higher quality breads on the menu enhances the perceived values of the breads and of the dishes they accompany.

Artisan Breads

Sullivan University

Figure 14-49. Artisan breads are being featured more often on menus than in the past.

Bread Classifications

A slice of bread has a hard or soft exterior crust and a moist interior. The crust is a result of starches and proteins coming in direct contact with heat and caramelizing. The moist interior or crumb of the bread is due to leavening. Breads are classified as either yeast breads or quick breads depending on the type of leavening agents used. A *leavening agent* is any ingredient that causes a baked product to rise by the action of air, steam, chemicals, or yeast. The most common leavening agents used in the preparation of breads are yeast, baking soda, and baking powder.

Yeast is a microscopic, single-celled, fungus that releases carbon dioxide and alcohol through a process called fermentation when provided with food (sugar) in a warm, moist environment. Yeast increases volume, improves flavor, and adds texture to dough.

Baking soda, also known as sodium bicarbonate, is an alkaline chemical leavening agent that reacts to an acidic dough or batter by releasing carbon dioxide without being heated. Baking soda is often used in conjunction with acidic liquids such as buttermilk, lemon juice, or molasses. Using too much baking soda can produce a yellow color, cause brown spots, and leave a bitter aftertaste.

Baking powder is a chemical leavening agent that is a combination of baking soda and cream of tartar or sodium aluminum sulfate. When mixed with liquids, the alkalinity of the baking soda reacts with the acidity of the other ingredients in the baking powder to produce carbon dioxide. Double-acting baking powder is a chemical leavening agent that reacts similarly to baking powder when it is mixed with liquids, releasing some carbon dioxide, but also reacts again when it is heated. Double-acting baking powder allows a batter to be prepared in advance.

Yeast Breads. Yeast breads include items such as Pullman loaves, baguettes, ciabattas, hard rolls, focaccias, pitas, challahs, doughnuts, and croissants. **See Figure 14-50.** The preparation of yeast breads takes longer than that of quick breads because

longer than that of quick breads because yeast is used as the leavening agent and requires time to ferment. *Fermentation* is the breakdown of carbohydrates due to the actions of bacteria, microorganisms, or yeast. Yeast is very sensitive to temperature and feeds on sugar to produce carbon dioxide gas and alcohol. **See Figure 14-51.** The carbon dioxide gas gives bread its airy texture. The alcohol burns off during baking but leaves behind flavor. Yeast breads may be prepared with active dry yeast, compressed yeast, instant yeast, or a yeast starter.

- *Active dry yeast* is yeast that has been dehydrated and looks like small granules. Compared to other types of yeasts, it produces the least amount of carbon dioxide. It is best activated in warm water between 105°F and 115°F.
- *Compressed yeast,* also known as fresh yeast, is yeast that has approximately 70% moisture and is available in 1 lb cakes or blocks. Compressed yeast produces the most carbon dioxide of any yeast. The best method for using compressed yeast is to dissolve it in double its weight of lukewarm water (between 90°F and 100°F) prior to adding it to a bread recipe.

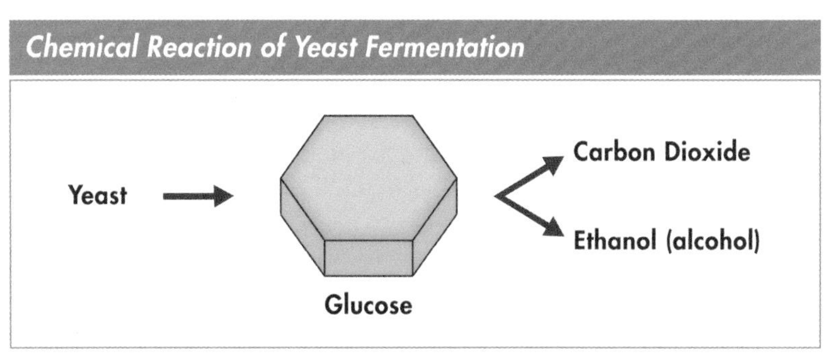

Figure 14-51. During fermentation, yeast feeds on sugar to produce carbon dioxide gas and alcohol.

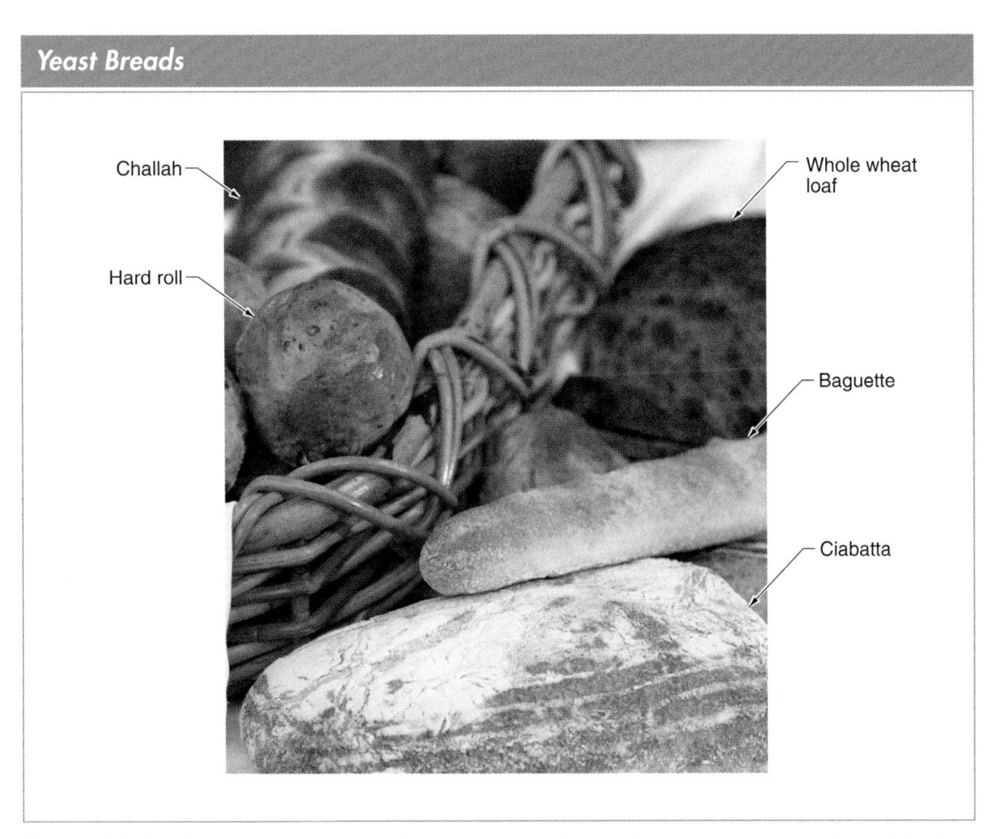

Figure 14-50. Yeast breads include items such as whole wheat loaves, baguettes, ciabattas, hard rolls, and challahs.

- *Instant yeast* is yeast that does not need to be hydrated and may be directly added to a warm flour mixture. It is often labeled as rapid-rise yeast and is available in both dried and vacuum-packaged forms.
- A *yeast starter* is a mixture of flour, yeast, sugar, and water. This mixture is set in a warm place until the yeast ferments, causing the mixture to foam. **See Figure 14-52.** A portion of a yeast starter can be used as a leavener and as a base for some breads. For example, a sourdough starter is used to make sourdough bread. The active yeast spores in the starter give sourdough bread its distinct tangy flavor. A yeast starter can be kept going for years if held in the right environment and if equal parts of flour and water are added each time a portion of the starter is removed.

Quick Breads. Quick breads include items such as pancakes, waffles, biscuits, scones, muffins, coffee cakes, cornbreads, flatbreads, banana nut bread, pumpkin bread, and zucchini bread. **See Figure 14-53.** Instead of yeast, quick breads use chemical leaveners to achieve volume during baking. The two most common chemical leaveners used for quick breads are baking soda and baking powder. Baking soda and baking powder create leavening in quick breads without the need for a long fermentation process.

Baking soda produces carbon dioxide when it comes in contact with an acidic ingredient. Some acidic ingredients include buttermilk, yogurt, lemon juice, and cocoa powder. The carbon dioxide produced by baking soda helps the final product rise when it is baked. Baking soda also darkens as it bakes, which produces a browned crust.

Yeast Activation

Figure 14-52. When yeast ferments, it begins to foam.

INGREDIENT SPOTLIGHT: Focaccia

Alpha Baking Co., Inc.

Unique Features

- Flat Italian bread commonly flavored with olive oil, herbs, and sea salt
- Holes poked in the dough before baking, giving it a distinct appearance
- Baked in round or rectangular loaves with a texture similar to pizza dough
- Low in saturated fat and cholesterol
- Source of niacin, folate, and iron

Menu Applications

- Serve as an accompaniment to soups and salads
- Top with vegetables, lean meats, and low-fat cheeses and serve as a featured menu item

Nutrition Facts
Serving Size 1 piece plain (57 g)

Amount Per Serving	
Calories 142	Calories from Fat 41
	% Daily Value*
Total Fat 4.5g	7%
Saturated Fat 0.5g	3%
Trans Fat	
Cholesterol 0mg	0%
Sodium 320mg	13%
Total Carbohydrate 20g	7%
Dietary Fiber 1g	4%
Sugars 1g	
Protein 5g	
Vitamin A 0% • Vitamin C	0%
Calcium 2% • Iron	10%

* Percent Daily Values (DV) are based on a 2,000 calorie diet.

Figure 14-53. Quick breads include items such as biscuits, muffins, and quick bread loaves.

Double-acting baking powder provides leavening in quick breads twice during the baking process. The first time is known as the fast-acting phase. During the fast-acting phase, double-acting baking powder creates carbon dioxide when it comes in contact with the liquid ingredients. The carbon dioxide gets trapped in the quick bread batter, creating a lighter consistency. The second phase, known as the slow-acting phase, occurs when the batter is heated. As the bread bakes, the double-acting baking powder causes a chemical reaction to slowly occur that also releases carbon dioxide and helps further leaven the bread.

Gluten-Free Breads. Gluten-free breads are made from the flours of gluten-free grains such as rice, millet, teff, quinoa, amaranth, and buckwheat. **See Figure 14-54.** Gluten-free breads can be classified as yeast breads or quick breads. Gluten-free breads vary in texture and consistency and are often more brittle than breads containing gluten. Gums such as xanthan gum and guar gum are often added to gluten-free breads to provide the structure that typically comes from gluten. Many chefs use blends of different flours and gums in gluten-free breads in order to produce the desired texture and flavor.

Breads Made with Gluten-Free Flours

- Amaranth breads
- Buckwheat breads
- Cornbreads
- Injera (made from teff)
- Millet breads
- Nut and seed breads
- Quinoa breads
- Rice breads
- Sorghum breads
- Sprouted breads (some varieties)
- Tapioca breads

Figure 14-54. Gluten-free breads are made from the flours of gluten-free grains as well as nut and seed flours.

Food Safety

Spelt is a type of wheat sometimes marketed as a wheat alternative. However, people on a gluten-free diet need to avoid spelt.

Knowledge Check 14-9

1. Explain why breads can have a high perceived value on the menu.
2. Name the two classifications of breads.
3. Define leavening agent.
4. Identify what gives yeast bread its airy texture.
5. Contrast four types of yeast.
6. List common types of quick breads.
7. Name the two common chemical leaveners used in quick breads.

PREPARING NUTRITIOUS BREADS

There are many ways to increase the nutrient density of breads. **See Figure 14-55.** For example, whole grain flours can be used instead of refined flours. Whole grain flours provide more protein, dietary fiber, vitamins, and minerals than refined flours. Whole grain flours also tend to add a darker color and more intense flavor. Adding sprouted grains such as quinoa or toasted grains like oats is another way to increase nutrient density while developing texture and flavor. Adding dried fruits, nuts, or seeds can also be a flavorful way to add essential nutrients to both yeast breads and quick breads.

Breads that use large amounts of butter or oil can be made more nutritious by substituting some of the fat with puréed fruits such as bananas, apples, or dates. Puréed fruits reduce fat and calories, add dietary fiber, vitamins, and minerals, and produce a final product with a pleasing texture and flavor.

Yeast Bread Preparation

The preparation of yeast breads involves five basic ingredients: flour, liquid, sugar, salt, and yeast. Wheat flour contains two proteins known as glutenin and gliadin. When wheat flour is mixed with a liquid, usually water or milk, glutenin and gliadin join together to form gluten. Gluten gives dough elasticity and the ability to capture carbon dioxide and provides structure to the final product. Different types of flours contain different amounts of protein, which directly affect gluten development. **See Figure 14-56.** For example, hard wheat has a high percentage of protein and develops more gluten than soft wheat, making it ideal for bread. In contrast, soft wheat has less protein and develops less gluten, making it ideal for pastries in which a more tender texture is desired.

Preparing Nutritious Breads

Culinary Application	Increased Benefits of Application
Using whole grain flours	• Protein • Dietary fiber • B vitamins and iron
Using sprouted grain flours	• Protein • Dietary fiber • Carotene, B vitamins, and vitamin C • Antioxidants • Nutrient absorption
Adding fruits and vegetables	• Dietary fiber • Vitamins and minerals • Color, texture, and flavor
Adding nuts and seeds	• Protein • Dietary fiber • Unsaturated fats • Vitamins and minerals • Color, texture, and flavor
Substituting puréed fruits for fats	• Dietary fiber • Vitamins and minerals • Flavor
Using natural sugars (agave nectar, honey, maple syrup, molasses)	• Vitamins and minerals • Flavor

Figure 14-55. There are many ways to increase the nutrient density of breads.

Protein Content in Flours

Type of Flour	Protein Percentage	Common Uses
Whole wheat	14%	• Yeast and quick breads • Blending with other flours
Bread	12%–13%	• Yeast and quick breads
All-purpose (a blend of hard and soft flours)	9%–12%	• Yeast and quick breads and pastries • Blending with whole wheat flours
Self-rising	9%–11%	• Quick breads, pastries, and cookies
Pastry	8%–9%	• Quick breads, pastries, and cookies
Cake	5%–8%	• Cakes

Figure 14-56. Different types of flours contain different amounts of protein, which directly affect gluten development.

Sugar acts as a stimulant to the yeast so that fermentation can occur. Salt brings out flavor and helps the bread retain moisture. Some yeast doughs also contain fat and eggs. Fat supplies tenderness and improves shelf life. Eggs supply structure to the dough and add color. Depending on the ingredients and proportions, yeast doughs can be lean doughs, enriched doughs, rich doughs, or rolled-in doughs.

- A *lean dough* is a yeast dough that is low in sugar and fat. Breads made from lean dough have a thin, crispy crust and a soft, tender crumb. Hard rolls, baguettes, and rye breads are made from lean doughs.

- An *enriched dough* is a yeast dough that is low in sugar and fat and typically includes dried milk solids. Enriched dough products have a softer crust and crumb than lean doughs and are popular for soft dinner rolls and hamburger buns.
- A *rich dough* is a yeast dough that incorporates a lot of fat, various amounts of sugar, and eggs into a heavy, soft structure. The finished product may be faintly yellow in color due to the large number of eggs used. Brioche, challah, cinnamon rolls, and doughnuts are made from rich doughs.
- A *rolled-in dough* is a yeast dough with a flaky texture that results from the incorporation of fat through a rolling and folding procedure. By alternating the layers of dough and fat, a very light and flaky texture is achieved. Rolled-in doughs can produce a sweet Danish or a flaky croissant.

Yeast bread preparation involves mixing ingredients to form a dough. Mixing is important for gluten development and uniform distribution of yeast throughout the dough. Improper mixing will prohibit even gluten formation. Once the dough is mixed, yeast dough preparation includes kneading, fermenting, punching, proofing, scoring or docking, and baking. **See Figure 14-57.**

Kneading Yeast Doughs. *Kneading* is the process of pushing and folding dough until it is smooth and elastic. Kneading aids gluten development and uniformly distributes yeast throughout the dough.

Fermenting Yeast Doughs. Once the dough has been properly kneaded, it is covered and allowed to rest so that the yeast can ferment and create carbon dioxide. Yeast dough should rest until it doubles in size or no longer springs back when pressed.

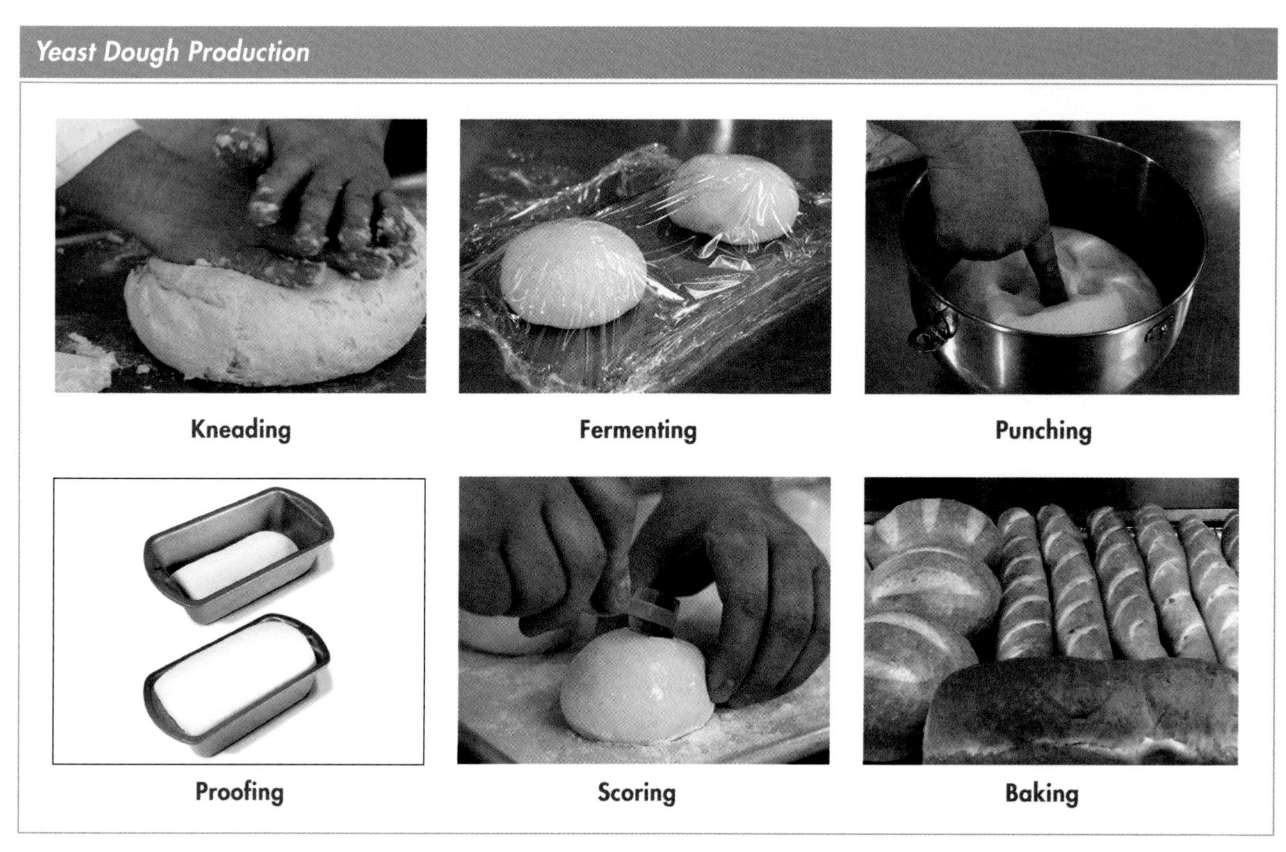

Figure 14-57. Yeast dough production involves kneading, fermenting, punching, proofing, scoring, and baking.

Punching Yeast Doughs. After the dough has fermented and doubled in size, it must be gently deflated. *Punching* is the process of deflating risen yeast dough to allow carbon dioxide to escape. Punching also relaxes gluten development and helps redistribute yeast.

Proofing Yeast Doughs. *Proofing* is the process of letting yeast dough rise in a warm (85°F) and moist (80% humidity) environment until the dough doubles in size. It is important not to overproof or underproof yeast doughs, as this will result in poor texture.

Scoring or Docking Yeast Doughs. After proofing, some yeast breads require scoring or docking. *Scoring* is the process of making shallow, angled cuts across the top of unbaked yeast breads to allow carbon dioxide to escape during baking and to promote even baking. *Docking* is the process of making small holes in dough before it is baked to allow steam to escape and to promote even baking. A dough docker or the tines of a fork can be used to dock yeast doughs.

Baking Yeast Doughs. During the baking process, the internal temperature of the dough rises and causes the dough to expand. *Oven spring* is the rapid expansion of yeast dough in the oven, resulting from the expansion of gases within the dough. When the temperature of the dough reaches 140°F, the yeast dies and there is no further expansion. Oven spring creates characteristic tiny holes in the interior of the bread.

FOOD SCIENCE EXPERIMENT: Yeast Activation 14-3

Objective

- To compare the effect of water temperature on yeast activation in terms of CO_2 production.

Materials and Equipment

- Masking tape
- Marker
- 3 bowls
- 3 sealable plastic bags, 4 gal. each
- Measuring cup
- 3 cups water
- Instant read thermometer
- 3 envelopes of active dry yeast
- 1½ cups sugar
- Spoon

Procedure

1. Use the masking tape and marker to label one bowl "Less than 105°F," the second bowl "Between 105°F and 115°F," and the third bowl "Greater than 115°F."
2. Use the marker to label one plastic bag "Less than 105°F," the second bag "Between 105°F and 115°F," and the third bag "Greater than 115°F."
3. Add 1 cup of water that is less than 105°F and one envelope of yeast to the bowl labeled "Less than 105°F."
4. Add 1 cup of water that is between 105°F and 115°F and one envelope of yeast to the bowl labeled "Between 105°F and 115°F."
5. Add 1 cup of water that is greater than 115°F and one envelope of yeast to the bowl labeled "Greater than 115°F."
6. Add ½ cup sugar to each bowl and stir.
7. Pour each yeast mixture into its corresponding bag.
8. Carefully push as much air as possible out of each bag and seal tightly.
9. Allow each bag to rest for one hour at room temperature.
10. After one hour, compare the amount of carbon dioxide (CO_2) gas in each bag.

Typical Results

There will be variances in the amount of CO_2 contained in each bag.

Quick Bread Preparation

Quick breads often include ingredients such as flour, liquid, sugar, salt, and chemical leaveners. When preparing quick breads, flour is mixed with a liquid, typically milk, to hydrate and disperse ingredients. Flour gives quick breads their bulk, but unlike yeast breads in which the dough is kneaded to create structure, quick breads are only mixed for a short amount of time. The texture of quick breads relies on the amount of mixing done to incorporate the ingredients. Overmixing ingredients causes unwanted gluten to develop and produces a tough final product.

Quick breads are often sweet and can be made with a variety of different sugars. Refined sugars such as granulated or confectioners' sugars are considered empty calories. Natural sugars such as agave nectar, honey, maple syrup, and molasses contribute both flavor and nutrients. **See Figure 14-58.**

Salt is often used to enhance flavor and bring out sweetness. Chemical leavening agents, such as baking soda and baking powder, create carbon dioxide without undergoing fermentation. The use of time-saving chemical leaveners gives quick breads their distinctive name. In addition to the basic ingredients, quick breads commonly include sweet spices such as cinnamon and cardamom, as well as flavor extracts such as vanilla and almond.

Flavor Development

The fermentation process is an important step in the development of flavor. For example, using a yeast starter to increase fermentation time heightens the complexity of flavor in the final product. The type of flour used also contributes flavor and can range from the hearty, earthy flavor of rye to the sweet, mellow flavor of corn. Toasted grains and flours add a rich, nutty flavor to breads.

Natural Sugars

Type of Sugar	Description	Nutrients
Agave nectar	• Nectar from an agave plant • High in fructose • Flavor comparable, but not identical, to honey • Also known as agave syrup	• Dietary fiber • Calcium • Magnesium • Iron
Honey	• Sweet, thick fluid made by honeybees from flower nectar • Approximately 1½ times as sweet as granulated sugar • Flavor depends on flowers from which nectar was derived; clover, buckwheat, and lavender each yield a different flavor	• Antioxidants
Maple syrup	• Sap from maple trees that has been boiled until most of the water has evaporated to produce a sweet product with a thick consistency • Includes light, medium, and dark amber varieties • Darker syrup has more robust flavor	• Zinc • Manganese
Molasses	• Dark syrup that is left after sugar cane has been processed • Cane syrup is processed three to four times, yielding a more concentrated grade of molasses each time • Blackstrap molasses is the strongest and most processed form	• Vitamin B_6 • Magnesium • Potassium • Manganese

Figure 14-58. Natural sugars such as agave nectar, honey, maple syrup, and molasses contribute both flavor and nutrients to quick breads.

In addition to the types of flours used and the fermentation process, a simple way to develop flavors is to add high-quality, flavorful ingredients such as vegetables, fruits, nuts, seeds, and cheeses to yeast breads and quick breads. **See Figure 14-59.** Flavor enhancers such as herbs, spices, citrus zest, and extracts also add unique flavors.

Bread Flavor Development

Rosemary
Thyme
Sun-dried tomatoes
National Honey Board

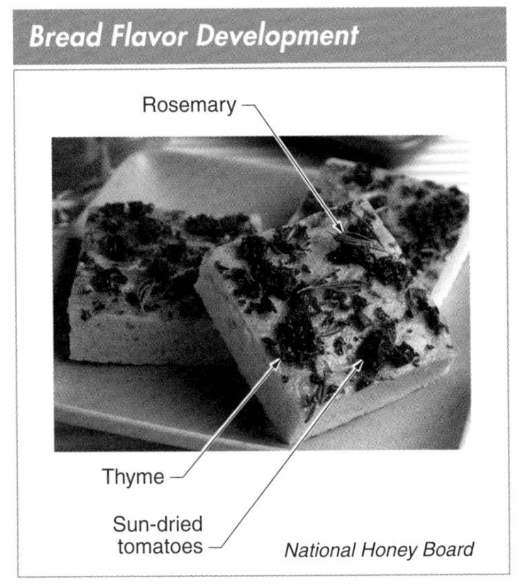

Figure 14-59. Adding high-quality, flavorful ingredients to breads helps develop flavor.

Plating Breads

There are virtually endless ways the various shapes and types of breads can be used to enhance plated presentations. Breads can be hollowed out and used as a bowl, stacked or layered to create height and dimension, or sliced to produce a unique shape. **See Figure 14-60.**

The many different forms of breads also add to their versatility in plated presentations. The nuttier and slightly chewy texture of whole grain breads complements sandwiches and is also ideal for breakfast items such as French toast. The neutral color of whole grain cornbread can bring balance to a colorful salad, or the cornbread can be turned into croutons to accent a spicy chili. A sandwich of thinly sliced turkey, lettuce, and tomato can be served in a pita pocket. That same pita bread can be grilled and served with scrambled eggs instead of toast or topped with tomato sauce, fresh mozzarella cheese, and basil as a flatbread pizza.

Plating Breads

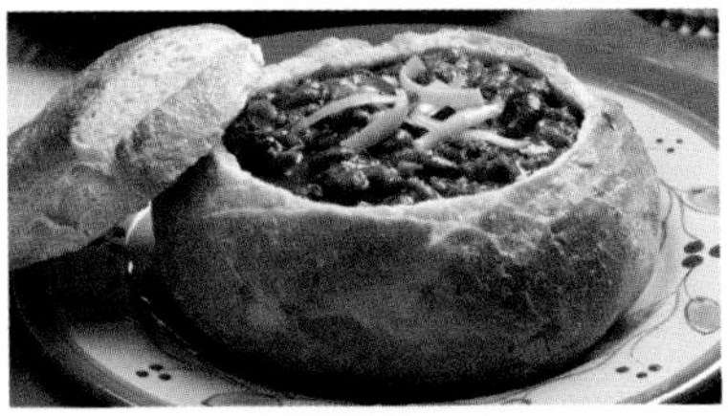

Basic American Foods

Whole Grain Bread Bowl with Chili

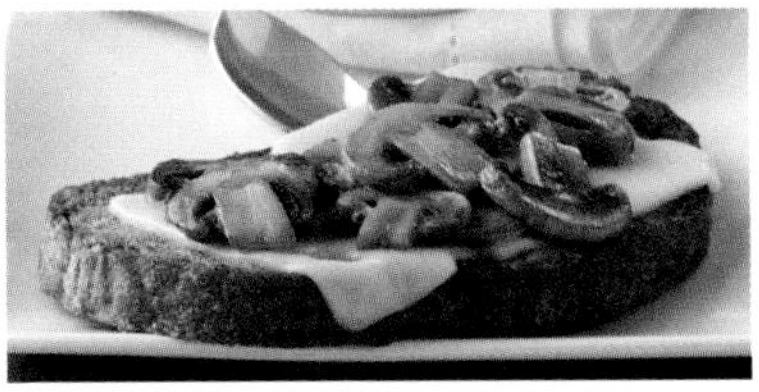

National Turkey Federation

Multigrain Toast with Cheese and Mushrooms

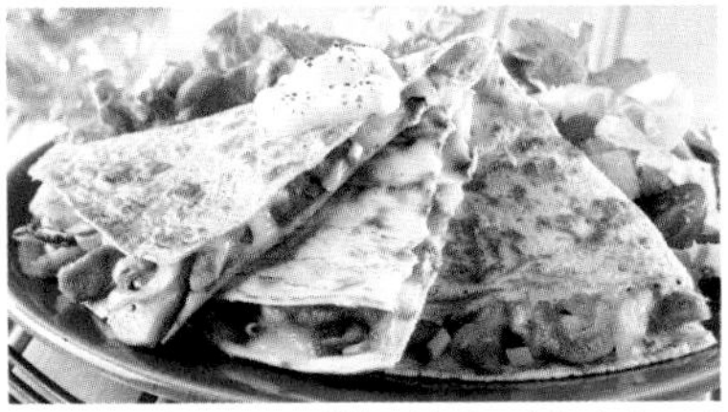

Courtesy of The National Pork Board

Whole Grain Tortilla with Pork and Vegetable Filling

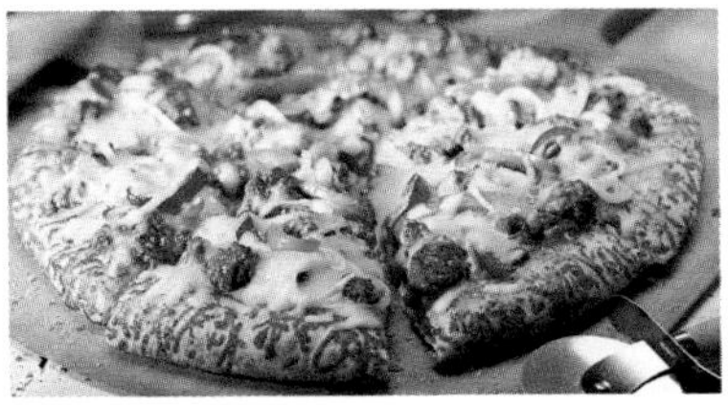

Photo Courtesy of the Beef Checkoff

Pizza with Beef and Peppers

Figure 14-60. Breads can be hollowed out and used as a bowl, stacked or layered to create height and dimension, or sliced to produce a unique shape.

CULINARY NUTRITION RECIPE-MODIFICATION PROCESS

Cinnamon Raisin Sunflower Buns

Cinnamon buns are popular quick breads that commonly include a sweet filling comprised of brown sugar and butter. This recipe also includes raisins and sunflower seeds as well as an icing made from confectioners' sugar and milk. While these buns offer a variety of textures and a rich flavor, they are high in saturated fat and sugar. The buns are also low in dietary fiber due to the use of refined flour.

Yield: 12 servings, 1 bun each

Ingredients

Dough

½ c	light brown sugar, packed
1 tsp	baking soda
½ tsp	salt
½ tsp	vanilla extract
1 ea	whole egg
1 c	buttermilk
⅓ c	butter, melted
3 c	all-purpose flour

Filling

4 tbsp	butter, room temperature
1 c	light brown sugar, packed
¾ tsp	ground cinnamon
⅓ c	raisins
⅓ c	sunflower seeds

Icing and garnish

1 c	confectioners' sugar
2 tbsp	whole milk
4 tbsp	sunflower seeds

Preparation

Dough

1. Combine brown sugar, baking soda, salt, vanilla, and egg in a mixing bowl and mix until well blended.
2. Add buttermilk and melted butter. Stir to combine.
3. Mix in flour until a dough forms and holds together.
4. Place the dough on a lightly floured surface and knead for 1–2 minutes.
5. Roll the dough into a 12″ × 24″ rectangle.

Filling

6. Spread room temperature butter evenly over the dough using a rubber spatula.
7. Sprinkle with brown sugar, cinnamon, raisins, and sunflower seeds.
8. Roll the dough into a log, starting from the short side of the rectangle.
9. Cut the dough into 12 one-inch slices and place in prepared wells of muffin pan.
10. Bake at 375°F for 15 minutes or until golden brown.
11. Place pan on wire rack and allow buns to cool for 5 minutes.
12. Remove buns from pan and place on wire rack to cool completely.

Icing and garnish

13. Combine confectioners' sugar and milk in a mixing bowl until and mix well blended.
14. Drizzle icing over buns.
15. Sprinkle sunflower seeds on top of each bun.

Evaluate original recipe for sensory and nutritional qualities

- Cinnamon raisin sunflower buns are a sweet, buttery pastry with varied textures.
- Buns are high in saturated fat and sugar and low in dietary fiber.

Establish goals for recipe modifications

- Reduce saturated fat and sugar content.
- Increase dietary fiber.
- Add texture and maintain flavor profile.

Identify modifications or substitutions

- Use a substitute for the butter.
- Reduce the amount of brown sugar.
- Use a substitute for a portion of the all-purpose flour.
- Increase the amount of raisins and sunflower seeds.
- Change the recipe from buns to muffins.

Determine functions of identified modifications or substitutions

- Butter adds a rich mouthfeel and creates flaky layers in the dough.
- Brown sugar provides sweetness.
- Flour provides structure and texture to the buns.
- Raisins and sunflower seeds add contrasting textures and flavors.
- Muffins are more nutrient dense than buns.

Select appropriate modifications or substitutions

- Substitute vegetable oil for the butter.
- Reduce brown sugar from 1½ cups to ⅔ cup.
- Substitute whole wheat flour for half of the all-purpose flour.
- Increase the amount of raisins to 1 cup and the sunflower seeds to ¾ cup.
- Incorporate nonfat milk for the liquid needed to prepare the muffins.

Test modified recipe to evaluate sensory and nutritional qualities

- Substituting oil for butter reduces saturated fat.
- Reducing the amount of brown sugar lowers overall sugar content.
- Incorporating whole wheat flour adds dietary fiber.
- Increasing the amount of raisins and sunflower seeds adds dietary fiber and protein while maintaining the flavor profile of the buns.
- Modifying the recipe from buns to muffins yields a heartier texture and adds both protein and calcium due to the incorporation of milk.

CULINARY NUTRITION RECIPE-MODIFICATION PROCESS

Modified Cinnamon Raisin Sunflower Muffins

While this modified recipe is still a quick bread, the form has been changed from buns to muffins. The flavor profile is similar, but the amount of sugar is reduced. Butter is replaced with oil and whole wheat flour is incorporated. These modifications result in muffins that offer sustained energy due to a reduction in simple carbohydrates and an increase in dietary fiber. The muffins are also lower in saturated fat, higher in protein, and contain more than twice the calcium of the cinnamon raisin sunflower buns.

Yield: 12 servings, 1 muffin each

Ingredients

2 c	all-purpose flour
2 c	whole wheat flour
⅔ c	light brown sugar, packed
1 tbsp	baking powder
¼ tsp	salt
1½ tsp	ground cinnamon
1½ c	nonfat milk
⅓ c	vegetable oil
2 ea	whole eggs
2 tsp	vanilla extract
1 c	raisins
¾ c	sunflower seeds

Preparation

1. Combine all-purpose flour, whole wheat flour, brown sugar, baking powder, salt, and cinnamon in a mixing bowl.
2. Whisk together milk, oil, eggs, and vanilla in a separate bowl.
3. Make a well in the center of the flour mixture. Add the milk mixture and stir until just combined. Do not overmix.
4. Add the raisins and sunflower seeds. Stir to incorporate. Do not overmix.
5. Spoon batter into prepared muffin pan.
6. Bake at 400°F for 15–20 minutes or until a toothpick inserted in the center of a muffin comes out clean.
7. Place pan on wire rack and allow muffins to cool for 5 minutes.
8. Remove muffins from pan and place on wire rack to cool completely.

Cinnamon Raisin Sunflower Buns/Muffins Nutritional Comparison

Nutrition Facts	Original	Modified
Calories	402.0	356.7
Total Fat	13.9 g	12.0 g
Saturated Fat	6.7 g	1.3 g
Dietary Fiber	1.8 g	4.6 g
Sugar	38.5 g	21.0 g
Protein	6.3 g	9.4 g
Calcium	71.3 mg	150.7 mg

Signature Cinnamon Raisin Nut Butter Muffins

In this signature muffin, a combination of sunflower nut butter and puréed bananas is used to elevate texture and flavor. Like the modified recipe, this recipe relies on whole wheat flour and oil to produce a flavorful, nutrient-dense muffin that is low in saturated fat and abundant in dietary fiber and protein.

Yield: 12 servings, 1 muffin each

Ingredients

2 c	all-purpose flour
2 c	whole wheat flour
⅔ c	brown sugar, packed
2 tsp	baking soda
½ tsp	salt
1½ tsp	ground cinnamon
2 ea	bananas, medium
⅓ c	vegetable oil
2 ea	whole eggs
½ c	sunflower nut butter
1 tsp	vanilla extract
½ c	raisins

Preparation

1. Combine all-purpose flour, whole wheat flour, brown sugar, baking soda, salt, and cinnamon in a mixing bowl.
2. Place bananas, oil, eggs, nut butter, and vanilla in a blender and purée until smooth.
3. Make a well in the center of the flour mixture. Add the banana mixture and stir until just combined. Do not overmix.
4. Add the raisins and stir to incorporate. Do not overmix.
5. Spoon batter into prepared muffin pan.
6. Bake at 375°F for 15–20 minutes or until a toothpick inserted in the center of a muffin comes out clean.
7. Place pan on wire rack and allow muffins to cool for 5 minutes.
8. Remove muffins from pan and place on wire rack to cool completely.

Cinnamon Raisin Sunflower Buns/Muffins Nutritional Comparison

Nutrition Facts	Original	Modified	Signature
Calories	402.0	356.7	354.6
Total Fat	13.9 g	12.0 g	12.7 g
Saturated Fat	6.7 g	1.3 g	1.4 g
Dietary Fiber	1.8 g	4.6 g	3.9 g
Sugar	38.5 g	21.0 g	18.0 g
Protein	6.3 g	9.4 g	8.5 g
Calcium	71.3 mg	150.7 mg	45.4 mg

Knowledge Check 14-10

1. Explain ways to increase the nutrient density of breads.
2. Explain the functions of yeast bread ingredients.
3. List production steps involved in the preparation of yeast breads.
4. Compare the purposes of kneading and punching doughs.
5. Describe the proofing process.
6. Define oven spring.
7. Explain why quick breads are mixed for only a short amount of time.
8. Describe factors that affect the flavor development of breads.
9. Give two examples of how to effectively plate breads.

BREAD MENU MIX

To respond to the rising interest in healthy menu options, chefs can offer nutrient-dense breads, with an emphasis on whole grain varieties, at every meal. From breakfast breads to breads served with soups, salads, and sandwiches at lunch to the assortment of breads used during dinner service, there is a place for nutrient-dense bread across all meal periods.

Menu descriptions that feature the unique qualities of breads can increase sales. **See Figure 14-61.** For example, menu descriptions that emphasize freshly baked breads, artisan breads, gluten-free breads, or breads made with whole grains are an effective way to draw attention to menu items.

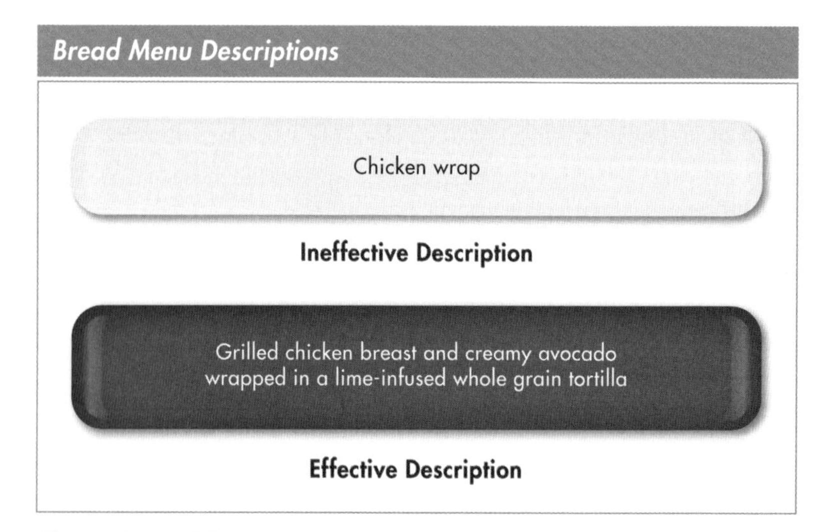

Figure 14-61. Effective menu descriptions that feature the unique qualities of breads can increase sales.

Breakfast Breads

Classic breakfast breads include quick breads such as coffee cake, banana bread, and blueberry muffins. Breakfast menus also include yeast breads such as bagels and English muffins, as well as calorie-laden doughnuts and cinnamon rolls. Day-old yeast breads, such as baguettes, can often be repurposed and used on the breakfast menu as French toast.

Many breakfast breads are high in sugar, fat, and calories. To produce nutritious breakfast breads, chefs can use puréed fruits as fat replacers, replace refined sugars with natural sugars, or incorporate whole grains. Topping French toast, pancakes, and waffles with fresh fruits instead of syrup and butter also reduces the sugar, fat, and calories of the items while increasing vitamins, minerals, dietary fiber, and antioxidants. **See Figure 14-62.**

Appetizer Breads

Breads provide complex carbohydrates, structure, and texture to appetizers. Crostini (small slices of toasted bread) form the foundation of many flavorful canapés with toppings such as beef tartare or olive tapenade. Bruschetta is another appetizer in which bread acts as the base for additional ingredients. **See Figure 14-63.** Pita chips and crackers are commonly served with dips, such as hummus, and toast points are traditionally served with caviar. Care should be taken to choose breads that complement the flavors of the other foods being highlighted on the plate.

Bread crumbs are used in stuffings and as binders for appetizers such as stuffed mushrooms, crab cakes, and meatballs. Bread crumbs are also used to coat foods that are fried.

Chef's Tip

Panko is a type of Japanese bread crumb that is lighter and crispier than a traditional bread crumb.

Nutritional Comparison of Pancakes and Toppings

Type	Calories	Total Fat	Saturated Fat	Sugar
Pancakes with syrup and butter*	368	14 g	6 g	34 g
Pancakes with fresh berries†	217	8 g	2 g	6 g

* based on a serving of two 4" pancakes, 2 oz maple syrup, and ¼ tbsp butter
† based on a serving of two 4" pancakes, ¼ cup blueberries, and ¼ cup sliced strawberries

Figure 14-62. Topping pancakes with fresh fruits instead of syrup and butter reduces the sugar, fat, and calories of the menu item.

Figure 14-63. Crostini, bruschetta, and pita chips are three popular ways to use breads in appetizers.

Sandwich Breads

Most guests make sandwich choices based on the fillings, but a well-crafted artisan bread can elevate the flavor profile of a sandwich to new heights. Yeast breads are the most common types of breads used for making sandwiches. Whole grains should be emphasized when using sandwich breads such as pitas, tortillas, and rolls. In addition to using whole grains, chefs should also have gluten-free options for sandwiches and flatbreads. **See Figure 14-64.**

Chef's Tip

Panini sandwiches are cooked on a panini grill, which creates grill marks on the bread that add visual appeal and flavor.

Gluten-Free Corn Tortillas

Courtesy of The National Pork Board

Figure 14-64. In addition to using whole grains, chefs should also have gluten-free options for sandwiches and flatbreads.

Sides of Bread

For many individuals, a meal would not be complete without bread. Dinner rolls, flatbreads, crackers, breadsticks, cornbreads, muffins, and croissants are commonly served with meals. The assortment of unique, flavorful, and nutritious breads that chefs can create by combining whole grains, vegetables, herbs, spices, oils, and cheeses is unlimited. **See Figure 14-65.**

Bread Basket

Figure 14-65. An assortment of breads can be created by combining whole grains, vegetables, herbs, spices, oils, and cheeses.

Dessert Breads

Breakfast breads such as coffee cakes can be offered on the dessert menu. Leftover breads can also be featured in desserts such as bread puddings. A healthy bread pudding known as summer bread pudding is an English dessert in which slices of bread are layered in a bowl along with fruits, usually mixed berries. The dessert rests several hours before it is turned onto a plate and served.

Although bread pudding can be a nutritious dessert, chefs should be mindful of how quickly the sugar, fat, and calories in bread puddings can escalate. For example, rich breads such as croissants or brioche are often incorporated with cream, eggs, and sugar to form the base of bread puddings. Bread puddings are also commonly topped with calorie-laden sauces, such as alcohol-laced caramel sauce.

To keep bread puddings more nutritionally balanced, whole grain breads can be used. The custard can be made with evaporated low-fat milk, and the sauce can be replaced with fresh fruit and frozen yogurt. **See Figure 14-66.** Modifications like these can result in a menu item that is lower in sugar, fat, and calories but is still a satisfying, flavorful dessert.

Dessert Breads

Figure 14-66. Bread pudding made with whole grain bread and low-fat dairy products, served with fresh fruit and frozen yogurt, is more nutrient dense than traditional bread pudding.

Knowledge Check 14-11

1. Explain how to raise the nutrient density of breakfast breads.
2. List ways breads are used in appetizers.
3. Identify breads commonly used in sandwiches and sides.
4. Explain how to prepare nutritious dessert breads.

PROMOTING BREADS ON THE MENU

Breads are an integral part of the menu at virtually all foodservice venues. Highlighting both ancient and whole grain breads is a promotional strategy that helps meet the rising demand for healthy menu options. **See Figure 14-67.** For example, a turkey sandwich served on whole grain bread is generally more enticing to guests than a turkey sandwich on white bread.

Promoting Whole Grain Breads

Photo Courtesy of McCain Foods USA

Figure 14-67. Using whole grain breads is a promotional strategy that helps meet the rising demand for healthy menu options.

Chef's Tip

Parbaked breads purchased from local bakeries and artisans allow a foodservice operation to promote freshly baked breads as well as local products.

Quick Service and Fast Casual Venues

Sandwiches are an essential carry-out menu item for quick service and fast casual venues. Some of these operations bake breads on site. Breads are also commonly served as a side with soups and salads. Flatbreads are used for sandwiches and pizzas and served with dips. Breakfast buffets typically feature a wide assortment of yeast bread and quick bread varieties. Sandwich bread options commonly include multigrain breads, pitas, flatbreads, and buns. **See Figure 14-68.**

Sandwich Breads

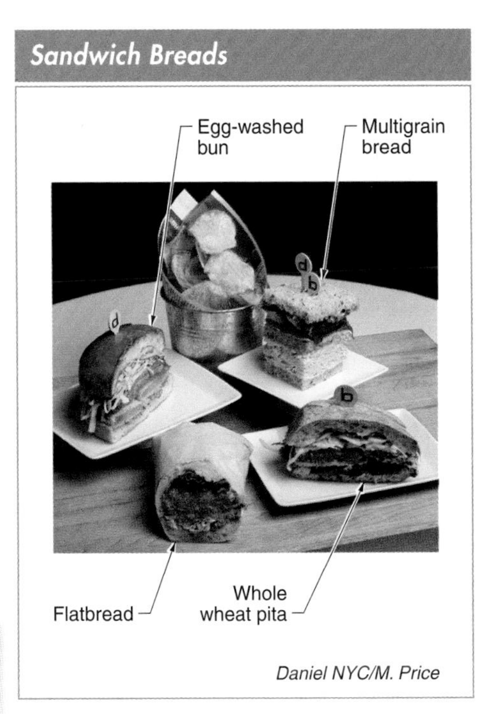

Daniel NYC/M. Price

Figure 14-68. Popular sandwich bread options include multigrain breads, pitas, flatbreads, and buns.

Fine Dining and Special Event Venues

House-made breads as well as artisan breads are often featured at fine dining venues. Some fine dining venues serve a different bread to complement each course. Depending on the occasion, special event venues may serve baskets of assorted specialty breads.

Tanimura & Antle®

Fine dining venues often serve house-made breads to complement each course.

Knowledge Check 14-12

1. Give an example of how breads can be used to meet the demand for healthy menu options.
2. Identify ways that flatbreads can be used on the menu.

Chapter Summary

Pastas are made from flour-based doughs and come in a variety of forms including ribbon, shaped, tube, and stuffed. Asian noodles are vegetable-based, egg-based, or grain-based ribbonlike strands. Gluten-free pastas are made from gluten-free flours. Nutrient-dense whole grain pastas can be enhanced with lean animal-based proteins, seasonal produce, aromatics, and unsaturated oils to create visually appealing, flavorful, and healthy menu items.

Service Ideas, Inc.

Grains can be classified as cereal grains and pseudocereals. Cereal grains include rice, wheat, corn, oats, barley, rye, triticale, millet, teff, farro, and spelt. Pseudocereals include quinoa, amaranth, and buckwheat. Rice, corn, millet, teff, quinoa, amaranth, and buckwheat are gluten-free. Ancient grains like millet, teff, farro, spelt, quinoa, and amaranth are becoming increasingly popular. Whole grains are more nutritious than refined grains because only the husk is removed. Grains have a mellow flavor that pairs well with savory and sweet ingredients.

The aroma of freshly baked yeast and quick breads is a marketing tool for many foodservice operations. Yeast breads take extended time to produce because they are leavened with yeast, which needs time to ferment. Quick breads do not require fermentation. Breads develop complex flavor by added ingredients such as fruits, vegetables, herbs, spices, nuts, and seeds. Whole grain breads provide more nutrients than refined breads.

Chapter Review

Digital Resources
ATPeResources.com/QuickLinks
Access Code: 267412

1. List factors that affect the perceived value of pastas.
2. Explain how to cook pastas.
3. Identify methods of developing flavors in pasta dishes.
4. Explain how to effectively plate pasta dishes.
5. Describe the ways in which pastas may be incorporated throughout the menu mix.
6. Give examples of how to promote pastas in various foodservice venues.
7. Compare the processing of whole and refined grains.
8. Describe the uses of common cereal grains and pseudocereals.
9. Describe common cooking methods for grains.
10. Identify methods of developing flavors in grain dishes.
11. Explain how to effectively plate grain dishes.
12. Describe the ways in which grains may be incorporated throughout the menu mix.
13. Give examples of how to promote grains in various foodservice venues.
14. Compare yeast breads and quick breads.
15. Identify methods of developing flavors in bread dishes.
16. Explain how to effectively plate bread dishes.
17. Describe the ways in which breads may be incorporated throughout the menu mix.
18. Give examples of how to promote breads in various foodservice venues.

Desserts on the Menu

All-Clad Metalcrafters

Sullivan University

While desserts are generally considered an indulgent way to end a meal, they can also be light and refreshing. Highlighting seasonal fresh fruits at their peak of ripeness, serving appropriate portion sizes, and offering guests a variety of flavors are key components of healthier dessert menus. Nutrient-dense ingredients as well as bite-sized portions of rich desserts allow guests to fulfill their desire for a sweet treat without overindulging in sugar, fat, and calories.

Chapter Objectives

1. Identify factors to consider when planning a healthier dessert menu.
2. Summarize the perceived value of desserts.
3. Describe the dessert classifications commonly found on menus.
4. Explain how desserts can be prepared to be healthy and flavorful.
5. Describe ways in which desserts can be plated.
6. Identify ways beverages, cheeses, fruits, nuts, seeds, and breads are used in desserts.
7. Summarize how foodservice venues promote desserts on the menu.

Wisconsin Milk Marketing Board, Inc.

DESSERTS

Many guests sacrifice dessert in an effort at healthy eating. However, dessert does not need to be avoided by guests if chefs keep in mind the concepts of balance, moderation, and variety. Desserts need to be given the same level of attention as other menu categories in terms of nutrient density, appropriate portion sizes, and flavor development. When these aspects are considered, guests can receive a nutrient-dense meal that is flavorful, satisfying, and healthy from appetizers through desserts. **See Figure 15-1.**

Nutrient-Dense Desserts

Whole grain crust
Seasonal fruits
Almonds

National Honey Board

Figure 15-1. Desserts made with fruits, nuts, and whole grains can provide guests with a nutrient-dense, flavorful option.

Many desserts contain large amounts of refined sugar, butter, eggs, and cream. This commonly leads to desserts that are high in sugar, fat, and calories. However, these ingredients can be used to create nutritious desserts when recipe modifications are made and portion control is followed. For example, smaller amounts of highly caloric ingredients can be incorporated with natural sugars, low-fat dairy products, fruits, nuts, seeds, and whole grains in order to create exceptional desserts that can be highlighted on the menu. **See Figure 15-2.**

Perceived Value of Desserts

When desserts complement nutrient-dense menu items, guests are more apt to indulge in them. Chefs can exceed guest expectations by showcasing both fruit-based desserts and small portions of decadent desserts. Having options that appeal to different tastes allows guests to choose a light and sweet ending to a meal or a sampling of something rich. A varied dessert menu can entice guests to stay longer and order more. This can lead to a larger sale of both food and beverages.

Dessert Classifications

A well-planned dessert menu offers a variety of desserts that complement nutrient-dense menu items while satisfying guest preferences. A varied dessert menu can include cakes and cookies, pies and tarts, pastries, crêpes, custards and creams, frozen desserts, and confections.

Cakes and Cookies. Cakes generally contain flour, sugar, fat, eggs, liquid, leaveners, and flavorings. Vanilla and chocolate are popular cake flavors, but hundreds of different flavor profiles are possible. Cakes are likely to include icing and are often elaborately decorated or transformed into tortes. A *torte* is a rich cake that is usually layered with creams, jams, or fruits and is iced with a sweet icing or glazed and garnished. **See Figure 15-3.**

Cake batter can also be used to make cupcakes. Cupcakes offer guests portion control and on-the-go convenience. Cupcakes may be simple or complex in terms of flavor and presentation. Cupcakes are often topped with large amounts of icing and are therefore laden with sugar, fat, and calories.

Creating Healthier Dessert Menus

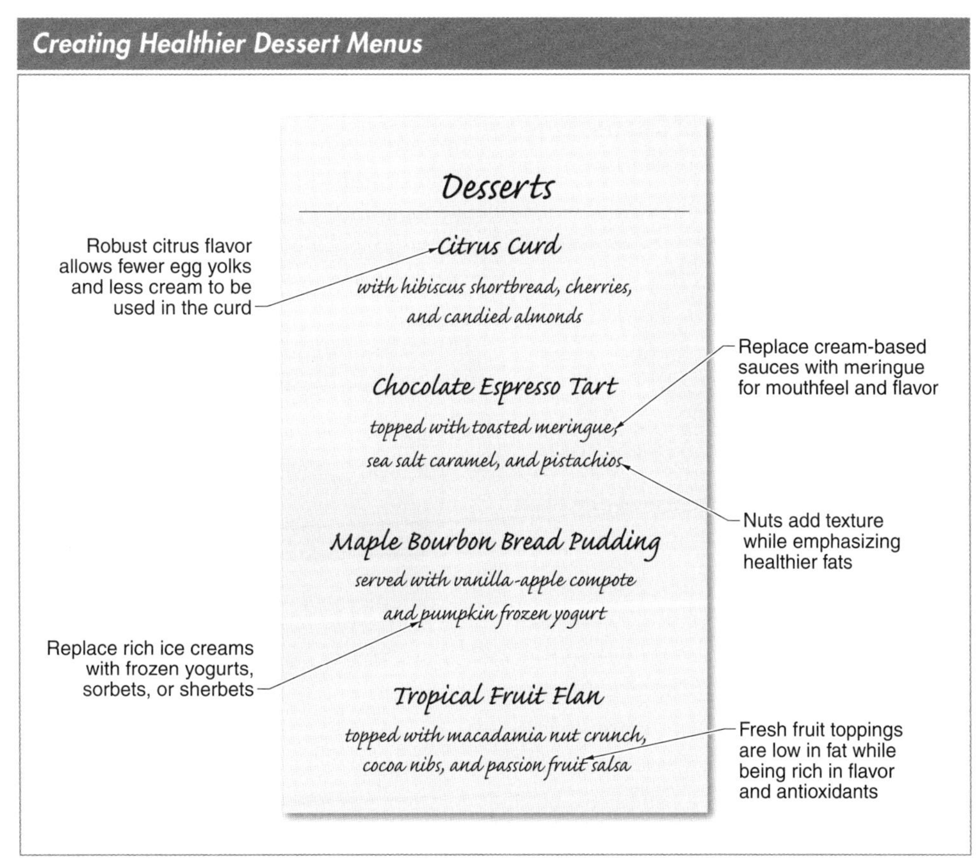

Figure 15-2. Smaller amounts of highly caloric ingredients can be incorporated with natural sugars, low-fat dairy products, fruits, nuts, seeds, and whole grains in order to create healthier dessert menus.

Tortes

Daniel NYC/T. Schauer

Figure 15-3. A torte is a rich cake that is usually layered with creams, jams, or fruits and is iced with a sweet icing or glazed and garnished.

Cookies are made from the same basic ingredients as cakes. The major difference between cookies and cakes is the ratio of wet to dry ingredients. Cookie dough typically contains more fat and less moisture than cake batter. When cookies are baked, they have a texture that is classified as soft or crisp. **See Figure 15-4.** A soft cookie is prepared from dough that contains a high percentage of moisture, while a crisp cookie contains a high percentage of sugar.

Chef's Tip

In order to lower saturated fat and develop a rich, nutty flavor in cookies that call for melted butter, brown a smaller amount of butter and add canola oil to equal the remaining amount of butter called for in the recipe.

Cookies

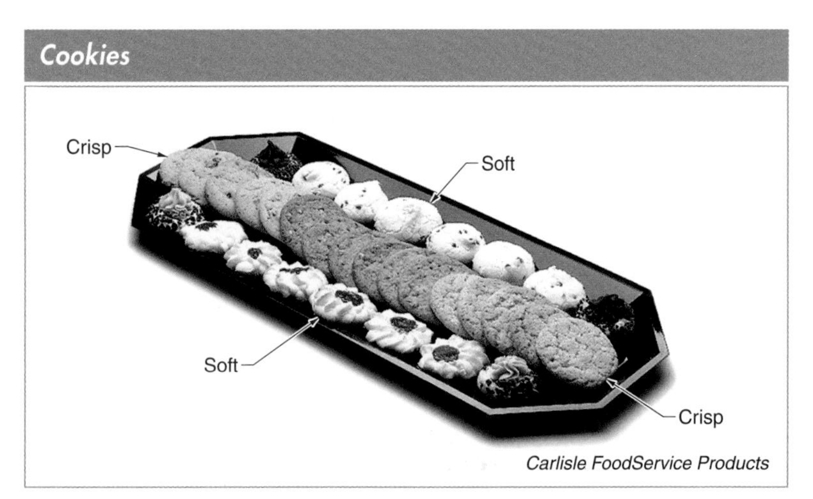

Figure 15-4. When cookies are baked, they have a texture that is classified as soft or crisp.

Pies and Tarts. Pies and tarts are similar and offer a wide range of flavor combinations. The main difference between pies and tarts is their depth. Pies are generally deeper than tarts. Both pies and tarts contain a single bottom crust made from a dough of flour, fat, and sometimes sugar. While tarts always have a single crust, pies can have a single or a double crust. A single crust keeps the amounts of fat and calories lower than a double crust.

A meringue can sometimes be used as an alternative to traditional piecrusts. A *meringue* is a mixture of egg whites and sugar. Meringues can be baked to a crisp consistency to make pavlova. *Pavlova* is a crisp meringue that is filled with whipped cream and fruit. Filling pavlovas with seasonal fresh fruit and a small dollop of whipped cream provides a low-fat dessert option. **See Figure 15-5.**

The filling options for pies and tarts are virtually limitless. Fruit fillings such as apple, cherry, blueberry, and peach are popular. However, pies and tarts also can be filled with custards, creams, nuts, and vegetables such as pumpkin or sweet potatoes.

Variations of pies and tarts include cobblers and crisps. Cobblers and crisps are both deep-dish desserts with fruit fillings, but their toppings differ. Cobblers are generally topped with a thick biscuit crust and sprinkled with sugar. Crisps are topped with a crumbly mixture of flour, fat, sugar, and sometimes oats. Cobblers and crisps do not have a bottom crust.

Pavlovas

Figure 15-5. Filling pavlovas with fresh fruit provides a low-fat dessert.

Pastries. Pastries can be made from a variety of different doughs including éclair paste, puff pastry, and phyllo dough. Pastry doughs are typically baked or deep-fried and filled with an assortment of sweet ingredients. Most pastry doughs are high in saturated fat and calories. When pastry doughs contain cream-based fillings, more fat and calories are added, so smaller portions should be served.

Crêpes. A *crêpe* is a French pancake that is light and very thin. Crêpes are made from a batter of eggs, milk, and flour and are often filled with fruits or pastry cream. Crêpes can be served hot or cold. For a nutrient-dense dessert, crêpes can be stuffed with fresh fruit and lightly sweetened Greek yogurt or low-fat ricotta cheese. **See Figure 15-6.**

Crêpes

Frieda's Specialty Produce

Figure 15-6. For a nutrient-dense dessert, crêpes can be stuffed with fresh fruit and lightly sweetened Greek yogurt or low-fat ricotta cheese.

Custards and Creams. There are many varieties of custards and creams including baked custards, soufflés, crème anglaise, sabayons, and mousses. Custards do not contain flour and instead rely on eggs, sugar, and cream for structure. To make lighter versions of custards, some of the egg yolks can be replaced with egg whites, and low-fat milk or milklike beverages can be substituted for the cream. Custards are often enhanced with additional ingredients such as fruits, chocolate, and sauces.

- A *cheesecake* is a baked custard that typically has a graham cracker or cookie crust. Cheesecake filling can have either a light or dense texture. The fat and calories in cheesecake filling can be reduced by substituting low-fat dairy products such as cream cheese, sour cream, or cottage cheese for full-fat varieties. The crust can be made with whole grain flour, and some of the butter in the crust can be replaced with canola oil to reduce the amount of saturated fat.
- A *sabayon,* also known as a zabaglione, is a light and foamy custard prepared by whisking egg yolks, sugar, and white wine over a double boiler into a silky foam. Any type of dry, sweet, or sparkling wine can be used to make sabayon. Fresh fruit can be enrobed in a small amount of sabayon for a light finish to a meal.
- A *mousse* is a rich and airy foam that has been lightened with whipped cream, whipped pasteurized egg whites, or both. When mousse is made without whipped cream and flavored with fruit purées or semisweet chocolate, it can be flavorful and lower in fat. **See Figure 15-7.**

Mousses

Cape Cod Cranberry Growers' Association

Figure 15-7. Mousse can be made without whipped cream and flavored with fruit purées to reduce fat content.

CULINARY NUTRITION RECIPE-MODIFICATION PROCESS

Tiramisu Trifle

Tiramisu trifle is an Italian dessert made from mascarpone cheese layered with espresso-soaked ladyfinger cookies. Tiramisu is prepared by incorporating egg yolks into the mascarpone cheese. This provides a smooth, creamy texture, but also contributes a large amount of calories, fat, and cholesterol to the dish. Calories and fat are also increased by the shaved chocolate garnish.

Yield: 9 servings, approximately 6 oz each

Ingredients

6 ea	pasteurized egg yolks
3 tbsp	granulated sugar
1 lb	mascarpone cheese
1½ c	espresso, chilled
2 tsp	coffee liqueur
24 ea	ladyfinger cookies
3 oz	bittersweet chocolate shavings

Preparation

1. Whip egg yolks and sugar until mixture becomes thick and pale.
2. Add mascarpone cheese and 1 tbsp espresso and beat until smooth.
3. Pour remaining espresso and coffee liqueur into a shallow dish.
4. Dip a ladyfinger cookie into the espresso mixture for 3–5 seconds and then place the cookie in the bottom of a 9″ × 13″ baking dish. Repeat this process until 12 cookies line the bottom of the dish.
5. Spread half of the mascarpone mixture over the cookies.
6. Dip the remaining cookies in the espresso mixture and arrange them on top of the mascarpone mixture.
7. Spread the remaining mascarpone mixture over the top layer of cookies.
8. Cover the baking dish and refrigerate until chilled.
9. Sprinkle chocolate shavings over the trifle before serving.

Chef's Tip: *Use pasteurized egg yolks in this recipe to reduce the risk of foodborne illness.*

Evaluate original recipe for sensory and nutritional qualities

- Tiramisu trifle is a creamy dessert that is rich and flavorful.
- Dish is high in calories, fat, and cholesterol.

Establish goals for recipe modifications

- Reduce calories, fat, and cholesterol.
- Maintain texture and flavor.

Identify modifications or substitutions

- Eliminate egg yolks.
- Use a substitute for a portion of the mascarpone cheese.
- Replace granulated sugar with a natural sweetener.
- Use a substitute for bittersweet chocolate shavings.

Determine functions of identified modifications or substitutions

- Egg yolks add richness to the mascarpone filling.
- Mascarpone provides structure to the filling.
- Granulated sugar adds sweetness.
- Chocolate shavings add color, texture, and flavor.

Select appropriate modifications or substitutions

- Omit egg yolks.
- Substitute low-fat cream cheese for half of the mascarpone cheese.
- Replace granulated sugar with agave nectar.
- Substitute Dutch cocoa powder for the chocolate shavings.

Test modified recipe to evaluate sensory and nutritional qualities

- Omitting the egg yolks from the filling reduces calories, fat, and cholesterol, and their absence is not detectable.
- Substituting low-fat cream cheese for half of the mascarpone reduces calories, fat, and cholesterol, helps stabilize the filling, and maintains a rich mouthfeel.
- Using agave nectar incorporates natural sugar and adds sweetness.
- Substituting Dutch cocoa powder for chocolate shavings reduces calories and fat.

Modified Tiramisu Trifle

This modified tiramisu trifle recipe does not contain egg yolks, thus reducing calories, fat, and cholesterol. Some of the mascarpone cheese is replaced with low-fat cream cheese, which helps stabilize the filling. The extra stability is needed because granulated sugar is replaced with natural agave nectar, which is a liquid. Shaved chocolate is replaced with a dusting of cocoa powder to maintain visual appeal and flavor while reducing calories and fat.

Yield: 9 servings, approximately 6 oz each

Ingredients

8 oz	low-fat cream cheese, softened
3 tbsp	agave nectar
8 oz	mascarpone cheese
1½ c	espresso, chilled
2 tsp	coffee liqueur
24 ea	ladyfinger cookies
⅓ c	Dutch cocoa powder

Wisconsin Milk Marketing Board, Inc.

Preparation

1. Whip cream cheese and agave nectar until mixture becomes thick and pale.
2. Add mascarpone cheese and 1 tbsp espresso and beat until smooth.
3. Pour remaining espresso and coffee liqueur into a shallow dish.
4. Dip a ladyfinger cookie into the espresso mixture for 3–5 seconds and then place the cookie in the bottom of a 9″ × 13″ baking dish. Repeat this process until 12 cookies line the bottom of the dish.
5. Spread half of the mascarpone mixture over the cookies.
6. Dip the remaining cookies into the espresso mixture and arrange them on top of the mascarpone mixture.
7. Spread the remaining mascarpone mixture over the top layer of cookies.
8. Cover the baking dish and refrigerate until chilled.
9. Dust the trifle with cocoa powder before serving.

Chef's Tip: *Lemon or orange zest can be added to the cheese filling for a burst of flavor.*

Tiramisu Trifle Nutritional Comparison

Nutrition Facts	Original	Modified
Calories	444.1	314.1
Total Fat	32.0 g	19.4 g
Saturated Fat	3.9 g	4.1 g
Cholesterol	313.8 mg	155.9 mg

CULINARY NUTRITION RECIPE-MODIFICATION PROCESS

Signature Orange Espresso Trifle

This signature trifle recipe maintains the layering of the trifle but changes the flavor to a lighter and more complex profile. A low-fat citrus pound cake accented with espresso syrup, chocolate yogurt, and oranges provides contrasting colors, textures, and flavors. This signature trifle shows that a rich dessert, like tiramisu, can be replaced with a more nutrient-dense yet flavorful and satisfying dessert.

Yield: 12 servings, approximately 6 oz each

Ingredients

Citrus pound cake

1 c	whole wheat flour
¾ c + 1 tbsp	all-purpose flour
2 tsp	baking powder
¼ tsp	salt
¾ c	granulated sugar
3 oz	low-fat cream cheese, softened
2 tbsp	butter, softened
1 tbsp	lemon zest, minced
1 tbsp	orange zest, minced
1 tbsp	poppy seeds
3 ea	egg whites
1 ea	whole egg
½ c	2% milk

Espresso syrup

3 oz	espresso, chilled
3 oz	agave nectar

Filling and garnish

3 c	nonfat chocolate yogurt
3 c	orange segments

Preparation

Citrus pound cake

1. Sift whole wheat flour, all-purpose flour, baking powder, and salt into a mixing bowl and set aside.
2. Beat the sugar, cream cheese, butter, lemon zest, orange zest, and poppy seeds until light and fluffy.
3. Add egg whites, one at a time, to cream cheese mixture, beating well after each addition.
4. Add the whole egg and beat to incorporate.
5. Reduce mixer speed to medium and add milk. *Note:* The mixture will look curdled.
6. Reduce mixer speed to low and add half the flour mixture until just combined.
7. Scrape down the sides and bottom of the bowl and then add remaining flour mixture until just combined.
8. Transfer batter to a prepared 5″ × 9″ loaf pan and smooth the top with a rubber spatula.
9. Bake at 350°F for approximately 45–50 minutes or until a toothpick inserted in the center of the cake comes out clean.
10. Set cake on a wire rack and allow to cool for 15 minutes.
11. Invert the cake onto a plate and allow to cool completely.

Espresso syrup

12. Add espresso and agave nectar to a bowl and stir to combine.

Assembly

13. Cut the cake into 12 equal slices and then cut each slice in half for a total of 24 slices.
14. Place one slice of cake in the bottom of a clear goblet.
15. Pour ¼ cup chocolate yogurt over the cake.
16. Place a second piece of cake over the yogurt-covered cake.
17. Spoon 1 tbsp of espresso syrup over the second layer of cake and garnish with ¼ cup orange segments.
18. Repeat steps 14–17 to fill a total of 12 goblets.

Trifle Nutritional Comparison

Nutrition Facts	Original	Modified	Signature
Calories	444.1	314.1	273.8
Total Fat	32.0 g	19.4 g	4.5 g
Saturated Fat	3.9 g	4.1 g	2.4 g
Cholesterol	313.8 mg	155.9 mg	28.1 mg
Calcium	34.9 mg	47.7 mg	153.3 mg

Icings and Sauces. An *icing* is a sugar-based coating often spread on the outside of and between the layers of a baked product. The consistency of icings ranges from thin and syrupy to thick and dense, with many possible flavor profiles. **See Figure 15-8.** All too often, icing is applied in large quantities, which adds calories. Excess icing can also create a dessert that is overly sweet and mask the flavor of other ingredients. An appropriate amount of icing provides balance and complements the dessert.

Sauces used on desserts can be made from ingredients such as fruits, chocolate, dairy products, wine, and even balsamic vinegar. Dessert sauces provide moisture and flavor and are also visually pleasing. For example, plating crème anglaise with a raspberry coulis enhances its presentation using contrasting colors. It is important not to use too much sauce, as this can increase the sugar, fat, and calories in the dessert.

Daniel NYC/N. Clutton

A small amount of sauce can elevate presentation and flavor.

Icings

Type	Preparation and Ingredients	Common Uses	Low in Fat
Whipped cream	Made by beating heavy cream until stiff peaks form	Piped onto fruit-based cakes and pies	No
Glaze	Made by thinning out jams, fruit purées, or chocolate	Poured over baked products	Generally yes
Foam (boiled)	Made by boiling a glucose and water mixture to 240°F and adding the resulting syrup to an egg white meringue	Applied to cakes, cupcakes, and cookies	Yes
Flat	Made by combining water, confectioners' sugar, corn syrup, and flavoring and heating the mixture to 100°F	Applied to sweet rolls, doughnuts, and Danishes	Yes
Royal	Made like flat icing but egg whites are added to produce a thicker icing that hardens to a brittle texture	Formed into decorations for cakes and sugar sculptures	Yes
Buttercream	Made by creaming together shortening or butter, confectioners' sugar, and vanilla	Applied to cakes, cupcakes, and cookies	No
Fondant	Made by cooking a sugar, glucose, and water mixture to 240°F, letting it cool to 150°F, and then mixing it until creamy and smooth; hardens when exposed to air	Rolled out and draped over cakes for a smooth appearance; heated and poured over cakes; sculpted into decorations	Yes
Fudge	Made by adding hot liquid or syrup to required ingredients and whipping the mixture until smooth	Applied to cakes, cupcakes, and cookies	Generally no
Ganache	Made by heating chocolate and whipping cream until the chocolate melts	Applied to cakes and pastries; used as filling for bon bons; rolled to make truffles	No

Figure 15-8. The consistency of icings ranges from thin and syrupy to thick and dense, with a wide variety of possible flavor profiles.

Frozen Desserts. Frozen desserts are served as stand-alone menu items or as accompaniments to other desserts and can be simple or elaborate in presentation. Ingredients such as fruits, nuts, and spices are often added to frozen desserts to create a variety of flavor profiles. Some frozen desserts are high in fat, while others are nearly fat-free but high in sugar. **See Figure 15-9.** Many frozen desserts are served with calorie-laden sauces such as chocolate or caramel sauce. In contrast, seasonal fruits are naturally sweet and a nutrient-dense complement to frozen desserts. Common frozen desserts include ice cream, frozen yogurt, baked Alaska, sorbets, and sherbets.

Nutritional Comparison of Common Frozen Desserts*

Dessert	Calories	Total Fat	Sugar
Ice cream	137	7 g	14 g
Frozen yogurt	110	3 g	17 g
Sherbet	107	1 g	18 g

* based on ½ cup serving size

Figure 15-9. Some frozen desserts are higher in fat, while others are nearly fat-free but high in sugar.

- *Ice cream* is a frozen dessert made from cream, butterfat, sugar, and sometimes eggs. The cream and butterfat in ice cream provide a velvety texture but also add saturated fat and calories. Low-fat milk and milklike beverages can be used in place of cream, but they produce a grainier texture due to crystallization. To compensate for the difference in texture, it is essential to develop the flavor using high-quality ingredients. For example, using vanilla beans as opposed to vanilla extract provides a richer flavor. Similarly, roasted nuts create depth of flavor and texture beyond plain nuts.
- *Frozen yogurt* is a frozen dessert made from yogurt, sugar, and sometimes gelatin. Frozen yogurt is lower in fat than ice cream but is high in sugar. Also, the freezing process destroys most of the beneficial live bacteria in yogurt.
- *Baked Alaska* is a composed frozen dessert made with a sponge cake soaked in a liqueur-flavored syrup that is topped with ice cream and covered with Italian meringue. Baked Alaska is commonly flambéed tableside. Replacing the ice cream with frozen yogurt and substituting angel food cake for the sponge cake creates a lower fat version of baked Alaska.
- A *sorbet* is a frozen dessert made by combining puréed fresh fruit and simple syrup. Simple syrup is made by heating a mixture of water and sugar until the sugar dissolves. Sorbet is a light and refreshing fat-free and dairy-free dessert that is relatively low in calories but high in sugar. **See Figure 15-10.**

Sorbets

Vita-Mix® Corporation

Figure 15-10. A sorbet is a light and refreshing fat-free and dairy-free dessert that is relatively low in calories but high in sugar.

- A *sherbet* is a frozen dessert made by combining puréed fresh fruit, simple syrup, and milk. Sherbets are similar to sorbets but have a creamier texture due to the presence of milk. Sherbets are lower in fat and calories than ice cream, with a high percentage of their calories coming from sugar.

Confections. A *confection* is a small sugar-based food item such as a candy or a chocolate. Hard and soft caramels, toffee, fudge, lollipops, marshmallows, truffles, and candied nuts are considered confections. Many of these items are high in empty calories due to their sugar content. Confections can be served as a plate of assorted items but are more commonly used to garnish plated desserts. For example, candied nuts or truffles often garnish a tray of cheeses and dried fruits.

Chef's Tip

Bite-sized, decorated cakes called petite fours are confections that make an ideal dessert for guests wanting just a bite of something sweet after a meal but not a lot of calories.

INGREDIENT SPOTLIGHT: Chocolate

Irinox USA

Nutrition Facts

Serving Size ½ oz, unsweetened

Amount Per Serving	
Calories 73	Calories from Fat 72
	% Daily Value*
Total Fat 8g	12%
Saturated Fat 5g	23%
Trans Fat	
Cholesterol 0mg	0%
Sodium 3mg	0%
Total Carbohydrate 4g	2%
Dietary Fiber 2g	10%
Sugars 0g	
Protein 2g	
Vitamin A 0% • Vitamin C	0%
Calcium 2% • Iron	14%

* Percent Daily Values (DV) are based on a 2,000 calorie diet.

Unique Features

- Product of the cacao bean, which grows in podlike fruits on tropical cacao trees
- Contains flavonoids, antioxidants that fight cellular damage caused by free radicals
- Studies suggest that it may help lower blood pressure, improve blood flow, and reduce the risk of blood clots
- Less processed forms such as dark or unsweetened varieties contain greater health benefits than more processed forms
- Low in cholesterol, sodium, and sugar
- Source of dietary fiber, magnesium, zinc, copper, and manganese

Menu Applications

- Use in sauces and garnishes
- Incorporate into cakes, cookies, pies, pastries, crêpes, custards, and frozen desserts
- Use as part of a well-rounded dessert menu

Knowledge Check 15-1

1. Identify three concepts that should be considered when developing healthier dessert menus.
2. Name four ingredients that can cause desserts to be high in sugar, fat, and calories.
3. Identify the importance of a strong dessert menu.
4. Explain the major difference between cakes and cookies.
5. Compare pies and tarts.
6. Describe how a meringue can be used to make a nutrient-dense dessert.
7. Differentiate between pastries and crêpes.
8. Give an example of how to prepare a custard or cream to be a lower-fat dessert option.
9. Name three frozen desserts that are generally lower in fat than ice cream.
10. Explain why confections are high in empty calories.

DESSERT PREPARATIONS

Many foodservice operations purchase desserts from vendors because of limited kitchen space or the cost, time, and skill level necessary to produce fine desserts. Whether desserts are purchased or house-made, they should be created from high-quality ingredients, contain minimal preservatives, and be appropriately portioned.

It is important to remember that dessert is most often served at the end of a meal when many calories have already been consumed. For this reason, it is best to highlight desserts that are light and flavorful as opposed to those that are overly rich and high in empty calories. Nutritionally balanced desserts should be lower in calories and focus on nutrient-dense ingredients that build flavor. **See Figure 15-11.** This can be achieved by emphasizing proper portion sizes, seasonal ingredients, and healthy cooking techniques.

Cooking Desserts

Baked items such as cakes, cookies, pies, tarts, and custards are commonly high in sugar, fat, and calories. Because baking does not add calories to a dish, the cooking technique does not need to be modified. Instead, emphasis should be placed on identifying ingredients that can be modified, such as granulated sugar, butter, cream, eggs, and refined flours. For example, a double-crust apple pie can be made with a single, whole wheat crust to lower its fat and calories and supply complex carbohydrates. The apples can be lightly sweetened with maple syrup to reduce refined sugars and increase nutrients. Whenever possible, oil should be substituted for butter in a recipe to reduce the amount of saturated fats.

Fruits can be a nutrient-dense dessert option, but care should be given to how they are prepared. Poaching or stewing fruit in a cooking liquid with added sugars increases the overall sugar content of the dessert. Instead, fruits can be grilled or roasted to caramelize their natural sugars. Caramelized fruit served with low-fat vanilla ice cream and roasted walnuts provides a nutrient-dense dessert with contrasting textures and depth of flavor.

Guidelines for Modifying Desserts and Enhancing Flavors

- Reduce sugars by one-half to two-thirds
- Replace refined sugars with natural sugars such as honey, maple syrup, and agave nectar
- Replace refined flours with whole grain varieties
- Substitute low-fat dairy products for full-fat varieties
- Replace butter and shortening with unsaturated oils
- Replace oil with applesauce, banana purée, or prune purée
- Use egg substitutes or egg whites instead of whole eggs
- Incorporate fresh, seasonal fruits; avoid fruits canned in syrup
- Use flavor enhancers such as citrus zest, vanilla beans, and spices
- Add liqueurs and wines to poaching liquids and sauces
- Roast nuts and use them sparingly
- Replace frying with healthier cooking techniques

Daniel NYC/J. Gripper

Figure 15-11. Nutritionally balanced desserts should be lower in calories and focus on nutrient-dense ingredients that build flavor.

FOOD SCIENCE EXPERIMENT: Baking Cookies with Natural, Refined, and Artificial Sweeteners

Objective

- To compare vanilla cookies made with honey, granulated sugar, and sucralose in terms of color, texture, and flavor.

Materials and Equipment

- Sifter
- Measuring cups
- Measuring spoons
- 2 mixing bowls
- Mixing machine
- Rubber spatula
- Parchment paper
- Baking sheet
- Nonstick cooking spray
- #16 scoop
- Cooling rack
- 3 plates
- Vanilla cookie ingredients

6 c	cake flour
¾ tsp	baking powder
¾ tsp	salt
12 oz	unsalted butter
⅓ c	honey
⅔ c	granulated sugar
⅔ c	sucralose
3 ea	whole eggs
¾ tsp	vanilla extract
12 oz	whole milk

Procedure

Allergen Alert

The materials used in this experiment may contain one or more food allergens.

1. Sift 2 cups cake flour, ¼ tsp baking powder, and ¼ tsp salt into a mixing bowl and set aside.
2. Place 4 oz of butter and ⅓ cup honey in a second mixing bowl and beat on medium speed until creamy.
3. Add 1 egg and ¼ tsp vanilla extract to the butter mixture and beat until incorporated.
4. Add half of the flour mixture and beat until incorporated.
5. Scrape down the sides and bottom of the bowl with a spatula, add 4 oz milk, and beat until incorporated.
6. Beat in the remaining flour mixture, scraping the sides and bottom of the bowl as needed.
7. Line a baking sheet with parchment paper and spray with nonstick cooking spray.
8. Place 12 mounds of dough onto the baking sheet using a #16 scoop and space 2 inches apart.
9. Moisten fingers with water and gently press each mound of dough into a disk 2½ inch wide and ⅜ inch thick.
10. Bake at 375°F for approximately 20 minutes or until the centers of the cookies are firm and the edges are just beginning to turn light golden brown.
11. Remove the cookies from the oven and allow them to cool on the baking sheet for 2 minutes before placing them on a wire rack to cool completely.
12. Repeat steps 1–11, substituting ⅔ cup granulated sugar for the honey.
13. Repeat steps 1–11, substituting ⅔ cup sucralose for the honey.
14. Place the cookies made with honey on the first plate, the cookies made with granulated sugar on the second plate, and the cookies made with sucralose on the third plate.
15. Evaluate the color, texture, and flavor of each plate of cookies.

Typical Results

There will be variances in the color, texture, and flavor of each plate of vanilla cookies.

Flavor Development

Choosing seasonal fruits at the peak of ripeness is one of the best ways to enhance flavors and increase nutrients in desserts. **See Figure 15-12.** Dry-heat cooking techniques such as grilling, roasting, and sautéing naturally concentrate the sugars in fruits to create depth of flavor.

Seasonal Fruit Tart

Sullivan University

Figure 15-12. Choosing seasonal fruits at the peak of ripeness is one of the best ways to enhance flavors and increase nutrients in desserts.

When creating nutrient-dense desserts or modifying existing dessert recipes, flavor is the key to success. Aromatic ingredients such as citrus, spices, vanilla, coffee, and tea can be used in virtually any type of dessert to build flavor without adding fat and calories. Because aromatic ingredients amplify flavor, the amount of sugar added to a recipe can often be reduced. For example, when lemon juice and ginger are added to a simple syrup, less sugar is needed to produce a flavorful product. Wine and liquor also enhance flavors and limit the overall fat content, but can also add calories to a dish.

Certain desserts develop flavor if they are prepared the day before service. Early preparation allows the ingredients to disperse and the flavors to intensify. Cake batters, poached and stewed fruits, and chutneys are commonly prepared using this technique.

Plating Desserts

An effective way to increase visual appeal and nutrients is to create colorful plated presentations. For example, a crêpe with fresh blackberries and raspberries served with a dollop of low-fat sour cream, slivered almonds, and a sprig of mint provides contrasting colors that enliven the presentation while providing a nutrient-dense dessert.

The plate used for a dessert should complement the overall presentation. Desserts can be positioned on the plate in a variety of ways. For example, main dessert components can be plated off-center and balanced with decorative patterns of sauces and garnishes. Dessert items can also be plated as duos or trios, layered, or elevated. **See Figure 15-13.**

However a dessert is plated, it is always important to serve appropriate portion sizes. When serving reduced portion sizes, the size of the plate should also be reduced. This practice makes the portion look abundant and does not leave the guest feeling deprived.

Chef's Tip

Shot glasses, demitasse cups, and wine glasses create a unique presentation when providing smaller dessert portions.

Plating Desserts

Daniel NYC/J. Bartelsman

Off-Center

Daniel NYC/T. Schauer

Trio

Daniel NYC/G. Russell

Layered

Daniel NYC/B. Milne

Elevated

Figure 15-13. Dessert components can be plated in a variety of ways to enhance presentation.

Knowledge Check 15-2

1. Identify ingredients that can be modified to prepare more nutrient-dense desserts.
2. Give four examples of aromatic ingredients that build flavor without adding calories.
3. Describe ways in which desserts can be plated to create visually appealing presentations.

CULINARY NUTRITION RECIPE-MODIFICATION PROCESS

Black Forest Cake

Black Forest cake is made by layering rich chocolate cake with whipped cream and a cherry filling that has been enhanced with cherry liqueur. There are different variations of Black Forest cake, but all feature a combination of chocolate cake and cherries, and most are high in calories, fat, cholesterol, and total carbohydrates.

Yield: 12 servings (8-inch cake cut into 12 equal slices)

Ingredients

Cake

2 tbsp + 2 tsp	unsalted butter
2 tbsp + 1 c	granulated sugar
3 oz	all-purpose flour
3 oz	unsweetened cocoa powder
1 tsp	baking powder
⅛ tsp	kosher salt
½ c	2% milk
8 ea	whole eggs
1 tsp	vanilla extract

Cream filling

6 tbsp	unsalted butter, softened
4 c	confectioners' sugar
1½ tsp	vanilla extract
4 tbsp	heavy cream

Cherry filling

2 c	granulated sugar
1 c	water
1¼ c	cherry liqueur
4 c	dark, sweet, pitted cherries in heavy syrup
2 tbsp	cornstarch

Icing

2 c	heavy cream
4 tbsp	confectioners' sugar
2 tsp	cherry liqueur

Preparation

Cake

1. Coat two 8-inch cake pans with 1 tbsp butter each. Sprinkle each pan with 1 tbsp sugar.
2. Sift flour, cocoa powder, baking powder, and salt into a mixing bowl and set aside.
3. Warm the milk and 2 tsp butter in a saucepan over low to medium heat.
4. Whip the eggs and 1 cup sugar on medium-high speed until the mixture is pale yellow and has tripled in volume.
5. Reduce mixer speed to low and add the warm milk mixture.
6. Add half the flour mixture to the egg and milk mixture and mix on low speed until batter is smooth.
7. Add remaining flour, scraping down the sides and bottom of the bowl, and mix until batter is smooth.
8. Add vanilla and mix.
9. Pour batter into prepared pans. Bake at 350°F until cakes spring back when touched (approximately 15–20 minutes).
10. Cool for a few minutes in pan on a wire rack. Gently flip cakes out onto a cooling rack and let cool completely.

Cream filling

11. Beat butter, confectioners' sugar, vanilla, and cream until ingredients are fully incorporated and mixture is smooth.

Cherry filling

12. Place sugar and water in a saucepan and bring to a boil over medium heat, stirring to dissolve the sugar.
13. Remove sugar mixture from heat and set aside to cool completely.
14. Add 1 cup cherry liqueur to sugar mixture. Stir to incorporate and set aside.
15. Place cherries, including heavy syrup, in a saucepan over medium heat and bring to a boil.
16. Combine cornstarch with remaining cherry liqueur to make a slurry.
17. Whisk the slurry into the cherries and stir until the mixture thickens.
18. Cook cherry mixture 1–2 minutes longer. Remove from heat to cool completely.

Icing

19. Whip cream, confectioners' sugar, and cherry liqueur until soft peaks form. Cover and refrigerate.

Assembly

20. Use a serrated knife to slice each layer of cake in half horizontally to form four equal layers.
21. Place a cake layer on a serving plate and spread evenly with one-third of the cream filling. Spread one-third of the cherry filling over the cream filling.
22. Place a second cake layer over the filling and spread evenly with one-third of the cream filling, followed by one-third of the cherry filling.
23. Place a third cake layer over the filling and spread evenly with the remaining cream filling, followed by the remaining cherry filling.
24. Place the final cake layer on top. Spread icing over the top and sides of the cake.
25. Cut cake into 12 equal slices and serve.

Chef's Tip: *To enhance presentation, garnish with chocolate shavings and maraschino cherries.*

Evaluate original recipe for sensory and nutritional qualities

- Black Forest cake is a moist, layered chocolate cake with sweet cherries and a rich cream icing.
- Cake is high in calories, saturated fat, cholesterol, and total carbohydrates.

Establish goals for recipe modifications

- Reduce the calories, saturated fat, cholesterol, and refined sugars.
- Maintain flavor and appearance.

Identify modifications or substitutions

- Modify portion size.
- Use a substitute for the milk, butter, and granulated sugar in the cake.
- Modify the amount of eggs, flour, and cocoa powder used in the cake.
- Eliminate the cream filling and reduce the amount of icing.
- Replace the cherry filling with a cherry compote.

Determine functions of identified modifications or substitutions

- Portion size is large.
- Milk, butter, and granulated sugar provide the cake with texture and flavor.
- Eggs, flour, and cocoa powder provide the cake with structure, texture, and flavor.
- Cream filling and icing add richness and sweetness.
- Cherry filling and cherries add color and sweetness that is essential to Black Forest cake.

Select appropriate modifications or substitutions

- Replace 8-inch cake pans with a 9″ × 13″ baking pan to eliminate layers and reduce portion size.
- Use applesauce and canola oil as substitutes for milk and butter, and replace granulated sugar in the cake with cherry juice concentrate.
- Reduce the amount of eggs and increase the amount of flour and cocoa powder.
- Omit the cream filling and use a smaller amount of icing.
- Replace cherry filling with cherry compote as a garnish.

Test modified recipe to evaluate sensory and nutritional qualities

- Reducing the portion size lowers calories, saturated fat, cholesterol, and total carbohydrates.
- Using applesauce and canola oil maintains moistness while lowering calories, saturated fat, and cholesterol, and using cherry juice concentrate reduces the amount of refined sugars and total carbohydrates while elevating flavor.
- Reducing the amount of eggs lowers calories, saturated fat, and cholesterol, and increasing the amount of flour and cocoa powder maintains structure, texture, and flavor.
- Omitting the cream filling and using less icing reduces calories, fat, and cholesterol.
- Replacing cherry filling with a garnish of cherry compote reduces refined sugars, elevates flavor, and maintains presentation.

CULINARY NUTRITION RECIPE-MODIFICATION PROCESS

Modified Black Forest Cake

This modified Black Forest cake recipe decreases the portion size, eliminates the cream filling, and reduces the amount of icing. Fats and refined sugars are reduced or replaced through the use of natural sugars such as cherry juice concentrate, applesauce, orange juice, and dried cherries. The result is a moist chocolate cake infused with cherry flavor that contains fewer calories, less saturated fat and cholesterol, and fewer total carbohydrates.

Yield: 12 servings (9″ × 13″ cake cut into 12 equal slices)

Ingredients

Cake

9.2 oz	all-purpose flour
1.5 oz	unsweetened cocoa powder
2 tsp	baking soda
1 tsp	kosher salt
2 c	cherry juice concentrate, thawed
½ c	unsweetened applesauce
¼ c	canola oil
2 ea	whole eggs
2 tbsp	apple cider vinegar
1 tsp	vanilla extract

Icing

1½ c	heavy cream
2 tbsp	confectioners' sugar
2 tsp	cherry liqueur

Cherry compote

½ c	granulated sugar
2 c	orange juice
¼ c	cherry liqueur
2 c	dried dark cherries
1 tbsp	cornstarch

Preparation

Cake

1. Sift flour, cocoa powder, baking soda, and salt into a mixing bowl and set aside.
2. Beat cherry juice, applesauce, oil, eggs, vinegar, and vanilla on low speed until well blended.
3. Gradually add flour mixture to cherry juice mixture, mixing on low speed until batter is smooth.
4. Pour batter into a prepared 9″ × 13″ pan. Bake at 350°F for approximately 35–40 minutes, or until a toothpick inserted in the center comes out clean.
5. Place pan on a wire rack and allow to cool completely.

Icing

6. Whip cream, confectioners' sugar, and cherry liqueur until soft peaks form. Cover and refrigerate.

Cherry compote

7. Place sugar, orange juice, cherry liqueur, dried cherries, and cornstarch in a saucepan and stir to combine.
8. Bring cherry mixture to a boil and continue boiling for 5 minutes.
9. Remove saucepan from heat and allow to cool completely.

Assembly

10. Spread icing evenly over top of cake.
11. Spoon cooled cherry compote evenly over icing.

Chef's Tip: *Finely chopped, toasted walnuts can be added as a garnish for additional texture and flavor with only a slight increase in calories and fat.*

Black Forest Cake Nutritional Comparison

Nutrition Facts	Original	Modified
Calories	760.6	549.5
Total Fat	29.4 g	17.5 g
Saturated Fat	17.3 g	7.8 g
Cholesterol	225.1 mg	76.0 mg
Total Carbohydrates	110.0 g	91.2 g

Signature Gluten-Free Black Forest Cupcakes

This signature recipe features the flavors of Black Forest cake in gluten-free cupcakes. Teff flour, rice flour, and hazelnut flour are combined with bananas and chocolate to produce a rich chocolate cake accented with a low-fat cream cheese frosting and fresh cherries. These cupcakes offer smaller portion sizes and are gluten-free.

Yield: 12 servings, 1 cupcake each

Ingredients

Cupcakes

4 oz	teff flour
2 oz	sweet rice flour
2 oz	hazelnut flour
½ oz	unsweetened cocoa powder
½ tsp	baking soda
½ tsp	kosher salt
3 ea	whole eggs
7 oz	granulated sugar
10 oz	ripe bananas, mashed
⅔ c	canola oil
2 tsp	vanilla extract
3½ oz	dark chocolate chips

Icing

8 oz	low-fat cream cheese, room temperature
1 tbsp	2% milk
1 tsp	vanilla extract
5 c	confectioners' sugar

Garnish

12 ea	fresh cherries, pitted

National Cherry Growers and Industries Foundation

Preparation

Cupcakes

1. Combine teff flour, rice flour, hazelnut flour, cocoa powder, baking soda, and salt in a mixing bowl.
2. Place eggs, sugar, bananas, oil, and vanilla in a blender and purée until smooth.
3. Add wet ingredients to the dry mixture and stir to combine.
4. Fold chocolate chips into the batter.
5. Divide the batter evenly to fill prepared muffin cups. Bake at 350°F for approximately 20 minutes, or until a toothpick inserted in the center of a cupcake comes out clean.
6. Remove cupcakes from oven and allow to cool completely on a wire rack.

Icing

7. Beat cream cheese, milk, and vanilla on medium speed until well combined.
8. Add confectioners' sugar one cup at a time to the cream cheese mixture, beating on low speed until all the sugar is incorporated.

Assembly

9. Spread icing on each cupcake and top with a fresh cherry.

Chef's Tip: *The cupcakes can be topped with a compote of fresh cherries that have been diced and mixed with 1 tbsp of agave nectar and 1 tsp of lemon zest for flavor and additional nutrients.*

Black Forest Cake Nutritional Comparison

Nutrition Facts	Original	Modified	Signature
Calories	760.6	549.5	548.2
Total Fat	29.4 g	17.5 g	22.7 g
Saturated Fat	17.3 g	7.8 g	5.2 g
Cholesterol	225.1 mg	76.0 mg	63.6 mg
Total Carbohydrates	110.0 g	91.2 g	83.5 g

DESSERT MENU MIX

Desserts are part of a satisfying dining experience. A well-designed dessert menu offers guests a variety of dessert options. Fruit, chocolate, and nut-based desserts as well as custards and frozen desserts are popular dessert menu items. Some guests opt for a plate of fine cheeses with dried fruits and nuts or a beverage such as a smoothie or cappuccino. Like menus for savory items, dessert menus should emphasize seasonal ingredients that are at their peak of flavor. By focusing on flavor profiles that are familiar to guests, chefs can create nutrient-dense desserts that guests will find hard to resist.

On a written dessert menu, the flavor profiles, cooking techniques, seasonal ingredients, and unique features of the desserts should be emphasized. Since guests often order elaborate desserts they would not prepare themselves, chefs should highlight complex preparations in menu descriptions. **See Figure 15-14.**

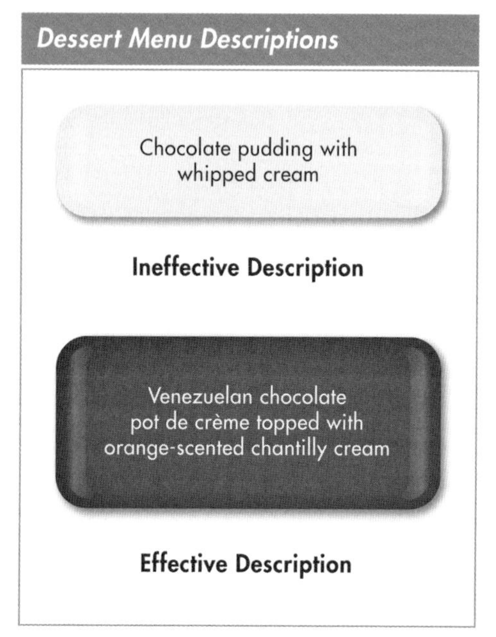

Figure 15-14. Elaborate preparations are highlighted in effective menu descriptions.

Beverages as Desserts

Coffee and tea are common beverage choices at the end of a meal. They can be elevated to dessert beverages when flavorings such as specialty syrups, alcohol, and whipped cream are added. Some beverage houses focus almost exclusively on specialty coffee and tea beverages, such as blended beverages with seasonally inspired flavors. In addition to coffee and tea, fruit-based smoothies, milkshakes, malts, and floats are popular dessert beverages. Large amounts of added sugar and fat are common in these dessert beverages.

Cheeses as Desserts

Cheese is a common dessert option at fine dining venues. A cheese plate generally includes three to five types of cheeses, each with different textures and flavor profiles. Imported and locally sourced artisan cheeses are often featured. Dessert cheeses are served at room temperature for optimum flavor and are frequently accompanied by honey, fruit preserves, dried fruits, nuts, or crackers. It is important to consider the fat and sodium content of the cheeses and the added sugars in the garnishes.

Fruits as Desserts

Fruits are a naturally sweet ending to a meal and are used in a variety of desserts. **See Figure 15-15.** Fruits such as poached pears, stewed apricots, and baked apples are often served whole. Poached and stewed fruits can also be made into compotes and chutneys to accompany other dessert items.

Fresh fruit salads are popular dessert options that are served alone or as a component of a dessert. For example, fresh berries topped with a dollop of mascarpone cheese and a drizzle of honey make a flavorful and nutritious fruit salad on their own. In contrast, a fruit salad of diced pineapples and mangos may serve as a garnish to rice pudding.

Fruits as Desserts

Wisconsin Milk Marketing Board, Inc.

Figure 15-15. Fruits are a naturally sweet ending to a meal and are used in a variety of desserts, such as blueberry pie.

Fruits not only add color, texture, and flavor to desserts, but they can also be used to contrast temperatures. For example, bananas Foster features bananas sautéed with brown sugar, rum, and banana liqueur and served over ice cream. Likewise, grilled watermelon served with frozen yogurt provides an appealing blend of warm and cold ingredients.

While fruits are the main ingredient in desserts such as pies and tarts, they can also serve as a garnish for virtually any type of dessert. Slices of peaches or nectarines can accompany cakes and tarts, while raspberries and blueberries can add a burst of color and flavor to a chocolate cake or bread pudding. Using fresh, seasonal fruits helps to ensure the highest level of color, flavor, and nutrients. In contrast, some forms of canned and frozen fruit are higher in calories due to added sugars.

Nuts and Seeds as Desserts

Nuts and seeds are commonly featured in desserts because they enhance texture, provide a rich mouthfeel, and add flavor. Nuts can be used raw or roasted for a deeper, more intense flavor. Candied nuts are often used as a garnish for added sweetness and crunch. Nuts and seeds can also be ground into flours to make baked goods such as cakes and cookies. Nuts and seeds supply healthy fats but are high in calories.

Breads as Desserts

Dessert menu items that showcase breads include bread puddings, coffee cakes, Danishes, and quick breads. Dessert breads offer a range of options because they can feature fruits, vegetables, nuts, chocolate, and cheeses. **See Figure 15-16.** When creating dessert breads, the emphasis should be on using whole grain flours for added dietary fiber and seasonal fruits for superior flavor and nutrients. In contrast, highly processed ingredients such as refined flours and jellies only add empty calories.

Breads as Desserts

Vita-Mix® Corporation

Figure 15-16. Dessert breads offer a range of options because they can feature a variety of ingredients, including fruits and nuts.

Quick breads can be a nutrient-dense dessert item because they can easily be made with whole grain flours, fruits, vegetables, nuts, and seeds. Popular quick breads include banana bread, pumpkin bread, and zucchini bread. Quick breads can be sliced into smaller portions that are visually appealing.

Plated Desserts

Plated desserts often contain four key components: a main item, a crunchy element, a sauce, and a garnish. **See Figure 15-17.** Common main items include cakes, mousses, tarts, puddings, and pastries. The main item is typically the largest in terms of size and calories. Some plated desserts also include a frozen element such as ice cream or sorbet. Frozen elements may be the main item or an accompaniment.

Crunchy elements often include nuts or cookies. For example, almonds and gingersnap cookies provide crunch when served whole, chopped, or processed into a crust. Although a crunchy element enhances plated desserts, it is not always present, especially when the main item contains flour, such as a cake or pie.

Dessert sauces such as fruit purées, chocolate, caramel, and cream-based sauces add color, moisture, and flavor to plated desserts. Dessert sauces can be a source of sugar and fat, but they can also increase the nutritional value of a dessert. For example, a fruit compote or coulis contains sugar but also adds vitamins and minerals to a plate.

Garnishes for desserts can be simple, such as strategically placed berries, or more elaborate, such as piped chocolate swirls. Like sauces, garnishes should enhance the nutritional profile of the dish rather than add only sugar, fat, and calories.

Plated Dessert Components

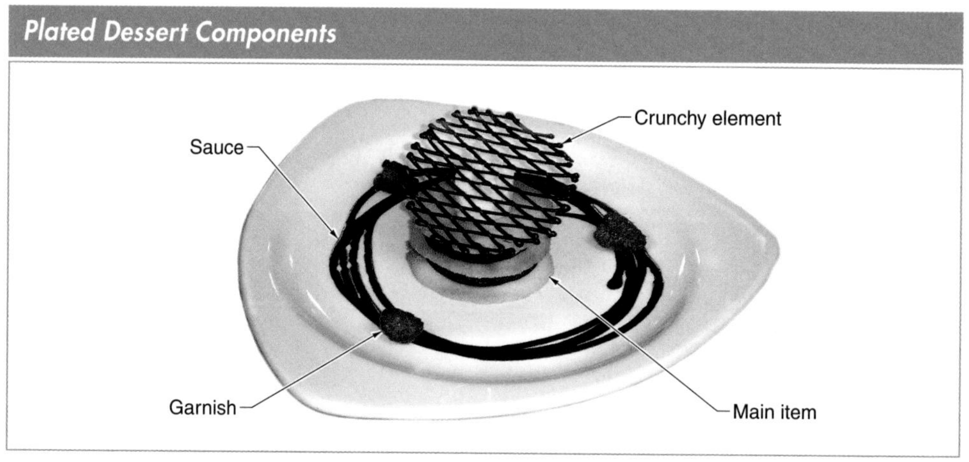

Figure 15-17. Plated desserts often contain four key components: a main item, a crunchy element, a sauce, and a garnish.

Knowledge Check 15-3

1. Identify beverages commonly served as desserts.
2. Summarize how cheeses are presented as desserts.
3. Explain why desserts should emphasize seasonal fruits.
4. Describe how desserts are enhanced by nuts and seeds.
5. Identify nutrient-dense ingredients that can be incorporated into quick breads.
6. List the four key components often included with a plated dessert.

PROMOTING DESSERTS ON THE MENU

Three effective strategies for promoting desserts on the menu include creating signature desserts, serving tasting portions of desserts, and replacing written menus with dessert trays. Signature desserts are enticing to guests because they are promoted as a unique dining experience. A signature dessert may be as simple as a seasonal pumpkin pie or as sophisticated as a trio of desserts featuring unique flavor profiles.

Many chefs offer tasting portions of desserts to allow guests to end the meal with something sweet without overindulging. **See Figure 15-18.** Dessert trays featuring a variety of attractively plated desserts can be presented tableside to guests. This is a powerful selling tool based on the concept that "we first eat with our eyes." Allowing guests to see what they will be ordering rather than read a written description often makes desserts hard to refuse.

Quick breads and pastries are common dessert options at quick service and fast casual venues. Soft-serve ice cream and frozen yogurt are also popular options for customers on the go. Although some casual dining and institutional venues prepare desserts in-house, many offer desserts that have been made off the premises. These desserts are often marketed as signature items.

Sweet specialty beverages are popular desserts at beverage houses and bars. House-made desserts or desserts purchased from local bakeries may also be featured on the menu. Fine dining and special event venues are likely to employ a full-time pastry chef to create elegantly plated desserts. **See Figure 15-19.** Dessert menus at fine dining venues often highlight seasonal ingredients. Special event venues such as wedding venues and banquet halls often feature desserts that reflect the theme of the event.

Dessert Tasting Portions

L'Auberge Carmel
C.S. White, photographer

Figure 15-18. Tasting portions of desserts allow guests to end the meal with something sweet without overindulging.

Desserts at Fine Dining Venues

Daniel NYC/C. Echavarria

Pear Cassis

Daniel NYC/T. Schauer

Chocolate Clafoutis

Figure 15-19. Chefs at fine dining venues often create elegantly plated desserts.

Knowledge Check 15-4

1. Describe three effective strategies for promoting desserts on the menu.
2. Identify ways in which desserts are featured at various types of foodservice venues.

Chapter Summary

Desserts enhance the dining experience. However, some guests avoid desserts that are high in refined sugars, saturated fat, and calories. Providing guests with a variety of nutrient-dense dessert options and smaller portion sizes of decadent menu items can encourage sales and increase revenue.

Cakes and cookies, pies and tarts, pastries, crêpes, custards and creams, frozen desserts, and confections are common dessert menu items. These menu items can be made healthier by incorporating seasonal fruits, natural sugars, unsaturated fats, low-fat dairy products, and whole grains into them. Signature desserts, tasting portions, and visually appealing presentations are effective strategies used in a variety of venues to promote dessert menu items.

National Honey Board

Chapter Review

Digital Resources
ATPeResources.com/QuickLinks
Access Code: 267412

1. Identify the ingredients that will make desserts more nutrient dense.
2. Explain how a strong dessert menu can increase revenue.
3. Describe desserts commonly found on a varied dessert menu.
4. Give an example of how to modify a dessert to make it more nutrient dense.
5. Explain how to enhance the flavor of nutrient-dense desserts.
6. Explain why it is important to serve reduced portions sizes on smaller plates.
7. Identify nutritional considerations for beverages, cheeses, nuts, and seeds served as desserts.
8. Describe ways in which fruits can be used in desserts.
9. Explain how to make dessert breads more nutrient dense.
10. Summarize how desserts are promoted across various foodservice venues.

Appendix

FLAVOR DEVELOPMENT INGREDIENTS AND TECHNIQUES

Herbs

Type	Description	Common Use
Basil	Leaf herb with pointy, green leaves; member of mint family; most common variety is sweet basil; some varieties have hints of lemon, garlic, cinnamon, clove, and chocolate flavor	Pasta, vegetable, seafood, chicken, and egg dishes; tomato sauces; pestos; infused oils
Cilantro	Leaf herb from stem and leaves of coriander plant; added just before serving, as heat will destroy its flavor	Salsas, soups, salads, and sandwiches
Chives	Stem herb with hollow, grass-shaped sprouts; mild onion flavor	Teas, soups, and Thai dishes
Dill (dill weed)	Leaf herb with feathery leaves; member of parsley family; slight licorice flavor	Salads, vegetables, meats, and sauces; pickling
Lemongrass (citronella)	Stem herb with long, thin, gray-green leaves and white scallionlike base; lemon flavor	Teas, soups, and Thai dishes
Mint	General term used to describe family of similar leaf herbs, such as peppermint and spearmint; cool, refreshing flavor	Sweet and savory dishes; beverages
Oregano (wild marjoram)	Leaf herb with small, dark-green, slightly curled leaves; member of mint family; pungent, peppery flavor	Italian and Greek dishes
Parsley	Leaf herb with curly or flat, dark-green leaves; tangy flavor (most predominant in stems)	Soups, salads, stews, sauces, and vegetable dishes; bouquets garnis; garnishing
Rosemary	Leaf herb with needlelike leaves; member of the evergreen family; slight mint flavor with scent resembling fresh pine	Grilled or roasted meats and poultry
Sage	Leaf herb with narrow, velvety leaves; member of mint family	Stews, sausages, and bean or tomato preparations
Tarragon	Leaf herb with smooth, slightly elongated leaves; slight licorice flavor	Seafood, poultry, tomato dishes, salads, salad dressings, béarnaise sauces
Thyme	Leaf herb with very small leaves; member of mint family; some varieties have hints of nutmeg, lemon, mint, and sage flavor	Stews, chowders, poultry, and vegetable dishes; bouquets garnis

FLAVOR DEVELOPMENT INGREDIENTS AND TECHNIQUES

Spices

Type	Description	Common Use
Capers	Flower spice from unopened flower bud of shrub; only used after being pickled in strongly salted white vinegar	Fish and poultry dishes; sauces
Cardamom	Berry spice from dried, immature fruit of tropical bush in ginger family; slight lemon flavor	Curries and pastries; pickling
Cassia (Chinese cinnamon)	Bark spice from bark of small evergreen tree; most spices labeled as cinnamon and sold in United States are actually cassia; bittersweet, warm flavor stronger than cinnamon	Sweet and savory dishes
Cinnamon	Bark spice from dried, thin, inner bark of small evergreen tree; bark is rolled into quills (cinnamon sticks); slightly sweet, warm flavor	Sweet and savory dishes
Coriander	Seed spice from light-brown, ridged seed of coriander plant; combination of lemon, sage, and caraway flavor	Savory dishes and baked goods; global cuisines, especially Caribbean, Indian, Mexican, and North African
Cumin	Seed spice from crescent-shaped seed of plant in parsley family; slightly bitter, warm flavor	Chili; soups; tamales; rice and cheese dishes; curry powders and chili powders
Fennel seeds	Seed spice from oval seed of fennel plant; slight licorice flavor and aroma	Sausages; roasted duck, chicken, and pork dishes
Ginger	Root spice from large, bumpy root of tropical plant; sold fresh, dried, powdered, crystallized, candied, or pickled; spicy, pungent, and warm flavor	Meat, poultry, and seafood dishes; fruits; desserts
Horseradish	Root spice from large, brown-skinned root of shrub related to radish; commonly grated; hot, spicy, and pungent flavor	Grilled or roasted meats; seafood dishes; sauces
Mustard seeds	Extremely tiny seed from mustard plant; hot, pungent flavor when ground and mixed with hot water	Cabbage, beets, sauerkraut, sauces, and salad dressings; pickling
Nutmeg	Seed spice from oval seed found in yellow, nectarine-shaped fruit of large tropical evergreen; sweet, spicy, warm flavor	Soups, potato dishes, sauces, and desserts
Paprika	Berry spice from dried, ground sweet red pepper; comes in many varieties including sweet, smoked, and hot	Goulashes; chicken paprika and egg dishes; sauces

FLAVOR DEVELOPMENT INGREDIENTS AND TECHNIQUES

Complementary Herb and Spice Blends

Name of Blend	Common Ingredients
Cajun seasoning	Zesty spice blend; ground chiles, fennel seeds, garlic, oregano, paprika, red and black pepper, salt, and thyme
Chili powder	Ground chiles, cloves, coriander, cumin, garlic, and oregano
Chinese five-spice powder	Equal proportions of ground cinnamon, cloves, fennel seeds, star anise, and Szechuan pepper
Curry powder	Mild to spicy blend; can consist of more than 20 spices including cardamom, cinnamon, cloves, coriander, cumin, fenugreek, ginger, mace, nutmeg, red and black pepper, and turmeric
Fines herbes	Chopped fresh chervil, chives, parsley, and tarragon; may include marjoram, savory, or thyme
Herbes de Provence	Dried basil, fennel seed, lavender, marjoram, rosemary, sage, savory, and thyme
Jerk seasoning	Spicy blend; ground allspice, chiles, cinnamon, cloves, garlic, and ginger
Pickling spice	Whole and coarsely ground allspice, bay leaves, cinnamon, cloves, dill, fennel seeds, ginger, mace, mustard, nutmeg, peppercorns, and red pepper
Poultry seasoning	Black pepper, marjoram, nutmeg, rosemary, sage, and thyme; may include celery salt, mustard powder, or oregano

FLAVOR DEVELOPMENT INGREDIENTS AND TECHNIQUES

Complementary Herb and Spice Food Pairings

Food	Flavoring	Food	Flavoring
Beef	Bay leaf, marjoram, nutmeg, sage, thyme	**Potatoes**	Dill, paprika, parsley, sage
Carrots	Cinnamon, cloves, marjoram, nutmeg, rosemary, sage	**Poultry**	Ginger, marjoram, oregano, paprika, poultry seasoning, rosemary, sage, tarragon, thyme
Corn	Cayenne, chervil, chives, cumin, curry powder, paprika, parsley	**Salads**	Basil, celery seed, chervil, chives, cilantro, dill, oregano, rosemary, sage, tarragon, thyme
Eggs	Basil, cilantro, cumin, savory, tarragon, turmeric	**Soups**	Bay leaves, cayenne, chervil, chili powder, cilantro, cumin, curry powder, dill, marjoram, nutmeg, oregano, rosemary, sage, savory, thyme
Fish	Curry powder, dill, dry mustard, lemon zest and juice, marjoram, paprika		
Fruits	Cinnamon, ground cloves, ginger, mace, mint	**Summer squash**	Cloves, curry powder, marjoram, nutmeg, rosemary, sage
Green beans	Dill, curry powder, lemon juice, marjoram, oregano, tarragon, thyme	**Tomatoes**	Basil, bay leaves, dill, marjoram
Lamb	Curry powder, rosemary, mint	**Veal**	Bay leaves, curry powder, ginger, marjoram, oregano
Peas	Ginger, marjoram, parsley, sage	**Winter squash**	Cinnamon, ginger, nutmeg
Pork	Sage, oregano		

FLAVOR DEVELOPMENT INGREDIENTS AND TECHNIQUES

Oils

Type	Description	Common Use
Canola oil (rapeseed oil)	Oil produced from rapeseeds; neutral flavor	Salad dressings, cooking, and baking
Corn oil	Oil produced from the germ of corn kernels; slight corn flavor	Pepper- or garlic-flavored salad dressings and dips
Grapeseed oil	Oil produced from grape seeds; delicate flavor	Dips and salsas; mixing with garlic and basil to drizzle on toasted bread
Olive oils	Oil produced from olives	
Extra-virgin olive oil	Produced from first pressing of olives without use of heat or chemicals; has an acid content of less than 1%; highest quality of olive oil and known for its rich flavor	Salad dressings and garnishing applications when stronger olive oil flavor is desired
Virgin olive oil	Produced from first pressing of olives without use of heat or chemicals; has an acid content of as much as 3%	Salad dressings and garnishing applications when milder olive oil flavor is desired
Pure olive oil	Produced using heat, and often chemicals, to extract additional oils from olive pulp after first pressing	Cooking and baking
Olive-pomace oil	Produced using heat, and often chemicals, to extract additional oils from olive pulp and olive pits after first pressing	Cooking and baking
Sesame oils	Oil produced from sesame seeds	Sausages; roasted duck, chicken, and pork dishes
Light sesame oil	Mild, nutty flavor	Salad dressings and spreads
Dark sesame oil	Strong in flavor because sesame seeds are toasted before oil is produced	Asian cuisine and garnishing applications
Soybean oil	Oil produced from soybeans	Salad dressings, mayonnaises, and spreads

FLAVOR DEVELOPMENT INGREDIENTS AND TECHNIQUES

Vinegars

Type	Description	Common Use
Balsamic vinegar	Made by aging trebbiano wine vinegar for many years in wooden casks until it darkens in color and becomes sweeter; sweet flavor mellows its high acidity content	Salads, dressings, sauces, and garnishing applications
Champagne vinegar	Made from Champagne grapes; mild flavor	Salads, dressings, and sauces
Cider vinegar	Made by fermenting unpasteurized apple juice or cider until sugars are converted into alcohol; subtle apple flavor and mild acidity	Salads, dressings, and sauces
Distilled vinegar	Made by fermenting diluted, distilled grain alcohol; strong flavor and high acidity	Pickling
Malt vinegar	Made from malted barley; strong, tangy flavor	Serving with fish and chips
Rice vinegar	Made from rice wine; slightly sweet and mildly acidic	Asian dishes and dipping sauces
Sherry vinegar	Made from sherry wine; strong sherry flavor	Salads, dressings, and sauces
Wine vinegars	Made from red or white wine; red wine vinegars are aged longer than white wine vinegars and are slightly more pungent	Salads, dressings, and sauces

FLAVOR DEVELOPMENT INGREDIENTS AND TECHNIQUES

Stocks and Bouillons

Type	Description
Brown stocks	Roasted meat, poultry, or game bones are simmered with mirepoix and an optional tomato product
White stocks	Poultry, veal, or fish bones are simmered in water with vegetables and herbs
Fish stocks	Fish bones or shellfish shells, vegetables, and a sachet d'épices are simmered in water
Fumets	Fish bones, shellfish shells, and vegetables are first sweated, then simmered in water to produce a concentrated flavor
Essences	Concentrated fish stock that includes a large amount of aromatic ingredients
Vegetable stocks	Vegetables, a white mirepoix, and a sachet are simmered in water
Glaces	Highly reduced stock with an intense flavor that, when cooled, takes on a rubbery texture that can be cut into cubes
Remouillages	Stock made from bones that have already been used once to make a stock
Bouillons	Liquid that is strained off after cooking vegetables, poultry, meat, or seafood in water
Court bouillons	Highly flavored, aromatic vegetable broth made from simmering vegetables with herbs and a small amount of an acidic liquid such as vinegar or wine

FLAVOR DEVELOPMENT INGREDIENTS AND TECHNIQUES

Balancing Ingredients

- **Balance sweet ingredients with salty, sour, bitter, or spicy ingredients**
- **Balance salty ingredients with sweet or sour ingredients**
- **Balance sour ingredients with sweet, salty, or bitter ingredients**
- **Balance bitter ingredients with sweet, salty, or sour ingredients**
- **Balance spicy ingredients with sweet ingredients**

Sweet Ingredients	Salty Ingredients	Sour Ingredients	Bitter Ingredients	Spicy Ingredients
Agave Cinnamon Fresh/dried fruits Fruit juices Honeys Maple syrup Sugars	Capers Celery Fish sauces Olives Salts Soy sauces Worcestershire sauces	Citrus juices Citrus zests Cranberries Pickles Vinegars	Arugula Cocoa Coffee Cumin Kale Spinach	Onions Garlic Ginger Horseradish Hot peppers Mustards Wasabi

FLAVOR DEVELOPMENT INGREDIENTS AND TECHNIQUES

Balanced Flavor Profiles

Cuisine	Common Ingredients
Asian	Anise, chile oil, chiles, coconut, garlic, ginger, green onions, lemongrass, sesame oil, soy sauce, star anise
Caribbean	Allspice, bay leaves, black pepper, cardamom, cayenne pepper, cilantro, cinnamon, cloves, coconut milk, coriander, cumin, curry, fenugreek, ginger, hot peppers, oregano, thyme, turmeric, vanilla
French	Bay leaves, black pepper, chervil, chives, fines herbes, garlic, green and pink peppercorns, marjoram, mustard, nutmeg, parsley, rosemary, shallots, tarragon, thyme
German	Allspice, caraway, cinnamon, dill, juniper berries, mustard, onions, vinegar, white pepper
Greek	Cinnamon, dill, garlic, lemon, mint, nutmeg, olives, onions, oregano, tomatoes, yogurt
Indian	Anise, cardamom, chiles, cilantro, cinnamon, coriander, cumin, curry, fennel, garlic, ginger, mint, mustard, nutmeg, saffron, turmeric, yogurt
Italian	Basil, bay leaves, fennel, garlic, marjoram, onions, oregano, parsley, pine nuts, rosemary, tomatoes
Mexican	Chiles, cilantro, cinnamon, cocoa, coriander, cumin, garlic, lime, onions, oregano, vanilla
North African	Cilantro, cinnamon, coriander, cumin, garlic, ginger, mint, red pepper, saffron, turmeric
Scandinavian	Cardamom, dill, lemons, mustard, nutmeg, white pepper

Estimated Calorie Needs per Day by Age, Gender, and Physical Activity Level

Estimated amounts of calories* needed to maintain calorie balance for various gender and age groups at three different levels of physical activity. The estimates are rounded to the nearest 200 calories. An individual's calorie needs may be higher or lower than these average estimates.

Gender/ Activity Level†	Male/ Sedentary	Male/ Moderately Active	Male/ Active	Female‡/ Sedentary	Female‡/ Moderately Active	Female‡/ Active
Age (years)						
2	1000	1000	1000	1000	1000	1000
3	1200	1400	1400	1000	1200	1400
4	1200	1400	1600	1200	1400	1400
5	1200	1400	1600	1200	1400	1600
6	1400	1600	1800	1200	1400	1600
7	1400	1600	1800	1200	1600	1800
8	1400	1600	2000	1400	1600	1800
9	1600	1800	2000	1400	1600	1800
10	1600	1800	2200	1400	1800	2000
11	1800	2000	2200	1600	1800	2000
12	1800	2200	2400	1600	2000	2200
13	2000	2200	2600	1600	2000	2200
14	2000	2400	2800	1800	2000	2400
15	2200	2600	3000	1800	2000	2400
16	2400	2800	3200	1800	2000	2400
17	2400	2800	3200	1800	2000	2400
18	2400	2800	3200	1800	2000	2400
19–20	2600	2800	3000	2000	2200	2400
21–25	2400	2800	3000	2000	2200	2400
26–30	2400	2600	3000	1800	2000	2400
31–35	2400	2600	3000	1800	2000	2200
36–40	2400	2600	2800	1800	2000	2200
41–45	2200	2600	2800	1800	2000	2200
46–50	2200	2400	2800	1800	2000	2200
51–55	2200	2400	2800	1600	1800	2200
56–60	2200	2400	2600	1600	1800	2200
61–65	2000	2400	2600	1600	1800	2000
66–70	2000	2200	2600	1600	1800	2000
71–75	2000	2200	2600	1600	1800	2000
76+	2000	2200	2400	1600	1800	2000

* Based on Estimated Energy Requirements (EER) equations, using reference heights (average) and reference weights (healthy) for each age-gender group. For children and adolescents, reference height and weight vary. For adults, the reference man is 5 feet 10 inches tall and weighs 154 pounds. The reference woman is 5 feet 4 inches tall and weighs 126 pounds. EER equations are from the Institute of Medicine. Dietary Reference Intakes for Energy, Carbohydrate, Fiber, Fat, Fatty Acids, Cholesterol, Protein, and Amino Acids. Washington (DC): The National Academies Press; 2002

† Sedentary means a lifestyle that includes only the light physical activity associated with typical day-to-day life. Moderately active means a lifestyle that includes physical activity equivalent to walking about 1.5 to 3 miles per day at 3 to 4 miles per hour, in addition to the light physical activity associated with typical day-to-day life. Active means a lifestyle that includes physical activity equivalent to walking more than 3 miles per day at 3 to 4 miles per hour, in addition to the light physical activity associated with typical day-to-day life.

‡ Estimates for females do not include women who are pregnant or breastfeeding.

Source: *Dietary Guidelines for Americans, 2010*

Macronutrient Calorie Contributions

1g of protein = 4 calories
1 g of carbohydrate = 4 calories
1 g of lipid (fat) = 9 calories

Nutritional Goals for Age-Gender Groups, Based on DRIs and Dietary Guidelines Recommendations

Nutrient (units)	Source of Goal*	Child 1–3	Female 4–8	Male 4–8	Female 9–13	Male 9–13	Female 14–18	Male 14–18	Female 19–30	Male 19–30	Female 31–50	Male 31–50	Female 51+	Male 51+
Macronutrients														
Protein (g)	RDA[†]	13	19	19	34	34	46	52	46	56	46	56	46	56
(% of calories)	AMDR[‡]	5–20	10–30	10–30	10–30	10–30	10–30	10–30	10–35	10–35	10–35	10–35	10–35	10–35
Carbohydrate (g)	RDA	130	130	130	130	130	130	130	130	130	130	130	130	130
(% of calories)	AMDR	45–65	45–65	45–65	45–65	45–65	45–65	45–65	45–65	45–65	45–65	45–65	45–65	45–65
Total fiber (g)	IOM[§]	14	17	20	22	25	25	31	28	34	25	31	22	28
Total fat (% of calories)	AMDR	30–40	25–35	25–35	25–35	25–35	25–35	25–35	20–35	20–35	20–35	20–35	20–35	20–35
Saturated fat (% of calories)	DG[‖]	<10	<10	<10	<10	<10	<10	<10	<10	<10	<10	<10	<10	<10
Linoleic acid (g)	AI[#]	7	10	10	10	12	11	16	12	17	12	17	11	14
(% of calories)	AMDR	5–10	5–10	5–10	5–10	5–10	5–10	5–10	5–10	5–10	5–10	5–10	5–10	5–10
Alpha-Linolenic acid (g)	AI	0.7	0.9	0.9	1.0	1.2	1.1	1.6	1.1	1.6	1.1	1.6	1.1	1.6
(% of calories)	AMDR	0.6–1.2	0.6–1.2	0.6–1.2	0.6–1.2	0.6–1.2	0.6–1.2	0.6–1.2	0.6–1.2	0.6–1.2	0.6–1.2	0.6–1.2	0.6–1.2	0.6–1.2
Cholesterol (mg)	DG	<300	<300	<300	<300	<300	<300	<300	<300	<300	<300	<300	<300	<300
Minerals														
Calcium (mg)	RDA	700	1000	1000	1300	1300	1300	1300	1000	1000	1000	1000	1200	1200
Iron (mg)	RDA	7	10	10	8	8	15	11	18	8	18	8	8	8
Magnesium (mg)	RDA	80	130	130	240	240	360	410	310	400	320	420	320	420
Phosphorus (mg)	RDA	460	500	500	1250	1250	1250	1250	700	700	700	700	700	700
Potassium (mg)	AI	3000	3800	3800	4500	4500	4700	4700	4700	4700	4700	4700	4700	4700
Sodium (mg)	UL**	<1500	<1900	<1900	<2200	<2200	<2300	<2300	<2300	<2300	<2300	<2300	<2300	<2300
Zinc (mg)	RDA	3	5	5	8	8	9	11	8	11	8	11	8	11
Copper (mcg)	RDA	340	440	440	700	700	890	890	900	900	900	900	900	900
Selenium (mcg)	RDA	20	30	30	40	40	55	55	55	55	55	55	55	55
Vitamins														
Vitamin A (mcg RAE)	RDA	300	400	400	600	600	700	900	700	900	700	900	700	900
Vitamin D[††] (mcg)	RDA	15	15	15	15	15	15	15	15	15	15	15	15	15
Vitamin E (mg AT)	RDA	6	7	7	11	11	15	15	15	15	15	15	15	15
Vitamin C (mg)	RDA	15	25	25	45	45	65	75	75	90	75	90	75	90
Thiamin (mg)	RDA	0.5	0.6	0.6	0.9	0.9	1.0	1.2	1.1	1.2	1.1	1.2	1.1	1.2
Riboflavin (mg)	RDA	0.5	0.6	0.6	0.9	0.9	1.0	1.3	1.1	1.3	1.1	1.3	1.1	1.3
Niacin (mg)	RDA	6	8	8	12	12	14	16	14	16	14	16	14	16
Folate (mcg)	RDA	150	200	200	300	300	400	400	400	400	400	400	400	400
Vitamin B_6 (mg)	RDA	0.5	0.6	0.6	1.0	1.0	1.2	1.3	1.3	1.3	1.3	1.3	1.5	1.7
Vitamin B_{12} (mcg)	RDA	0.9	1.2	1.2	1.8	1.8	2.4	2.4	2.4	2.4	2.4	2.4	2.4	2.4
Choline (mg)	AI	200	250	250	375	375	400	550	425	550	425	550	425	550
Vitamin K (mcg)	AI	30	55	55	60	60	75	75	90	120	90	120	90	120

* Dietary Guidelines recommendations are used when no quantitative Dietary Reference Intake value is available; apply to ages 2 years and older.
† Recommended Dietary Allowance, IOM
‡ Acceptable Macronutrient Distribution Range, IOM
§ 14 grams per 1000 calories, IOM
‖ Dietary Guidelines recommendation
\# Adequate Intake, IOM
** Upper Limit, IOM
†† 1 mcg of vitamin D is equivalent to 40 IU
AT = alpha-tocopherol; DFE = dietary folate equivalents; RAE = retinol activity equivalents

Source: *Dietary Guidelines for Americans, 2010*

USDA Food Patterns

For each food group or subgroup,[*] recommended average daily intake amounts[†] at all calorie levels. Recommended intakes from vegetable and protein foods subgroups are per week.

Calorie level of pattern[‡]	1000	1200	1400	1600	1800	2000	2200	2400	2600	2800	3000	3200
Fruits	1 c	1 c	1½ c	1½ c	1½ c	2 c	2 c	2 c	2 c	2½ c	2½ c	2½ c
Vegetables[§]	1 c	1½ c	1½ c	2 c	2½ c	2½ c	3 c	3 c	3½ c	3½ c	4 c	4 c
Dark-green vegetables	½ c/wk	1 c/wk	1 c/wk	1½ c/wk	1½ c/wk	1½ c/wk	2 c/wk	2 c/wk	2½ c/wk	2½ c/wk	2½ c/wk	2½ c/wk
Red and orange vegetables	2½ c/wk	3 c/wk	3 c/wk	4 c/wk	5½ c/wk	5½ c/wk	6 c/wk	6 c/wk	7 c/wk	7 c/wk	7½ c/wk	7½ c/wk
Beans and peas (legumes)	½ c/wk	½ c/wk	½ c/wk	1 c/wk	1½ c/wk	1½ c/wk	2 c/wk	2 c/wk	2½ c/wk	2½ c/wk	3 c/wk	3 c/wk
Starchy vegetables	2 c/wk	3½ c/wk	3½ c/wk	4 c/wk	5 c/wk	5 c/wk	6 c/wk	6 c/wk	7 c/wk	7 c/wk	8 c/wk	8 c/wk
Other vegetables	1½ c/wk	2½ c/wk	2½ c/wk	3½ c/wk	4 c/wk	4 c/wk	5 c/wk	5 c/wk	5½ c/wk	5½ c/wk	7 c/wk	7 c/wk
Grains[‖]	3 oz-eq	4 oz-eq	5 oz-eq	5 oz-eq	6 oz-eq	6 oz-eq	7 oz-eq	8 oz-eq	9 oz-eq	10 oz-eq	10 oz-eq	10 oz-eq
Whole grains	1½ oz-eq	2 oz-eq	2½ oz-eq	3 oz-eq	3 oz-eq	3 oz-eq	3½ oz-eq	4 oz-eq	4½ oz-eq	5 oz-eq	5 oz-eq	5 oz-eq
Enriched grains	1½ oz-eq	2 oz-eq	2½ oz-eq	2 oz-eq	3 oz-eq	3 oz-eq	3½ oz-eq	4 oz-eq	4½ oz-eq	5 oz-eq	5 oz-eq	5 oz-eq
Protein foods[§]	2 oz-eq	3 oz-eq	4 oz-eq	5 oz-eq	5 oz-eq	5½ oz-eq	6 oz-eq	6½ oz-eq	6½ oz-eq	7 oz-eq	7 oz-eq	7 oz-eq
Seafood	3 oz/wk	5 oz/wk	6 oz/wk	8 oz/wk	8 oz/wk	8 oz/wk	9 oz/wk	10 oz/wk	10 oz/wk	11 oz/wk	11 oz/wk	11 oz/wk
Meat, poultry, eggs	10 oz/wk	14 oz/wk	19 oz/wk	24 oz/wk	24 oz/wk	26 oz/wk	29 oz/wk	31 oz/wk	31 oz/wk	34 oz/wk	34 oz/wk	34 oz/wk
Nuts, seeds, soy products	1 oz/wk	2 oz/wk	3 oz/wk	4 oz/wk	4 oz/wk	4 oz/wk	4 oz/wk	5 oz/wk	5 oz/wk	5 oz/wk	5 oz/wk	5 oz/wk
Dairy[#]	2 c	2½ c	2½ c	3 c	3 c	3 c	3 c	3 c	3 c	3 c	3 c	3 c
Oils[**]	15 g	17 g	17 g	22 g	24 g	27 g	29 g	31 g	34 g	36 g	44 g	51 g
Maximum SoFAS[††] limit, calories (% of calories)	137 (14%)	121 (10%)	121 (9%)	121 (8%)	161 (9%)	258 (13%)	266 (12%)	330 (14%)	362 (14%)	395 (14%)	459 (15%)	596 (19%)

[†] Food group amounts are shown in cup (c) or ounce-equivalents (oz-eq). Oils are shown in grams (g). Quantity equivalents for each food group are:
- Grains, 1 ounce-equivalent is: 1 one-ounce slice bread; 1 ounce uncooked pasta or rice; ½ cup cooked rice, pasta, or cereal; 1 tortilla (6″ diameter); 1 pancake (5″ diameter); 1 ounce ready-to-eat cereal (about 1 cup cereal flakes)
- Vegetables and fruits, 1 cup equivalent is: 1 cup raw or cooked vegetable or fruit; ½ cup dried vegetable or fruit; 1 cup vegetable or fruit juice; 2 cups leafy salad greens
- Protein foods, 1 ounce-equivalent is: 1 ounce lean meat, poultry, seafood; 1 egg; 1 tbsp peanut butter; ½ ounce nuts or seeds. Also, ¼ cup cooked beans or peas may also be counted as 1 ounce-equivalent
- Dairy, 1 cup equivalent is: 1 cup milk, fortified soy beverage, or yogurt; 1½ ounces natural cheese (e.g., cheddar); 2 ounces of processed cheese (e.g., American)

[‡] See Estimated Calorie Needs per Day by Age, Gender, and Physical Activity Level table on page 581. Food intake patterns at 1000, 1200, and 1400 calories meet the nutritional needs of children ages 2 to 8 years. Patterns from 1600 to 3200 calories meet the nutritional needs of children ages 9 years and older and adults. If a child ages 4 to 8 years needs more calories and, therefore, is following a pattern at 1600 calories or more, the recommended amount from the dairy group can be 2½ cups per day. Children ages 9 years and older and adults should not use the 1000, 1200, or 1400 calorie patterns

[§] Vegetable and protein foods subgroup amounts are shown in this table as weekly amounts, because it would be difficult for consumers to select foods from all subgroups daily.

[‖] Whole-grain subgroup amounts shown in this table are minimums. More whole grains up to all of the grains recommended may be selected, with offsetting decreases in the amounts of enriched refined grains.

[#] The amount of dairy foods in the 1200 and 1400 calorie patterns have increased to reflect new RDAs for calcium that are higher than previous recommendations for children ages 4 to 8 years.

[**] Oils and soft margarines include vegetable, nut, and fish oils and soft vegetable oil table spreads that have no *trans* fats.

[††] SoFAS are calories from solid fats and added sugars. The limit for SoFAS is the remaining amount of calories in each food pattern after selecting the specified amounts in each food group in nutrient-dense forms (forms that are fat-free or low-fat and with no added sugars). The number of SoFAS is lower in the 1200, 1400, and 1600 calorie patterns than in the 1000 calorie pattern. The nutrient goals for the 1200 to 1600 calorie patterns are higher and require that more calories be used for nutrient-dense foods from the food groups.

Note: For further information regarding USDA Food Patterns, see page 583.

Source: *Dietary Guidelines for Americans, 2010*

Note for USDA Food Patterns

***All foods are assumed to be in nutrient-dense forms, lean or low-fat and prepared without added fats, sugars, or salt. Solid fats and added sugars may be included up to the daily maximum limit identified in the table. Food items in each group and subgroup are:**

Fruits	All fresh, frozen, canned, and dried fruits and fruit juices: for example, oranges and orange juice, apples and apple juice, bananas, grapes, melons, berries, and raisins
Vegetables	
Dark-green vegetables	All fresh, frozen, and canned dark-green leafy vegetables and broccoli, cooked or raw: for example, broccoli; spinach; romaine; and collard, turnip, and mustard greens
Red and orange vegetables	All fresh, frozen, and canned red and orange vegetables, cooked or raw: for example, tomatoes, red peppers, carrots, sweet potatoes, winter squash, and pumpkin
Beans and peas (legumes)	All cooked beans and peas: for example, kidney beans, lentils, chickpeas, and pinto beans. Does not include green beans or green peas (See additional comment under protein foods group)
Starchy vegetables	All fresh, frozen, and canned starchy vegetables: for example, white potatoes, corn, and green peas
Other vegetables	All fresh, frozen, and canned other vegetables, cooked or raw: for example, iceberg lettuce, green beans, and onions
Grains	
Whole grains	All whole-grain products and whole grains used as ingredients: for example, whole-wheat bread, whole-grain cereals and crackers, oatmeal, and brown rice
Enriched grains	All enriched refined-grain products and enriched refined grains used as ingredients: for example, white breads, enriched grain cereals and crackers, enriched pasta, and white rice
Protein foods	All meat, poultry, seafood, eggs, nuts, seeds, and processed soy products. Meat and poultry should be lean or low-fat and nuts should be unsalted. Beans and peas are considered part of this group as well as the vegetable group, but should be counted in one group only
Dairy	All milks, including lactose-free and lactose-reduced products and fortified soy beverages, yogurts, frozen yogurts, dairy desserts, and cheeses. Most choices should be fat-free or low-fat. Cream, sour cream, and cream cheese are not included due to their low calcium content

Source: *Dietary Guidelines for Americans, 2010*

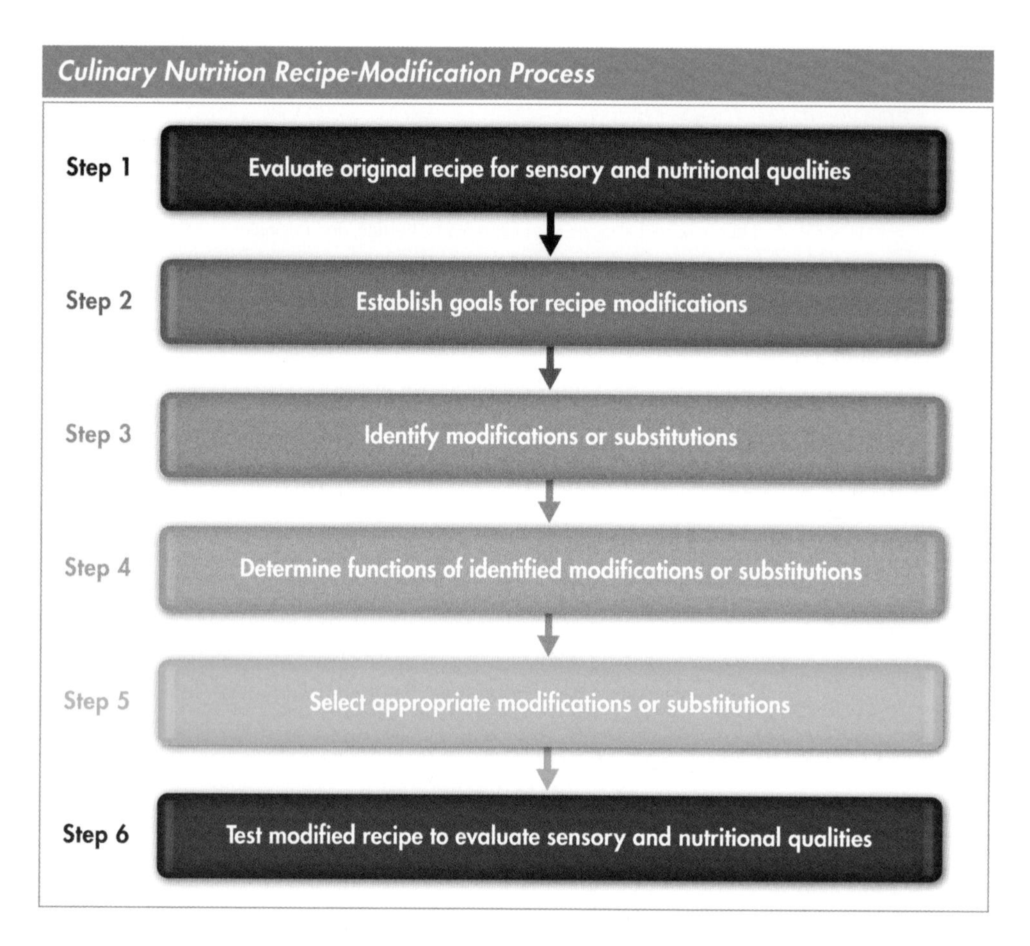

Recipe Goals and Possible Modifications

Goals	Modifications
Reduce refined sugars	• Reduce the amount of refined sugars used in the recipe • Replace refined sugars with sweet spices and natural sugars (fresh fruit, fruit juices and nectars, honey, maple syrup, and/or molasses)
Reduce fats	• Reduce the amount of fat used in the recipe • Replace solid fats and oils with puréed fruits and vegetables, juices, vinegars, stocks, alcohol, or yogurt • Replace whole eggs with egg whites or egg substitutes • Replace dairy products with nonfat or low-fat varieties • Remove skin from poultry • Use lean meats and trim away visible fat • Replace some of the meat with vegetables, legumes, and/or whole grains *Note:* Replace solid fats with heart-healthy oils to reduce saturated fats
Reduce sodium	• Replace salt with fresh herbs, spices, vinegars, citrus juice, and/or citrus zest • Use low-sodium convenience products (canned items, stocks, soy sauce, and Worcestershire sauce)
Convert to gluten-free	• Replace gluten-containing grains (barley, bulgur, oats, rye, farro, spelt, and wheat) with gluten-free grains (amaranth, buckwheat, corn, millet, quinoa, rice, teff, and wild rice) • Use flours made with gluten-free grains, nut flours, soy flour, and potato flour • Replace gluten-containing spice blends with fresh herbs and spices • Replace soy sauce with wheat-free varieties
Convert to lactose-free	• Replace dairy products with soy-, grain-, nut-, and seed-based milklike products
Convert to vegetarian	• Replace animal-based proteins (poultry, meats, seafood, and sometimes eggs and dairy products) with plant-based foods (soy products, vegetables, legumes, fruits, nuts, seeds, and whole grains) • Replace chicken and beef stock with vegetable stock

Recommended Nutrient Daily Values

Food Component	Daily Values
Total fat	65 g
Saturated fat	20 g
Cholesterol	300 mg
Sodium	2400 mg
Potassium	3500 mg
Total carbohydrate	300 g
Dietary fiber	25 g
Protein	50 g
Vitamin A	500 IU
Vitamin C	60 mg
Calcium	1000 mg
Iron	18 mg
Vitamin D	400 IU
Vitamin E	30 IU
Vitamin K	80 mcg
Thiamin	1.5 mg
Riboflavin	1.7 mg
Niacin	20 mg
Vitamin B_6	2 mg
Folate	400 mcg
Vitamin B_{12}	6 mcg
Biotin	300 mcg
Pantothenic acid	10 mg
Phosphorus	1000 mg
Iodine	150 mcg
Magnesium	400 mg
Zinc	15 mg
Selenium	70 mcg
Copper	2 mg
Manganese	2 mg
Chromium	120 mcg
Molybdenum	75 mcg
Chloride	3400 mg

Source: www.fda.gov/Food/GuidanceRegulation/GuidanceDocumentsRegulatoryInformation/LabelingNutrition/ucm064928.htm

Sources of Sugars, Fats, and Sodium

Sugars	Fats	Sodium
• Anhydrous dextrose	• Animal fat	• Baking powder
• Brown rice sugar	• Butter	• Disodium phosphate
• Brown sugar	• Cocoa butter	• Iodized salt
• Corn syrup	• Coconut oil	• Kosher salt
• Corn syrup solids	• Cream	• Monosodium glutamate
• Dextrin	• Hydrogenated oils	• Rock salt
• Dextrose	• Lard	• Salt
• Evaporated cane juice	• Margarine	• Sea salt
• Fructose	• Milk solids	• Sodium benzoate
• Glucose	• Oils (vegetable oils, olive oil, nut and seed oils)	• Sodium bicarbonate (baking soda)
• High-fructose corn syrup	• Palm kernel oil	• Sodium caseinate
• Honey	• Palm oil	• Sodium citrate
• Invert sugar	• Partially hydrogenated oils	• Sodium lactate
• Lactose	• Shortening	• Sodium nitrate
• Maltose	• Suet	• Sodium nitrite
• Malt syrup	• Tallow	• Sodium phosphates
• Maple syrup		• Sodium propionate
• Molasses		• Sodium saccharin
• Nectars		• Sodium sulfite
• Raw sugar		
• Rice syrup		
• Sorghum syrup		
• Sucrose		
• Sugar		
• Sugar alcohol		
• Treacle		

Sources of Gluten

Latin Terms for Gluten on Food Labels	Items Containing Gluten	Items Typically Containing Gluten
• Hordeum vulgare (barley) • Secale cereale (rye) • Triticum (cross of wheat and rye) • Triticum spelta (spelt) • Triticum vulgare (wheat)	• Barley • Bran • Bulgur • Durum • Farina • Farro • Graham flour • Malt (made from barley) • Rye • Seitan • Semolina • Spelt • Triticale • Wheat • Wheat berries • Wheat flour (e.g., all-purpose flour, bread flour, cake flour, bleached and unbleached flour) • Wheat germ/wheat germ oil • Wheat grass • Wheat protein/hydrolyzed wheat protein • Wheat starch/hydrolyzed wheat starch	• Ales and beers • Artificial flavorings (may come from barley) • Bouillon cubes • Bread crumbs and croutons • Breads • Cakes and cookies • Candies • Cereals • Condiments (e.g., ketchup, malt vinegar) • Dry roasted nuts • Flour tortillas • Gravies • Ground spice blends • Hot dogs • Instant coffees and teas • Lunch meats • Matzo • Meat analogs • Oats • Pastas • Pies and pastries • Pretzels • Salad dressings • Seasoned potato and tortilla chips • Seasoned rice mixes • Surimi

Sources of Whole Grains

Gluten-Containing Whole Grains	Dietary Fiber	Gluten-Free Whole Grains	Dietary Fiber
Bulgur (cracked wheat) (1 cup, cooked)	8.2 g	**Amaranth (1 cup, cooked)**	5.2 g
Dark rye flour (¼ cup, dry)	7.6 g	**Brown rice (1 cup, cooked)**	3.5 g
Rolled oats (1 cup, cooked)	5.3 g	**Buckwheat (1 cup, cooked)**	4.5 g
Spelt (1 cup, cooked)	7.6 g	**Millet (1 cup, cooked)**	2.3 g
Whole-grain barley (1 cup, cooked)	6.0 g	**Quinoa (1 cup, cooked)**	5.2 g
Whole-grain triticale flour (¼ cup, dry)	4.7 g	**Whole-grain cornmeal (¼ cup, dry)**	2.2 g
Whole-grain wheat flour (¼ cup, dry)	3.2 g	**Wild rice (1 cup, cooked)**	3.0 g
		Teff (1 cup, cooked)	7.1 g

Sources of Omega-3 Fatty Acids

Plant Sources	Omega-3 Fatty Acids
Black walnuts (1 oz)	0.562 g
Canola oil (1 tbsp)	0.812 g
English walnuts (1 oz)	2.542 g
Flaxseed oil (1 tbsp)	7.196 g
Flaxseeds, ground (1 tbsp)	1.597 g
Olive oil (1 tbsp)	0.103 g
Soybean oil (1 tbsp)	0.917 g
Tofu (3 oz)	0.489 g
Wheat germ oil (1 tbsp)	0.932 g
Seafood Sources*	**Omega-3 Fatty Acids**
Atlantic salmon (farmed)	2.130 g
Atlantic mackerel	2.270 g
Blue crab	0.272 g
Blue mussels	0.411 g
Catfish (farmed)	0.391 g
Clams	0.168 g
Coho salmon (farmed)	1.089 g
Dungeness crab	0.269 g
Flounder (sole)	0.215 g
Oysters	0.564 g
Pacific cod	0.188 g
Pacific halibut	0.444 g
Rainbow trout (farmed)	0.838 g
Sablefish	1.410 g
Sardines (canned in oil)	1.242 g
Scallops	0.183 g
Shrimp	0.459 g
Spiny lobster	0.353 g
Squid	0.422 g
Striped bass	0.654 g
Swordfish	0.701 g
Tilapia	0.185 g
White tuna (canned in water)	0.808 g
Yellowfin tuna	0.207 g

* based on 3 oz serving size

Common Units of Measure

Volume Units

Customary System	
Unit	**Abbreviation**
teaspoon	tsp
tablespoon	tbsp
fluid ounce	fl oz
cup	c
pint	pt
quart	qt
gallon	gal.
Metric System	
Unit	**Abbreviation**
liter	L
milliliter	mL

Temperature Units

Customary System	
Unit	**Abbreviation**
degrees Fahrenheit	°F
Metric System	
Unit	**Abbreviation**
degrees Celsius	°C

Weight Units

Customary System	
Unit	**Abbreviation**
ounce	oz
pound	lb or #
Metric System	
Unit	**Abbreviation**
milligram	mg
gram	g
kilogram	kg

Distance Units

Customary System	
Unit	**Abbreviation**
inch	in.
foot	ft
Metric System	
Unit	**Abbreviation**
millimeter	mm
centimeter	cm
meter	m

Volume Measurement Equivalents

Customary System Equivalents

1 gal. = 4 qt = 8 pt = 16 cups = 128 fl oz = 256 tbsp = 768 tsp
1 qt = 2 pt = 4 cups = 32 fl oz = 64 tbsp = 192 tsp
1 pt = 2 cups = 16 fl oz = 32 tbsp = 96 tsp
1 cup = 8 fl oz = 16 tbsp = 48 tsp
1 fl oz = 2 tbsp = 6 tsp
1 tbsp = 3 tsp

Customary—Metric Equivalents

1 gal. = 3.79 L
1 qt = 0.95 L
1 cup = 236.6 mL
1 fl oz = 29.6 mL
1 tsp = 5 mL
1 L = 1.06 qt
1 L = 33.8 fl oz

Metric System Equivalents

1 kiloliter (kL) = 1000 L
1 hectoliter (hL) = 100 L
1 dekaliter (daL) = 10 L
1 deciliter (dL) = 0.1 L
1 centiliter (cL) = 0.01 L
1 milliliter (mL) = 0.001 L

Weight Measurement Equivalents

Customary System Equivalents

1 lb = 16 oz

Customary—Metric Equivalents

1 lb = 0.454 kg
1 lb = 454 g
1 oz = 28.4 g
1 kg = 2.2 lb

Metric System Equivalents

1 kilogram (kg) = 1000 g
1 hectogram (hg) = 100 g
1 dekagram (dag) = 10 g
1 decigram (dg) = 0.1 g
1 centigram (cg) = 0.01 g
1 milligram (mg) = 0.001 g

Recommended Safe Minimum Internal Cooking Temperatures

Food	Degrees Fahrenheit (°F)	Rest Time
Ground meat and meat mixtures		
Beef, pork, veal, and lamb	160	None
Turkey and chicken	165	None
Fresh beef, veal, and lamb		
Steaks, roasts, and chops	145	3 minutes
Poultry		
Chicken and turkey, whole	165	None
Poultry breasts, roasts	165	None
Poultry thighs, legs, and wings	165	None
Duck and goose	165	None
Stuffing (cooked alone or in bird)	165	None
Pork and ham		
Fresh pork	145	3 minutes
Fresh ham	145	3 minutes
Eggs and egg dishes		
Eggs	Cook until yolk and white are firm.	None
Egg dishes	160	None
Seafood		
Fish	145 or cook fish until it is opaque (milky white) and flakes with a fork.	None
Shellfish		
Shrimp, lobster, and crabs	Cook until the flesh is pearly and opaque.	None
Clams, oysters, and mussels	Cook until shells open.	None
Scallops	Cook until flesh is milky white or opaque and firm.	None

Source: FoodSafety.gov

A

acid: A compound that releases hydrogen ions and produces a solution with a pH less than 7.

acid-base balance: The state of equilibrium between acids and bases in body fluids.

acidophilus milk: A fermented liquid dairy product made by adding lactobacillus acidophilus bacteria to milk.

acidosis: A harmful condition in which the blood contains excess acids.

active dry yeast: Yeast that has been dehydrated and looks like small granules.

added sugar: Any sugar not naturally occurring in a food and added during processing or preparation.

additive: A substance added to a food to maintain or improve nutritional quality, maintain quality and freshness, aid in processing or preparation, or make food more appealing.

adenosine triphosphate (ATP): A molecule that captures the released energy generated during catabolism and that uses that energy to power all cellular activity.

adequate intake (AI): The estimated daily intake level for a nutrient for which no RDA has been set but nutrient intake appears to be adequate to meet the needs of a healthy population.

adipose cell: A cell with the primary function of storing excess fat for later energy use.

aging: The period of rest that occurs for a length of time after an animal has been slaughtered.

à la carte menu: A menu that offers separately priced food and beverages.

albumen: *See* egg white.

alcohol: A liquid substance produced from fermented fruits or grains; it is commonly used as a flavoring or beverage.

ale: A beer, such as a porter or a stout, that is fermented rapidly.

alimentary canal: *See* digestive tract.

alkalosis: A harmful condition in which the blood contains excess bases.

almond: A teardrop-shaped fruit that grows on a small almond tree.

alpha-linolenic acid: An essential polyunsaturated fatty acid that is part of the omega-3 fatty acid family. Commonly referred to as omega-3 fatty acid.

amaranth: A gluten-free pseudocereal with a small, yellow, black-flecked seed that is high in protein.

amino acid: A chemical compound consisting of carbon and hydrogen with a nitrogen-containing amino group at one end, an acid group at the other end, and a distinct side chain.

amino acid pool: A collection of amino acids broken down from their original protein structure and distributed through the body to provide the material for cells to build new proteins.

amylase: An enzyme in saliva that breaks down carbohydrates into simple sugars.

amylopectin: A soluble component of starch that forms a branched chain.

amylose: An insoluble component of starch that forms an unbranched chain.

anabolism: The metabolic reaction by which the body absorbs the energy released during catabolism to build complex molecules from simpler molecules.

anaphylaxis: A severe allergic reaction in which breathing is prohibited due to narrowing of the airway.

anemia: A blood condition in which red blood cells do not supply an adequate amount of oxygen to the body.

antibody: A type of protein produced by the immune system that destroys or deactivates antigens.

antidiuretic hormone (ADH): A hormone secreted by the body that signals the kidneys to conserve water rather than eliminate it.

antigen: A foreign substance the body considers harmful.

antioxidant: A substance that may protect cells from damage caused by free radicals.

anus: The opening located at the end of the digestive tract where waste is excreted from the body.

appetizer: A food that is larger than a single bite and that is typically served as the first course of a meal.

aroma: A scent detected by the sense of smell.

aromatic: An ingredient added to a food to enhance its natural aromas and flavors.

artery: A blood vessel that carries oxygen-containing blood from the heart to the rest of the body.

artesian water: Bottled still water obtained by tapping an underground water source, which causes the water to rise into a well.

artificial sweetener: An intensely sweet, nonnutritive synthetic substance with zero to almost no calories.

ascorbic acid: *See* vitamin C.

atherosclerosis: A condition characterized by excess cholesterol that causes plaque to build up in artery walls and restrict blood flow.

B

baked Alaska: A composed frozen dessert made with a sponge cake soaked in a liqueur-flavored syrup that is topped with ice cream and covered with Italian meringue.

baking powder: A chemical leavening agent that is a combination of baking soda and cream of tartar or sodium aluminum sulfate.

baking soda: An alkaline chemical leavening agent that reacts to an acidic dough or batter by releasing carbon dioxide without being heated. Also known as sodium bicarbonate.

barbequing: A dry-heat cooking technique in which food is slowly cooked over hot coals or smoldering hardwoods.

barley: A cereal grain that resembles brown rice in shape, yet takes longer to cook.

basal metabolic rate (BMR): The minimum amount of calories the body needs to support life.

base: A compound that receives hydrogen ions and produces a solution with a pH greater than 7.

basting: The process of using a brush or a ladle to place pan drippings or sauces over an item during the cooking process to help retain moisture and enhance flavor.

bean: The oval or kidney-shaped edible seed of various plants in the legume family.

bean curd: *See* tofu.

beef: Meat from domesticated cattle.

beer: An alcoholic beverage that is brewed from a mash composed of water and malted barley or other cereal grains, such as corn, rye, or wheat, flavored with hops, and fermented with yeast.

beriberi: A disease caused by a thiamin deficiency with symptoms that include muscle weakness, fatigue, nerve damage, edema, and heart damage.

berry: A type of fruit that is small and has many tiny, edible seeds.

bile: A yellow-green fluid produced by the liver and stored in the gallbladder until it passes into the small intestine where it enables fats to mix with water for digestion.

bioavailability: The extent to which a substance is absorbed by the body.

biotin: A water-soluble vitamin made in small amounts by the body that is necessary for energy production, the synthesis of fatty acids, proper nervous system function, and healthy skin, hair, and nails. Also known as vitamin B_7.

bivalve: A mollusk that has a top shell and a bottom shell connected by a central hinge.

blood glucose level: The amount of glucose circulating in the blood.

blood pressure: The pressure of blood within the arteries.

blue-veined cheese: A cheese produced by inserting live mold spores into the center of the ripening cheese with a needle.

body mass index (BMI): A calculation based on height and weight to determine body fat composition.

bolus: A compacted mass of food that has been mixed with saliva and is ready to be swallowed.

bouquet garni: A bundle of herbs and vegetables tied together with twine that is used to flavor stocks and soups.

bran: The tough outer layer of grain that covers the endosperm.

Brazil nut: A 1–2 inch long, richly flavored, white seed of the very large fruit of the Brazil nut tree.

buckwheat: A gluten-free pseudocereal with a dark, three-cornered seed.

bulb vegetable: A strongly flavored vegetable that grows underground and consists of a short stem base with one or more buds that are enclosed in overlapping membranes or leaves.

butter: A solid dairy product made by churning butterfat until it reaches a solid state.

butterfat: The natural fat found in milk.

butterflied fillet: Two single fillets from a dressed fish that are held together by the uncut back or belly of the fish.

buttermilk: A fermented liquid dairy product made by adding bacterial cultures to low-fat or skim milk to create a thickened texture and tangy flavor.

B vitamins: A group of vitamins that include thiamin, riboflavin, niacin, pantothenic acid, vitamin B_6, biotin, folate, and vitamin B_{12}.

C

calcium: A major mineral required for blood clotting and the development of healthy teeth and bones.

California menu: A menu that offers all food and beverage items for breakfast, lunch, and dinner throughout the entire day.

cancer: A group of diseases characterized by abnormal cell growth.

canning: The process of preserving foods in an airtight container to prevent spoilage.

caramelization: The reaction that occurs when sugars are exposed to high heat and that results in browning and a change in flavor.

carbohydrate: An energy-providing nutrient in the form of sugar, starch, or dietary fiber; it is the main source of energy for the body.

carbonated water: Water that has been infused with carbon dioxide.

cardiovascular disease: A class of medical conditions that affects the heart and blood vessels. Also known as heart disease.

carotenoid: An organic pigment found in orange or yellow vegetables.

carpaccio: Thin slices of meat or seafood that are served raw.

cartilaginous fish: A fish that has a skeleton composed of cartilage instead of bones.

casein: The main group of proteins found in milk.

cashew: A nut extracted from the kidney-shaped kernel inside the fruit of the cashew tree; it has a buttery flavor.

catabolism: The metabolic reaction by which the body breaks down the complex molecules in nutrients into simpler molecules in order to release useable energy.

celiac disease: A condition in which gluten damages the ability of the small intestine to absorb nutrients.

cellulose: An insoluble fiber that is the main component of plant cell walls.

cephalopod: A mollusk that does not have an external shell.

cereal grain: A grain that is derived from plants in the grass family.

ceviche: Extremely thin slices of raw fish that have been marinated in lemon or lime juice.

chalazae: Twisted cordlike strands of egg white that anchor the yolk to the center of the egg white.

cheese: A solid dairy product made from milk curds.

cheesecake: A baked custard that typically has a graham cracker or cookie crust.

chestnut: A nut from a rounded-off, triangular-shaped kernel found inside the burrlike fruit of the chestnut tree.

chia seed: A small brown seed from a plant in the mint family.

chloride: A major mineral that helps maintain the acid-base balance of the body as well as proper blood volume and blood pressure.

chlorophyll: The green pigment plants use to capture sunlight for photosynthesis.

cholesterol: A type of sterol in the lipid family that is a soft, waxy substance made by the body and found in every cell and in foods of animal origin.

chromium: A trace mineral that is needed to regulate blood glucose levels.

chutney: A sauce typically made by cooking fruits with flavorings and seasonings.

chylomicron: A lipoprotein that transports large fragments of partially digested triglycerides and cholesterol from the small intestine to the liver and other tissues for absorption.

chyme: A thick, semifluid mass of partly digested food and gastric juices that originates in the stomach and is passed to the small intestine.

citrus fruit: A fruit with a brightly colored, thick rind and pulpy, segmented flesh that grows on trees in warm climates.

coagulation: The process of a protein changing from a liquid to a semisolid or a solid state when heat or friction is applied.

coenzyme: A compound that attaches to an enzyme to activate or enhance the functions of the enzyme.

collagen: A soft, white connective tissue that breaks down into gelatin when heated.

combination cooking: Any cooking technique that uses both moist and dry heat.

complementary protein: A combination of two or more incomplete proteins that supplies sufficient amounts of all the essential amino acids.

complete protein: A dietary source of protein that contains all the essential amino acids in the amounts needed by the body.

complex carbohydrate: A type of carbohydrate made from more than two attached sugar units.

compound butter: A flavorful butter made by mixing cold, softened butter with flavoring ingredients such as fresh herbs, garlic, vegetable purées, dried fruits, preserves, or wine reductions.

compressed yeast: Yeast that has approximately 70% moisture and is available in 1 lb cakes or blocks. Also known as fresh yeast.

condiment: A savory, sweet, spicy, or salty accompaniment added to or served with a food to impart a particular flavor that will complement the dish.

conditionally essential amino acid: A nonessential amino acid that the body cannot make in sufficient quantities.

conduction: A type of heat transfer method in which heat passes from one object to another through direct contact.

confection: A small sugar-based food item such as a candy or a chocolate.

convection: A type of heat transfer method that occurs due to the circular movement of a fluid or gas.

cooking: The process of heating foods in order to make them taste better, make them easier to digest, and kill harmful microorganisms that may be present in the food.

copper: A trace mineral that works with iron to form red blood cells.

corn: A gluten-free cereal grain cultivated from an annual grass that bears kernels on large woody cobs called ears. Also known as maize.

coulis: A sauce typically made from either raw or cooked puréed fruits or vegetables.

cracked grain: A whole kernel of grain that has been cracked by being placed between rollers.

craft beer: A beer produced by a small, independently owned brewer.

cream: A liquid dairy product made from the butterfat that separates from non-homogenized milk.

crème fraîche: A soured cream containing approximately 28% butterfat.

crêpe: A French pancake that is light and very thin.

crustacean: A shellfish that has a hard, segmented shell protecting its soft flesh and that lacks an internal bone structure.

crystallization: A process in which crystals precipitate from a solution to form a solid in the presence of heat.

curd: The thick, casein-rich part of coagulated milk.

curing: The process of using salt and sodium nitrite alone or with flavorings or sugar to preserve a food item.

cycle menu: A menu written for a specific period and repeated once that period ends.

D

deficiency: The lack of a substance necessary to maintain health.

dehydration: 1. A condition in which the body does not have an adequate amount of water to function properly. **2.** The process of removing moisture from food.

denaturing: The process of structurally changing protein from its natural state.

dental caries: *See* tooth decay.

dextrose: *See* glucose.

diabetes: A chronic disease characterized by elevated blood sugar levels and ineffective insulin production or absorption. Also known as diabetes mellitus.

diabetes mellitus: *See* diabetes.

diastolic pressure: The pressure of blood within the arteries before the heart contracts.

dietary fiber: An indigestible form of carbohydrate found in vegetables, fruits, and grains.

dietary reference intake (DRI): A set of nutrient reference standards that include estimated average requirements, recommended dietary allowances, adequate intakes, tolerable upper intake levels, and estimated energy requirements.

digestion: The process the human body uses to break down food into a form that can be absorbed and used or excreted.

digestive tract: A hollow tube that serves as the passageway for food to move from the mouth through the esophagus, stomach, small and large intestines, rectum, and anus. Also known as the gastrointestinal (GI) tract or alimentary canal.

dipeptide: A protein fragment consisting of two peptide bonds.

disaccharide: A simple carbohydrate made of two attached sugar units. Also known as double sugar.

distilled water: Purified water that has been boiled until it steams and then cooled until it condenses back to liquid form.

diuretic: A substance that causes an increased output of urine.

DNA: The genetic building blocks of the body.

docking: The process of making small holes in dough before it is baked to allow steam to escape and to promote even baking.

double sugar: *See* disaccharide.

drawn fish: A fish that has had only the viscera (internal organs) removed.

dressed fish: A fish that has been scaled and has had the viscera, gills, and fins removed.

drupe: A fruit that contains one hard seed or pit. Also known as a stone fruit.

dry aging: The process of aging larger cuts of meat by hanging them in a well-controlled environment.

dry-heat cooking: Any cooking technique that uses hot air, hot metal, a flame, or hot fat to conduct heat and brown food.

dry-rind cheese: A semisoft cheese that is allowed to ripen through exposure to air.

dry rub: A rub made by grinding herbs and spices into a fine powder.

E

edamame: Green soybeans.

edema: A condition in which body tissues swell due to fluid retention.

eggless egg substitute: A yellow-colored liquid composed of soy, vegetable gums, and starches derived from corn or flour.

eggshell: The brittle, porous covering of an egg that protects the fragile yolk and white.

egg substitute: An egg product made from pasteurized eggs that have had the yolks removed and vegetable gum, coloring, and flavoring added.

egg white: The clear portion of a raw egg, which makes up two-thirds of the egg and consists mostly of ovalbumin protein. Also known as the albumen.

egg yolk: The yellow portion of an egg.

electrolyte: A mineral compound that forms positively or negatively charged ions when dissolved in body fluids.

empty-calorie food: A food that is high in calories and low in nutrients.

emulsifier: A substance that mixes with two unlike liquids to produce an emulsion.

emulsion: A combination of two unlike liquids that have been forced to bond with each other.

endosperm: The largest component of a grain kernel; it consists of carbohydrates and a small amount of protein.

enriched dough: A yeast dough that is low in sugar and fat and typically includes dried milk solids.

enriched food: A food that has had nutrients added that were originally present in the food but were lost during processing.

enzymatic browning: The oxidation of fruits and vegetables, resulting in a brown surface color.

enzyme: A type of protein that acts as a catalyst to accelerate chemical reactions.

ergosterol: A type of sterol present in fungi.

esophagus: A muscular tube that moves food from the mouth to the stomach.

espresso: An intensely flavored coffee made from beans that have been roasted to the very dark, or espresso-roasted, stage.

essential amino acid: An amino acid that the body cannot make in sufficient quantities and that must be obtained from dietary sources.

essential fatty acid: A fatty acid that cannot be made by the body and that must be obtained from dietary sources.

essential nutrient: A nutrient the body cannot make in sufficient amounts to meet its needs and that must be obtained from food.

estimated average requirement (EAR): The daily nutrient intake level that is sufficient enough to meet the needs of half a healthy population group based on age and gender.

estimated energy requirement (EER): The energy intake (calories) needed by healthy individuals to maintain their weight based on age, gender, weight, height, and activity level.

exotic fruit: Any fruit grown in the moderately humid subtropics where the coldest months average 50°F and summers are hotter than winters.

extracellular fluid: Fluid that surrounds the cells of the body.

extract: Flavorful oil that has been combined with alcohol.

F

fabricated cut: A ready-to-cook cut of meat that is packaged to certain size and weight specifications.

farm-raised game bird: A game bird that is raised for legal sale.

farro: A hearty ancient cereal grain that tastes similar to wheat yet resembles brown rice.

fat: A triglyceride that remains solid at room temperature.

fat-soluble vitamin: A vitamin that dissolves in fat.

fatty acid: An organic compound consisting of a chain of carbon and hydrogen atoms attached to an acid group on one end.

fermentation: The breakdown of carbohydrates due to the actions of bacteria, microorganisms, or yeast.

fibrin: A protein fiber that forms a clot to stop bleeding.

filbert: *See* hazelnut.

fish: The classification of animals that have fins, gills, and an internal skeleton made of bones or cartilage.

fish fillet: A lengthwise piece of flesh cut away from the backbone of a fish.

fish loin: A lengthwise cut from either side of the backbone of a large roundfish.

fish steak: A cross section of a dressed fish.

fixed menu: A menu that remains the same or rarely changes.

flaked grain: A refined grain that has been rolled to produce a flake. Also known as a rolled grain.

flatfish: A thin, wide fish with both eyes located on one side of the head and a backbone that runs from head to tail through the lateral line of the body.

flavonoid: An organic pigment found in purple, dark red, and white vegetables.

flavor: The combined sensory experience of taste and smell along with visual, textural, and temperature sensations.

flavoring: An item that alters or enhances the natural flavor of food.

flavor perception: A preconceived idea about the taste of a food item.

flavor profile: The combination of ingredients, such as herbs, spices, and seasonings, associated with a specific cuisine.

flaxseed: The dark-amber-colored seed of the flowering flax plant.

flexitarian: An individual who primarily consumes plant-based foods and consumes animal-based foods, such as meat, dairy, and eggs, only occasionally or in small amounts.

flower vegetable: The edible flower of a nonwoody plant that is prepared as a vegetable.

fluoride: A trace mineral that provides protection against tooth decay and aids the formation of healthy teeth and bones.

fluorosis: A condition caused by an excess of fluoride that results in white or brown marks on tooth enamel and damage to bones.

foie gras: The fattened liver of a duck or goose.

folate: A water-soluble vitamin that supports red blood cell production, the nervous system, and the formation of DNA. Also known as vitamin B_9.

folic acid: A synthetic form of folate.

food additive: A substance added to food to preserve or improve its appearance, texture, flavor, and/or nutritional qualities.

food allergy: An adverse reaction to a specific food that involves the immune system.

food intolerance: An adverse reaction to a specific food that does not involve the immune system.

fortified food: A food that has nutrients added in order to increase nutritive value.

free radical: An unstable compound produced as a byproduct of metabolism.

fresh cheese: A cheese that is not aged or allowed to ripen.

fresh yeast: *See* compressed yeast.

frozen yogurt: A frozen dessert made from yogurt, sugar, and sometimes gelatin.

fructose: A monosaccharide that is the sweetest of all sugars.

fruit: The edible, ripened ovary of a flowering plant that usually contains one or more seeds.

fruit-vegetable: A botanical fruit that is sold, prepared, and served as a vegetable.

functional food: A food that provides potential health benefits when consumed on a regular basis as part of a varied diet.

G

galactose: A monosaccharide that is most often attached to glucose to form lactose (milk sugar).

gastric juice: An acidic digestive fluid secreted by glands in the stomach that contains water, enzymes, and hydrochloric acid.

gastrointestinal (GI) tract: *See* digestive tract.

gastropod: *See* univalve.

gelatinization: The process in which a heated starch absorbs moisture and swells, which thickens a liquid.

germ: The smallest part of a grain kernel; it contains a small amount of natural oils as well as vitamins and minerals.

giblets: The name for the grouping of the neck, gizzard, liver, and heart of a bird.

glucagon: A pancreatic hormone sent to the liver to trigger the conversion of glycogen to glucose for energy.

glucose: A monosaccharide that is the primary source of energy in plants and animals. Also known as dextrose.

glutamate: A substance that is derived from glutamic acid and that is associated with the taste of umami.

gluten: A type of protein found in grains such as wheat, rye, barley, and some varieties of oats.

glycemic index (GI): A measure indicating the rate at which an ingested food causes blood sugar levels to rise.

glycemic response: A measure indicating the rate at which blood sugar rises after eating.

glycerol: An organic compound that contains three carbon atoms.

glycogen: A polysaccharide made and used by the liver and muscles to store long chains of glucose.

goiter: An enlarged thyroid gland that appears as a protruding growth in the neck.

gout: A condition in which an excess of uric acid builds up within the joints and results in joint pain and swelling.

grain: The edible fruit, in the form of a kernel or seed, of a grass.

grain-fed beef: Meat from cattle that were grain-fed in confined feeding operations for 90 days to 1 year.

grape: An oval fruit that has a smooth skin and grows on woody vines in large clusters.

grass-fed beef: Meat from cattle that were raised on grass with little or no special feed.

grating cheese: A hard, crumbly, dry cheese that is commonly grated or shaved and placed on top of food.

greens: *See* leaf vegetables.

groundnut: *See* peanut.

gum: A thickening agent derived from plant flours.

H

hard cheese: A firm, somewhat pliable, and supple cheese with a slightly dry texture and buttery flavor.

hazelnut: A grape-sized nut found in the fuzzy outer husks of the hazel tree. Also known as a filbert.

headed and gutted fish: *See* pan-dressed fish.

health claim: A statement declaring a relationship between certain foods or nutrients and disease.

heart disease: *See* cardiovascular disease.

heat: Energy that is transferred between two objects or substances of different temperatures.

heat capacity: The amount of energy needed to raise 1 g of a substance 1°C.

heme iron: A dietary iron found only in animal-based foods that is easily absorbed by the body.

hemoglobin: A protein in red blood cells that transports oxygen throughout the body.

herb: A flavoring derived from the leaves or stem of a very aromatic plant.

herbivore: An animal that feeds on grass and other plant-based foods.

high-density lipoprotein (HDL): A type of lipoprotein used to collect cholesterol and bring it to the liver so that it can be dismantled and recycled in bile or excreted.

high-fructose corn syrup (HFCS): Corn syrup that has been treated with an enzyme to convert part of its glucose to fructose.

homogenization: The process of emulsifying the fat particles in milk in order to provide a uniform consistency and prevent fat separation.

hormone: A chemical messenger secreted by an organ and sent to another destination in the body to deliver a message.

hull: *See* husk.

husk: The inedible, protective outer covering of grain. Also known as a hull.

hybrid fruit: A fruit that is the result of crossbreeding two or more fruits of different species to obtain a completely new fruit.

hydrogenation: The chemical process of using heat to force hydrogen atoms into an unsaturated fatty acid to make it similar in structure to a saturated fat.

hydrolysis: The process of splitting a substance into smaller parts by the addition of water.

hyperkalemia: A condition characterized by an elevated concentration of potassium in the blood.

hypernatremia: A condition characterized by abnormally high levels of sodium in the body usually due to excessive water loss.

hypertension: A condition characterized by high blood pressure.

hypervitaminosis A: An adverse health condition caused by the excessive intake of preformed vitamin A.

hypoglycemia: A condition that results from low blood glucose levels.

hypokalemia: A condition characterized by a low concentration of potassium in the blood.

hyponatremia: A sodium deficiency characterized by muscle cramps, nausea, dizziness, seizures, and coma.

I

ice cream: A frozen dessert made from cream, butterfat, sugar, and sometimes eggs.

icing: A sugar-based coating often spread on the outside of and between the layers of a baked product.

ileum: A one-way valve located at the base of the small intestine that enables chyme to pass into the large intestine.

incomplete protein: A dietary source of protein that lacks one or more of the essential amino acids.

ingredient alternative: An ingredient that replaces an item of different characteristics.

ingredient substitution: An ingredient that replaces an item of similar characteristics.

inorganic nutrient: A nutrient that does not contain carbon.

insoluble fiber: Dietary fiber that does not dissolve in water.

instant yeast: Yeast that does not need to be hydrated and may be directly added to a warm flour mixture.

insulin: A hormone produced by the pancreas that is necessary for regulating blood sugar levels.

international unit: A globally accepted amount of a substance, such as a vitamin, that is necessary to facilitate a specific body response.

intra-abdominal fat: *See* visceral fat.

intracellular fluid: Fluid that is contained within the cells of the body.

intrinsic factor: A protein produced in the stomach that promotes vitamin B_{12} absorption.

iodine: A trace mineral that is required to make hormones in the thyroid gland.

iron: A trace mineral that is required to make the proteins that distribute oxygen throughout the body.

isoflavone: An organic compound found in legumes and considered a phytonutrient.

K

kasha: Roasted buckwheat.

kecap: A type of Asian sauce made from fermented soybeans, palm sugar, and usually star anise and garlic.

kefir: A semisolid dairy product that has been fermented by adding strains of active bacteria and yeast to milk.

Keshan disease: A disease associated with selenium deficiency that results in an enlarged heart that works inefficiently.

ketone: An acidic substance that forms in the blood when lipids are metabolized for energy instead of carbohydrates.

ketosis: A condition in which ketones build up in the body because lipids are improperly metabolized for energy.

kilocalorie: The amount of heat required to raise the temperature of 1 g of water by 1°C.

kneading: The process of pushing and folding dough until it is smooth and elastic.

kwashiorkor: A form of PEM characterized by a diet deficient in proteins but adequate in caloric intake.

L

lactic acid: A type of organic, colorless acid made by red blood cells and muscle tissues.

lactose: A disaccharide comprised of glucose and galactose. Also known as milk sugar.

lager: A beer, such as a pilsner or a malt liquor, that is fermented slowly using cold temperatures.

lamb: Meat from slaughtered sheep that are less than a year old.

large intestine: The wide, lower section of the intestinal tract approximately 5 feet in length where water is extracted from chyme.

leaf vegetables: Edible plant leaves that are often accompanied by edible stems and shoots. Also known as greens.

lean dough: A yeast dough that is low in sugar and fat.

leavening agent: Any ingredient that causes a baked product to rise by the action of air, steam, chemicals, or yeast.

legume: The edible seed of a nonwoody plant; grows in multiples within a pod.

lentil: A very small pulse that has been split in half.

liaison: A thickening agent that is a mixture of egg yolks and heavy cream used to thicken sauces.

lignin: A nonpolysaccharide form of fiber that provides rigidity and support for the woody cell walls of plants.

limiting amino acid: An amino acid that is missing or unavailable in an adequate amount.

lingual lipase: An enzyme in saliva that helps break down the lipids found in milk for digestion.

linoleic acid: An essential polyunsaturated fatty acid that is part of the omega-6 fatty acid family. Commonly referred to as omega-6 fatty acid.

lipid: An energy-providing nutrient made from fatty acids that includes solid fats and oils.

lipoprotein: A protein-coated lipid that acts as an emulsifier for larger lipid fragments and cholesterol to facilitate their transport into the lymphatic system and bloodstream for absorption.

lipoprotein lipase: An enzyme that breaks down the triglycerides transported by a chylomicron into fatty acids and glycerol that can be absorbed.

liqueur: An alcoholic beverage made from distilled alcohol and flavored with fruits, spices, herbs, flowers, nuts, or cream, and bottled with added sugar.

liver: A large organ located in the upper abdomen that produces bile for digestion, stores and filters blood, and helps convert nutrients into usable energy.

low-density lipoprotein (LDL): A type of lipoprotein used to transport cholesterol to the cells and tissues of the body.

lymph: A type of fluid drained from tissues that can transport material from the bloodstream into tissues.

lymphatic system: A system for transporting fat-soluble nutrients from the small intestine into the bloodstream.

M

macadamia nut: A tan, marble-sized nut with a hard shell grown on the macadamia tree.

macronutrient: A nutrient needed by the body in large amounts.

magnesium: A major mineral that is needed to keep bones strong and that helps facilitate hundreds of chemical reactions in the body.

Maillard reaction: A reaction that occurs when the proteins and sugars in a food are exposed to heat and merge together to form a brown exterior.

maize: *See* corn.

major mineral: A mineral with a recommended dietary allowance that exceeds 100 mg per day.

malnutrition: An adverse health condition caused by an excess or deficiency of nutrients.

maltose: A disaccharide comprised of two units of glucose. Also known as malt sugar.

malt sugar: *See* maltose.

manganese: A trace mineral that facilitates protein and carbohydrate metabolism and bone formation.

marasmus: A form of PEM characterized by a diet deficient in both proteins and caloric intake.

marinade: A flavorful liquid used to soak uncooked foods, such as meat, poultry, and fish, to impart flavor and sometimes to tenderize.

market menu: A menu that changes frequently to coincide with changes in product availability.

meat analog: A meatless product made from soybean by-products that is formed into the familiar shape of a processed meat product.

Mediterranean diet: A diet that emphasizes a high consumption of vegetables, fruits, whole grains, olives, olive oils, nuts, and seeds; a moderate consumption of lean meats, fish, and wine; and a limited consumption of red meats and processed foods.

melon: A type of fruit with a hard outer rind (skin) and a soft inner flesh that typically contains many seeds.

menu: A document that markets a foodservice operation to guests.

menu mix: The assortment of items that may be ordered from a given menu.

meringue: A mixture of egg whites and sugar.

metabolism: The sum of all chemical reactions that occur within the body.

micronutrient: A nutrient needed by the body in small amounts.

microvillus: A hairlike projection on the villus of the small intestine that increases the surface area available for nutrient absorption.

milk: A liquid dairy product produced by the mammary glands of cows, goats, sheep, or water buffalo.

milk sugar: *See* lactose.

milled grain: A refined grain that has been ground into a fine meal or powder.

millet: A small, round, butter-colored ancient cereal grain that is gluten-free.

mineral: An inorganic nutrient that is required in small amounts to help regulate body processes.

mineral water: Bottled still water obtained from an underground water source that contains not less than 250 ppm (parts per million) of total dissolved solids such as calcium and magnesium.

mirepoix: A roughly cut mixture consisting of 50% onions, 25% celery, and 25% carrots.

miso: A type of Asian flavoring made from fermented soybeans, starch such as barley or rice, salt, and water.

mixologist: An individual formally trained in the preparation of mixed drinks, or cocktails.

modified starch: A starch that has been chemically or physically altered from its original state.

moist-heat cooking: Any cooking technique that uses liquid or steam as a cooking medium.

mollusk: A shellfish with a soft, nonsegmented body.

monoglyceride: A large lipid fragment consisting of a glycerol molecule with one fatty acid chain attached.

monosaccharide: A simple carbohydrate made of one sugar unit.

monounsaturated fat: A type of triglyceride made primarily of monounsaturated fatty acids.

monounsaturated fatty acid: A type of fatty acid that contains one point of unsaturation.

mousse: A rich and airy foam that has been lightened with whipped cream, whipped pasteurized egg whites, or both.

mouthfeel: The way a food or beverage feels in the mouth.

mushroom: The fleshy, spore-bearing body of an edible fungus that grows above the ground.

myoglobin: An iron-containing protein that binds oxygen in muscle tissues.

N

natto: A type of Asian sauce made from fermented soybeans that has a distinct aroma and cheese-like flavor.

niacin: A water-soluble vitamin necessary for skin, nerve, and digestive system functions. Also known as vitamin B_3.

nitrogen balance: The difference between nitrogen intake and nitrogen excretion.

nonessential amino acid: An amino acid that can be made by the body in sufficient quantities.

nonheme iron: A dietary iron found primarily in plant-based foods that is not easily absorbed by the body.

nut: A dry, hard-shelled fruit or seed that contains an inner kernel.

nutrient: A substance that provides nourishment to the body.

nutrient claim: A statement that uses approved terminology to describe the nutrient content of a food.

nutrient-dense food: A food that is high in nutrients and low in calories.

nutrition: The science of how the body receives and uses the substances found in food.

O

oat groat: An oat grain from which only the husk has been removed.

oats: A cereal grain derived from the berries of oat grass.

obesity: A medical condition characterized by an excess of body fat.

offal: The edible part of an animal that is not part of a primal cut.

oil: A triglyceride that remains liquid at room temperature.

oilseed: An oil-rich seed grown primarily for its oil.

okara: The ground soybean pulp left over from the production of soy milk.

old-fashioned oats: *See* rolled oats.

oligosaccharide: A complex carbohydrate with a chain that generally ranges from three to eleven sugar units.

omega-3 fatty acid: A polyunsaturated fatty acid with a final double bond that is three carbons from the end of its carbon chain.

omega-6 fatty acid: A polyunsaturated fatty acid with a final double bond that is six carbons from the end of its carbon chain.

organic food: Food produced without the use of chemically formulated fertilizers, growth stimulants, antibiotics, pesticides, or spoilage-inhibiting radiation.

organic nutrient: A nutrient that contains carbon.

osmosis: The movement of water across cell membranes from a dilute solution to a more concentrated solution.

osteomalacia: An adulthood disease caused by a vitamin D deficiency characterized by bones that soften, resulting in bone pain and muscle weakness.

osteoporosis: A condition characterized by a loss of bone density that results in porous, brittle bones that are susceptible to breakage.

oven spring: The rapid expansion of yeast dough in the oven, resulting from the expansion of gases within the dough.

P

pancreas: A long organ located near the stomach that secretes digestive juices into the small intestine.

pancreatic juice: An alkaline digestive fluid secreted by the pancreas that contains enzymes.

pancreatic lipase: An enzyme released by the pancreas that breaks lipids into glycerol, fatty acids, and monoglycerides.

pan-dressed fish: A dressed fish that has had its head removed. Also known as a headed and gutted fish.

pantothenic acid: A water-soluble vitamin necessary for producing hormones and metabolizing carbohydrates and lipids. Also known as vitamin B_5.

par stock: The amount of a particular product that should be kept in inventory to ensure that an adequate supply is on hand for regular food production within a given operation.

pasta: A rolled or extruded dough product composed of flour, water, salt, oil, and sometimes eggs.

pasteurization: The process of destroying harmful microorganisms by heating and then quick-cooling a food product.

pavlova: A crisp meringue that is filled with whipped cream and fruit.

pea: The round edible seed of various plants in the legume family.

peanut: A legume found within a thin, netted, tan-colored pod that grows underground. Also known as a groundnut.

pearled grain: A refined grain with a pearl-like appearance that results from having been scrubbed and tumbled to remove the bran.

pecan: A light-brown nut from the fruit of the pecan tree.

pectin: A soluble fiber found in and around plant cell walls.

pellagra: A disease caused by a niacin deficiency with symptoms referred to as "the three Ds": dementia (mental impairment), dermatitis (inflamed skin), and diarrhea.

pepita: *See* pumpkin seed.

peppercorn: The dried berry of a climbing vine known as the Piper nigrum; used whole, ground, or crushed.

pepsin: A digestive enzyme in the stomach that breaks the peptide bonds connecting amino acids.

peptide bond: A bond that connects a nitrogen-containing amino group of one amino acid to the acid group of another amino acid to form a link in a protein chain.

peristalsis: The involuntary and wavelike contraction of muscles that transports food through the digestive tract.

pesto: A sauce made from fresh ingredients that have been crushed with a mortar and pestle or finely chopped in a food processor before being mixed with oil.

pH level: The measurement of the acidity or alkalinity found in a substance based on a scale of 1 to 14.

phospholipid: A type of lipid consisting of two fatty acids and a phosphate molecule.

phosphorus: A major mineral that works with calcium to build strong teeth and bones, helps synthesize proteins, plays an essential role in energy production and storage, filters waste from the kidneys, and assists in the formation of DNA.

photosynthesis: The process by which plants turn carbon dioxide and water into carbohydrates by using chlorophyll to capture energy from sunlight.

phytochemicals: *See* phytonutrients.

phytonutrients: Naturally occurring substances found in plants that have been found to protect against disease. Also known as phytochemicals.

pignoli: *See* pine nut.

pine nut: An ivory-colored, torpedo-shaped nut from the pine cone of various types of pine trees. Also known as a pignoli.

pistachio: A green, bean-shaped nut from the pistachio tree.

plaque: The accumulation of bacteria and food residue on teeth that can cause tooth decay and gum disease.

plating: The process of arranging food in an appealing manner for presentation.

point of unsaturation: The point on a fatty acid chain where a missing hydrogen atom is replaced with a double carbon bond.

polyol: *See* sugar replacer.

polypeptide: A protein fragment consisting of more than 10 peptide bonds.

polyphenol: An antioxidant that can cause curding.

polysaccharide: A complex carbohydrate with a chain usually consisting of hundreds to thousands of sugar units.

polyunsaturated fat: A type of triglyceride made primarily of polyunsaturated fatty acids.

polyunsaturated fatty acid: A type of fatty acid that contains two or more points of unsaturation.

pome: A fleshy fruit that contains a core of seeds and an edible skin.

poppy seed: A tiny, blue-gray seed from the poppy plant.

pork: Meat from slaughtered hogs that are less than a year old.

portal system: A system for transporting water-soluble nutrients from the small intestine to the liver and then into the bloodstream.

portion size: The amount of a food or beverage item served to an individual.

potassium: A major mineral that helps regulate normal heart functioning and muscle contraction and that is necessary for energy production and protein synthesis.

poultry: The collective term for various kinds of birds that are raised for human consumption.

preformed retinoid: A collective term for the most active forms of vitamin A, including retinol, retinal, and retinoic acid.

primal cut: A large cut of meat from a whole or a partial carcass.

primary structure: A long, straight chain of amino acids linked by peptide bonds.

prix fixe menu: A menu that offers limited choices within a collection of specific items for a multicourse meal at a set price.

probiotic: A live microorganism that has been found to be beneficial to the digestive tract.

processed food: Food that has been altered by processes such as canning, cooking, freezing, dehydration, or milling.

proofing: The process of letting yeast dough rise in a warm (85°F) and moist (80% humidity) environment until the dough doubles in size.

protease: A digestive enzyme in the small intestine that breaks the peptide bonds connecting amino acids.

protein: An energy-providing nutrient made of carbon, hydrogen, oxygen, and nitrogen assembled in chains of amino acids.

protein-calorie malnutrition: *See* protein energy malnutrition (PEM).

protein energy malnutrition (PEM): A general term used to describe various forms of malnutrition that result from a body deficient of protein. Also known as protein-calorie malnutrition.

protein sparing: The process by which carbohydrates and lipids supply energy so that the main functions of proteins can be fulfilled.

protein turnover: The continuous breakdown, rebuilding, and recycling of amino acids.

provitamin carotenoid: A collective term for compounds that are converted to vitamin A, including beta-carotene, alpha-carotene, and beta-cryptoxanthin.

pseudocereal: A seed that is classified as a grain but is derived from broadleaf plants instead of grasses.

pulse: The dried seed of a legume.

pumpkin seed: A green seed that comes from a pumpkin plant. Also known as a pepita.

punching: The process of deflating risen yeast dough to allow carbon dioxide to escape.

purchasing specification: A written form listing the specific characteristics of a product that is to be purchased from a supplier.

purified water: Bottled still water that has been distilled or filtered to remove minerals and other impurities.

pyloric sphincter: A muscular ring located at the base of the stomach that enables chyme to pass from the stomach to the small intestine.

Q

quaternary structure: A complex protein chain that results from globular structures combining with each other.

quinoa: A gluten-free pseudocereal with a small, round seed that is classified as a complete protein.

R

radiation: A type of heat transfer method in which electromagnetic waves transfer energy.

ratite: A flightless bird that has a flat breastbone and small wings in relation to its body size.

raw bar: A presentation of a variety of raw and steamed seafood presented and served on a bed of ice.

recommended dietary allowance (RDA): The daily nutrient intake level that is sufficient enough to meet the nutrient needs of approximately 98% of a healthy population group based on age and gender.

rectum: The lowest section of the large intestine.

reduction: A thick liquid with a flavor that has become concentrated as a result of being gently simmered until reduced in volume.

refined grain: A grain that has been processed to remove the bran, germ, or both.

registered dietician (RD): An individual who has specialized in the study of nutrition and met a stringent set of academic and professional standards.

retrogradation: A process in which a gelatinized starch causes a thickened liquid to turn into a gel as it cools.

ribbon pasta: Pasta that is cut into thin, round strands or flat, ribbonlike strands.

riboflavin: A water-soluble vitamin necessary for energy production, growth, red cell production, and the activation of vitamin B_6. Also known as vitamin B_2.

rich dough: A yeast dough that incorporates a lot of fat, various amounts of sugar, and eggs into a heavy, soft structure.

rickets: A childhood disease caused by a vitamin D deficiency characterized by bones that soften and bend.

rind-ripened cheese: *See* soft cheese.

rolled grain: *See* flaked grain.

rolled-in dough: A yeast dough with a flaky texture that results from the incorporation of fat through a rolling and folding procedure.

rolled oats: Oats that have been steamed and flattened into small flakes. Also known as old-fashioned oats.

root vegetable: An earthy-flavored vegetable that grows underground and has leaves that extend above ground.

roundfish: A fish with a cylindrical body, an eye located on each side of the head, and a backbone that runs from head to tail in the center of the body.

roux: A thickening agent made by cooking a mixture of equal amounts, by weight, of flour and fat and used to thicken soups and sauces.

rub: A blend of ingredients that is pressed onto the surface of uncooked foods, such as meat, poultry, and fish, to impart flavor and sometimes to tenderize.

rye: A hearty cereal grain with dark-brown berries that are long and thin.

S

sabayon: A light and foamy custard prepared by whisking egg yolks, sugar, and white wine over a double boiler into a silky foam. Also known as a zabaglione.

saccharide: A unit of sugar.

sachet d'épices: A mixture of spices and herbs placed in a piece of cheesecloth and tied with butcher's twine.

sake: A Japanese wine that is made from rice and that is not aged.

saliva: The watery fluid secreted by the salivary glands in the mouth that moistens food and contains digestive enzymes.

salsa: A sauce made by mixing or puréeing diced vegetables or fruits, flavorings, and seasonings together.

salt: A crystalline solid composed mainly of sodium chloride and used as a seasoning and a preservative.

sashimi: Very thin slices of raw fish served with garnishes and sauces.

satiety: The state of feeling full.

saturated fat: A triglyceride made primarily of saturated fatty acids.

saturated fatty acid: A fatty acid chain in which each carbon atom is filled to capacity with hydrogen atoms.

sauce: An accompaniment that is served with food to complete a dish or enhance the flavor and/or moistness of a dish.

scoring: The process of making shallow, angled cuts across the top of unbaked yeast breads to allow carbon dioxide to escape during baking and to promote even baking.

scurvy: A disease caused by a vitamin C deficiency with symptoms that include swollen or bleeding gums, loosening of the teeth, impaired wound healing, and general weakness.

searing: The process of using high heat to quickly brown the surface of a food.

seasoning: An item that intensifies or improves the flavor of food.

sea vegetable: An edible saltwater plant that contains high amounts of dietary fiber, vitamins, and minerals.

secondary structure: A coiled protein chain that has the appearance of a spring.

seed: A fruit or unripened ovule of a nonwoody plant.

seed vegetable: The edible seed of a nonwoody plant.

selenium: A trace mineral that helps maintain the function of thyroid hormones and a healthy immune system.

selenosis: A toxicity of selenium with symptoms that include brittle hair and nails, nerve damage, and breath with a garlic odor.

semolina: The granular product that results from milling the endosperm of durum wheat.

serving size: A specific amount of food expressed using a unit of measurement such as a tablespoon, ounce, or cup.

sesame seed: The small teardrop-shaped seed of the sesame plant.

shaped pasta: Pasta that has been extruded into complex shapes such as corkscrews, bowties, shells, flowers, or stars.

shellfish: The classification of aquatic invertebrates that may or may not have a hard, external shell.

shell membrane: The thin, skinlike material located directly under an eggshell.

sherbet: A frozen dessert made by combining puréed fresh fruit, simple syrup, and milk.

shrinkage: The loss of volume and weight of food as it cooks.

silverskin: A tough, rubbery, silver-white connective tissue that does not break down when heated.

simple carbohydrate: A type of carbohydrate made from one or two units of sugar. Also known as a simple sugar.

simple sugar: *See* simple carbohydrate.

small intestine: The narrow, upper section of the intestinal tract approximately 20 feet in length where most nutrients are absorbed.

smoke point: The temperature at which oil begins to smoke and emit an odor.

smoking: A dry-heat cooking technique in which food is slowly cooked over smoldering hardwoods in a vented enclosure.

sodium: A major mineral that helps maintain the volume of fluid surrounding the cells of the body and regulate the acid-base balance of the body.

sodium bicarbonate: *See* baking soda.

soft cheese: A cheese that has been sprayed with a harmless live mold to produce a thin skin or rind. Also known as a rind-ripened cheese.

soft drink: A nonalcoholic beverage made from carbonated water, flavorings, sweeteners, colors, acids, and preservatives.

solanine: A bitter, poisonous alkaloid that most commonly develops in potatoes and tomatoes.

soluble fiber: Dietary fiber that dissolves in water.

solvent: A substance that has the ability to dissolve other substances.

sommelier: An individual formally trained in ordering, storing, and serving wines to complement a wide variety of menu items.

sorbet: A frozen dessert made by combining puréed fresh fruit and simple syrup.

sour cream: A semisolid dairy product that has been fermented by adding strains of active bacteria to cream.

soybean: The oil-rich seed from the soybean plant; it is a complete protein.

soybean curd: *See* tofu.

soy milk: The liquid expressed from soybeans that have been soaked and finely ground.

soy nut: A soybean that has been soaked in water and roasted or baked.

soy sauce: A type of Asian sauce made from fermented soybeans, wheat, salt, and water.

spelt: An ancient cereal grain with a nutty flavor and high protein content.

spice: A flavoring derived from the bark, seeds, roots, flowers, berries, or beans of an aromatic plant.

spirit: An alcoholic beverage made from distilled grains, fruits, or vegetables.

spit-roasting: The process of cooking food by skewering it and suspending and rotating it above or next to a heat source.

spring water: Bottled still water obtained from a spring that flows naturally to the surface.

sprout: An edible strand of a germinated bean or seed.

stabilizer: A substance added to food to help maintain the suspension of one liquid in another.

starch: A polysaccharide made by plants to store long chains of glucose molecules.

steel-cut oats: Oat groats that have been toasted and cut into small pieces.

stem vegetable: The main trunk of a plant that develops edible buds and shoots instead of roots.

sterol: A type of lipid that is a waxy, insoluble substance composed of natural steroid alcohols derived from plants or animals.

stevia: A natural sweetener derived from a plant native to South America.

still water: Bottled water that does not contain carbonation.

stock: An unthickened liquid that is flavored by simmering seasonings with vegetables, and often, the bones of meat, poultry, or fish.

stomach: A sac-shaped organ located between the esophagus and small intestine where food is partially digested.

stone fruit: *See* drupe.

stuffed pasta: Pasta that has been formed by hand or machine to hold a filling.

subcutaneous fat: Fat that is stored just below the surface of the skin.

sucrose: A disaccharide comprised of glucose and fructose.

sugar alcohol: *See* sugar replacer.

sugar replacer: A naturally occurring carbohydrate that is only partially digestible and has a chemical structure that resembles both sugar and alcohol. Also known as sugar alcohol or polyol.

sunflower seed: A tan seed that comes from the sunflower plant.

supersaturated solution: A solution in which a substance has been dissolved into another substance in an amount greater than what is usually possible as a result of the substances being heated and cooled.

surimi: Imitation shellfish made from fish that is colored, flavored, and shaped to resemble a particular type of shellfish.

sushi: A vinegar-seasoned rice dish garnished with raw fish, cooked seafood, eggs, or vegetables.

sustainable agriculture: A method of food production that seeks to conserve land and water, reduce energy consumption, avoid the use of harmful fertilizers and pesticides, and limit air pollution.

sweating: The process of slowly cooking food to soften its texture.

syneresis: The separation of a liquid from a gel. Also known as weeping.

systolic pressure: The pressure of blood within the arteries after the heart contracts.

T

table d'hôte menu: A menu that identifies specific items that will be served for each course at a set price.

tannin: An astringent compound found in tea, the bark of some trees, and the skins, seeds, and stems of grapes.

tartare: Freshly ground or chopped raw meat or seafood.

taste: A sense that is activated by receptor cells that make up the taste buds.

teff: A gluten-free ancient cereal grain; it is the smallest grain in the world.

tempeh: A flat, dense cake made from fermented soybeans.

tempering: A process in which a hot liquid is gradually added to eggs in order to slowly raise the temperature of the eggs, which stabilizes their texture and prevents curdling.

tertiary structure: A protein chain that folds back in an irregular pattern to form a compact spherical or globular shape.

texture: The appearance and feel of an item.

textured soy protein (TSP): A granular, meatless protein product made from soy flour.

thiamin: A water-soluble vitamin necessary for metabolizing carbohydrates, regulating nerve impulses, and the proper functioning of the heart and muscles. Also known as vitamin B_1.

thickening agent: A substance that adds body to a hot liquid.

thyroid gland: A gland found in the neck that secretes hormones necessary for regulating metabolism.

tisane: An herbal beverage created by steeping herbs, spices, flowers, dried fruits, or roots in boiling water.

tofu: The curd of ground, cooked soybeans that has been pressed into a cake. Also known as soybean curd or bean curd.

tolerable upper intake level (UL): The maximum daily intake level of a nutrient before the body experiences an adverse effect.

tooth decay: A condition in which teeth decay due to acids produced by bacterial growth and improper dental care. Also known as dental caries.

torte: A rich cake that is usually layered with creams, jams, or fruits and iced with a sweet icing or glazed and garnished.

toxicity: The degree to which a substance can harm living organisms.

trace mineral: A mineral with a recommended dietary allowance of less than 100 mg per day.

trans fat: An unsaturated fatty acid with hydrogen atoms on opposite sides of a double carbon bond, which make its structure similar to a saturated fat. Also known as a trans-fatty acid.

trans-fatty acid: *See* trans fat.

triglyceride: A lipid that consists of a glycerol backbone with three fatty acids attached and is the main form of fat in the body and foods.

tripeptide: A protein fragment consisting of three peptide bonds.

triticale: A hybrid cereal grain made by crossing rye and wheat.

tropical fruit: A type of fruit grown in the humid tropics where temperatures average 80°F with little temperature variation between winter and summer.

tube pasta: Pasta that has been pushed through an extruder to form a tube shape and then fed through a cutter that cuts the tube to the desired length.

tuber: A short, fleshy vegetable that grows underground and bears buds capable of producing new plants.

type 1 diabetes: A form of diabetes in which the pancreas does not produce enough insulin to normalize blood sugar levels.

type 2 diabetes: A form of diabetes in which the body resists absorbing enough insulin to normalize blood sugar levels.

U

umami: The Japanese word for "deliciousness"; it is a savory taste.

univalve: A mollusk that has a single solid shell and a single foot. Also known as a gastropod.

unsaturated fat: A type of triglyceride made primarily of unsaturated fatty acids.

unsaturated fatty acid: A fatty acid chain that lacks some hydrogen atoms.

uric acid: A product of nitrogen waste found in the urine and blood.

V

variety fruit: A fruit that is the result of breeding two or more fruits of the same species that have different characteristics.

veal: Meat from calves, which are young cattle.

vegetable: An edible root, tuber, bulb, stem, leaf, flower, or seed of a nonwoody plant.

vegetarian: An individual who avoids the consumption of animal-based foods and consumes plant-based foods, such as vegetables, fruits, beans, grains, nuts, and seeds, and sometimes dairy products and/or eggs.

very-low-density lipoprotein (VLDL): A type of lipoprotein used to transport triglycerides and cholesterol made by the liver through the body for absorption.

villus: A fingerlike projection located in the small intestine that facilitates nutrient absorption.

vinegar: A sour, acidic liquid made from fermented alcohol.

visceral fat: Fat that is stored within the abdomen and surrounds the internal organs. Also called intra-abdominal fat.

vitamin: An organic nutrient that is required in small amounts to help regulate body processes.

vitamin A: A fat-soluble vitamin that plays a key role in promoting healthy vision and proper growth and development and that is vital for a healthy immune system.

vitamin B_1: *See* thiamin.

vitamin B_2: *See* riboflavin.

vitamin B_3: *See* niacin.

vitamin B_5: *See* pantothenic acid.

vitamin B_6: A water-soluble vitamin necessary for metabolic functions, nervous system and immune system functions, and the formation of hemoglobin, which is the protein in red blood cells that transports oxygen throughout the body.

vitamin B_7: *See* biotin.

vitamin B_9: *See* folate.

vitamin B_{12}: A water-soluble vitamin that helps support the formation of red blood cells and DNA, promotes proper nerve development, and plays a role in energy production.

vitamin C: A water-soluble vitamin with antioxidant properties that protect the body from disease by improving immune functions. Also known as ascorbic acid.

vitamin D: A fat-soluble vitamin that is required for proper kidney functioning, the health of teeth and bones, and for the absorption of vitamin A, calcium, phosphorus, magnesium, iron, and zinc.

vitamin E: A fat-soluble vitamin that strengthens the immune system and widens blood vessels, which reduces the risk of blood clots.

vitamin K: A fat-soluble vitamin that plays an essential role in blood clotting and bone formation.

W

walnut: A nut from the fruit of the walnut tree.

washed-rind cheese: A semisoft cheese with an exterior rind that is washed with a brine, wine, olive oil, nut oil, or fruit juice.

water-soluble vitamin: A vitamin that dissolves in water.

weeping: *See* syneresis.

wet aging: The process of aging meat in vacuum-sealed plastic.

wet rub: A rub made by incorporating wet ingredients, such as Dijon mustard, flavored oils, puréed garlic, or honey, into a dry rub mixture.

wheat: A light-yellow cereal grain cultivated from an annual grass that yields the flour used in many pastas and baked goods.

wheel: The round center cut of a large fish from which steaks are cut.

whey: The watery part of milk.

whole fish: The market form of a fish that has not been treated or had any parts removed.

whole food: A food in its natural state such as fresh vegetables and fruit.

whole grain: A grain that has only had the husk removed.

Wilson's disease: A genetic disorder characterized by the accumulation of copper in various organs.

wine: An alcoholic beverage commonly made from fermented fruit juice that has been aged for a period of time.

Y

yeast: A microscopic, single-celled, fungus that releases carbon dioxide and alcohol through a process called fermentation when provided with food (sugar) in a warm, moist environment.

yeast starter: A mixture of flour, yeast, sugar, and water.

yogurt: A semisolid dairy product that has been fermented by adding bacterial cultures to milk.

young goose: A goose that is usually less than six months of age and weighs approximately 4–10 pounds.

yuba: The protein skin that forms when soy milk is heated.

Z

zabaglione: *See* sabayon.

zest: The colored, outermost layer of a citrus fruit peel and contains a high concentration of oil.

zinc: A trace mineral that is considered a primary protector of the immune system, plays an essential role in regulating genetic material, and assists in wound healing, protein synthesis, energy metabolism, and insulin storage.

Index

Page numbers in italic refer to figures.

D

E

F

G

N

T

DIGITAL RESOURCES

Digital resources for *Culinary Nutrition Principles and Applications* expand the learning experience by providing supplemental learning materials that focus on content retention and application. Digital resources include Quick Quizzes®, an Illustrated Glossary, Flash Cards, Knowledge Checks and Reviews, Culinary Nutrition Recipe Modifications, a Menu Planning Challenge, the Digestion and Absorption Process, a Media Library, and access to ATPeResources.com. Digital resources can be accessed from a computer with an Internet connection.

Location of Digital Resources

Digital resources for *Culinary Nutrition Principles and Applications* can be accessed by visiting ATPeResources.com/QuickLinks and entering access code 267412 or by scanning the QR Codes located in the textbook.

System Requirements

- Windows or Mac OS
- A modern JavaScript-enabled web browser:
    - Google Chrome™
    - Mozilla Firefox
    - Internet Explorer® 9 or later
    - Safari®
- Adobe® Reader™ (software application)
- Adobe® Flash® Player (browser plugin)
- Microsoft® Word (viewer or full-software application)
- Microsoft® PowerPoint® (viewer or full-software application)

Opening Files

Digital resource content will open in a web browser window when a web browser plugin is available for that specific file type. Files types that are not supported by a web browser plugin (such as PDF and PowerPoint® files) will need to be downloaded to the computer and opened using the appropriate software.